ANOXIA

Cellular Origin, Life in Extreme Habitats and Astrobiology

Volume 21

Series Editor:

Joseph Seckbach
The Hebrew University of Jerusalem, Israel

For further volumes:
http://www.springer.com/series/5775

Anoxia

Evidence for Eukaryote Survival and Paleontological Strategies

Edited by

Alexander V. Altenbach
Ludwig-Maximilians-University, Munich, Germany

Joan M. Bernhard
Wood Hole Oceanographic Institution, MA, USA

and

Joseph Seckbach
The Hebrew University of Jerusalem, Israel

Editors
Alexander V. Altenbach
Department for Earth and Environmental
Science, and GeoBio-Center
Ludwig-Maximilians-University
Richard-Wagner-Str. 10
80333 Munich
Germany
a.altenbach@lrz.uni-muenchen.de

Joan M. Bernhard
Geology and Geophysics Department
Wood Hole Oceanographic Institution
MS52, Woods Hole, MA 02543
USA
jbernhard@whoi.edu

Joseph Seckbach
P.O. Box 1132
90435 Efrat
Israel
seckbach@cc.huji.ac.il

ISSN 1566-0400
ISBN 978-94-007-1895-1   e-ISBN 978-94-007-1896-8
DOI 10.1007/978-94-007-1896-8
Springer Dordrecht Heidelberg London New York

Library of Congress Control Number: 2011935457

© Springer Science+Business Media B.V. 2012
No part of this work may be reproduced, stored in a retrieval system, or transmitted in any form or by any means, electronic, mechanical, photocopying, microfilming, recording or otherwise, without written permission from the Publisher, with the exception of any material supplied specifically for the purpose of being entered and executed on a computer system, for exclusive use by the purchaser of the work.

Printed on acid-free paper

Springer is part of Springer Science+Business Media (www.springer.com)

TABLE OF CONTENTS

Introduction/**Joseph Seckbach**	ix
Stepping into the Book of Anoxia and Eukaryotes/**The Editors**	xi
List of Authors and their Addresses	xxi
List of External Reviewers and Referees	xxix
Acknowledgment to Authors, Reviewers, and any Special People Who Assisted	xxxiii

PART 1:
GENERAL INTRODUCTION

Anaerobic Eukaryotes [**Fenchel, T.**]	3
Biogeochemical Reactions in Marine Sediments Underlying Anoxic Water Bodies [**Treude, T.**]	17
Diversity of Anaerobic Prokaryotes and Eukaryotes: Breaking Long-Established Dogmas [**Oren, A.**]	39

PART 2:
FUNCTIONAL BIOCHEMISTRY

The Biochemical Adaptations of Mitochondrion-Related Organelles of Parasitic and Free-Living Microbial Eukaryotes to Low Oxygen Environments [**Tsaousis, A.D. et al.**]	51
Hydrogenosomes and Mitosomes: Mitochondrial Adaptations to Life in Anaerobic Environments [**de Graaf, R.M. and Hackstein, J.H.P.**]	83
Adapting to Hypoxia: Lessons from Vascular Endothelial Growth Factor [**Levy, N.S. and Levy, A.P.**]	113

PART 3: MANAGING ANOXIA

Magnetotactic Protists at the Oxic–Anoxic Transition Zones of Coastal Aquatic Environments [**Bazylinski, D.A. et al.**] 131

A Novel Ciliate (Ciliophora: Hypotrichida) Isolated from Bathyal Anoxic Sediments [**Beaudoin, D.J. et al.**] 145

The Wood-Eating Termite Hindgut: Diverse Cellular Symbioses in a Microoxic to Anoxic Environment [**Dolan, M.F.**] 155

Ecological and Experimental Exposure of Insects to Anoxia Reveals Surprising Tolerance [**Hoback, W.W.**] 167

The Unusual Response of Encysted Embryos of the Animal Extremophile, *Artemia franciscana*, to Prolonged Anoxia [**Clegg, J.S.**] 189

Survival of Tardigrades in Extreme Environments: A Model Animal for Astrobiology [**Horikawa, D.D.**] 205

Long-Term Anoxia Tolerance in Flowering Plants [**Crawford, R.M.M.**] 219

PART 4: FORAMINIFERA

Benthic Foraminifera: Inhabitants of Low-Oxygen Environments [**Koho, K.A. and Piña-Ochoa, E.**] 249

Ecological and Biological Response of Benthic Foraminifera Under Oxygen-Depleted Conditions: Evidence from Laboratory Approaches [**Heinz, P. and Geslin, E.**] 287

The Response of Benthic Foraminifera to Low-Oxygen Conditions of the Peruvian Oxygen Minimum Zone [**Mallon, J. et al.**] 305

Benthic Foraminiferal Communities and Microhabitat Selection on the Continental Shelf Off Central Peru [**Cardich, J. et al.**] 323

PART 5: ZONES AND REGIONS

Living Assemblages from the "Dead Zone" and Naturally Occurring Hypoxic Zones [**Buck, K.R. et al.**] 343

The Return of Shallow Shelf Seas as Extreme Environments:
  Anoxia and Macrofauna Reactions in the Northern
  Adriatic Sea [**Stachowitsch, M. et al.**]... 353
Meiobenthos of the Oxic/Anoxic Interface in the Southwestern
  Region of the Black Sea: Abundance and Taxonomic
  Composition [**Sergeeva, N.G. et al.**] 369
The Role of Eukaryotes in the Anaerobic Food Web
  of Stratified Lakes [**Saccà, A.**] 403
The Anoxic Framvaren Fjord as a Model System to Study
  Protistan Diversity and Evolution
  [**Stoeck, T. and Behnke, A.**] 421
Characterizing an Anoxic Habitat: Sulfur Bacteria in a Meromictic
  Alpine Lake [**Fritz, G.B. et al.**].. 449
Ophel, the Newly Discovered Hypoxic Chemolithotrophic
  Groundwater Biome: A Window to Ancient Animal Life
  [**Por, F.D.**].. 463
Microbial Eukaryotes in the Marine Subsurface?
  [**Edgcomb, V.P. and Biddle, J.F.**].. 479

PART 6:
MODERN ANALOGS AND TEMPLATES
FOR EARTH HISTORY

On The Use of Stable Nitrogen Isotopes in Present and Past
  Anoxic Environments [**Struck, U.**]... 497
Carbon and Nitrogen Isotopic Fractionation in Foraminifera:
  Possible Signatures from Anoxia
  [**Altenbach, A.V. et al.**]... 515
The Functionality of Pores in Benthic Foraminifera
  in View of Bottom Water Oxygenation: A Review
  [**Glock, N. et al.**].. 537
Anoxia-Dysoxia at the Sediment-Water Interface of the Southern
  Tethys in the Late Cretaceous: Mishash Formation,
  Southern Israel [**Almogi-Labin, A. et al.**] 553
Styles of Agglutination in Benthic Foraminifera from Modern
  Santa Barbara Basin Sediments and the Implications
  of Finding Fossil Analogs in Devonian and Mississippian
  Black Shales [**Schieber, J.**].. 573
Did Redox Conditions Trigger Test Templates in Proterozoic
  Foraminifera? [**Altenbach, A.V. and Gaulke, M.**]........................... 591
The Relevance of Anoxic and Agglutinated Benthic Foraminifera
  to the Possible Archean Evolution of Eukaryotes
  [**Altermann, W. et al.**]... 615

Organism Index 631

Subject Index 639

Author Index 647

INTRODUCTION TO ANOXIA: EVIDENCE FOR EUKARYOTE SURVIVAL AND PALEONTOLOGICAL STRATEGIES

Research in anoxic environments is a relatively new and rapidly growing branch of science that is of general interest to many students of diverse microbial communities. The term *anoxia* means absence of atmospheric oxygen, while the term *hypoxia* refers to O_2 depletion or to an extreme form of "low oxygen." Both terms anoxia and hypoxia are used in various contexts.

It is accepted that the initial microorganisms evolved anaerobically and thrived in an atmosphere without oxygen. The rise of atmospheric oxygen occurred ~2.3 bya through the photosynthesis process of cyanobacteria which "poisoned" the environment by the release of toxic O_2. Microorganisms that could adapt to the oxygenated environment survived and some of them evolved further to the eukaryotic kingdom in an aerobic atmosphere, while others vanished or escaped to specific anaerobic niches where they were protected. Most of the anaerobes are prokaryotes, while some are also within the Eukaryan kingdom. Those latter organisms are the focus of this new volume.

Anaerobic areas of marine or fresh water that are depleted of dissolved oxygen have restricted water exchange. In most cases, oxygen is prevented from reaching the deeper levels by a physical barrier (e.g., silt or mud) as well as by temperature or concentration stratification, such as in denser hypersaline waters. Anoxic conditions will occur if the rate of oxidation of organic matter is greater than the supply of dissolved oxygen. Anoxic waters are a natural phenomenon, and have occurred throughout the geological history. At present, for example, anoxic basins exist in the Baltic and Mediterranean Seas and elsewhere. Eutrophication of freshwater lakes and marine environments often causes increase in the extent of the anoxic areas. Decay of phytoplankton blooms also intensifies the anoxic conditions in a water body. Although algae produce oxygen in the daytime via photosynthesis, during the night hours they continue to undergo cellular respiration and can therefore deplete available oxygen. In addition, when algal blooms die off, oxygen is further used during bacterial decomposition of the dead algal cells. Both of these processes can result in a significant depletion of dissolved oxygen in the water, creating hypoxic conditions or a dead zone (low-oxygen areas).

Among the eukaryotic anaerobes one could find protists that live in hypersaline environments (up to 365 g/l NaCl), for example, the groups of ciliates, dinoflagellates, choanoflagellates, and other marine protozoa. We are aware of some eukaryotes that act in anaerobic conditions such as the yeast that ferments sugars to ethanol and CO_2, wine fermentation, and in the baking process. Second, the protozoa (e.g., ciliates) in the rumen of cows and other ruminant animals act in anaerobic conditions. In some anoxic single eukaryotic cells, the mitochondria are replaced by hydrogenosomes, or the mitochondrion is adapted as an unusual organelle structure for the anaerobic metabolism.

Lately a group of metazoa was detected living in a permanently anoxic environment in the sediments of the deep hypersaline basin 3.5 km below the surface of the Mediterranean Sea. Others have detected eukaryotes in anoxic areas of the Black Sea and near Costa Rica. Some Foraminifera are found living in oxygen-free zones, such as in Swedish Fjords, in the Cariaco and Santa Barbara Basin, the Black Sea, or off Namibia.

In the severely cold winters of the Northern Arctic zones, there are plants that can survive under a covering of ice which completely prevents access to oxygen. Any remaining oxygen in the soil atmosphere is consumed by microbial activity. There is therefore a total cessation of aerobic metabolism for several months in the overwintering organs, such as tubers and underground stems. The ability of these perennial organs to maintain viable buds throughout an anoxic winter enables the plants to grow new roots and shoots when aerobic metabolism is resumed on thawing in spring (see Crawford in this volume). We know also that in certain species seed germination can take place in anaerobic conditions. Similarly, the tolerance of insects to anoxia has also been recorded in this volume (Hoback, in this volume).

Tardigrades (segmented polyextremophilic eukaryotic animals, less than 1 mm in length) can survive and exhibit extraordinary tolerance to several extreme environments. The results with anhydrobiotic tardigrades strongly suggest that these invertebrate animals can survive even in anoxic environments in outer space. It seems that oxygen supply to the tardigrades causes critical damage to these anhydrobiotic animals under such conditions (Horikawa in this volume).

The present topic of *ANOXIA: Evidence for Eukaryote Survival and Paleontological Strategies* is timely and exciting and we now present it in this volume, which is aimed at biological researchers of ecology and biodiversity, to astrobiologists, to readers interested in extreme environments, and also paleoecologists and paleontologists (and some sedimentologists). This volume is number 21 of the *Cellular Origin, Life in Extreme Habitats and Astrobiology* [COLE] series, [www.springer.com/series/5775]. It contains 32 chapters contributed by 71 authors from 13 countries (given here in alphabetical order): Austria, Canada, Denmark, France, Germany, Israel, Italy, Japan, Peru, the Netherlands, Ukraine, the United Kingdom, and the USA. We availed ourselves of 25 external referees in addition to our peer reviewers to evaluate the chapters. It is our hope that our readers will enjoy this book in which we invested so much enthusiasm and effort.

The author thanks Professors Aharon Oren and David Chapman for their constructive suggestions to improve this Introduction.

Joseph Seckbach The Hebrew University of Jerusalem
Jerusalem, Israel

STEPPING INTO THE BOOK OF ANOXIA AND EUKARYOTES

With this book, the editors, authors, and reviewers cooperated in promoting the debate on the persistence of eukaryotes in anoxic environments and newly discovered adaptations of eukaryotes in oxygen-depleted habitats. Also with this book, we wish to attract scientists and students from all types of science to conduct research in low-level oxygen to truly anoxic environments. We not only seek to provide overviews and basics that lead to a better understanding, but also want to communicate the endeavor and fascination involved in this research. The six parts of the book span a broad range from molecular biology to field research, from environmental monitoring to paleoecology. Hopefully, this may also enhance interest and cooperation on interdisciplinary grounds. Most of the questions raised are under discussion at present, a positive sign for frontier research where rapid developments transpire.

Thriving eukaryotes and anoxic environments were considered quite incompatible for a long time. In Part I, basics on eukaryotes recovered from anoxic environments are summarized (Fenchel), and principles of biogeochemical activities near the redoxcline are outlined (Treude). The comparison of common former considerations about anoxic life and our present knowledge offers insight into the possible revision of some dogmatic views (Oren).

Part II exemplifies the biochemical pathways required for eukaryotes under oxygen stress or absence of molecular oxygen. This section covers the biochemical adaptation to low-oxygen environments (Tsaousis), and an overview on the specific function of hydrogenosomes and mitosomes in anaerobes (De Graaf and Hackstein). The present debate about eukaryotic cell evolution is ultimately linked to the issue of how mitochondria originated and evolved. In the context of a classical view, the Archaea and the Eukarya have a common ancestor. Alternative views propose that the Eukarya evolved directly from the archaeal lineage. The definition of modern anaerobic eukaryotes as remnants of the one or other lineage is an as-yet-unresolved question. One possible implication in this context is of utmost importance for evolutionary biology: anaerobiosis in extant eukaryotes would be either a late adaptation developed by obligate aerobic eukaryotes, or an omnipresent ability since the most ancestral lineage. A comprehensive overview of pathways for the adaptation to anoxic conditions are explained and discussed by Levy and Levy.

Part III presents contributions on the surprising tolerance and diversity of eukaryotes to hypoxia and anoxia, demonstrating that anoxic life is not strictly anaerobic microbes able to cope with the reducing chemical habitat of their substrate. All kinds of biota may attune to anoxic conditions following the demands of hosted symbionts, for prolonging the survival and success of their offspring and encystments, for enhancement of their competitiveness, and/or for successful survival and rapid repopulation after sporadic oxygen deficiency. Very different eukaryotes employ considerable and sometimes decisive advantage by coping and managing anoxic conditions; and all this for quite varied reasons. These chapters cover magnetotactic protists (Bazylinski et al.), ciliates (Beaudoin et al.), and protistan symbionts hosted by termites (Dolan). In addition, a number of experimental works involve insects (Hoback), brine shrimp (Clegg), and the superstar specialists in surviving super-stressors, the tardigrades (Horikawa). Even flowering plants face driving forces to acquire specific capabilities for coping with pulsed or sporadic anoxia (Crawford).

A specialized part of the book, Part IV, presents work on Foraminifera, which are a unique taxon in that most extant forms easily fossilize (vs. metazoans and other protists common to anoxic habitats) and because foraminifera have been shown to perform complete denitrification (Koho and Pina-Ochoa). Thus, this group could be considered a key taxon with respect to facultative eukaryotic anaerobiosis (Heinz and Geslin). Distribution-oriented studies (Mallon et al.; Cardich et al.) illustrate how abundant this group can be in certain oxygen-depleted settings.

Part V focuses on community responses in specific oxygen-stressed habitats. Our planet faces increasing surface temperatures, record-breaking heat in summers, catastrophic storms and rain falls, and the most rapid melting of ice sheets and mountain glaciers ever observed by humans. Declining densities of surface waters reduce mixing rates with deeper water masses, the intensity of downwelling, and the supply of well-oxygenated bottom water masses. Increased surface water temperatures as well as the enhanced inflow of freshwater from melting ice shields cause such density drops. Marine realms with enhanced degradation of organic carbon fluxes and oxygen consumption, called oxygen minimum zones (OMZ), seem highly sensitive to these perturbations. These regions are actively becoming more and more depleted in oxygen. Their annual reduction of dissolved oxygen ranges from about 0.1 to 0.4 µmol per liter of seawater at mid-water depths, expanding the area where larger metazoa start to suffer from hypoxia by 4.5 million km^2 during the last decades (see Stramma et al. 2010 in Table 1). As the inner core of such OMZ's may reach anoxia, the expansion of ocean-wide "death zones" is forewarned (Gewin 2010, Table 1), with hypoxia and anoxia as prime stressors (Buck et al.). Inevitably, hypoxia and anoxia must be monitored more carefully in the future, in order to follow environmental change (Stachowitsch et al.). Well-investigated hypoxic to anoxic regions, such as the Black Sea (Sergeeva et al.), stratified basins (Sacca), silled fjord basins (Stoecke and Behnke), and meromictic lakes (Fritz et al.) still offer new insights about community responses after close examination. More recently, chemolithotrophic groundwaters (Por) and the deep sedimentary habitats (Edgcomb and Biddle) have started to attract a growing number of scientists, and are expected to deliver a wealth of new insights, novel biota, and fascinating biogeochemical dynamics.

Table 1. Thresholds, ranges, and technical terms in use for the definition of dissolved oxygen concentrations, their biotic response, and observed environmental impacts and repercussions.

Range	≈ μmol/kg	Term	Indication	Reference
>8–2 ml/l	400–100	Oxic	Geochemists relate this term primarily to redox conditions (Eh), biologists to availability of O_2	Tyson and Pearson (1991)
2–0.2 ml/l	100–10	Dysoxic	Seasonal dysoxic conditions occur in stratified estuarine or pro-delta settings, more extensively on open shelves at water depths deeper than 60 m, and in near bottom water	Tyson and Pearson (1991)
120–60 μmol/kg	120–60	Hypoxic	Lethal or stressful to specific mobile macro-organisms	Stramma et al. (2008)
<70 μmol/kg	70		Some large mobile macro-organisms are unable to abide	Stramma et al. (2010)
<2 mg/l	70		Reduction of meiofaunal abundance and diversity	Wetzel et al. (2001)
<1.5 ml/l	68		Critical for larger fish	Gewin (2010)
<63 μmol	63		Range for the definition of specific biotic and biogeochemical consequences of coastal "hypoxia"	Helly and Levin (2004), Middelburg and Levin (2009)
1.42 ml/l	62.5		Threshold for coastal seafloor hypoxia, near 30% oxygen saturation	Levin et al. (2009)
<5 kPa O_2	50		Onset for specific physiological adaptations required for certain transition of ecosystems	Seibel (2011)
>1 ml/l	45	Oxic	Technical term	Bernhard and Sen Gupta (1999), Levin (2004)
1–0.1 ml/l	45–4.5	Dysoxic	Technical term	Bernhard and Sen Gupta (1999), Levin (2004)
<0.5 ml/l	22		Threshold for contour lining "hypoxic" oxygen depletion on shelves and bathyal sea floors	Helly and Levin (2004)
20 μmol	20	Suboxia	Upper threshold of the transition layer from O_2 to NO_3^- respiration, (0.7–20 μmol), termed "suboxia" by biologists and biogeochemists	Helly and Levin (2004), Middelburg and Levin (2009)
<20 μmol	20		Global definition of the most intense oxygen minimum zones (OMZ)	Helly and Levin (2004), Middelburg and Levin (2009)
<1 kPa O_2	10		Threshold for a certain transition of ecosystems; represents a limit to evolved oxygen extraction capacity	Seibel (2011)

(continued)

Table 1. (continued)

Range	≈ μmol/kg	Term	Indication	Reference
0.2–0.0 ml/l	10	Suboxic	Formation of laminated sediments without macrofauna, but with in situ microfauna	Tyson and Pearson (1991)
<10 μmol/kg	10	Suboxic	Nitrate becomes involved in respiration if present	Stramma et al. (2008)
<0.15 ml/l	10		Bioturbation is reduced, chemosynthesis becomes important	Levin (2004)
<10 μM	10		Accuracy of common O_2 probes in field research	Paulmier and Ruiz-Pino (2009)
10–2 μM	10–2		Reproducibility of O_2 measures in the field	Paulmier and Ruiz-Pino (2009)
>0.2 ml/l	9		No effect on midwater biomass, low effect on biodiversity	Childress and Seibel (1998)
<0.2 ml/l	9		Threshold for contour lining a more strict definition of "hypoxia"	Helly and Levin (2004)
<0.15 ml/l	7		Significant drop in zooplankton biomass	Childress and Seibel (1998)
2–6 μmol/l	2–6	C – layer	Coexistence of H_2S and O_2	Sorokin (2005)
<0.1 ml/l	4.5	Suboxic	Marks a distinct change in the environment, rise of nitrate removal	Karstensen et al. (2008)
>0–0.1 ml/l	4.5	Microxic	Technical term	Bernhard and Sen Gupta (1999), Levin (2004)
μ50 Ciliate	2–4		Half-saturation for larger ciliates (≈ 1–2% atm. pressure)	Fenchel and Finlay (2008)
>10^{-6} mol	1		Accuracy of high level O_2 detection methodologies (< ≈ 0.5% saturation)	Berner (1981), Paulmier and Ruiz-Pino (2009)
1 μmol	1		Minimum level reached in the core of OMZs	Helly and Levin (2004), Middelburg and Levin (2009)
0.7 μmol	0.7	Suboxia	Lower threshold of the transition layer from O_2 to NO_3^- respiration (0.7–20 μmol/kg)	Helly and Levin (2004), Middelburg and Levin (2009)
μ50 Amoeba	0.4		Half-saturation for an amoeba (≈ 0.2% of atmospheric pressure)	Fenchel and Finlay (2008)
μ50 Yeast	0.15		Half-saturation for yeast cells (≈ 0.07% of atmospheric pressure)	Fenchel and Finlay (2008)
μ50 Bacteria	0.1		Half-saturation for bacteria and mitochondria (≈ 0.05% of atmospheric pressure)	Fenchel and Finlay (2008)
0 ml/l	0	Postoxic	Neither free oxygen nor reducing conditions (e.g., production of hydrogen sulfide)	Berner (1981), Baernhard and Sen Gupta (1999)
0 ml/l	0	Anoxic	No dissolved oxygen	Baernhard and Sen Gupta (1999), Levin (2004), Stramma et al. (2008), and Tyson and Pearson (1991)

Part VI turns back in time in search for signals, tracers and evidence from modern anoxic environments that can be applied to the reconstruction, and understanding of the fossil record. Stable isotopes are commonly used in biogeochemistry, but rarely scaled for their specific behavior under anoxia (Struck; Altenbach et al.). Test porosity in Foraminifera channels diffusional gradients between the cell and the environment. New insights indicate that nitrate utilization and denitrification might be deduced using modern and fossil test porosity (Glock et al.). Also, the correlation between modern and Mesozoic upwelling systems is discussed (Almogi et al.), as are analogs of foraminiferal test structures in Devonian black shales and modern Foraminifera from the Santa Barbara Basin (Schieber). Many recent insights into modern facultative anaerobic Foraminifera support a hypothesis on the basic reasons for foraminiferal test construction (Altenbach and Gaulke), and even more far-reaching speculations about the evolutionary path in these rhizarians (Altermann et al.).

Common terms in use for distinguishing levels of oxygen deficiency in environments, for biota, in physiology, in clinical research, and in geosciences may be identical, but their definitions may be quite variable in different disciplines of natural sciences. The terms "hypoxia" and "hypoxic" are basically clinical terminologies that define a pathological condition of an organism or its tissues when deprived of appropriate oxygen availability, as opposed to stress-free "normoxia" or "normoxic" conditions. However, these terms are meanwhile broadly used in environmental research, indicating depleted oxygen conditions irrespective of stress impact on biota. But as hypoxia might be reached for different biota at very different oxygen concentrations, this term should not be linked to an explicit level or range of dissolved oxygen available in the environment. In physiology, the half-saturation constant "$\mu 50$" is a common measure. It marks the substrate concentration at which "μ" equals half of the maximum rate of growth/turnover/consumption "μmax" of an organism or a cellular structure. The term "euxinia" was coined by considerations on sedimentary facies in geosciences; it defines either stagnant water exchange or reduced solubility of oxygen, provoking oxygen depletion down to anoxia rather than a specific level of dissolved oxygen.

Sedimentologists often condense conflicting redox conditions to the reconstruction of either "euxinic" conditions, which means that sulfide was present and molecular oxygen practically absent in the water column, or they consider "suboxic" conditions, which defines residual molecular oxygen in the bottom water but also sulfide production within or at the surface of the sediment column.

Table 1 provides a guide for the different units, ranges, and terms used in this book and an example of the wide range of literature employing them. The first column quotes the threshold or range exemplified in a source reference quoted in column five. The second column unifies this unit to µmol, either as given in the respective source or roughly calculated without specific corrections (e.g., for temperature, density, media, etc.). The third column refers to the specific terminology in use, and the fourth column describes its basic usage.

Last but not the least, it may be noted that the definition of "anoxia" itself – as complete absence of dissolved oxygen – is easily defined, but impossible to measure in the field. Most field methodologies available are not able to detect concentrations below 1 µmol, and their reproducibility may range well above 2 µmol (Table 1). This range of methodological restriction can interfere with a number low level thresholds presented in Table 1. When approaching anoxia, methods must necessarily be more and more sophisticated. In addition, Eh profiling and specific biochemical analyses should extend oxygen probe measures. Fine-tuning between suboxic, postoxic, and anoxic conditions is a meticulous task under debate (see Sorokin 2007). However, whether chemically aggressive, free radicals are present or not may be more decisive for the prevailing eukaryotes than small-scale drops in O_2 concentrations already near zero. For a multitude of research topics, there is much future work to define what anoxia is, and what eukaryotes actually do during exposure to anoxia. We hope many readers of this book will dedicate studies to these unknowns.

References

Berner RA (1981) A new geochemical classification of sedimentary environments. J Sed Petrol 51:359–365

Bernhard JM, Sen Gupta B (1999) Foraminifera of oxygen-depleted environments. In: Sen Gupta B (ed) Modern Foraminifera. Kluwer Academic, Dordrecht, pp 200–216

Childress JJ, Seibel BA (1998) Life at stable low oxygen levels: adaptations of animals to oceanic oxygen minimum layers. J Exp Biol 201:1223–1232

Fenchel T, Finlay B (2008) Oxygen and the spatial structure of microbial communities. Biol Rev 83:553–569

Gewin V (2010) Dead in the water. Nature, 466:812–814

Helly JJ, Levin LA (2004) Global distribution of naturally occurring marine hypoxia on continental margins. Deep-Sea Res I 51:1159–1168

Karstensen J, Stramma L, Visbeck M (2008) Oxygen minimum zones in the eastern tropical Atlantic and Pacific oceans. Prog Oceanogr 77:331–350

Levin LA (2004) Oxygen minimum zone benthos: adaptation and community response to hypoxia. In: Gibson RN, Atkinson RJA (ed) Oceanography and marine biology, an annual review, vol 41. CRC Press, Boca Raton, pp 1–45

Levin LA, Ekau W, Gooday AJ, Jorissen F, Middelburg JJ, Naqvi SWA, Neira C, Rabalais NN, Zhang J (2009) Effects of natural and human-induced hypoxia on coastal benthos. Biogeosciences 6:2063–2098

Middelburg JJ, Levin LA (2009) Coastal hypoxia and sediment biogeochemistry. Biogeosciences 6:1273–1293

Paulmier A, Ruiz-Pino D (2009) Oxygen minimum Zones (OMZs) in the modern ocean. Prog Oceanogr 80:113–128

Seibel BA (2011) Critical oxygen levels and metabolic suppression in oceanic oxygen minimum zones. J Exp Biol 214:326–336

Sorokin YI (2005) On the structure of the Black Sea redox zone. Oceanology [Engl Transl AGU, Oceanologiya], Interperiodica 45:S51–S60

Sorokin YI (2007) Suboxic zone in the Black Sea: real fact or analytical artefact? Mar Biol Res 3:265–271

Stramma L, Johnson GC, Sprintall J, Mohrholz V (2008) Expanding oxygen-minimum zones in the tropical oceans. Science 320:655–658

Stramma L, Schmidtko S, Levin LA, Johnson GC (2010) Ocean oxygen minima expansions and their biological impacts. Deep Sea Res I 57:587–595

Tyson RV, Pearson TH (1991) Modern and ancient continental shelf anoxia: an overview. Geol Soc Lond Spec Publ 58:1–24

Wetzel MA, Fleeger JW, Powers SP (2001) Effects of hypoxia and anoxia on meiofauna: a review with new data from the Gulf of Mexico. Coast Est Stud 58:165–184

Alexander V. Altenbach, Joan M. Bernhard, and Joseph Seckbach
The Editors. Munich, Woods Hole, Jerusalem

Biodata of **Alexander V. Altenbach**, **Joan M. Bernhard**, and **Joseph Seckbach**, the editors of this volume.

Dr. Alexander Volker Altenbach earned his Ph.D. from Kiel University in 1985, followed by activities as a reader, team leader, and chief scientists in paleoceanography and micropaleontology in Kiel. Cooperation with multidisciplinary joint research units and the Geomar Research Center convinced him that mainly interdisciplinary geobiochemical approaches pave the way for understanding system earth. External affairs sum up from mudlogging and engineering geology to cofounding a software company in Hamburg.

Since 1994, he is Professor for micropaleontology at the Dept. of Earth and Environmental Science of the Ludwig-Maximilians-University in Munich (Germany). Times in Munich include activities as Chaperon of the Bavarian States Collection for Micropaleontology, Speaker of the Research Center for Geobiology and Biodiversity, and Dean of the Faculty of Geosciences.

All in all, years were spent on research vessels, boats, scuba diving, and during field excursions on five oceans and continents, among them very fruitful sabbaticals at the Australian National University in Canberra, and at the Huinay Research Station in Patagonia. Publications mainly deal with Foraminifera, but some do also cover the development of laboratory equipment and software, tectonics, fossils, the ecology of lizards and snakes, and a natural field guide to Australia. Most investigations center on ecology, biomass, food webs, and stable isotope fractionation.

Reasonable for getting involved in this book was the rejection of an early manuscript on foraminifera thriving under sulfidic conditions by two internationally recommended journals. As noted by a critical reviewer, this was because "anoxic foraminifera don't seem reasonable."

E-mail: **a.altenbach@lrz.uni-muenchen.de**

Dr. Joan M. Bernhard, who is a Senior Scientist at Woods Hole Oceanographic Institution (Massachusetts, USA), is a biogeochemist with a major focus on the adaptations and ecology of protists living in the chemocline. Another focus of her work uses experimental approaches to investigate the controls on geochemical proxies recorded in calcareous foraminiferal tests (shells), as well as other aspects of foraminiferal biology. Her work largely involves the bathyal to abyssal deep sea and recently has concentrated on modern environments and organisms, but her career began with interpretation of the fossil record. She continues to do paleoecologically and paleoceanographically relevant research.

Bernhard has degrees in geology (B.A. 1982, Colgate University; M.S. 1984, University of California Davis) and biological oceanography (Ph.D. 1990, Scripps Institution of Oceanography, University of California San Diego), and did post-doctoral work in cell biology at the Wadsworth Center (New York State Department of Health, Albany New York). She also worked in the Department of Environmental Health Sciences at the University of South Carolina's Arnold School of Public Health from 1997 to 2004. She has served as Chief Scientist on 18 research cruises and participant on 33 others. Her research has included submersible and Remotely Operated Vehicle (ROV) work. Her earlier career included nine field seasons totaling 23 months in the Antarctic, performing nearly 200 SCUBA dives under ice.

Her multidisciplinary training and diverse experience gives her a unique perspective into anoxic habitats.

E-mail: **jbernhard@whoi.edu**

Biodata of **Joseph Seckbach**, editor of this volume, and author of *"Introduction to Anoxia: Evidence for Eukaryote Survival and Paleontological Strategies"*.

Professor Joseph Seckbach is the founder and chief editor of *Cellular Origins, Life in Extreme Habitats and Astrobiology* ("COLE") book series (the present volume is number 21 of this series, see: www.springer.com/series/5775). He has coedited other volumes, such as the Proceeding of Endocytobiology VII Conference (Freiburg, Germany) and the Proceedings of Algae and Extreme Environments Meeting (Trebon, Czech Republic). See http://www.schweizerbart.de/pubs/books/bo/novahedwig-051012300-desc.ht). His recent volume (with coeditor Richard Gordon) entitled Divine Action and Natural Selection: Science, Faith, and Evolution was published by World Scientific Publishing Company.

Dr. Seckbach earned his Ph.D. from the University of Chicago (1965) and did a postdoctoral training in the Division of Biology at Caltech, in Pasadena, CA. He was appointed to the faculty of the Hebrew University (Jerusalem, Israel). He spent sabbaticals at UCLA and Harvard University and DAAD-sponsored periods in Tübingen, Germany, and at LMU, Munich. Dr. Seckbach served at Louisiana State University, Baton Rouge, as the first selected occupant of the Endowed Chair for the Louisiana Sea Grant and Technology transfer.

Beyond editing academic volumes, he has published scientific articles on plant ferritin–phytoferritin, cellular evolution, acidothermophilic algae, and life in extreme environments. He also edited and translated several popular books. Professor Seckbach is the coauthor, with R. Ikan, of the Hebrew language *Chemistry Lexicon* (DeVeer publisher, Tel Aviv, Israel). His recent interest is in the field of enigmatic microorganisms and life in extreme environments.

E-mail: **seckbach@huji.ac.il**

LIST OF AUTHORS AND THEIR ADDRESSES

ALMOGI-LABIN, AHUVA
GEOLOGICAL SURVEY OF ISRAEL, 30 MALKHE ISRAEL ST., JERUSALEM 95501, ISRAEL

ALTENBACH, ALEXANDER VOLKER
DEPARTMENT FOR EARTH AND ENVIRONMENTAL SCIENCE, AND GEOBIO-CENTER, LUDWIG-MAXIMILIANS-UNIVERSITÄT, MUNICH, AND RICHARD-WAGNER-STR. 10, 80333 MUNICH, GERMANY

ALTERMANN, WLADYSLAW
DEPARTMENT OF GEOLOGY, UNIVERSITY OF PRETORIA, PRETORIA 0002, SOUTH AFRICA

ANIKEEVA, OKSANA V.
INSTITUTE OF BIOLOGY OF THE SOUTHERN SEAS NASU, SEVASTOPOL, UKRAINE

ASHCKENAZI-POLIVODA, SARIT
GEOLOGICAL SURVEY OF ISRAEL, 30 MALKHE ISRAEL ST., JERUSALEM 95501, ISRAEL
DEPARTMENT OF GEOLOGICAL AND ENVIRONMENTAL SCIENCES, BEN GURION UNIVERSITY OF THE NEGEV, BEER SHEVA 84105, ISRAEL

BARRY, JAMES P.
MONTEREY BAY AQUARIUM RESEARCH INSTITUTE, 7700 SANDHOLDT ROAD, MOSS LANDING, CA 95039, USA

BAZYLINSKI, DENNIS A.
SCHOOL OF LIFE SCIENCES, UNIVERSITY OF NEVADA AT LAS VEGAS, 4505 MARYLAND PARKWAY, LAS VEGAS, NV 89154-4004, USA

BEAUDOIN, DAVID J.
BIOLOGY DEPARTMENT, WOODS HOLE OCEANOGRAPHIC INSTITUTION, WOODS HOLE, MA 02543, USA

BEHNKE, ANKE
DEPARTMENT OF ECOLOGY, UNIVERSITY OF KAISERSLAUTERN, ERWIN SCHROEDINGER STR. 14, KAISERSLAUTERN D-67663, GERMANY

BENJAMINI, CHAIM
DEPARTMENT OF GEOLOGICAL AND ENVIRONMENTAL SCIENCES, BEN GURION UNIVERSITY OF THE NEGEV, BEER SHEVA 84105, ISRAEL
RAMON SCIENCE CENTER, MIZPE RAMON 80600, ISRAEL

BERNHARD, JOAN M.
GEOLOGY AND GEOPHYSICS DEPARTMENT, WOODS HOLE OCEANOGRAPHIC INSTITUTION, MS #52, WOODS HOLE, MA 02543, USA

BIDDLE, JENNIFER F.
COLLEGE OF EARTH, OCEAN AND THE ENVIRONMENT, UNIVERSITY OF DELAWARE, LEWES, DE 19958, USA

BRÜMMER, FRANZ
DEPARTMENT OF ZOOLOGY, BIOLOGICAL INSTITUTE, UNIVERSITY OF STUTTGART, 70569 STUTTGART, GERMANY

BUCK, KURT R.
MONTEREY BAY AQUARIUM RESEARCH INSTITUTE, 7700 SANDHOLDT ROAD, MOSS LANDING, CA 95039, USA

CARDICH, JORGE
FACULTAD DE CIENCIAS Y FILOSOFÍA, PROGRAMA MAESTRÍA EN CIENCIAS DEL MAR, UNIVERSIDAD PERUANA CAYETANO HEREDIA, AV. HONORIO DELGADO 430, LIMA 31, PERU
DIRECCIÓN DE INVESTIGACIONES OCEANOGRÁFICAS, INSTITUTO DEL, MAR DEL PERÚ (IMARPE), AV. GAMARRA Y GRAL. VALLE, S/N, CHUCUITO, CALLAO, PERU
JOINT INTERNATIONAL LABORATORY 'DYNAMICS OF THE HUMBOLDT CURRENT SYSTEM' (LMI DISCOH), LIMA, PERU

CLEGG, JAMES S.
BODEGA MARINE, LABORATORY, SECTION OF MOLECULAR AND CELLULAR BIOLOGY, UNIVERSITY OF CALIFORNIA, DAVIS, BODEGA BAY, CA 94923, USA

CRAWFORD, ROBERT M.M.
SCHOOL OF BIOLOGY, THE UNIVERSITY OF ST ANDREWS,
ST ANDREWS KY16 AL, SCOTLAND, UK

DE GRAAF, ROB M.
DEPARTMENT OF EVOLUTIONARY MICROBIOLOGY, FACULTY
OF SCIENCE, RADBOUD UNIVERSITY NIJMEGEN,
HEYENDAALSEWEG 135, 6525AJ NIJMEGEN, THE NETHERLANDS

DOLAN, MICHAEL F.
GEOSCIENCES DEPARTMENT, UNIVERSITY OF MASSACHUSETTS,
AMHERST, MA 01003, USA

EDELMAN-FURSTENBERG, YAEL
GEOLOGICAL SURVEY OF ISRAEL, 30 MALKHE ISRAEL ST.,
JERUSALEM 95501, ISRAEL

EDGCOMB, VIRGINIA P.
GEOLOGY AND GEOPHYSICS DEPARTMENT, WOODS HOLE
OCEANOGRAPHIC INSTITUTION, WOODS HOLE, MA 02543, USA

FENCHEL, TOM
MARINE BIOLOGICAL LABORATORY, UNIVERSITY
OF COPENHAGEN, STRANDPROMENADEN 5, DK-3000
HELSINGØR, DENMARK

FRANKEL, RICHARD B.
DEPARTMENT OF PHYSICS, CALIFORNIA POLYTECHNIC STATE
UNIVERSITY, SAN LUIS OBISPO, CA 93407, USA

FRITZ, GISELA B.
DEPARTMENT OF ZOOLOGY, BIOLOGICAL INSTITUTE,
UNIVERSITY OF STUTTGART, 70569 STUTTGART, GERMANY

GAULKE, MAREN
GEOBIO-CENTER, LUDWIG-MAXIMILIANS-UNIVERSITY, RICHARD-
WAGNER-STR. 10, 80333 MUNICH, GERMANY

GESLIN, EMMANUELLE
LABORATOIRE D'ETUDE DES BIO-INDICATEURS ACTUELS
ET FOSSILES (BIAF) AND LEBIM, UNIVERSITY OF ANGERS,
2 BOULEVARD LAVOISIER, ANGERS CEDEX 49045, FRANCE

GLOCK, NICOLAAS
CHRISTIAN-ALBRECHTS-UNIVERSITY KIEL,
SONDERFORSCHUNGSBEREICH 754, KIEL, GERMANY
LEIBNIZ INSTITUTE OF MARINE SCIENCES,
IFM-GEOMAR, WISCHHOFSTRASSE 1-3, D-24148 KIEL, GERMANY

GOODAY, ANDREW J.
NATIONAL OCEANOGRAPHY CENTRE, SOUTHAMPTON SO14 3ZH, UK

GUTIÉRREZ, DIMITRI
FACULTAD DE CIENCIAS Y FILOSOFÍA, PROGRAMA MAESTRÍA
EN CIENCIAS DEL MAR, UNIVERSIDAD PERUANA CAYETANO
HEREDIA, AV. HONORIO DELGADO 430, LIMA 31, PERU
DIRECCIÓN DE INVESTIGACIONES OCEANOGRÁFICAS,
INSTITUTO DEL, MAR DEL PERÚ (IMARPE), AV. GAMARRA Y
GRAL. VALLE, S/N, CHUCUITO, CALLAO, PERU
JOINT INTERNATIONAL LABORATORY 'DYNAMICS OF THE
HUMBOLDT CURRENT SYSTEM' (LMI DISCOH), LIMA, PERU

HACKSTEIN, JOHANNES H.P.
DEPARTMENT OF EVOLUTIONARY MICROBIOLOGY, FACULTY
OF SCIENCE, RADBOUD UNIVERSITY NIJMEGEN,
HEYENDAALSEWEG 135, 6525AJ NIJMEGEN, THE NETHERLANDS

HEINZ, PETRA
DEPARTMENT FOR GEOSCIENCES, UNIVERSITY OF TÜBINGEN,
HÖLDERLINSTR. 12, TÜBINGEN 72074, GERMANY

HENGHERR, STEFFEN
DEPARTMENT OF ZOOLOGY, BIOLOGICAL INSTITUTE,
UNIVERSITY OF STUTTGART, 70569 STUTTGART, GERMANY

HISS, MARTIN
GEOLOGICAL SURVEY NRW, 1080, 47710 KREFELD, GERMANY

HOBACK, WILLIAM WYATT
DEPARTMENT OF BIOLOGY, UNIVERSITY OF NEBRASKA
AT KEARNEY, 905 WEST 25TH STREET, KEARNEY 68849, NE, USA

HORIKAWA, DAIKI D.
UNIVERSETY PARIS DESCARTES-SITE NECKER, INSERM U 1001,
75751, PARIS CEDEX 15, FRANCE
MEDITERRANEAN INSTITUTE FOR LIFE SCIENCES, 21000 SPLIT,
CROATIA

KOHO, KAROLIINA A.
DEPARTMENT OF EARTH SCIENCES, FACULTY
OF GEOSCIENCES, UTRECHT UNIVERSITY, BUDAPESTLAAN 4,
3584 CD UTRECHT, THE NETHERLANDS

KOLESNIKOVA, ELENA A.
INSTITUTE OF BIOLOGY OF THE SOUTHERN SEAS NASU,
SEVASTOPOL, UKRAINE

KOSHELEVA, TETIANA N.
INSTITUTE OF BIOLOGY OF THE SOUTHERN SEAS NASU,
SEVASTOPOL, UKRAINE

LEFÈVRE, CHRISTOPHER T.
SCHOOL OF LIFE SCIENCES, UNIVERSITY OF NEVADA
AT LAS VEGAS, 4505 MARYLAND PARKWAY, LAS VEGAS, NV
89154-4004, USA

LEGER, MICHELLE M.
DEPARTMENT OF BIOCHEMISTRY AND MOLECULAR BIOLOGY,
CENTRE FOR COMPARATIVE GENOMICS AND EVOLUTIONARY
BIOINFORMATICS, DALHOUSIE UNIVERSITY, HALIFAX, CANADA
B3H 4R2

LEITER, CAROLA
DEPARTMENT FOR EARTH AND ENVIRONMENTAL SCIENCE, LMU
MUNICH, RICHARD-WAGNER-STR. 10, 80333 MUNICH, GERMANY
GEOBIOCENTER LUDWIG-MAXIMILIANS-UNIVERSITY,
RICHARD-WAGNER-STR. 10, 80333 MUNICH, GERMANY

LEVY, ANDREW P.
DEPARTMENT OF ANATOMY, TECHNION SCHOOL OF MEDICINE,
TECHNION INSTITUTE OF TECHNOLOGY, HAIFA, ISRAEL

LEVY, NINA S.
RAPPAPORT INSTITUTE, TECHNION INSTITUTE OF TECHNOLOGY,
HAIFA, ISRAEL

LICHTSCHLAG, ANNA
MAX-PLANCK-INSTITUTE FOR MARINE MICROBIOLOGY,
BREMEN, GERMANY

MALLON, JÜRGEN
CHRISTIAN-ALBRECHTS-UNIVERSITY KIEL,
SONDERFORSCHUNGSBEREICH 754, KIEL, GERMANY
LEIBNIZ INSTITUTE OF MARINE SCIENCES,
IFM-GEOMAR, WISCHHOFSTRASSE 1-3, D-24148 KIEL, GERMANY

MAYR, CHRISTOPH
GEOBIO-CENTERLMU, RICHARD-WAGNER-STR. 10, 80333 MUNICH,
GERMANY AND INSTITUTE FOR GEOGRAPHY, FAU NÜRNBERG/
ERLANGEN, KOCHSTR. 4/4, 91054 ERLANGEN, GERMANY

MAZLUMYAN, SOFIA A.
INSTITUTE OF BIOLOGY OF THE SOUTHERN SEAS NASU,
SEVASTOPOL, UKRAINE

MORALES, MARÍA
LABORATORIO DE MICROPALEONTOLOGÍA, INSTITUTO
GEOLÓGICO MINERO METALÚRGICO (INGEMMET), AV. CANADÁ
1470, LIMA 41, PERU

OREN, AHARON
DEPARTMENT OF PLANT AND ENVIRONMENTAL SCIENCES,
THE INSTITUTE OF LIFE SCIENCES, AND THE MOSHE SHILO
MINERVA CENTER FOR MARINE BIOGEOCHEMISTRY, THE
HEBREW UNIVERSITY OF JERUSALEM, 91904 JERUSALEM, ISRAEL

PIÑA-OCHOA, ELISA
CENTER FOR GEOMICROBIOLOGY, INSTITUTE OF BIOLOGICAL
SCIENCES, AARHUS UNIVERSITY, DK-8000 AARHUS, DENMARK

PFANNKUCHEN, MARTIN
RUĐER BOŠKOVIŠINSTITUTE, ROVINJ, CROATIA

POR, FRANCIS DOV
DEPARTMENT OF EVOLUTION, ECOLOGY AND BEHAVIOUR,
NATIONAL COLLECTIONS OF NATURAL HISTORY, THE HEBREW
UNIVERSITY OF JERUSALEM, GIVAT RAM, 91904 JERUSALEM,
ISRAEL

QUIPÚZCOA, LUIS
DIRECCIÓN DE INVESTIGACIONES OCEANOGRÁFICAS, INSTITUTO
DEL, MAR DEL PERÚ (IMARPE), AV. GAMARRA Y GRAL. VALLE,
S/N, CHUCUITO, CALLAO, PERU

RABALAIS, NANCY N.
LOUISIANA UNIVERSITIES MARINE CONSORTIUM, 8124 HWY. 56, CHAUVIN, LA 70344, USA

RADIC, ANTONIO
DEPARTMENT FOR EARTH AND ENVIRONMENTAL SCIENCE, LMU MUNICH, RICHARD-WAGNER-STR. 10, 80333 MUNICH, GERMANY

RIEDEL, BETTINA
DEPARTMENT OF MARINE BIOLOGY, UNIVERSITY OF VIENNA, ALTHANSTRASSE 14, 1090 VIENNA, AUSTRIA

ROGER, ANDREW J.
DEPARTMENT OF BIOCHEMISTRY AND MOLECULAR BIOLOGY, CENTRE FOR COMPARATIVE GENOMICS AND EVOLUTIONARY BIOINFORMATICS, DALHOUSIE UNIVERSITY, HALIFAX, CANADA B3H 4R2

SACCÀ, ALESSANDRO
DEPARTMENT OF ANIMAL BIOLOGY AND MARINE ECOLOGY, UNIVERSITY OF MESSINA, VIALE FERDINANDO STAGNO D'ALCONTRES 33, 98166 MESSINA, ITALY

SCHIEBER, JÜRGEN
DEPARTMENT OF GEOLOGICAL SCIENCES, INDIANA UNIVERSITY, 1001 E. 10TH STR., BLOOMINGTON, IN 47405, USA

SCHÖNFELD, JOACHIM
LEIBNIZ INSTITUTE OF MARINE SCIENCES, IFM-GEOMAR, WISCHHOFSTRASSE 1-3, D-24148 KIEL, GERMANY

SECKBACH, JOSEPH
HEBREW UNIVERSITY OF JERUSALEM, HOME ADDRESS: MEVO HADAS 20, P.O. BOX 1132, EFRAT 90435, ISRAEL

SERGEEVA, NELLI G.
INSTITUTE OF BIOLOGY OF THE SOUTHERN SEAS NANU, SEVASTOPOL, UKRAINE

SIFEDDINE, ABDELFETTAH
CENTRE IRD FRANCE-NORD, LOCEAN, UMR 7159, 32 AVENUE HENRI VARAGNAT, 93143 BONDY CEDEX, FRANCE AND DEPARTAMENTO DE GEOQUIMICA, UNIVERSIDADE FEDERAL FLUMINENSE, LMI PALEOTRACES, NITEROI, RJ, BRAZIL

LIST OF AUTHORS AND THEIR ADDRESSES

STACHOWITSCH, MICHAEL
DEPARTMENT OF MARINE BIOLOGY, UNIVERSITY OF VIENNA, ALTHANSTRASSE 14, 1090 VIENNA, AUSTRIA

STAIRS, COURTNEY A.W.
DEPARTMENT OF BIOCHEMISTRY AND MOLECULAR BIOLOGY, CENTRE FOR COMPARATIVE GENOMICS AND EVOLUTIONARY BIOINFORMATICS, DALHOUSIE UNIVERSITY, HALIFAX, CANADA B3H 4R2

STOECK, THORSTEN
DEPARTMENT OF ECOLOGY, UNIVERSITY OF KAISERSLAUTERN, ERWIN SCHROEDINGER STR. 14, KAISERSLAUTERN D-67663, GERMANY

STROHMEIER, STEPHAN
DEPARTMENT OF ZOOLOGY, BIOLOGICAL INSTITUTE, UNIVERSITY OF STUTTGART, 70569 STUTTGART, GERMANY

STRUCK, ULRICH
GEOBIO-CENTERLMU, RICHARD-WAGNER-STR. 10, 80333 MUNICH, GERMANY AND MUSEUM FÜR NATURKUNDE, LEIBNIZ-INSTITUT FÜR EVOLUTIONS- UND BIODIVERSITÄTSFORSCHUNG, HUMBOLDT UNIVERSITÄT ZU BERLIN, INVALIDENSTRAßE 43, 10115 BERLIN, GERMANY

TREUDE, TINA
DEPARTMENT OF MARINE BIOGEOCHEMISTRY, LEIBNIZ INSTITUTE OF MARINE SCIENCES (IFM-GEOMAR), 24148 KIEL, GERMANY

TSAOUSIS, ANASTASIOS D.
DEPARTMENT OF BIOCHEMISTRY AND MOLECULAR BIOLOGY, CENTRE FOR COMPARATIVE GENOMICS AND EVOLUTIONARY BIOINFORMATICS, DALHOUSIE UNIVERSITY, HALIFAX, CANADA B3H B3H 4R2

ZUSCHIN, MARTIN
DEPARTMENT OF PALEONTOLOGY, UNIVERSITY OF VIENNA, ALTHANSTRASSE 14, 1090 VIENNA, AUSTRIA

LIST OF EXTERNAL REVIEWERS AND REFEREES

BARNHART, MILES CHRISTOPHER
DEPARTMENT OF BIOLOGY, MISSOURI STATE UNIVERSITY,
SPRINGFIELD, MO 65897, USA

BONSDORFF, ERIK
DEPT. OF BIOSCIENCES, ENVIRONMENTAL AND MARINE BIOLOGY,
ÅBO AKADEMI UNIVERSITY, BIOCITY, FI-20520 TURKU/ÅBO,
FINLAND

BRUST, MATHEW L.
DEPARTMENT OF BIOLOGY, CHADRON STATE COLLEGE,
1000 MAIN STREET, CHADRON, NE 69337, 308-432-6446, USA

BUNGE, JOHN A.
DEPT. STATISTICAL SCIENCES, CORNELL UNIVERSITY,
COMSTOCK HALL-ACADEMIC II, ROOM 1198, ITHACA, USA

DE TULLIO, MARIO C.
DIPART. DI BIOLOGIA E PATOLOGIA VEGETALE,
UNIVERSITÀ DI BARI, VIA E, ORABONA 4, BARI, I-70125, ITALY

DYER, BETSEY D.
WHEATON COLLEGE, 26 EAST MAIN STREET, NORTON,
MA 02766–2322, USA

FRENZEL, PETER
FRIEDRICH SCHILLER UNIVERSITY OF JENA, INSTITUTE
OF EARTH SCIENCES, BURGWEG 11, D-07749 JENA, GERMANY

VAN DER GIEZEN, MARK
CENTRE FOR EUKARYOTIC EVOLUTIONARY MICROBIOLOGY,
COLLEGE OF LIFE AND ENVIRONMENTAL SCIENCES,
UNIVERSITY OF EXETER, STOCKER ROAD, EXETER, EX4 4QD,
UNITED KINGDOM

HASEMANN, CHRISTIANE
DEEP SEA ECOLOGY AND TECHNOLOGY, ALFRED WEGENER INSTITUTE, AM HANDELSHAFEN 12, D-27570, BREMERHAVEN, GERMANY

HEIDELBERG, KARLA
DEPT. OF BIOLOGICAL SCIENCES, UNIVERSITY OF SOUTHERN CALIFORNIA, PO BOX 5069 AVALON, CA 90704, USA

HROMIC, TATIANA
INSTITUTO DE LA PATAGONIA, UNIVERSIDAD DE MAGALLANES, CASILLA 113-D, PUNTA ARENAS, CHILE

KORMAS, KONSTANTINOS AR.
AQUATIC MICROBIAL ECOLOGY, DEPARTMENT OF ICHTHYOLOGY & AQUATIC ENVIRONMENT, UNIVERSITY OF THESSALY, 384 46 NEA IONIA, GREECE

KUHNT, WOLFGANG
MARINE MICROPALEONTOLOGY, CHRISTINA-ALBERTINA-UNIVERSITY, LUDEWIG-MEYN-STR. 14, D-24118 KIEL, GERMANY

LEADBETTER, JARED
CALIFORNIA INSTITUTE OF TECHNOLOGY, ENVIRONMENTAL SCIENCE & ENGINEERING, 137 KECK LABORATORIES, MAILCODE 138–78, PASADENA, CA 91125–7800, USA

LOPEZ-GARCIA, PURIFICATION
UNITÉ D'ECOLOGIE, SYSTÉMATIQUE ET EVOLUTION, UMR CNRS 8079, BÂTIMENT 360, UNIVERSITE PARIS-SUD, 91405 ORSAY CEDEX, FRANCE

MCLENNAN, ALEXANDER
INSTITUTE OF INTEGRATIVE BIOLOGY, CELL REGULATION AND SIGNALLING DIVISION, UNIVERSITY OF LIVERPOOL, LIVERPOOL L69 7ZB, UNITED KINGDOM

MEYSMAN, FILIP
NETHERLANDS INSTITUTE OF ECOLOGY, CENTRE FOR ESTUARINE AND MARINE ECOLOGY, KORRINGAWEG 7, 4401 NT YERSEKE, THE NETHERLANDS

LIST OF EXTERNAL REVIEWERS AND REFEREES

NIELSEN, LARS PETER
DEPT. OF BIOLOGICAL SCIENCES, AARHUS UNIVERSITY,
BYNING 1540, NY MUNKEGADE 116, 8000 ARHUS, DENMARK

PAWLOWSKI, JAN
DEPT. OF ZOOLOGY AND ANIMAL BIOLOGY, UNIVERSITY
OF GENEVA, CH-1211 GENEVA 4, SWITZERLAND

POLLEHNE, FALK
BIOLOGICAL OCEANOGRAPHY, LEIBNIZ INSTITUTE FOR BALTIC
SEA RESEARCH, SEESTRASSE 15, D-18119 ROSTOCK-WARNEMÜNDE,
GERMANY

SCHÜLER, DIRK
DEPT. BIOLOGY I, LUDWIG-MAXIMILIANS-UNIVERSITÄT
MÜNCHEN, LMU BIOZENTRUM ZI. E 01.028, GROSSHADERNER STR.
2, 82152 PLANEGG-MARTINSRIED, GERMANY

SCHWEIKERT, MICHAEL
BIOL. INSTITUTE, UNIVERSITÄT STUTTGART, PFAFFENWALDRING 57,
70550 STUTTGART, GERMANY

SIMPSON, ALASTAIR
DEPT. OF BIOLOGY, DALHOUSIE UNIVERSITY, HALIFAX,
NOVA SCOTIA, CANADA B3H 4J1

TIELENS, LOUIS
DEPT. OF ORGANIC CHEMISTRY, RIJKSUNIVERSITEIT VAN
UTRECHT, CROESESTRAAT 79, UTRECHT, THE NETHERLANDS

WETZEL, MARKUS
INSTITUT FÜR INTEGRIERTE NATURWISSENSCHAFTEN,
ABTEILUNG BIOLOGIE, UNIVERSITY KOBLENZ-LANDAU,
UNIVERSITÄTSSTRASSE 1, 56070 KOBLENZ, GERMANY

ACKNOWLEDGMENTS

We heartily thank all those involved with this book. Of course, the book would not be possible without the dedicated efforts of our expert authors. We explicitly selected a mixture of World authorities and beginning investigators to be our contributors because we feel that beginning investigators and students will be the leaders in the field in the years to come. Also, we selected contributors with differing opinions about the topics at hand. In this way, we feel the book reveals the spectrum of present activities in the field, presents new and emerging insights, and provides a sound foundation for the reader.

However, aside from these expert authors, we must also thank our collaborators who suggested potential author names; their suggestions improved the breadth and scope of this volume. In addition, the experienced reviewers, peers, and externals are also heartily thanked. Their efforts, sometimes within inconvenient time constraints, often helped to clarify passages, aiding many authors.

Finally, we thank those technicians, laboratory helpers, and administrative professionals who may not be named as authors because science often requires the collaboration of many, sometimes too many, to name formally. And, of course, we thank our families for their patience in our undertaking of this effort. Their support and encouragement made the project more enjoyable.

Alex Altenbach, Joan Bernhard & Joseph Seckbach (eds.)

PART I:
GENERAL INTRODUCTION

Fenchel
Treude
Oren

Biodata of **Tom Fenchel**, author of "*Anaerobic Eukaryotes.*"

Tom Fenchel is an Emeritus Professor of Marine Biology at the University of Copenhagen. He received his Ph.D. and his D.Sc. from the University of Copenhagen in 1964 and 1969, respectively. In the period from 1970 to 1987, he was a Full Professor at the University of Aarhus of zoology and ecology and then became Professor and Director of the Marine Biological Laboratory, University of Copenhagen 1987–2010. His research has included the ecology and physiology of microorganisms and evolutionary biology.

E-mail: **tfenchel@bio.ku.dk**

ANAEROBIC EUKARYOTES

TOM FENCHEL
Marine Biological Laboratory, University of Copenhagen,
Strandpromenaden 5, DK-3000 Helsingør, Denmark

1. Anoxia: An Extreme Environment?

Among the prokaryotes, anoxic habitats can hardly be considered an "extreme environment" – many major groups such as the archaebacterial methanogens, various types of sulfate reducers, clostridia, certain phototrophic bacteria, etc., are all obligate anaerobes with a variable tolerance to the presence of low O_2 tensions. It is generally believed that life arose in an anoxic world. During perhaps the first two billion years of life's history, only prokaryotes existed, and during that period, probably most major groups of prokaryotes diverged. Some prokaryotes were also the first organisms to apply free O_2 in their energy metabolism. The first eukaryotes arose maybe some two billion years ago in a world where organisms with oxygenic photosynthesis in the form of cyanobacteria had already evolved. While the atmospheric pO_2 may have remained low for a long period of time and the bulk of the oceans may have remained anaerobic throughout most of the Proterozoic (Canfield et al. 2000), patches of high O_2 concentrations around cyanobacterial mats and in surface waters with unicellular planktonic cyanobacteria exposed to light are likely to have existed much earlier (Fenchel and Finlay 1995).

As discussed below, evidence suggests that eukaryotes arose as aerobic organisms. Patches of anaerobic habitats have always existed and still occur in the biosphere such as stratified water columns, aquatic sediments below a certain depth, and the intestine of larger animals – in a sense, these represent "extreme" conditions to which some eukaryotes have adapted.

2. Anaerobic Multicellular Eukaryotes

It is well established that many aquatic and parasitic animals live for longer or shorter periods under strict anaerobic conditions. Intertidal animals such as bivalves and barnacles must, at low tide, depend on anaerobic energy metabolism as may some burrowing shallow-water invertebrates if the water column immediately above the sediment surface becomes anoxic. Likewise, many freshwater animals can sustain life in anaerobic water for longer or shorter periods, e.g., under ice cover or during periodic events of anoxia in the deeper parts of stratified lakes. Certain intestinal parasites depend on anaerobic energy metabolism throughout most of their life cycle, e.g., the nematode *Ascaris*. Likewise, there are many reports on the presence of representatives of the meiofauna including nematodes, gastrotrichs, and gnathostomulids in anaerobic and sulfidic sediments (Bryant 1991; Fenchel and Finlay 1995).

There are also reports in the literature suggesting the existence of animals that maintain their entire life cycle under anaerobic conditions including certain oligochaetes and nematodes. Much of this literature seems anecdotal or relies on experiments in which a complete removal of O_2 can be questioned – small aerobic organisms can maintain aerobic metabolism even under a very low pO_2 (for discussion, see Fenchel and Finlay 1995). More recently, Danovaro et al. (2010) presented evidence to show that a representative of the meiofaunal group Loricifera lives in a deep permanently anaerobic and sulfidic deep water basin in the Mediterranean Sea. It was also suggested that its mitochondria, which lack cristae, are hydrogenosomes (see below), but no strong evidence for this has been provided.

The general consensus is that while many metazoans can manage for periods or even permanently on the basis of anaerobic energy metabolism, so far no animal has been found without the capacity of oxidative phosphorylation. All metazoans also have an absolute requirement for exposure to oxygen at least during part of their lifetime. This is due to the absolute requirement of O_2 for synthetic pathways leading to sterols, collagen, and quinone tanning (Barrett 1991).

The energy metabolism of animals under anaerobic conditions is fermentative and depends either partly or exclusively on the glycolytic pathway. The resulting metabolites are lactate and other volatile organic acids, ethanol, or, in the case of co-metabolism with amino acids, so-called opines. Increased energetic yield may be obtained from fumarate that is reduced to succinate and further transformed into propionate as the major metabolite. This process takes place in mitochondria and involves parts of the TCA cycle and the cytochrome chain (Bryant 1991).

Eukaryotes that are capable of oxygenic photosynthesis (plants, various types of multi- or unicellular algae) have the ability of oxidative phosphorylation, but some unicellular forms (*Chlamydomonas*) are capable of sustaining life heterotrophically in the absence of oxygen (Mus et al. 2007). By far, most fungi are aerobes; exception are some chytridiomycete fungi (see below).

3. Anaerobic Unicellular Eukaryotes

Obligate or facultative anaerobes occur in many protist groups including several taxa among the flagellates, ciliates, amoebae, and foraminifera, and they occur in many different sorts of anoxic habitats: stratified lakes and marine environments, aquatic sediments and anaerobic sewage treatment plants, or they live as intestinal parasites and commensals. Some major taxa – e.g., the diplomonad and trichomonad flagellates and the pelobiont amoebas – exclusively include obligate anaerobes, but in most cases, the anaerobic species (or genera, families) are phylogenetically embedded in higher level groups that include aerobic forms.

3.1. MITOCHONDRIA, HYDROGENOSOMES, AND MITOSOMES

For a long time, it was believed that some groups of anaerobic protists including the free-living amoeboid pelebionts, the parasitic amoeba *Entamoeba*, and representatives of free-living or parasitic flagellates (diplomonads, retortamonads) were examples of primarily amitochondriate eukaryotes, i.e., they were considered as relics of primitive eukaryotes from before they acquired mitochondria. This view was supported by the fact that some of these groups appear to branch off early in phylogenetic tree of the eukaryotes based on sequencing of rRNA genes.

It is now recognized that all extant eukaryotes possess either mitochondria or organelles derived from mitochondria whether anaerobes or not (Hjort et al. 2010). Two mitochondria-like organelles have been named hydrogenosomes and mitosomes, respectively. To these can be added that some obligate anaerobic protozoa have mitochondria-like organelles, but these do neither have cytochrome oxidase and usually no cristae, and they do not generate hydrogen. Presumably, they function like mitochondria in some animals, that is, generating ATP through fermentation processes in mitochondria in which parts of the TCA cycle are used to ferment malate into propionate.

All mitochondria and mitochondria-derived organelles are characterized by being surrounded by a double membrane, but cristae are usually absent in hydrogenosomes and always absent in mitosomes. They also all seem to carry out some mitochondrial functions that are not directly related to energy metabolism such as Ca^{2+} storage and the synthesis of Fe/S proteins (Biagini et al. 1997; Hjort et al. 2010).

Hydrogenosomes were first discovered and characterized by Lindmark and Müller (1973) in the parasitic flagellates *Trichomonas*; since then, they have been found in anaerobic chytridiomycetes, in the free-living heterolobosid flagellate *Psalteriomonas*, in the *Trichomonas*-related hypermastigidid flagellates, and especially in a variety of free-living and commensal ciliates belonging to different taxa (Fenchel and Finlay 1995; Müller 1993).

It was first suggested that hydrogenosomes are modified mitochondria on the basis of structural evidence (Cavalier-Smith 1987; Finlay and Fenchel 1989). Also, the two types of organelles share much of their biochemistry and a number of functional roles of typical mitochondrial functions (Hjort et al. 2010). At least one type of studied hydrogenosomes possesses a mitochondria-like chromosome, but in other cases it seems to have been lost, and all necessary proteins are coded by nuclear genes (Hjort et al. 2010). However, sequencing the chromosome of the anaerobic ciliate *Nyctotherus* (Boxma et al. 2005) gave support for the mitochondrial origin of hydrogenosomes (see Hjort et al. 2010 and Hackstein et al. 2006). The fact that hydrogenosomes apparently have evolved independently within a variety of ciliate taxa (Embley and Finlay 1994) also provides evidence that hydrogenosomes are modified mitochondria and that they have evolved independently within several different lineages. Hydrogenosomes do not include cytochrome oxidase. It remains unresolved exactly how these organelles obtained the enzymes ferredoxin oxidoreductase and hydrogenase (Hjort et al. 2010).

Hydrogenosomes can be identified from the fact that the organisms in question produce H_2 as a metabolite – or CH_4 in the case when organisms harbor methanogenic symbionts – and by the cytochemical demonstration of hydrogenase (Fenchel and Finlay 1991a; Zwart et al. 1988).

The primary function of hydrogenosomes is that malate or pyruvate deriving from the glycolytic pathways is imported into the organelle; here pyruvate is fermented in a clostridium-like manner with the principle end products being acetate + H_2. The key mechanism is that pyruvate reduces ferredoxin catalyzed by the enzyme ferredoxin oxidoreductase. Reduced ferredoxin then reduces protons to hydrogen catalyzed by a hydrogenase.

ATP is generated through substrate-level phosphorylation, but some electron transport phosphorylation may also be possible in that an electrochemical gradient over the membrane of functioning hydrogenosomes has been demonstrated. This type of fermentation yields about twice the amount of ATP relative to the glycolytic pathway alone per unit of dissimilated glucose. There are some deviations in the fermentative pathways when hydrogenosomes from different organisms are compared – which could be expected given that these organelles have evolved independently from mitochondria at several occasions.

Mitosomes do not play any role in ATP generation. Mitosomes are considerably smaller than typical mitochondria or hydrogenosomes. Mitosomes are found in some anaerobic flagellates (*Giardia*) and amoebas (*Entamoeba*). In these organisms, the energy metabolism is cytosolic and is based on fermentative processes yielding different low molecular organic acids and ethanol, but not H_2. However, it has recently been found that some diplomonads actually produce H_2 by some not entirely clarified mechanism (Lloyd et al. 2002; Millet et al. 2010). Mitosomes like hydrogenosomes still function like mitochondria in terms of synthesis of Fe/S clusters and storage of Ca^{2+} (Biagini et al. 1997; Hjort et al. 2010).

3.2. FACULTATIVE ANAEROBIC PROTISTS

Like many bacteria, there are also protists that are capable of sustained balanced growth under aerobic conditions on the basis of oxidative phosphorylation as well as under strict anaerobic conditions. This has especially been shown for some marine ciliates (Bernard and Fenchel 1996). The growth rate constant as well as growth yield was significantly lowered under anaerobic conditions. These forms possess cytochrome oxidase, and studies of their motile chemosensory behavior showed a preference for microaerobic conditions. These organisms could also sustain growth rates comparable to anaerobic growth when exposed to up to 0.5 mM CN^- and, in some cases, up to 5 mM HS^-. Presumably, these organisms depend on some sort of fermentative energy metabolism in the absence of O_2 (or poisoned with CN^- or HS^-), but this was not investigated. A survey of heterotrophic flagellates in marine anaerobic and sulfidic sediments showed a mixture of established anaerobes as well as species that are well known from aerobic marine habitats (Bernard et al. 2000).

Different prokaryotes are capable of using an array of different inorganic electron acceptors (such as oxidized N, S, Fe, and Mn compounds) in respiratory processes, either exclusively or as an alternative to O_2. Among protists, nitrate respiration (denitrification) was first discovered in the freshwater ciliate *Loxodes* (Finlay et al. 1983; Finlay 1985). Under microaerobic conditions it respires oxygen, but under anaerobic conditions it instead reduces NO_3^- to NO_2^-, a process that probably takes place in the mitochondria and involves the electron carrier chain. It has previously been shown that many benthic foraminifera occur in and on anoxic marine sediments (Bernhard 1989; Bernhard and Reimers 1991). Recently, it was shown that a variety of marine benthic foraminifera and gromiids store nitrate, and under anaerobic conditions, they reduce this in a respiratory process. All these foraminifera are capable of oxygen respiration when O_2 is available, and the study indicated that the process may be of quantitative significance in terms of denitrification in marine sediments (Piña-Ochoa et al. 2010).

It is plausible that denitrification is a more widespread phenomenon among the many protists that occur in the microaerobic chemocline of aquatic sediments or in the stratified water column.

3.3. COPING WITH ANOXIA USING ENDOSYMBIOTIC PHOTOTROPHS

Many protists, especially ciliates, harbor unicellular algae as endosymbionts. Among freshwater forms, the symbionts are mostly the green alga *Chlorella*, and among marine forms, they are usually dinoflagellates or cryptomonads. Interest in this type of symbiosis has mainly centered on the transfer of photosynthate in the form of carbohydrates from symbiont to the host cell. However, there are also several examples showing that such symbiont-carrying ciliates can use the O_2 production of the symbionts for respiration and thus can exploit anaerobic habitats in the light (Esteban et al. 2010).

This reliance of endosymbiotic phototrophs plays a role especially for stratified lakes and ponds in which the zone beneath the chemocline is still exposed to light. Here, some common ciliates that harbor symbiotic *Chlorella* such as species of *Euplotes*, *Frontonia*, and *Loxodes* exploit this.

4. Sensitivity to Oxygen

Like in the case of anaerobic prokaryotes, the sensitivity to oxygen exposure varies among eukaryotic anaerobes. Some are killed or inactivated relatively rapidly at O_2 tensions exceeding about 2% atm. saturation while some show extended survival at even higher O_2 tensions. In some anaerobes, growth is enhanced when exposed to microaerobic conditions – this probably means that O_2 can act as an electron sink in conjunction with a fermentative pathway (Paget and Lloyd 1990). In general, the sensitivity of the individual species is an expression of the energetic cost of detoxification mechanisms and the probability that the cells become exposed to O_2 in their natural habitats.

Several studied species of anaerobic ciliates all showed a variable negative response to O_2 tensions exceeding about 2% atm. sat. The cells seem to possess superoxide dismutase, but they lack catalase. An additional protection mechanism against O_2 toxicity is a sort of O_2 respiration that allows the interior of the cells to remain anaerobic. This O_2 uptake is not coupled to ATP generation. The O_2 uptake can be measured directly and is of the same magnitude as that of an aerobic protist of similar size. That this mechanism works can be seen from the fact that the endosymbiotic methanogens remain active up to an ambient O_2 tension of 1–2% atm. sat. The ciliates also show a chemosensory motile response to O_2 and always accumulate wherever the O_2 tension is lowest (Fenchel and Finlay 1990b). Parasitic anaerobic flagellates and amoebas appear to be considerably more tolerant to O_2 – presumably because they will be exposed to oxygen when transferring from one host to another (e.g., Williams and Lloyd 1993).

5. Sterol Synthesis

It was previously mentioned that one reason that there exist no obligate anaerobic animals is that – in contrast to prokaryotes – they depend on O_2 for certain synthetic processes. Among them, the problem of sterol synthesis is relevant in the present context since sterols constitute an essential component of the eukaryote cell membrane.

In natural habitats and in most applied culture methods, there are probably available sterols deriving from the lysis of other eukaryotic cells. There is evidence to show that monoxenic culture of an anaerobic ciliate fed with bacteria in a chemically defined medium required the addition of sterols to the medium. It has, however, been found that some ciliates are capable of synthesizing a certain cell membrane sterol, tetrahymenol, that is synthesized without access to free O_2 (Wagener and Pfennig 1987). This aspect warrants further studies.

6. Prokaryote Symbionts

Endo- as well as ectosymbiotic prokaryotes represent a common phenomenon among protists, but are especially common among anaerobic protists (Fenchel et al. 1977). The consistent presence of large numbers of bacterial cells in or on anaerobic ciliates indicated that there is some sort of relationship between the presence of symbiotic bacteria and the anaerobic life of the host. When the host has hydrogenosomes, the functional significance is probably always one of syntrophic hydrogen transfer. Fermentation involving H_2 generation is sensitive to the ambient H_2 tension. It is a well-established fact that fermenting bacteria that generate H_2 often are associated with other bacteria that can consume H_2 under anaerobic conditions such as methanogens and sulfate reducers. A form of mutualistic relationship is established – so-called syntrophic hydrogen transfer. The fermentative reactions will become less energetically efficient and end products, including more reduced metabolites than acetate, will be produced when the ambient H_2 tension increases (Fenchel et al. 1998).

Harboring H_2-consuming symbionts will then result in the maintenance of a lower intracellular H_2 tension. However, some protists that apparently do not excrete H_2 also harbor intense lawns of symbiotic bacteria on their cell surface (Fenchel and Finlay 1992). Three types of such symbionts have been established.

There are methanogens that occur (with one exception, see below) exclusively as endosymbionts in H_2-producing protists. Sulfate reducers have been identified as ectosymbionts of two hydrogenosome-containing marine anaerobic ciliates, but ectosymbiotic bacteria are widespread among several other anaerobic marine ciliates but absent in their freshwater counterparts – suggesting that these bacterial symbionts are also sulfate reducers. Finally, one example of endosymbiotic purple non-sulfur bacterium is known.

6.1. METHANOGENS

Van Bruggen et al. (1983) were the first to identify the endosymbiotic bacteria of some ciliates as methanogens based on the in vivo fluorescence in violet light caused by the coenzyme F_{420} characteristic of methanogens. Practically all obligate anaerobic freshwater ciliates have endosymbiotic methanogens, and marine anaerobic ciliates often also harbor methanogens.

The number of methanogens per cell is approximately similar to the number of hydrogenosomes and can vary between several hundreds to several thousands per cell according to the ciliate species. Often the bacteria and the organelles are juxtaposed or together form clusters in accordance with the need to minimize diffusional limitation in the transfer of H_2 from the hydrogenosomes to the methanogens. In one species, *Plagiopyla frontata*, the methanogens and the hydrogenosomes are both disk shaped and arranged like rolls of coins with alternating positions of bacteria and hydrogenosomes. This arrangement appears almost like a separate organelle. The hydrogenosomes and the methanogens undergo division simultaneously with the cell division of the host without losing the integrity of the complexes (Fenchel and Finlay 1991c). In some species, the methanogens are polymorphic and undergo a transformation in which the cell wall appears to degrade, and the consequently irregularly shaped cells then attach to hydrogenosomes (Finlay and Fenchel 1991).

The endosymbiotic methanogens can be inactivated by bromoethane sulfonic acid, and so aposymbiotic ciliates can be made. These are viable, but their growth rate constant and cell yield are reduced by about 30% relative to the symbiont-containing cells (Fenchel and Finlay 1991b). The conversion of H_2 to CH_4 is almost complete in normal cells, but aposymbiotic cells excrete H_2. That H_2 production is partly inhibited in the absence of methanogenic activity could be shown by comparing the H_2 production of aposymbiotic cells with the CH_4 production of the normal cells and the fact that in H_2/CO_2 methanogenesis, it takes four H_2 to produce one CH_4. Calculations also showed that in a ciliate cell, outward diffusion of H_2 would not be able to prevent the buildup of an inhibitory level of pH_2 inside the ciliate host at typical rates of H_2 production unless the H_2 was consumed by H_2-scavenging endosymbionts (Fenchel and Finlay 1992).

Molecular analysis has shown that each ciliate species harbor their own specific species of methanogens and that they cannot be identified with any free-living genotype. The fact that symbiotic methanogens belong to different groups of methanogens even within related anaerobic ciliates indicates that the evolution of endosymbiotic methanogens in anaerobic ciliates has taken place independently many times during geological time (Embley and Finlay 1994).

Anaerobic ciliates are not the only anaerobic protists with endosymbiotic methanogens. They also occur, e.g., in the excavate flagellate *Psalteriomonas* and in the hypermastigine flagellates of the termite intestine. The most intriguing case is the anaerobic freshwater amoebas *Pelomyxa* and *Mastigella* that live in the bottom ooze of lakes and ponds. They do not possess hydrogenosomes, but still they harbor methanogenic bacteria. *Pelomyxa* also harbors another type of bacterium – its physiology is unknown, but it is not a methanogen. It has been suggested that this bacterium is responsible for H_2 production, but solid evidence for this is not present, and more detailed studies on this organism are required. Unfortunately, it is difficult to grow *Pelomyxa* under controlled conditions (van Bruggen et al. 1985, 1988).

6.2. ECTOSYMBIOTIC BACTERIA

As mentioned previously, marine anaerobic ciliates are often covered by a dense layer of symbiotic bacteria. In two cases (the marine anaerobic ciliates, *Metopus contortus* and *Caenomorpha levanderi*), it was possible to demonstrate, using FISH, that these bacteria are sulfate reducers. This stands to reason as they would be able to maintain a low intracellular pH_2 in the ciliates in a similar way as do methanogens (Fenchel and Ramsing 1992). This may also apply in some other cases. But many anaerobic marine ciliates that do not have hydrogenosomes also show a spectacular cover of ectosymbiotic bacteria such as species of *Parablepharsma* and *Sonderia*. The nature of this relationship remains to be understood.

Ectosymbiotic bacteria also characterize the euglenoid flagellate *Postgaardi* and its relatives. They occur in marine anaerobic sediments and have mainly been found at greater depths (Buck and Bernhard 2002; Simpson et al. 1996; Yubuki et al. 2009). The functional significance of this association is also not known.

6.3. PHOTOTROPHIC ENDOSYMBIONTS

Only one example is known. The oligotrich ciliate, *Strombidium purpureum*, occurs among masses of purple sulfur in shallow marine habitats. It harbors many cells that appear to be a non-sulfur purple bacterium – morphologically, it closely resembles *Rhodopseudomonas*. The ciliate can be grown strictly anaerobically under constant light exposure; it shows photosensory motile behavior, and the action spectrum of the motile response is identical to the absorption spectrum of bacteriochlorophyll *a*. In darkness it becomes a microaerobe that prefers an O_2 tension of about 4% atm. sat. So that it is the symbiotic bacteria that determine the motile behavior of their host cell (Fenchel and Bernard 1993a, b).

7. Energy Metabolism and the Structure of Anaerobic Communities

One general aspect of anaerobic energy metabolism is that it is much less efficient than aerobic energy metabolism. In anaerobic eukaryotes – excluding denitrification – this means different types of fermentative pathways and will yield 2–4 mol of ATP per mole dissimilated glucose. In contrast, oxidative phosphorylation yields 32 mol of ATP per mole glucose.

The consequence of this is that aerobes have maximum growth rate constants and cell yields that are about four times higher than for similarly sized anaerobes. This result can be obtained both through theoretical considerations and empirical results (Fenchel and Finlay 1990a, 1995).

Anaerobic microbial communities are dominated by prokaryotes, and the biotic interactions that structure the communities are mainly competition for common resources and syntrophic interactions. Predation is limited to the present anaerobic protists that are almost exclusively feeding on bacteria although there may be a few ciliates that also feed on other smaller protists such as heterotrophic flagellates. In contrast, in aerobic microbial communities, predation plays a much larger role and includes not only protists feeding on bacteria but also longer food chains that include several trophic levels. Furthermore, it could be expected that the biomass ratios between bacteria and their protozoan predators are much higher than is the case for aerobic microbial communities.

Consequently, anaerobic habitats offer a limited scope for phagotrophic organisms, and phagotrophy is a fundamental property of eukaryotes. First, the advent of O_2 in the biosphere and of oxidative phosphorylation provided conditions that allowed for biotic communities characterized by complex food webs, longer food chains, and eventually the evolution of larger organisms.

8. The Origin of Anaerobic Eukaryotes

We have seen that many groups of protists have evolved representatives with an anaerobic lifestyle. Most of these anaerobes are embedded in groups that mostly include aerobes – this is, in particular, evident among the ciliates: most ciliate orders are dominated by aerobes – or are related to other orders dominated by aerobic species. Molecular studies also show that the evolution of hydrogenosomes has taken place independently several times. Some anaerobic species are congeners with aerobic species, e.g., *Cyclidium porcatum* that has hydrogenosomes and methanogenic symbionts (Esteban et al. 1993).

A few protist groups include only anaerobes – such as the diplomonad flagellates and the pelebiont amoeba. Among them some appear to branch off early in the phylogenetic tree and were for some time considered Precambrian relicts. Among them the mentioned pelebionts were included, but it is now known that they represent a branch on the phylogenetic tree for amoebae (Edgcomb et al. 2002; Minge et al. 2009). The microsporidians were for some time considered to be such very primitive representatives of Precambrian origin – we know now that they derive from fungi that simply have adapted to life as intracellular parasites of other eukaryotes. The existence of facultative anaerobes among different protists – that is, they can undergo balanced growth under both aerobic and anaerobic conditions – also indicates that the transformation from aerobic to anaerobic life is not so complicated for at least some types of unicellular eukaryotes. It is also clear that many examples of anaerobic intestinal parasites of animals have secondarily evolved from aerobic ancestors, e.g., the ciliates of the rumen and the intestine of some other mammals – they would seem to have evolved as recently as the Tertiary.

Some still subscribe to the idea that some extant anaerobic protists somehow represent a relict of the first anaerobic life (e.g., Mentel and Martin 2008). But all extant anaerobic eukaryotes are characterized by having "degenerate mitochondria" (van der Giezen and Tovar 2005). Apparently, all extant eukaryotes – anaerobes as well as aerobes – harbor mitochondria or organelles that are somehow derived from mitochondria.

Many problems remain unresolved with respect to the origin of the eukaryotic cell from some sort of prokaryote progenitor, and it is a topic that is characterized by speculative ideas. One of the more established aspects is that mitochondria (and related organelles) derive from an endosymbiotic alpha proteobacterium. It is also likely that aerobic life is a fundamental property of eukaryotes that evolved very early and that we have not (yet) found an extant organism that can throw light on the properties of an anaerobic pre-eukaryote progenitor.

9. References

Barrett J (1991) Parasitic helminths. In: Bryant C (ed) Metazoan life without oxygen. Chapman & Hall, London, pp 146–164

Bernard C, Fenchel T (1996) Some microaerophilic ciliates are facultative anaerobes. Eur J Prostitol 32:293–297

Bernard C, Simpson AGB, Patterson DJ (2000) Some free-living flagellates (Protista) from anoxic habitats. Ophelia 52:113–142

Bernhard JM (1989) The distribution of benthic Foraminifera with respect to oxygen concentration and organic carbon levels in shallow-water Antarctic sediments. Limnol Oceanogr 34:1131–1141

Bernhard JM, Reimers CE (1991) Benthic foraminiferal populations related to anoxia: Santa Barbara Basin. Biogeochemistry 15:127–149

Biagini GA, van der Giezen M, Hill B et al (1997) Ca^{2+} accumulation in the hydrogenosomes of *Neocallimastix frontalis* L2: a mitochondrial-like physiological role. FEMS Microbiol Lett 149:227–232

Boxma B, Graf RM, van der Staay G et al (2005) An anaerobic mitochondrion that produces hydrogen. Nature 434:74–97

Bryant C (ed) (1991) Metazoan life without oxygen. Chapman & Hall, London

Buck KR, Bernhard JM (2002) Symbiosis in deep-sea sulfidic sediments. In: Sekbach J (ed) Symbiosis: mechanisms and model-systems. Kluwer, Dordrecht, pp 213–225

Canfield DE et al (2000) The Archean sulfur cycle and the early history of atmospheric oxygen. Science 288:658–661

Cavalier-Smith T (1987) The simultaneous symbiotic origin of mitochondria, chloroplasts and microbodies. Ann N Y Acad Sci 503:55–71

Danovaro R, Dell'Anno A, Pusceddu A et al (2010) The first metazoa living in permanently anoxic conditions. BMC Biol 8:30

Edgcomb VP, Simpson AGB, Zettler AL et al (2002) Pelobionts are degenerate protists: insights from molecules and morphology. Mol Biol Ecol 19:978–982

Embley TM, Finlay BJ (1994) The use of small subunit rRNA sequences to unravel the relationships between anaerobic ciliates and their methanogen endosymbionts. Microbiology 140:225–235

Esteban G, Guhl BE, Clarke KJ et al (1993) *Cyclidium porcatum* n.sp.: a free-living anaerobic scuticociliate containing a stable complex of hydrogenosomes, eubacteria and archaebacteria. Eur J Protistol 29:262–270

Esteban T, Fenchel T, Finlay BJ (2010) Mixotrophy in ciliates. Protist 161:621–641

Fenchel T, Bernard C (1993a) A purple protist. Nature 362:300

Fenchel T, Bernard C (1993b) Endosymbiotic purple non-sulphur bacteria in an anaerobic ciliated protozoon. FEMS Microbiol Lett 119:21–25

Fenchel T, Finlay BJ (1990a) Anaerobic free living protozoa: growth efficiencies and the structure of anaerobic communities. FEMS Microbiol Ecol 74:269–276

Fenchel T, Finlay BJ (1990b) Oxygen toxicity, respiration and behavioural responses to oxygen in freeliving anaerobic ciliates. J Gen Microbiol 136:1953–1959

Fenchel T, Finlay BJ (1991a) The biology of free living anaerobic ciliates. Eur J Protistol 26:201–215

Fenchel T, Finlay BJ (1991b) Endosymbiont methanogenic bacteria in anaerobic ciliates: significance for the growth efficiency of the host. J Protozool 38:22–28

Fenchel T, Finlay BJ (1991c) Synchronous division of an endosymbiotic methanogenic bacterium in the anaerobic ciliate *Plagiopyla frontata* Kahl. J Protozool 38:22–28

Fenchel T, Finlay BJ (1992) Production of methane and hydrogen by anaerobic ciliates containing symbiotic methanogens. Arch Microbiol 157:475–480

Fenchel T, Finlay BJ (1995) Ecology and evolution in anoxic worlds. Oxford University Press, Oxford

Fenchel T, Ramsing NB (1992) Identification of sulphate-reducing ectosymbiotic bacteria from anaerobic ciliates using 16S rRNA binding oligonucleotide probes. Arch Microbiol 158:394–397

Fenchel T, Perry T, Thane A (1977) Anaerobiosis and symbiosis with bacteria in free-living ciliates. J Protozool 24:154–163

Fenchel T, King GM, Blackburn TH (1998) Bacterial biogeochemistry. Academic, San Diego

Finlay BJ (1985) Nitrate respiration by protozoa (*Loxodes* spp.) in the hypolimnetic nitrite maximum of a freshwater pond. Freshw Biol 15:333–346

Finlay BJ, Fenchel T (1989) Hydrogenosomes in some anaerobic protozoa resemble mitochondria. FEMS Microbiol Lett 65:311–314

Finlay BJ, Fenchel T (1991) Polymorphic bacterial symbionts in the anaerobic protozoon *Metopus*. FEMS Microbiol Lett 79:187–190

Finlay BJ, Span ASW, Harman JMP (1983) Nitrate respiration in primitive eukaryotes. Nature 303:333–336
Hackstein JHP, Tjuden J, Huynen M (2006) Mitochondria, hydrogenosomes and mitosomes: products of evolutionary tinkering! Curr Genet 50:225–245
Hjort K, Goldberg AV, Anastasios D et al (2010) Diversity and reductive evolution of mitochondria among microbial eukaryotes. Philos Trans R Soc B 365:713–727
Lindmark D, Müller M (1973) Hydrogenosomes a cytoplasmic organelle of the anaerobic flagellate *Trichomonas foetus*, and its role in pyruvate metabolism. J Biol Chem 248:7724–7728
Lloyd D, Ralphs JR, Harris JC (2002) *Giardia intestinalis*, a eukaryote without hydrogenosomes, produces hydrogen. Microbiology 148:727–733
Mentel M, Martin W (2008) Energy metabolism among eukaryotic anaerobes in light of Proterozoic ocean chemistry. Philos Trans R Soc B 363:2717–2729
Millet COM, Cable J, Lloyd D (2010) The diplomonad fish parasite *Spironucleus vortens* produces hydrogen. J Eukaryot Microbiol 57:400–404
Minge COM, Silberman JD, Orr JS, Cavalier-Smith T, Koran S-T, Fabian B, Skjæveland Å, Jakobsen KS (2009) Evolutionary position of breviate amoebae and the primary eukaryote divergence. Proc R Soc B 276:597–604
Müller M (1993) The hydrogenosome. J Gen Microbiol 139:2879–2889
Mus F, Dubini A, Sebert M et al (2007) Anaerobic acclimation in *Chlamydomonas* – anoxic gene expression, hydrogenase induction, and metabloc pathway. J Biol Chem 282:25475–25486
Paget TA, Lloyd D (1990) *Trichomonas vaginalis* requires traces of oxygen and high concentrations of carbon dioxide for optimal growth. Mol Biochem Parasitol 41:65–72
Piña-Ochoa E, Høgslund S, Geslin E et al (2010) Widespread occurrence of nitrate storage and denitrification among foraminifera and Gromiida. Proc Natl Acad Sci USA 107:1148–1153
Simpson AGB, Hoff J, Bernard C et al (1996) The ultrastructure and systematic position of the Euglenozoon *Postgaardi mariagergensis* Fenchel et al. Arch Protistenk 147:213–225
van Bruggen JJA, Stumm CK, Vogels GD (1983) Symbiosis of methanogenic bacteria and sapropelic protozoa. Arch Microbiol 136:89–96
van Bruggen JJA, Stumm CK, Zwart et al (1985) Endosymbiotic methanogenic bacteria of the sapropelic amoeba *Mastigella*. FEMS Microbiol Ecol 31:187–192
van Bruggen JJA, van Rens GLM, Geertman EJM et al (1988) Isolation of a methanogenic endosymbiont of the sapropelic amoeba *Pelomyxa palustris* Greef. J Protozool 35:20–23
van der Giezen M, Tovar J (2005) Degenerate mitochondria. EMBO Rep 6:525–530
Wagener S, Pfennig N (1987) Monoxenic culture of the anaerobic ciliate *Trimyema compressum* Lackey. Arch Microbiol 149:4–11
Williams AG, Lloyd D (1993) Biological activities of symbiotic and parasitic protists in low oxygen environments. Adv Microb Ecol 13:211–262
Yubuki N, Edgcomb V, Bernhard JM et al (2009) Ultrastructure and molecular phylogeny of *Calkinsia aureus*: cellular identity of a novel clade of deep-sea euglenozoans with epibiotic bacteria. BMC Microbiol 9:16
Zwart KB, Goosen NK, von Schijndel MW et al (1988) Cytochemical localization of hydrogenase activity in the anaerobic protozoa *Trichomonas vaginalis*, *Plagiopyla nasuta* and *Trimyema compressum*. J Gen Microbiol 134:2165–2170

Biodata of **Tina Treude**, author of *"Biogeochemical Reactions in Marine Sediments Underlying Anoxic Water Bodies."*

Tina Treude is a professor at the Leibniz Institute of Marine Sciences (IFM-GEOMAR) and at the Christian-Albrecht University in Kiel, Germany. She is team leader of the Research Group "Marine Geobiology" and member of the Cluster of Excellence "The Future Ocean." She studied biological oceanography in Kiel and did her Ph.D. thesis with the title "Anaerobic oxidation of methane in marine sediments" at the Max Planck Institute for Marine Microbiology in Bremen, Germany. She earned her Ph.D. from the University of Bremen in 2004. Her current interests include biogeochemical reactions at cold-seep systems, submarine gas hydrates, benthic processes in oxygen minimum zones, and microbe–mineral interactions in marine sediments.

E-mail: **ttreude@ifm-geomar.de**

BIOGEOCHEMICAL REACTIONS IN MARINE SEDIMENTS UNDERLYING ANOXIC WATER BODIES

TINA TREUDE
Department of Marine Biogeochemistry, Leibniz Institute of Marine Sciences (IFM-GEOMAR), 24148 Kiel, Germany

1. Introduction

This chapter provides an overview of biogeochemical reactions in marine sediments underlying temporal or permanent hypoxic and anoxic water bodies in modern and past oceans. The aim of this review is to describe the chemical environment that organisms inhabiting surface sediments encounter during oxygen depletion or deficiency. It also introduces important metabolic processes that govern or are governed by different redox settings. In Sect. 2, biogeochemical processes in sediments underlying fully oxygenated water bodies are elucidated. Section 3 explains differences in biogeochemical reactions in hypoxic and anoxic environments as opposed to oxygenated environments. Modern oxygen minimum zones and permanent anoxic environments are introduced. In Sect. 4, biogeochemical processes during past anoxic events and during the era before the first rise of oxygen are reviewed.

2. Sediment Biogeochemistry in Modern Oxygenated Oceans

In modern oxygenated oceans, a cascade of electron acceptors (oxygen, nitrate, manganese, iron, and sulfate – in this order) is utilized by prokaryotes to degrade organic matter on the ocean floor (Canfield et al. 2005; Jørgensen 2000) (Fig. 1). Of this repertoire, oxygen, and in some rare cases nitrate (Risgaard-Petersen et al. 2006) are so far the only known electron acceptors used by eukaryotes. The order of electron acceptor utilization, and hence the order of depletion in the sediment, is governed by the Gibbs free energy gain of the reaction (Fig. 2): the higher the energy gain (kJ mol^{-1}), the more successful the consumer is in competition with other organisms for food, i.e., the more attractive the electron acceptor is (Canfield et al. 2005; Jørgensen 2000). Organic matter degradation with oxygen generates the highest energy output, making it the most attractive and first consumed electron acceptor. Furthermore, free oxygen radicals, such as the superoxide anion, hydrogen peroxide, and hydroxyl radicals, are very powerful tools for the breakdown of refractory organic compounds,

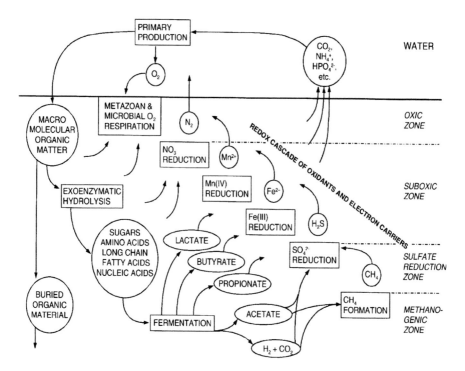

Figure 1. Biogeochemical processes in marine sediments. Organic matter, produced on the basis of photosynthesis, is degraded by microorganisms in a redox cascade of processes utilizing different electron acceptors (O_2 respiration, NO_3^- reduction, Mn^{4+} reduction, Fe^{3+} reduction, SO_4^{2-} reduction, CH_4 formation). The reduced compounds (Mn^{2+}, Fe^{2+}, H_2S, CH_4) are reoxidized by abiotic and biotic (chemoautotrophic) processes (From Jørgensen (2000). With permission).

allowing aerobic organisms to use a wide spectrum of food sources (Hedges and Keil 1995; Canfield 1994).

After aerobic organisms, nitrate reducers are still very versatile in substrate utilization, whereas others (manganese, iron, and sulfate reducers and methanogens) use a smaller spectrum of substrates, such as volatile fatty acids, provided by fermentation (Jørgensen 2000). Ideally, organic matter degradation leads to the complete oxidation of labile organic carbon to CO_2 as electron acceptors are reduced. In the final step of this degradation process, CO_2 (along with other short-chained organic molecules) is converted to methane. Reduced compounds such as Mn^{2+}, Fe^{2+}, sulfide, methane, and ammonium are finally reoxidized by chemical and biological processes at the sediment–water interface. Total oxygen uptake of sediments is therefore used as a measure to determine the sum of all respiration processes (Smith and Hinga 1983; Wenzhöfer and Glud 2002).

Pathway and stoichiometry of reaction	ΔG^0 (kJ mol-1)
Oxic respiration:	
$CH_2O + O_2 \rightarrow CO_2 + H_2O$	-479
Denitrification:	
$5CH_2O + 4NO_3^- \rightarrow 2N_2 + 4HCO_3^- + CO_2 + 3H_2O$	-453
Mn(IV) reduction:	
$CH_2O + 3CO_2 + H_2O + 2MnO_2 \rightarrow 2Mn^{2+} + 4HCO_3^-$	-349
Fe(III) reduction:	
$CH_2O + 7CO_2 + 4Fe(OH)_3 \rightarrow 4Fe^{2+} + 8HCO_3^- + 3H_2O$	-114
Sulfate reduction:	
$2CH_2O + SO_4^{2-} \rightarrow H_2S + 2HCO_3^-$	-77
$4H_2 + SO_4^{2-} + H^+ \rightarrow HS^- + 4H_2O$	-152
$CH_3COO^- + SO_4^{2-} + 2H^+ \rightarrow 2CO_2 + HS^- + 2H_2O$	-41
Methane production:	
$4H_2 + HCO_3^- + H^+ \rightarrow CH_4 + 3H_2O$	-136
$CH_3COO^- + H^+ \rightarrow CH_4 + CO_2$	-28
Acetogenesis:	
$4H_2 + 2CO_3^- + H^+ \rightarrow CH_3COO^- + 4H_2O$	-105
Fermentation:	
$CH_3CH_2OH + H_2O \rightarrow CH_3COO^- + 2H_2 + H^+$	10
$CH_3CH_2COO^- + 3H_2O \rightarrow CH_3COO^- + HCO_3^- + 3H_2 + H^+$	77

Figure 2. Microbial redox processes during organic matter degradation in marine sediments and the Gibbs free energy gain under standard conditions (From Jørgensen (2000). With permission).

2.1. RANGE AND VARIATIONS OF OXYGEN AVAILABILITY IN SEDIMENTS

Oxygen availability in seawater is relatively low in comparison to other electron acceptors such as sulfate (~300 µM O_2 compared to 28 mM sulfate). Penetration depth of oxygen in sediments is often restricted to a few millimeters (Revsbech et al. 1980; Glud 2008) because it is quickly consumed by biological and chemical processes (Cai and Sayles 1996; Wenzhöfer and Glud 2002). Global studies on oxygen availability revealed an exponential increase of oxygen penetration into sediments with water depth (Wenzhöfer and Glud 2002; Glud 2008) (Fig. 3). The relationship behind this link is the decrease in oxygen consumption rates in sediments with decreasing sedimentary organic carbon flux (Suess 1980). The larger the water depth, the less organic matter reaches the seafloor, and the less car-

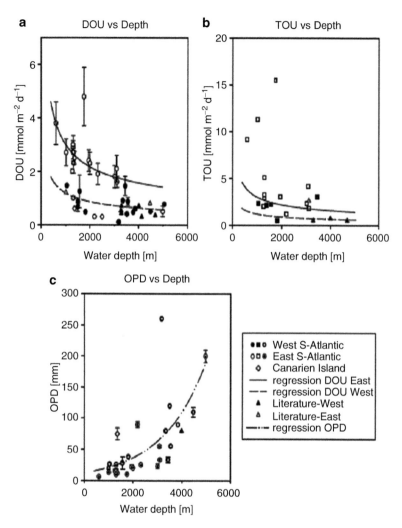

Figure 3. In situ diffusive oxygen uptake (DOU, **a**), total oxygen uptake (TOU, **b**), and oxygen penetration depth (OPD, **c**) of sediments of the Atlantic Ocean as a function of water depth (From Wenzhöfer and Glud (2002). With permission).

bon is degraded in the sediments. In fine-grained organic-rich sediments within coastal areas, oxygen penetration is only a few millimeters (e.g., Jørgensen et al. 2005) compared to several centimeters to meters in the deep sea (Wenzhöfer and Glud 2002; Fischer et al. 2009).

Besides water depth, productivity in the euphotic zone controls particle flux: high productivity results in more export of organic matter to the seafloor (Lampitt and Antia 1997). Therefore, it is not surprising that within the same

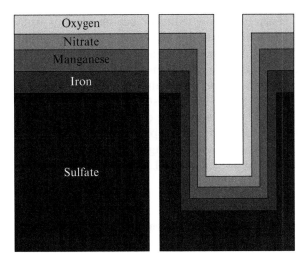

Figure 4. Ideal laminated sediment (*left*) without bioturbation/bioirrigation and sediment inhabited by burro-forming fauna (*right*). Electron acceptors penetrate deeper into the sediment through the burrow walls (From Canfield et al. (2005). With permission).

depth range, deep-sea sediments underneath productive upwelling regions reveal a much shallower penetration depth of oxygen (e.g., 20 cm at 4,986 m water depth, Glud et al. 1994) than sediments below the oligotrophic pacific gyres (>8 m around 5,000 m water depth, Fischer et al. 2009).

Another important factor affecting oxygen penetration is sediment permeability. Permeable sandy sediments in coastal areas are highly dynamic systems, especially in wave-impacted areas featuring ripple formation (Huettel and Gust 1992; Ziebis et al. 1996; Precht et al. 2004). Pressure differences around the ripples caused by wave action, combined with high permeability, enables fast advective exchange processes between the sediment and the water column. As a consequence, oxygen penetrations >1 cm are detected even in intertidal mud flat sediment from productive coastal areas (Precht et al. 2004). In addition, other important factors controlling oxygen availability in sediments are photosynthetic microorganisms and bottom-dwelling fauna. During daylight, oxygenic benthic microorganisms, such as cyanobacteria and microalgae, increase oxygen concentrations in their environment to levels above seawater concentration (Revsbech et al. 1986). Bottom-dwelling fauna facilitate deep oxygen penetration into sediments (Kristensen 2000; Kristensen et al. 2005). Here, two separate processes need to be distinguished: (1) bioturbation, which is the displacement and mixing of sediment particles by benthic fauna (and flora), and (2) bioirrigation, which refers to the process of benthic organisms flushing their burrows with overlying seawater (Meysman et al. 2006). Animal burrows basically represent invaginations of the sediment surface that increase the exchange area between sediment and water (Fig. 4). Oxygenated surface water that is pumped through the burrows enables oxygen penetration into sediment depths where it would be normally depleted (Revsbech et al. 1980), i.e., providing oxic microhabitats embedded in anoxic layers. Bioturbation and bioirrigation can in some extreme cases extend up to several meters into sediments (Pemberton et al. 1976). The processes are very complex, depending on sediment and organism type, and have been summarized elsewhere (Kristensen 2000; Kristensen et al. 2005). For this chapter, it is important to remember that the presence or absence of bioirrigating infauna is an important key factor controlling oxygen penetration and organic matter turnover in sediments as well as element cycling across the sediment–water interface (e.g., Fossing et al. 2000; Jørgensen et al. 2005; Kostka et al. 2002; Bertics et al. 2010).

3. Changes in Sedimentary Biogeochemical Reactions Under Hypoxia and Anoxia

A comprehensive summary of biogeochemical processes in sediments and at the benthic boundary under hypoxia and anoxia has been published by Middelburg and Levin (2009). With respect to the focus of this book, I will concentrate on factors that could directly affect the lifestyle of eukaryotes: oxygen/nitrate availability and release of toxic substances. In the literature, definitions are not always consistent with respect to the terms "suboxic" or "hypoxic." In order to strive for more clarity, it was recently suggested to define conditions after metabolic zones, i.e., after the thermodynamically most attractive electron acceptor available (Canfield and Thamdrup 2009). In this chapter, we will use the definitions of oxic (>63 μM O_2), hypoxic (<63 μM O_2), anoxic (oxygen below detection limit of microsensors <1 μM), and sulfidic (oxygen below detection limit and measureable free sulfide) conditions provided by Middelburg and Levin (2009). According to these authors, sediments underlying oxic water bodies feature aerobic respiration by prokaryotes and eukaryotes and reoxidation of reduced compounds (see Sect. 2.1). Under hypoxic conditions in the water column, processes correlated with oxygen are diminished in the sediment but reoxidation would still continue with nitrate, sulfate, and metal oxides. In permanently anoxic or sulfidic conditions, most oxidants have been exhausted and processes in the sediments are dominated by sulfate reduction, methanogenesis, and anaerobic oxidation of methane.

3.1. BIOGEOCHEMICAL PROCESSES IN NATURALLY OCCURRING OXYGEN MINIMUM ZONES ON CONTINENTAL MARGINS

In modern oceans, the largest naturally occurring oxygen minimum zones (OMZ) are often located in coastal upwelling regions featuring intense primary productivity and high deposition rates of organic material (Strauss 2005; Paulmier and

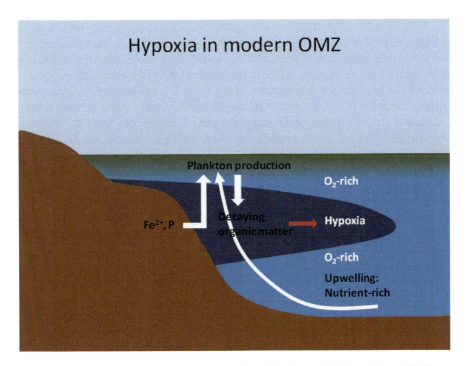

Figure 5. Simplified scheme of important processes in modern Oxygen Minimum Zones (OMZ).

Ruiz-Pino 2009). Sinking organic particles are degraded in the water column by heterotrophic organisms, causing strong oxygen depletions (Fig. 5). Global ocean area and volume occupied by OMZ featuring oxygen concentrations <20 μM are estimated to be in the range of 30 Mio km^2 and 100 Mio km^3, respectively, accounting for 8% and 7% of the global ocean (Paulmier and Ruiz-Pino 2009).

The most extensive and permanent OMZ are located in the Eastern Pacific Ocean (e.g., off Peru/Chile, off Central America), in the Northern Indian Ocean (e.g., Arabian Sea, Bay of Bengal), and in the Southeast Atlantic (e.g., off Namibia). Seasonal OMZ have been observed also, e.g., in the West Bering Sea and the Gulf of Alaska during autumn-winter-spring seasons. Commonly, oxygen minima intercept the continental margin at the shelf and continental slope. According to Helly and Levin (2004), world-wide OMZ with oxygen concentrations <20 μM cover ~1.2 Mio km^2 of seafloor (31% occur in the Eastern Pacific, 59% in the Indian Ocean, 10% in the Southeast Atlantic). Due to high organic matter accumulation rates (>30 gC m^{-2} year^{-1}, Henrichs and Farrington 1984; Müller and Suess 1979) generating sediment carbon contents >10% (Levin et al. 2002; Henrichs and Farrington 1984; Müller and Suess 1979), oxygen is depleted quickly or is completely absent in sediments, leaving organic matter degradation to mostly anaerobic processes (Canfield 1989).

With 28 mM concentrations in seawater, sulfate is often the most abundant electron acceptor in sediments, and hence, a majority of organic matter in oxygen minimum zone sediments is degraded via sulfate reduction. The importance of sulfate reduction in organic carbon respiration increases with sedimentation rates and can account for nearly 100% of the total C_{org} oxidation rates when sedimentation reaches 0.1 g cm^{-2} year^{-1} (Canfield 1989). Areal sulfate reduction rates in surface sediments within OMZ range between 1 and 60 mmol m^{-2} d^{-1} (Fossing 1990; Brüchert et al. 2003; Ferdelman et al. 1997; Fossing et al. 2000). The buildup of sometimes high levels of hydrogen sulfide in the sediments supports chemosynthetic communities. OMZ sediments commonly feature characteristic sulfur bacteria, which harvest chemical energy from the oxidation of sulfide. Sulfidic sediments in the OMZ region off Peru/Chile are dominated by the giant filamentous sulfide-oxidizing bacteria *Thioploca* (Fossing et al. 1995). Their ability to store high concentrations of nitrate for sulfide oxidation in vacuoles and to migrate in the sediment through sheaths enables them to oxidize sulfide even if it is spatially separated from the electron acceptor.

The OMZ sediments off Namibia are dominated by a different kind of sulfur bacteria, *Thiomargarita*, which are also giant in size and able to store high concentrations of nitrate, but are nonfilamentous and immobile (Schulz et al. 1999). Their lifestyle, although still enigmatic, has been suggested to be connected with sediment resuspension processes that enable them to pick up nitrate from the water column when suspended and to oxidize sulfide after resettling on the seafloor. Other, very ubiquitous inhabitants of sulfide-rich OMZ sediments are representatives of the genus *Beggiatoa* (Rosenberg and Diaz 1993) – a filamentous sulfur bacterium. Their lifestyle is often bound to very steep sulfide and oxygen gradients at the sediment–water interface (Nelson et al. 1986); however, some species are also reported to vertically transport nitrate in the sediment (Preisler et al. 2007). All afore mentioned sulfur bacteria are able to store particulate elemental sulfur as an intermediate in their cells (Jørgensen and Nelson 2004). Their ability to reduce sulfide levels in sediments is thought to be advantageous with respect to detoxification for certain metazoans and protozoans that can live under very low oxygen concentrations (Levin et al. 2002, and references therein). Recently, pelagic sulfur bacteria have been found to play an important role in detoxification of gas-driven sulfide eruptions into the water column of OMZ (Lavik et al. 2009). Such gas eruptions develop when methanogenesis, the final step in organic matter degradation, causes oversaturation of methane in the sediment porewater, building up a high gas pressure that first lifts and finally ejects the upper sediment burden. This process has been described to form temporal "islands" in coastal areas caused by lifted seafloor. Such gas eruption can be so intensive that they transport and release large amounts of sulfide into the water column (Weeks et al. 2002; 2004). If the sulfide is not quickly oxidized, it can become toxic to many marine organisms already at micromolar levels (Bagarinao 1992; Grieshaber and Völkel 1998), forcing, e.g., mobile animals, such as lobsters, to literally walk out of the water onto the beaches (Weeks et al. 2004).

With respect to processes at the sediment–water interface of OMZ, some elements (e.g., manganese/iron and phosphorous) show remarkable differences in their behavior under a hypoxic environment compared to oxic or anoxic-sulfidic settings. When oxygen concentrations decline in the bottom waters, the zone of reoxidation processes in sediments thins, allowing higher fluxes of soluble reduced iron/manganese from the sediment into the water column (Sundby and Silverberg 1985; Konovalov et al. 2007; Middelburg and Levin 2009). Because of dissimilar thermodynamic and kinetic behavior, manganese and iron show different responses to certain oxygen and sulfide levels in the water column. In short, reduced manganese is liberated at higher bottom-water oxygen levels than iron, whereas reactive iron compounds in the anoxic water react easily with sulfide, and are stripped out of the water due to sulfide-mineral precipitation (Middelburg and Levin 2009, and references therein). In consequence, iron effluxes are maximal under low-oxic, sulfide-free condition. While manganese and iron are leached from sediments under hypoxic conditions into the water column, they resettle as metal oxides at the hypoxic–oxic interface leading to enrichment in the sediments. Phosphorous cycling is closely connected to iron cycling because phosphorous is bound to iron oxides in the sediment. Hence, phosphorous fluxes are usually accelerated in conjunction with iron fluxes (Sundby et al. 1992). Because both iron and phosphorous are important nutrients for primary production, their faster release from sediments in OMZ ultimately feeds primary productivity at the surface. This positive feedback helps to sustain the OMZ. Nitrate, although it might first accumulate in hypoxic zones because of initially enhanced nitrification processes, is usually depleted in the water column and sediment under prolonged oxygen deficient situations (Middelburg and Levin 2009; Brüchert et al. 2003).

3.2. BIOGEOCHEMICAL PROCESSES IN SEDIMENTS UNDERLYING EUXINIC WATER BODIES

Euxinic comes from the old Greek name for the Black Sea "Euxinus." The term euxinic basically refers to environments with restricted circulation, caused by topography and/or pycnoclines, creating stagnant or anoxic conditions. Euxinic conditions are at the same time both anoxic and sulfidic. In the modern oceans, we find permanently anoxic water bodies in the Black Sea, the Baltic Sea (in deeper basins), and the Cariaco Basin. In these environments, stagnant conditions combined with high primary productivity in the euphotic zone, i.e., high export rates of organic carbon to the seafloor, cause complete oxygen consumption in deeper water layers (Neretin 2006) (Fig. 6). The permanent anoxia has dramatic consequences with respect to biogeochemical processes and sediment characteristics, which distinguishes these environments from oxygen minimum zones (OMZ) that exhibit only hypoxic or temporarily anoxic conditions (Middelburg and Levin 2009). One of the most remarkable characteristics of euxinic environments is probably the permanent lack of macroscopic benthic fauna. In the absence of bottom-dwelling organisms responsible for bioturbation and bioirrigation, sediments accumulate

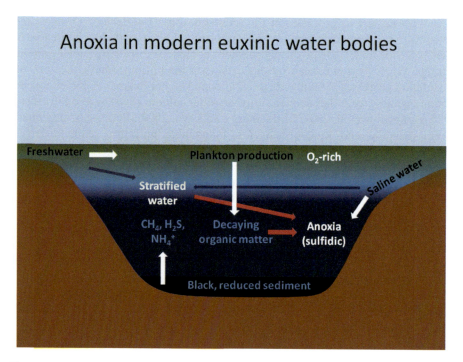

Figure 6. Simplified scheme of important biogeochemical reactions in modern euxinic water bodies, such as the Black Sea.

without vertical or horizontal biological mixing. The lack of mixing commonly results in the formation of sediment laminae due to the seasonal changes in plankton blooms depositing on the seafloor (Kemp et al. 1999). The complete lack of seafloor ventilation and particle mixing by metazoans has major consequences for element cycling, such as the accumulation of end products (e.g., sulfide, methane, ammonium) in sediments (Bianchi et al. 2000), reduced exchange rates (e.g., of nutrients) between sediment and water column (Krantzberg 1985), and long-term burial of authigenic minerals (Raiswell and Berner 1985). Although the issue is still a matter of debate (Hedges and Keil 1995; Canfield 1994; Kristensen 2000), organic matter degradation in sediments appears to be less efficient under permanently anoxic waters, leading to higher preservation rates (Middelburg and Levin 2009; Moodley et al. 2005; Kristensen et al. 1995). Two factors that are thought to mainly control burial efficiency are exposure time to oxygen and sediment-accumulation rates. Oxygen is often needed to break down refractory compounds that otherwise would not become accessible to anaerobic processes (Hedges and Keil 1995; Canfield 1994; Kristensen et al. 1995).

Hence, without oxygen some organic compounds can remain untouched by biological degradation processes. High sediment-accumulation rates can shift organic matter relatively "fast" out of the diffusive range of electron acceptors on geological time scales, enabling long-term burial until potential reexposure to oxic conditions (e.g., in turbidites, Wilson et al. 1985). Under such conditions, so-called sapropels (a contraction of the ancient Greek words sapros and pelos, meaning putrefaction and mud) with organic carbon contents >2% in weight form (Kidd et al. 1978; Howell and Thunell 1992). Sapropelic deposits can form important oil source rocks such as the ones deposited during the Ocean Anoxic Events (OAE, see below, Irving et al. 1974). Under permanent anoxic conditions, reduced end products of organic matter degradation, which would usually be reoxidized at the sediment–water interface in oxygenated oceans, tend to finally accumulate in the anoxic water column. While Fe^{2+} and phosphorous fluxes from the sediments into the water column are highest under hypoxic conditions (see above and Middelburg and Levin 2009), methane, ammonium, and sulfide fluxes are highest at permanent anoxia, causing accumulation in the water column until they are consumed by reoxidation processes at pelagic redoxclines (Kuypers et al. 2003; Reeburgh et al. 1991).

4. Anoxia Through Time

4.1. BIOGEOCHEMICAL PROCESSES IN SEDIMENTS DURING OCEAN ANOXIC EVENTS

Ocean anoxic events (OAE) have probably occurred several times in the Earth's history, e.g., in the Ordovician-Silurian (443 Ma, Armstrong et al. 2009), late Devonian (374 Ma, Bond and Wignall 2008), Permian-Triassic (251 Ma, Wignall and Twitchett 1996), Triassic-Jurassic (201 Ma, Morante and Hallam 1996), Early Jurassic (199.6 Ma Jenkyns 1988; Schootbrugge et al. 2005), and the Late Cretaceous (120 and 93 Ma Schlanger and Jenkyns 1976; Jones and Jenkyns 2001), as well as during the Paleocene-Eocene Thermal Maximum (PETM, 55.8 Ma, Jiang et al. 2006). Most, but not all, of the OAE were connected to global mass extinctions, as well as the deposition of organic-rich sediments (sapropel and black shale). Discussion surrounding the causes of OAE is still quite controversial; however, the causes are most likely diverse. In some studies, global warming induced by strong volcanic and hydrothermal activity (Jones and Jenkyns 2001; Turgeon and Creaser 2008; Bralower 2008) (Fig. 7) is suggested as the initial trigger, resulting in pronounced ocean water stratification. When combined with high productivity at the surface, organic matter degradation drive oxygen concentrations at depth towards zero. Once certain tipping points were reached, additional factors such as massive methane releases from submarine gas hydrates (Kennett et al. 2003) or sulfide eruptions from organic-rich sediments (Kump et al. 2005) might have boosted the catastrophe. In some studies, mass extinctions and OAE are brought into context with sea-level changes (Hallam and Wignall 1999;

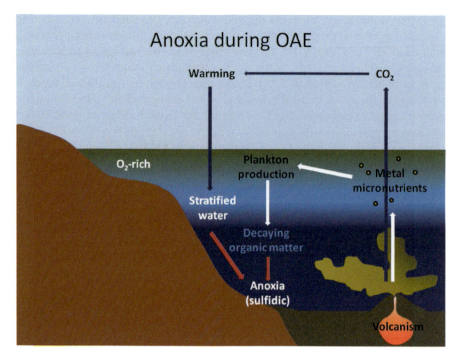

Figure 7. Scheme illustrating a potential scenario during Ocean Anoxic Events (OAE) in cases where volcanism was probably the initial driving force of OAE development (Modified after Turgeon and Creaser (2008) and Bralower (2008)).

Sandberg et al. 2002). In other cases, researchers suggest that the thermohaline circulation during the late Permian was very different from today given the palaeogeography with one supercontinent (Pangaea) and one superocean (Panthalassa) (Petsch 2005). Coupled to higher CO_2 levels (Berner 1994; Berner and Kothavala 2001), the climate was generally warmer, with weaker wind stress, and low equator to pole temperature gradients. As a consequence of salinity-driven bottom-water formation in low latitudes – as opposed to today's low-temperature-driven formation at high latitudes – salty warm water could have limited oxygenation of the deep oceans, i.e., facilitating anoxia development. OAE typically last less than 0.5 Ma before full recovery (e.g., Jenkyns 1988).

Irrespective of the cause of the OAE's, which in many cases are not fully understood, conditions during anoxia might have been very similar to today's anoxic/euxinic environments, as shown by several fossil records: (1) absence of macrofauna, i.e., lack of bioturbation and bioirrigation (Twitchett 1999), facilitating laminae formation (Kemp et al. 1999; Armstrong et al. 2009); (2) higher carbon preservation and burial rates (Schlanger and Jenkyns 1976; Jenkyns 1988); (3) sedimentary degradation processes dominated by sulfate reduction (Canfield

1989), creating sulfidic sediment with pyritization processes (Passier et al. 1996; Raiswell and Berner 1985); and (4) eventually the accumulation of sulfide in the anoxic water column (Jenkyns 1988; Kump et al. 2005).

4.2. THE INITIAL EARLY EARTH ANOXIA AND CHANGES DURING THE FIRST RISE OF OXYGEN

The evidences, theories, and controversial discussions that exist concerning the beginning of life, the rise of oxygen, as well as the evolution of the Earth's atmosphere and oceanic redox states are so complex that it would fill books to give credit to all studies. Here, I would like to summarize only some hypotheses about the development of the most important chemical substances and metabolisms before and after the first accumulation of free oxygen. Today's consensus is that the early oceans and atmosphere were anoxic (e.g., Kasting 1993). It is still under debate to what extent reduced compounds (methane, ammonia, and sulfide) were present in the early atmosphere. There seems, however, no doubt that nitrogen, CO_2, hydrogen, and water vapor were important constituents of the early atmosphere (for a review of Early Earth atmosphere development, see Catling and Claire 2005). Oxygen, if present at all, was very short-lived as a by-product of photolysis of water (via H_2O_2, McKay and Hartman 1991). Molecular studies suggest that the small amounts of chemically available oxygen might have been enough to support the evolution of aerobic respiration in microniches already before the biologically induced rise of oxygen by oxygenic photosynthesis (Castresana and Saraste 1995). The basis for this "respiration-early hypothesis" is that cytochrome oxidase, which is part of the aerobic respiration machinery, evolved prior to the water-splitting enzymes needed for oxygenic photosynthesis. The origin of aerobic respiration is probably denitrification because the key enzyme of denitrification (NO reductase) is homologous to cytochrome oxidase. Besides the debate about the beginning of aerobic respiration, good indications exist that the very first metabolic processes on Early Earth included methanogenesis (earliest record 3.5 Ga, Ueno et al. 2006) and sulfate reduction (earliest record 3.47 Ga, Shen et al. 2001). While the beginning of oxygenic photosynthesis by cyanobacteria is still unclear, ancient stromatolites, which are conventionally thought to be fossil remains of cyanobacterial mats, date back 3.5 Ga (Awramik 1992; Allwood et al. 2006). However, evidences that ancient stromatolite formation was ultimately connected to biological processes are not always consistent in the literature (e.g., Walter 1983; Grotzinger and Knoll 1999) and restrict the clearly biogenic origin to "only" 2.7 Ga old fossils (e.g., Buick 1992; Des Marais 2000; Lepot et al. 2008).

Because of its inherent complexity, it is hard to imagine how photosynthesis once evolved through intermediate steps. Nisbet et al. (1995) provided a model in which photosynthesis evolved from thermophilic bacteria using phototaxis of near-infrared light to locate hydrothermal systems. Phototaxis was then later turned into an energy-saving process in combination with the development

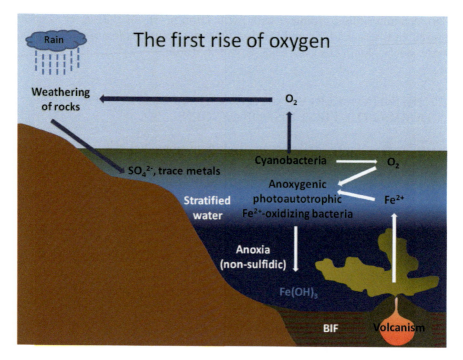

Figure 8. Illustration of potential key processes during the first rise of oxygen at the end of the Archaean era before the development of sulfidic conditions (see text).

of pigments, which originally served as protection against UV-light in shallow hydrothermal systems. Recently, a phototrophic green sulfur bacterium was isolated from deep-sea hydrothermal systems at the East Pacific Rise that utilizes geothermal radiation (Beatty et al. 2005), which supports the hypothesis of Nisbet et al. (1995). Imagining the diversity of early biomes that once occupied the ocean seafloors and water columns is still difficult given the limited access to fossil records. Nisbet and Sleep (2001) introduced their perspective on microbial communities during the late Archaean, which consisted of (1) anoxygenic photosynthetic mats on shallow sediments and rocks in the "red to infrared light"–penetrated zone, (2) oxygenic photosynthetic mats and stromatolites on shallow sediments in the "visible light"–penetrated zone, (3) hyperthermophilic biofilms and mats on rocks at hydrothermal or volcanic systems, and (4) a separation into planktonic cyanobacteria and benthic decomposers in the open water areas.

Before the final accumulation of free oxygen about 2.4–2.3 Ga ago (Catling and Claire 2005), the oceans went through some radical changes in chemical composition and redox states. The developments can be best described by following three key elements: sulfur, iron, and molybdenum (Fig. 8). The Early Archaean anoxic ocean was probably non-sulfidic because sulfate, the electron acceptor of sulfate reduction, has not been weathered from rocks and was therefore available only in low concentrations (Shen et al. 2003). The same situation applied for molybdenum. Lacking a highly oxidizing reagent such as hydrogen sulfide, Fe^{2+}, which was expelled from volcanic activity, accumulated in the anoxic ocean (Anbar and Knoll 2002). When oxygenic photosynthesis evolved, the first biologically produced oxygen was only short-lived because it instantly reacted with reduced iron. Such a reaction could have been purely chemically or, as suggested by recent studies, mediated by anoxygenic Fe^{2+}-oxidizing phototrophs at the redoxcline (Kappler et al. 2005).

Because the produced iron oxides were insoluble, they sedimented on the seafloor, generating the legendary banded iron formations (BIF's; Beukes and Klein 1992). As a result of this precipitation process, surface oceans became depleted in iron. The Archaean biosphere was basically the inverse of the modern situation (Walker 1987); today, the oxidized partner in photosynthesis is usually free oxygen, which accumulates in the atmosphere while the reduced partner, organic matter, settles on the seafloor in a more stagnant pool compared to the atmosphere. When oxygenic photosynthesis developed in the Archaean surface oceans, the ultimate oxidized partner was probably insoluble iron, which settled faster in a stagnant pool in sediments than the comparably more volatile reduced partner, organic matter. Hence, the Archaean biosphere might have been reducing at the top and oxidizing at the bottom.

After continuous oxygen production, surface oceans finally reached oxygenated levels and oxygen accumulated in the atmosphere approximately 2.32 Ga ago (Bekker et al. 2004; Anbar and Knoll 2002). Atmospheric oxygen promoted weathering of rocks. Rainfall and rivers flushed sulfate and molybdenum, as well as other trace metals, from weathered terrestrial rocks into the oceans where they slowly increased in concentration. Sulfate-reducing organisms turned sulfate into sulfide, which accumulated in the deep anoxic oceans. Under these sulfidic conditions, Fe^{2+}, which was still abundant in the anoxic zone, precipitated with sulfur to form iron sulfides and eventually pyrite (Canfield 1998). At that time (about 1.8–1.2 Ga ago), iron was probably low both at the oxygenated surface and in the deep sulfidic ocean (Anbar and Knoll 2002). Overall, we can assume that biologically important trace metals (Fe, Mb, Zn, Cu, Cd) would have been scarce in most marine environments due to infant weathering processes and sulfidic conditions, potentially restricting the nitrogen cycle, affecting primary productivity, and limiting the ecological distribution of eukaryotic algae. Mid-proterozoic eurkaryotic algae might have found sufficient conditions in the proximity of riverine metal sources in regions with upwelling ammonium. In the Phanerozoic, the oceans were finally fully oxygenated, similar to today's conditions, limiting sulfide concentrations and hence precipitation of trace metals. However, there is some debate about the timing of the oxygenation of the deep oceans. Box model studies suggest that oxygenation did not happen until the Neoproterozoic era because sufficient oxygen was needed to remove sulfide from the deep waters (Canfield 1998). Other indications come from several biological and geochemical innovations that occurred during that time, including the late Proterozoic rise of animals (Cloud 1972; Knoll 1992) and the emergence of colorless sulfur bacteria (Canfield and Teske 1996).

5. References

Allwood AC, Walter MR, Kamber BS, Marshall CP, Burch IW (2006) Stromatolite reef from the early Archaean era of Australia. Nature 441:714–718

Anbar AD, Knoll AH (2002) Proterozoic ocean chemistry and evolution: a bioinorganic bridge? Science 297:1137–1142

Armstrong HA, Abbott GD, Turner BR, Makhlouf IM, Muhammad AB, Pedentchouk N, Peters H (2009) Black shale deposition in an upper Ordovician-Silurian permanently stratified, peri-göacial basin, southern Jordan. Palaeogeogr Paleoclimatol Palaeoecol 273:368–377

Awramik SM (1992) The oldest record of photosynthesis. Photosynth Res 33:75–89

Bagarinao T (1992) Sulfide as an environmental factor and toxicant: tolerance and adaptations in aquatic organisms. Aquat Toxicol 24:21–62

Beatty TJ, Overmann J, Lince MT, Manske AK, Lang AS, Blankenship RE, Van Dover CL, Martinson TA, Plumley FG (2005) An obligately photosynthetic bacterial anaerobe from a deep-sea hydrothermal vent. Proc Natl Acad Sci 102(26):9306–9310

Bekker A, Holland HD, Wang P-L, Rumble D III, Stein DL, Hannah JL, Coetzee LL, Beukes NJ (2004) Dating the rise of atmospheric oxygen. Nature 427:117–120

Berner RA (1994) GEOCARB II: a revised model of atmospheric CO2 over Phanerozoic time. Am J Sci 294(56):56–91

Berner RA, Kothavala Z (2001) GEOCARB III: a revised model of atmospheric CO2 over Phanerozoic time. Am J Sci 301:182–204

Bertics VJ, Sohm JA, Treude T, Chow C-ET, Capone DG, Fuhrman JA, Ziebis W (2010) Burrowing deeper into the benthic nitrogen fixation: the impact of bioturbation on nitrogen fixation coupled to sulfate reduction. Mar Ecol Prog Ser 409:1–15

Beukes NJ, Klein C (1992) Models for iron-formation deposition. In: Schopf JW, Klein C (eds) The Proterozoic biosphere. Cambridge University Press, Cambridge, pp 147–151

Bianchi CN, Johansson B, Elmgren R (2000) Breakdown of phytoplankton in Baltic sediments: effects of anoxia and loss of deposit-feeding macrofauna. J Exp Mar Biol Ecol 251:161–183

Bond PG, Wignall PB (2008) The role of sea-level change and marine anoxia in the Frasnian-Famennian (Late Devonian) mass extinction. Palaeogeogr Paleoclimatol Palaeoecol 263:107–118

Bralower TJ (2008) Volcanic cause of catastrophe. Nature 454:285–287

Brüchert V, Jørgensen BB, Neumann K, Riechmann D, Schlösser M, Schulz H (2003) Regulation of bacterial sulfate reduction and hydrogen sulfide fluxes in the central Namibian coastal upwelling zone. Geochim Cosmochim Ac 67(23):4505–4518

Buick R (1992) The antiquity of oxygenic photosynthesis: evidence from Stromatolites in sulphate-deficient Archaean lakes. Science 255:74–77

Cai W-J, Sayles FL (1996) Oxygen penetration depths and fluxes in marine sediments. Mar Chem 52:123–131

Canfield DE (1989) Sulfate reduction and oxic respiration in marine sediments: implications for organic carbon preservation in euxinic environments. Deep-Sea Res 36(1):121–138

Canfield DE (1994) Factors influencing organic carbon preservation in marine sediments. Chem Geol 114:315–329

Canfield DE (1998) A new model for Proterozoic ocean chemistry. Nature 396:450–453

Canfield DE, Teske A (1996) Late Proterozoic rise in atmospheric oxygen concentration inferred from phylogenetic and sulfur-isotope studies. Nature 382:127–132

Canfield DE, Thamdrup B (2009) Towards a consistent classification scheme for geochemical environments, or, why we wish the term "suboxic" would go away. Geobiology 7:385–392

Canfield DE, Thamdrup B, Kristensen E (2005) Aquatic geomicrobiology. Advances in marine biology. Elsevier, San Diego

Castresana J, Saraste M (1995) Evolution of energetic metabolism: the respiration-early hypothesis. Trends Biochem Sci 20:443–448

Catling DC, Claire MW (2005) How Earth' atmosphere evolved to an oxic state: a status report. Earth Planet Sci Lett 237:1–20

Cloud PE Jr (1972) A working model of the primitive Earth. Am J Sci 272:537–548

Des Marais DJ (2000) When did photosynthesis emerge on Earth? Science 289:1703–1705

Ferdelman TG, Lee C, Pantoja S, Harder J, Bebout BM, Fossing H (1997) Sulfate reduction and methanogenesis in a Thioploca-dominated sediment off the coast of Chile. Geochim Cosmochim Acta 61(15):3065–3079

Fischer JP, Ferdelman TG, D'Hondt SD, Roey H, Wenzhöfer F (2009) Oxygen penetration deep into the sediment of the South Pacific gyre. Biogeosciences 6:1467–1478

Fossing H (1990) Sulfate reduction in shelf sediments in the upwelling region off Central Peru. Continent Shelf Res 10(4):355–367

Fossing H, Gallardo VA, Jørgensen BB, Hüttel M, Nielsen LP, Schulz H, Canfield DE et al (1995) Concentration and transport of nitrate by the mat-forming sulphur bacterium *Thioploca*. Nature 374:713–715

Fossing H, Ferdelman TG, Berg P (2000) Sulphate reduction and methane oxidation in continental sediments influenced by irrigation (South-East Atlantic off Namibia). Geochim Cosmochim Acta 64(5):897–910

Glud RN (2008) Oxygen dynamics of marine sediments. Mar Biol Res 4:243–289

Glud RN, Gundersen JK, Jørgensen BB, Revsbech NP, Schulz HD (1994) Diffusive and total oxygen uptake of deep-sea sediments in the eastern South Atlantic Ocean: in situ and laboratory measurements. Deep-Sea Res I 41(11/12):1767–1788

Grieshaber MK, Völkel S (1998) Animal adaptations for tolerance and exploitation of poisonous sulfide. Annu Rev Physiol 60:33–53

Grotzinger JP, Knoll AH (1999) Stromatolites in precambrian carbonates: evolutionary mileposts or environmental dipsticks? Annu Rev Earth Planet Sci 27:313–358

Hallam A, Wignall PB (1999) Mass extinctions and sea-level changes. Earth Sci Rev 48:217–250

Hedges JI, Keil RG (1995) Sedimentary organic matter preservation: an assessment and speculative synthesis. Mar Chem 49:81–115

Helly JJ, Levin LA (2004) Global distribution of naturally occurring marine hypoxia on continental margins. Deep Sea Res I 51:1159–1168

Henrichs SM, Farrington JW (1984) Peru upwelling region sediments near 15°S. 1. Remineralization and accumulation of organic matter. Limnol Oceanogr 29(1):1–19

Howell MW, Thunell RC (1992) Organic carbon accumulation in Bannock Basin: evaluating the role of productivity in the formation of eastern Mediterranean sapropels. Mar Geol 103:461–471

Huettel M, Gust G (1992) Impact of bioroughness on intertidal solute exchange in permeable sediments. Mar Ecol Prog Ser 89(2–3):253–267

Irving E, North FK, Couillard R (1974) Oil, climate and tectonics. Can J Earth Sci 11(1):1–17

Jenkyns HC (1988) The early Toarcian (Jurassic) anoxic event: Stratigraphic, sedimentary, and geochemical evidences. Am J Sci 288:101–151

Jiang G, Shi X, Zhang S (2006) Methane seeps, methane hydrate destabilization, and the late Neoproterozoic postglacial cap carbonates. Chin Sci Bull 51(10):1152–1173

Jones CE, Jenkyns HC (2001) Seawater strontium isotopes, oceanic anoxic events, and seafloor hydrothermal activity in the Jurassic and Cretaceous. Am J Sci 301:112–149

Jørgensen BB (2000) Bacteria and marine biogeochemistry. In: Schulz HD, Zabel M (eds) Marine biogeochemistry. Springer, Berlin, pp 173–201

Jørgensen BB, Nelson DC (2004) Sulfide oxidation in marine sediments: geochemistry meets microbiology. Geol Soc Am Spec Pap 379:63–81

Jørgensen BB, Glud RN, Holby O (2005) Oxygen distribution and bioirrigation in Arctic fjord sediments (Svalbard, Barents Sea). Mar Ecol Prog Ser 292:85–95

Kappler A, Pasquero C, Konhauser KO, Newman DK (2005) Deposition of banded iron formations by anoxygenic phototrophic Fe(II)-oxidizing bacteria. Geology 33(11):865–863

Kasting JF (1993) Earth's early atmosphere. Science 259:920–926

Kemp AES, Pearce RB, Koizumi I, Pike J, Rance SJ (1999) The role of mat-forming diatoms in the formation of Mediterranean sapropels. Nature 398:57–61

Kennett JP, Cannariato KG, Hendy IL, Behl RJ (2003) Methane hydrates in quaternary climate change: the clathrate gun hypothesis. American Geophysical Union, Washington, DC

Kidd RB, Cita MB, Ryan WBF (1978) Stratigraphy of eastern Mediterranean sapropel sequences recovered during DSDP Leg 42A and their paleoenvironmental significance. Init Rep DSDP 42:421–443

Knoll AH (1992) Biological and biogeochemical preludes to the Ediacardian radiation. In: Lipps JH, Signor PW (eds) Origin and early evolution of the metazoa. Plenum, New York, pp 53–84

Konovalov SK III, Luther G, Yücel M (2007) Porewater redox and processes in the Black Sea sediments. Chem Geol 245:254–274

Kostka JE, Gribsholt B, Petrie E, Dalton D, Skelton H (2002) The rates and pathways of carbon oxidation in bioturbated saltmarsh sediments. Limnol Oceanogr 47(1):230–240

Krantzberg G (1985) The influence of bioturbation on physical, chemical, and biological parameters in aquatic environments: a review. Environ Pollut 39(Series A):99–122

Kristensen E (2000) Organic matter diagenesis at the oxic/anoxic interface in coastal marine sediments, with emphasis on the role of burrowing animals. Hydrobiologia 426:1–24

Kristensen E, Ahmed SI, Devol AH (1995) Aerobic and anaerobic decomposition of organic matter in marine sediments: which is faster? Limnol Oceanogr 40(3):1430–1437

Kristensen E, Haese RR, Kostka JE (2005) Interactions between macro- and microorganisms in marine sediments. Coastal and estuarine studies. American Geophysical Union, Washington, DC

Kump LR, Pavlov AA, Arthur MA (2005) Massive release of hydrogen sulfide to the surface ocean and atmosphere during intervals of ocean anoxia. Geology 33(5):397–400

Kuypers MMM, Sliekers AO, Lavik G, Schmid M, Jørgensen BB, Kuenen JG, Damsté JSS, Strous M, Jetten MSM (2003) Anaerobic ammonium oxidation by anammox bacteria in the Black Sea. Nature 422:608–611

Lampitt RS, Antia AN (1997) Particle flux in deep seas: regional characteristics and temporal variability. Deep-Sea Res I 44(8):1377–1403

Lavik G, Stührmann T, Brüchert V, Van der Plas A, Mohrholz V, Lam P, Mußmann M et al (2009) Detoxification of sulphidic African shelf waters by blooming chemolithotrophs. Nature 457:581–584

Lepot K, Benzerara K, Brown GE Jr, Philippot P (2008) Microbially influenced formation of 2,724-million-year-old stromatolites. Nature Geosci 1:118–121

Levin LA, Gutierrez D, Rathburn AE, Neira C, Sellanes J, Munoz P, Gallardo VA, Salamance M (2002) Benthic processes on the Peru margin: a transect across the oxygen minimum zone during the 1997-98 El Niño. Prog Oceanogr 53:1–27

McKay CP, Hartman H (1991) Hydrogen peroxide and the evolution of oxygenic photosynthesis. Orig Life Evol Biosph 21:157–163

Meysman FJR, Middelburg JJ, Heip CHR (2006) New insights into Darwin's last idea: bioturbation. Trends Ecol Evol 21:688–695

Middelburg JJ, Levin LA (2009) Coastal hypoxia and sediment biogeochemistry. Biogeosciences 6:1273–1293

Moodley L, Middelburg JJ, Herman PMJ, Soetaert K, de Lange GJ (2005) Oxygenation and organic-matter preservation in marine sediments: direct experimental evidence from ancient organic carbon-rich deposits. Geology 33(11):889–892

Morante R, Hallam A (1996) Organic carbon isotopic record across the Triassic-Jurassic boundary in Austria and its bearing on the cause of the mass extinction. Geology 24(5):391–394

Müller PJ, Suess E (1979) Productivity, sedimentation rate, and sedimentary organic matter in the oceans- I. Organic carbon preservation. Deep Sea Res 26A:1347–1367

Nelson DC, Jørgensen BB, Revsbech NP (1986) Growth pattern and yield of a chemoautotrophic *Beggiatoa* sp. in oxygen-sulfide microgradients. Appl Environ Microbiol 52(2):225–233

Neretin LN (2006) Past and present water column anoxia, Nato Science Series, IV. Earth and environmental sciences, vol 64. Springer, Dordrecht

Nisbet EG, Sleep NH (2001) The habitat and nature of early life. Nature 409:1083–1091

Nisbet EG, Cann JR, Van Dover CL (1995) Origins of photosynthesis. Nature 373:479–480

Passier HF, Middelburg JJ, van Os BJH, De Lange GJ (1996) Diagenetic pyritisation under eastern Mediterranean sapropels caused by downward sulphide diffusion. Geochim Cosmochim Acta 60(5):751–763

Paulmier A, Ruiz-Pino D (2009) Oxygen minimum zones (OMZs) in modern ocean. Prog Oceanogr 80:113–128

Pemberton GS, Risk MJ, Buckley D (1976) Supershrimp: deep bioturbation in the Strait of Canso, Nova Scotia. Science 192:790–791

Petsch ST (2005) The global oxygen cycle. In: Schlesinger WH (ed) Biogeochemistry. Elsevier, Amsterdam, pp 515–556

Precht E, Franke U, Polerecky L, Huettel M (2004) Oxygen dynamics in permeable sediments with wave-driven pore water exchange. Limnol Oceanogr 49(3):693–705

Preisler A, De Beer D, Lichtschlag A, Lavik G, Boetius A, Jørgensen BB (2007) Biological and chemical sulfide oxidation in a Beggiatoa inhabited marine sediment. ISME J, 1, 341–353

Raiswell R, Berner RA (1985) Pyrite formation in euxinic and semi-euxinic sediments. Am J Sci 285:710–724

Reeburgh WS, Ward BB, Whalen SC, Sandbeck KA, Kilpatrick KA, Kerkhof LJ (1991) Black Sea methane geochemistry. Deep-Sea Res I 38(2):1189–1210

Revsbech NP, Soerensen J, Blackburn TH (1980) Distribution of oxygen in marine sediments measured with microelectrodes. Limnol Oceanogr 25(3):403–411

Revsbech NP, Madsen B, Jørgensen BB (1986) Oxygen production and consumption in sediments determined at high spatial resolution by computer simulation of oxygen microelectrode data. Limnol Oceanogr 31(2):293–304

Risgaard-Petersen N, Langezaal AM, Ingvardsen S, Schmid MC, Jetten MSM, Op den Camp HJM, Derksen JWM et al (2006) Evidence for complete denitrification in a benthic foraminifer. Nature 443:93–96

Rosenberg R, Diaz RJ (1993) Sulfur bacteria (Beggiatoa spp.) mats indicate hypoxic condition in the inner Stockholm Archipelago. Ambio 22(1):32–36

Sandberg CA, Morrow JR, Ziegler W (2002) Late Devonian sea-level changes, catastrophic events, and mass extinctions. Geol Soc Am Spec Pap 356:473–487

Schlanger SO, Jenkyns HC (1976) Cretaceous oceanic anoxic events: causes and consequences. Geol En Mijnbouw 55(3–4):179–184

Schulz HN, Brinkhoff T, Ferdelman TG, Hernández Mariné M, Teske A, Jørgensen BB (1999) Dense populations of a giant sulfur bacterium in Namibian shelf sediments. Science 284:493–495

Shen Y, Buick R, Canfield DE (2001) Isotopic evidence for microbial sulphate reduction in the early Archaean era. Nature 410:77–81

Shen Y, Knoll AH, Walter MR (2003) Evidence for low sulphate and anoxia in a mid-Proterozoic marine basin. Nature 423:632–635

Smith KL, Hinga KR (1983) Sediment community respiration in the deep sea. In: Rowe GT (ed) Deep Sea Biol. Wiley, New York

Strauss H, Strauss H (2005) Anoxia through time. In: Neretin LN (ed) Past and present water column anoxia, NATO science series: IV: Earth and environmental sciences. Springer, Dordrecht, p 541

Suess E (1980) Particulate organic carbon flux in the oceans – surface productivity and oxygen utilization. Nature 288:260–288

Sundby B, Silverberg N (1985) Manganese fluxes in the benthic boundary layer. Limnol Oceanogr 30(2):372–381

Sundby B, Gobeil C, Silverberg N, Mucci A (1992) The phosphorous cycle in coastal marine sediments. Limnol Oceanogr 37(6):1129–1145

Turgeon SC, Creaser RA (2008) Cretaceous oceanic anoxic event 2 triggered by a massive magmatic episode. Nature 454:323–327

Twitchett RJ (1999) Palaeoenvironments and faunal recovery after the end-Permian mass extinction. Palaeogeogr Paleoclimatol Palaeoecol 154:27–37

Ueno Y, Yamada K, Yoshida N, Maruyama S, Isozaki Y (2006) Evidence from fluid inclusions for microbial methanogenesis in the early Archaean era. Nature 440:516–519

van de Schootbrugge B, McArthur JM, Baily TR, Rosenthal Y, Wright JD, Miller KG (2005) Toarcian oceanic anoxic event: an assessment of global causes using belemnite C isotope records. Palaeogeography 20:1–10

Walker JCG (1987) Was the Archaean biosphere upside down? Nature 329:710–712

Walter MR (1983) Archean stromatolites - evidence of the earth's earliest benthos. In: Schopf JW (ed) Earth's earliest biosphere: its origin and evolution. Princeton University Press, Princeton, pp 187–213

Weeks SJ, Currie B, Bakun A (2002) Massive emissions of toxic gas in the Atlantic. Nature 415:493–494

Weeks SJ, Currie B, Bakun A, Kathleen RP (2004) Hydrogen sulphide eruptions in the Atlantic Ocean off southern Africa: implications of a new view based on Sea WiFS satellite imagery. Deep Sea Res I 51:153–172

Wenzhöfer F, Glud RN (2002) Benthic carbon mineralization in the Atlantic: a synthesis based on in situ data from the last decade. Deep Sea Res I 49:1255–1279

Wignall PB, Twitchett RJ (1996) Oceanic anoxia and the end Permian mass extinction. Science 272:1155–1158

Wilson TRS, Thomsen L, Colley S, Hydes DJ, Higgs NC (1985) Early organic diagenesis: the significance of progressive subsurface oxidation fronts in pelagic sediments. Geochim Cosmochim Ac 49:811–822

Ziebis W, Huettel M, Forster S (1996) The impact of biogenic sediment topography on oxygen fluxes in permeable sediments. Mar Ecol Prog Ser 140:227–237

Biodata of **Aharon Oren**, author of *"Diversity of Anaerobic Prokaryotes and Eukaryotes: Breaking Long-Established Dogmas."*

Aharon Oren is professor of microbial ecology at the Institute of Life Sciences, the Hebrew University of Jerusalem, Israel. He earned his M.Sc. degree from the University of Groningen, The Netherlands (1972) and his Ph.D. from the Hebrew University in 1978, and spent a postdoctoral period at the University of Illinois at Urbana-Champaign. In 2010, he received an honorary doctorate from the University of Osnabrück, Germany. His research centers on the ecology, physiology, biochemistry, and taxonomy of halophilic microorganisms and the microbiology of hypersaline and other extreme environments.

E-mail: **orena@cc.huji.ac.il**

DIVERSITY OF ANAEROBIC PROKARYOTES AND EUKARYOTES: BREAKING LONG-ESTABLISHED DOGMAS

AHARON OREN
Department of Plant and Environmental Sciences, The Institute of Life Sciences, and the Moshe Shilo Minerva Center for Marine Biogeochemistry, The Hebrew University of Jerusalem, 91904 Jerusalem, Israel

1. Introduction

We are long familiar with the existence of anaerobic eukaryotic life. It is known for centuries that yeasts can grow anaerobically while generating energy by fermentation of sugars to ethanol and CO_2. The existence of anaerobic protozoa in the rumen of cows and other ruminant animals was documented already in the first half of the nineteenth century. The rumen is populated not only by a variety of anaerobic bacteria but also contains a great diversity of ciliates, including holotrichs (e.g., *Dasytricha* and *Isotricha*) and entodiniomorphs (oligotrichs) (e.g., *Diplodinium*). These protozoa make a living in a way very similar to that of some of the fermentative bacteria in the rumen ecosystem: They ferment sugars to products such as H_2, CO_2, acetate, butyrate, lactate, and sometimes propionate as well (Hungate 1966; Wolin 1979).

Until recently, it was assumed that anaerobic life of animals is restricted to such unicellular organisms as are found in the digestive system of ruminants, in anaerobic sediments, and in some other ecosystems as well. All metazoa were considered obligate aerobes. This dogma was broken recently with the publication of a report documenting the existence of a diverse community of metazoans living in permanently anoxic conditions in the sediments of the deep anoxic hypersaline L'Atalante basin, 3.5 km below the surface of the Mediterranean Sea (Danovaro et al. 2010). The sediments of the L'Atalante basin are inhabited by three newly discovered species of the relatively recently discovered animal phylum Loricifera (*Spinoloricus* sp., *Rugiloricus* sp., and *Pliciloricus* sp.). These meiofauna (animals of <1 mm in size) organisms are metabolically active and show specific adaptations to the extreme conditions of the deep basin: a high pressure, high salinity, and high sulfide environment. They lack mitochondria and contain hydrogenosome-like organelles, associated with rod-shaped structures that may be endosymbiotic prokaryotes. Hydrogenosomes are membrane-enclosed organelles found in some anaerobic ciliates, trichomonads, and fungi, and they mediate the production of acetate, CO_2, H_2, and ATP from pyruvate (Martin and Müller 1998; Müller 1993). It has been speculated that the rod-shaped structures in close proximity to the hydrogenosome-like organelles may belong to the domain Archaea, similar to the methanogenic archaea often found inside anaerobic ciliates in association with hydrogenosomes (Levin 2010).

2. Novel Types of Anaerobic Metabolism in the Prokaryotic World

Not only in eukaryotic world dogmas were broken in recent years relating to their ability to live in the absence of molecular oxygen, our concepts on the ways anaerobic life is possible in the prokaryotes have greatly changed as well.

Much of the prokaryote diversity originated before oxygenic photosynthesis had evolved before cyanobacteria and later eukaryotic algae started to produce molecular oxygen as waste product of their photoautotrophic mode of life. As a result, we find a tremendous diversity of anaerobic metabolic processes in the prokaryotes, bacteria as well as archaea: anaerobic respiration, fermentation, anoxygenic photosynthesis, and others (Oren 2007; Schmitz et al. 2006; Zehnder and Svensson 1986).

Energy generation both in chemoheterotrophic (using organic compounds as carbon and energy source) and in chemolithotrophic metabolism (obtaining energy from the oxidation of inorganic compounds with the use of CO_2 as carbon source) is based on oxidation processes. Any oxidation has to be coupled with a reduction, and in aerobic respiration, molecular oxygen is used as electron acceptor. Many of the problems with which the anaerobic microbial world has to cope are related to the "quest for electron acceptors" (Oren 2007), the diverse ways how to get rid of excess electrons. In the prokaryote world, many compounds can serve as electron acceptor in "anaerobic respiration" processes. These include nitrate, nitrite, sulfate, elemental sulfur, Fe(III), Mn(IV), arsenate, selenate, dimethyl sulfoxide, trimethylamine-N-oxide, fumarate, and also CO_2 (if we consider oxidation of hydrogen by methanogenic and homoacetogenic prokaryotes as forms of anaerobic respiration) (Oren 2009).

In microbial metabolism, molecular oxygen serves not only as electron acceptor in respiration processes but also as a substrate in certain reactions catalyzed by monooxygenases and dioxygenases – enzymes that incorporate respectively one or two of the oxygen atoms from O_2 into the oxidation product. Here, oxygen can obviously not be replaced by other compounds. Two major aerobic microbial processes depend on such oxygenase reactions: the first step of nitrification in which ammonia is oxidized to nitrite and the oxidation of methane by aerobic methanotrophs. The mode of action of the key enzymes, ammonia monooxygenase and methane monooxygenase, is very similar, yielding hydroxylamine and methanol, respectively:

$$NH_3 + O_2 + 2 \text{ electrons} + 2H^+ \rightarrow NH_2OH + H_2O$$

$$CH_4 + O_2 + 2 \text{ electrons} + 2H^+ \rightarrow CH_3OH + H_2O$$

Because of the absolute requirement for oxygen of these two reactions, it was in the past claimed that ammonia and methane cannot be oxidized under anaerobic conditions. This dogma has been broken in recent years with the discovery of anaerobic ammonia oxidation (the "anammox" process) and no less than two completely different ways in which methane can be oxidized in anoxic environments.

The history of the discovery of the anammox process – the anaerobic oxidation of ammonia with nitrite as oxidant to yield molecular nitrogen – is an interesting one. The thermodynamic feasibility of the reaction was predicted by Broda (1977):

$$NH_4^+ + NO_2^- \to N_2 + 2H_2O \quad \Delta G_o' = -358 \text{ kJ}$$

It is little known that the same equation featured as early as 1930 in a textbook on bacterial metabolism, not as a mode of energy generation but in the reverse direction as a possible mechanism to explain the fixation of gaseous nitrogen in a period in which the mode of action of nitrogenase was not yet known (Stephenson 1930).

It was generally assumed that the energy-yielding reaction cannot occur in nature as no molecular oxygen-independent ways of oxidizing ammonia were known. Therefore, the discovery that the same process predicted by Broda was indeed operative in a denitrifying fluidized bed reactor (Mulder et al. 1995) came as a great surprise. The new chemolithotroph was identified to belong to the phylum *Planctomycetes* (Strous et al. 1999). Elucidation of the genome of one of the newly discovered anammox bacteria showed ammonia monooxygenase to be absent, and hydroxylamine does not serve as an intermediate. Instead, NO (nitric oxide) and – highly unexpectedly – N_2H_4 (hydrazine) are involved in the biochemical pathway of energy generation (Strous et al. 2006). When it was also shown that the process is widespread in nature, and may even contribute significantly to the nitrogen cycle in environments such as the Black Sea (Kuypers et al. 2003) and anoxic waters near Costa Rica (Dalsgaard et al. 2003), it had become clear that the existing concepts about the lack of dissimilatory anaerobic metabolism of ammonia had to be modified.

Strous et al. (2002) presented the anammox case as "a new experimental manifesto for microbiological ecophysiology." Using a modern approach based on the time-tested strategy of selective enrichment pioneered in the last decades of the nineteenth century and the first decades of the twentieth century by Sergei Winogradsky and Martinus Beijerinck, the old dogma of that ammonia oxidation required molecular oxygen was broken following postulation of an ecological niche based on thermodynamic considerations and macroecological field data, engineering of this niche into a laboratory bioreactor for enrichment culture, physiological characterization of the enrichment culture, phylogenetic characterization of the enriched community using molecular tools, and verification of the *in situ* importance of these species in the actual ecosystems.

Until a decade ago, the only way known to oxidize methane was based on the methane monooxygenase reaction shown above. However, already in the 1970s evidence accumulated to show that also in the absence of molecular oxygen, methane disappeared from anaerobic sediments, and in most cases the process appeared to be linked to the reduction of sulfate:

$$CH_4 + SO_4^{2-} + H^+ \rightarrow CO_2 + HS^- + 2H_2O \quad \Delta G_0' = -22 \text{ kJ}$$

In spite of the very small free energy change associated with this reaction, the process occurs in many places on the sea bottom where methane is evolved from methane seeps and dissolution of methane hydrates.

A consortium consisting of an archaeon which is the putative methane oxidizer and a bacterial sulfate reducer mediates the anaerobic oxidation of methane (Boetius et al. 2000). Attempts to cultivate the consortium in the laboratory were thus far unsuccessful, and the nature of the substance(s) transferred between the partners was never unequivocally ascertained. To obtain information about the possible mechanism of the reactions involved in the anaerobic oxidation of methane, a metagenomic approach was used. Starting with material collected from marine sediments near methane seeps, the analysis of the genetic material recovered from the site suggested that methane may be oxidized in a series of reactions of "reverse methanogenesis," reversing the process of methane formation from $H_2 + CO_2$. Nearly all genes needed for the process were detected (Hallam et al. 2004).

An even greater surprise was obtained during the recent analysis of another anaerobic methane oxidation process, this time coupled not with sulfate in a marine environment, but with nitrate. Following the successful enrichment of a microbial consortium that couples anaerobic methane oxidation to denitrification (Raghoebarsing et al. 2006), the methane oxidizing organism from the consortium was studied by metagenomic sequencing. The bacterium, provisionally named "*Candidatus* Methylomirabilis oxyfera," makes a living by performing the following reaction:

$$3CH_4 + 8NO_2^- + 8H^+ \rightarrow 3CO_2 + 4N_2 + 10H_2O \quad \Delta G_0' = -928 \text{ kJ/mol } CH_4$$

Genomic analysis showed that methane oxidation in this organism is based on the conventional methane monooxygenase characteristic of the aerobic methane oxidizers. Therefore, the organism has a special strategy to function in an anaerobic environment: if there is no molecular oxygen around and you still need it, then you should make your own. Instead of nitric oxide reductase and nitric oxide reductase,

$$NO_3^- \rightarrow NO_2^- \rightarrow NO \rightarrow N_2O \rightarrow N_2$$

two enzymes present in the classic denitrification pathway, "*Candidatus* Methylomirabilis oxyfera" possesses a novel enzyme, nitric oxide dismutase (NO dismutase) which forms oxygen and nitrogen gas from nitric oxide:

$$(NO_3^- \rightarrow) NO_2^- \rightarrow NO \rightarrow N_2 + O_2$$

The reduction of nitrate to nitrite is not performed by the methanotroph but by other microbes in the enrichment culture (Ettwig et al. 2010).

"*Candidatus* Methylomirabilis oxyfera" is thus a cryptic anaerobe. Oremland (2010) summarized its metabolism as follows: "Essentially, it contains its own little scuba tank that allows it to 'breathe' oxygen while immersed in methane-rich anoxic muck, thereby accessing the methane that conventional methanotrophs cannot reach."

There are more substrates that under aerobic conditions are metabolized in a chain of reactions initiated by a monooxygenase or a dioxygenase enzyme. Well-known examples are the saturated and aromatic hydrocarbons. But this does not imply that such compounds cannot be degraded under anaerobic conditions. Breakdown of aliphatic and aromatic hydrocarbons in the absence of molecular oxygen can be coupled with denitrification or with sulfate reduction. Anaerobic degradation of saturated and aromatic hydrocarbons can be initiated by a radical reaction with fumarate, yielding substituted succinates; in other cases, dehydrogenations with the formation of a secondary alcohol or carboxylation reactions appear to be involved (Widdel and Rabus 2001). Anaerobic degradation of aromatic compounds often involves a reductive attach as the first step, leading to the formation of benzoyl-CoA as intermediate (Harwood et al. 1999). There are other anaerobic pathways as well (Schmitz et al. 2006).

Finally, it is interesting to note that the second of the "two kinds of lithotrophs missing in nature" identified in the theoretical study by Broda (1977), namely an anoxygenic phototroph that derives its electrons from the oxidation of ammonia, has not been isolated yet. However, a highly similar process in which nitrite supplies the electrons and is oxidized to nitrate was recently discovered (Griffin et al. 2007).

3. Anaerobic Metabolism in the Eukaryotic World: Revisited

It is generally assumed that evolution of the eukaryotes largely occurred during times when most of the biosphere was already oxygenated by the photosynthetic action of cyanobacteria. If this is indeed the case, it is not surprising that the diversity of anaerobic metabolism among the eukaryotes is much more restricted than the wealth of different processes that enable prokaryotes to grow in the absence of molecular oxygen.

Still, we are from time to time surprised by new findings showing that the eukaryotic world knows some unexpected options to survive and even grow under anaerobic conditions. Anaerobic respiration using oxidized compounds of nitrogen, sulfur, and other elements, is indeed a process typical of the prokaryote world, and modes of anaerobic respiration, using alternative electron acceptors when oxygen is not available, are very rare among the eukaryotes. After it was discovered that foraminifera may proliferate in anoxic environments (Bernhard et al. 2006), it was found that a globobuliminid foraminiferan that lives in the oxygen-free zone of a Swedish fjord can use nitrate as electron acceptor and performs true denitrification, forming N_2 as end product. The organism can accumulate so much nitrate that sustained anaerobic growth is possible for more than a month (Risgaard-Petersen et al. 2006). It now appears that this mode of metabolism may occur in many other benthic foraminifera (Piña-Ochoa et al. 2010).

Most anaerobic eukaryotes make a living by fermentation, based on substrate-level phosphorylation. This is true for the yeasts producing alcohol as well as for the rumen protozoa. It is yet unknown what type of metabolism may drive the growth of the newly discovered Loricifera in the meiofauna of the sediments of L'Atalante basin. However, the presence of hydrogenosome-like organelles and associated prokaryote-like rods suggests that the process may be similar to that known from several types of anaerobic protozoa.

The finding of the anaerobic multicellular animals in sediment below the anoxic brine in the depths of the Mediterranean Sea may be not too surprising, as in general, when one searches in unusual environments, one finds unusual organisms. The discovery by Danovaro et al. (2010) provides the first evidence of a metazoan life cycle in permanently anoxic sediments. "Are there more metazoan taxa out there that can live without oxygen? Almost certainly" (Levin 2010).

4. References

Bernhard JM, Habura A, Bowser SS (2006) An endobiont-bearing allogromiid from the Santa Barbara Basin: implications for the early diversification of foraminifera. J Geophys Res 111:G03002. doi:10.1029/2005JG000158

Boetius A, Ravenschlag K, Schubert CJ, Rickert D, Widdel F, Gieseke A, Amann R, Jørgensen BB, Witte U, Pfannkuche O (2000) A marine microbial consortium apparently mediating anaerobic oxidation of methane. Nature 407:623–626

Broda E (1977) Two kinds of lithotrophs missing in nature. Z Allg Mikrobiol 17:491–493

Dalsgaard T, Canfield DE, Petersen J, Thamdrup B, Acuña-González J (2003) N_2 production by the anammox reaction in the anoxic water column of Golfo Dulce, Costa Rica. Nature 422:606–608

Danovaro R, Dell'Anno A, Pusceddu A, Gambi C, Heiner I, Møjberg Kristensen R (2010) The first metazoan living in permanently anoxic conditions. BMC Biol 8:30

Ettwig KF, Butler MK, Le Paslier D, Pelletier E, Mangenot S, Kuypers MMM, Schreiber F, Dulith BE, Zedelius J, de Beer D, Gloerich J, Wessels HJCT, van Alen T, Luesken F, Wu ML, van de Pas-Schoonen KT, Op den Camp HJM, Janssen-Megens EM, Francoijs K-J, Stunnenberg H, Weissenbach J, Jetten MSM, Strous M (2010) Nitrite-driven anaerobic methane oxidation by oxygenic bacteria. Nature 464:543–548

Griffin BM, Schott J, Schink B (2007) Nitrite, an electron donor for anoxygenic photosynthesis. Science 316:1870

Hallam SJ, Putnam N, Preston CM, Detter JC, Rokhsar D, Richardson PM, DeLong EF (2004) Reverse methanogenesis: testing the hypothesis with environmental genomics. Science 305:1457–1462

Harwood CS, Burchhardt G, Herrmann H, Fuchs G (1999) Anaerobic metabolism of aromatic compounds via the benzoyl-CoA pathway. FEMS Microbiol Rev 22:439–458

Hungate RE (1966) The rumen and its microbes. Academic, New York

Kuypers MMM, Sliekers AO, Lavik G, Schmid M, Jørgensen BB, Kuenen JG, Sinninghe Damsté JS, Strous M, Jetten MSM (2003) Anaerobic ammonium oxidation by anammox bacteria in the Black Sea. Nature 422:608–611

Levin LA (2010) Anaerobic metazoans: no longer an oxymoron. BMC Biol 8:31

Martin W, Müller M (1998) The hydrogen hypothesis for the first eukaryote. Nature 392:37–41

Mulder A, van de Graaf AA, Robertson LA, Kuenen JG (1995) Anaerobic ammonium oxidation discovered in a denitrifying fluidized bed reactor. FEMS Microbiol Ecol 16:177–184

Müller M (1993) The hydrogenosome. J Gen Microbiol 139:2879–2889

Oremland RS (2010) NO connection with methane. Nature 464:500–501

Oren A (2007) Anaerobes. In: Encyclopedia of life sciences. Wiley, Chichester. doi:10.1002/9780470015902.a0020369

Oren A (2009) Anaerobic respiration. In: Encyclopedia of life sciences. Wiley, Chichester. doi:10.1002/9780470015902.a0001414.pub2

Piña-Ochoa E, Høgslund S, Geslin E, Cedhagen T, Revsbech NP, Nielsen LP, Schweizer M, Jorissen F, Rysdaard S, Risgaard-Petersen N (2010) Widespread occurrence of nitrate storage and denitrification among Foraminifera and *Gromiida*. Proc Natl Acad Sci USA 107:1148–1153

Raghoebarsing AA, Pol A, van de Pas-Schoonen KT, Smolders AJP, Ettwig KF, Rijpstra WIC, Schouten S, Sinninghe Damsté JS, Op den Camp HJM, Jetten MSM, Strous M (2006) A microbial consortium couples anaerobic methane oxidation to denitrification. Nature 440:918–921

Risgaard-Petersen N, Langezaal AM, Ingvardsen S, Schmid MC, Jetten MSM, Op den Camp HJM, Derksen JWM, Piña-Ochoa E, Eriksson SP, Nielsen LP, Revsbech NP, Cedhagen T, van der Zwaan GJ (2006) Evidence for complete denitrification in a benthic foraminifer. Nature 443:93–96

Schmitz R, Daniel R, Deppenmeier U, Gottschalk G (2006) The anaerobic way of life. In: Dworkin M, Falkow S, Rosenberg E, Schleifer K-H, Stackebrandt E (eds) The prokaryotes. A handbook on the biology of bacteria: ecophysiology and biochemistry, vol 2. Springer, New York, pp 86–101

Stephenson M (1930) Bacterial metabolism, 1st edn. Longmans/Green & Co, London

Strous M, Fuerst JA, Kramer EHM, Logemann S, Muyzer G, van de Pas-Schoonen KT, Webb R, Kuenen JG, Jetten MSM (1999) Missing lithotroph identified as new planctomycete. Nature 400:446–449

Strous M, Kuenen JG, Fuerst JA, Wagner M, Jetten MSM (2002) The anammox case – a new experimental manifesto for microbiological eco-physiology. Antonie Van Leeuwenhoek 81:693–702

Strous M, Pelletier E, Mangenot S, Rattei T, Lehner A, Taylor MW, Horn M, Daims H, Bartol-Mavel D, Wincker P, Barbe V, Fonknechten N, Vallenet D, Segurens B, Schenowitz-Truong C, Médigue C, Collingro A, Snel B, Dulith BE, Op den Camp HJM, van der Drift C, Cirpus I, van de Pas-Schoonen KT, Harhangi HR, van Niftrik L, Schmid M, Keltjens J, van de Vossenberg J, Kartal B, Meier H, Frishman D, Huynen MA, Mewes H-W, Weissenbach J, Jetten MSM, Wagner M, Le Paslier D (2006) Deciphering the evolution and metabolism of an anammox bacterium from a community genome. Nature 440:790–794

Widdel F, Rabus R (2001) Anaerobic biodegradation of saturated and aromatic hydrocarbons. Curr Opin Biotechnol 12:259–276

Wolin MJ (1979) The rumen fermentation: a model for microbial interactions in anaerobic ecosystems. Adv Microb Ecol 3:49–77

Zehnder AJB, Svensson BH (1986) Life without oxygen: what can and what cannot? Experientia 42:1197–1205

PART II:
FUNCTIONAL BIOCHEMISTRY

Tsaousis
Leger
Stairs
Roger
De Graaf
Hackstein
Levy
Levy

Biodata of **Anastasios D. Tsaousis**, **Michelle M. Leger**, **Courtney A.W. Stairs**, and **Andrew J. Roger**, authors of "*The Biochemical Adaptations of Mitochondrion-Related Organelles of Parasitic and Free-Living Microbial Eukaryotes to Low Oxygen Environments.*"

Dr. Anastasios D. Tsaousis is currently a post-doctoral fellow at the Dalhousie University, Canada. He obtained his B.Sc. from University of Crete, Greece, in 2003 and his Ph.D. from Newcastle University, United Kingdom, in 2007. His scientific interests include evolution and function of mitochondrion-related organelles, anaerobic metabolic adaptations of protozoa, metabolic exchange between organelles and their hosts, and early eukaryotic evolution.

E-mail: **tsaousis.anastasios@gmail.com**

Michelle M. Leger is currently studying for a Ph.D. at Dalhousie University, Canada. She obtained her B.Sc. (Hons.) from the University of York, UK, in 2005, and her M.Sc. from the University of British Columbia, Canada, in 2007. Her scientific interests are in the areas of early eukaryotic evolution, mitochondrion-related organelles, organellar reduction, and adaptations to anaerobic environments in microbial eukaryotes.

E-mail: **m.leger@dal.ca**

Anastasios D. Tsaousis   **Michelle M. Leger**

Courtney A.W. Stairs is currently a Ph.D. candidate at Dalhousie University in Halifax, Nova Scotia, Canada. She obtained her B.Sc. (Hons.) from Dalhousie University in 2009. Courtney's areas of interest include the evolution and biochemistry of mitochondrion-related organelles and their associated anaerobic metabolism.

E-mail: **courtney.stairs@dal.ca**

Professor Andrew J. Roger is currently Professor of Biochemistry and Molecular Biology and Canada Research Chair in Comparative Genomics and Evolutionary Bioinformatics at Dalhousie University in Halifax, Nova Scotia, Canada. He obtained his Ph.D. from Dalhousie University in 1997 and completed his post-doctoral studies at the Marine Biological Laboratory in Woods Hole, Massachusetts, USA. Professor Roger's scientific interests include the early evolutionary history of eukaryotic cells, their organelles and genomes; microbial eukaryote biodiversity; biochemical adaptations to low oxygen conditions; as well as statistical modeling of molecular evolution.

E-mail: **andrew.roger@dal.ca**

Courtney A.W. Stairs   **Andrew J. Roger**

THE BIOCHEMICAL ADAPTATIONS OF MITOCHONDRION-RELATED ORGANELLES OF PARASITIC AND FREE-LIVING MICROBIAL EUKARYOTES TO LOW OXYGEN ENVIRONMENTS

ANASTASIOS D. TSAOUSIS, MICHELLE M. LEGER*,
COURTNEY A.W. STAIRS*, AND ANDREW J. ROGER
Department of Biochemistry and Molecular Biology, Centre for Comparative Genomics and Evolutionary Bioinformatics, Dalhousie University, Halifax, Canada B3H 4R2

1. Introduction

The majority of cellular and biochemical diversity within the eukaryotic realm is found within microbial lineages (Fig. 1). Over the last 1–2 billion years of evolution, unicellular eukaryotes have invaded a wide spectrum of habitats on the Earth through the evolutionary differentiation and specialization of their external morphology, cellular ultrastructure, biochemistry, and metabolism, and the compartmentation of these processes in organelles within their cells. Low oxygen environments are particularly interesting in this respect because they represent a variety of habitats ranging from freshwater and marine sediments to oxygen-depleted zones in the water column to the animal intestinal tract. The unicellular eukaryotes that occupy these environments are similarly diverse, displaying a variety of trophic types ranging from heterotrophic free-living flagellated and amoeboid forms to specialized obligate intracellular parasites. Furthermore, all of these unicellular eukaryotes have adapted to low oxygen environments by the loss of aerobic respiration and by retailoring their mitochondria into one of a number of types of mitochondrion-related organelles (e.g., hydrogenosomes and mitosomes). These organelles are found in single-celled anaerobic parasites such as *Giardia*, *Cryptosporidium*, *Entamoeba histolytica*, *Blastocystis* sp., *Trichomonas vaginalis*, and microsporidia, but also in free-living protists such as *Mastigamoeba balamuthi*, *Andalucia incarcerata*, *Trimastix pyriformis*, *Sawyeria marylandensis*, and *Psalteriomonas lanterna*. Understanding the functions of these organelles in the aforementioned organisms is of key importance to understanding their various adaptations to anoxic environments. Moreover, it is of great interest to clarify how mitochondria, energy-generating organelles that rely on oxygen to function in most eukaryotes, have changed during adaptation of the host organisms to anaerobic and, sometimes, parasitic lifestyles. Since mitochondria and mitochondrion-related organelles (MROs) are apparently vital to all eukaryotic

---

*These authors contributed equally in this work.

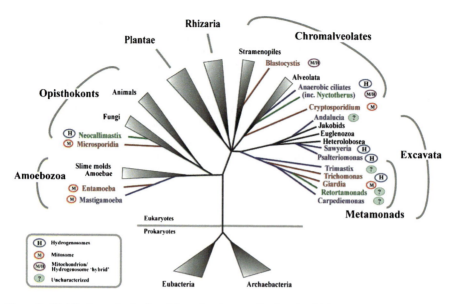

Figure 1. Distribution of mitochondrion-related organelles in microbial eukaryotes across the six super groups of eukaryotes. The six major eukaryotic super groups are labeled in *bold* and *larger fonts*. Anaerobic protozoan lineages are shown in *red* (parasites), *green* (commensals), and *blue* (free living). Their mitochondrion-related organelles are shown in *ovals* next to the species names: *H* hydrogenosome, *M* mitosome, *M/H* mitochondrion/hydrogenosome "hybrid," *?* uncharacterized organelle (The figure was modified from an original version that was kindly provided by Alastair G. B. Simpson).

cells, and transitions to anaerobic and parasitic lifestyles have happened in at least a dozen eukaryotic lineages independently (Fig. 1), understanding this process is of fundamental importance to evolutionary and biomedical science.

1.1. MITOCHONDRIA AND MITOCHONDRION-RELATED ORGANELLES

All living eukaryotes appear to have diverged after the endosymbiotic integration of the α-proteobacterial symbiont that gave rise to mitochondria and mitochondrion-related organelles (MROs) (Roger and Silberman 2002; Embley and Martin 2006). Aerobic respiration is one of the key functions of mitochondria of aerobic eukaryotes that provides the majority of ATP required for these cells. Energy metabolism in these organisms typically utilizes pyruvate produced by glycolysis that is decarboxylated by the pyruvate dehydrogenase complex (PDC) generating acetyl-CoA, which is subsequently directed into the TCA cycle, generating one molecule of ATP (via GTP) by substrate level phosphorylation and reducing NAD^+ to NADH and FAD to $FADH_2$. The four NADH and one $FADH_2$ generated in the catabolism of one molecule of pyruvate are oxidized by complexes of the electron transport chain (ETC), where they ultimately reduce

O_2 to produce H_2O. The movement of these electrons through the ETC generates a proton gradient across the inner mitochondrial membrane that drives ATP production by the F_1F_o ATP synthase; the ATP produced is then exported to the cytosol via an ATP/ADP transporter that belongs to the mitochondrial carrier family (MCF). However, ATP generation is not the only key process in the aerobic mitochondrion, since only 15% of mitochondrial proteins in model organisms such as *Saccharomyces cerevisiae* are associated with energy metabolism (Sickmann et al. 2003). Other important processes within mitochondria include heme biosynthesis, β-oxidation of fatty acids, amino acid metabolism, mitochondrial genome maintenance and gene expression, protein folding and translocation, mitochondrial biogenesis (fusion and fission), and apoptosis. One of these functions, the biosynthesis of iron–sulfur (Fe–S) clusters, has been suggested to be a core function that may be common to all mitochondria (Lill 2009).

In contrast to aerobes, anaerobic microbial eukaryotes have a variety of functionally distinct MROs, many of which have been classified as either hydrogenosomes or mitosomes. The hydrogenosomes of parasitic protozoa such as trichomonads and commensals such as rumen chytrid fungi and ciliates (Fig. 1) represent the best-studied MROs. Putative hydrogenosomes have also been described in free-living organisms such as *Psalteriomonas lanterna* (de Graaf et al. 2009) and *Sawyeria marylandensis* (Barbera et al. 2010), suggesting that this kind of functional adaptation is unlikely to be restricted to parasitic organisms. In addition, a hydrogenosome-like structure has been shown to exist in three free-living anoxic metazoan species of the animal phylum Loricifera, although no biochemical or molecular evidence is available to support this designation (Danovaro et al. 2010). Like aerobic mitochondria, hydrogenosomes are double-membrane-bounded organelles; some of them have well-defined cristae (Finlay and Fenchel 1989) while others do not (e.g., trichomonad hydrogenosomes), and most of them lack an organellar genome [with the exceptions of *Blastocystis* and *Nyctotherus* discussed below (Boxma et al. 2005; Perez-Brocal and Clark 2008)]. Some hydrogenosomes have been demonstrated to possess homologues of mitochondrial proteins, including chaperones (Bui et al. 1996; van der Giezen et al. 2003), proteins of the mitochondrial carrier family (Dyall et al. 2000; van der Giezen et al. 2002; Tjaden et al. 2004), conserved proteins of the Fe–S cluster biosynthesis pathway (Sutak et al. 2004), and proteins homologous to the 51-kDa and 24-kDa subunits of the mitochondrial NADH dehydrogenase module in complex I (Hrdy et al. 2004). The main defining feature of hydrogenosomes is the oxidative decarboxylation of malate or pyruvate to ultimately produce acetate, carbon dioxide (CO_2), and molecular hydrogen (H_2), with the simultaneous production of ATP (Muller 1993, 2007), which is then exported to the host cell using an ATP/ADP transporter (van der Giezen et al. 2002; Tjaden et al. 2004). In contrast to classical aerobic mitochondria, which contain pyruvate dehydrogenase, hydrogenosomes contain pyruvate:ferredoxin oxidoreductase (PFO) and [FeFe]-hydrogenase, enzymes that are common among anaerobic bacteria and which do not appear (so far) to be of α-proteobacterial origin (Horner et al. 2000; Hug et al. 2010). PFO activity has been demonstrated in the hydrogenosomes of the chytrid fungi *Neocallimastix patriciarum*

(Yarlett et al. 1986). Furthermore, *Neocallimastix* sp. and *Piromyces* sp. E2 use the enzyme pyruvate formate lyase (PFL) to convert pyruvate into acetyl-CoA and formate (Akhmanova et al. 1999; Boxma et al. 2004).

Recent genomic/transcriptomic investigations have revealed two anaerobic protists: the ciliate *Nyctotherus ovalis* (Boxma et al. 2005) and the stramenopile *Blastocystis* sp. (Stechmann et al. 2008) contain MROs with hydrogenosomal properties (i.e., [FeFe]-hydrogenase and associated enzymes). Although they seem to lack the capacity for aerobic respiration, MROs from both organisms possess mitochondrial genomes and are predicted to contain a large number of mitochondrial proteins that presumably carry out canonical mitochondrial functions. These organelles have been described as mitochondrion–hydrogenosome (M/H) intermediates or "hybrid" organelles (Fig. 1). Further investigations of these organelles are likely to yield insights into the early stages of adaptation of mitochondria to anoxic conditions (Tsaousis et al. 2010).

In addition to these MROs, double-membrane-bounded organelles, collectively referred to as mitosomes, have been identified in other parasitic (e.g., *Entamoeba, Giardia, Cryptosporidium*, and microsporidia) and free-living protozoa (e.g., *Mastigamoeba*). The biochemical functions of these organelles are mostly unknown. Mitosomes are significantly smaller than typical mitochondria and do not contain cristae or an organellar genome. Mitochondrial proteins, including chaperones (Mai et al. 1999; Tovar et al. 1999; Williams et al. 2002; Gill et al. 2007; Tsaousis et al. 2008), and conserved proteins of Fe–S cluster biosynthesis (Tovar et al. 2003; Goldberg et al. 2008) have been localized to the mitosomes in some of these organisms. Although the exact function(s) of these organelles remain(s) unclear, they are provisionally defined as organelles that consume, but do not produce, ATP (Chan et al. 2005; Tsaousis et al. 2008), and most mitosomes are thought to function in Fe–S cluster synthesis (Embley and Martin 2006). The mitosomes of *Entamoeba histolytica* may represent a possible exception as there is controversy concerning the localization of the unusual Fe–S cluster assembly machinery of this organism (Mi-ichi et al. 2009; Maralikova et al. 2010). Recent work suggests that the sulfate activation pathway, typically confined to the cytoplasm or plastids in other eukaryotes, is present in these organelles and may represent their primary function (Mi-ichi et al. 2009). Curiously, these mitosomes appear to constitute 1.4–1.9% of the total cell volume (Maralikova et al. 2010), a surprisingly large volume (canonical mitochondria occupy 15–20% of the volume of the cell) considering that these organelles do not appear to be involved in energy metabolism.

In the following sections, we will concentrate on a number of functions associated with MROs of anaerobic organisms that have been studied to date, focusing mainly on the better-studied pathways that were acquired or maintained as a result of adaptation to low oxygen habitats. These pathways include alternative modes of ATP production and pyruvate metabolism, as well as conserved pathways found in most MROs such as Fe–S cluster assembly, metabolic exchange, and protein import that are related directly or indirectly to the anaerobic adaptations of these organelles.

2. Protein Import into Mitochondrion-Related Organelles

With the exceptions of *Blastocystis* and *Nyctotherus*, MROs from all other anaerobic unicellular eukaryotes that have been studied to date appear to lack a genome as well as associated transcriptional and translational machineries. In order for these anaerobic organelles to function in the absence of their own protein synthesis apparatus, they must import proteins from the cytosol; the mechanism by which this is accomplished is apparently shared between all MROs studied to date (Embley and Martin 2006). The mitochondrial protein import machinery of model system eukaryotes consists of five oligomeric complexes: the Translocase of the Outer Membrane (TOM) complex, the Translocase of the Inner Membrane 22 and 23 (TIM22 and TIM23) complexes, the Sorting and Assembling Machinery (SAM), and the Mitochondrial Intermembrane space Assembly (MIA) [for review see Lithgow and Schneider (2010)]. The TOM complex is responsible for the recognition and translocation of mitochondrial-targeted proteins through the outer mitochondrial membrane, whereas after passing through the intermembrane space via the MIA complex, they are then translocated through the mitochondrial matrix or inner mitochondrial membrane by the TIM22 or TIM23, respectively. Mitochondrial proteins that are targeted to the outer membrane of the organelle are first transported through the TOM complex to the *trans* side of the outer membrane, where the TIM chaperone complexes guide the hydrophobic precursors to the SAM complex (Pfanner et al. 2004).

Essential proteins of the mitochondrial protein import machinery have been detected in the genomes of all anaerobic protozoa to date, suggesting that this machinery is indispensable for the maintenance of these organelles. Nonetheless, some orthologues of the yeast mitochondrial protein import apparatus that have been deemed "essential" are apparently absent in completely sequenced genomes of some anaerobic protozoa (Dolezal et al. 2006). This suggests that either the sensitivity of the current bioinformatics methods is not sufficient to detect these highly divergent transmembrane proteins or that the proteins have been lost through reductive evolution (Dolezal et al. 2006; Embley and Martin 2006). Interestingly, some hydrogenosomal and mitosomal proteins can be imported into yeast and/or *Trichomonas* mitochondria or hydrogenosomes (Burri et al. 2006; Dolezal et al. 2010; Tsaousis et al. 2011) by heterologous expression of their native sequences, suggesting the conservation of import signals between mitosomes, hydrogenosomes, and yeast mitochondria. The ancestral state of the mitochondrial protein import machinery of the last common ancestor of eukaryotes is still unclear. However, new genomic and transcriptomic data, along with functional studies of mitochondrial protein import machineries from diverse aerobic and anaerobic lineages, should allow us to determine the presence, absence, and functions of various components of the TIM, TOM, MIA, and SAM systems in different lineages of eukaryotes and clarify their origins (Tsaousis et al. 2011).

3. Organellar Substrate Exchange: The Mitochondrial Carrier Family

Members of the mitochondrial carrier family (MCF) of proteins mediate the exchange of molecules between metabolic pathways in the cytosol and the mitochondrial matrix (Arco and Satrustegui 2005). Model system aerobic eukaryotes typically have between 30 and 60 different MCF proteins, which are specialized to transport different molecules (Kunji 2004). These carriers are divided in several classes, which include the nucleotide carrier class (e.g., ATP/ADP exchange carrier), the amino acid carrier class (e.g., aspartate/glutamate), the keto acid class, the inorganic phosphate (Pi) carrier, and the iron carrier (Kunji and Robinson 2006). By contrast, the genomes of the malaria parasite *Plasmodium falciparum* (Gardner et al. 2002), the anaerobic urogenital parasite *Trichomonas vaginalis* (Carlton et al. 2007), and the anaerobic intestinal parasite *Cryptosporidium parvum* (Abrahamsen et al. 2004) encode only 9, 5, and 5 MCF proteins, respectively. Although the precise function of many of these detected MCFs remains to be experimentally tested, one member of this family of transporters in *T. vaginalis* is known to function as an ATP/ADP carrier that is used to export anaerobically produced ATP in hydrogenosomes to the cytosol (Tjaden et al. 2004). The tendency of these parasites to import host metabolites rather than synthesizing metabolites *de novo* allows for reduction or elimination of many metabolic pathways and transport mechanisms related to these pathways in their MROs (Chan et al. 2005). An example of extreme reduction is found in the genome of the parasitic amoeba *Entamoeba histolytica* (Loftus et al. 2005) and the microsporidian *Antonospora locustae*, which both apparently encode only a single mitochondrial carrier protein, localized in their mitosomes and specialized to transport ATP in exchange for ADP (Chan et al. 2005; Williams et al. 2008). Even stranger, the genomes of the microsporidian *Encephalitozoon cuniculi* and the diplomonad *Giardia intestinalis* (Katinka et al. 2001; Morrison et al. 2007) do not encode any recognizable MCFs (Boxma et al. 2007), raising the question of how their mitosomes acquire ATP (although it is possible that MCFs in these organism are so divergent as to be unrecognizable on the basis of sequence similarity). In the case of *E. cuniculi*, a non-mitochondrial *Rickettsia/Chlamydia*-like nucleotide transporter (NTT) has been characterized and localized in the mitosome, potentially explaining how the organelle acquires ATP (Tsaousis et al. 2008).

4. Iron–Sulfur Cluster Assembly

Proteins that contain iron–sulfur (Fe–S) clusters (referred as Fe–S proteins) have pivotal roles in the function of all cells; for example, in *Escherichia coli*, almost 80 proteins depend on Fe–S clusters in order to function (Py and Barras 2010). In eukaryotic cells, Fe–S proteins are present in mitochondria, the cytoplasm, the nucleus (Gerber and Lill 2002), and in plastids (Lill and Kispal 2000). Fe–S proteins are responsible for central functions in enzymatic catalysis, electron transport, and regulation of gene expression (Beinert and Kiley 1999; Lill and Muhlenhoff 2006). In aerobic mitochondria, Fe–S proteins are involved in the respiratory chain (subunits of complexes I, II, and III) and in the citric acid cycle (aconitase) (Lill and Kispal 2000). Characteristic proteins of anaerobic MROs, such as [FeFe]-hydrogenase and PFO, also contain Fe–S clusters, explaining the necessity of this machinery in the organelle. The mitochondrial Fe–S cluster assembly machinery is required not only for production of mitochondrial Fe–S proteins, but it is also essential for the biosynthesis of Fe–S proteins in the cytosol and nucleus (Lill and Kispal 2000).

In bacteria and eukaryotes, the assembly of the Fe–S clusters can be achieved by three different systems (Lill and Muhlenhoff 2005). These include the iron–sulfur cluster (ISC) assembly and sulfur mobilization (SUF) machineries that are responsible for general assembly of Fe–S proteins, and the nitrogen fixation (NIF) system of nitrogen fixing bacteria which specializes in the assembly of the Fe–S clusters of the nitrogenase complex metalloprotein (Lill and Muhlenhoff 2006). In eukaryotes, the formation and export of Fe–S clusters for mitochondria, cytosolic, and nuclear proteins is an important function of the mitochondrial ISC system; indeed, it is the only known mitochondrial biosynthetic pathway that is essential for cell viability in yeast (Lill and Kispal 2000). In contrast, the SUF machinery is found in plastids, in some Bacteria and Archaebacteria and is usually expressed under oxygen stress or iron limitation [for reviews see Lill (2009) and/or Py and Barras (2010)].

The common mechanism of the assembly of Fe–S clusters depends on the enzymatic transfer of molecular sulfur from cysteine by a cysteine desulfurase and of molecular iron to a scaffold iron-binding protein to form the Fe–S clusters, which by a subsequent mechanism are transferred to the candidate apoproteins (Fig. 2a). In mitochondria, this reaction can be achieved by several proteins: the IscS(Nfs)/Isd11 complex is the cysteine desulfurase, frataxin (Yfh1p) is the iron donor (or iron storage protein), and the IscU (Isu) is the scaffold protein (Fig. 2b). Several other proteins were shown to be directly or indirectly involved with the formation and export of Fe–S clusters [for further reading, see (Lill 2009)]. The IscS and IscU proteins are encoded by all of the eukaryotic genomes sequenced so far, with the exception of *Entamoeba* and *Mastigamoeba* (see below). While studies on these proteins in all other anaerobic protozoa that have been investigated have demonstrated their localization and function inside the MROs (LaGier et al. 2003; Tovar et al. 2003; Sutak et al. 2004; Goldberg et al. 2008), in the microsporidian *Trachipleistophora hominis*, the main pools of IscU and frataxin were localized in the cytosol of the parasite (Goldberg et al. 2008) while IscS was in mitosomes. Although it is possible that a minor pool of IscU and frataxin exists in the *T. hominis* mitosome (and likewise that some IscS functions in the cytosol), no evidence was obtained for this from the immunofluorescence microscopy data. This disjunct distribution of these three proteins raises intriguing questions as to how the separated ISC components can perform their usually tightly coordinated function, and consequently it suggests that the Fe–S cluster assembly machinery

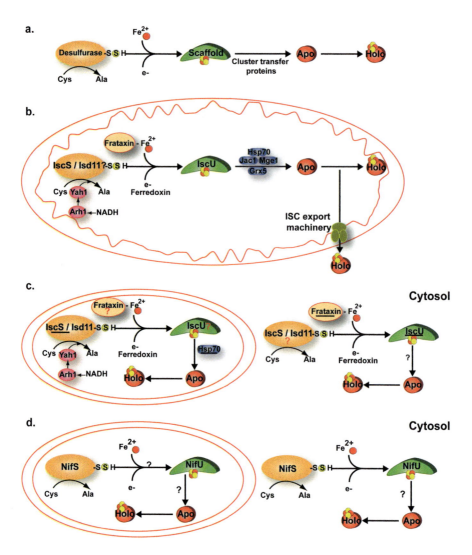

Figure 2. Models for Fe–S cluster biosynthesis in mitochondria and MROs. (**a**) Schematic representation of Fe–S protein biosynthesis. Enzymatic transfer of molecular sulfur from cysteine by a cysteine desulfurase and of molecular iron to a scaffold iron-binding protein to form the Fe–S clusters, which by a subsequent mechanism are transferred to the candidate apoproteins. (**b**) A model of Fe–S cluster biosynthesis in mitochondrion-related organelles. The biogenesis of Fe–S cluster first requires iron that binds to IscU (*green*), which serves as a scaffold for synthesis of the ISC. Metal delivery to IscU is assisted by frataxin (Yfh1), which directly binds to the IscU proteins. ISC synthesis on IscU further requires the cysteine desulfurase complex (IscS–Isd11 or Nfs1–Isd11), which mediates the release of sulfur from cysteine, and the electron-transfer chain consisting of NADH, ferredoxin reductase (Arh1), and ferredoxin (Yah1). The chaperone system, which consists of Hsp70 (Ssq1), Jac1, and Mge1, and the glutaredoxin Grx5 are required after ISC assembly on the IscU proteins; e.g., in ISC transfer to the apoproteins. Several other proteins (e.g., Isa1/Isa2 and Nfu1) are involved in the biogenesis,

in *T. hominis* may be in the process of relocating from the mitosome to the cytosol (Goldberg et al. 2008) (Fig. 2c).

The ISC system in most eukaryotes is derived from an ancient endosymbiotic gene transfer from the original α-proteobacteria that gave rise to mitochondria (Lill and Muhlenhoff 2006). The one exception is the ε-proteobacterial-type NIF Fe–S cluster biosynthesis system (and the apparent absence of the ISC system) in *Entamoeba* (Ali et al. 2004; van der Giezen et al. 2004) and its free-living relative *Mastigamoeba*. One localization study on the NIF system in *Entamoeba histolytica* suggested that it was mainly located in the cytosol (Mi-ichi et al. 2009) (Fig. 2d). In contrast, subsequent studies have claimed a dual localization in both cytosol and the mitosome, with a 10-fold higher concentration of the NIF proteins within mitosomes (Maralikova et al. 2010). Since there appears to be at least one Fe–S protein in the mitosomes of *Entamoeba* organelle, rubrerythrin (Maralikova et al. 2010), there must be a mitosomal Fe–S cluster biosynthesis machinery or a mechanism of Fe–S cluster import into the organelle. In the case of *Mastigamoeba*, several Fe–S proteins have been predicted to be present in the organelle (e.g., succinate dehydrogenase, aconitase, ferredoxin, PFO) (Gill et al. 2007), suggesting a dual localization of the NIF system (cytosol and mitosome), an alternative mitosomal Fe–S cluster biosynthesis pathway, or the presence of Fe–S cluster import system (Fig. 2d).

Although Fe–S clusters can be found as rhombic, [2Fe–2S], or cubic, [4Fe–4S] forms, the iron can change oxidation status from Fe^{2+} to Fe^{3+}, while sulfur is always present in the S^{2-} oxidation state. Under the presence of molecular oxygen, some Fe–S clusters are unstable. For example, in proteins harboring an [4Fe–4S] cluster, under the presence of excess oxygen, an exposed Fe^{2+} atom can be oxidized and released, resulting in an unstable cluster $[3Fe-4S]^+$, which is subsequently degraded to free Fe^{2+}, Fe^{3+}, and S^{2-} ions. To avoid this, organisms have developed multicomponent systems that promote the biogenesis of Fe–S proteins while protecting the cellular surroundings from the potentially detrimental effects of free Fe^{2+}, Fe^{3+}, and S^{2-} ions [for review see Py and Barras (2010)]. Sequence analyses

Figure 2. (continued) but their function and role is not well established. The ISC export machinery, which consists of an ABC transporter Atm1 of the inner membrane, exports an unknown compound to the cytosol for the need of the extramitochondrial Fe–S proteins. (**c**) Schematic hypothetical model for Fe–S cluster biosynthesis in the microsporidian *Trachipleistophora hominis*. Different components of the Fe–S biosynthesis pathway in *T. hominis* are not localized together (underlined proteins). The mtHsp70 and the essential sulfur donor (IscS) are localized in the *T. hominis* mitosome, whereas the main pools of the scaffold protein (IscU) and frataxin (Yfh1) were shown to be cytosolic (Goldberg et al. 2008). These results suggest that in *T. hominis* the Fe–S cluster assembly machinery may be relocating from the mitosome to the cytosol. Proteins with "?" have not been shown to localize in the compartments. (**d**) Schematic model of dual localization of NIF Fe–S cluster biosynthesis in *Entamoeba*. Both components (NifS and NifU) of the Fe–S cluster machinery of *Entamoeba* have been localized in both the cytosol and in the mitosomes of the parasite (Maralikova et al. 2010).

of whole genomes have discovered that these systems are highly conserved in Bacteria and eukaryotes (Py and Barras 2010), although the situation in Archaebacteria is still unclear. One such system is the SUF mobilization, which is highly expressed under oxygen stress conditions in bacteria and plastids (Py and Barras 2010). The majority of parasitic protozoa do not have plastids containing the SUF biogenesis machinery, and thus do not appear to harbor a system for recovery of the Fe–S clusters under oxygen stress conditions. It is therefore of interest to understand whether (and how) these organisms have adapted to overcome the toxic effects of Fe–S cluster oxidation that could occur when they are exposed transiently to oxygen.

5. Pyruvate Metabolism

The canonical pyruvate dehydrogenase complex (PDC), which is composed of three enzymes: pyruvate dehydrogenase (E1; EC 1.2.4.1), dihydrolipoyl transacetylase (E2; EC 2.3.1.12), and dihydrolipoyl dehydrogenase (E3, EC 1.8.1.4) catalyzes the oxidative decarboxylation of glycolysis-generated pyruvate to acetyl-CoA with concomitant reduction of NAD^+ to NADH. In most aerobic organisms, the PDC complex functions in mitochondria. The acetyl-CoA generated by this complex is utilized by the TCA cycle and respiratory chain to ultimately generate ATP as discussed earlier. In low oxygen conditions, this enzyme complex has been shown to participate in anaerobic acetyl-CoA generation in organisms such as the liver fluke *Fasciola hepatica* (Tielens et al. 2002). However, the majority of MRO-containing organisms do not use PDC for acetyl-CoA generation; instead, these facultative or obligate anaerobic unicellular eukaryotes use the oxygen-sensitive enzymes pyruvate:ferredoxin oxidoreductase (PFO), pyruvate:$NADP^+$ oxidoreductase (PNO), or pyruvate formate lyase (PFL).

PFO utilizes coenzyme A (CoA) and ferredoxin in the decarboxylation of pyruvate to form acetyl-CoA, CO_2, and reduced ferredoxin. The presence of PFO only in anaerobic/microaerophilic eukaryotic microbes may be a direct consequence of the oxygen sensitivity of this enzyme's Fe–S clusters. Catalysis in this case differs slightly from the PDC system, as PFO-mediated pyruvate decarboxylation only requires one enzyme. Eukaryotic PFO activity was first described by Reeves and colleagues as early as 1977 in *Entamoeba histolytica* under the name of pyruvate synthase (Reeves et al. 1977). Since then, PFO has been studied in detail in the hydrogenosomes of *Trichomonas vaginalis* (Steinbuchel and Muller 1986; Hrdy and Muller 1995) and *Neocallimastix patriciarum* (Yarlett et al. 1986), in the cytosol of the mitosome-bearing organisms *Entamoeba histolytica* (Rodriguez et al. 1996) and *Giardia intestinalis* (Ellis et al. 1993) and also in the chloroplasts of *Chlamydomonas reinhardtii* (Mus et al. 2007; Dubini et al. 2009). PFO homologues have been identified in other eukaryotes that transiently or permanently experience anoxic conditions including *Blastocystis* sp., *Mastigamoeba balamuthi*, *Acanthamoeba castellanii*, *Peranema trichophorum*, *Andalucia incarcerata*, *Spironucleus barkhanus*, *Oxyrrhis marina*, and *Perkinsus marinus*, but their enzymatic properties remain uncharacterized to date (Hug et al. 2010) (See Fig. 6 for a complete list).

PNO (EC 1.2.1.51) is a fusion protein of the aforementioned eubacterial PFO domain and a NADPH-cytochrome P450 reductase domain that converts $NADP^+$, CoA, and pyruvate to NADPH, acetyl-CoA, and CO_2. This enzyme has also been identified and characterized in the mitochondria of *Euglena gracilis* (Buetow 1989), the cytosol and crystalloid body of the apicomplexan *Cryptosporidium parvum* (Ctrnacta et al. 2006), and in an MRO-enriched subcellular fraction of *Blastocystis* sp. (Lantsman et al. 2008). Studies of *Euglena* metabolism hypothesize that the NADPH and acetyl-CoA generated in this reaction are used in processes such as wax ester formation, and not energy generation under anaerobic conditions (Inui et al. 1985). Recent studies performed by Hug and colleagues report additional putative PNO sequences in *Rhizopus oryzae*, *Astasia longa*, and *Thalassiosira pseudonana* (Hug et al. 2010). Phylogenetic analyses of the PFO domain of these PNO sequences combined with available eukaryotic PFO sequences revealed monophyly of the eukaryotic sequences indicating a single common origin for these enzymes in eukaryotes (Horner et al. 1999; Rotte et al. 2001; Embley 2006; Hug et al. 2010).

The final enzyme to be discussed with respect to pyruvate metabolism is PFL (EC 2.3.1.54), which catalyzes the non-oxidative formation of formate and acetyl-CoA from pyruvate and CoA. While reactions catalyzed by PDC, PFO, and PNO depend on the use of redox cofactors ($NAD(P)^+$ or ferredoxins) to convert pyruvate into acetyl-CoA, PFL does not. Instead, the reaction hinges on the activation of the enzyme by PFL activating enzyme (PFLA; EC 1.97.1.4) which uses S-adenosyl methionine (S-Ado-Met) and reduced flavodoxin or ferredoxin to generate a glycyl radical (Wagner et al. 1992). PFL is a strictly anaerobic enzyme due to its oxygen sensitivity and regulation of activation. The active radical form of PFL upon oxygen exposure is irreversibly cleaved at the radical storage site (Wagner et al. 1992) explaining why these enzymes have only been identified in organisms that are either obligatory or facultative anaerobes (Thauer et al. 1972). PFL and PFLA protein levels are similar under aerobic and anaerobic conditions, but activation of PFL by PFLA is only induced under anaerobiosis (Sawers and Watson 1998). PFL activity in eukaryotes has been described in the cell-free extracts (Marvin-Sikkema et al. 1993) and hydrogenosomes of the chytrid fungus *Necallimastix* sp. (Akhmanova et al. 1999) as well as its close relative *Piromyces* sp. E2 (Boxma et al. 2004), and the chloroplasts and mitochondria of the chlorophyte alga *Chlamydomonas reinhardtii* (Atteia et al. 2006; Hemschemeier et al. 2008). Additional PFL and PFLA homologues were identified in archaeplastids (*Scenedesmus obliquus*, *Chlorella* sp., *Acetabularia acetabulum*, *Haematococcus pluvialis*, *Volvox carteri*, *Ostreococcus tauri*, *Ostreococcus lucimarinus*, *Micromonas pusilla*, *Micromonas* sp., *Porphyra haitanensis*, and *Cyanophora paradoxa*), an opisthokont (*Amoebidium parasiticum*), an amoebozoan (*Mastigamoeba balamuthi*), a stramenopile (*Thalassiosira pseudonana*), and a haptophyte (*Prymnesium parvum*) by Stairs and colleagues (Stairs et al. 2011).

Atteia and colleagues propose that in the chloroplasts and mitochondria of *Chlamydomonas reinhardtii*, PFL-derived acetyl-CoA is converted to acetyl-phosphate and then acetate by the enzymes phosphotransacetylase and acetate kinase yielding one molecule of ATP (Atteia et al. 2006). Initial activation of PFL by the addition of a glycyl radical requires S-Ado-Met whose synthesis consumes ATP (S-Ado-Met synthetase). Therefore, in order for this reaction to be energy generative, the radical must be protected or else the PFL system is not as efficient as a PFO-based system for energy generation (Stairs et al. 2011). Phylogenetic analysis of available PFL and PFLA sequences from bacterial and eukaryotic sources revealed monophyly of the eukaryotes with strong affinity to the firmicute bacteria (Stairs et al. 2011) suggesting a common origin of the eukaryotic orthologues. The authors provide evidence against an endosymbiotic origin of these enzymes and propose that an operon containing PFL–PFLA was transferred laterally to a eukaryotic lineage and subsequently transferred within eukaryotes via multiple eukaryote-to-eukaryote lateral gene transfer events.

The reason why eukaryotes have evolved four different mechanisms for acetyl-CoA generation from pyruvate is unknown, especially when organisms such as *Thalassiosira pseudonana* and *Chlamydomonas reinhardtii* appear to possess PDC, PFO, and PFL. One hypothesis is redox coping mechanisms: under anaerobic conditions, where acetyl-CoA is required and high reducing equivalents (NADH) are inhibitory of PDC, using an enzyme that is unaffected by NADH levels such as PFO (or PFL) is advantageous. Furthermore, if the organisms inhabit a CO_2-rich environment, decarboxylating enzymes (e.g., PDC, PFO, or PNO) might be affected by product inhibition. Therefore, having an acetyl-CoA generating enzyme that does not create CO_2 such as PFL may be beneficial. However, the fates of acetyl-CoA may differ between PDC, PFO, PNO, and PFL due to the metabolic state of the cell or the location of the enzyme within the cell.

5.1. MALATE DECARBOXYLATION

In *Trichomonas vaginalis*, two subunits (24 kDa/NuoE and 51 kDa/NuoF) of complex I of the electron transport complex, also known as NADH:ubiquinone oxidoreductase, were found to localize to the hydrogenosome (Dyall et al. 2004; Hrdy et al. 2004). This represented the first instance of MROs retaining elements of the respiratory chain. However, with the apparent absence of other complex I subunits as well as the downstream complexes, the authors proposed a redox-balancing role for the Nuo proteins, whereby NADH is reoxidized to NAD^+ for use by the hydrogenosomal malic enzyme that produces pyruvate (outlined in Fig. 3). In addition to being capable of reducing ubiquinone, this unique complex I can also reduce ferredoxin, the principal redox molecule responsible for hydrogen production (Muller 1993), and is thought to have ferredoxin-reducing activity in *Trichomonas* (Dyall et al. 2004; Hrdy et al. 2004; Do et al. 2009). Interestingly, two more recent studies suggest that a similar pathway may be present in the

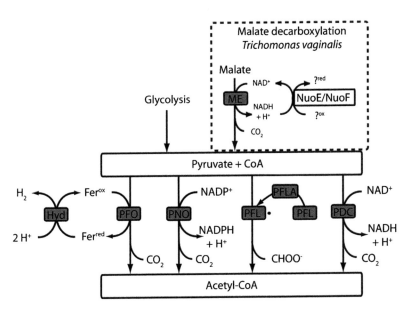

Figure 3. Pyruvate is metabolized by a variety of different enzyme systems. Pyruvate is generated either from malate by the malic enzyme (e.g., *Trichomonas vaginalis*) and/or by glycolysis. Pyruvate can then be metabolized to acetyl-CoA by four different enzymes each with unique chemistry and cofactors and phylogenetic distribution among eukaryotes. Unknown participating molecules are indicated (?). Abbreviations are as follows: malic enzyme (*ME*), NADH:ubiquinone oxidoreductase (*Nuo*), coenzyme A (*CoA*), [FeFe]-hydrogenase (*Hyd*), ferredoxin (*Fer*), pyruvate:ferredoxin oxidoreductase (*PFO*), pyruvate:NADP$^+$ oxidoreductase (*PNO*), pyruvate formate lyase (*PFL*), pyruvate formate lyase activating enzyme (*PFLA*), and pyruvate dehydrogenase complex (PDC).

free-living protists *Psalteriomonas lanterna* (in which only the 51 kDa subunit has been found so far (de Graaf et al. 2009)) and *Sawyeria marylandensis* [in which homologues of both subunits are present (Barbera et al. 2010)]. As in *Trichomonas*, there is no sign in either organism of the downstream enzymes that would be expected to be present if these subunits functioned in conjunction with TCA cycle activity. The 51 kDa subunits found in all three organisms form a clade with each other, eukaryotic complex I subunits, and α-proteobacterial homologues. Meanwhile, in the ciliate *Nyctotherus ovalis*, laterally acquired bacterial homologues of these subunits are found fused to an [FeFe]-hydrogenase (see below) and may perform the same function (Boxma et al. 2007).

6. [FeFe]-Hydrogenase

The first report of hydrogenase activity in a non-photosynthetic eukaryote (Lindmark and Muller 1973), *Tritrichomonas foetus*, also marked the discovery that hydrogenase activity in this organism occurred within an organelle for which

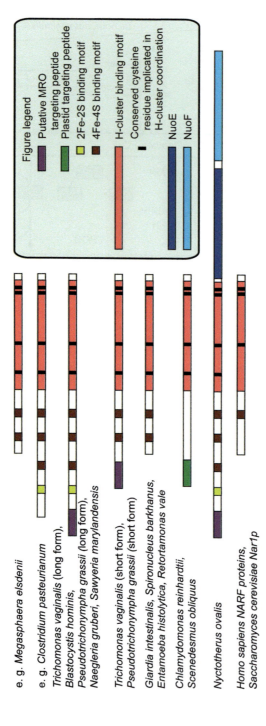

Figure 4. Schematic diagram of the domain structure of selected [FeFe]-hydrogenases, based on data presented in Horner et al. (2002) and Vignais and Billoud (2007), or inferred from full-length sequences using InterProScan (http://www.ebi.ac.uk/Tools/InterProScan/; Quevillon et al. 2005); the domains are not drawn to scale.

authors proposed the name hydrogenosome. Since then, [FeFe]-hydrogenase has been cloned from a number of diverse, unrelated eukaryotes, and while it is not always localized to the MRO, it has come to be seen as a characteristic enzyme of anaerobic microbial eukaryotes.

As discussed previously, [FeFe]-hydrogenase acts to remove reducing equivalents from pyruvate decarboxylation, catalyzing the transfer of electrons from ferredoxin to protons to produce molecular hydrogen. [FeFe]-hydrogenases are a major class of hydrogenases, enzymes capable of either reducing protons to produce molecular hydrogen or performing the reverse reaction in the uptake of H_2. These enzymes are classified according to the metal ions present in their catalytic sites; other classes include the [NiFe]-hydrogenases, [NiFeSe]-hydrogenases, and [Fe]-hydrogenases (Vignais and Billoud 2007). While all of the aforementioned classes of hydrogenases are found in prokaryotes, only [FeFe]-hydrogenases have been described to date in eukaryotes, and further discussion here will therefore be limited to this class of enzymes.

Even though [FeFe]-hydrogenases may be multimeric, only monomeric enzymes have been reported in eukaryotes. [FeFe]-hydrogenases vary greatly in length; however, all possess a conserved active site, the H-cluster, comprised of a dimeric iron ([FeFe]) center bound to a [4Fe–4S] cluster, both joined to each other and to the protein by conserved cysteine residues in the H-domain of the enzyme (Fig. 4). The iron atoms of the [FeFe] center are bound by CN^-, CO, and di(thiomethyl)amine ligands. [FeFe]-hydrogenases generally possess two accessory [4Fe–4S] clusters, bound by N-terminal domains, that are involved in electron transfer from the electron donor to the active site; they may also possess additional [4Fe–4S] or [2Fe–2S] clusters, accounting for the variation in overall size (Vignais and Billoud 2007). Upon exposure to oxygen, [FeFe]-hydrogenases are irreversibly inactivated as a result of oxidation of their [FeFe] center and Fe–S clusters (Vincent et al. 2005). Three Fe–S cluster-containing maturases – the radical S-adenosylmethionine enzymes HydE and HydG and the small GTPase HydF – are required for the assembly and insertion of the [4Fe–S] cluster into the enzyme (Mulder et al. 2010; Posewitz et al. 2004). HydF acts as a scaffold on which the [FeFe] center is assembled, possibly in a process requiring HydE (Shepard et al. 2010); meanwhile, HydG synthesizes the CN^-, CO, and di(thiomethyl)amine ligands (Pilet et al. 2009; Shepard et al. 2010).

6.1. EVOLUTIONARY RELATIONSHIPS AMONG [FeFe]-HYDROGENASES

The presence of [FeFe]-hydrogenases in MROs raised questions about the origin of this enzyme. As a result, in 1998, a modified version of the endosymbiotic theory for the origin of mitochondria was proposed (Martin and Muller 1998). The hydrogen hypothesis holds that the original endosymbiont was a facultative anaerobe possessing an [FeFe]-hydrogenase, which was retained by a methanogenic, hydrogen-dependent host, not for its ability to carry out oxidative phosphorylation, but instead for the hydrogen it generated as a waste product (Martin and Muller 1998). The presence of enzymes associated with anaerobic ATP production (e.g., [FeFe]-hydrogenase, PFO etc.) in anaerobic microbial eukaryotes (and the widespread distribution among eukaryotes of distant homologues of [FeFe]-hydrogenases known as *Nuclear prelamin A Recognition Factor* (NARF/Nar) proteins (Horner et al. 2002)) has been cited as possible evidence supporting the hydrogen hypothesis. Examining the distribution and phylogenetic relationships of eukaryotic [FeFe]-hydrogenases is therefore of great interest.

Evidence of hydrogen production in eukaryotes was first discovered in green algae of the genus *Scenedesmus* (Gaffron and Rubin 1942), which were found to produce hydrogen gas under anaerobic conditions. Hydrogen production was later described in another green alga, *Chlamydomonas reinhardtii*, and hydrogenase genes have since been cloned from both organisms. In each organism, two hydrogenases were discovered, localized to the chloroplast, and found to be expressed only under anaerobic conditions (Florin et al. 2001; Wunschiers et al. 2001; Happe and Kaminski 2002; Forestier et al. 2003). In addition to their role in carbon metabolism, these enzymes function in anaerobic photosynthesis, using hydrolysis by photosystem II as a source of electrons. The green algal enzymes are also notable for their lack of accessory [4Fe–4S] or [2Fe–2S] clusters. *C. reinhardtii* was found to possess two maturases required for [FeFe]-hydrogenase assembly: HydG and HydEF, the latter a fusion of HydE and HydF (Posewitz et al. 2004).

Enzyme activity and/or genes encoding a [FeFe]-hydrogenase-like product have been discovered in a wide range of non-photosynthetic anaerobic eukaryotes possessing MROs (Fig. 6), such as the parabasalid *Trichomonas vaginalis* (Bui and Johnson 1996; Horner et al. 2000), the chytrid fungus *Neocallimastix frontalis* (Davidson et al. 2002; Voncken et al. 2002), the preaxostylan excavate *Trimastix pyriformis* (Hampl et al. 2008), the parabasalids *Pseudotrichonympha grassii* and *Histomonas meleagridis* (Mazet et al. 2008), the stramenopile *Blastocystis* sp. (Stechmann et al. 2008), and the heteroloboseans *Psalteriomonas lanterna* (de Graaf et al. 2009) and *Sawyeria marylandensis* (Barbera et al. 2010). The ciliate *Nyctotherus ovalis* (Boxma et al. 2007) was found to possess a chimeric enzyme, consisting of a fusion of an [FeFe]-hydrogenase with homologues of bacterial NuoE and NuoF; these subunits may function in reoxidization of NADH (see above). Although [FeFe]-hydrogenase maturases have been less investigated in eukaryotes, homologues have been discovered in *Trichomonas vaginalis* (Putz et al. 2006), *Trimastix pyriformis* (Hampl et al. 2008), *Mastigamoeba balamuthi, Andalucia incarcerata*, and *Acanthamoeba castellanii* (Hug et al. 2010).

[FeFe]-hydrogenase genes and/or activity have also been found in the mitosome-bearing organisms *Giardia lamblia* (Lloyd et al. 2002a) and *Entamoeba histolytica* (Horner et al. 2000; Nixon et al. 2003) as well as the MRO-bearing *Spironucleus barkhanus* (Horner et al. 2000); in these organisms, the enzyme lacks an organellar targeting peptide. No maturases have been found in the genomes of *Giardia* or *Entamoeba*, suggesting that either their H-clusters are assembled by a novel mechanism, or their maturases have diverged beyond the limit of detection with current bioinformatic methods. Nevertheless, it has been suggested (Meyer 2007) that less efficient H-cluster assembly – resulting in lower levels of fully functional [FeFe]-hydrogenase – may account for the reduced levels of hydrogen production in *Giardia* compared with those found in *Trichomonas vaginalis* (Lloyd et al. 2002a, b; Nixon et al. 2003). Most recently, genes encoding an [FeFe]-hydrogenase and all three maturases have been found in the genome of the free-living heterolobosean *Naegleria gruberi* (Fritz-Laylin et al. 2010; Hug et al. 2010), while sequences encoding a [FeFe]-hydrogenase, two maturases, and PFO have been found in EST data from the amoebozoan *Acanthamoeba castellanii* (Hug et al. 2010). These last examples are notable because these organisms are non-photosynthetic aerobes, possessing mitochondria capable of oxidative phosphorylation. While the *A. castellanii* sequences are incomplete, the *N. gruberi* enzymes are predicted to be targeted to their mitochondria. This raises the possibility that mitochondria in these organisms may be capable of switching between aerobic and anaerobic modes of energy generation, presenting a tantalizing glimpse of how hydrogenosomes might have arisen.

While the number of eukaryotes in which [FeFe]-hydrogenases are known is still small, several studies have analyzed the phylogenetic relationships among these enzymes (Horner et al. 2002; Hackstein 2005; Meyer 2007; Vignais and Billoud 2007; Hug et al. 2010). Despite the small number of eukaryotic taxa represented in the earlier studies, and the low bootstrap support plaguing all of the studies, monophyly of eukaryotic [FeFe]-hydrogenases could not be recovered, suggesting at least two independent origins of these enzymes in eukaryotes. Furthermore, these studies indicated that an α-proteobacterial ancestry for eukaryotic [FeFe]-hydrogenase was rejected in topology tests (Hug et al. 2010). If the original endosymbiont, which gave rise to the mitochondrion, did possess an [FeFe]-hydrogenase, then it has long been replaced by more recent gene transfers in modern anaerobic eukaryotes. One of the studies (Hug et al. 2010) also examined the relationships between [FeFe]-maturases. In each case, eukaryote monophyly was recovered with a closest affinity to the firmicute bacteria, although the number of eukaryotic taxa represented was small (between 2 and 6).

7. ATP-Generation from Acetyl-CoA in Microbial Eukaryotes

In most aerobic eukaryotes, acetyl-CoA generated by PDC is used to fuel ATP synthesis via the TCA cycle, electron transport chain and oxidative phosphorylation in mitochondria as outlined above. While some anaerobes have evolved mechanisms of maintaining this oxygen-dependent pathway, such as using alternative terminal electron donors to oxygen, many eukaryotes use different pathways that ultimately end in acetate production. These reactions may generate ATP either directly, or as a result of ATP generation from one of the products (Fig. 5). In the mitosome-bearing organisms such as *Giardia lamblia* (Sanchez and Muller

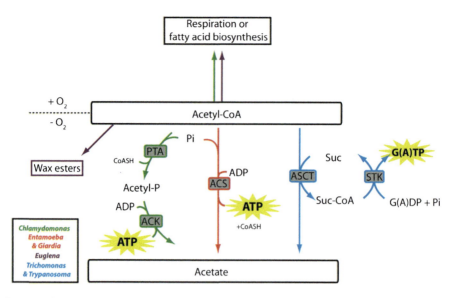

Figure 5. Acetyl-CoA metabolism in anaerobic protists. Aerobically generated acetyl-CoA can be metabolized aerobically using respiration or fatty acid biosynthesis. Anaerobically, PNO-generated acetyl-CoA is metabolized to wax esters (*Euglena*), while other organisms are hypothesized to generate acetate and ATP. Organisms where the pathway has been identified and characterized are labeled as follows: green, *Chlamydomonas reinhardtii* mitochondria and chloroplasts; red, *Entamoeba histolytica* and *Giardia* cytosol; purple, *Euglena gracilis* mitochondria; and blue, *Trichomonas vaginalis* hydrogenosome and *Trypanosoma brucei* mitochondria. Abbreviations are as follows: coenzyme A (*CoA*), phosphoacetyl-transferase (*PTA*), acetate kinase (*ACK*), acetyl-CoA synthetase (*ACS*), acetate:succinate-CoA transferase (*ASCT*), and succinate thiokinase (*STK*).

1996) and *Entamoeba histolytica* (Field et al. 2000), the PFO-derived acetyl-CoA is converted to acetate with the concomitant production of ATP by the enzyme acetyl-CoA synthetase (or acetate thiokinase; ACS; EC 6.2.1.13), directly generating ATP, a pathway that is also found in some Archaebacteria. Genes encoding ACS-like sequences have also been identified in *Blastocystis* sp. (Genbank accession number: EC64851) (Stechmann et al. 2008) and *Plasmodium falciparum* (accession number XM_001348495) (Sanchez et al. 2000).

A second ATP generative mechanism using acetyl-CoA involves a two-step reaction whereby acetyl-CoA is converted to acetyl-phosphate and then acetate by phosphotransacetylase (PTA; EC 2.3.1.8) and acetate kinase (ACK; EC 2.7.2.1), respectively. While this system is primarily found in eubacteria, PTA and ACK have been identified in the chlorophyte alga *Chlamydomonas reinhardtii* and the oomycete stramenopile *Phytophthora sojae*, and only ACK was found in *Entamoeba histolytica* (Atteia et al. 2006). It has been hypothesized that this system might also be important in acetate assimilation to generate trioses that can be incorporated into the glyoxylate cycle in the absence of another carbon source (Heifetz et al. 2000).

A third ATP generative mechanism using acetyl-CoA involves the transfer of CoA from acetyl-CoA to succinate, generating succinyl-CoA and acetate, catalyzed by an acetate:succinate CoA transferase (ASCT; EC 2.3.1.99) using a ping-pong bi-bi mechanism (Tielens et al. 2010). The succinyl-CoA is then converted back to succinate by the TCA cycle enzyme succinyl-CoA synthetase (SCS; or succinate thiokinase, STK; EC 6.2.4.1) generating ATP by substrate-level phosphorylation (Hrdy et al. 2004). ASCTs in eukaryotes have a number of different evolutionary origins. Tielens et al. (Tielens et al. 2010) have proposed a scheme for classifying these enzymes into three subfamilies, found in the mitochondria of trypanosomatids (subfamily IA), in the anaerobic mitochondria of helminths (subfamily IB), and in hydrogenosomes (subfamily IC).

The first eukaryotic ASCT (subfamily 1A) was discovered in procyclic *Trypanosoma brucei*, and is a homologue of the branched-chain fatty acid degradation pathway enzyme succinyl-CoA:3-ketoacid transferase (SCOT; EC 2.8.3.5) (Riviere et al. 2004). Another member of this subfamily has since been found in *Nyctotherus ovalis* (AJ871320) (Boxma et al. 2005).

The anaerobic mitochondria of the liver fluke *Fasciola hepatica* contain examples of the second subfamily of ASCT (1B; Fig. 5). In this organism, phosphoenolpyruvate from the glycolytic pathway is converted to malate, which is subsequently reduced to fumarate and then succinate (a process known as malate dismutation). Succinate is then converted to succinyl-CoA and then methylmalonyl-CoA, which is converted to propionyl-CoA in a reaction that produces ATP. Coenzyme A is recycled by propionyl-CoA, which transfers it back to succinate. Alternatively, malate is oxidized by malic enzyme and pyruvate dehydrogenase, producing acetate via pyruvate and acetyl-CoA. It is the final step in this process of acetate production that is catalyzed by ASCT. The subfamily B enzyme is a short-chain acyl-CoA transferase that is also capable of catalyzing the transfer of CoA to acetate, propionate, and butyrate; it is only distantly related to the *Trypanosoma* enzyme, and more closely related to a CoA transferase from the brine shrimp *Artemia franciscana* (van Grinsven et al. 2008). *Blastocystis* also appears to possess a member of this subfamily (Tielens et al. 2010), in addition to the previously identified subfamily 1C enzyme (Stechmann et al. 2008; see below).

Meanwhile, *Trichomonas vaginalis* possesses an ASCT homologous to one found in *Saccharomyces cerevisiae* (subfamily 1C); the *T. vaginalis* enzyme appears to function in the hydrogenosome (van Grinsven et al. 2008). A homologue of this enzyme was later found (Stechmann et al. 2008) and characterized (Lantsman et al. 2008) in *Blastocystis* sp. MROs. Acetate production has been detected in the hydrogenosomes of *Psalteriomonas lanterna* and *Neocallimastix frontalis;* however, the enzymes responsible for its generation in these organisms have not been identified.

Similar to the aforementioned diversity of pyruvate metabolism, eukaryotes have evolved multiple ways to generate ATP from acetyl-CoA with the end product of acetate. It is remarkable that multiple enzyme systems have been acquired by many distantly related lineages, creating organisms that have convergently evolved

Species	PDC	P(F/N)O	PFL	PFLA	Hyd.	Matur.	Fe-S
Other Metazoans (84)	+	-	-	-	-	-	ISC
Amoebidium parasiticum *	+	?	+	?	+	?	ISC
Capitella capitata	+	N	-	-	?	?	ISC
other ascomycotes (78)	+	-SR	-	-	-	-	ISC
Neurospora crassa	+	-SR	-	-	-	-	ISC
Basidiomycota (18)	+	-SR	-	-	-	-	ISC
Rhizopus oryzae	+	N	-	-	-	-	ISC
Neocallimastix frontalis *	?	?	+	+	+(m)	?	ISC
Piromyces sp. *	?	?	+	?	+	?	ISC
Batrachochytrium dend.	+	-SR	-	-	?	?	ISC
Spizellomyces punctatus *	?	-SR	?	?	?	?	ISC
Encephalitozoon cuniculi	+	-	-	-	-	-	ISC
Entamoeba histolytica	-	F	-	-	+(2; c)	-	NIF
Mastigamoeba balamuthi *	?	F	+	+	+	G	NIF
Dictyostelium discoideum	+	-	-	-	-	-	ISC
Acanthamoeba castellanii *	+	F	?	?	+	E,G	ISC
Polysphondylium pallidum	+	-	-	-	-	-	ISC
Trichomonas vaginalis	-	F	-	-	+(3; m)	E,F,G	ISC
Histomonas meleagridis *	?	F	?	?	+(m)	?	ISC
Pseudotrichonympha pall. *	?	?	?	?	+(2)	?	ISC
Giardia intestinalis	-	F	-	-	+(c)	-	ISC
Spironucleus (2) *	?	F	-	-	+	?	ISC
Retortamonas sp. *	?	F	?	?	+(2)	?	ISC
Euglena gracilis *	+	N	?	?	?	?	ISC?/SUF?
Astasia longa *	?	N	?	?	?	?	ISC
Peranema trichophorum *	?	N/F	?	?	?	?	ISC
Kinetoplastida (5)	+	-	-	-	-	-	ISC
Psalteriomonas lanterna *	-	N/F	-	-	+(m)	-	ISC
Sawyeria marylandensis *	?	F	?	?	+(2)	F,G	ISC
Naegleria gruberi	+	-	-	-	+	E,F,G	ISC
Andalucia incarcerata *	?	N/F	?	?	+	G	ISC
Trimastix pyriformis *	?	F	?	?	+	E,G	ISC?
Monocercomonoides sp. *	?	F	?	?	+	?	ISC
Oxyrrhis marina *	?	N/F	?	?	?	?	ISC
Perkinsus marinus **	+	N/F	?	?	?	?	ISC
Other Apicomplexa (13)	+	-	-	-	-	-	ISC/SUF#
Cryptosporidium parvum	+	F	-	-	-	-	ISC
Other ciliates (2)	+	-	-	-	-	-	ISC
Metopus contortus *	?	?	?	?	+	?	ISC
Nyctotherus ovalis *	+	?	?	?	+(m)	?	ISC
Caenomorphidae gen. sp. *	?	?	?	?	+	?	ISC
Blastocystis sp. *	+	N/F	?	?	+(m)	?	ISC/SUF?
Phytophthora ramorum	+	-SR	-	-	-	-	ISC/SUF
Thalassiosira pseudonana	+	N	+	+	+	?	ISC/SUF
Prymnesium parvum	+	?	+	?	?	?	ISC/SUF
Emiliania huxleyi	+	-	-	-	?	?	ISC/SUF
Chlamydomonas reinhardtii	+	F	+	+	+(2; chl)	E/F,G	ISC/SUF
Scenedesmus obliquus *	?	?	+	?	+(2; chl)	?	ISC/SUF
Chlorella sp. *	+	F	+	?	+	?	ISC/SUF
Acetabularia acetabulum *	+	?	+	?	?	?	ISC/SUF
Haematococcus pluvialis *	+	?	+	?	?	?	ISC/SUF
Volvox carteri *	+	?	+	+	+	G	ISC/SUF
Streptophytes (11)	+	-	-	-	-	-	ISC/SUF
Ostreococcus (2)	+	-	+	+	-	-	ISC/SUF
Micromonas (2)	+	-	+#	-	?	?	ISC/SUF
Porphyra haitanensis *	+	?	+	?	?	?	ISC/SUF
Cyanidioschyzon merolae	+	-	-	-	?	?	ISC/SUF
Galdieria sulphuraria	+	?	?	?	?	?	ISC/SUF
Cyanophora paradoxa*	+	?	+	?	?	?	ISC/SUF

to possess near-identical anaerobic metabolisms. Understanding the origin of these pathways, in addition to the mechanism(s) by which they were obtained by eukaryotes, will be monumental in determining the ancestral state of the mitochondrion.

8. Conclusions

In adaptation to anoxia, mitochondria have been converted into a spectrum of mitochondrion-related organelles with diverse functions (Fig. 6), and these conversions have occurred independently in a wide range of eukaryotic lineages. The mitochondrial functions believed to be almost universally retained by these organelles include Fe–S cluster biosynthesis, and those functions required for maintenance of the organelle and its involvement in this pathway, namely protein import and substrate exchange. Convergent evolution is also apparent in the newly acquired functions of the MROs of distinct eukaryotic lineages. Organisms adapting to permanently anaerobic lifestyles are no longer under selection pressure to maintain oxidative phosphorylation, and so many MROs have independently lost some or all components of the TCA cycle, the electron transport complexes, and the F_1F_0 ATP synthase. In addition, anaerobic lineages have independently acquired the means of generating ATP under anaerobic conditions, and the pathways involved share a number of common features, such as the conversion of pyruvate to acetyl-CoA, the subsequent formation of acetate in the course of ATP synthesis, and the involvement of [FeFe]-hydrogenase as a means of removing reducing equivalents. However, the independent nature of these adaptations is also apparent in their diversity. The newly acquired pyruvate metabolism may take place in the MRO, in the cytosol, or even in the chloroplast. The enzymes that catalyze different steps vary between organisms; for example, acetyl-CoA may be generated by the PDC, PFL, PFO, or PNO, and acetate may likewise be generated by any one of four different pathways. These differences have downstream effects on other pathways that are retained or lost by the MROs.

Figure 6. Distribution of homologues of the enzymes discussed in this chapter among publicly available genome or expressed sequence tag data from eukaryote taxa (Figure adapted and modified from Stairs et al. 2011). *, taxa for whom only EST data are available; +, homologue present; −, no homologue yet found in whole genome data; ?, no homologue found to date, although complete genome data are not yet available. F, PFO; N, PNO; N/F, incomplete sequence that might encode either PNO or PFO; SR, sulfate reductase. For [FeFe]-hydrogenase, the number of homologues believed to be present, where greater than one, is shown in brackets. Where one or more of these enzymes has been localized to a specific subcellular compartment, that compartment is indicated in *brackets*: (*m*), mitochondrial; (*c*), cytosolic; (*chl*), chloroplast. E, HydE; F, HydF; G, HydG; #, present in some of the species of the group.

This is the case for ATP/ADP transporters, which are required to export ATP from organelles that retain a role in ATP production; in mitosomes, which do not perform this function, they instead import ATP into the organelle (Chan et al. 2005; Tsaousis et al. 2008).

In the years since hydrogen production in eukaryotes was first described, many features that were initially described as curiosities of a single organism have been found to be widespread among anaerobic protists. With the study of an ever-wider range of anaerobic protists, many more comparisons between adaptations to anoxia are possible. Green algae such as *Chlamydomonas reinhardtii* are among the best-studied organisms that are capable of surviving in anoxic as well as oxygen-rich habitats; they are unique in that they do so with the aid of metabolic pathways present in their chloroplasts. The recent discovery of anaerobic metabolic enzymes in *Acanthamoeba* and *Naegleria* will permit the study of similar capabilities in organisms lacking chloroplasts, and possibly allow a glimpse into the first steps of mitochondrial adaptation to anoxia. Indeed, it may turn out that a large number of eukaryotic microbes have both aerobic and anaerobic energy metabolic capabilities (see Fig. 6).

Until recently, investigations of protists living under anoxic conditions focused mainly on pathogens, such as *Trichomonas*, *Entamoeba*, *Giardia*, and microsporidia. This has led to difficulties in determining which adaptations in these organisms might be responses to anoxia, as opposed to the adoption of a parasitic lifestyle. With an increasing number of studies choosing to focus on free-living protists, the influences of these two factors may be teased apart. This will be instrumental in elucidating the influence of anoxic conditions in the course of the evolutionary history of eukaryotes.

9. Acknowledgments

This work was supported from grant MOP-62809 from the Canadian Institutes of Health Research awarded to AJR. ADT was supported by a Marie Curie International Outgoing fellowship. AJR was supported by the Integrated Microbial Biodiversity program of the Canadian Institute for Advanced Research and the Canada Research Chairs program. MML was supported by an Aide à la Formation-Recherche awarded by the Fonds National de la Recherche (Luxembourg). CWS was supported by scholarships from the Natural Sciences and Engineering Research Council of Canada and Killam Trusts.

10. References

Abrahamsen MS, Templeton TJ, Enomoto S, Abrahante JE, Zhu G, Lancto CA, Deng M, Liu C, Widmer G, Tzipori S, Buck GA, Xu P, Bankier AT, Dear PH, Konfortov BA, Spriggs HF, Iyer L, Anantharaman V, Aravind L, Kapur V (2004) Complete genome sequence of the apicomplexan, *Cryptosporidium parvum*. Science 304(5669):441–445

Akhmanova A, Voncken FG, Hosea KM, Harhangi H, Keltjens JT, op den Camp HJ, Vogels GD, Hackstein JH (1999) A hydrogenosome with pyruvate formate-lyase: anaerobic chytrid fungi use an alternative route for pyruvate catabolism. Mol Microbiol 32(5):1103–1114

Ali V, Shigeta Y, Tokumoto U, Takahashi Y, Nozaki T (2004) An intestinal parasitic protist, *Entamoeba histolytica*, possesses a non-redundant nitrogen fixation-like system for iron-sulfur cluster assembly under anaerobic conditions. J Biol Chem 279(16):16863–16874

Arco AD, Satrustegui J (2005) New mitochondrial carriers: an overview. Cell Mol Life Sci 62(19–20): 2204–2227

Atteia A, van Lis R, Gelius-Dietrich G, Adrait A, Garin J, Joyard J, Rolland N, Martin W (2006) Pyruvate formate-lyase and a novel route of eu-karyotic ATP synthesis in *Chlamydomonas* mitochondria. J Biol Chem 281(15):9909–9918

Barbera MJ, Ruiz-Trillo I, TJY A, Bery A, Silberman JD, Roger AJ (2010) *Sawyeria marylandensis* (Hetetolobosea) has a hydrogensome with novel metabolic properties. Eukaryot Cell 9(12): 1913–1924

Beinert H, Kiley PJ (1999) Fe-S proteins in sensing and regulatory functions. Curr Opin Chem Biol 3(2):152–157

Boxma B, Voncken F, Jannink S, van Alen T, Akhmanova A, van Weelden SW, van Hellemond JJ, Ricard G, Huynen M, Tielens AG, Hackstein JH (2004) The anaerobic chytridiomycete fungus *Piromyces* sp. E2 produces ethanol via pyruvate:formate lyase and an alcohol dehydrogenase E. Mol Microbiol 51(5):1389–1399

Boxma B, de Graaf RM, van der Staay GW, van Alen TA, Ricard G, Gabaldon T, van Hoek AH, Moon-van der Staay SY, Koopman WJ, van Hellemond JJ, Tielens AG, Friedrich T, Veenhuis M, Huynen MA, Hackstein JH (2005) An anaerobic mitochondrion that produces hydrogen. Nature 434(7029):74–79

Boxma B, Ricard G, van Hoek AH, Severing E, Moon-van der Staay SY, van der Staay GW, van Alen TA, de Graaf RM, Cremers G, Kwantes M, McEwan NR, Newbold CJ, Jouany JP, Michalowski T, Pristas P, Huynen MA, Hackstein JH (2007) The [FeFe] hydrogenase of *Nyctotherus ovalis* has a chimeric origin. BMC Evol Biol 7:230

Buetow DE (1989) The mitochondrion. In: The Biology of *Euglena*. Vol. 4: Subcellular Biochemistry and Molecular Biology. Academic Press, San Diego, pp 247–314

Bui ET, Johnson PJ (1996) Identification and characterization of [Fe]-hydrogenases in the hydrogenosome of *Trichomonas vaginalis*. Mol Biochem Parasitol 76(1–2):305–310

Bui ETN, Bradley PJ, Johnson PJ (1996) A common evolutionary origin for mitochondria and hydrogenosomes. Proc Natl Acad Sci U S A 93:9651–9656

Burri L, Williams BA, Bursac D, Lithgow T, Keeling PJ (2006) Microsporidian mitosomes retain elements of the general mitochondrial targeting system. Proc Natl Acad Sci U S A 103(43): 15916–15920

Carlton JM, Hirt RP, Silva JC, Delcher AL, Schatz M, Zhao Q, Wortman JR, Bidwell SL, Alsmark UC, Besteiro S, Sicheritz-Ponten T, Noel CJ, Dacks JB, Foster PG, Simillion C, Van de Peer Y, Miranda-Saavedra D, Barton GJ, Westrop GD, Muller S, Dessi D, Fiori PL, Ren Q, Paulsen I, Zhang H, Bastida-Corcuera FD, Simoes-Barbosa A, Brown MT, Hayes RD, Mukherjee M, Okumura CY, Schneider R, Smith AJ, Vanacova S, Villalvazo M, Haas BJ, Pertea M, Feldblyum TV, Utterback TR, Shu CL, Osoegawa K, de Jong PJ, Hrdy I, Horvathova L, Zubacova Z, Dolezal P, Malik SB, Logsdon JM Jr, Henze K, Gupta A, Wang CC, Dunne RL, Upcroft JA, Upcroft P, White O, Salzberg SL, Tang P, Chiu CH, Lee YS, Embley TM, Coombs GH, Mottram JC, Tachezy J, Fraser-Liggett CM, Johnson PJ (2007) Draft genome sequence of the sexually transmitted pathogen *Trichomonas vaginalis*. Science 315(5809):207–212

Chan KW, Slotboom DJ, Cox S, Embley TM, Fabre O, van der Giezen M, Harding M, Horner DS, Kunji ER, Leon-Avila G, Tovar J (2005) A novel ADP/ATP transporter in the mitosome of the microaerophilic human parasite *Entamoeba histolytica*. Curr Biol 15(8):737–742

Ctrnacta V, Ault JG, Stejskal F, Keithly JS (2006) Localization of pyruvate:NADP+ oxidoreductase in sporozoites of *Cryptosporidium parvum*. J Eukaryot Microbiol 53(4):225–231

Danovaro R, Dell'Anno A, Pusceddu A, Gambi C, Heiner I, Kristensen RM (2010) The first metazoa living in permanently anoxic conditions. BMC Biol 8:30

Davidson EA, van der Giezen M, Horner DS, Embley TM, Howe CJ (2002) An [Fe] hydrogenase from the anaerobic hydrogenosome-containing fungus *Neocallimastix frontalis* L2. Gene 296(1–2):45–52

de Graaf RM, Duarte I, van Alen TA, Kuiper JW, Schotanus K, Rosenberg J, Huynen MA, Hackstein JH (2009) The hydrogenosomes of *Psalteriomonas lanterna*. BMC Evol Biol 9:287

Do PM, Angerhofer A, Hrdy I, Bardonova L, Ingram LO, Shanmugam KT (2009) Engineering *Escherichia coli* for fermentative dihydrogen production: potential role of NADH-ferredoxin oxidoreductase from the hydrogenosome of anaerobic protozoa. Appl Biochem Biotechnol 153(1–3):21–33

Dolezal P, Likic V, Tachezy J, Lithgow T (2006) Evolution of the molecular machines for protein import into mitochondria. Science 313(5785):314–318

Dolezal P, Dagley MJ, Kono M, Wolynec P, Likic VA, Foo JH, Sedinova M, Tachezy J, Bachmann A, Bruchhaus I, Lithgow T (2010) The essentials of protein import in the degenerate mitochondrion of *Entamoeba histolytica*. PLoS Pathog 6(3):e1000812

Dubini A, Mus F, Seibert M, Grossman AR, Posewitz MC (2009) Flexibility in anaerobic metabolism as revealed in a mutant of *Chlamydomonas reinhardtii* lacking hydrogenase activity. J Biol Chem 284(11):7201–7213

Dyall SD, Koehler CM, Delgadillo-Correa MG, Bradley PJ, Plumper E, Leuenberger D, Turck CW, Johnson PJ (2000) Presence of a member of the mitochondrial carrier family in hydrogenosomes: conservation of membrane-targeting pathways between hydrogenosomes and mitochondria. Mol Cell Biol 20(7):2488–2497

Dyall SD, Yan W, Delgadillo-Correa MG, Lunceford A, Loo JA, Clarke CF, Johnson PJ (2004) Non-mitochondrial complex I proteins in a hydrogenosomal oxidoreductase complex. Nature 431(7012):1103–1107

Ellis JE, Williams R, Cole D, Cammack R, Lloyd D (1993) Electron transport components of the parasitic protozoon *Giardia lamblia*. FEBS Lett 325(3):196–200

Embley TM (2006) Multiple secondary origins of the anaerobic lifestyle in eukaryotes. Philos Trans R Soc Lond B Biol Sci 361(1470):1055–1067

Embley TM, Martin W (2006) Eukaryotic evolution, changes and challenges. Nature 440(7084):623–630

Field J, Rosenthal B, Samuelson J (2000) Early lateral transfer of genes encoding malic enzyme, acetyl-CoA synthetase and alcohol dehydrogenases from anaerobic prokaryotes to *Entamoeba histolytica*. Mol Microbiol 38(3):446–455

Finlay B, Fenchel T (1989) Hydrogenosomes in some anaerobic protozoa resemble mitochondria. FEMS Microbiol Lett 65:311–314

Florin L, Tsokoglou A, Happe T (2001) A novel type of iron hydrogenase in the green alga *Scenedesmus obliquus* is linked to the photosynthetic electron transport chain. J Biol Chem 276(9):6125–6132

Forestier M, King P, Zhang L, Posewitz M, Schwarzer S, Happe T, Ghirardi ML, Seibert M (2003) Expression of two [Fe]-hydrogenases in *Chlamydomonas reinhardtii* under anaerobic conditions. Eur J Biochem 270(13):2750–2758

Fritz-Laylin LK, Prochnik SE, Ginger ML, Dacks JB, Carpenter ML, Field MC, Kuo A, Paredez A, Chapman J, Pham J, Shu S, Neupane R, Cipriano M, Mancuso J, Tu H, Salamov A, Lindquist E, Shapiro H, Lucas S, Grigoriev IV, Cande WZ, Fulton C, Rokhsar DS, Dawson SC (2010) The genome of *Naegleria gruberi* illuminates early eukaryotic versatility. Cell 140(5):631–642

Gaffron H, Rubin J (1942) Fermentative and photochemical production of hydrogen in algae. J Gen Physiol 26(2):219–240

Gardner MJ, Hall N, Fung E, White O, Berriman M, Hyman RW, Carlton JM, Pain A, Nelson KE, Bowman S, Paulsen IT, James K, Eisen JA, Ruth-erford K, Salzberg SL, Craig A, Kyes S, Chan MS, Nene V, Shallom SJ, Suh B, Peterson J, Angiuoli S, Pertea M, Allen J, Selengut J, Haft D, Mather MW, Vaidya AB, Martin DM, Fairlamb AH, Fraunholz MJ, Roos DS, Ralph SA,

McFadden GI, Cummings LM, Subramanian GM, Mungall C, Venter JC, Carucci DJ, Hoffman SL, Newbold C, Davis RW, Fraser CM, Barrell B (2002) Genome sequence of the human malaria parasite *Plasmodium falciparum*. Nature 419(6906):498–511

Gerber J, Lill R (2002) Biogenesis of iron-sulfur proteins in eukaryotes: components, mechanism and pathology. Mitochondrion 2(1–2):71–86

Gill EE, Diaz-Trivino S, Barbera MJ, Silberman JD, Stechmann A, Gaston D, Tamas I, Roger AJ (2007) Novel mitochondrion-related organelles in the anaerobic amoeba *Mastigamoeba balamuthi*. Mol Microbiol 66(6):1306–1320

Goldberg AV, Molik S, Tsaousis AD, Neumann K, Kuhnke G, Delbac F, Vivares CP, Hirt RP, Lill R, Embley TM (2008) Localization and functionality of microsporidian iron-sulphur cluster assembly proteins. Nature 452(7187):624–628

Hackstein JH (2005) Eukaryotic Fe-hydrogenases – old eukaryotic heritage or adaptive acquisitions? Biochem Soc Trans 33(Pt 1):47–50

Hampl V, Silberman JD, Stechmann A, Diaz-Trivino S, Johnson PJ, Roger AJ (2008) Genetic evidence for a mitochondriate ancestry in the 'amitochondriate' flagellate *Trimastix pyriformis*. PLoS One 3(1):e1383

Happe T, Kaminski A (2002) Differential regulation of the Fe-hydrogenase during anaerobic adaptation in the green alga *Chlamydomonas reinhardtii*. Eur J Biochem 269(3):1022–1032

Heifetz PB, Forster B, Osmond CB, Giles LJ, Boynton JE (2000) Effects of acetate on facultative autotrophy in *Chlamydomonas reinhardtii* assessed by photosynthetic measurements and stable isotope analyses. Plant Physiol 122(4):1439–1445

Hemschemeier A, Jacobs J, Happe T (2008) Biochemical and physiological characterization of the pyruvate formate-lyase Pfl1 of *Chlamydomonas reinhardtii*, a typically bacterial enzyme in a eukaryotic alga. Eukaryot Cell 7(3):518–526

Horner DS, Hirt RP, Embley TM (1999) A single eubacterial origin of eukaryotic pyruvate: ferredoxin oxidoreductase genes: implications for the evolution of anaerobic eukaryotes. Mol Biol Evol 16(9):1280–1291

Horner DS, Foster PG, Embley TM (2000) Iron hydrogenases and the evolution of anaerobic eukaryotes. Mol Biol Evol 17(11):1695–1709

Horner DS, Heil B, Happe T, Embley TM (2002) Iron hydrogenases–ancient enzymes in modern eukaryotes. Trends Biochem Sci 27(3):148–153

Hrdy I, Muller M (1995) Primary structure and eubacterial relationships of the pyruvate:ferredoxin oxidoreductase of the amitochondriate eukaryote *Trichomonas vaginalis*. J Mol Evol 41(3):388–396

Hrdy I, Hirt RP, Dolezal P, Bardonova L, Foster PG, Tachezy J, Embley TM (2004) Trichomonas hydrogenosomes contain the NADH dehydrogenase module of mitochondrial complex I. Nature 432(7017):618–622

Hug LA, Stechmann A, Roger AJ (2010) Phylogenetic distributions and histories of proteins involved in anaerobic pyruvate metabolism in eukaryotes. Mol Biol Evol 27(2):311–324

Inui H, Miyatake K, Nakano Y, Kitaoka S (1985) The physiological role of oxygen-sensitive pyruvate dehydrogenase in mitochondrial fatty acid synthesis in *Euglena gracilis*. Arch Biochem Biophys 237(2):423–429

Katinka MD, Duprat S, Cornillot E, Metenier G, Thomarat F, Prensier G, Barbe V, Peyretaillade E, Brottier P, Wincker P, Delbac F, El Alaoui H, Peyret P, Saurin W, Gouy M, Weissenbach J, Vivares CP (2001) Genome sequence and gene compaction of the eukaryote parasite *Encephalitozoon cuniculi*. Nature 414(6862):450–453

Kunji ER (2004) The role and structure of mitochondrial carriers. FEBS Lett 564(3):239–244

Kunji ER, Robinson AJ (2006) The conserved substrate binding site of mitochondrial carriers. Biochim Biophys Acta 1757(9–10):1237–1248

LaGier MJ, Tachezy J, Stejskal F, Kutisova K, Keithly JS (2003) Mitochondrial-type iron-sulfur cluster biosynthesis genes (IscS and IscU) in the api-complexan *Cryptosporidium parvum*. Microbiology 149(Pt 12):3519–3530

Lantsman Y, Tan KS, Morada M, Yarlett N (2008) Biochemical characterization of a mitochondrial-like organelle from *Blastocystis* sp. subtype 7. Microbiology 154(Pt 9):2757–2766

Lill R (2009) Function and biogenesis of iron-sulphur proteins. Nature 460(7257):831–838

Lill R, Kispal G (2000) Maturation of cellular Fe-S proteins: an essential function of mitochondria. Trends Biochem Sci 25(8):352–356

Lill R, Muhlenhoff U (2005) Iron-sulfur-protein biogenesis in eukaryotes. Trends Biochem Sci 30(3):133–141

Lill R, Muhlenhoff U (2006) Iron-sulfur protein biogenesis in eukaryotes: components and mechanisms. Annu Rev Cell Dev Biol 22:457–486

Lindmark DG, Muller M (1973) Hydrogenosome, a cytoplasmic organelle of the anaerobic flagellate *Trichomonas foetus*, and its role in pyruvate metabolism. J Biol Chem 248(22):7724–7728

Lithgow T, Schneider A (2010) Evolution of macromolecular import pathways in mitochondria, hydrogenosomes and mitosomes. Philos Trans R Soc Lond B Biol Sci 365(1541):799–817

Lloyd D, Ralphs JR, Harris JC (2002a) Giardia intestinalis, a eukaryote without hydrogenosomes, produces hydrogen. Microbiology 148(Pt 3):727–733

Lloyd D, Ralphs JR, Harris JC (2002b) Hydrogen production in *Giardia intestinalis*, a eukaryote with no hydrogenosomes. Trends Parasitol 18(4):155–156

Loftus B, Anderson I, Davies R, Alsmark UC, Samuelson J, Amedeo P, Roncaglia P, Berriman M, Hirt RP, Mann BJ, Nozaki T, Suh B, Pop M, Duchene M, Ackers J, Tannich E, Leippe M, Hofer M, Bruchhaus I, Willhoeft U, Bhattacharya A, Chillingworth T, Churcher C, Hance Z, Harris B, Harris D, Jagels K, Moule S, Mungall K, Ormond D, Squares R, Whitehead S, Quail MA, Rabbinowitsch E, Norbertczak H, Price C, Wang Z, Guillen N, Gilchrist C, Stroup SE, Bhattacharya S, Lohia A, Foster PG, Sicheritz-Ponten T, Weber C, Singh U, Mukherjee C, El-Sayed NM, Petri WA Jr, Clark CG, Embley TM, Barrell B, Fraser CM, Hall N (2005) The genome of the protist parasite Entamoeba histolytica. Nature 433(7028):865–868

Mai Z, Ghosh S, Frisardi M, Rosenthal B, Rogers R, Samuelson J (1999) Hsp60 is targeted to a cryptic mitochondrion-derived organelle ("crypton") in the microaerophilic protozoan parasite Entamoeba histolytica. Mol Cell Biol 19(3):2198–2205

Maralikova B, Ali V, Nakada-Tsukui K, Nozaki T, van der Giezen M, Henze K, Tovar J (2010) Bacterial-type oxygen detoxification and iron-sulfur cluster assembly in amoebal relict mitochondria. Cell Microbiol 12(3):331–342

Martin W, Muller M (1998) The hydrogen hypothesis for the first eukaryote. Nature 392(6671):37–41

Marvin-Sikkema FD, Pedro Gomes TM, Grivet JP, Gottschal JC, Prins RA (1993) Characterization of hydrogenosomes and their role in glucose metabolism of *Neocallimastix* sp. L2. Arch Microbiol 160(5):388–396

Mazet M, Diogon M, Alderete JF, Vivares CP, Delbac F (2008) First molecular characterisation of hydrogenosomes in the protozoan parasite *Histomonas meleagridis*. Int J Parasitol 38(2):177–190

Meyer J (2007) [FeFe] hydrogenases and their evolution: a genomic perspective. Cell Mol Life Sci 64(9):1063–1084

Mi-ichi F, Abu Yousuf M, Nakada-Tsukui K, Nozaki T (2009) Mitosomes in *Entamoeba histolytica* contain a sulfate activation pathway. Proc Natl Acad Sci U S A 106(51):21731–21736

Morrison HG, McArthur AG, Gillin FD, Aley SB, Adam RD, Olsen GJ, Best AA, Cande WZ, Chen F, Cipriano MJ, Davids BJ, Dawson SC, Elmen-dorf HG, Hehl AB, Holder ME, Huse SM, Kim UU, Lasek-Nesselquist E, Manning G, Nigam A, Nixon JE, Palm D, Passamaneck NE, Prabhu A, Reich CI, Reiner DS, Samuelson J, Svard SG, Sogin ML (2007) Genomic minimalism in the early diverging intestinal parasite *Giardia lamblia*. Science 317(5846):1921–1926

Mulder DW, Boyd ES, Sarma R, Lange RK, Endrizzi JA, Broderick JB, Peters JW (2010) Stepwise [FeFe]-hydrogenase H-cluster assembly revealed in the structure of HydA(DeltaEFG). Nature 465(7295):248–251

Muller M (1993) The hydrogenosome. J Gen Microbiol 139(12):2879–2889

Muller M (2007) The road to hydrogenosomes. In: Martin W, Muller M (eds) Origin of mitochondria and hydrogenosomes. Springer, Berlin Heidelberg, pp 1–11

Mus F, Dubini A, Seibert M, Posewitz MC, Grossman AR (2007) Anaerobic acclimation in *Chlamydomonas reinhardtii*: anoxic gene expression, hydrogenase induction, and metabolic pathways. J Biol Chem 282(35):25475–25486

Nixon JE, Field J, McArthur AG, Sogin ML, Yarlett N, Loftus BJ, Samuelson J (2003) Iron-dependent hydrogenases of *Entamoeba histolytica* and *Giardia lamblia*: activity of the recombinant entamoebic enzyme and evidence for lateral gene transfer. Biol Bull 204(1):1–9

Perez-Brocal V, Clark CG (2008) Analysis of two genomes from the mitochondrion-like organelle of the intestinal parasite *Blastocystis*: complete sequences, gene content, and genome organization. Mol Biol Evol 25(11):2475–2482

Pfanner N, Wiedemann N, Meisinger C, Lithgow T (2004) Assembling the mitochondrial outer membrane. Nat Struct Mol Biol 11(11):1044–1048

Pilet E, Nicolet Y, Mathevon C, Douki T, Fontecilla-Camps JC, Fontecave M (2009) The role of the maturase HydG in [FeFe]-hydrogenase active site synthesis and assembly. FEBS Lett 583(3):506–511

Posewitz MC, Smolinski SL, Kanakagiri S, Melis A, Seibert M, Ghirardi ML (2004) Hydrogen photoproduction is attenuated by disruption of an isoamylase gene in *Chlamydomonas reinhardtii*. Plant Cell 16(8):2151–2163

Putz S, Dolezal P, Gelius-Dietrich G, Bohacova L, Tachezy J, Henze K (2006) Fe-hydrogenase maturases in the hydrogenosomes of *Trichomonas vaginalis*. Eukaryot Cell 5(3):579–586

Py B, Barras F (2010) Building Fe-S proteins: bacterial strategies. Nat Rev Microbiol 8(6):436–446

Quevillon E, Silventoinen V, Pillai S, Harte N, Mulder N, Apweiler R, Lopez R (2005) InterProScan: protein domains identifier. Nucleic Acids Res 33(Web Server issue):W116–W120

Reeves RE, Warren LG, Susskind B, Lo HS (1977) An energy-conserving pyruvate-to-acetate pathway in *Entamoeba histolytica*. Pyruvate synthase and a new acetate thiokinase. J Biol Chem 252(2):726–731

Riviere L, van Weelden SW, Glass P, Vegh P, Coustou V, Biran M, van Hellemond JJ, Bringaud F, Tielens AG, Boshart M (2004) Acetyl:succinate CoA-transferase in procyclic Trypanosoma brucei. Gene identification and role in carbohydrate metabolism. J Biol Chem 279(44):45337–45346

Rodriguez MA, Hidalgo ME, Sanchez T, Orozco E (1996) Cloning and characterization of the *Entamoeba histolytica* pyruvate: ferredoxin oxidoreductase gene. Mol Biochem Parasitol 78(1–2):273–277

Roger AJ, Silberman JD (2002) Cell evolution: mitochondria in hiding. Nature 418(6900):827–829

Rotte C, Stejskal F, Zhu G, Keithly JS, Martin W (2001) Pyruvate: NADP+ oxidoreductase from the mitochondrion of Euglena gracilis and from the apicomplexan *Cryptosporidium parvum*: a biochemical relic linking pyruvate metabolism in mitochondriate and amitochondriate protists. Mol Biol Evol 18(5):710–720

Sanchez LB, Muller M (1996) Purification and characterization of the acetate forming enzyme, acetyl-CoA synthetase (ADP-forming) from the amitochondriate protist, *Giardia lamblia*. FEBS Lett 378(3):240–244

Sanchez LB, Galperin MY, Muller M (2000) Acetyl-CoA synthetase from the amitochondriate eukaryote *Giardia lamblia* belongs to the newly recognized superfamily of acyl-CoA synthetases (Nucleoside diphosphate-forming). J Biol Chem 275(8):5794–5803

Sawers G, Watson G (1998) A glycyl radical solution: oxygen-dependent interconversion of pyruvate formate-lyase. Mol Microbiol 29(4):945–954

Shepard EM, McGlynn SE, Bueling AL, Grady-Smith CS, George SJ, Winslow MA, Cramer SP, Peters JW, Broderick JB (2010) Synthesis of the 2Fe subcluster of the [FeFe]-hydrogenase H cluster on the HydF scaffold. Proc Natl Acad Sci U S A 107(23):10448–10453

Sickmann A, Reinders J, Wagner Y, Joppich C, Zahedi R, Meyer HE, Schonfisch B, Perschil I, Chacinska A, Guiard B, Rehling P, Pfanner N, Meis-inger C (2003) The proteome of *Saccharomyces cerevisiae* mitochondria. Proc Natl Acad Sci U S A 100(23):13207–13212

Stairs CW, Roger AJ, Hampl V (2011) Eukaryotic pyruvate formate lyase and its activating enzyme were acquired laterally from a firmicute. Mol Biol Evol 28:2087–2099

Stechmann A, Hamblin K, Perez-Brocal V, Gaston D, Richmond GS, van der Giezen M, Clark CG, Roger AJ (2008) Organelles in *Blastocystis* that blur the distinction between mitochondria and hydrogenosomes. Curr Biol 18(8):580–585

Steinbuchel A, Muller M (1986) Anaerobic pyruvate metabolism of *Tritrichomonas foetus* and Trichomonas vaginalis hydrogenosomes. Mol Biochem Parasitol 20(1):57–65

Sutak R, Dolezal P, Fiumera HL, Hrdy I, Dancis A, Delgadillo-Correa M, Johnson PJ, Muller M, Tachezy J (2004) Mitochondrial-type assembly of FeS centers in the hydrogenosomes of the amitochondriate eukaryote *Trichomonas vaginalis*. Proc Natl Acad Sci U S A 101(28):10368–10373

Thauer RK, Kirchniawy FH, Jungermann KA (1972) Properties and function of the pyruvate-formate-lyase reaction in clostridiae. Eur J Biochem 27(2):282–290

Tielens AG, Rotte C, van Hellemond JJ, Martin W (2002) Mitochondria as we don't know them. Trends Biochem Sci 27(11):564–572

Tielens AG, van Grinsven KW, Henze K, van Hellemond JJ, Martin W (2010) Acetate formation in the energy metabolism of parasitic helminths and protists. Int J Parasitol 40(4):387–397

Tjaden J, Haferkamp I, Boxma B, Tielens AG, Huynen M, Hackstein JH (2004) A divergent ADP/ATP carrier in the hydrogenosomes of *Trichomonas gallinae* argues for an independent origin of these organelles. Mol Microbiol 51(5):1439–1446

Tovar J, Fischer A, Clark CG (1999) The mitosome, a novel organelle related to mitochondria in the amitochondrial parasite *Entamoeba histolytica*. Mol Microbiol 32(5):1013–1021

Tovar J, Leon-Avila G, Sanchez LB, Sutak R, Tachezy J, van der Giezen M, Hernandez M, Muller M, Lucocq JM (2003) Mitochondrial remnant organelles of *Giardia* function in iron-sulphur protein maturation. Nature 426(6963):172–176

Tsaousis AD, Kunji ER, Goldberg AV, Lucocq JM, Hirt RP, Embley TM (2008) A novel route for ATP acquisition by the remnant mitochondria of *Encephalitozoon cuniculi*. Nature 453(7194): 553–556

Tsaousis AD, Stechmann A, Perez-Brocal V, Hamblin KA, van der Giezen M, Clark CG (2010) The *Blastocystis* mitochondrion-like organelles. In Anaerobic Parasitic Protozoa: Genomics and Molecular Biology, C.G. Clark, R.D. Adam, and P.J. Johnson, eds. (Horizon Scientific Press)

Tsaousis AD, Gaston D, Stechmann A, Walker PB, Lithgow T, Roger AJ (2011) A functional Tom70 in the human parasite *Blastocystis* sp.: implications for the evolution of the mitochondrial import apparatus. Mol Biol Evol 28(1):781–791

van der Giezen M, Slotboom DJ, Horner DS, Dyal PL, Harding M, Xue GP, Embley TM, Kunji ER (2002) Conserved properties of hydrogenosomal and mitochondrial ADP/ATP carriers: a common origin for both organelles. EMBO 21(4):572–579

van der Giezen M, Birdsey GM, Horner DS, Lucocq J, Dyal PL, Benchimol M, Danpure CJ, Embley TM (2003) Fungal hydrogenosomes contain mitochondrial heat-shock proteins. Mol Biol Evol 20(7):1051–1061

van der Giezen M, Cox S, Tovar J (2004) The iron-sulfur cluster assembly genes iscS and iscU of *Entamoeba histolytica* were acquired by horizontal gene transfer. BMC Evol Biol 4:7

van Grinsven KW, Rosnowsky S, van Weelden SW, Putz S, van der Giezen M, Martin W, van Hellemond JJ, Tielens AG, Henze K (2008) Acetate:succinate CoA-transferase in the hydrogenosomes of *Trichomonas vaginalis*: identification and characterization. J Biol Chem 283(3):1411–1418

Vignais PM, Billoud B (2007) Occurrence, classification, and biological function of hydrogenases: an overview. Chem Rev 107(10):4206–4272

Vincent KA, Parkin A, Lenz O, Albracht SP, Fontecilla-Camps JC, Cammack R, Friedrich B, Armstrong FA (2005) Electrochemical definitions of O2 sensitivity and oxidative inactivation in hydrogenases. J Am Chem Soc 127(51):18179–18189

Voncken FG, Boxma B, van Hoek AH, Akhmanova AS, Vogels GD, Huynen M, Veenhuis M, Hackstein JH (2002) A hydrogenosomal [Fe]-hydrogenase from the anaerobic chytrid *Neocallimastix* sp. L2. Gene 284(1–2):103–112

Wagner AF, Frey M, Neugebauer FA, Schafer W, Knappe J (1992) The free radical in pyruvate formate-lyase is located on glycine-734. Proc Natl Acad Sci U S A 89(3):996–1000

Williams BA, Hirt RP, Lucocq JM, Embley TM (2002) A mitochondrial remnant in the microsporidian *Trachipleistophora hominis*. Nature 418(6900):865–869

Williams BA, Haferkamp I, Keeling PJ (2008) An ADP/ATP-specific mitochondrial carrier protein in the microsporidian *Antonospora locustae*. J Mol Biol 375(5):1249–1257

Wunschiers R, Stangier K, Senger H, Schulz R (2001) Molecular evidence for a Fe-hydrogenase in the green alga *Scenedesmus obliquus*. Curr Microbiol 42(5):353–360

Yarlett N, Orpin CG, Munn EA, Yarlett NC, Greenwood CA (1986) Hydrogenosomes in the rumen fungus *Neocallimastix patriciarum*. Biochem J 236(3):729–739

Biodata of **Rob M. de Graaf** and **Johannes H.P. Hackstein**, authors of *"Hydrogenosomes and Mitosomes: Mitochondrial Adaptations to Life in Anaerobic Environments."*

Dr. Rob M. de Graaf is a scientist at the Institute of Water and Wetland Research at the Radboud University of Nijmegen, the Netherlands. Dr. de Graaf obtained his Ph.D. in 2010 from the Radboud University of Nijmegen. His main interests are Evolutionary Microbiology, in particular of anaerobic protists, and Exobiology.

E-mail: **exorest@science.ru.nl**

Dr. Johannes H.P. Hackstein is associate Professor at the Institute of Water and Wetland Research at the Radboud University of Nijmegen, the Netherlands. He received his Ph.D. from the University of Cologne in 1976. Subsequently, he studied the genetics of spermatogenesis in *Drosophila melanogaster* and *D. hydei* at the Friedrich-Miescher-Laboratorium of the Max-Planck-Gesellschaft (Tübingen, Germany) and at the Department of Genetics at the University of Nijmegen. Since 1999 he is head of the research group "Evolutionary Microbiology." His main interests are the symbioses between animals and methanogenic archaea and the evolution of anaerobic protists.

E-mail: **J.hackstein@science.ru.nl**

Rob M. de Graaf   **Johannes H.P. Hackstein**

HYDROGENOSOMES AND MITOSOMES: MITOCHONDRIAL ADAPTATIONS TO LIFE IN ANAEROBIC ENVIRONMENTS

ROB M. DE GRAAF AND JOHANNES H.P. HACKSTEIN
Department of Evolutionary Microbiology, Faculty of Science, Radboud University Nijmegen, Heyendaalseweg 135, 6525AJ Nijmegen, The Netherlands

1. Introduction

The origin of the eukaryotic cell was a major breakthrough in life's history. The textbook eukaryotic cell as we know it today is very efficient in producing large amounts of ATP by oxidative phosphorylation in their mitochondria under aerobic conditions. Since Lynn Margulis published in 1967 the endosymbiosis theory (Sagan 1967) that for the first time provided a plausible explanation for the origin of the eukaryotic cell by endosymbiosis, a lot of molecular data accumulated that clearly show that alpha proteobacteria are the precursors of mitochondria. The origin of the host is still less clear. For some time, it was thought that ancestral eukaryotic cells ("archezoa," (Cavalier-Smith 1987)) existed that became the host for an alpha proteobacterium that eventually evolved into the mitochondrion. This archezoa hypothesis was supported by the observation that there were (anaerobic) eukaryotic cells that obviously did not contain mitochondria. However, upon closer inspection, these cells appeared to possess vestigial double membrane–bound organelles. Notably, in contrast to mitochondria, these organelles did not contain DNA and did not generate ATP with the aid of an electron transport chain. Some of these organelles produced hydrogen and were named hydrogenosomes. They were for the first time described in the parasitic flagellate *Trichomonas vaginalis* where they produced ATP by substrate level phosphorylation (Lindmark et al. 1975). In other unicellular anaerobes, these organelles were rather inconspicuous and did neither produce hydrogen nor ATP. These organelles were named "mitosomes" since they were assumed to be remnants of mitochondria. Nowadays, it is clear that all cells with mitosomes or hydrogenosomes possess genes in their nuclear DNA that code for proteins from mitochondrial origin. Consequently, there is no eukaryotic cell known so far that does not contain genes of mitochondrial origin. This means there is no evidence that archezoa ever existed, and consequently, the nature of the ancestral host for the mitochondria remains uncertain (Embley and Martin 2006). On the other hand, with the discovery of a genome in the hydrogenosomes of the anaerobic ciliate *Nyctotherus ovalis* that looks in general like an ordinary mitochondrial genome of an aerobic ciliate, the final proof was delivered that hydrogenosomes are the result of an adaptation of mitochondria to an anoxic environment (Boxma et al. 2005). This was further strengthened by the discovery of another "mitochondrial" genome in the elusive double membrane–bound organelle of the stramenopile *Blastocystis* sp. Thus, the adaptation to anaerobic environments can involve dramatic changes in the most important organelles of the eukaryotic cell.

All eukaryotic cells can adapt to (transient) anaerobiosis by using cytoplasmic glycolysis instead of using the mitochondrial metabolism. They cope with the accumulation of reduced metabolites such as NAD(P)H by the reoxidation of these substances by fermentation, i.e., for example, the formation of ethanol or lactate. Since these metabolites accumulate inside or around the cell, extended anaerobiosis creates problems for the cell, which still possesses functional, but quiescent mitochondria. Therefore, adaptation to permanent anaerobiosis requires substantial modifications of the mitochondria. For example, mitochondria can use alternative electron acceptors such as nitrate, or the citric acid cycle can be modified to allow the use of endogenous fumarate as electron acceptor, which leads to the formation of succinate that is excreted (Tielens et al. 2002). As a further consequence, the electron transport chain can be reduced and finally lost completely. In hydrogenosomes, hydrogenases allow the use of protons as electron acceptors, and the citric acid cycle enzyme succinyl-CoA synthetase allows the generation of ATP by substrate level phosphorylation. Eventually, even the formation of ATP in the organelles can be lost.

In this chapter, we will give an overview of the different types of mitochondrial-related organelles found in the recent years. Furthermore, we will discuss the different kinds of hydrogenosomes and mitosomes and show that differences between these organelles become smaller with the discovery of intermediate types of organelles. It will become clear that mitochondria under permanently anoxic conditions can evolve reductively to hydrogenosomes and mitosomes. Finally, we will present a potential scenario for the evolution of mitochondria to hydrogenosomes and mitosomes.

2. Anaerobic Protists: Diversity and Distribution of Hydrogenosomes and Related Organelles

Adaptations to anaerobic environments imply the evolution of anaerobically functioning organelles. It is assumed that, in particular, the mitochondrion was a principally aerobic (or facultatively anaerobic) organelle that evolved from a symbiotic α-proteobacterium as shown in Fig. 1 under a substantial gene loss and the transfer of the majority of the residual genes into the nucleus. The adaptation of this (proto)-mitochondrion to life under anaerobic circumstances led to its transformation into a hydrogenosome or a mitosome (see below). This adaptation, in most cases, led to a complete loss of the organellar genome and the electron transport chain besides several changes of the organellar metabolism, which are described in more detail below.

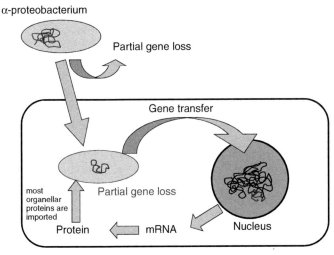

Figure 1. According to the endosymbiosis theory, an ancestor of the α-proteobacteria was adopted by a host cell. During this process, a large number of genes were lost, a part stayed in the organelle, and another part was transferred to the nucleus. The organellar genes that were transferred to the nucleus are transcribed, subsequently translated in the cytosol, and finally, the protein is imported into the organelle. In the case of almost all hydrogenosomes, all genes were lost from the organelle, requiring all organellar proteins to be imported.

Most aerobic eukaryotes possess classical mitochondria. It is well known that certain eukaryotes with classical mitochondria can adapt to anaerobic/microaerobic conditions by a modification of their mitochondrial metabolism, for example, by the use of alternative electron acceptors such as (environmental) nitrate or (endogenous) fumarate (Tielens et al. 2002). Other adaptations imply the loss of mitochondrial genes that are not useful anymore, and the acquisition of new genes necessary for the anaerobic metabolism by horizontal gene transfer. In recent years at least 16 distinct lineages of anaerobic, unicellular eukaryotes have been identified that contain mitochondrion-related organelles (Embley et al. 2003; van der Giezen and Tovar 2005; Hackstein et al. 2006; Barbera et al. 2007; van der Giezen 2009; Hjort et al. 2010). These organelles can be divided into two groups: hydrogenosomes and mitosomes.

3. Mitochondrial Genomes

Textbook mitochondria are organelles that are surrounded by a double-membrane and use oxidative phosphorylation to produce ATP. However, there are many (anaerobic) mitochondria that can produce ATP without using oxygen as terminal electron acceptor. Besides in ATP production by oxidative phosphorylation, mitochondria play an important role in Fe-S cluster assembly, ion homeostasis, apoptosis and a broad spectrum of metabolic pathways. All mitochondria possess a genome that shows that mitochondria are without any doubt related to the alpha-proteobacteria as described above. In contrast to nuclear genomes that usually possess several different chromosomes, mitochondrial genomes in general consist of a single molecule. However, eukaryotic cells almost always contain more than one mitochondrion per cell (average 10–100) and often more than a dozen copies of mtDNA per mitochondrion. This high multiplicity of mtDNA within the cell implies that mitochondrial mutations that can be caused by free radicals produced as a by-product of oxidative phosphorylation can accumulate gradually without immediate deleterious impact. At the moment (February 2010), 3,229 complete mitochondrial genomes of 2,090 distinct organisms have been sequenced according to GOBASE (http://gobase.bcm.umontreal.ca/). These 2,090 different organisms consist of 1,296 vertebrates, 581 invertebrates, 60 plants, 78 fungi, and 75 protists. The mitochondrial genomes vary in size between 6 kb in the genome of the human malaria parasite *Plasmodium falciparum* (Feagin 1994) and more than (estimated) 2,200 kb in land plants. The largest sequenced mitochondrial genome is that of maize (569,630 bp); most of the mtDNA consist of intergenic regions. These intergenic regions often consist of tandem repeat arrays (Lunt et al. 1998) or stem–loop motifs (Paquin and Lang 1996). Also, introns can play a role in the extension of the mitochondrial genome. For example, in the filamentous fungus *Podospora anserina*, introns make up 75% of the total size of the mitochondrial DNA (Cummings et al. 1990). The size of most other (non-plant) eukaryotic mitochondrial genomes ranges between 15 and 60 kb. In contrast to the enormous variance in mitochondrial genome size, the coding capacity of mtDNA ranges from 97 genes in the flagellate *Reclinomonas americana* (Lang et al. 1997) to only 5 in *Plasmodium falciparum* (Feagin 1994), with an average of 40–50 genes.

The majority of the sequenced mitochondrial genomes, especially those of animals, are circular. However, also linear mitochondrial genomes are found in very diverse groups of unrelated organisms i.e., chlorophycean green algae (like *Chlamydomonas*), fungi, cnidarian animals (jellyfish), apicomplexa (*Plasmodium* and relatives), and ciliates. These linear molecules contain telomere-like repeats of varying length that in some cases can be different at both ends (Morin and Cech 1988b; Nosek et al. 1998). In some cases mitochondrial genomes consist of multiple circular molecules like in *Spizellomyces punctutatus* (Burger and Lang 2003) or of a large number of linear molecules of different sizes with telomeres as in the fungus like organism *Amoebidium parasiticum* (Lang et al. 2002). Probably, the most complex mitochondrial genomes are found in a group of unicellular algae: the dinoflagellates. They belong, together with the apicomplexa and ciliates, to the clade Alveolata and possess a highly complex genome structure with pseudogenes, partial gene fragments, transcript editing, and a dense pattern of inverted repeats (Nash et al. 2008; Waller and Jackson 2009). Ciliate mitochondrial genomes are much larger and contain many more genes than the apicomplexan genomes (~50). At the moment, the complete mitochondrial genomes of *Paramecium aurelia* and of 5 different *Tetrahymena* species (*T. pyriformis*, *T. thermophila*, *T. pigmentosa*, *T. paravorax* and *T. malaccensis*) have been sequenced completely (Pritchard et al. 1990; Burger et al. 2000; Brunk et al. 2003; Moradian et al. 2007). Recently, the nearly complete mitochondrial genome of *Euplotes minuta* and a large part of the mitochondrial genome of *Euplotes crassus* have been described (de Graaf et al. 2009b). We can conclude that, in general, ciliate mitochondrial genomes have a size between 40 and 50 kb. So far, all have a linear mitochondrial genome capped with telomere like repeats that vary substantially in length (Morin and Cech 1988b; Morin and Cech 1988a). The ciliate mitochondrial genomes harbor tightly packed genes, only 4% of the genomes consist of intergenic spacers. Comparing the *Tetrahymena* and *Paramecium* mitochondrial genomes the gene-order is conserved but when compared to the *Euplotes* species, it is very different. In general, it can be concluded that mitochondrial genomes – even among ciliates – can be very different.

4. Hydrogenosomes

Hydrogenosomes are double membrane–bound organelles that produce hydrogen and ATP (Müller 1993). They are found in some anaerobic (microaerophilic) unicellular eukaryotes that represent a broad spectrum of species. Within the Excavata, three unrelated species with hydrogenosomes have been identified (Fig. 2): the heterolobosean amoeboflagellate *Psalteriomonas lanterna*, the preaxostylid flagellate *Trimastix pyriformis*, and the parabasalid flagellate *Trichomonas vaginalis* with its relatives *Tritrichomonas foetus*, *Monocercomonas* sp., and *Histomonas meleagridis*. Within the Chromalveolata, several anaerobic ciliates with hydrogenosomes evolved from different aerobic ancestors. One of them, *Nyctotherus ovalis*, which thrives in the hindgut of cockroaches, (Fig. 3) possesses a large organellar genome. (Akhmanova et al. 1998; Boxma et al. 2005; de Graaf et al. 2011). Recently, in the stramenopile *Blastocystis*, another member of the Chromalveolata with hydrogenosomes that is rather unrelated to the ciliates, an organellar genome was identified (Perez-Brocal and Clark 2008; Stechmann et al. 2008; Wawrzyniak et al. 2008), which is a striking example of convergent evolution. In the anaerobic chytridiomycete fungus *Neocallimastix* sp, (and its relative *Piromyces* sp. as well) hydrogenosomes were identified (Yarlett et al. 1986; Müller 1993; Boxma et al. 2004). These hydrogenosomes lack a genome (van der Giezen et al. 1997). These chytrids are phylogenetically related to aerobic fungi with "classical mitochondria" and to fungi-related anaerobic organisms (Microsporidia) that possess mitosomes. So far, hydrogenosomes were not identified in multicellular organisms. However, very recently, organelles resembling hydrogenosomes were identified in an anaerobic multicellular organism belonging to the animal phylum Loricefera, which lives under anaerobic conditions in the deep sea. If this interpretation is correct, this would be the first metazoan with hydrogenosomes (Danovaro et al. 2010). Here, we will first describe the different hydrogenosomes of unicellular organisms and later discuss the mitosomes.

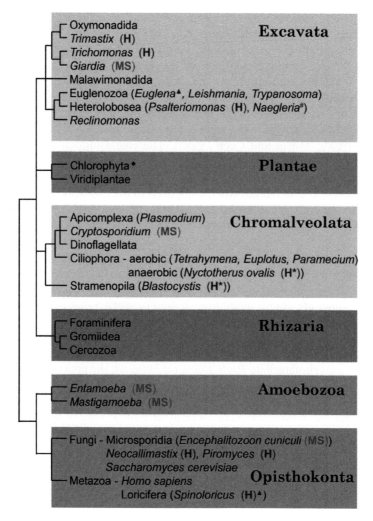

Figure 2. Schematic cladogram of eukaryotes indicating that hydrogenosomes (H) and mitosomes (MS) are widespread in the empire of eukaryotes and associated (in most cases) with related eukaryotes with "textbook" mitochondria. H* indicates hydrogenosomes with a mitochondrial genome, ♦ some single-celled algae, like *Scenedesmus*, possess hydrogenases in their chloroplasts (Ghirardi et al. 2007), # *Naegleria* contains a hydrogenase. ▲ potentially hydrogenosome containing species.

4.1. *TRICHOMONAS*

The name hydrogenosome was introduced by (Lindmark and Müller 1973) for subcellular particles resembling microbodies that play a central role in pyruvate degradation and generate electrons for the production of hydrogen from protons in *Tritrichomonas foetus*. This organism is a close relative of the human parasite *Trichomonas vaginalis* that was later intensively investigated and of which recently the complete genome has been sequenced (Carlton et al. 2007) and the hydrogenosomal proteome analyzed (Henze 2007). The combination of these data made it possible to reconstruct the hydrogenosomal metabolism (Table 1). About 200 proteins could be identified that are involved in the hydrogenosomal metabolism of *Trichomonas*, (Henze 2007) approximately 13% of what is found in the human mitochondrial proteome consisting of up to 1500 proteins (Meisinger et al. 2008). The trichomonad hydrogenosomes import pyruvate and malate (Hrdý et al. 2007). Malate is decarboxylated to pyruvate by a NAD-dependent malic enzyme inside the hydrogenosome. The initial step in the catabolism of pyruvate is the oxidative decarboxylation by the pyruvate:ferredoxin oxidoreductase (PFO) to acetyl-CoA and CO_2. The reduced ferredoxin is reoxidized by a [FeFe] hydrogenase. The genome of *T. vaginalis* encodes 5 [FeFe] hydrogenases with hydrogenosomal targeting signals. It is assumed that one or the other hydrogenase reacts directly with NADH, potentially involving ferredoxin and the 24-kD and 51-kD proteins that are orthologous to the corresponding subunits of a mitochondrial complex I (Hrdy et al. 2004). The next step in the catabolism of pyruvate is the formation of acetate from acetyl-CoA

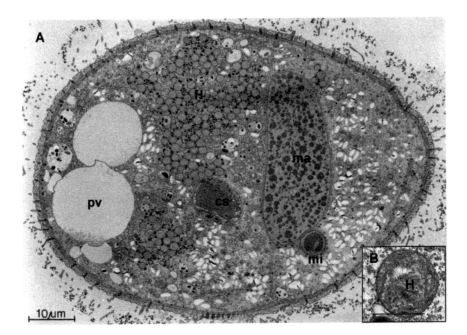

Figure 3. Electron micrograph of *Nyctotherus ovalis* (**A**). Insert (**B**) a hydrogenosome, scale bar represents 0.5 μm. *H* hydrogenosomes, *ma* macronucleus, *mi* micronucleus, *cs* cytostome, *pv*: pulsating vacuole. (After Hackstein et al. 2008, modified.)

with the simultaneous transfer of the CoA moiety to succinate. This reaction is catalyzed by the enzyme acetate:succinate CoA-transferase (ASCT). The corresponding gene was not identified in the draft genome version. However, recently, the gene and the enzyme have been identified and characterized in detail (van Grinsven et al. 2008). The succinyl-CoA synthetase uses the energy-rich CoA bond for the generation of ATP/GTP from ADP/GDP. It is the only enzyme of *Trichomonas* also known from the TCA cycle of aerobic mitochondria. It regenerates succinate for the reaction with ASCT. The acetate that is formed by the action of ASCT is excreted by a so far unidentified transporter.

Some components of other mitochondria-related metabolic pathways have been identified. For example, Fe–S cluster synthesizing proteins, glycine cleavage system components, components of a defense system against reactive oxygen species (ROS), and a number of genes encoding organellar import proteins (see Table 1).

4.2. *TRIMASTIX PYRIFORMIS*

Trimastix pyriformis is, just like the trichomonads, another member of the Excavata, but rather unrelated to the trichomonads. This organism belongs to the Preaxostyla, a poorly investigated subgroup of the Excavata that was believed to represent a primitive pre-mitochondrial lineage of eukaryotes (Hampl et al. 2008). However, a recent EST study (Hampl et al. 2008) definitively revealed a mitochondrial ancestry for the hydrogenosomes in this organism (Table 1). Hampl et al. identified a highly expressed [FeFe] hydrogenase in combination with two enzymes involved in the maturation of [FeFe] hydrogenases. Also, a pyruvate:ferredoxin oxidoreductase (PFO) was found in this EST study. The localization of both enzymes has not yet been studied. Thus, it remains unclear as to whether the hydrogenase and the PFO are localized in the putative hydrogenosomes or in the cytoplasm. Notwithstanding, the presence of highly expressed hydrogenase and PFO genes suggests that the double-membrane-bound organelles in *Trimastix* are hydrogenosomes that have a metabolism similar to trichomonads. Notably, four proteins belonging to the glycine cleavage system (GCS) have been identified; the GCS is characteristic for mitochondria and is also present in the hydrogenosomes of *Trichomonas*, *Blastocystis*, and *Nyctotherus*, and the mitosomes of *Mastigamoeba balamuthi*. Further support for a mitochondrial/hydrogenosomal nature of the organelles comes from the presence of three genes encoding mitochondrial carriers, three genes encoding components of organellar (mitochondrial) import machinery (TOM40, MPP, and Hsp60), a lipoyl transferase, and a pyridine nucleotide transhydrogenase alpha. In addition, a gene encoding the TCA cycle enzyme aconitase was found that contrasts with the hydrogenosomes of *Trichomonas*, which does not possess TCA cycle enzymes except SCS (succinyl-CoA synthetase) (Carlton et al. 2007; Hampl et al. 2008). All the data discussed

Table 1. Distribution of mitochondrial-derived genes in organisms with hydrogenosomes or mitosomes.

Protistan group	Species	Organelle	Genome	Mitochondrial-derived genes	Anaerobic metabolism
Entamoeba	*Entamoeba histolytica* (A)	M	No	Cpn60, Cpn10, Hsp70, MDH, NifS*, NifU*, PNT, ME, ACS	PFO1, PFO2, hydrogenase
Pelobionts	*Mastigamoeba balamuthi* (A)	M	No	Cpn60, Cpn10, Mge1p, MDH, Aconitase, ICDH, SDHb, SDHc PC, NifS*, NifU*, AKL, SHMT, GCS-H, GCS-L, GCS-T, GCS-P, IVDH, Fd, PNT, ME, ACS	PFO1, PFO2, hydrogenase
Microsporidia	*Encephalitozoon cuniculi* (B)	M	No	PDH E1α, PDH E1β, ATM 1p, IscU, IscS, Fd, frataxin, Hsp70, ERV 1p, Tim22, Tom70, IMP-2, NADPH-FOR, MnSOD, NTT3	
	Antonospora loustae (B)	M	–	PDH E1β, Hsp70, TIM22, Fd, G3PH, frataxin, IscU, NADPH-FOR, IscS, pyruvate importer, IMP2, ATM1, MnSOD, modified AAC (C)	
Apicomplexa	*Cryptosporidium parvum* (D)	M$_R$	No	AOX, PNT, IscU, IscS, Fd, MDH, Hsp60, Hsp70	PNO, Narf-like hydrogenase
Diplomonads	*Giardia intestinalis*	M	No	IscU, IscS, Cpn60, Hsp70, Fd, MPP, Pam18, TOM40 (E)	PFO, hydrogenase
	Spironucleus	M	No	GroE1(Hsp60/Cpn60), NifS* (F)	
Parabasalids	*Trichomonas vaginalis*	H	No	Hmp31, Hmp35, Cpn10, Hsp20 Hsp60, Hsp70, IscU, IscS, 51 kDa, 24 kDa, MPP, Pam18, Sam50, Tim23, Tim44, Fd, ScoAα, ScoAβ, ME, ASCT, AK PDH E2, GCS-H, MHF, ALAT, STK, SHMT, Argdeiminase, IscA, IscU, IscS, NifU, frataxin, MnSOD, Rr, Px, thiroedoxin, GLO, GK (I, J)	PFO, hydrogenase
Chydtrid fungi	*Neocallimastix* spp.	H	No	Hsp70, Hsp60, AAC, ScoA(α+β)	PFL, hydrogenase
	Piromyces spp.	H	No	Hsp60, Hsp90, AAC, ME, ScoA(α+β), 51 kDa, SHMT, GDH (New results presented in this chapter)	PFL, hydrogenase

(continued)

Table 1. (continued)

Protistan group	Species	Organelle	Genome	Mitochondrial-derived genes	Anaerobic metabolism
Ciliophora	*Nyctotherus ovalis*	H	Yes	PDH-E1α/E1β/E2/E3, AK, LDH, ME, mtC1, mtC2, mtRP, mttRNA, mt-rns, mt-rnl, MMCoA, PcoA, SCOT, LCFA-CoA, GK, GCS-L, GCS-H, GCS-T, SHMT, OIDα, OIDβ, BC, BCAA, ALAT, GDH, CS, SP, MHF, ACS, adrenodoxin, mtPepα, FtsJ, FtsH, PPI, Hsp10, Hsp60, Hsp70, Hsp90, AAC, PET8, OGCP, ROS, uracil-DNA glycosylase, EfTu, cyt-C, PC, G3PDH, FUS3, MPV17, and less common mt genes	Hydrogenase
Heterokonts	*Blastocystis* sp.	H	Yes	PDH-E1α/E2/E3, mtC2, mtRP, mttRNA, mt-rns, mt-rnl, AOX, PNT, frataxin, Fd, Grx5, IscS, Isca2, ABC transporters, Mmt1, MDH, FUM, Scoα/β, DTA, ALAT, SHMT, GCS-H, GCS-L, GCS-T, GCS-P, SerDH, ThrDH, BCCA, ICoADH, PCCa, 3-HBDH, MUT, Hsp70, TOM70, TIM50, TIM21, TIM17, TIM9, MP1, OXAI, ASCT, ACS, glyoxalasel and II, PC, enoyl-CoA hydratase, trans-2-enoyl-CoA reductase, ALDH, long-chain fatty acid-CoA ligase, ACC, MAT, MSD1, LARS, EfTu, EfG, IF-2, Tfa, MRF1, MSS1, MTO1, OGCP, G3PDH, Mcart1, Aralar, several other carriers, and less common mt genes	PFO, hydrogenase
Carpediemonas	*Carpediemonas membranifer*	H	–	Hsp90 (H)	
Trimastix	*Trimastix pyriformis* (G)	H	–	Aconitase, GCS-L, GCS-H, GCS-T, GCS-P, lipoyl-transferase, PNT, TOM40, MPP, Cpn60, mt carriers	PFO, hydrogenase
Heterolobsea	*Psalteriomonas lanterna*	H	No	Hsp60, AAC, 51 kDa, PCCB,GDH, EfTuα	PFO, hydrogenase

*NifS and NifU are bacterial genes, acquired by horizontal gene transfer
(A): (Gill et al. 2007); (B): (Williams and Keeling 2005); (C): (Williams et al. 2008); (D): (Henriquez et al. 2005); (E): (Dagley et al. 2009); (F): (Andersson et al. 2007); (G): (Hampl et al. 2008); (H): (Kolisko et al. 2008); (I): (Barbera et al. 2007); (J): (Carlton et al. 2007)

Cpn60: chaperonin 60, *Cpn10*: chaperonin 10, *Hsp*: heat shock protein, *Mge1p*: nucleotide exchange factor for Hsp70, *MDH*: malate dehydrogenase, *LDH*: lactate dehydrogenase, *AK*: adenylate kinase, *ICDH*: Isocitrate dehydrogenase, *SDH*: succinate dehydrogenase, *PC*: pyruvate carboxylase, *Nif*: Fe-S cluster assembly, *Isc*: Fe-S cluster assembly, *AKL*: α-amino-β-ketobutyrate, *SHMT*: serine hydroxymethyltransferase, *GCS*: glycine cleavage system, *IVDH*: isovaleryl-/CoA-dehydrogenase, *Fd*: ferredoxin, *PNT*: pyridine nucleotide transhydrogenase, *ME*: malic enzyme, *ACS*: acetyl-CoA synthetase, *PFO*: pyruvate:ferredoxin oxidoreductase, *Hyd*: hydrogenase, *mtC1*: mitochondrial complex 1, *mtC2*: mitochondrial complex 2, *mtRP*: set of mitochondrial proteins, *mtRNA*: set of mitochondrial tRNAs, *MMCoA*: methylmalonyl-CoA mutase, *PCoA*: propionyl CoA carboxylase, *SCOT*: succinyl-CoA:ketoacid-CoA transferase, *LCFA-CoA*: longchain-fatty-acid CoA ligase, *GK*: glycerolkinase, *OID*: 2-oxoisovalerate dehydrogenase, *BC*: branched chain α-keto acid dihydrolipoyl acyltransferase, *BCCA*: branched chain aa aminotransferase, *ALAT*: alanine aminotransferase, *GDH*: glutamate dehydrogenase, *CS*: cystathione β-synthetase, *SP*: saccharopepsin, *MHF*: methenyltetrahydrofolate, *mtPepa*: mitochondrial processing peptidase-α, *PPI*: peptidyl-prolyl cis-trans isomerase, *AAC*: ADP-ATP carier, *OGCP*: 2-oxoglutarate-malate carier, *ROS*: components of ROS defense system, *EfTu*: elongation factor Tu, *Cyt-C*: cytochrome C, *G3PDH*: glycerol-3-phosphate dehydrogenase, *IMP2*: inner mitochondrial membrane peptidase2, *ATMI*: mitochondrial ABC transporter, *NADPH-FOR*: NADPH adrenodoxin oxidoreductase, *MnSOD*: superoxide dismutase Mn, *MPP*: mitochondrial processing protease, *AOX*: alternative oxidase, *STK*: succinate thiokinase, *Grx5*: glutaredoxin, *MmtI*: mitochondrial metal transporter 1, *FUM*: fumarate hydratase, *ScoA*: succinyl-CoA synthetase, *DTA*: aspartate aminotransferase, *SerDH*: serine dehydrogenase, *ThrDH*: threonine dehydrogenase, *ICoADH*: isovaleryl-CoA dehydrogenase, *PCCα*: propionyl CoA carboxylase chain, *3-HBDH*: 3-hydroxyisobutyrate dehydrogenase, *MUT*: methylmalonyl-CoA mutase, *MPI*: metalloprotease 1, *OXAI*: oxidase assembly like protein, *ASCT*: acetate:succinate CoA transferase, *ALDH*: aldehyde dehydrogenase, *ACC*: acetyl-CoA-carboxylase, *MAT*: malonyl-CoA carboxylase, *MSDI*: aspartate tRNA ligase, *LARS*: leucyl tRNA synthase, *EfG*: translation initiation factor G, *IF-2*: translation initiation factor 2, *tfa*: Transcription factor A, *Aralar*: aspartate/malate carier, MPP matrix processing peptidase, *PAM18*: presequence translocase-associated motor, *Sam50*: sorting and assembling machinery. *Rr*: rubrerythrin, *Px*: thioperoxidase, *GLO*: glyoxylase, *NTT3*: nucleotide transporter.

above indicate that without doubt the organelles in *Trimastix pyriformis* are related to mitochondria that retained a unique set of mitochondrial genes. The available data are not sufficient to definitively identify these organelles as hydrogenosomes as that depends on the localization of the hydrogenase protein and hydrogen production.

Another eukaryotic member of the Excavata that harbors hydrogenosomes is the microaerophilic amoeboflagellate *Psalteriomonas lanterna* (Broers et al. 1990; Broers 1992). This free-living freshwater eukaryote is related to the aerobic heterolobosean amoeboflagellates *Naegleria gruberi* and *Naegleria fowleri* (see Fig. 2). The latter is a pathogen that can cause severe meningitis. Genes identified related to the organellar metabolism can be found in Table 1. The analysis of these genes suggests that the hydrogenosomal metabolism of *Psalteriomonas lanterna* is similar to that of *Trichomonas vaginalis*. These genes and the ultrastructure of the hydrogenosomes are discussed in detail by (de Graaf et al. 2009a).

4.3. *NEOCALLIMASTIX* SP. AND *PIROMYCES E2*

In the very diverse group of fungi, most of the investigated aerobic fungi possess mitochondria that can perform oxidative phosphorylation. However in the anaerobically functioning chytridiomycete fungi there are some that harbor hydrogenosomes. They cluster phylogenetically clearly with the aerobic chytridiomycete fungi. The anaerobic chytrids are found in the intestinal tract of many herbivorous mammals like cattle, deer, sheep etc. but also in marsupials (Trinci et al. 1994) The best investigated microaerophilic chytridiomycete fungi are *Neocallimastix sp.* (Yarlett et al. 1986; Marvin-Sikkema et al. 1993) and *Piromyces E2*. Their hydrogenosomes produce ATP by substrate level phosphorylation and hydrogen together with CO_2, formate and acetate as end products of their organellar metabolism. The fungus produces succinate, lactate and ethanol when growing on cellulose, glucose or fructose (Marvin-Sikkema et al. 1993). These hydrogenosomes differ in metabolic properties from other hydrogenosomes by the presence of a PFL as key enzyme, and not PDH (as in *Nyctotherus ovalis*) or PFO (as in *Trichomonas*) (Table 1; (Akhmanova et al. 1999). Experiments using [6-^{14}C]-glucose and [U-^{14}C]-glucose indicated that an incomplete TCA cycle operates in the reductive mode allowing the formation of succinate from oxaloacetate via a malate intermediate. Since the formation of significant amounts of labeled CO_2 could be excluded while formate and acetate plus ethanol were formed in a 1:1 ratio (Boxma et al. 2004), it must be concluded that PFL and not pyruvate:ferredoxin oxidoreductase (PFO) or pyruvate dehydrogenase (PDH) play the central role in the hydrogenosomal metabolism. The activity of the latter enzymes would have generated one molecule of labeled carbon dioxide per molecule of pyruvate degraded. However, less than 1% of the expected amount could be measured (Boxma et al. 2004). The observation that the hydrogenosomal PFL and the cytoplasmic ADHE are the key enzymes in the degradation of carbohydrates by anaerobic chytrids reveals that the metabolism of these hydrogenosomes is fundamentally different from the hydrogenosomal metabolism in both trichomonads and *N. ovalis*-like ciliates. Hydrogen formation depends solely on the import of malate into the hydrogenosome, where malate is decarboxylated by malic enzyme, which provides the electrons for the reduction of H^+ to hydrogen. Recently, a large number of sequences (>17000) from an EST study of *Piromyces E2* have become available on Genbank by the DOE Joint Genome Institute. A preliminary screen showed that many clones with sequence similarity to NADP-dependent malic enzyme could be identified. Other hydrogenosomal genes identified in this screen are: succinyl-CoA synthetase α and β subunit, elongation factor 1-alpha and ADP/ATP carrier; all were already identified in *Neocallimastix* sp. Notably, in contrast to Trichomonas the anaerobic chytrids host a mitochondrial type ADP/ATP carrier (AAC) similar to that of Nyctotherus (van der Giezen et al. 2002; Voncken et al. 2002a). Additional mitochondrial genes found were: the 51kD complex I subunit, serine hydroxymethyltransferase, NADP-specific glutamate dehydrogenase and Hsp 90. Many clones were identified with sequence similarity to the previously identified [FeFe] hydrogenase (Voncken et al. 2002b) and PFL genes. Among the Piromyces ESTs there are no sequences with significant similarity to either a PFO gene or any of the PDH subunit encoding genes. It can therefore be concluded that the hydrogenosomes of the chytrid fungi have a fungal mitochondrial ancestor and a different organellar metabolism than other hydrogenosomes.

4.4. *BLASTOCYSTIS* SP.

In the eukaryotic supergroup Chromalveolata many organisms with hydrogenosomes have been identified so far. Almost al of them belong to the ciliates but one of them is a gut parasite: the stramenopile *Blastocystis* sp. that is not related to ciliates. This organism harbours organelles with a mitochondrial genome that lacks genes encoding proteins of mitochondrial Complex III, IV and V of the respiratory chain (Perez-Brocal and Clark 2008; Stechmann et al. 2008; Wawrzyniak et al. 2008). The hydrogenosomal key enzyme, a [FeFe] hydrogenase, could be identified in an EST study and localized in the organelle by cytohistochemistry (Stechmann et al. 2008). However, hydrogenase activity has not been detected so far (Lantsman et al. 2008). Also, the EST analysis provided evidence for the presence of a pyruvate:ferredoxin oxidoreductase (PFO) (Table 1) (Stechmann et al. 2008). Enzymatic studies, however, did reveal pyruvate:NADP oxidoreductase (PNO) activity instead of PFO activity. PNO is as PFO a strictly anaerobic enzyme that decarboxylates pyruvate. It has been found also in the mitochondrial remnants of *Cryptosporidium* sp. and the mitochondria of *Euglena gracilis* (Rotte et al. 2001). In *Blastocystis*, PNO decarboxylates pyruvate to acetyl-CoA and CO_2; acetyl-CoA is metabolized by the "hydrogenosomal" enzyme ASCT (acetate:succinate CoA-transferase) to acetate. The CoA moiety is transferred to succinate that is recycled via a SCS. This is a typical hydrogenosomal pathway, and some of the corresponding genes have significant sequence similarity to the homologous genes of *Trichomonas*. Also the malic enzyme activity is a hydrogenosomal trait. A corresponding EST, however, has not been identified (Stechmann et al. 2008). The organelles host an incomplete TCA cycle. The EST studies provided evidence for SCS, succinate dehydrogenase (SDH), fumarase (Fum) and malate dehydrogenase (MDH) genes, but enzymatic studies revealed enzymatic activity of aconitase, isocitrate dehydrogenase, and α-ketoglutarate dehydrogenase in addition to the STK/SCS activity. A SDH activity was not observed, although the EST studies revealed the presence of all four subunits of a mitochondrial Complex II. The EST study in combination with sequence analyses of the organellar genome revealed 16 mitochondrial Complex I genes, 10 of which were encoded on the organellar genome. Remarkably, mitochondrial Complex III, IV and V genes were not found. Because an alternative oxidase was identified, it is likely that *Blastocystis* has an incomplete electron transport chain consisting of Complex I, II and the alternative oxidase (Perez-Brocal and Clark 2008; Stechmann et al. 2008). The metabolism of *Blastocystis* possesses a combination of mitochondrial and hydrogenosomal characteristics, but the presence of a mitochondrial genome makes it, just as for *Nyctotherus ovalis*, a missing link between mitochondria and hydrogenosomes (Table 1).

4.5. ANAEROBIC CILIATES

Ciliates form a very species-rich monophyletic group of unicellular eukaryotes that belongs to the Chromalveolata, just like the rather unrelated stramenopile *Blastocystis*. The genomes of the aerobic model organisms *Tetrahymena* sp. and *Paramecium* sp. (discovered by Anthony van Leeuwenhoek) have been sequenced completely. Ciliates possess a very complex genetic system with two nuclei: a micronucleus that represents the germline with high molecular weight DNA and a macronucleus with lower molecular weight DNA that is transcriptionally active during somatic development. The macronucleus contains many copies of subchromosomal fragments that in *Tetrahymena* sp. and *Paramecium* sp. have a size between 50 and 1,500 kb (Eisen et al. 2006). In other ciliates like *Euplotes* sp., *Oxytricha*, and *Nyctotherus*, they can be present as "gene-sized" pieces (Prescott 1994). Macronuclei are generated from the post conjugation micronuclei in a complex process that involves DNA elimination and large-scale genomic rearrangements (Prescott 1994; Nowacki et al. 2008). Another group of anaerobic ciliates that forms a monophyletic group is the rumen ciliates, such as, for example, *Dasytricha*, *Isotricha*, *Epidinium*, and *Eudiplodinium*, but not all of them possess hydrogenosomes (Yarlett et al. 1985; Strüder-Kypke et al. 2006). The main end products of the metabolism of exogenously added glucose as well as of intracellular amylopectine of rumen ciliates with hydrogenosomes are hydrogen, acetate, lactate, butyrate, and CO_2 (Yarlett et al. 1985). The investigated rumen ciliates are able to use oxygen as terminal electron acceptor. The nature of the terminal oxidase is still unknown, but cytochromes appear not to be involved. *Dasytricha ruminantium* is the best-studied rumen ciliate, but even the knowledge of the metabolism of this rumen ciliate is still far from complete. The enzyme used for the degradation of pyruvate to acetyl-CoA in this protist is suggested to be PFO, which has been identified tentatively in the hydrogenosomal fraction of cellular extracts (Yarlett et al. 1981, 1982, 1985). A hypothetical scheme for the hydrogenosomal metabolism of the holotrich rumen ciliate *Dasytricha* is based on various studies. The scheme is remarkable as it requires the export of acetyl-CoA from the hydrogenosome for the formation of butyryl-CoA. This butyryl-CoA is then used for the production of butyrate, which is accompanied by the production of ATP (Yarlett et al. 1985; Ellis et al. 1991a, b). These aspects make this hydrogenosome of rumen ciliates very different from that of *Nyctotherus* and also different from the hydrogenosomes of *Trichomonas* and the anaerobic chytrids.

An organism that has to be mentioned briefly is *Trimyema compressum*, a free-living plagiopylid ciliate that possesses hydrogenosomes and consumes oxygen under microaerobic conditions. A metabolic study showed that *T. compressum* produces formate as the major end product with minor amounts of acetate and lactate (Goosen et al. 1990). Under these microaerobic conditions, hydrogen and ethanol are not produced. Under strictly anaerobic conditions, however, ethanol is the main end product, while acetate, lactate, formate, and hydrogen are then formed in minor amounts (Goosen et al. 1990). This pattern of anaerobic fermentation products resembles the one found in anaerobic chytridiomycete fungi. These fungi perform a bacterial-type mixed-acids fermentation, using PFL for the degradation of pyruvate, instead of PDH or PFO, which is used by *N. ovalis* and *Trichomonas*, respectively (see above). Albeit that no additional biochemical or molecular biological data are available and that no cell fractionation studies have been performed, it is likely that the plagiopylids evolved a type of hydrogenosome that is clearly different from those of *Nyctotherus* and *Dasytricha*.

4.6. *NYCTOTHERUS OVALIS*

Nyctotherus species are anaerobic, heterotrichous ciliates with hydrogenosomes that thrive in the intestinal tract of cockroaches, millipedes, frogs and reptiles (van Hoek et al. 1999; van Hoek et al. 2000). Notably, the presence of a "mitochondrial" genome has been demonstrated in the hydrogenosomes of *N. ovalis* (Akhmanova et al. 1998; van Hoek et al. 2000; Boxma et al. 2005). This genome was shown to be a typical mitochondrial genome of ciliate origin (Boxma et al. 2005; de Graaf et al. 2011). This ciliate origin is reinforced by the analysis of many nuclear genes that encode "mitochondrial" proteins (Boxma et al. 2005; Ricard 2008; Hackstein and Tielens 2010; de Graaf et al. 2011).

Metabolic studies revealed that less than a third of the supplemented glucose was degraded to typical end products of a glycolytic fermentation: approximately 24% of the degraded glucose was excreted as lactate and 5% as ethanol

(Boxma et al. 2005). Notably, the major part of the glucose was degraded via the hydrogenosomes to acetate and succinate. These studies also revealed that *N. ovalis* does not operate a complete TCA cycle for the degradation of glucose. It also does not use pyruvate formate lyase (PFL) activity in its pyruvate metabolism, as is the case in hydrogenosomes of anaerobic chytrids. The product of glycolysis in the cytosol, pyruvate, is apparently either converted into lactate or ethanol, or transported into the hydrogenosome to be converted into acetate or succinate. For the production of acetate, this pyruvate is decarboxylated by a pyruvate dehydrogenase complex (PDH) and not by a pyruvate:ferredoxin oxidoreductase (PFO) as in *Trichomonas* and other hydrogenosome-bearing protists (Boxma et al. 2005). The excretion of significant amounts of succinate indicated that endogenously produced fumarate is used as an electron acceptor. Protons act as another hydrogenosomal electron acceptor, which results in the formation of hydrogen. Fumarate reduction is most likely catalyzed by a membrane-bound fumarate reductase (an anaerobically functioning variant of complex II), coupled to complex I of the electron transport chain via quinones.

The significance of these experimental data might be circumstantial without molecular support (Boxma et al. 2005; Ricard 2008; de Graaf et al. 2011). As expected, genes for all four subunits of a PDH are present and are expressed. In addition, a gene was detected for acetyl-CoA synthase, as well as the enzyme ASCT (acetate:succinate CoA-transferase), enzymes for the production of acetate from acetyl-CoA, and also several genes, which are predicted to encode enzymes of the TCA cycle, i.e., malate dehydrogenase, succinate dehydrogenase (2 subunits), and succinyl-CoA synthetase. Thus, basically, the core energy metabolism of a typical ciliate mitochondrion was detected, albeit in an anaerobic version. In fact, *N. ovalis* contains hydrogen-producing mitochondria (Boxma et al. 2005).

There is no evidence for genes encoding components of mitochondrial complexes III and IV (de Graaf et al. 2011). Notably, these complexes are also absent in the electron transport chains of anaerobic mitochondria and the hydrogenosomes of *Blastocystis* (Tielens et al. 2002; Perez-Brocal and Clark 2008; Stechmann et al. 2008; Wawrzyniak et al. 2008). Therefore, it is unlikely that these hydrogenosomes gain their energy by the generation of a PMF. Of the subunits of a mitochondrial complex I, 12 out of the 14 subunits that form the core of a bacterial complex I were cloned and sequenced until now (de Graaf et al. 2011). Accordingly, imaging studies using inhibitors and fluorescent dyes not only demonstrated the presence of a functional complex I in these hydrogenosomes but also indicated the absence of functional complexes III and IV and the absence of a plant-like alternative terminal oxidase (Boxma et al. 2005).

Also, no homologues of an F_1F_0-ATP synthase have been discovered so far, as in the organelles of *Blastocystis* and *Trichomonas* (Carlton et al. 2007; Perez-Brocal and Clark 2008; Stechmann et al. 2008; Wawrzyniak et al. 2008).

In addition, components of a mitochondrial amino acid metabolism were identified, including a glycine cleavage system. Moreover, components of fatty acid metabolism, several AACs, several members of the mitochondrial solute carriers family, a malate-oxoglutarate translocator, components of a mitochondrial protein import and processing machinery, components of a protein synthesizing machinery, and proteins belonging to ROS defense systems were found. Several proteins originated from lateral gene transfer (Ricard 2008). Thus, the hydrogenosome of *N. ovalis* is not simply a rudimentary mitochondrion. It is a highly specialized organelle of considerable complexity.

5. Mitosomes

5.1 *ENTAMOEBA HISTOLYTICA*

Mitosomes were for the first time identified in *Entamoeba histolytica*, a human parasite, by the presence of a mitochondrial-type Cpn60 that could be localized immunologically to an hitherto unknown organelle (Mai et al. 1999; Tovar, Fischer, and Clark 1999; Loftus and Hall 2005). The sequence of the complete genome revealed a set of "mitochondrial" genes, but the absence of other mitochondrial genes like those for oxidative phosphorylation (Loftus and Hall 2005). Interestingly, *Entamoeba* lacks a typical ADP/ATP carrier (AAC) with similarity to mitochondrial carriers. Instead, it hosts an alternative AAC that clusters with the brittle-like carrier of the mitochondriate relative *Dictyostelium* (Chan et al. 2005; Hackstein et al. 2006). Obviously, the mitosomes of *Entamoeba* are not engaged in the energy metabolism. It seems that energy conversion is cytosolic and ATP is generated only by substrate level phosphorylation in the cytoplasm (Müller 1998). The only known function of the mitosomes seemed to be Fe-S cluster synthesis. These genes (NifS and NifU), however, are not of mitochondrial origin. They were acquired by horizontal gene transfer from epsilon-proteobacteria (Ali et al. 2004; van der Giezen et al. 2004). Unexpectedly, these proteins did not contain mitochondrial targeting signals indicating that they might be located in the cytoplasm. However, recently Maralikova et al. (Maralikova et al. 2010) showed by immunomicroscopy and biochemical methods that the proteins are present in a 10× higher concentration in the mitosomes than in the cytosol. It seems that, although the original mitochondrial Fe-S cluster genes were replaced by bacterial ones in these organisms, the mitosomes still play an important role in Fe-S cluster syntheses. Another example of a hydrogenosomal relation of these mitosomes is the presence of two [FeFe] hydrogenases (Nixon et al. 2003) (Table 1). It has not been proven yet that hydrogen production occurs in vivo and that this protein is localized in the mitosomes of *Entamoeba histolytica*. The hydrogenase is probably cytoplasmatic because no mitochondrial targeting signal could be identified. The presence of functional N-terminal extensions with sequence similarity to mitochondrial targeting peptides has been shown unequivocally for Cpn60 (Mai et al. 1999; Tovar et al. 1999). However, a PNT (Pyridine nucleotide transhydrogenase) with such a N-terminal extension is not imported into the mitosomes (Yousuf et al. 2010).

Recently, it has been shown that the mitosomes of *Entamoeba* possess a minimal "mitochondrial" import machinery. In particular, a phosphate carrier and the membrane proteins Tom 40 and Sam 50 have been identified (Dolezal et al. 2010).

5.2. *MASTIGAMOEBA BALAMUTHI*

Most eukaryotes with mitosomes are parasites, and in general, mitosomes are abundant in the cells. However, also the free-living amoeba *Mastigamoeba balamuthi*, which is phylogenetically related to *Entamoeba*, hosts mitosomes. In an EST project from *Mastigamoeba balamuthi* (Gill et al. 2007) a much larger amount of mitochondrion- related genes were identified than in *Entamoeba histolytica* although the ESTs are far from representing the whole genome (Table 1). For example, all components of a mitochondrial glycine cleavage system (GCS) have been identified, furthermore malate dehydrogenase, SDHb and SDHc, several components of a TCA cycle and acetyl-CoA synthetase. In addition, a [FeFe] hydrogenase is well represented among the ESTs. If the [FeFe] hydrogenase of *M. balamuthi* can be located in the mitosomes and hydrogenase activity measured, then these organelles must be classified as hydrogenosomes. This indicates in any case that *Entamoeba histolytica* possesses a more reduced mitosomal metabolism and thus represents a further step in the mitochondrial adaptation to anaerobic environments than *M. balamuthi*. Notably, as in *Entamoeba*, the Fe-S cluster metabolism in *M. balamuthi* is represented by bacterial NifS and NifU genes (Gill et al. 2007).

5.3. *ENCEPHALITOZOON CUNICULI, ANTONOSPORA LOCUSTAE, TRACHIPLEISTOPHORA HOMINIS*

Within the protistan group of microsporidia, which contains thousands of species of unicellular, intracellular parasitic eukaryotes, the best investigated organisms are *Encephalitozoon cuniculi* (Katinka et al. 2001), *Antonospora locustae* (Williams and Keeling 2005) and *Trachipleistophora hominis* (Williams et al. 2002).

From *Encephalitozoon cuniculi*, the complete genome is known, and as early as 2001 the presence of a mitochondrion-derived organelle was predicted on the basis of 22 "mitochondrial" proteins (Katinka et al. 2001). In 2008, the immunofluorescent localization of a ferredoxin involved in Fe-S cluster synthesis confirmed the presence of tiny, punctuate organelles (Williams et al. 2002). With *Antonospora locustae*, an extended EST study was performed (Williams and Keeling 2005). The microsporidian metabolism is highly reduced, but the glycolytic pathway is conserved whereas TCA cycle enzymes are completely absent. Another major difference with respect to other eukaryotes with mitosomes is the absence of a hydrogenase and PFO. Instead of PFO, two PDH-E1 subunits (α and β) were identified but the other PDH subunits seem to be absent (Table 1). It is unclear how pyruvate decarboxylation can take place without the lacking PDH-subunits. In *Antonospora locustae* a transporter for pyruvate into the organelle is found that could be an indication that the last steps of glucose metabolism could be compartmentalized. Other mitochondrial transporters were identified, for example TIM 22 in both species and TOM 70 only in *Encephalitozoon cuniculi*. Because of incompleteness of the genome data of *Antonospora locustae* it is difficult to say something about similarity and differences of the mitosomal metabolism of these two microsporidia, but the similarity is considerable and without any doubt a mitochondrial relation exists. However, in contrast to *Encephalitozoon*, *Antonospora* possesses an alternative ADP/ATP carrier; its DNA sequence does not cluster with the mitochondrial-type AACs. *Encephalitozoon* does not contain any AAC. The necessary ATP transport is performed with that aid of unusual ATP transporters (Tsaousis et al. 2008). Notably, the Fe-S cluster biosynthesis and the localization of the proteins in *Encephalitozoon* and *Trachipleistophora* is different (Goldberg et al. 2008).

Studies with both *Encephalitozoon* and *Antonospora* revealed the presence of a rudimentary mitochondrion-type protein import system (Burri et al. 2006).

The mitosomes of *Trachipleistophora hominis* were identified with the aid of homologous antibodies against Hsp70 as tiny, double-membrane bounded organelles under light and electron microscopy (Williams et al. 2002).

5.4. *CRYPTOSPORIDIUM PARVUM*

Cryptosporidium parvum belongs to another group of eukaryotes in which most organisms possess aerobic mitochondria: the apicomplexa. These protists share a common ancestor with dinoflagellates and ciliates that together constitute the clade Alveolata. *C. parvum* is a parasite that can infect humans and contains only a single "mitochondrial remnant" organelle per cell (Keithly 2008). Although the complete genome of *C. parvum* has been sequenced (Abrahamsen et al. 2004), the organellar metabolism is not completely understood. Glycolysis seems to be the major pathway of the energy metabolism because of lack of a functional TCA cycle and electron transport chain. Notwithstanding, some components of the oxidative phosphorylation pathway were identified (Table 1). Pyruvate is converted into acetyl-CoA by a PNO (an pyruvate:NADPH oxidoreductase, (Rotte et al. 2001)) that possess an N-terminal PFO domain that is fused with a C-terminal NADPH-cytochrome P450 reductase domain (Abrahamsen et al. 2004). In contrast to Cpn60, which targets to the mitosome (Riordan et al. 2003), it is unlikely that PNO is located in the organelle of *C. parvum* (Ctrnacta et al. 2006). The reduced respiratory chain (the identified genes have N-terminal mitochondrial targeting signals) possesses an alternative oxidase (AOX) as terminal electron acceptor (Roberts et al. 2004). How the electrons enter the respiratory chain is unknown because no genes for mitochondrial Complexes I or II were found.

Alternative routes have been described in other organisms, but none of them seem to be plausible for *C. parvum*. Remarkably, two subunits of Complex V (α and β subunits of the F_1F_0 ATP synthase) were identified in *C. parvum*; both contain an N terminal targeting sequence (Abrahamsen et al. 2004; Putignani et al. 2004). The presence of this ATPase leaves the possibility for a proton gradient coupled to oxidative phosphorylation. The identification of a PNT (pyridine nucleotide transhydrogenase), that can create a proton gradient, a Narf-like protein with similarity to a [FeFe] hydrogenase, Cpn60, Hsp70, a number of carriers belonging to the mitochondrial carrier family (Abrahamsen et al. 2004; Hackstein et al. 2006) and several mitochondrial transporters show us another example of a rather complex mitochondrion-related organelle instead of the degenerated organelles that they were supposed to be some years ago (Keithly 2008). Of course, there is evidence for the presence of enzymes involved in Fe-S cluster biosynthesis (LaGier et al. 2003; Keithly 2008). It is obvious that the mitosome of *Cryptosporidium* is a very complex organelle with a complex morphology (Keithly 2008).

5.5. *GIARDIA SP.*

The last group of organisms with mitosomes that will be discussed here are the diplomonads, a group of microaerophilic unicellular protists. Because of lack of classical mitochondria and their deep branching position in some phylogenies they had been considered as basal eukaryotes. However, it has been shown by confocal immunofluorescence microscopy and transmission electron immunomicroscopy using specific antibodies against the Fe-S cluster synthesizing proteins IscS and IscU that *Giardia* possesses tiny double-membrane bounded organelles that host mitochondrial enzymes (Tovar et al. 2003). The data show that also these organelles are mitosomes that are the product of adaptation of an aerobic, mitochondrial ancestor to an anaerobic niche. The best studied member of this group is *Giardia intestinalis*, a human parasite, but recently a large amount of molecular data became available for *Spironucleus salmonicida* a fish parasite. The presence of a MPP (mitochondrial processing peptidase) in *G. intestinalis* that processes targeting sequences after import into the organelle is one of the hallmarks of a mitochondrial relationship. Also the recent discovery of TOM40 (Dagley et al. 2009), an important compound of the mitochondrial TOM import complex, and of Pam18, an important compound of the translocase of the mitochondrial inner membrane, provides evidence for the mitochondrial ancestry of these mitosomes. However, the mitochondrial import system is very reduced (Dolezal et al. 2005), and while the import of IscU and ferredoxin is still dependent on their amino-terminal presequences, targeting of *Giardia* Cpn60, IscS, or mtHsp70 into mitosomes does no longer require cleavable presequences (Regoes et al. 2005). *G. lamblia* completely lacks the aerobic respiratory components of aerobic mitochondria and generates ATP exclusively by substrate-level phosphorylation in the cytosol. Pyruvate in G. lamblia is oxidized by a cytosolic PFO. There is also evidence for hydrogenase activity in *G. lamblia*. The expression of a short type of an [FeFe] hydrogenase (Nixon et al. 2003) that looks like the one found in *Spironucleus barkhanus* (Horner et al. 2000), and even the production of a small amount of hydrogen could be shown (Lloyd et al. 2002). It is obvious that the difference between mitosomes and hydrogenosomes might become smaller with the investigation of more mitosome/hydrogenosome harboring organisms. A interesting candidate could be the deep branching relative of diplomonads *Hicanonectes teleskopos* of which almost no molecular data are present, but of which transmission electron micrographical observations show mitochondria-like organelles that lack cristae and look more like the hydrogenosomes of the parabasalids than the mitosomes of their closest relatives, the diplomonads (Park et al. 2009).

6. Conclusions and Discussion

Some of the intriguing questions that remain are: Why did *N. ovalis* and *Blastocystis* spp. retain a genome? Why is this not the case in other eukaryotic organisms with hydrogenosomes like *Trichomonas vaginalis* or *Psalteriomonas lanternae*? Why did mitosomes loose much more organellar proteins in addition to the organellar genome?

A popular hypothesis that gives a possible explanation for the presence of a mitochondrial genome is that some of its genes encode highly hydrophobic proteins that are difficult to import across the mitochondrial membranes. The two protein encoding genes in all sequenced mitochondrial genomes that are always present are *cox* I and *cob*, and exactly these are the most hydrophobic proteins present in mitochondria (Claros et al. 1995). Direct experimental evidence showed that reduction of hydrophobicity was essential for the rare transfer event for the cytochrome c oxidase subunit 2 in legumes (Daley et al. 2002). Another example of hydrophobicity reduction is found in *Chlamydomonas* were transfer of *cox2*, *cox3*, and *atp6* genes occurred in combination with reduced hydrophobicity in transmembrane regions (Perez-Martinez et al. 2000, 2001; Funes et al. 2002). A second hypothesis that must be mentioned here is the possible toxicity of some mitochondrial-encoded proteins when present in the cytosol (Martin and Schnarrenberger 1997). This also could be a good reason for keeping some genes encoded on the organellar genome as exemplified by the mitochondrial genome of *Plasmodium falciparum* where the *cob* gene and two *cox* genes are retained as the only protein-encoding genes of the tiny organelle genome. Both of the above mentioned hypotheses are not valid in the case of *N. ovalis* or *Blastocystis* because both do not possess mitochondrial complex III, IV, and V genes anymore. Does this mean that we are dealing here with a beginning transfer of the complete mitochondrial genome to the nucleus? If the hypotheses discussed above cannot explain why mitochondrial genomes still exist, then it is possible that the remaining mitochondrial genome in *N. ovalis* and *Blastocystis* will disappear eventually.

However, another hypothesis that has to be mentioned here is the CORR hypothesis (colocation for redox regulation (Allen 2003)). This hypothesis states that chloroplasts and mitochondria retain those genes (located on an organellar genome) whose expression is required to be under direct regulatory control of the redox state of their gene products, or the electron carriers with which their gene products interact. In *N. ovalis* and *Blastocystis*, the presence of an active mitochondrial complex I might still require redox regulation notwithstanding that complexes III and IV are lacking. Therefore, an organellar genome is retained in both organisms according to the CORR hypothesis.

The research presented here indicates that one of the first steps of adaptation of aerobic protists to an anaerobic environment might be the evolution of a fumarate respiration by a reversion of action of the succinate dehydrogenase and the acquisition of a hydrogenase followed by the loss of complexes III, IV, and V genes. That such an event is not specific for the ciliate *N. ovalis* is underlined by the study of the stramenopile *Blastocystis* which (most likely) also performs a fumarate respiration, just as quite a number of anaerobic mitochondria of certain unicellular and multicellular organisms (Tielens et al. 2002). The analysis of the complete genome of *Naegleria gruberi*, which has recently been sequenced (Fritz-Laylin et al. 2010), showed that this could be another example of the first step to anaerobic adaptation. *Neagleria* possesses an organelle that can function like a "normal" mitochondrion under aerobic conditions with the potential for a fumarate respiration under anaerobic conditions. Notably, this organism also possesses a FeFe hydrogenase, closely related to that of *Psalteriomonas*, which functions under anaerobic/microaerobic conditions. The presence of a set of genes obtained by horizontal gene transfer in *N. ovalis* seems to be a logical consequence to obtain an organelle with adaptations to anaerobic environments. A possible scenario for eukaryotes with a textbook mitochondrion genome and metabolism to evolve into eukaryotes with hydrogenosomes/mitosomes could be as follows:

1. Reverse action of the TCA cycle (fumarate respiration) (as shown in *N. gruberi* and a number of other anaerobic mitochondria (Tielens et al. 2002)).
2. Recruitment of a hydrogenase (like shown in *N. gruberi* and *N. ovalis*).
3. Changing to a more anaerobic metabolism by obtaining suitable genes by HGT (Ricard et al. 2006).
4. Loss of complexes III, IV, and V of the electron transport chain (as shown in *N. ovalis* and *Blastocystis*).
5. Loss of the remaining part of the organellar genome (as shown, for example in *Trichomonas*, *Psalteriomonas* and many other organisms with hydrogenosomes and mitosomes (Palmer 1997)).

The hydrogenosomes of *Nyctotherus ovalis* form an evolutionary link between ciliate mitochondria and hydrogenosomes. It is obvious that hydrogenosomes evolved several times from different aerobic progenitors. The eukaryotic cell is clearly characterized by the presence of a mitochondrion that has the capacity to adapt to various anaerobic environments by reductive evolution to yield a rainbow of hydrogenosomes and mitosomes.

7. References

Abrahamsen MS, Templeton TJ, Enomoto S, Abrahante JE, Zhu G, Lancto CA, Deng M, Liu C, Widmer G, Tzipori S, Buck GA, Xu P, Bankier AT, Dear PH, Konfortov BA, Spriggs HF, Iyer L, Anantharaman V, Aravind L, Kapur V (2004) Complete genome sequence of the apicomplexan, Cryptosporidium parvum. Science 304:441–445

Akhmanova A, Voncken F, van Alen T, van Hoek A, Boxma B, Vogels G, Veenhuis M, Hackstein JH (1998) A hydrogenosome with a genome. Nature 396:527–528

Akhmanova, Voncken FG, Hosea KM, Harhangi H, Keltjens JT, op den Camp HJ, Vogels GD, Hackstein JHP (1999) A hydrogenosome with pyruvate formate-lyase: anaerobic chytrid fungi use an alternative route for pyruvate catabolism. Mol Microbiol 32:1103–1114

Ali V, Shigeta Y, Tokumoto U, Takahash Y, Nozaki T (2004) An intestinal parasitic protist, Entamoeba histolytica, possesses a non-redundant nitrogen fixation-like system for iron-sulfur cluster assembly under anaerobic conditions. J Biol Chem 279:16863–16874

Allen JF (2003) The function of genomes in bioenergetic organelles. Philos Trans R Soc Lond B Biol Sci 358:19–37, discussion 37–18

Andersson JO, Sjogren AM, Horner DS, Murphy CA, Dyal PL, Svard SG, Logsdon JM Jr, Ragan MA, Hirt RP, Roger AJ (2007) A genomic survey of the fish parasite *Spironucleus salmonicida* indicates genomic plasticity among diplomonads and significant lateral gene transfer in eukaryote genome evolution. BMC Genomics 8:51

Barbera MJ, Ruiz-Trillo I, Leigh J, Hug LA, Roger AJ (2007) The diversity of mitochondrion-related organelles amongst eukaryotic microbes. In: Martin WF, Müller M (eds) Origin of mitochondria and hydrogenosomes. Springer, Berlin/Heidelberg, pp 239–275

Boxma B, Voncken F, Jannink S, van Alen T, Akhmanova A, van Weelden S, van Hellemond J, Ricard G, Huynen M, Tielens A, Hackstein J (2004) The anaerobic chytridiomycete fungus Piromyces sp. E2 produces ethanol via pyruvate:formate lyase and an alcohol dehydrogenase E. Mol Microbiol 51:1389–1399

Boxma B, de Graaf RM, van der Staay GW, van Alen TA, Ricard G, Gabaldon T, van Hoek AHAM, Moon-van der Staay SY, Koopman WJ, van Hellemond JJ, Tielens AG, Friedrich T, Veenhuis M, Huynen MA, Hackstein JHP (2005) An anaerobic mitochondrion that produces hydrogen. Nature 434:74–79

Broers CAM (1992) Anaerobic psalteriomonad amoeboflagellates. Thesis, Catholic University Nijmegen, Nijmegen

Broers CAM, Stumm CK, Vogels GD, Brugerolle G (1990) Psalteriomonas lanterna gen. nov., sp. nov., a free-living amoeboflagellate isolated from fresh-water anaerobic sediments. Eur J Protistol 25:369–380

Brunk CF, Lee LC, Tran AB, Li J (2003) Complete sequence of the mitochondrial genome of Tetrahymena thermophila and comparative methods for identifying highly divergent genes. Nucleic Acids Res 31:1673–1682

Burger G, Lang BF (2003) Parallels in genome evolution in mitochondria and bacterial symbionts. IUBMB Life 55:205–212

Burger G, Zhu Y, Littlejohn TG, Greenwood SJ, Schnare MN, Lang BF, Gray MW (2000) Complete sequence of the mitochondrial genome of Tetrahymena pyriformis and comparison with Paramecium aurelia mitochondrial DNA. J Mol Biol 297:365–380

Burri L, Williams BA, Bursac D, Lithgow T, Keeling PJ (2006). Microsporidian mitosomes retain elements of the general mitochondrial targeting system. Proc Natl Acad Sci USA 103:15916–15920

Carlton JM, Hirt RP, Silva JC, Delcher AL, Schatz M, Zhao Q, Wortman JR, Bidwell SL, Alsmark UC, Besteiro S, Sicheritz-Ponten T, Noel CJ, Dacks JB, Foster PG, Simillion C, Van de Peer Y, Miranda-Saavedra D, Barton GJ, Westrop GD, Muller S, Dessi D, Fiori PL, Ren Q, Paulsen I, Zhang H, Bastida-Corcuera FD, Simoes-Barbosa A, Brown MT, Hayes RD, Mukherjee M, Okumura CY, Schneider R, Smith AJ, Vanacova S, Villalvazo M, Haas BJ, Pertea M, Feldblyum TV, Utterback TR, Shu CL, Osoegawa K, de Jong PJ, Hrdy I, Horvathova L, Zubacova Z, Dolezal P, Malik SB, Logsdon JM Jr, Henze K, Gupta A, Wang CC, Dunne RL, Upcroft JA, Upcroft

P, White O, Salzberg SL, Tang P, Chiu CH, Lee YS, Embley TM, Coombs GH, Mottram JC, Tachezy J, Fraser-Liggett CM, Johnson PJ (2007) Draft genome sequence of the sexually transmitted pathogen Trichomonas vaginalis. Science 315:207–212

Cavalier-Smith T (1987) The origin of eukaryotic and archaebacterial cells. Ann N Y Acad Sci 503:17–54

Chan KW, Slotboom DJ, Cox S, Embley TM, Fabre O, van der Giezen M, Harding M, Horner DS, Kunji ER, Leon-Avila G, Tovar J (2005) A novel ADP/ATP transporter in the mitosome of the microaerophilic human parasite Entamoeba histolytica. Curr Biol 15:737–742

Claros MG, Perea J, Shu Y, Samatey FA, Popot JL, Jacq C (1995) Limitations to in vivo import of hydrophobic proteins into yeast mitochondria. The case of a cytoplasmically synthesized apocytochrome b. Eur J Biochem 228:762–771

Ctrnacta V, Ault JG, Stejskal F, Keithly JS (2006) Localization of pyruvate:NADP+ oxidoreductase in sporozoites of Cryptosporidium parvum. J Eukaryot Microbiol 53:225–231

Cummings DJ, McNally KL, Domenico JM, Matsuura ET (1990) The complete DNA sequence of the mitochondrial genome of Podospora anserina. Curr Genet 17:375–402

Dagley MJ, Dolezal P, Likic VA, Smid O, Purcell AW, Buchanan SK, Tachezy J, Lithgow T (2009) The protein import channel in the outer mitosomal membrane of Giardia intestinalis. Mol Biol Evol 26:1941–1947

Daley DO, Clifton R, Whelan J (2002) Intracellular gene transfer: reduced hydrophobicity facilitates gene transfer for subunit 2 of cytochrome c oxidase. Proc Natl Acad Sci USA 99:10510–10515

Danovaro R, Dell'anno A, Pusceddu A, Gambi C, Heiner I, Mobjerg Kristensen R (2010) The first metazoa living in permanently anoxic conditions. BMC Biol 8:30

de Graaf RM, Duarte I, van Alen TA, Kuiper JW, Schotanus K, Rosenberg J, Huynen MA, Hackstein JH (2009a) The hydrogenosomes of Psalteriomonas lanterna. BMC Evol Biol 9:287

de Graaf RM, van Alen TA, Dutilh BE, Kuiper JW, van Zoggel HJ, Huynh MB, Görtz HD, Huynen MA, Hackstein JH (2009b) The mitochondrial genomes of the ciliates Euplotes minuta and Euplotes crassus. BMC Genomics 10:514

de Graaf RM, Ricard G, van Alen TA, Duarte I, Dutilh BE, Burgtorf C, Kuiper JW, van der Staay GW, Tielens AGM, Huynen MA, Hackstein JHP (2011) The organellar genome and metabolic potential of the hydrogen-producing mitochondrion of Nyctotherus ovalis. Mol Biol Evol. doi 10.1093/molbev/msr059

Dolezal P, Smid O, Rada P, Zubacova Z, Bursac D, Sutak R, Nebesarova J, Lithgow T, Tachezy J (2005) Giardia mitosomes and trichomonad hydrogenosomes share a common mode of protein targeting. Proc Natl Acad Sci USA 102:10924–10929

Dolezal P, Dagley MJ, Kono M, Wolynec P, Likic VA, Foo JH, Sedinova M, Tachezy J, Bachmann A, Bruchhaus I, Lithgow T (2010) The essentials of protein import in the degenerate mitochondrion of Entamoeba histolytica. PLoS Pathog 6:e1000812

Eisen JA, Coyne RS, Wu M, Wu D, Thiagarajan M, Wortman JR, Badger JH, Ren Q, Amedeo P, Jones KM, Tallon LJ, Delcher AL, Salzberg SL, Silva JC, Haas BJ, Majoros WH, Farzad M, Carlton JM, Smith RK Jr, Garg J, Pearlman RE, Karrer KM, Sun L, Manning G, Elde NC, Turkewitz AP, Asai DJ, Wilkes DE, Wang Y, Cai H, Collins K, Stewart BA, Lee SR, Wilamowska K, Weinberg Z, Ruzzo WL, Wloga D, Gaertig J, Frankel J, Tsao CC, Gorovsky MA, Keeling PJ, Waller RF, Patron NJ, Cherry JM, Stover NA, Krieger CJ, del Toro C, Ryder HF, Williamson SC, Barbeau RA, Hamilton EP, Orias E (2006) Macronuclear genome sequence of the ciliate Tetrahymena thermophila, a model eukaryote. PLoS Biol 4:e286

Ellis JE, McIntyre PS, Saleh M, Williams AG, Lloyd D (1991a) Influence of CO_2 and low concentrations of O_2 on fermentative metabolism of the ruminal ciliate Polyplastron multivesiculatum. Appl Environ Microbiol 57:1400–1407

Ellis JE, McIntyre PS, Saleh M, Williams AG, Lloyd D (1991b) Influence of CO_2 and low concentrations of O_2 on fermentative metabolism of the rumen ciliate Dasytricha ruminantium. J Gen Microbiol 137:1409–1417

Embley TM, Martin W (2006) Eukaryotic evolution, changes and challenges. Nature 440:623–630

Embley TM, van der Giezen M, Horner DS, Dyal PL, Bell S, Foster PG (2003) Hydrogenosomes, mitochondria and early eukaryotic evolution. IUBMB Life 55:387–395

Feagin JE (1994) The extrachromosomal DNAs of apicomplexan parasites. Annu Rev Microbiol 48:81–104

Fritz-Laylin LK, Prochnik SE, Ginger ML, Dacks JB, Carpenter ML, Field MC, Kuo A, Paredez A, Chapman J, Pham J, Shu S, Neupane R, Cipriano M, Mancuso J, Tu H, Salamov A, Lindquist E, Shapiro H, Lucas S, Grigoriev IV, Cande WZ, Fulton C, Rokhsar DS, Dawson SC (2010) The genome of Naegleria gruberi illuminates early eukaryotic versatility. Cell 140:631–642

Funes S, Davidson E, Claros MG, van Lis R, Perez-Martinez X, Vazquez-Acevedo M, King MP, Gonzalez-Halphen D (2002) The typically mitochondrial DNA-encoded ATP6 subunit of F1F0-ATPase is encoded by a nuclear gene in Chlamydomonas reinhardtii. J Biol Chem 277:6051–6058

Ghirardi ML, Posewitz MC, Maness PC, Dubini A, Yu J, Seibert M (2007) Hydrogenases and hydrogen photoproduction in oxygenic photosynthetic organisms. Annu Rev Plant Biol 58:71–91

Gill EE, Diaz-Trivino S, Barbera MJ, Silberman JD, Stechmann A, Gaston D, Tamas I, Roger AJ (2007) Novel mitochondrion-related organelles in the anaerobic amoeba Mastigamoeba balamuthi. Mol Microbiol 66:1306–1320

Goldberg AV, Molik S, Tsaousis AD, Neumann K, Kuhnke G, Delbac F, Vivares CP, Hirt RP, Lill R, Embley TM (2008) Localization and functionality of microsporidian iron-sulphur cluster assembly proteins. Nature 452:624–628

Goosen NK, van der Drift C, Stumm C, Vogels GD (1990) End products of metabolism in the anaerobic ciliate Trimyema compressum. FEMS Microbiol Lett 69:171–175

Hackstein JHP, Tielens AGM (2010) Hydrogenosomes. In: Hackstein JHP (ed) (Endo)symbiotic methanogenic archea. Springer, Heidelberg

Hackstein JHP, Tjaden J, Huynen MA (2006) Mitochondria, hydrogenosomes and mitosomes: products of evolutionary tinkering! Curr Genet 50:225–245

Hackstein JHP, de Graaf RM, van Hellemond JJ, Tielens AG (2008) Hydrogenosomes of anaerobic ciliates. In: Tachezy J (ed) Hydrogenosomes and mitosomes: mitochondria of anaerobic eukaryotes. Springer, Berlin/Heidelberg, pp 97–112

Hampl V, Silberman JD, Stechmann A, Diaz-Trivino S, Johnson PJ, Roger AJ (2008) Genetic evidence for a mitochondriate ancestry in the 'amitochondriate' flagellate Trimastix pyriformis. PLoS One 3:e1383

Henriquez FL, Richards TA, Roberts F, McLeod R, Roberts CW (2005) The unusual mitochondrial compartment of Cryptosporidium parvum. Trends Parasitol 21:68–74

Henze K (2007) The proteome of T.vaginalis hydrogenosomes. In: Tachezy J (ed) Hydrogensomes and mitosomes: mitochondria of anaerobic eukaryotes. Springer, Berlin/Heidelberg, pp 163–178

Hjort KA, Goldberg AV, Tsaousis AD, Hirt RP, Embley TM (2010) Diversity and reductive evolution of mitochondria among microbial eukaryotes. Philos Trans R Soc Lond B Biol Sci 365:713–727

Horner DS, Foster PG, Embley TM (2000) Iron hydrogenases and the evolution of anaerobic eukaryotes. Mol Biol Evol 17:1695–1709

Hrdy I, Hirt RP, Dolezal P, Bardonova L, Foster PG, Tachezy J, Embley TM (2004) Trichomonas hydrogenosomes contain the NADH dehydrogenase module of mitochondrial complex I. Nature 432:618–622

Hrdý I, Tachezy J, Müller M (2007) Metabolism of trichomonads hydrogenosomes. In: Tachezy J (ed) Hydrogenosomes and mitosomes: mitochondria of anaerobic eukaryotes. Springer, Berlin/Heidelberg, pp 113–145

Katinka MD, Duprat S, Cornillot E, Metenier G, Thomarat F, Prensier G, Barbe V, Peyretaillade E, Brottier P, Wincker P, Delbac F, El Alaoui H, Peyret P, Saurin W, Gouy M, Weissenbach J, Vivares CP (2001) Genome sequence and gene compaction of the eukaryote parasite Encephalitozoon cuniculi. Nature 414:450–453

Keithly J (2008) The mitochondrion related organelle of Cryptosporidium parvum. In: Tachezy J (ed) Hydrogenosomes and mitosomes: mitochondria of anaerobic eukaryotes. Springer, Berlin/Heidelberg, pp 231–253

Kolisko M, Cepicka I, Hampl V, Leigh J, Roger AJ, Kulda J, Simpson AG, Flegr J (2008) Molecular phylogeny of diplomonads and enteromonads based on SSU rRNA, alpha-tubulin and HSP90 genes: implications for the evolutionary history of the double karyomastigont of diplomonads. BMC Evol Biol 8:205

LaGier MJ, Tachezy J, Stejskal F, Kutisova K, Keithly JS (2003) Mitochondrial-type iron-sulfur cluster biosynthesis genes (IscS and IscU) in the apicomplexan Cryptosporidium parvum. Microbiology 149:3519–3530

Lang BF, Burger G, O'Kelly CJ, Cedergren R, Golding GB, Lemieux C, Sankoff D, Turmel M, Gray MW (1997) An ancestral mitochondrial DNA resembling a eubacterial genome in miniature. Nature 387:493–497

Lang BF, O'Kelly C, Nerad T, Gray MW, Burger G (2002) The closest unicellular relatives of animals. Curr Biol 12:1773–1778

Lantsman Y, Tan KS, Morada M, Yarlett N (2008) Biochemical characterization of a mitochondrial-like organelle from Blastocystis sp. subtype 7. Microbiology 154:2757–2766

Lindmark DG, Müller M (1973) Hydrogenosome, a cytoplasmic organelle of the anaerobic flagellate Tritrichomonas foetus, and its role in pyruvate metabolism. J Biol Chem 248:7724–7728

Lindmark DG, Muller M, Shio H (1975) Hydrogenosomes in trichomonas-vaginalis. J Parasitol 61:552–554

Lloyd D, Ralphs JR, Harris JC (2002) Giardia intestinalis, a eukaryote without hydrogenosomes, produces hydrogen. Microbiology 148:727–733

Loftus BJ, Hall N (2005) Entamoeba: still more to be learned from the genome. Trends Parasitol 21:453

Lunt DH, Whipple LE, Hyman BC (1998) Mitochondrial DNA variable number tandem repeats (VNTRs): utility and problems in molecular ecology. Mol Ecol 7:1441–1455

Mai Z, Ghosh S, Frisardi M, Rosenthal B, Rogers R, Samuelson J (1999) Hsp60 is targeted to a cryptic mitochondrion-derived organelle ("crypton") in the microaerophilic protozoan parasite Entamoeba histolytica. Mol Cell Biol 19:2198–2205

Maralikova B, Ali V, Nakada-Tsukui K, Nozaki T, van der Giezen M, Henze K, Tovar J (2010) Bacterial-type oxygen detoxification and iron–sulfur cluster assembly in amoebal relict mitochondria. Cell Microbiol 12(3):331–342

Martin W, Schnarrenberger C (1997) The evolution of the Calvin cycle from prokaryotic to eukaryotic chromosomes: a case study of functional redundancy in ancient pathways through endosymbiosis. Curr Genet 32:1–18

Marvin-Sikkema F, Gomes T, Grivet J, Gottschall J, Prins RA (1993) Characterization of hydrogenosomes and their role in glucose metabolism of Neocallimastix sp L2. Arch of Microbiol 160:388–396

Meisinger C, Sickmann A, Pfanner N (2008) The mitochondrial proteome: from inventory to function. Cell 134:22–24

Moradian MM, Beglaryan D, Skozylas JM, Kerikorian V (2007) Complete mitochondrial genome sequence of three Tetrahymena species reveals mutation hot spots and accelerated nonsynonymous substitutions in Ymf genes. PLoS One 2(7):e650

Morin GB, Cech TR (1988a) Telomeric repeats of Tetrahymena malaccensis mitochondrial DNA: a multimodal distribution that fluctuates erratically during growth. Mol Cell Biol 8:4450–4458

Morin GB, Cech TR (1988b) Mitochondrial telomeres: surprising diversity of repeated telomeric DNA sequences among six species of Tetrahymena. Cell 52:367–374

Müller M (1993) The hydrogenosome. J Gen Microbiol 139:2879–2889

Müller M (1998) Enzymes and compartmentation of core energy metabolism of anaerobic protists – a special case in eukaryotic evolution? In: Coombs GH, Vickerman K, Sleigh MA, Warren A (eds) Evolutionary relationships among protozoa. Kluwer, Dordrecht, pp 109–132

Nash EA, Nisbet RE, Barbrook AC, Howe CJ (2008) Dinoflagellates: a mitochondrial genome all at sea. Trends Genet 24:328–335

Nixon JE, Field J, McArthur AG, Sogin ML, Yarlett N, Loftus BJ, Samuelson J (2003) Iron-dependent hydrogenases of Entamoeba histolytica and Giardia lamblia: activity of the recombinant entamoebic enzyme and evidence for lateral gene transfer. Biol Bull 204:1–9

Nosek J, Tomaska L, Fukuhara H, Suyama Y, Kovac L (1998) Linear mitochondrial genomes: 30 years down the line. Trends Genet 14:184–188

Nowacki M, Vijayan V, Zhou Y, Schotanus K, Doak TG, Landweber LF (2008) RNA-mediated epigenetic programming of a genome-rearrangement pathway. Nature 451:153–158

Palmer JD (1997) Organelle genomes: going, going, gone! Science 275:790–791

Paquin B, Lang BF (1996) The mitochondrial DNA of Allomyces macrogynus: the complete genomic sequence from an ancestral fungus. J Mol Biol 255:688–701

Park JS, Kolisko M, Heiss AA, Simpson AG (2009) Light microscopic observations, ultrastructure, and molecular phylogeny of Hicanonectes teleskopos n. g., n. sp., a deep-branching relative of diplomonads. J Eukaryot Microbiol 56:373–384

Perez-Brocal V, Clark CG (2008) Analysis of two genomes from the mitochondrion-like organelle of the intestinal parasite Blastocystis: complete sequences, gene content, and genome organization. Mol Biol Evol 25:2475–2482

Perez-Martinez X, Vazquez-Acevedo M, Tolkunova E, Funes S, Claros MG, Davidson E, King MP, Gonzalez-Halphen D (2000) Unusual location of a mitochondrial gene. Subunit III of cytochrome C oxidase is encoded in the nucleus of Chlamydomonad algae. J Biol Chem 275:30144–30152

Perez-Martinez X, Antaramian A, Vazquez-Acevedo M, Funes S, Tolkunova E, d'Alayer J, Claros MG, Davidson E, King MP, Gonzalez-Halphen D (2001) Subunit II of cytochrome c oxidase in Chlamydomonad algae is a heterodimer encoded by two independent nuclear genes. J Biol Chem 276:11302–11309

Prescott DM (1994) The DNA of ciliated protozoa. Microbiol Rev 58:233–267

Pritchard AE, Seilhamer JJ, Mahalingam R, Sable CL, Venuti SE, Cummings DJ (1990) Nucleotide sequence of the mitochondrial genome of Paramecium. Nucleic Acids Res 18:173–180

Putignani L, Tait A, Smith HV, Horner D, Tovar J, Tetley L, Wastling JM (2004) Characterization of a mitochondrion-like organelle in Cryptosporidium parvum. Parasitology 129:1–18

Regoes A, Zourmpanou D, Leon-Avila G, van der Giezen M, Tovar J, Hehl AB (2005) Protein import, replication, and inheritance of a vestigial mitochondrion. J Biol Chem 280:30557–30563

Ricard G (2008) Evolution and genome structure of anaerobic ciliates. Center for Molecular and Biomolecular informatics. Thesis, Radboud University, Nijmegen

Ricard G, McEwan NR, Dutilh BE, Jouany JP, Macheboeuf D, Mitsumori M, McIntosh FM, Michalowski T, Nagamine T, Nelson N, Newbold CJ, Nsabimana E, Takenaka A, Thomas NA, Ushida K, Hackstein JHP, Huynen MA (2006) Horizontal gene transfer from Bacteria to rumen Ciliates indicates adaptation to their anaerobic, carbohydrates-rich environment. BMC Genomics 7:22

Riordan CE, Ault JG, Langreth SG, Keithly JS (2003) Cryptosporidium parvum Cpn60 targets a relict organelle. Curr Genet 44:138–147

Roberts CW, Roberts F, Henriquez FL, Akiyoshi D, Samuel BU, Richards TA, Milhous W, Kyle D, McIntosh L, Hill GC, Chaudhuri M, Tzipori S, McLeod R (2004) Evidence for mitochondrial-derived alternative oxidase in the apicomplexan parasite Cryptosporidium parvum: a potential anti-microbial agent target. Int J Parasitol 34:297–308

Rotte C, Stejskal F, Zhu G, Keithly JS, Martin W (2001) Pyruvate : NADP+ oxidoreductase from the mitochondrion of Euglena gracilis and from the apicomplexan Cryptosporidium parvum: a biochemical relic linking pyruvate metabolism in mitochondriate and amitochondriate protists. Mol Biol Evol 18:710–720

Sagan L (1967) On the origin of mitosing cells. J Theor Biol 14:255–274

Stechmann A, Hamblin K, Perez-Brocal V, Gaston D, Richmond GS, van der Giezen M, Clark CG, Roger AJ (2008) Organelles in blastocystis that blur the distinction between mitochondria and hydrogenosomes. Curr Biol 18:580–585

Strüder-Kypke MC, Wright AD, Foissner W, Chatzinotas A, Lynn DH (2006) Molecular phylogeny of litostome ciliates (Ciliophora, Litostomatea) with emphasis on free-living haptorian genera. Protist 157:261–278

Tielens AG, Rotte C, van Hellemond JJ, Martin W (2002) Mitochondria as we don't know them. Trends Biochem Sci 27:564–572

Tovar J, Fischer A, Clark CG (1999) The mitosome, a novel organelle related to mitochondria in the amitochondrial parasite Entamoeba histolytica. Mol Microbiol 32:1013–1021

Tovar J, Leon-Avila G, Sanchez LB, Sutak R, Tachezy J, van der Giezen M, Hernandez M, Müller M, Lucocq JM (2003) Mitochondrial remnant organelles of Giardia function in iron-sulphur protein maturation. Nature 426:172–176

Trinci A, Davies D, Gull K, Lawrence M, Nielsen B, Rickers A, Theodorou M (1994) Anaerobic fungi and herbivorous animals. Mycol Res 98:129–152

Tsaousis AD, Kunji ER, Goldberg AV, Lucocq JM, Hirt RP, Embley TM (2008) A novel route for ATP acquisition by the remnant mitochondria of Encephalitozoon cuniculi. Nature 453:553–556

van der Giezen M (2009) Hydrogenosomes and mitosomes: conservation and evolution of functions. J Eukaryot Microbiol 56:221–231

van der Giezen M, Cox S, Tovar J (2004) The iron-sulfur cluster assembly genes IscS and IscU of Entamoeba histolytica were acquired by horizontal gene transfer. BMC Evol Biol 4:7

van der Giezen M, Sjollema KA, Artz RR, Alkema W, Prins RA (1997) Hydrogenosomes in the anaerobic fungus Neocallimastix frontalis have a double membrane but lack an associated organelle genome. FEBS Lett 408:147–150

van der Giezen M, Slotboom DJ, Horner DS, Dyal PL, Harding M, Xue GP, Embley TM, Kunji ER (2002) Conserved properties of hydrogenosomal and mitochondrial ADP/ATP carriers: a common origin for both organelles. Embo J 21:572–579

van der Giezen M, Tovar J (2005) Degenerate mitochondria. EMBO Rep 6:525–530

van Grinsven KW, Rosnowsky S, van Weelden SW, Putz S, van der Giezen M, Martin W, van Hellemond JJ, Tielens AG, Henze K (2008) Acetate:succinate CoA-transferase in the hydrogenosomes of Trichomonas vaginalis: identification and characterization. J Biol Chem 283:1411–1418

van Hoek AH, Sprakel VS, van Alen TA, Theuvenet AP, Vogels GD, Hackstein JH (1999) Voltage-dependent reversal of anodic galvanotaxis in Nyctotherus ovalis. J Eukaryot Microbiol 46:427–433

van Hoek AH, Akhmanova AS, Huynen MA, Hackstein JH (2000) A mitochondrial ancestry of the hydrogenosomes of Nyctotherus ovalis. Mol Biol Evol 17:202–206

Voncken F, Boxma B, Tjaden J, Akhmanova A, Huynen M, Verbeek F, Tielens AGM, Haferkamp I, Neuhaus HE, Vogels G, Veenhuis M, Hackstein JHP (2002a) Multiple origins of hydrogenosomes: functional and phylogenetic evidence from the ADP/ATP carrier of the anaerobic chytrid Neocallimastix sp. Mol Microbiol 44:1441–1454

Voncken FGJ, Boxma B, van Hoek AHAM, Akhmanova AS, Vogels GD, Huynen MA, Veenhuis M, Hackstein JHP (2002b) A hydrogenosomal [Fe]-hydrogenase from the anaerobic chytrid Neocallimastix sp. L2. Gene 284:103–112

Waller RF, Jackson CJ (2009) Dinoflagellate mitochondrial genomes: stretching the rules of molecular biology. Bioessays 31:237–245

Wawrzyniak I, Roussel M, Diogon M, Couloux A, Texier C, Tan KS, Vivares CP, Delbac F, Wincker P, El Alaoui H (2008) Complete circular DNA in the mitochondria-like organelles of Blastocystis hominis. Int J Parasitol 38:1377–1382

Williams BA, Hirt RP, Lucocq JM, Embley TM (2002) A mitochondrial remnant in the microsporidian Trachipleistophora hominis. Nature 418:865–869

Williams BA, Keeling PJ (2005) Microsporidian mitochondrial proteins: expression in Antonospora locustae spores and identification of genes coding for two further proteins. J Eukaryot Microbiol 52:271–276

Williams BA, Haferkamp I, Keeling PJ (2008) An ADP/ATP-specific mitochondrial carrier protein in the microsporidian Antonospora locustae. J Mol Biol 375:1249–1257

Yarlett N, Hann AC, Lloyd D, Williams A (1981) Hydrogenosomes in the rumen protozoon Dasytricha ruminantium Schuberg. Biochem J 200:365–372

Yarlett N, Lloyd D, Williams AG (1982) Respiration of the rumen ciliate Dasytricha ruminantium Schuberg. Biochem J 206:259–266

Yarlett N, Lloyd D, Williams AG (1985) Butyrate formation from glucose by the rumen protozoon Dasytricha ruminantium. Biochem J 228:187–192

Yarlett N, Orpin CG, Munn EA, Yarlett NC, Greenwood CA (1986) Hydrogenosomes in the rumen fungus Neocallimastix patriciarum. Biochem J 236:729–739

Yousuf MA, Mi-ichi F, Nakada-Tsukui K, Nozaki T (2010) Localization and targeting of an unusual pyridine nucleotide transhydrogenase in Entamoeba histolytica. Eukaryot Cell 9:926–933

Biodata of **Nina S. Levy** and **Andrew P. Levy**, authors of *"Adapting to Hypoxia: Lessons from Vascular Endothelial Growth Factor."*

Dr. Nina S. Levy obtained her Ph.D. in 1989 from Johns Hopkins University. After postdoctoral work at Johns Hopkins University, she began to work together with her husband, Andrew P. Levy, on the regulation of VEGF while working in the laboratories of Dr. Joe Loscalzo and Dr. Mark Goldberg at the Brigham and Woman's Hospital in Boston, MA. Nina continued this work in Andrew Levy's laboratory at Georgetown University in Washington, D.C., and later at the Technion Institute of Technology in Haifa, Israel, where she is currently working as a research associate.

E-mail: **ninal@tx.technion.ac.il**

Andrew P. Levy is currently a Professor of Medicine at theTechnion Institute of Technology in Haifa, Israel. He received his M.D. and Ph.D. degrees in 1990 from Johns Hopkins University where he co-discovered VEGF. Following an internship in internal medicine at Johns Hopkins Hospital and a residency in cardiology at the Brigham and Woman's Hospital in Boston, MA, Andrew continued his research on the hypoxic regulation of VEGF in Boston in the laboratories of Dr. Joe Loscalzo and Dr. Mark Goldberg. After spending time at Georgetown University in D.C. as an assistant professor, he relocated to the Technion Institute of Technology where he continued work on VEGF as well as a new area of research involving the role of haptoglobin genotype on vascular disease in diabetic individuals.

E-mail: **alevy@tx.technion.ac.il**

Nina S. Levy   **Andrew P. Levy**

ADAPTING TO HYPOXIA: LESSONS FROM VASCULAR ENDOTHELIAL GROWTH FACTOR

NINA S. LEVY[1] AND ANDREW P. LEVY[2]

[1]*Rappaport Institute, Technion Institute of Technology, Haifa, Israel*
[2]*Department of Anatomy, Technion School of Medicine,*
Technion Institute of Technology, Haifa, Israel

1. Introduction

Planet Earth is unique among member planets of its solar system in that its atmosphere contains a whopping 21% oxygen, more than one hundred times the amount of atmospheric oxygen of its next closest competitor Mars (0.13%). Accordingly, the vast majority of life forms on the Earth have evolved not only to utilize oxygen but to depend on oxygen for their continued existence. However, the Earth's atmosphere was not always rich in oxygen. Most geologists believe that the Earth's atmosphere started out as mostly nitrogen and carbon dioxide. About 2 billion years ago, photosynthesizing bacteria evolved and began releasing oxygen into the atmosphere. The level of oxygen has steadily risen since then from zero to its current level of approximately 21%. Today, life in the absence of oxygen (anoxia) is deadly for the vast majority of living organisms. The human brain can withstand only 4–5 min in the absence of oxygen before irreversible brain damage occurs. However, given the evolutionary history of our planet, it is perhaps not surprising that periods of low oxygen can be tolerated and are even inherent in such processes as normal embryonic development.

Hypoxia can be generally defined as a pathological condition in which the body as a whole or a region of the body is deprived of adequate oxygen supply. Hypoxia can result from a change in atmospheric oxygen such as might occur at high altitude, underground, or in outer space. Within seconds, the carotid body of the lungs senses the decrease in oxygen and releases neurotransmitters which activate the autonomic nervous system and cause an increase in respiration. Blood vessels in the lung become constricted (in an effort to shunt blood to more highly oxygenated areas of the lung), while blood vessels in the rest of the body dilate in order to increase the amount of oxygen delivered to the tissues. These mechanisms are activated within seconds, occur in specialized cells in the body, and do not require new protein synthesis. Prolonged hypoxia, however, results in an intricate adaptive response which can be activated in virtually all cells in the body and causes extensive changes in gene expression including the stimulation as well as the downregulation of hundreds of genes.

Hypoxia can occur during repeated, strenuous exercise when oxygen levels in the muscle become depleted faster than the body can replenish them. During normal embryogenesis, growing tissues become hypoxic due to the lack of supporting vasculature. Other pathological situations which result in hypoxia include insufficient oxygen supply to the heart or brain due to cholesterol buildup in the arteries, pulmonary insufficiency due to lung deposits, and necrosis of the extremities in diabetic patients due to insufficient vascularization. The genetic programs which are activated in the different tissues experiencing prolonged hypoxia are distinct and varied. However, in virtually all cases, hypoxic tissues require an increase in blood supply in order to alleviate hypoxic stress. Accordingly, the gene for vascular endothelial growth factor (VEGF) is induced by many cell types following prolonged hypoxia. VEGF is a 46-Kd secreted growth factor which binds to cognate receptors on the surface of endothelial cells and stimulates the formation of new blood vessels, thereby increasing oxygen delivery to hypoxic tissues. Due to the essential nature of this growth factor, many different mechanisms have evolved to up-regulate VEGF, making it a good model gene for the discussion of mechanisms of adaptive hypoxia, as will be discussed below.

2. Increased Gene Transcription

A master regulator of gene transcription under hypoxia is hypoxia-inducible factor (HIF). HIF was originally discovered during studies aimed at elucidating the regulation of erythropoietin (EPO), a hormone glycoprotein responsible for stimulating new red blood cell production. EPO was one of the first proteins shown to increase in response to hypoxia. Up-regulation of EPO in hepatic Hep3B cells was shown to be largely due to an increase in transcription initiation (Goldberg et al. 1991). Extracts from Hep3B cells showed hypoxia-inducible protein binding to a region in the 3'UTR of the EPO gene (Semenza et al. 1991). More importantly, extracts from other cell types exposed to hypoxia revealed the presence of inducible protein binding to the same region, indicating the universality of the hypoxia-inducible factor involved. Subsequent purification and characterization of the factor showed that HIF is comprised of alpha and beta chains which contain basic helix–loop–helix (bHLH) DNA-binding domains and PAS (PER, ARNT, SIM) protein-binding domains and are part of a large family of proteins with these features (Wang and Semenza 1995). Under stimulatory conditions, HIF alpha chains enter the nucleus and heterodimerize with HIF beta. Active complexes bind DNA at the consensus sequence RCGTG and can recruit cAMP responsive element-binding protein (CEBP/p300) and signal transducer and activator of transcription 3 (STAT3) to the transcriptional complex, thereby increasing transcription of downstream genes (see Fig. 1).

Similar to EPO, VEGF was shown to contain sequences responsible for hypoxic stimulation of transcription. In the case of VEGF, it was determined that a 47-bp sequence located approximately one kilobase upstream of the

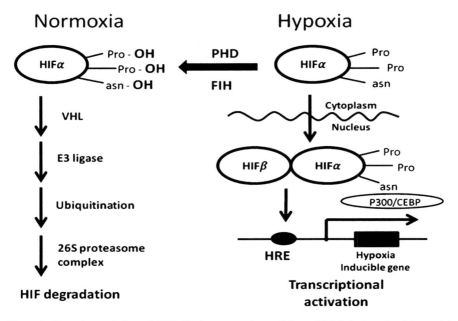

Figure 1. Hypoxic regulation of HIF. Under normoxic conditions, HIF is hydroxylated by prolyl hydroxylase (PHD) and factor inhibiting HIF (FIH). In its hydroxylated form, HIF alpha is bound by the von Hippel–Lindau tumor suppressor protein (VHL) which then recruits E3 ubiquitin ligase to the complex. Following ubiquination, HIF is directed to the 26S proteasome complex where it undergoes rapid degradation. Under hypoxic conditions, HIF is not hydroxylated and remains stable, entering the nucleus and dimerizing with HIF beta where it binds to HRE elements and promotes transcriptional activation of VEGF and other hypoxia-inducible genes.

transcriptional start site could mediate hypoxia-inducible expression of a luciferase reporter construct in Hep3B cells (Forsythe et al. 1996). Mutations in the hypoxia responsive element (HRE) sequence located in this region abolished the hypoxic response. Co-transfection of cells with a dominant negative form of HIF abolished the hypoxic response as well as transfection into cells which did not express HIF beta. HIF alpha and beta synthesized by in vitro translation were shown to bind specifically to the EPO and VEGF HREs, each being competitively inhibited from binding to one HRE by an excess of the other HRE. These studies extended the role of HIF as a ubiquitous regulator of hypoxia-inducible gene transcription. It is now apparent that HIF can up-regulate more than one hundred genes in response to hypoxia, including those which control angiogenesis, glycolysis, glucose transport, mitochondrial function, vascular tone, and cell proliferation.

A tremendous amount of research has been dedicated in recent years to elucidating the regulation of HIF and has provided a good understanding of this system (for review, see Webb et al. 2009; see Fig. 1). It is known that the protein levels of HIF alpha chain are greatly increased under hypoxia, while the beta chain is constitutive. The level of the alpha chain protein is increased under hypoxia by virtue of an increase in protein stabilization. Under normoxia, the HIF alpha chain is hydroxylated at two critical prolyl residues by the enzyme prolyl hydroxylase (PHD) of which there are several isoforms. The hydroxylated protein is recognized and bound by the von Hippel-Lindau tumor suppressor protein (pVHL), and this complex is then recognized by the E3 ubiquitin-protein ligase. Once ubiquinated, the alpha chain is rapidly degraded by the 26S proteasome complex. A second prolyl hydroxylase, named factor inhibiting HIF (FIH), specifically hydroxylates an asparagyl residue found in the C-terminus of HIF alpha. Hydroxylation at the latter site inhibits recruitment of the transcriptional coactivators p300/CBP to the HIF heterodimer. The activity of the HIF hydroxylases is dependent on oxygen, iron, and 2-oxyglutarate. The Michaelis-Menten constant (Km) of these enzymes is higher than the concentration of oxygen generally found in cells under normoxia, thereby making them very sensitive oxygen sensors. Under hypoxia, the activity of PHD and FIH is decreased, leading to an accumulation of the HIF alpha chain and an increase in transcriptional activity of the HIF heterodimer.

An additional pathway which becomes activated by hypoxia is the unfolded protein response (UPR). The trigger for this response is endoplasmic reticulum (ER) overload, in which protein traffic through the ER becomes backed up with unfolded or misfolded proteins due to the lack of ATP stores required for proper protein folding. The main effect of this response is to inhibit global protein synthesis which requires 25–30% of ATP stores and stimulate up-regulation of molecular chaperones and protein processing enzymes which act to keep proteins moving through the ER. In addition, protein degradation enzymes are synthesized to relieve the ER workload. Three of the main regulators of this response are inositol requiring 1 (IRE1), PKR-like ER kinase (PERK), and activating transcription factor 6 (ATF6) (Ghosh et al. 2010).

Cells exposed to conventional ER stress inducers (thapsigargin or tunicamycin) were shown to have threefold to fivefold increased levels of VEGF. This effect was independent of HIF, as shown by the lack of inhibition of VEGF stimulation in the presence of ER stress and antisense HIF alpha. Actinomycin D experiments indicated that VEGF mRNA was not being stabilized. Gene rescue studies as well as chromatin immunoprecipitation and luciferase reporter construct analysis showed that VEGF expression was increased following ER stress by several independent mechanisms. First, IRE1 alpha increases transcription of VEGF by promoting splicing of X-box-binding protein 1 (XBP-1) mRNA followed by binding of active transcription factor XBP-1p to the VEGF 5′UTR. Second, PERK increases translation of activating transcription factor 4 (ATF4) which can bind to the first intron in the VEGF gene and stimulate gene expression. Third, overexpression of cleaved ATF6 (the transcriptionally active form) can activate downstream expression of a reporter construct via interaction with the VEGF promoter, possibly through its activation of XBP-1. These results indicate that activation of the UPR leads to increased VEGF transcription via multiple transcription factors, some of which are independent of HIF.

3. Increased mRNA Stability

A second mechanism for increasing the level of hypoxia-inducible gene expression is by mRNA stabilization. Under hypoxia, it was found that the steady state level of VEGF mRNA was increased 12-fold in pheochromocytoma (PC12) cells (Levy et al. 1995). Nuclear runoff experiments showed an increase in transcription rate (mediated by HIF) of threefold only. In order to account for the remaining hypoxic increase in VEGF mRNA, subsequent studies were designed to measure the half-life of VEGF mRNA. Under normoxic conditions, VEGF has a short half-life of approximately 15 min. Two regions containing consensus AUUUA sequences associated with mRNA instability were found that mediated the rapid degradation of VEGF mRNA in an in vitro mRNA degradation assay. Under hypoxia, there was a 2.5-fold increase in mRNA half-life (Levy et al. 1996a). Cis-acting sequences in the 3'UTR of VEGF, termed VEGF regulatory sequences (VRS), were identified using a reporter gene with deletion constructs. Three distinct regions were capable of binding similar hypoxia-inducible protein complexes. Even though there was no clear primary consensus sequence or secondary structure, each sequence was able to compete with the others for binding of the hypoxia-inducible complex (Levy et al. 1996b).

One of the hypoxia-inducible binding proteins which mediates hypoxic mRNA stabilization is the human antigen R (HuR) protein. Several lines of evidence support this assertion. First, incubation of radiolabeled VRS with purified HuR showed specific binding to a 45-bp fragment with high affinity (9 mM) (Levy et al. 1998). Second, in cells expressing HuR antisense mRNA, there was no hypoxic increase in VEGF mRNA stability, while cells overexpressing HuR showed more efficient hypoxic induction of VEGF. Third, purified HuR was able to confer an increase in VEGF mRNA half-life in an in vitro mRNA degradation assay (Goldberg-Cohen et al. 2002). These results have since been extended to other hypoxia-inducible genes, such as HIF-1, and have established HuR as a general mRNA stabilizing element.

Because the HuR-binding site was found to be four nucleotides away from a consensus AUUUA instability site, it was hypothesized that binding of HuR may change the conformation of the mRNA or hinder binding of degradation enzymes. It was shown that removal of the 16 bp which make contact with HuR from the 40 bp HuR-binding region resulted in a segment which now conferred a destabilizing effect in a heterologous reporter system (Goldberg-Cohen et al. 2002). This suggested that stabilizing and destabilizing regions may be overlapping, and the overall effect depends on how well mRNA-binding factors can compete for the same space. It is known that HuR levels do not change appreciably in the cell following hypoxia; however, more HuR shuttles from the cytoplasm to the nucleus, suggesting that HuR may act to bind to the mRNA and protect it from being recognized by ribonucleoprotein degradation elements after it exits the nucleus (see Fig. 2).

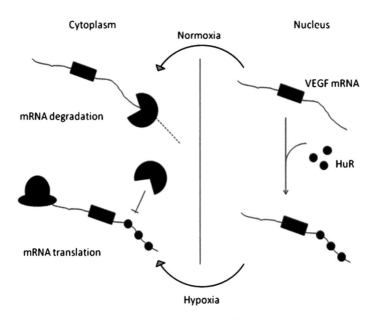

Figure 2. One hypothetical mechanism for the hypoxic stabilization of VEGF mRNA transcripts is shown diagrammatically above. VEGF mRNA transcripts, depicted by a filled-in rectangle flanked by 5′ and 3′ untranslated sequences (gray lines), become bound by HuR at specific site in the 3′UTR in the nucleus and are exported to the cytoplasm. Unbound cytoplasmic transcripts become rapidly degraded by RNAses (depicted by the pac-man). Transcripts bound by HuR as well as other stabilizing proteins are protected from degradation and are preferentially bound to polysomes (depicted by the hat shape) where they are actively translated.

Recent studies have identified poly(A)-binding protein interacting protein 2 (PAIP-2) as a VEGF mRNA stabilizing factor. Previous work showed that PAIP-2 prevents poly(A)-binding protein from binding to the poly(A) tail of messenger RNAs and recruiting deadenylation enzymes. In the case of VEGF, it was shown by RNA electromobility shift assays that PAIP-2 binds to at least two sites on the VEGF 3′UTR (Onesto et al. 2004). Overexpression of PAIP-2 led to an increase in VEGF mRNA half-life. It is not known if PAIP-2 is induced by hypoxia. However, co-immunoprecipitation and immunoblotting experiments showed that HuR and PAIP-2 interact with one another. This allows for the possibility that under hypoxic conditions, when HuR levels are increased in the cytoplasm, HuR and PAIP-2 may alter the mRNA conformation by binding their respective mRNA target sequences and each other. This type of change in tertiary structure may also contribute to preventing the recruitment of mRNA degradation enzymes.

Two other proteins shown to be involved in hypoxic VEGF mRNA stabilization are heteronuclear ribonucleoprotein L (hnRNP L) and double-stranded RNA-binding protein 76/NF90 (DRBP76/NF90), a specific isoform of the DRBP family (Claffey et al. 1998; Vumbaca et al. 2008). Both of these proteins were isolated by virtue of their binding to a 125-bp region in the 3′UTR of VEGF (upstream of the HuR-binding site) which confers increased mRNA stability in an in vitro luciferase reporter assay. This region was predicted to form a stem loop structure which may account for the binding of DRBP76. HnRNP L was shown to bind to a minimal 20-bp sequence at the 5′ end of the 125-bp segment, while the minimal binding site of DRBP76 was not investigated. Inhibition of DRBP76 expression or transfection with an antisense oligonucleotide to the 20 bp binding site for hnRNP L led to a partial decrease in VEGF mRNA stability. DRBP76 may act to keep transient stem loop structures more stable. Collectively, these results further support the notion that multiple mRNA stabilizing proteins act by promoting specific secondary structures which likely influence the binding of RNA degradation enzymes.

4. Translational Regulation

Messenger RNA translation is a high energy requiring process and is therefore repressed in cells experiencing a depletion in energy stores such as occurs during hypoxia. In general, mRNA translation is most tightly regulated at the initiation stage and involves reversible phosphorylation (for review, see van den Beucken et al. 2006). Specifically, the alpha subunit of elongation initiation factor 2 (eIF2), which normally recruits the aminoacylated tRNA to the 40S ribosomal subunit in its GTP bound form, becomes phosphorylated by the endoplasmic reticulum kinase PERK. The exchange of GDP for GTP, which is mediated by the guanine nucleotide exchange factor eIF2B, is inhibited by phosphorylated eIF2 alpha, resulting in a decrease in translational initiation rate.

The inhibition of translation via eIF2 is transient and appears to occur only under more severe conditions of hypoxia. A second mechanism of translational inhibition, which appears to function independently of eIF2 phosphorylation, is disruption of the cap-binding protein complex eIF4F. This complex consists of a number of proteins which act in concert to connect between the cap structure at the 5′ terminus of the mRNA and the 40S ribosomal subunit. A number of binding proteins (4E-BPs), when underphosphorylated, can obstruct formation of an active eIF4F initiation complex and prevent transcript recruitment. Upstream regulators of eIF4F include mammalian target of rapamyacin (mTOR) kinase, which phosphorylates 4E-BP1 under normoxia and prevents its binding to the eIF4F complex. Hypophosphorylation of 4E-BP1 is found under mild hypoxia as well as acute, severe hypoxia.

Global translation is repressed during hypoxia. It therefore became necessary to elucidate how transcripts encoding proteins needed to overcome hypoxic stress could escape this repression. Investigations aimed at solving this dilemma led to the discovery of internal ribosomal entry sites (IRESes), in which conserved sequences in the 5′UTR of specific genes are able to direct ribosome binding independent of eIF4F complex formation at the cap. IRESes have been previously

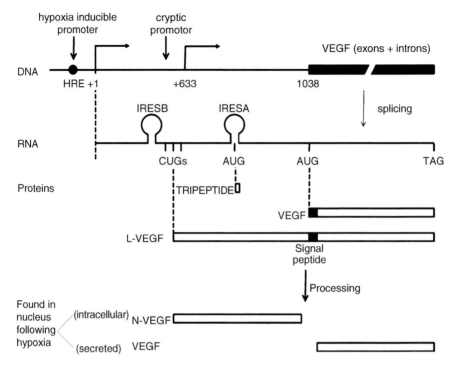

Figure 3. Genomic, mRNA, and protein structure of VEGF. At the DNA level, VEGF is transcriptionally activated by binding of HIF to the HRE approximately one kilobase upstream of the VEGF starting AUG codon. A cryptic promoter has been found downstream of the hypoxia-inducible promoter. Two IRES elements (A + B) have been described in the 5'UTR. VEGF contains one uORF starting with AUG located in IRES A. It encodes a tripeptide. Several uORFs beginning with CUG are in frame with VEGF and result in a longer protein called L-VEGF. Processing of this isoform leads to the production of intracellular N-VEGF and secreted VEGF. Both of these isoforms have been found in the nucleus under hypoxic conditions (This diagram is not drawn to scale).

described as a mechanism for translation of viral mRNAs in infected cells where host protein translation had been shut down through blockade of eIF4F. In humans, the VEGF 5'UTR is 1,038 nucleotides long and contains two IRESes (see Fig. 3). IRES A is located within the 300 nucleotides upstream of the AUG start codon, and IRES B is located approximately 500 nucleotides upstream of the AUG start codon. VEGF as well as a number of other hypoxia-inducible genes were reported to have functional IRESes, as determined by transfection studies using bicistronic plasmid constructs in which one reporter gene containing a functional promoter is placed upstream of a promoterless test gene containing 5'UTR sequences next to a second reporter gene (Huez et al. 1998). Activity of the second reporter gene is considered to be indicative of IRES activity.

A recent study which used bicistronic constructs containing no promoter as a control showed that VEGF IRES activity appears to be the result of a cryptic promoter in the 5′UTR sequences of VEGF, glucose transporter-1, and HIF, as well as other genes (Young et al. 2008). VEGF is now known to contain an additional promoter in between the two IRES sites, approximately 400 bp upstream of the VEGF AUG start codon. In order to eliminate cryptic promoter activity, synthetic VEGF capped and polyadenylated transcripts were transfected into NIH 3T3 cells and luciferase activity was measured. The results showed that only the cytomegalovirus IRES was capable of directing significant levels of translation of a downstream luciferase gene (Young et al. 2008). Incubation of the cells under hypoxia did not stimulate activity of any of the mammalian IRESs studied. Despite lack of evidence for IRES function, VEGF transcripts were found to be associated with polysomes following hypoxia, indicating that other mechanisms were responsible for the observed hypoxia-inducible translational upregulation (Young et al. 2008).

Additional studies continue to report that VEGF is translated by an IRES-dependent mechanism using methods which attempt to control for cryptic promoter activity. It was shown for example that antisense RNA to eIF4E did not affect VEGF IRES directed expression of a luciferase reporter gene (Bastide et al. 2008). This indicated that translation was occurring by a 5′cap-independent mechanism, which would preclude translation from transcripts being initiated by a cryptic promoter. Another study showed that expression of VEGF is up-regulated fivefold in ischemic skeletal muscle of transgenic mice carrying a luciferase reporter construct downstream of VEGF IRES A or IRES B (Bornes et al. 2007). In these experiments, the ratio of control luciferase mRNA to IRES directed luciferase mRNA was found to be 1, confirming the integrity of the bicistronic transcript and precluding any activity that might arise from a cryptic promoter. The authors of this study concluded that IRES-dependent translation is a rapid (3 h) and efficient mechanism for inducing protein synthesis in response to hypoxia.

The mechanisms by which cellular IRES elements mediate ribosome recruitment are unclear. It appears that a new group of proteins known as IRES trans-acting factors (ITAFs) are necessary for IRES function (for review, see Graber and Holcik 2007). ITAFs have been proposed to function as RNA chaperones, allowing for direct binding of the 40S ribosomal subunit through modification of RNA secondary and/or tertiary structure. Alternatively, ITAFs could function as adaptor proteins that facilitate RNA–ribosome interaction. As yet, there appears to be no "universal" ITAF that modifies the activity of all IRESes. However, it was recently found that pyrimidine tract–binding protein (PTB) positively regulates the activity of several IRESes following apoptotic stress. It could be that some of the discrepancies seen in the study of IRES-dependent translation of VEGF under hypoxia are due to the absence or presence of critical ITAFs.

Other mechanisms of increasing gene specific translation following cellular stress have been reported. Phosphorylation of eIF2 has been described so far as a mechanism for promoting global translational arrest. However, it was shown originally in yeast and subsequently in mammals that eIF2 phosphorylation can promote the translation of specific genes containing multiple upstream open reading frames (Lu et al. 2004). ATF-4, an important transcription factor up-regulated in the UPR response (noted above), contains two uORFs in its 5′UTR. Following the translation of the first uORF, the ribosome complex requires a second ternary complex containing eIF2-GTP and Met-tRNA. Phosphorylated eIF2 alpha delays the formation of this ternary complex and allows the ribosome to continue scanning past the second uORF until it reaches the ATF-4 ORF. When eIF2 alpha is not phosphorylated, translation reinitiates at the second uORF, which overlaps the ATF-4 ORF and precludes its translation. Other genes activated by the UPR, such as C/EBP homology protein (CHOP) and growth arrest DNA damage 34 (GADD34), also contain uORFs but are not up-regulated by phosphorylated eIF2, indicating that additional sequences or proteins are likely to be involved.

VEGF contains four uORFs, three beginning with CUG, that are in frame with VEGF and one that starts with AUG and terminates upstream of the VEGF start codon (see Fig. 3). The CUGs, which are clustered approximately 540 nucleotides upstream of the VEGF start codon, direct the translation of a long form of VEGF (L-VEGF) which becomes cleaved to produce secreted VEGF and N-VEGF, a 180 amino acid, nonsecreted protein. Under hypoxia, it was found that N-VEGF but not VEGF could translocate to the nucleus. (Rosenbaum-Dekel et al. 2005). A second study found that VEGF is able to translocate to the nucleus following hypoxia (Lejbkowicz et al. 2005); however, the function of either of these isoforms in the nucleus is not known. The AUG uORF maps to the IRES A site. This ORF encodes a tripeptide. Under normoxia, cap-independent translation was found to occur from this ORF (Bastide et al. 2008). Interestingly, translation of this ORF appeared to control specific VEGF isoform expression. The hypoxic regulation of VEGF's uORFs has not been investigated but has the potential to reveal additional mechanisms of translational upregulation.

A relatively new player in the game of gene regulation has been discovered in the form of micro RNAs (miRNAs). These ribosomal RNA molecules are 22 nucleotides in length and repress translation by binding to sites in the 3′UTR of an mRNA with partial complementarity. One possible mechanism of this repression suggests that miRNA associated proteins can bind to the cap structure of the mRNA and compete for binding by eIF4E, thus repressing cap-dependent translation. As the number of miRNAs appears to be in the hundreds and still growing, it is now accepted that most genes are regulated by miRNAs. Recent studies showed that groups of microRNAs can be controlled by transcription factors involved in the regulation of a particular pathway. For example, the c-myc oncogenic transcription factor was shown to up-regulate a group of miRNAs which are important for tumor formation. In the case of hypoxia, it is known that many genes undergo translational repression. Therefore, it is plausible that hypoxia-inducible transcription factors might regulate the expression of a group of miRNAs involved in translational repression. A recent study identified 23 hypoxia responsive miRNAs (HRMs) (Kulshreshtha et al. 2007). This analysis has since been extended to include 42 HRMs that have been reported to be up-regulated and 27 that were downregulated. (Kulshreshtha et al. 2008).

A logical regulator of HRM expression is HIF. A computerized search of miRNA promoters revealed a highly significant enrichment of the HIF binding consensus sequences in a group of 23 up-regulated HRMs. Transfection studies with reporter constructs containing HRM promoter regions adjacent to a luciferase reporter construct in cells constitutively expressing HIF showed an increase in HRM expression. Chromatin immunoprecipitation experiments showed a direct interaction of HIF with several HRM promoter regions. It follows from this discussion that HIF plays an even more extensive role than previously appreciated. Not only can HIF up-regulate hundreds of genes by virtue of direct binding to HREs, it can also downregulate perhaps thousands of genes through stimulation of miRNA expression, each of which can act to repress ten or more proteins.

Two studies have shown that VEGF can be regulated by miRNAs. One study showed that miRNA 92–1 can target the 3′UTR of the tumor suppressor gene VHL in chronic lymphocytic leukemia B (CLL-B) cells (Ghosh et al. 2009). This interaction led to a decrease in the VHL protein and an increase in HIF. As described above, VHL normally binds hydroxylated HIF, recruits E3 ubiquitin ligase, and promotes degradation of HIF (see regulation of HIF above). In CLL-B cells, miRNA 92–1 is expressed at a high level leading to a decrease in VHL and an increase in HIF. These studies suggest that miRNAs may act to decrease the expression of repressors of important growth factors such as VEGF.

A second study showed that miRNA 20b, which was downregulated by hypoxia in MCF-7 breast cancer cells, was correlated with an increase in VEGF mRNA. This suggested that miRNA 20b normally represses a protein or group of proteins which act to up-regulate VEGF. (Cascio et al. 2010). Further experiments showed that miRNA 20b targets HIF-1 and STAT3, two hypoxia-inducible transcription factors that bind to the promoter of VEGF and promote increased transcription (Lei et al. 2009). A decrease in miRNA 20b allows for accumulation of HIF and STAT3 and an increase in VEGF. These studies suggest the possible existence of a HIF autocrine pathway, whereby HIF stimulates the increase of miRNAs which subsequently act to decrease the level of HIF. This type of interaction may allow for fine tuning of the cellular response to hypoxia.

5. Summary and Future Perspectives

The regulation of gene expression by hypoxia is a multifaceted process. Many parameters influence this system such as the cell or tissue type being affected, the degree and duration of hypoxia, and the developmental or disease status of the various cells or tissues involved. In order to allow for the appropriate expression of genes involved in hypoxic adaptation, numerous mechanisms have evolved

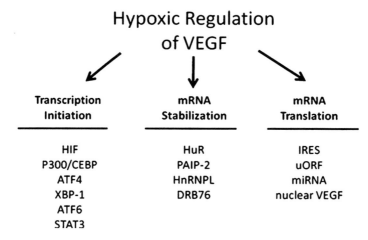

Figure 4. Summary of the factors involved in the hypoxic up-regulation of VEGF. Multiple factors act at multiple levels, allowing for a robust and highly sensitive response to hypoxia.

to ensure this process including increased transcription, mRNA stabilization, and mRNA translation. VEGF represents a good model gene for studying hypoxic upregulation as it is induced in virtually all cell types in response to hypoxia. Although many elements of this system have been discovered (see Fig. 4), there are still many questions to be answered. For example, how does hypoxia stimulate the translocation of the mRNA stabilizing protein HuR into the nucleus? What is the role of IRES-dependent translation in the hypoxic up-regulation of VEGF? What effect does hypoxia have on the translation of uORFs? What is the role of hypoxia-inducible nuclear localization of VEGF and N-VEGF? Is HIF regulated by miRNAs via an autocrine pathway?

Genetic polymorphisms are another area of research that is likely to provide new insights into the hypoxic regulation of VEGF. It was shown that monocytes from diabetic patients who have developed retinopathy induce VEGF to a lower degree following hypoxia than diabetic patients without retinopathy (Marsh et al. 2000). Similarly, monocytes from patients with a high degree of coronary collaterals are able to induce VEGF under hypoxia to a greater degree than monocytes from patients with fewer coronary collaterals (Schultz et al. 1999). Recently, lower levels of VEGF have been found in the cerebrospinal fluid from hypoxemic amyotrophic lateral sclerosis patients compared to hypoxemic controls, suggesting an abnormal response to hypoxia (Just et al. 2007). All of these studies suggest that genetic polymorphisms in genes involved in the hypoxic induction of VEGF exist, which can lead to vascular or neurological disorders. The continued isolation of important factors involved in response to hypoxia in general and in the control of VEGF in particular should present new therapeutic targets for treating these diseases.

6. References

Bastide A, Karaa Z, Bornes S, Hieblot C, Lacazetter E, Prats H, Touriol C (2008) An upstream open reading frame within an IRES controls expression of a specific VEGF-A isoform. Nucleic Acids Res 36:2434–2445

Bornes S, Prado-Lourenco L, Bastide A, Zanibellato C, Iacovoni JS, Lacazette E, Prats A-C, Touriol C, Prats H (2007) Translational induction of VEGF internal ribosome entry site elements during the early response to ischemic stress. Circulation 100:305–308

Cascio S, D'andrea A, Ferla R, Surmacz E, Gulotta E, Amodeo V, Bazan V, Gebbia N, Russo A (2010) miR-20b modulates VEGF expression by targeting HIF-1a and STAT3 in MCF-7 breast cancer cells. J Cell Physiol 224:242–249

Claffey KP, Shih S-C, Mullen A, Dziennis S, Cusick JL, Abrams KR, Lee SW, Detmar M (1998) Identification of a human VPF/VEGF 3′ untranslated region mediating hypoxia-induced mRNA stability. Mol Biol Cell 9:469–481

Forsythe JA, Jiang BH, Iyer NV, Agani F, Leung SW, Koos RD, Semenza GL (1996) Activation of vascular endothelial growth factor gene transcription by hypoxia-inducible factor 1. Mol Cell Biol 16:4604–13

Ghosh AK, Shanafelt TD, Cimmino A, Taccioli C, Volinia S, Liu C-g, Calin GA, Croce CM, Chan DA, Giaccia AJ, Secreto C, Wellik LE, Lee YK, Mukhopadhyay D, Kay NE (2009) Aberrant regulation of pVHL levels by microRNA promotes the HIF/VEGF axis in CLL B cells. Blood 2009(113):5568–5574

Ghosh R, Lipson KL, Sargent KE, Mercurio AM, Hunt JS, Ron D, Urano F (2010) Transcriptional regulation of VEGF-A by the unfolded protein response pathway. PLoS One 5:e9575

Goldberg MA, Gaut CC, Bunn HF (1991) Erythropoietin mRNA levels are governed by both the rate of gene transcription and posttranscriptional events. Blood 77:271–277

Goldberg-Cohen I, Furneaux H, Levy AP (2002) A 40 bp RNA element that mediates stabilization of VEGF mRNA by HuR. J Biol Chem 277:13635–40

Graber TE, Holcik M (2007) Cap-independent regulation of gene expression in apoptosis. Mol Biosyst 3:825–34

Huez I, Creancier L, Audigier S, Gensac MC, Prats AC, Prats H (1998) Two independent internal ribosome entry sites are involved in translation initiation of vascular endothelial growth factor mRNA. Mol Cell Biol 18:6178–90

Just N, Moreau C, Lassalle P, Gosset P, Perez T, Brunaud-Danel V, Wallaert B, Destee A, Defebvre L, Tonnel AB, Devos D (2007) High erythropoietin and low vascular endothelial growth factor levels in cerebrospinal fluid from hypoxemic ALS patients suggesting an abnormal response to hypoxia. Neuromuscul Disord 17:169–173

Kulshreshtha R, Ferracin M, Wojcik SE, Garzon R, Alder H, Agosto-Perez FJ, Davuluri R, Liu CG, Croce CM, Negrini M, Calin GA, Ivan M (2007) A microRNA signature of hypoxia. Mol Cell Biol 27:1859–1867

Kulshreshtha R, Davuluri RV, Calin GA, Ivan M (2008) A microRNA component of the hypoxic response. Cell Death Differ 15:667–671

Lei Z, Li B, Yang Z, Fang H, Zhang G-M, Feng Z-H, Huang B (2009) Regulation of HIF-1a and VEGF by miR-20b tunes tumor cells to adapt to the alteration of oxygen concentration. PLoS One 4:e7629

Lejbkowicz F, Goldberg-Cohen I, Levy AP (2005) New horizons for VEGF. Is there a role for nuclear localization? Acta Histochem 106:405–411

Levy AP, Levy NS, Wegner S, Goldberg MA (1995) Transcriptional regulation of the rat vascular endothelial growth factor gene by hypoxia. J Biol Chem 270:13333–13340

Levy AP, Levy NS, Goldberg MA (1996a) Post-transcriptional regulation of vascular endothelial growth factor by hypoxia. J Biol Chem 271:2746–2753

Levy AP, Levy NS, Goldberg MA (1996b) Hypoxia-inducible protein binding to vascular endothelial growth factor mRNA and its modulation by the von Hippel-Lindau protein. J Biol Chem 271:25492–25497

Levy NS, Chung S, Furneaux H, Levy AP (1998) Hypoxic stabilization of vascular endothelial growth factor mRNA by the RNA-binding protein HuR. J Biol Chem 273:6417–23

Lu PD, Harding HP, Ron D (2004) Translation reinitiation at alternative open reading frames regulates gene expression in an integrated stress response. J Cell Biol 167:27–33

Marsh S, Nakhoul FM, Skorecki K, Rubin A, Miller BP, Leibu R, Levy NS, Levy AP (2000) Hypoxic induction of vascular endothelial growth factor is markedly decreased in diabetic individuals who do not develop retinopathy. Diabetes Care 23:1375–1380

Onesto C, Berra E, Grépin R, Pagès G (2004) Poly(A)-binding protein-interacting protein 2, a strong regulator of vascular endothelial growth factor mRNA. J Biol Chem 279:34217–34226

Rosenbaum-Dekel Y, Fuchs A, Yakirevich E, Azriel A, Mazareb S, Resnick MB, Levi B-Z (2005) Nuclear localization of long-VEGF is associated with hypoxia and tumor angiogenesis. Biochem Biophys Res Commun 332:271–278

Schultz A, Lavi L, Hochberg I, Beyar R, Stone T, Skorecki K, Lavie P, Roguin A, Levy AP (1999) Interindividual heterogeneity in the hypoxic regulation of VEGF: significance for the development of the coronary artery collateral circulation. Circulation 100:547–552

Semenza GL, Nejfelt MK, Chi SM, Antonarakis SE (1991) Hypoxia inducible nuclear factors bind to an enhancer element located 30 to the human erythropoietin gene. Proc Natl Acad Sci USA 88:5680–5684

van den Beucken T, Koritzinsky M, Wouters BG (2006) Translational control of gene expression during hypoxia. Cancer Biol Ther 5:749–755

Vumbaca F, Phoenix KN, Rodriguez-Pinto D, Han DK, Claffey KP (2008) Double-stranded RNA-binding protein regulates vascular endothelial growth factor mRNA stability, translation, and breast cancer angiogenesis. Mol Cell Biol 28:772–783

Wang GL, Semenza GL (1995) Purification and characterization of hypoxia-inducible factor 1. J Biol Chem 270:1230–1237

Webb JD, Coleman ML, Pugh CW (2009) Hypoxia, hypoxia-inducible factors (HIF), HIF hydroxylases and oxygen sensing. Mol Life Sci 66:3539–3554

Young RM, Wang SJ, Gordan JD, Ji X, Liebhaber SA, Simon MC (2008) Hypoxia-mediated selective mRNA translation by an internal ribosome entry site-independent mechanism. J Biol Chem 283:16309–16319

PART III:
MANAGING ANOXIA

Bazylinski
Lefèvre
Frankel
Beaudoin
Bernhard
Edgcomb
Dolan
Hoback
Clegg
Horikawa
Crawford

Biodata of **Dennis A. Bazylinski**, author of "*Magnetotactic Protists at the Oxic–Anoxic Transition Zones of Coastal Aquatic Environments*," with coauthors **Christopher T. Lefèvre** and **Richard B. Frankel**.

Dennis A. Bazylinski received his Ph.D. in Microbiology from the University of New Hampshire in 1984. He is currently Director of and a Professor in the School of Life Sciences at the University of Nevada at Las Vegas. He joined this Department after spending ten years in the Department of Microbiology at Iowa State University. His main research interests are in Microbial Geochemistry and Microbial Ecophysiology with a focus on Biomineralization. His organisms of study are the magnetotactic bacteria, prokaryotes that biomineralize intracellular magnetic crystals, which he has been working on for over 25 years after being introduced to them during his Ph.D. work.

E-mail: **dennis.bazylinski@unlv.edu**

Christopher T. Lefèvre received his Ph.D. in Marine Biology from the Centre d'Océanologie de Marseille, Université Aix-Marseille II, France in 2008. He joined Professor Bazylinski's lab as a Postdoctoral Associate in 2008. He recently started a second Postdoctoral at the Comissariat à l'Energie Atomique of Cadarache in France. His research interest is the ecophysiology and evolution of the magnetotactic bacteria.

E-mail: **lefevrechristopher@hotmail.com**

Dennis A. Bazylinski   **Christopher T. Lefèvre**

Richard B. Frankel received his Ph.D. from the University of California, Berkeley in 1965. He is currently Emeritus Professor of Physics at the California Polytechnic State University, San Luis Obispo. He came to Cal Poly in 1988 after 23 years at the F. Bitter National Magnet Laboratory, MIT. His main research interests are in biophysics of magnetotaxis and biomineralization of magnetic iron minerals. He started working on magnetotactic bacteria in 1978.

E-mail: **rfrankel@calpoly.edu**

MAGNETOTACTIC PROTISTS AT THE OXIC–ANOXIC TRANSITION ZONES OF COASTAL AQUATIC ENVIRONMENTS

DENNIS A. BAZYLINSKI[1], CHRISTOPHER T. LEFÈVRE[1], AND RICHARD B. FRANKEL[2]

[1]School of Life Sciences, University of Nevada at Las Vegas, 4505 Maryland Parkway, Las Vegas, NV 89154-4004, USA
[2]Department of Physics, California Polytechnic State University, San Luis Obispo, CA 93407, USA

1. Introduction

Magnetotactic bacteria are a diverse group of motile prokaryotes that biomineralize intracellular, membrane-bounded, tens-of-nanometer-sized crystals of a magnetic mineral, either magnetite (Fe_3O_4) or greigite (Fe_3S_4) (Bazylinski and Frankel 2004). These structures, called magnetosomes, cause cells to align along the Earth's geomagnetic field lines as they swim, a trait called magnetotaxis (Frankel et al. 1997). Magnetotactic bacteria are known to mainly inhabit the oxic–anoxic transition zone (OATZ) of aquatic habitats (Bazylinski and Frankel 2004), and it is currently thought that the magnetosomes function as a means of making chemotaxis more efficient in locating and maintaining an optimal position for growth and survival at the OATZ (Frankel et al. 1997). In addition to magnetotactic bacteria, there have been a few reports of the presence of magnetotactic protists in similar environments (Bazylinski et al. 2000). Unfortunately, this discovery raised more questions than it answered! For example, what is the origin of the magnetic crystals in these organisms and do they consist of the same magnetic minerals as the magnetotactic bacteria? Do these organisms contribute to iron cycling in their respective habitats? etc.

While the magnetotactic bacteria have been relatively well-studied as to their phylogeny, ecology, physiology, and genetics (Bazylinski and Frankel 2004), little to nothing is known regarding magnetotactic protists. The purpose of this chapter is to present what is known about magnetotactic protists and to discuss some of the questions and problems regarding these organisms that remain to be addressed in future studies.

2. Discovery of Magnetotactic Protists

The first magnetotactic protist, a euglenoid alga, was discovered in brackish mud and water samples collected from a coastal mangrove swamp near Fortaleza, Brazil (Fig. 1) (Torres de Araujo et al. 1985). This organism was tentatively

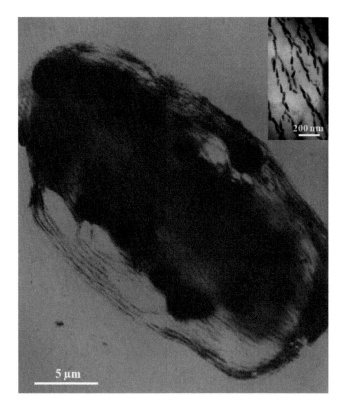

Figure 1. Transmission electron micrograph (TEM) image of a cell of the magnetotactic euglenoid alga described by Torres de Araujo et al. (1985). The parallel "lines" that traverse the cell along its long axis represent chains of bullet-shaped magnetite magnetosomes. *Upper inset* shows higher magnification electron micrograph image of the magnetosomes.

identified as *Anisonema platysomum* and contained numerous chains of bullet-shaped magnetite crystals ranging from 80 to 180 nm long by 40 to 50 nm wide. Chemical concentrations of the environment where this organism was located were not determined, but it seems likely from the presence of magnetotactic coccoid and spirillar bacteria that conditions were microaerobic/microoxic.

While studying magnetotactic bacteria populations related to the chemistry and magnetic properties of the water column of a chemically stratified coastal salt pond, Bazylinski et al. (2000) and later Simmons and Edwards (2007) reported the presence of numerous, different magnetotactic protists whose presence in the water column appeared to be correlated with depth and specific chemical parameters.

3. Magnetotactic Protists at Salt Pond

Salt Pond is a semi-anoxic eutrophic marine basin in Woods Hole (MA, USA) that becomes chemically stratified during the summer and fall (Wakeham et al. 1984, 1987). It has a maximum depth of about 5.5 m and receives significant freshwater input resulting in a well-defined pycnocline (density gradient), which helps to stabilize the chemical gradients (Fig. 2).

Hydrogen sulfide, resulting from the activity of sulfate-reducing bacteria in the anoxic hypolimnion, diffuses toward the surface causing the formation of steep opposing gradients of hydrogen sulfide and oxygen diffusing from the surface. The OATZ rises to within 2–3.5 m of the surface in summer. Dense populations of prokaryotes exist in the OATZ in Salt Pond, including magnetotactic and purple photosynthetic bacteria, the latter of which cause the water to appear pink when collected.

3.1. TYPES OF MAGNETOTACTIC PROTISTS

Magnetotactic protistan types observed in Salt Pond included dinoflagellates, biflagellates, and one ciliate (Fig. 3). The sizes of these organisms were typical of eukaryotes rather than prokaryotes and ranged from 3.8–9.3 to 15–28 μm.

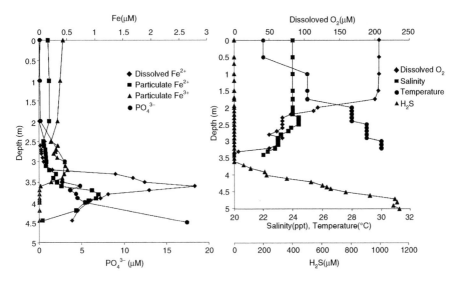

Figure 2. Chemical concentration profiles versus depth in Salt Pond determined during July 1995 in the summer when this semi-anaerobic basin becomes chemically stratified. The OATZ occurred from about 3.3 to 3.7 m. Magnetite-producing magnetotactic bacteria are generally found at the OATZ proper, while greigite-producing types are found below the OATZ in the anaerobic sulfidic zone. The distribution of magnetotactic protists is shown in Table 1 with depths relating to this figure.

Table 1. Magnetotactic protists present in Salt Pond water column samples collected in July 1995[a].

Protist	Type	Tentative identification	Depth distribution (m)	Zone
A[b]	Dinoflagellate	Dinoflagellate	3.3–3.6	OATZ to anoxic
B	Biflagellate	Cryptomonad	3.5–3.7	Anoxic
C	Biflagellate	Cryptomonad	3.7–4.1	Anoxic
D	Ciliate	*Cyclidium* species	3.8–3.9	Anoxic

[a]Data from Bazylinski et al. (2000)
[b]Letter denotes protists shown in Fig. 3

All types had clearly identifiable eukaryotic cell structures, including nuclei, flagella, and/or cilia.

Magnetotactic protists were associated with specific depths in Salt Pond, as shown in Table 1, and, in general, most types were more abundant in the anoxic zone. It is presently unclear whether they follow magnetotactic or other bacteria to graze on or whether they require specific chemical and/or anoxic conditions for growth and survival.

Apparently, the types of magnetotactic protists present in Salt Pond change, as Simmons and Edwards (2007) show images of three types, two of which are clearly different in morphology than those described here. What controls the types of magnetotactic protists present in Salt Pond is not currently known but may be related to changes or disruptions in water column chemical concentration profiles which have been shown to occur (Moskowitz et al. 2008; Bazylinski et al. unpublished data).

3.2. BEHAVIOR OF MAGNETOTACTIC PROTISTS

Several different types of magnetotactic protists were found in Salt Pond, all of which displayed magnetotaxis but differed in how they swam. Like magnetotactic bacteria, they migrated and accumulated at the edge of a hanging water droplet in a magnetic field (Fig. 4). A dinoflagellate (Fig. 3a) and a specific biflagellate (Fig. 3c) migrated more or less directly along magnetic field lines and responded to a reversal of the magnetic field by rotating 180° and continued to swim in the same direction relative to the field direction. In contrast, another biflagellate (Fig. 3b) did not migrate directly along the magnetic field lines; instead rather it swam with frequent, spontaneous changes in direction, only gradually migrating along magnetic field lines. The ciliate shown in Fig. 3d exhibited the most unusual behavior. It frequently attached to the microscope slide and/or cover slip, making rapid excursions before attaching again. Cells that apparently died and became free-floating responded to a reversal of the magnetic field, as described for the dinoflagellate. Unlike magnetotactic bacteria, it took the protists a significantly

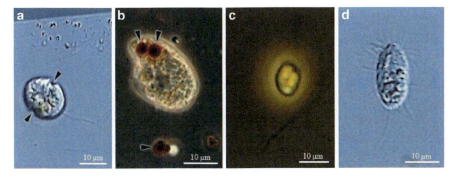

Figure 3. Light micrographs of four types of magnetotactic protozoa. (**a**) Differential interference contrast (DIC) micrograph of a dinoflagellate. *Arrows* denote the girdle or annulus surrounding the test typical of the dinoflagellate group. One flagellum has separated from the cell. (**b**) Phase-contrast (PC) micrograph of a biflagellate. Neither flagellum is visible. *Arrows* denote structures that are sometimes extruded from the cell and that orient in a magnetic field. (**c**) PC micrograph of another biflagellate showing the two flagella typical of the group. (**d**) DIC micrograph of a ciliate probably belonging to the genus *Cyclidium* (Figure adapted from Bazylinski et al. 2000).

longer period of time to accumulate near the edge of the drop and, in addition, they never actually reached the very edge of the drop (Fig. 4).

In the original report on these organisms (Bazylinski et al. 2000), they were described as magnetic (meaning that they are magnetically responsive) rather than magnetotactic. "Magnetotactic" may be more appropriate as all those described from natural environments showed a polar preference in their swimming direction like polar magnetotactic bacteria (Frankel et al. 1997; Simmons and Edwards 2007).

3.3. "MAGNETOSOMES" IN MAGNETOTACTIC PROTISTS?

Electron microscopy in conjunction with selected area electron diffraction and energy dispersive x-ray analyses was consistent with the presence of magnetite crystals in cells of the dinoflagellate and the biflagellate shown in Fig. 3a–c (Bazylinski et al. 2000) (Fig. 5). Although crystals were similar in size to (about 55–75 nm in diameter) and morphologically resembled those of magnetotactic bacteria, precise crystal morphologies could not be determined due to the thickness of the cells. Interestingly, despite the fact that greigite is a magnetic component in a number of magnetotactic bacteria in Salt Pond, it was never identified in any protistan cell. It is possible that this mineral dissolves rapidly in protistan cells, as discussed in Sect. 4.

Whether the magnetite crystals in these magnetotactic protists can be considered as "magnetosomes" depends on whether they are enclosed in a lipid bilayer membrane as they are in their bacterial counterparts. This is presently

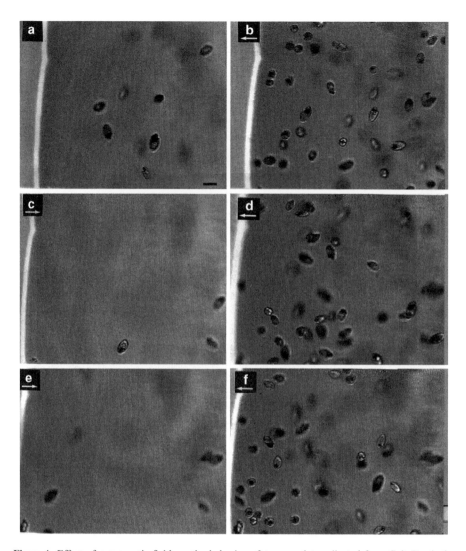

Figure 4. Effect of a magnetic field on the behavior of two protists collected from Salt Pond, the dinoflagellate shown in Fig. 3a and the biflagellate shown in Fig. 3b. The microscope was focused on a point at the edge of the water drop closest to the south pole of a bar magnet. (**a**) shows the organisms in that region before the magnet was placed on the microscope stage. (**b**) shows the effect of the magnet with the south pole closest to the drop, producing a local field direction indicated by the *arrow* in the *upper left corner*. (**c**) shows the effect of reversing the bar magnet so that the north magnetic pole is closest to the edge of the drop. (**d**) shows the reversal of the magnet again, with the same orientation as in (**b**). (**e**) and (**f**) are a repeat of (**c**) and (**d**). Images were taken approximately 45–60 s after the field was reversed. *Bar in panel A* represents 20 µm.

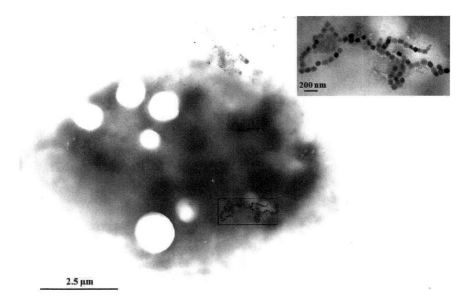

Figure 5. TEM image of the biflagellate shown in Fig. 3c. Dark, electron-dense structures in chains are the mineral crystals identified as magnetite. *Inset* depicts a high-magnification TEM of the magnetite crystals that are bracketed in the image of the whole cell. Note that the magnetite crystals are not organized as they are in the organism shown in Fig. 1.

unknown. Nonetheless, the presence of these crystals explains the magnetotactic behavior of these organisms.

4. Origin of Magnetite in Magnetotactic Protists

An important question that needs to be answered deals with the origin of the putative "magnetosomes" in magnetotactic protists. Two possibilities have been raised: (1) Do the protists biomineralize the magnetite crystals themselves? and (2) Do the protists ingest magnetotactic bacteria and/or bacterial magnetosomes from lysed cells and incorporate them either temporarily or permanently in the cell? Most researchers feel that both possibilities occur in nature and that what occurs is species dependent despite the fact that there is only direct evidence of the second possibility.

Because the arrangement of magnetosomes appears to be so precisely structured in the euglenoid alga described by Torres de Araujo et al. (1985) (Fig. 1), it seems unlikely that this arrangement could occur after the ingestion of what would have to be significant numbers of magnetotactic bacteria. Instead, it seems more likely that this organism biomineralizes and arranges endogenous magnetite crystals in a highly controlled fashion within the cell where intracellular structural elements play a significant role in magnetosome, as has been shown for magnetotactic prokaryotes (Komeili et al. 2006; Scheffel et al. 2006).

On the other hand, other protists seem to have magnetite crystals that might be in partial chains but not very organized within the cell (Fig. 5). One biflagellate was observed to extrude magnetic, dark-orange, roughly spherical inclusions (Fig. 3b) after which the cell was no longer magnetically responsive. It is possible that these inclusions represent indigestible remains of ingested magnetotactic bacteria in vacuoles. In the study of Bazylinski et al. (2000), none of the magnetic protistan cells were observed to be engulfing significant amounts of magnetotactic bacteria although the bacteria were abundant at the same depths in the water column and present in the water droplets examined. However, Simmons and Edwards (2007) observed direct feeding of protists on magnetotactic bacteria and subsequent egestion of magnetosomes.

Unfortunately, to our knowledge, there is only one published laboratory study on the ingestion of magnetotactic bacteria by a protozoan (Martins et al. 2007). In this study, the filter-feeding ciliate *Euplotes vannus* (*E. vannus*) was fed units of the multicellular magnetotactic prokaryote *Candidatus* Magnetoglobus multicellularis (*Ca.* M. multicellularis). However, cells of *E. vannus* did not respond to a magnetic field after confirmed ingestion of *Ca.* M. multicellularis. The reason for this is unclear although this organism biomineralizes greigite rather than magnetite in its magnetosomes, and it was shown that most of the greigite crystals within ingested cells were dissolved within 30–120 min of when the bacteria were present in vacuoles.

5. Role of Magnetotactic Protists in Iron Cycling

Simmons and Edwards (2007) observed up to $2.9 \pm 0.6 \times 10^3$ magnetotactic protists per ml in the OATZ of Salt Pond that, as shown here and previously (Bazylinski et al. 2000), can contain a large number of magnetosomes. Thus, these organisms clearly have a great potential for iron cycling in aquatic environments like Salt Pond. Iron is well recognized as a limiting factor in primary production in some oceanic environments and is often present in seawater in particulate and colloidal forms (Barbeau et al. 1996). Barbeau et al. (1996) and later Pernthaler (2005) showed that digestion of colloidal iron in the food vacuoles of protozoans during grazing of particulate and colloidal matter might generate more bioavailable iron for other species, such as phytoplankton. The observations of Simmons and Edwards (2007) together with the work of Martins et al. (2007) suggest that protists that ingest magnetotactic bacteria could play an important role in iron cycling by solubilizing iron in magnetosomes. If this is true of those that ingest magnetite-producing magnetotactic bacteria in habitats like Salt Pond, this would contribute to the high ferrous iron concentration at the OATZ and the high microbial concentrations present there. A key point here is whether magnetite is dissolved either partially or completely in the acidic environment of the digestive vacuoles (Ramoino et al. 1996) of the protists, as has been shown for greigite (Martins et al. 2007). Alternatively, magnetotactic protists that take up soluble forms of iron and biomineralize magnetic mineral crystals might do the opposite, making potentially significant amounts of iron unavailable to other organisms.

6. Future Research Directions

As stated early in this brief chapter, the interesting discovery of magnetotactic protists raises many important questions that should be addressed in future work. These questions range from the simple to the complex. The protists obviously need to be precisely identified. Are magnetotactic protists distributed widely in aquatic environments? It is sometimes difficult to determine the presence of magnetotactic protists because they are overlooked when samples contain large numbers of magnetotactic bacteria. In addition, it takes much more time for them to swim and accumulate at the edge of a water drop than magnetotactic bacteria. However, we have observed: magnetotactic protists in samples collected from the chemically stratified Pettaquamscutt Estuary (Donaghay et al. 1992); magnetotactic Gymnodinioid dinoflagellates in samples collected from salt marsh pools at the Ebro Delta, Spain; and magnetotactic biflagellates similar to that shown in Fig. 3b in samples collected from the Mediterranean Sea in Marseille, France [Lefèvre and Wu unpublished data]. Do the magnetite crystals in protists have a magnetosome membrane? Do any of the magnetotactic protists biomineralize their own magnetosomes? Axenic cultures of these organisms would certainly help in this regard. Clearly, more magnetotactic bacteria grazing experiments involving different types of protists are necessary to understand how these organisms behave after ingestion of the bacteria. These and more environmental studies would help to ascertain the role and estimate the impact of magnetotactic protists in iron cycling in natural habitats. Regardless of the origin of the magnetite crystals, the protists described here exhibit a magnetotactic response albeit a seemingly weak one. Does their magnetic dipole moment help them in any way as it is thought to do for the magnetotactic bacteria in locating and maintaining position at the OATZ? Many types of protists appear to prefer microoxic conditions and are known to be distributed in the OATZ (Fenchel 1969; Fenchel and Finlay 1984; Fenchel et al. 1989). Some appear to use geotactic mechanisms that involve mineral mechanoreceptors containing barium (Finlay et al. 1983) and strontium (Rieder et al. 1982), and aerotaxis to locate and maintain an optimal position in vertical oxygen concentration gradients (Fenchel and Finlay 1984).

What is the link between a specific protist and the chemistry in stratified aquatic environments? Many were found in the anoxic zone below the OATZ. Are these organisms following specific types of magnetotactic bacteria for grazing or are the chemical conditions necessary for their growth and survival? Do the protists possess any type of anaerobic metabolism? We hope this chapter and the studies cited herein stimulate the research studies necessary to answer these intriguing questions.

7. Acknowledgments

We are grateful to K.J. Edwards, S.S. Epstein, M. Pósfai, and S.L. Simmons for their collaboration in this work. DAB and CTL are supported by U.S. National Science Foundation grant EAR-0920718.

8. References

Barbeau K, Moffett JW, Caron DA, Croot PL, Erdner DL (1996) Role of protozoan grazing in relieving iron limitation of phytoplankton. Nature 380:61–64

Bazylinski DA, Frankel RB (2004) Magnetosome formation in prokaryotes. Nat Rev Microbiol 2:217–230

Bazylinski DA, Schlezinger DR, Howes BL, Frankel RB, Epstein SS (2000) Occurrence and distribution of diverse populations of magnetic protists in a chemically stratified coastal salt pond. Chem Geol 169:319–328

Donaghay PL, Rines HM, Sieburth JM (1992) Simultaneous sampling of fine scale biological, chemical and physical structure in stratified waters. Arch Hydrobiol Beih Ergeb Limnol 36:97–108

Fenchel T (1969) The ecology of marine microbenthos: IV. Structure and function of the benthic ecosystem, its chemical and physical factors and the microfauna communities with special reference to the ciliated protozoa. Ophelia 6:1–182

Fenchel T, Finlay BJ (1984) Geotaxis in the ciliated protozoan, Loxodes. J Exp Biol 110:17–33

Fenchel T, Finlay BJ, Gianni A (1989) Microaerophily in ciliates: responses of an *Euplotes* species (Hypotrichida) to oxygen tension. Arch Protistenkd 137:317–330

Finlay BJ, Hetherington NB, Davison W (1983) Active biological participation in lacustrine barium chemistry. Geochim Cosmochim Acta 47:1325–1329

Frankel RB, Bazylinski DA, Johnson MS, Taylor BL (1997) Magneto-aerotaxis in marine coccoid bacteria. Biophys J 73:994–1000

Komeili A, Li Z, Newman DK, Jensen GJ (2006) Magnetosomes are cell membrane invaginations organized by the actin-like protein MamK. Science 311:242–245

Martins JL, Silveira TS, Abreu F, Silva KT, da Silva-Neto ID, Lins U (2007) Grazing protozoa and magnetosome dissolution in magnetotactic bacteria. Environ Microbiol 9:2775–2781

Moskowitz BM, Bazylinski DA, Egli R, Frankel RB, Edwards KJ (2008) Magnetic properties of marine magnetotactic bacteria in a seasonally stratified coastal salt pond (Salt Pond, MA, USA). Geophys J Int 174:75–92

Pernthaler J (2005) Predation on prokaryotes in the water column and its ecological implications. Nat Rev Microbiol 3:537–546

Ramoino P, Beltrame F, Diaspro A, Fato M (1996) Time-variant analysis of organelle and vesicle movement during phagocytosis in *Paramecium primaurelia* by means of fluorescence confocal laser scanning microscopy. Microsc Res Tech 35:377–384

Rieder N, Ott HA, Pfundstein P, Schoch R (1982) X-ray microanalysis of the mineral contents of some protozoa. J Protozool 29:15–18

Scheffel A, Gruska M, Faivre D, Linaroudis A, Plitzko JM, Schüler D (2006) An acidic protein aligns magnetosomes along a filamentous structure in magnetotactic bacteria. Nature 440:110–114

Simmons SL, Edwards KJ (2007) Geobiology of magnetotactic bacteria. In: Schüler D (ed) Magnetoreception and magnetosomes in bacteria. Springer, Heidelberg, pp 77–102

Torres de Araujo FF, Pires MA, Frankel RB, Bicudo CEM (1985) Magnetite and magnetotaxis in algae. Biophys J 50:375–378

Wakeham SG, Howes BL, Dacey JWH (1984) Dimethyl sulphide in a stratified coastal salt pond. Nature 310:770–772

Wakeham SG, Howes BL, Dacey JWH, Schwarzenbach RP, Zeyer J (1987) Biogeochemistry of dimethylsulfide in a seasonally stratified coastal salt pond. Geochim Cosmochim Acta 51:1675–1684

Biodata of **David J. Beaudoin**, **Joan M. Bernhard**, and **Virginia P. Edgcomb**, authors of the chapter "*A Novel Ciliate (Ciliophora: Hypotrichida) Isolated from Bathyal Anoxic Sediments.*"

Mr. David J. Beaudoin is a Research Associate in the Biology Department of the Woods Hole Oceanographic Institution (Woods Hole, Massachusetts, USA). He received his B.S. in 1993 at the University of Rhode Island and his M.S. in 1998 at the University of Maine. He is a molecular ecologist studying microbial biodiversity.

E-mail: **dbeaudoin@whoi.edu**

Dr. Joan M. Bernhard, who is a Senior Scientist at Woods Hole Oceanographic Institution, is a biogeochemist with a major focus on the adaptations and ecology of protists living in the chemocline. Her work is largely in the bathyal to abyssal deep sea but also in high latitudes. Bernhard has degrees in geology and biological oceanography, and did postdoctoral work in cell biology. Her multidisciplinary training gives her a unique perspective into anoxic habitats. For more information, see Biodata of the Editors.

E-mail: **jbernhard@whoi.edu**

David J. Beaudoin   **Joan M. Bernhard**

Dr. Virginia P. Edgcomb is a Research Specialist in the Department of Geology and Geophysics at the Woods Hole Oceanographic Institution (WHOI) (Woods Hole, MA). She received her Ph.D. from the University of Delaware (Department of Biology) in 1997. As postdoctoral researcher, she spent 3 years at the Marine Biological Laboratory where she was involved in studies of early eukaryotic evolution and of microbial diversity at hydrothermal vents, followed by 2 years at WHOI where she studied the tolerance of several marine prokaryotes to extreme conditions found at hydrothermal vents. Her current research interests include the diversity and evolution of protists, the microbial ecology of microoxic and anoxic marine environments, and symbioses between protists and prokaryotes in extreme environments, including hypersaline anoxic basins, anoxic and sulfidic marine water column and sedimentary environments, and subsurface marine sediments.

E-mail: **vedgcomb@whoi.edu**

A NOVEL CILIATE (CILIOPHORA: HYPOTRICHIDA) ISOLATED FROM BATHYAL ANOXIC SEDIMENTS

DAVID J. BEAUDOIN[1], JOAN M. BERNHARD[2], AND VIRGINIA P. EDGCOMB[2]

[1]*Biology Department, Woods Hole Oceanographic Institution, Woods Hole, MA 02543, USA*
[2]*Geology and Geophysics Department, Woods Hole Oceanographic Institution, Woods Hole, MA 02543, USA*

1. Introduction

Oxygen-depleted to anoxic regions of marine environments similar to those found in the Cariaco Trench and the deep Black Sea occur globally and have likely persisted throughout the Earth's history. Such anaerobic environments have no doubt played an important role in the early formation and evolution of the known biosphere. It is likely that eukaryotes originated prior to the formation of an oxygenated atmosphere (Fenchel and Finlay 1995) and that similar extant anoxic environments may still harbor unknown ancestral lineages. Recent rRNA gene-based surveys of modern anoxic habitats have uncovered diverse communities of anaerobic and aerotolerant organisms that include many novel microscopic eukaryotes (e.g., Dawson and Pace 2002; Stoeck et al. 2003).

The Santa Barbara Basin (SBB) is an oxygen-depleted silled borderland basin situated between the northern Channel Islands and the Southern California (USA) mainland. Unlike many other deep sea environments, the SBB is largely devoid of larger macrofauna, resulting in relatively low sedimentary bioturbation rates. The relatively undisturbed state of the central basin (maximum depth ~600 m) combined with high sedimentation rates and nearly anoxic bottom waters ([O_2] up to ~5 µM; Reimers et al. 1996) results in varved sediments that have been extensively studied, particularly for paleoclimate and paleoceanographic reconstructions (e.g., Kennett and Ingram 1995).

Symbioses between prokaryotes and protists are well known in the deep sea at oxic–anoxic boundaries (e.g., Fenchel and Finlay 1995; Stewart et al. 2005). Previous microscopic evaluations of the SBB eukaryotic benthic community revealed many different taxa including ciliates, flagellates, foraminifera, and metazoa, most of which were associated with prokaryotic symbionts (Bernhard et al. 2000, 2010; Edgcomb et al. 2011). Of the 15 different ciliate morphotypes examined from SBB sediments, including *Metopus* and *Parablepharisma* sp., 13 were observed to harbor prokaryotic partners (Bernhard et al. 2000).

It is inferred that both partners benefit metabolically from these symbioses via chemosynthesis in the prokaryotic partner. Here, using culture-independent rRNA-targeted methods, we present data on the taxonomic position of a ciliate living in the anoxic upper sediments of the SBB. We also describe our initial efforts to phylogenetically characterize the putative prokaryotic symbionts observed on the ciliate exterior.

2. Materials and Methods

Sediments for this study were collected from a 585-m deep site in the SBB (34°17.5′N, 120°02.0′W) using a Soutar box corer in September 2007 aboard *RV Robert Gordon Sproul*. Bulk samples were taken from the upper ~1–2 cm of undisturbed box core casts and placed in high-density polyethylene bottles. Bottles were topped off with chilled bottom water, and live samples were maintained in the lab at in situ temperature (~7°C) for 2–3 months.

Individual ciliates were handpicked from gently sieved (63-μm) sediments with the aid of a dissecting microscope. Ciliates were determined to be alive if they were observed to be actively swimming after sieving. Samples were picked into chilled seawater and then rinsed three times in sterile seawater to reduce the likelihood of contaminating microorganisms stuck to the outside of the ciliate. Individuals were then either frozen at −80°C for later nucleic acid extraction or observed with DIC light microscopy. Light micrographs of live SBB ciliates were taken using a Zeiss Axioplan 2 imaging microscope equipped with a Zeiss AxioCam camera.

Twenty individually rinsed ciliates were combined together for DNA extraction. Cells were centrifuged briefly to remove residual seawater, and DNA was extracted using the DNeasy Plant Mini Kit (Qiagen) according to the manufacturer's instructions. Following DNA extraction, DNA was further purified and concentrated by ethanol precipitation. We attempted to amplify 18S rRNA gene fragments from ciliate DNA extracts using several primer sets targeting eukaryotes. A positive result was achieved with the 360F (Medlin et al. 1988) and U1492R (Longnecker and Reysenbach 2001) primer sets. PCR conditions were: 95°C for 5 min, followed by 40 cycles of 95°C for 1 min, 45°C for 1 min, and 72°C for 90 s, with a final incubation of 72°C for 10 min. Additionally, in an attempt to determine the origin of the putative prokaryotic symbiont, bacterial and archaeal PCR amplifications were run. The bacterial primer sets 515F (Lane 1991)/U1492R and archaeal primer 1100F (Reysenbach and Pace 1994)/U1492R both resulted in positive amplifications. Touchdown PCR conditions for prokaryote amplifications were: 10 cycles of 95°C for 1 min, 50°C for 1 min (−0.5°C/cycle), and 72°C for 90 s, followed by 30 cycles of 95°C for 1 min, 45°C for 1 min, and 72°C for 90 s.

PCR products were visualized by agarose gel electrophoresis and later excised and purified from the gels using the Qiaquick Gel Extraction Kit (Qiagen). Gel-purified fragments were then used to construct clone libraries using the pCR4 vector in the TOPO TA cloning kit for sequencing (Invitrogen) according to the manufacturer's instructions. Plasmids were recovered from overnight cultures using a MWG Biotech RoboPrep2500 at the Josephine Bay Paul Center Keck Facility (Marine Biological Laboratory [MBL]). Selected clones were sequenced bidirectionally with the primers M13F, M13R, 570F, 570R, 1055F, and 1055R (Elwood et al. 1985) using an Applied Biosystems 3730XL capillary sequencer also at the MBL Keck facility. Sequences were then edited and assembled into contigs using Sequencher (Gene Codes Corporation). Chimeric sequences were removed from further analyses by visual inspection and the CHECK_CHIMERA program available through the Ribosomal Database Project (Maidak et al. 2001). The approximate phylogenetic position of the gene sequences was then determined by comparison with other database sequences via BLASTN (Zhang et al. 2000).

Clone sequences were aligned using ARB (Ludwig et al. 2004). Sequence alignments were manually refined in ARB using secondary structure information. The white ciliate sequences and representative sequences from the major groups within the phylum Ciliophora as well as the top BLAST hits to our sequences were exported from ARB for phylogenetic analysis. Only sites that were reliably aligned were included in a phylogenetic analysis under Maximum Likelihood inference using RAxML (Stamatakis et al. 2008). All phylogenetic analyses were performed on the CIPRES portal (www.phylo.org) under the GTR+I+Gamma model using 1,000 bootstrap replicates and estimation of the proportion of invariable sites.

Catalyzed Reporter Deposition FISH (CARD-FISH) was performed according to the methods of Pernthaler et al. (2002) with slight modifications described in Edgcomb et al. (2011). Briefly, individual cells were first handpicked and rinsed in sterile seawater and fixed in 2% (final concentration) paraformaldehyde for 1 h, then rinsed three times with 5 ml sterile phosphate-buffered saline (PBS) by filtration onto a 0.2-μm pore size, 25-mm Isopore GTTP filter (Millipore, USA). Prior to CARD-FISH, the filters were overlaid with 37°C 0.2% (w/v) Metaphor agarose and were dried at 50°C. The Alexa488-labeled probes used include DELTA495a, b, and c and the corresponding competitor probes for each, cDELTA495a, b, and c (Lucker et al. 2007).

3. Results and Discussion

Bottom-water oxygen concentrations where this ciliate was living in SBB are typically <2 μM (e.g., Bernhard et al. 1997; JMB unpublished data). Notably, this ciliate persisted in our enrichment bottles, most of which were sulfidic and presumably anoxic, for many months after collection.

The long, slender body of approximately 350×90 μm is white-colored with both ends rounded (Fig. 1a). A number of putative rod-shaped ectobionts can be seen interspersed with pellicle cilia (Fig. 1b). Additionally, frontal cirri are present (Fig. 1c) and locomotion is exhibited by slow crawling.

To further characterize this white ciliate, we PCR-amplified rRNA gene fragments (1,369 bp) using primers that target the genes of eukaryotes. Clone

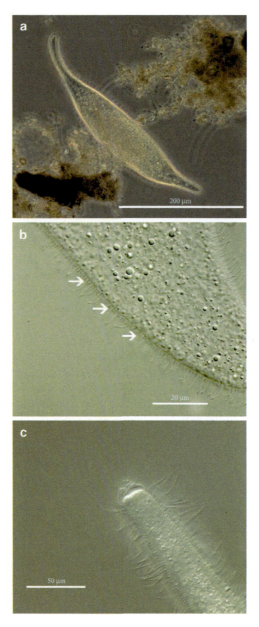

Figure 1. Photomicrographs of a typical white-colored ciliate specimen (**a**) showing the putative symbiotic prokaryotes (**b**) and frontal cirri (**c**).

sequencing resulted in numerous clones ($N=5$) corresponding to phylotypes of the ciliate order Hypotrichida (Levine et al. 1980), as determined by BLAST. No other ciliate phylotypes were detected in our analysis. Two clones were selected for bidirectional sequencing and were determined to share >99% sequence identity.

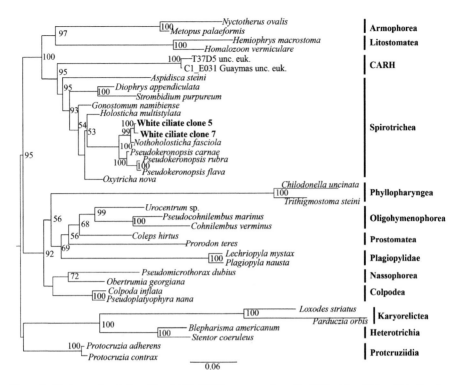

Figure 2. Phylogenetic relationship of 18S rDNA sequence from the SBB white ciliate within the phylum Ciliophora. The tree was constructed by maximum likelihood (RAxML) using an alignment of 925 unambiguous positions under the GTR+I+Gamma model of sequence evolution. Bootstrap values greater than 50% are shown at the nodes. Scale is given as substitutions/site.

These sequences are deposited in GenBank under accession numbers HQ259070 and HQ259071.

The phylogenetic analysis presented here (Fig. 2) demonstrates that our two SBB white ciliate sequences cluster together with a sequence from a marine ciliate isolated from coastal China, *Noviholosticha fasciola* (FJ377548.1; Yi et al. 2010). Other members of the ciliate family Pseudokeronopsidae, including *Pseudokeronopsis carna, P. rubra*, and *P. flava* (AY881633.1, DQ640314.1; DQ227798.1 Yi et al. 2008) are also highly similar to the SBB sequences as determined by BLAST and appear to form their own clade within the Spirotrichea. The rRNA evolutionary similarity between the *Noviholosticha* and *Pseudokeronopsis* genera ranges between 95.6% and 96.1% (Yi et al. 2010).

Ciliates are important components of the protistan community of many anoxic environments including freshwater lakes (Guhl et al. 1996; Finlay et al. 1996; Stoeck et al. 2007), wetland soils (Schwarz and Frenzel 2003), the rumen of herbivorous animals (Fenchel and Finlay 1995), and marine sediments (Dolan and Coats 1991; Hayward et al. 2003; Stoeck et al. 2003). In many cases, prokaryotes are associated with these ciliate hosts (Bernhard et al. 2000; Fenchel and Finlay 1995), suggesting the importance of symbiotic interactions to the survival of ciliates in oxygen-depleted environments.

Like many other protists from the SBB sediments, this ciliate appears to harbor prokaryotes, presumably symbiotic, on the outer pellicle. The exact nature of the ciliate/prokaryote relationship, like most others observed from this site, remains unknown. Other ciliates living in anaerobic habitats frequently harbor sulfate-reducing and sulfur-oxidizing bacteria as well as methanogenic Archaea as ecto- or endobionts (Görtz 2006).

Finally, we attempted to identify the symbionts observed on the ciliate pellicle by PCR amplification of 16S rDNA. Both bacterial and archaeal amplifications yielded positive results. The primary bacterial sequences that fit the observed morphology included a sulfate-reducing delta proteobacteria ($N=5$) of the *Desulfobulbaceae* isolated from deep tidal flat sediments (AM774317; Gittel et al. 2008) and an uncultured bacterium ($N=2$) from a cave sulfur-oxidizing biofilm (EF467539).

The major Archaeal sequence identified appears to be an uncultured relative of *Ferroplasma acidiphilum* ($N=4$), which does not have the observed morphology. The observed epibiont morphology makes it probable that this putative symbiont is a member of the delta proteobacteria. Preliminary CARD-FISH analyses ($N=4$ individuals) using the general probe to sulfate-reducing bacteria suggest that the epibiont of this ciliate is a sulfate-reducing bacterium (Fig. 3).

Future studies using additional specimens and more specific probes for taxonomic groups within the sulfate-reducing bacteria are needed to confirm

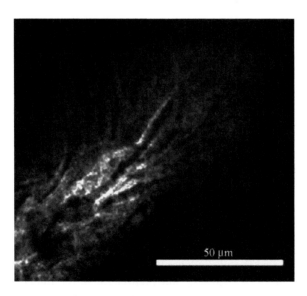

Figure 3. CARD-FISH image of the anterior half of a white ciliate cell showing a positive signal with a general probe to sulfate-reducing bacteria (DELTA495a, b, c).

this result prior to depositing the putative symbiont sequence into GenBank. Such studies will also begin to address the metabolic nature of the host–symbiont relationship.

4. Conclusion

In summary, we describe a hypotrich ciliate related to *Noviholosticha* isolated from sediments of the SBB. Phylogenetic analysis of 18S rDNA genes places this ciliate within the family *Pseudokeronopsidae*, a placement that is supported by morphological observations of living specimens. Although hypotrichous ciliates are found abundantly in freshwater and marine habitats, relatively few species are known to be obligate anaerobes (Fenchel and Finlay 1995). It is possible that the species described here was experiencing microoxic conditions at the time of capture or that it associates with another organism (i.e., ectobionts), enabling it to exist anaerobically. Further analyses are required to fully understand the ecology and physiology of this ectobiont-bearing hypotrich.

5. References

Bernhard JM, Sen Gupta BK, Borne PF (1997) Benthic foraminiferal proxy to estimate dysoxic bottom-water oxygen concentrations: Santa Barbara Basin, U.S. Pacific Continental Margin. J Foraminifer Res 27:301–310

Bernhard JM, Buck KR, Farmer MA, Bowser SS (2000) The Santa Barbara Basin is a symbiosis oasis. Nature 403:77–80

Bernhard JM, Goldstein ST, Bowser SS (2010) An ectobiont-bearing foraminiferan, *Bolivina pacifica*, that inhabits microxic pore waters: cell-biological and paleoceanographic insights. Environ Microbiol 12:2107–2119

Dawson SC, Pace NR (2002) Novel kingdom-level eukaryotic diversity in anoxic environments. Proc Natl Acad Sci USA 99:8324–8329

Dolan JR, Coats DW (1991) Changes in fine-scale vertical distributions of ciliate microzooplankton related to anoxia in Chesapeake Bay water. Mar Microb Food Webs 5:81–93

Edgcomb VE, Breglia SA, Yubuki N, Beaudoin DJ, Patterson DJ, Leander BS, Bernhard JM (2011) Identity of epibiotic bacteria on symbiontid euglenozoans in O_2-depleted marine sediments: evidence for symbiont and host co-evolution. ISME J 5:231–243. doi:10.1038/ismej.2010.121

Elwood HJ, Olsen GJ, Sogin ML (1985) The small-subunit ribosomal RNA gene sequences from the hypotrichous ciliates *Oxytricha nova* and *Stylonychia pustulata*. Mol Biol Evol 2:399–410

Fenchel T, Finlay BJ (1995) Ecology and evolution in anoxic worlds. Oxford University Press, New York

Finlay BJ, Maberly SC, Esteban GF (1996) Spectacular abundance of ciliates in anoxic pond water: contribution of symbiont photosynthesis to host respiratory oxygen requirements. FEMS Microbiol Ecol 20:229–235

Gittel A, Mussmann M, Sass H, Cypionka H, Konneke M (2008) Identity and abundance of active sulfate-reducing bacteria in deep tidal flat sediments determined by directed cultivation and CARD-FISH analysis. Environ Microbiol 10:2645–2658

Görtz H-D (2006) Symbiotic associations between ciliates and prokaryotes. Prokaryotes 1:364–402. doi:10.1007/0-387-30741-9_15

Guhl BE, Finlay BJ, Schink B (1996) Comparison of ciliate communities in the anoxic hypolimnia of three lakes: general features and the influence of lake characteristics. J Plankton Res 18:335–353

Hayward BH, Droste R, Epstein SS (2003) Interstitial ciliates: benthic microaerophiles or planktonic anaerobes? J Eukaryot Microbiol 50:356–359

Kennett JP, Ingram BL (1995) A 20,000-year record of ocean circulation and climate change from the Santa Barbara Basin. Nature 377:510–514

Lane DJ (1991) 16 S/23S rRNA sequencing. In: Stackebrandt E, Goodfellow M (eds) Nucleic acid techniques in bacterial systematics. Wiley, New York, pp 115–175

Levine ND, Corliss JO, Cox FEG, Deroux G et al (1980) A newly revised classification of the Protozoa. J Protozool 27:37–59

Longnecker K, Reysenbach A-L (2001) Expansion of the geographic distribution of a novel lineage of the epsilon Proteobacteria to a hydrothermal vent on the Southern east Pacific Rise. FEMS Microbiol Ecol 35:287–293

Lucker S, Steger D, Kjeldsen KU, MacGregor BJ, Wagner M, Loy A (2007) Improved 16 S rRNA-targeted probe set for analysis of sulfate-reducing bacteria by fluorescence in situ hybridization. J Microbiol Methods 69:523–528

Ludwig W, Strunk O, Westram R, Richter L, Meier H, Yadhukumar et al (2004) ARB: a software environment for sequence data. Nucleic Acids Res 32:1363–1371

Maidak BL, Cole JR, Lilburn TG, Parker CT Jr, Saxman PR, Farris RJ, Garrity GM, Olsen GJ, Schmidt TM, Tiedje JM (2001) The RDP-II (Ribosomal Database Project). Nucleic Acids Res 29:173–174

Medlin LK, Elwood HJ, Stickel S, Sogin ML (1988) The characterization of enzymatically amplified eukaryotic 16 S-like rRNA-coding regions. Gene 71:491–499

Pernthaler A, Pernthaler J, Amann R (2002) Fluorescence in situ hybridization and catalyzed reporter deposition for the identification of marine bacteria. Appl Environ Microbiol 68:3094–3101

Reimers CE, Ruttenberg KC, Canfield DE, Christiansen MB, Martin JB (1996) Porewater pH and authigenic phases formed in the uppermost sediments of the Santa Barbara Basin. Geochim Cosmochim Acta 60:4037–4057

Reysenbach A-L, Pace NR (1994) Reliable amplification of hyperthermophilic Archaeal 16 S rRNA genes by the polymerase chain reaction. In: Robb FT, Place AR (eds) Archaea: a laboratory manual: thermophiles. Cold Spring Harbor Laboratory Press, Plainview

Schwarz MV, Frenzel P (2003) Population dynamics and ecology of ciliates (Protozoa, Ciliophora) in an anoxic rice field soil. Biol Fertil Soils 38:245–252

Stamatakis A, Hoover P, Rougemont J (2008) A rapid bootstrap algorithm for the RAxML web servers. Syst Biol 57:758–771

Stewart FJ, Newton ILG, Cavanaugh CM (2005) Chemosynthetic endosymbioses: adaptations to oxic-anoxic interfaces. Trends Microbiol 13:439–448

Stoeck T, Taylor GT, Epstein SS (2003) Novel eukaryotes from a permanently anoxic deep-sea basin (Cariaco, Caribbean Sea). Appl Environ Microbiol 69:5656–5663

Stoeck T, Bruemmer F, Foissner W (2007) Evidence for local ciliate endemism in an alpine anoxic lake. Microb Ecol 54:478–486

Yi Z, Chen Z, Warren A, Roberts D, Al-Rasheid KAS, Miao M, Gao S, Shao C, Song W (2008) Molecular phylogeny of *Pseudokeronopsis* (Protozoa, Ciliophora, Urostylida), with reconsideration of three closely related species at inter- and intra-specific levels inferred from the small subunit ribosomal RNA gene and the ITS1-5.8 S-ITS2 region sequences. J Zool 275:268–275

Yi Z, Lin X, Warren A, Al-Rasheid KAS, Song W (2010) Molecular phylogeny of *Nothoholosticha* (Protozoa, Ciliophora, Urostylida) and systematic relationships of the *Holosticha*-complex. Syst Biodivers 8:149–155

Zhang Z, Schwartz S, Wagner L, Miller W (2000) A greedy algorithm for aligning DNA sequences. J Comp Biol 7:203–214

Biodata of **Michael F. Dolan**, author of *"The Wood-Eating Termite Hindgut: Diverse Cellular Symbioses in a Microoxic to Anoxic Environment."*

Dr. Michael F. Dolan is currently a postdoctoral researcher at the Marine Biological Laboratory, Woods Hole, Massachusetts, USA, and an adjunct professor and lecturer in the Department of Geosciences, University of Massachusetts, Amherst. He obtained his Ph.D. from the University of Massachusetts, Amherst in 1999. Dr. Dolan's scientific interests are in the areas of the taxonomy and cell biology of parabasalid protists, and in the history of protistology.

E-mail: **mdolan@geo.umass.edu**

THE WOOD-EATING TERMITE HINDGUT: DIVERSE CELLULAR SYMBIOSES IN A MICROOXIC TO ANOXIC ENVIRONMENT

MICHAEL F. DOLAN
Geosciences Department, University of Massachusetts, Amherst, MA 01003, USA

1. Introduction

The wood-eating termite hindgut is an unusual site of anoxia in that it is based on the breakdown of lignocelluloses, the fermentation of the sugar monomers, and the consumption of waste CO_2 and H_2 by chemolithoautotrophs. It evolved 140 million years ago, so it is a unique evolutionary experiment that is not homologous to the early anoxic environments where life originated or where eukaryotes evolved, presumably as a symbiogenetic fusion of archaeal and eubacterial lineages. But it is a well-known fact that eukaryotes evolved through symbiogenesis, at least in terms of the mitochondria and chloroplasts, that makes the anaerobic hindgut community so interesting. There may well be hundreds if not thousands of unique cases of ecto- and endosymbioses involving prokaryotes living on and in protists within the wood-eating termites' hindguts. The importance of these prokaryote associations is reflected in the names given the flagellates by early twentieth century protozoologists, such as *Devescovina striata* whose striations are ectosymbiotic bacteria or *Oxymonas pediculosa*, which is "lousy" with surface bacteria (Dolan 2001). With genomic approaches, those morphologically defined entities can be identified, and in some cases their complete genomes can be determined.

The guts of wood-eating termites are not relics of an ancient anoxic environment, but they host an extremely diverse array of microbial symbionts and cellular symbioses, perhaps the most diverse cases of symbiosis in nature, in an anoxic to low oxic environment (Breznak 1975). The hindguts of these insects are filled not only with the cellulose-digesting protists that are packed in by the thousands but also the numerous, often unique prokaryotes that live on their cell surfaces, in their cytoplasm, and in their nuclei. Further, diverse prokaryotes live unattached in the gut liquid. It is a specialized microbial community that is passed from one termite to another, from anus to mouth, as a way of inoculating the young or those that have recently molted and evacuated their gut contents.

This chapter will review the evolution of the wood-eating termites' gut, the microoxic to anoxic environment within the gut, the diversity of amitochondriate protists and prokaryotes that evolved there, the use of the cellular symbioses of prokaryotes living in and on the protists as analogs of early eukaryote evolution, recent discoveries in the field, particularly through genomics research, and the use of genomic vs. morphological data in studying this microbial community. This chapter only deals with findings from the families that contain the wood-eating termites (e.g., Kalotermitidae, Hodotermitidae, Archotermopsidae, and Rhinotermitidae), not the soil-feeding Termitidae.

The wood-eating termite's hindgut is a microcosm of anaerobic life walking around on six legs. The spectacle of thousands of protists flowing out of the burst walls of an excised hindgut, "reminding one of the turning out of a multitude of persons from the door of a crowded meeting house," as Joseph Leidy puts it in 1877, 1880, has amazed student and teacher alike for at least 130 years. It defies reason to see an organ, whose purpose is to digest food that flows through it, be jam-packed instead with a wreathing mass of protists and bacteria. But the big fleas have little fleas, bacteria that cover the protists or inhabit their cytoplasm. While this does not go on ad infinitum, the symbioses within symbioses present a complexity unlike any other microbial community.

2. Evolution of the Termite Hindgut

Termites and the related wood-feeding cockroach *Cryptocercus* are cryptic social or subsocial insects that transfer hindgut contents between individuals by feeding from the anus. Their lineage appeared by 140 million years ago and exploited an unoccupied niche – animals eating wood, spreading through tropical and temperate environments, and becoming the most ecologically important group of insects besides bees (Grimaldi and Engel 2005). Termites and *Cryptocercus* share a common ancestor that had the distinct anaerobic protist symbionts. The wood-feeding roach is now found in limited temperate environments. One of the major differences in its microbial community compared to that in termites is that the protists encyst and have a sexual stage that is triggered by the insect's molting (Cleveland 1947). The oldest family of termites, the Mastotermitidae, has only one species left, *M. darwiniensis* from northern Australia. The order and hypothesized dates of divergence of the termites after the Mastotermitidae are Hodotermitidae [145 mya], Archotermopsidae (including *Zootermopsis*) [135 mya], the drywood-eating Kalotermitidae (including *Cryptotermes* and *Incisitermes*) [125 mya], and the Rhinotermitidae (including *Coptotermes* and *Reticulitermes*) [115 mya] (Grimaldi and Engel 2005; Engel et al. 2009).

3. The Microoxic to Anoxic Gut Environment

The wood-eating termite's hindgut was assumed to be an anoxic chamber until measurements were taken with microelectrodes, and it was discovered that only the center of the gut contents was anoxic, while the periphery of the gut lumen, 50–200 μm in from the gut wall, is microoxic (see Brune and Friedrich 2000 for an excellent review of the hindgut environment). In *Reticulitermes flavipes*, whose

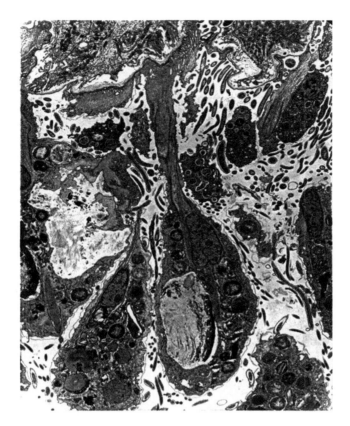

Figure 1. Microbiota on hindgut of the drywood-eating termite *Incisitermes minor* from Newbury Park, California, USA. The gut epithelial layer is at the *top*. A prominent *Oxymonas* sp. is seen in the center, attached to the gut wall. It is attached to the wall by a microtubule-based rostellum. It appears to be feeding both on wood and bacteria. A diversity of spirochetes is seen in cross section. This region of the hindgut is microoxic, while the center of the lumen is anoxic. Image is 50-μm wide (Transmission electron micrograph by David Chase).

hindgut is one of the most studied, only 40% of the hindgut contents, that in the center of the gut lumen, was anoxic, and it was determined that 21% of the insect's oxygen consumption was carried out by the gut wall and gut microbes (Brune et al. 1995). Some protists are attached to the gut wall (*Oxymonas*, *Pyrsonympha*), suggesting they may be more tolerant of oxygen (see Figs. 1 and 2).

In contrast to the gut's oxygen profile which has been measured at 5 kPa at the gut edge and at 0 in the lumen, H_2, which is produced in the fermentation of sugars, accumulates at partial pressure of 5 kPa in the lumen and 0 near the gut wall (Ebert and Brune 1997). This H_2 is consumed by methanogens and acetogens. It is thought that the termites feed on acetate that flows across the gut epithelium (Hungate 1955).

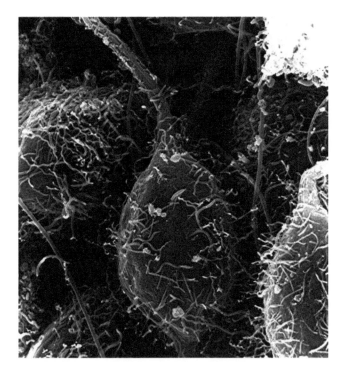

Figure 2. Microbiota attached to hindgut of *Pterotermes occidentis* from Arizona, USA. This is a comparable scene as in Fig. 1, as seen with the scanning electron microscope. Note the attached *Oxymonas* and the numerous rods, spirochetes, and other bacteria. Width of the image is 45 µm (Scanning electron micrograph by David Chase).

4. Diversity of Organisms in the Gut

The termite hindgut microbial community is based on fixed carbon input in the lignocellulose of wood whose sugar monomers are fermented, with chemolithoautotrophs consuming the waste CO_2 and H_2. Its inhabitants are a menagerie of protists including *Trichonympha* and *Pyrsonympha*, an oxymonad with strange intracellular motility, and many gigantic and multinucleate relatives of *Trichomonas*. This microcosm teams with exotic spirochetes, including those that are acetogenic or fix nitrogen. There are many more groups of prokaryotes found in the gut, particularly from the Bacteroidales. The ecto- and endosymbionts will be discussed in the section on recent discoveries.

4.1. PROTISTS

The anaerobic protists are of two main groups – the Parabasalia, which as the name suggests have a parabasal body (a densely packed Golgi complex) attached to its basal bodies or kinetosomes, and the Oxymonadida of the phylum Preaxostyla, some of which are not motile, but attach to the gut wall (Fig. 1), and others with striking internal microtubule motility. Traditionally, they are known as flagellates and are not related to the ciliates, for example of the cattle rumen. The parabasalids are relatives of the common parasitic protist *Trichomonas vaginalis*, but in many cases, they have become relatively gigantic, perhaps due to selection pressure to develop more flagella to swim faster and to have larger cells so that they may ingest larger pieces of wood. The parabasalids have historically been put into two groups: the trichomonads, which have their nuclei and flagella arranged into a single organellar system called the karyomastigont; and the hypermastigotes, like *Trichonympha*, that have numerous flagella arranged in elaborate patterns on their surface and a large central nucleus unattached to the flagella. The new order Cristamonadida was recently erected and includes several families taken from either the trichomonads or hypermastigotes (Devescovinidae, Calonymphidae, and Joenidae), mostly termite gut flagellates (Brugerolle and Patterson 2001). However, the parabasalids have recently been reclassified as it was recognized that the older morphology-based taxonomies were not supported by molecular phylogenetic analysis (Cepicka et al. 2010). These three families were suppressed, and all cristamonads were placed into the single family Lophomonadidae. The oxymonads are widespread in termite guts in the genus *Oxymonas*. Another oxymonad genus, *Pyrsonympha*, which exists exclusively in the subterranean termite genus *Reticulitermes*, was one of the first gut protists described by Joseph Leidy in 1877.

4.2. SPIROCHETES

Prior to the development of molecular phylogenetics and fluorescent in situ hybridization, the diversity of spirochetes in the termite gut was recognized by ultrastructural studies of spirochete cross sections (Gharagozlou 1968; Hollande and Gharagozlou 1967). Several families and genera of spirochetes were erected in this way (Bermudes et al. 1988). Because no ssuDNA sequences were presented for these organisms, they were considered "bacteria of unknown affiliation," although they were clearly spirochetes (Margulis and Hinkle 1991). A "termite cluster" of *Treponema* was established that was distinct from all previously known treponemes (Lilburn et al. 1999). As molecular phylogenetic data were accumulated, the termite gut spirochetes all nestled with *Treponema*, so the diversity of termite gut spirochetes are now listed as *Treponema* clusters I, II, and III. Unfortunately, 20 years later, the gap between the morphological descriptions and the gene sequence data has not been bridged.

Molecular phylogenetic analysis found a great diversity of spirochetes in termites, with each species of insect commonly having 10–20 phylotypes of these bacteria (Berlanga et al. 2007). It is common for a gut-dwelling protist to harbor multiple lineages of spirochetes, for example, *Holomastigotoides mirabile* in *Coptotermes formosanus* (Inoue et al. 2008). The first spirochete cultured from the wood-eating termite gut, *Treponema primitia*, is capable of acetogenic growth (Graber and Breznak 2004). Acetogens generate 20–30% of the hindgut acetate (Breznak and Switzer 1986). Molecular phylogenetic analysis suggests that the spirochetes have coevolved with their termite host (Berlanga et al. 2007). While the termites are related to *Cryptocercus*, less is known of the wood-feeding roach's hindgut physiology. There is a great diversity of protists, but they differ from the termite protist symbionts in that they are often covered by one type of bacteria and they generally lack endosymbiotic bacteria. Some spirochetes in *Cryptocercus* also differ from those found in the termites, while some are shared by both insects (Dolan and Melnitsky 2005; Berlanga et al. 2009).

5. Early Cell Evolution Analogs

The gut microbes have consistently led investigators to think in evolutionary terms. Cleveland claimed to have discovered the origin of mitosis in *Barbulanympha's* curious one-step reductive division. Margulis and coauthors have written extensively in this tradition from a morphological perspective, particularly regarding the hypothesized role of spirochetes in the evolution of the eukaryotic flagellum (e.g., Margulis 1993). There remains no biochemical or molecular phylogenetic evidence to support this hypothesis.

However, in studying this microbial community, one can't help but think that it holds clues by analogy, not homology, to the early evolution of life on the Earth. As one of the long time workers in the field puts it:

> The repeated postulation of intracellular symbionts in the protozoa and the numerous microboid bodies which they contain make them particularly favorable subjects for investigations into the interrelationships between cytoplasm and nucleus. Particularly with the rumen and termite protozoa, if techniques can be developed for growing them in pure or well-controlled cultures, it should be possible to replace speculation by experimentation as the basis of our understanding of the scope and nature of intracellular symbiosis. Results of experiments on intracellular symbiosis may suggest new experimental approaches to some of the fundamental problems in cell physiology (Hungate 1955).

While a spirochete contribution to the eukaryote lineage has not been found, the termite gut community has provided a wealth of examples of intimate, developmental associations between spirochetes and protist hosts (Wier et al. 2010).

6. Recent Discoveries

Culture-independent methods of study have greatly expanded our knowledge of the gut microbes in recent years because many of them have proven difficult to culture.

6.1. CO-EVOLUTION

From the early work of Kirby, Emerson, Krishna, and others, it has been known that many of the flagellate taxa have coevolved with their termite hosts (compiled in Yamin 1979). The oxymonad species *Pyrsonympha* is only found in *Reticulitermes*. The old families Devescovinidae and Calonymophidae were found only in the drywood-eating family Kalotermitidae. In contrast, some flagellate taxa were widespread in termites such as *Trichonympha* that was found in the Kalotermitidae, the Archotermopsidae, the Rhinotermitidae, as well as in the wood-feeding cockroach *Cryptocercus*. Recent work has found that in many cases, the bacterial ecto- and endosymbionts have also coevolved with their gut flagellate hosts (Desai et al. 2010; Ohkuma et al. 2007). "*Candidatus* Endomicrobium trichonymphae," the endosymbiont of *Trichonympha* species found in *Zootermopsis* and the Rhinotermitidae, have cospeciated with their hosts and are inherited by vertical transmission (Ikeda-Ohtsubo and Brune 2008).

6.2. WHOLE GENOME OF ENDOSYMBIONTS

Researchers at the RIKEN Genomic Sciences Center in Japan, one of the most productive laboratories in molecular phylogenetic studies of the termite gut microbiota, have recently published the complete genomes of two endosymbionts of gut flagellates: the Termite Group 1 endobiont of *Trichonympha agilis* in the gut of *Reticulitermes speratus*, and a Bacteroidales endobiont of *Pseudotrichonympha grassii* from *Coptotermes formosanus* (Hongoh et al. 2008a, b).

7. Genetic Diversity Versus Morphological Complexity

Most molecular phylogenetic studies of the termite gut community have found a greater diversity of populations and species than had been described by morphological investigations. For example, there are two distinct populations of the protist *Joenia annectens* in the European drywood-eating termite *Kalotermes flavicollis* (Strassert et al. 2010). There are many more phylotypes of spirochetes in the eastern subterranean termite of North America, *Reticulitermes flavipes* than were visually apparent (Lilburn et al. 1999).

Some studies using genomics techniques have found less diversity than previously described. For example, *Incisitermes marginipennis* reportedly harbored ten protist species, but a molecular phylogenetic analysis found only one protist species in the termite's hindgut (Strassert et al. 2009). One study that has found two morphology-based genera with the same molecular phylogeny of the ssurRNA gene has concluded that morphology-based taxa should be suppressed if they are not consistent with the ssurRNA gene phylogeny (Harper et al. 2009).

The drywood termite *Incisitermes harbors* no two genera of multinucleate parabasalids, *Coronympha octonaria* or *C. clevelandi* and *Metacoronympha senta*. *Coronympha* has a ring of karyomastionts (nuclei attached to the mastigont or group of four flagella). *C. octonaria* has eight. *C. clevelandi* has 16. In *Metacoronympha*, the karyomastigonts are arranged in a spiral complex and vary from 30 to 1,000 per cell. The two genera are also distinguished by details of their karyomastigonts. *Coronympha* has pyriform nuclei arranged in a ring, with each nucleus extending to the anterior tip of the cell where the mastigonts are arranged in a ring with each nucleus attached to a mastigont. *Metacoronympha* nuclei are more rounded, with the mastigont adjacent to the nucleus in a typical trichomonad fashion. In their study, Harper et al. obtained the ssurRNA gene for these two protist genera from four species of *Incisitermes*. They found that the one gene sequence was identical or nearly identical for the two genera, and that the sequence identity was greatest between the two within each of the four termite species rather than between termite species. They conclude that "Metacoronympha" is a life stage of the genus *Coronympha*, and that *Coronympha octonaria* differs in gene sequence between the four termite families so that it should be divided into four species.

To reach this conclusion, the authors (1) ignore all morphological data other than gross cell morphology and (2) develop an ad hoc explanation that one genus is a life stage of the other. They present no evidence to support the life stage hypothesis. The two genera have many subcellular morphological differences in the formation of their karyomastigonts that are regular inherited characters of the cells. There must then be genetic bases for these morphological characters, whether due to the DNA sequences of the cells or positional information of the cytoskeleton similar to the "cortical inheritance" found in *Paramecium* (Beisson and Sonnebon 1965). Furthermore, a lateral transfer of the gene in question from *Coronympha* to *Metacoronympha* and *Metacoronympha's* loss of its endogenous version is the most parsimonious explanation. As seen in this example and in many other cases, the anaerobic termite gut flagellates are a rich source of material to explore evolutionary questions and elucidate the genetic makeup of eukaryotic cells that contain numerous prokaryotic symbionts.

8. References

Beisson J, Sonnebon TM (1965) Cytoplasmic inheritance of organization of cell cortex in *Paramecium aurelia*. Proc Natl Acad Sci USA 53:275

Berlanga M, Paster BJ, Guerrero R (2007) Coevolution of symbiotic spirochete diversity in lower termites. Int Microbiol 10:133–139

Berlanga M, Paster BJ, Guerrero R (2009) The taxophysiological paradox: changes in the intestinal microbiota of the xylophagous cockroach *Cryptocercus punctulatus* depending on the physiological state of the host. Int Microbiol 12:227–236

Bermudes D, Chase D, Margulis L (1988) Morphology as a basis for taxonomy of large spirochetes symbiotic in wood-eating cockroaches and termites: *Pillotina* gen. nov., nom. rev., *Pillotina calotermitidis* sp. nov., nom. rev., *Dilpoclayx* gen. nov., nom. rev., *Diploclayx calotermitidis* sp. nov., nom. rev., *Hollandina* gen. nov., nom. rev., *Hollandina pterotermitidis* sp. nov., nom. rev. and *Clevelandina reticulitermitidis* gen. nov., sp. nov. Int J Syst Bacteriol 38:291–302

Breznak JA (1975) Symbiotic relationships between termites and their intestinal microbiota. In: Symbiosis. Symposia of the society for experimental biology No. 29. Cambridge University Press, Cambridge, pp 559–580

Breznak JA, Switzer JM (1986) Acetate synthesis from H_2 plus CO_2 by termite gut microbes. Appl Environ Microbiol 52:623–630

Brugerolle G, Patterson D (2001) Ultrastructure of Joenia pulchella Grassi, 1917 (Protista, Parabasalia), a reassessment of evolutionary trends in the parabasalids, and a new order Cristamonadida. Org Div Evol 1:147–160

Brune A, Friedrich M (2000) Microecology of the termite gut: structure and function on a microscale. Curr Opin Microbiol 3:263–269

Brune A, Emerson D, Breznak JA (1995) The termite gut microflora as an oxygen sink: microelectrode determination of oxygen and pH gradients in guts of lower and higher termites. Appl Environ Microbiol 61:2681–2687

Cepicka I, Hampl V, Kulda J (2010) Critical taxonomic revision of parabasalids with description of one new genus and three new species. Protist 161:400–433

Cleveland LR (1947) Sex produced in the protozoa of Cryptocercus by molting. Science 105:16–18

Desai MS, Strassert JFH, Meuser K, Hertel H, Ikeda-Ohtsubo W, Radek R, Brune A (2010) Strict cospeciation of devescovinid flagellates and Bacteroidales ectosymbionts in the gut of dry-wood termites (Kalotermitidae). Environ Microbiol 12:2120–2132

Dolan MF (2001) Speciation of termite gut protists: the role of bacterial symbionts. Int Microbiol 4:203–208

Dolan MF, Melnitsky H (2005) Patterns of protist-bacteria associations in the gut of the wood-feeding cockroack Cryptocercus. J N C Acad Sci 121:56–60

Ebert A, Brune A (1997) Hydrogen concentration profiles at the oxic-anoxic interface: a microsensor study of the hindgut of the wood-feeding lower termite Reticulitermes flavipes (Kollar). Appl Env Microbiol 63:4039–4046

Engel MS, Grimaldi DA, Krishna K (2009) Termites (Isoptera): their phylogeny, classification, and rise to ecological dominance. Am Mus Novit 3650:1–27

Gharagozlou ID (1968) Aspect infrastructural de Diplocalyx calotermitidis nov. gen nov. sp., spirochaetale de l'intestin de Calotermes flavicollis. C R Acad Sci (Paris) 266:494–496

Graber JR, Breznak JA (2004) Physiology and nutrition of Treponema primitia, and H_2/CO_2-acetogenic spirochete from termite hindguts. Appl Environ Microbiol 70:1307–1314

Grimaldi D, Engel MS (2005) Evolution of the insects. Cambridge University Press, Cambridge

Harper JT, Gile GH, James ER, Carpenter KJ, Keeling PJ (2009) The inadequacy of morphology for species and genus delineation in microbial eukaryotes: an example from the parabasalian termite symbiont Coronympha. PLoS One 4:e6577

Hollande AC, Gharagozlou I (1967) Morphologie infrastructurale de Pillotina calotermitidis nov. gen nov. sp., spirochaetale de l'intestin de Calotermes praecox. C R Acad Sci (Paris) 265:1309–1312

Hongoh Y, Sharma VK, Prakash T, Noda S, Taylor TD, Kudo T, Sakaki Y, Toyoda A, Hattori M, Ohkuma M (2008a) Complete genome of the uncultured Termite Group 1 bacteria in a single host protist cell. Proc Natl Acad Sci USA 105:5555–5560

Hongoh Y, Sharma VK, Prakash T, Noda S, Toh H, Taylor TD, Kudo T, Sakaki Y, Toyoda A, Hattori M, Ohkuma M (2008b) Genome of an endosymbiont coupling N_2 fixation to cellulolysis within protist cells in termite gut. Science 322:1108–1109

Hungate RE (1955) Speculations on the role of symbiosis in evolution. In: Hutner SH, Lwoff A (eds) Biochemistry and physiology of protozoa. Academic, New York, pp 194–195

Ikeda-Ohtsubo W, Brune A (2008) Cospeciation of termite gut flagellates and their bacterial endosymbionts: Trichonympha species and 'Candidatus Endomicrobium trichonymphae'. Mol Ecol 18:332–342

Inoue J-I, Noda S, Hongoh Y, Ui S, Ohkuma M (2008) Identification of endosymbiotic methanogen and ectosymbiotic spirochetes of gut protists of the termite Coptotermes formosanus. Microbes Environ 23:94–97

Leidy J (1877) On intestinal parasites of Termes flavipes. Proc Acad Nat Sci Phila 29:146–149

Leidy J (1880) Parasites of the termites. J Acad Nat Sci Phila 8:425–447

Lilburn TG, Schmidt TM, Breznak JA (1999) Phylogenetic diversity of termite gut spirochetes. Environ Microbiol 1:331–345

Margulis L (1993) Symbiosis in cell evolution, 2nd edn. Freeman, New York

Margulis L, Hinkle G (1991) Large symbiotic spirochetes: Clevelandina, Cristispira, Diplocalyx, Hollandina, and Pillotina. In: Balows A, Trüper HG, Dworkin M, Harder W, Schleifer KH (eds) The prokaryotes, vol IV, 2nd edn. Springer, Berlin, pp 3965–3978

Ohkuma M, Sato T, Noda S, Ui S, Kudo T, Hongoh Y (2007) The candidate phylum 'Termite Group 1' of bacteria: phylogenetic diversity, distribution, and endosymbiont members of various gut flagellated protists. FEMS Microbiol Ecol 60:467–476

Strassert JFH, Desai MS, Brune A, Radek R (2009) The true diversity of devescovinid flagellates in the termite Incisitermes marginipennis. Protist 160:522–535

Strassert JFH, Desai MS, Radek R, Brune A (2010) Identification and localization of the multiple bacterial symbionts of the termite gut flagellate Joenia annectens. Microbiology 156:2068–2079

Wier AM, Sacchi L, Dolan MF, Bandi C, MacAllister J, Margulis L (2010) Spirochete attachment ultrastructure: implications for the origin and evolution of cilia. Biol Bull 218:25–35

Yamin M (1979) Flagellates of the orders Trichomonadida Kirby, Oxymonadida Grassé, and Hypermastigida Grassi & Foà reported from lower termites (Isoptera families Mastotermitidae, Kalotermitidae, Hodotermitidae, Termopsidae, Rhinotermitidae, and Serritermitidae) and from the wood-feeding roach Cryptocercus (Dictyoptera: Cryptocercidae). Sociobiology 4:1–120

Biodata of **William Wyatt Hoback**, author of *"Ecological and Experimental Exposure of Insects to Anoxia Reveals Surprising Tolerance."*

Professor William Wyatt Hoback is currently at the University of Nebraska at Kearney, Nebraska. He obtained his Ph.D. in entomology from the University of Nebraska-Lincoln in 1999 and has continued his research in the areas of ecological physiology since joining the faculty at the University of Nebraska at Kearney in the same year. He has served as graduate chair and director of the distance MS degree in Biology while teaching classes in entomology, evolution, professionalism, freshwater biology, and herpetology. Professor Hoback's scientific interests lie in understanding the community structure of closely related organisms as it relates to biotic and abiotic factors. He has published more than 60 peer-reviewed papers on topics ranging from insect resistance to anoxia associated with flooding to physiology, behavior, and ecology or burying beetles (Coleoptera: Silphidae). He has obtained more than $2 million in grant funding and has been awarded the Pratt-Heins Foundation Faculty Award for scholarship and research at the University of Nebraska at Kearney. He has also received the North Central Branch of the Entomological Society Award in Teaching.

E-mail: **hobackww@unk.edu**

ECOLOGICAL AND EXPERIMENTAL EXPOSURE OF INSECTS TO ANOXIA REVEALS SURPRISING TOLERANCE

WILLIAM WYATT HOBACK
Department of Biology, University of Nebraska at Kearney, 905 West 25th Street, Kearney 68849, NE, USA

1. Introduction

1.1. EVOLUTION OF INSECTS AND EARLY TERRESTRIAL CONDITIONS

At sea level, oxygen accounts for about 20.94% of the molecules in air or about 209,460 ppm of volume. In oxygen saturated pure water at 15°C, oxygen accounts for only about 0.0006% of the molecules or about 10 ppm of mass. Aquatic insects in water are able to obtain their oxygen needs through gas exchange, while terrestrial insects submerged in water quickly become hypoxic because their oxygen demands are not met. Among other selective pressures, such a dramatic difference in availability of oxygen may have allowed the first invertebrates to venture onto land where they encountered roughly 30× more oxygen.

Hexapods first colonized the terrestrial environment from the ocean approximately 420 million years before present (Ward et al. 2006). On land, they presumably encountered reduced competition and predation pressures, and comparatively abundant oxygen. These early insects also faced challenges from gravity and desiccation. In response, these organisms evolved increasing specialization of the integument including the development of a waxy impermeable cuticle and an internal respiratory system in the form of a tracheal system. Diversity remained sparse for approximately 100 million years with only three hexapod orders (Rhyniognatha, Collembola, and Archaeognatha) known (Ward et al. 2006). Additional orders of insects including all modern forms evolved much later, beginning around 320 million years before present. Recent research suggests that "Romer's Gap," a 25-million-year period of few terrestrial animal fossils, is best explained by low (<15%) atmospheric oxygen levels (Ward et al. 2006) which limited further colonization and diversification of terrestrial groups.

With support of their efficient tracheal systems, most insects are capable of intense aerobic energy metabolism, and the majority of research interest has focused on the physiology that allows sustained, energetically expensive activity such as flight. However, freshwater habitats, which today house approximately

30,000 insect species, remained virtually uninhabited for nearly 200 million years after insects colonized land (Miller and Labandeira 2002). Insects colonized freshwater habitats from terrestrial environments during the Carboniferous.

Colonization of freshwater habitats, which are hypoxic compared to terrestrial environments, required the transformation of structures into gills along with behavioral and physiological adaptations. Aquatic insect larvae that inhabit burrows in the substrate were the last to evolve, occurring sometime in the late Paleozoic (Miller and Labandeira 2002). Despite their late arrival, some of these aquatic burrowing insect species possess the greatest known animal resistance to severe hypoxia and anoxia, being able to survive more than half a year without oxygen.

1.2. AQUATIC INSECTS AND ANOXIA

Generally, aquatic insect species are sensitive to hypoxia, and species assemblages can often be defined in terms of water quality as a function of dissolved oxygen (Hoback and Stanley 2001). The close association between degree of oxygenation and the presence or absence of particular insect groups allows aquatic insects to serve as biological indicator species (Gaufin 1973; Rosenberg and Resh 1993). Among the most widely used aquatic insect biological indices is the "EPT," which stands for the Ephemeroptera, Plecoptera, and Trichoptera, which are generally sensitive to hypoxia. Despite sensitivities among these orders and others, a number of aquatic insects are adapted to survive severe hypoxia (van der Geest 2007), and the most anoxia-tolerant insects currently known are aquatic.

The insects that are most resistant to anoxia are larval flies belonging to the family Chironomidae (Nagell and Landahl 1978). In anoxic water at 4°C, half of the individuals of *Chironomus plumosus* survived 205 days (more than 4,900 h). Over half of the tested larvae of a related species *C. anthracinus* survived 100 days (2,400 h). Survival times did not differ for these larvae when held in aerated water suggesting that death was a result of starvation rather than the absence of oxygen (Nagell and Landahl 1978). Other aquatic insects are exposed to anoxic conditions during winter when the shallow ponds they inhabit become covered by ice and snow. One such insect is the sediment-dwelling larvae of the mayfly, *Cloeon dipterum*, which survives an average of 130 days (Nagell 1977) at temperatures near 0°C. Starved larvae under the same conditions survived only 16 days (Nagell 1977). Under similar conditions, the stonefly, *Nemoura cinerea*, survived less than 6 days but maintained much greater behavioral responses than those of the mayfly (Nagell and Fagerstrom 1978).

Aquatic insect species, including chironomids and mayflies, that form burrows in the benthic sediments face severe hypoxia especially when water temperatures are warmer and they are active. In the absence of biotic perturbation, these sediments become anoxic within 3 mm of the interface with the overlying waters (Wang et al. 2001). During active periods when water temperatures are warmer, burrowing aquatic insects utilize a number of behavioral mechanisms to aerate their burrows, and resistance to hypoxia appears to vary among groups.

For example, the mayfly, *Hexagenia limbata*, maintains nearly constant water movement through its burrow and is exposed to oxygen levels >75% of overlying waters, while the alderfly, *Sialis velata*, does not constantly pump water and withstands periodic hypoxia (Wang et al. 2001; Gallon et al. 2008). Neither species has been investigated for resistance to hypoxia.

Taken in total, this very brief review of events in the adaptation of insects to first terrestrial and then subsequently to freshwater habitats suggests that oxygen availability may have been a key challenge for colonization of new habitats once adaptations were in place to efficiently utilize atmospheric oxygen. It is with this background that I explore insect respiration before highlighting the survival capabilities of terrestrial insects exposed to severe hypoxia and anoxia in various ecological and, more recently, experimental conditions.

2. Insect Respiration Patterns and Hypoxia/Hyperoxia

2.1. DISCONTINUOUS GAS EXCHANGE

Insects are the most diverse and abundant organisms on the planet and possess a unique combination of characteristics that allow for this success. Their small size, durable exoskeleton, and ability to fly have allowed them to achieve success in a variety of niches especially in the terrestrial environment. Among the insects are species that have the highest recorded mass-specific rates of metabolic activity in the animal kingdom (Bradley 2000) achieving this through delivery of oxygen directly to the tissues without the use of a circulatory system or oxygen binding molecules (Chapman 1982).

Most insects use a system of tracheae to acquire oxygen from the atmosphere and to expel carbon dioxide. Diffusion of molecules through the air is faster than through liquid, and the insect tracheal system provides advantages as long as the insect remains relatively small (Chapman 1982). The tracheal system is a series of cuticle-lined tubes that connect to spiracles on the outer surface of the insect's body. The spiracles can be opened and closed and are used by the insect to control the flow of atmospheric air into the body (Chown and Nicholson 2004). The tracheal system develops during the embryonic stage for most insects, and it increases in size and complexity as the insect grows. During each molt of the insect, the cuticular lining of the tracheal system is also discarded and reformed allowing insects to change the organization of the tracheal system and even modify the number of spiracles for each stage and during pupation for insects that possess complete metamorphosis (Chapman 1982). For example, some fly larvae have a single pair of spiracles on the abdomen during the first instar, and by the fourth instar, they have eight pairs of spiracles (Chapman 1982). The number of spiracles appears to loosely correspond to insect size and oxygen needs with a maximum of ten pairs of spiracles observed among insects. The size and number of tracheae generally scale with the insect's body size and need for oxygen, and numerous experiments have been conducted recently to examine this relationship which are reviewed by Cetanin et al. (2010).

Respiratory gases have long been thought to be transported to and from cells of the insect's tissues through convection and passive diffusion (Chown and Nicholson 2004). However, recent advances in technology using synchrotron x-ray imaging has revealed that many diverse groups of insects use rapid cycles of tracheal compression and expansion to create tidal flows for gas exchange (Westneat et al. 2003).

In many insects that are active, the spiracles close for only brief periods if at all and oxygen and carbon dioxide are exchanged relatively continuously. However, among a diversity of insect orders, a pattern termed "discontinuous gas exchange" has been observed and studied (Marais et al. 2005, Quinlan and Gibbs 2006). Discontinuous gas exchange is a pattern of opening and closing of the spiracles while the insect is at rest, which results in a cyclical pattern of oxygen uptake and carbon dioxide release from the insect (Chown and Nicholson 2004). According to a comprehensive analysis by Marais et al. (2005), this pattern of gas exchange has evolved independently in at three holometabolous insect orders (Lepidoptera, Hymenoptera, and Coleoptera) and two hemimetabolous insect orders (Orthoptera and Blattodea).

At least four hypotheses relating to respiration have been proposed to explain the adaptive value of discontinuous gas exchange, while a fifth suggests that the cycle is an adaptation to avoid parasites such as tracheal mites and is not presented in this review. The oldest hypothesis suggested that discontinuous gas exchange evolved in order for insects to conserve water (Marais et al. 2005), and recent experiments on *Drosophila melanogaster* reared under different moisture regimes have provided some experimental support (Williams et al. 1997). However, alternate hypotheses have been proposed because water savings only occurs if the insects use active ventilation of the tracheal system and is not different in species that rely on only passive diffusion.

A correlation between insects that possess discontinuous gas exchange and insects that dwell underground has been observed. Hypoxia and high concentrations of carbon dioxide in these environments led Lighton (1996, 2007) to propose that hypoxia and the buildup of carbon dioxide were strong selective forces in the evolution of discontinuous gas exchange among the arthropods. This hypothesis suggests that insects use discontinuous gas exchange to create favorable diffusion gradients in hypoxic underground environments. Because insects can also gain oxygen in these environments by keeping their spiracles open for a longer period as long as the environment is moist, debate continues as to whether hypoxic environments or water savings are the selective force for the evolution of discontinuous gas exchange (Chown et al. 2006; White et al. 2007). Research on dung beetles in severely hypoxic environments (Holter 1991) has provided additional insights into the discontinuous gas exchange observed in this group. Some genera of dung beetles dwell within the dung pat itself, and others excavate a portion of the pat and reside in the ground beneath it where they rear their young.

Chown and Holter (2000) tested oxygen uptake by *Aphodius fossor* under conditions of increasingly severe hypoxia. For this species, they found that discontinuous gas exchange ceased as the beetle was exposed to more severe hypoxia (2.84%) and that the beetles held their spiracles open switching to a mode of continuous diffusion (Chown and Holter 2000). However, by dwelling in fresh dung, this species has abundant access to water vapor and thus respiratory water loss while the spiracles are open would be negligible. Further examination of the role of severe hypoxia/anoxia as a selection pressure in the evolution of discontinuous gas exchange for species in drier environments may be warranted.

An alternative hypothesis for discontinuous gas exchange is that it evolved not in response to hypoxia but rather as a defense mechanism to prevent exposure to too much oxygen during periods of low metabolic activity (Hetz and Bradley 2005). Contreras and Bradley (2009) furthered this hypothesis with tests on the blood-feeding assassin bug, *Rhodnius prolixus*. They conclude that the oxidative damage hypothesis best explains the patterns associated with discontinuous gas exchange. If confirmed, this observation could suggest that the efficiency of the insect tracheole system is also a detriment. Although insect tolerance of severe hypoxia and anoxia at first seems counterintuitive, it may actually be a consequence of adaptations to avoid oxidative damage. More recently, Mathews and White (2011) suggested that discontinuous respiration is correlated with brain size and represents a method of energy conservation in species with larger brains. These authors show experimentally that discontinuous gas exchange can be induced in insects that do not normally exhibit this pattern through anesthetization or decapitation and suggest that this respiratory pattern represents a sleeplike state. Regardless of the selection pressures that drove the evolution of discontinuous gas exchange, insects that possess this pattern have tissues that are periodically exposed to hypoxia. This situation is exacerbated when insects are placed into environments that have reduced oxygen availability.

3. Insects Exposed to Hypoxic/Anoxic Conditions

One of the earliest observations of terrestrial insects surviving prolonged periods of submersion was made by J. Hamilton in 1885. He noted that larval tiger beetles in their burrows were underwater for approximately 1 week and resumed activity after the waters receded (Hamilton 1885). More than 100 years later, controlled experiments tested the survival of a related species of terrestrial tiger beetle larvae in severely hypoxic conditions (Hoback et al. 1998). Although still very limited in the number of species examined, research on insects exposed to hypoxia associated with immersion, anoxic environments, and experimental anoxia have yielded surprisingly long survival times for most species across several insect orders (Table 1). At freezing temperatures, some beetles survive more than a year in anoxia (Sømme 1974). Even at warm temperatures (29°C), some tiger beetle larvae survive more than 9 days exposure to anoxia (Zerm et al. 2004a). Despite limitations of

Table 1. Published survival times as lethal time to 50% mortality (LT50s) for insects exposed to severely hypoxic or anoxic conditions though submersion in hypoxic water or treatment with nitrogen or argon atmosphere.

Family	Species	Stage	Temperature (°C)	Survival time (h)	Reference
Carabidae (Cicindelinae)	Cicindela togata	Larva	10, 15, 20, 25, 35	132	Hoback et al. (1998)
Carabidae (Cicindelinae)	Amblycheila cylindriformis	Larva	25	34	Hoback et al. (2000)
Carabidae (Cicindelinae)	Cicindela denverensis	Larva	20	124	Brust and Hoback (2009)
Carabidae (Cicindelinae)	Cicindela formosa	Larva	20	65	Brust and Hoback (2009)
Carabidae (Cicindelinae)	Cicindela nevadica	Larva	20	136	Brust and Hoback (2009)
Carabidae (Cicindelinae)	Cicindela punctulata	Larva	20	120	Brust and Hoback (2009)
Carabidae (Cicindelinae)	Cicindela repanda	Larva	20	137	Brust and Hoback (2009)
Carabidae (Cicindelinae)	Cicindela tranquebarica	Larva	20	54	Brust and Hoback (2009)
Carabidae (Cicindelinae)	Cicindela hirticollis	Larva	16.5	108	Brust et al. (2005)
Carabidae (Cicindelinae)	Cicindela hirticollis	Larva	16.5	79	Brust et al. (2005)
Carabidae (Cicindelinae)	Phaeoxantha klugii	Larva	29	137	Zerm and Adis (2003)
Carabidae (Cicindelinae)	Phaeoxantha klugii	Larva	29	216	Zerm et al. (2004a, b)
Chrysomelidae	Diabrotica balteata	Larva	10, 15, 20, 25	22	Hoback et al. (2002)
Chrysomelidae	Diabrotica undecimpunctata	Larva	10, 15, 20, 25	10	Hoback et al. (2002)
Chrysomelidae	Diabrotica virgifera	Larva	10, 15, 20, 25	26	Hoback et al. (2002)
Staphylinidae	Bledius spectabilis	Larva	20	16	Wyatt (1986)
Tenebrionidae	Tenebrio molitor	Larva	15, 20, 25, 35	31	Hoback et al. (1998)
Calliphoridae	Callitroga macellaria	Larva	23	24	Meyer (1977)
Tephritidae	Eurosta solidaginis	Larva	0	50	Joanisse and Storey (1998)
Olethreutidae	Epiblema scudderiana	Larva	0	50	Joanisse and Storey (1998)
Acrididae	Ageneotettix deorum	Larva	20	5	Brust et al. (2007)
Acrididae	Arphia xanthoptera	Larva	20	3	Brust et al. (2007)
Acrididae	Chortophaga viridifasciata	Larva	20	13	Brust et al. (2007)
Acrididae	Melanoplus femurrubrum	Larva	20	3	Brust et al. (2007)

(continued)

Table 1. (continued)

Family	Species	Stage	Temperature (°C)	Survival time (h)	Reference
Acrididae	*Schistocerca gregaria*	Larva	23	4	Hochachka et al. (1993)
Byrrhidae	*Byrrhus pilula*	Adult	0	2,880	Sømme (1974)
Carabidae	*Oopterus soledadinus*	Adult	0	96	Block and Sømme (1983)
Carabidae	*Pelophila borealis*	Adult	0	3,048	Conradi-Larsen and Sømme (1973)
Carabidae	*Pelophila borealis*	Adult	0	3,744	Sømme (1974)
Carabidae	*Amara alpina*	Adult	0	1,224	Sømme (1974)
Carabidae (Cicindelinae)	*Cicindela hirticollis*	Adult	16.5	35	Brust et al. (2005)
Carabidae (Cicindelinae)	*Cicindela denverensis*	Adult	20	10	Brust and Hoback (2009)
Carabidae (Cicindelinae)	*Cicindela formosa*	Adult	20	45	Brust and Hoback (2009)
Carabidae (Cicindelinae)	*Cicindela nevadica*	Adult	20	15	Brust and Hoback (2009)
Carabidae (Cicindelinae)	*Cicindela punctulata*	Adult	20	11	Brust and Hoback (2009)
Carabidae (Cicindelinae)	*Cicindela repanda*	Adult	20	11	Brust and Hoback (2009)
Carabidae (Cicindelinae)	*Cicindela tranquebarica*	Adult	20	20	Brust and Hoback (2009)
Carabidae (Cicindelinae)	*Phaeoxantha klugii*	Adult	29	6	Zerm and Adis (2003)
Chrysomelidae	*Melasoma collaris*	Adult	0	2,880	Meidell (1983)
Chrysomelidae	*Agelastica alni*	Adult	20	10	Kölsch et al. (2002)
Curculionidae	*Lepyrus arcticus*	Adult	0	1,392	Sømme (1974)
Curculionidae	*Cosmopolites sordidus*	Adult	20	216	Kölsch (2001)
Curculionidae	*Temnoschoita nigroplagiata*	Adult	20	24	Kölsch (2001)
Curculionidae	*Otiorhynchus dubius*	Adult	0	2,880	Sømme (1974)
Perimylopidae	*Hydromedion sparsutum*	Adult	0	48	Block and Sømme (1983)
Perimylopidae	*Perimylops antarcticus*	Adult	0	48	Block and Sømme (1983)
Staphylinidae	*Bledius spectabilis*	Adult	20	43	Wyatt (1986)
	Cryptopygus antarcticus	Adult	0	672	Sømme and Block (1982)
	Parisotoma octoculata	Adult	0	96	Somme and Block (1982)
	Xenylla maritima	Adult	0	2,160	Leinaas and Sømme (1984)
	Anurophorus laricis	Adult	0	2,160	Leinaas and Sømme (1984)

(continued)

Table 1. (continued)

Family	Species	Stage	Temperature (°C)	Survival time (h)	Reference
	Tetracanthella wahlgreni	Adult	0	2,304	Sømme and Conradi-Larsen (1977)
	Isotoma violacea	Adult	0	192	Sømme and Conradi-Larsen (1977)
	Onychiurus vontoernei	Adult	0	480	Sømme (1979)
	Tetracanthella afurcata	Adult	0	2,160	Sømme (1979)
	Hypogastrura viatica	Adult	5	>864	Hodkinson and Bird (2004)
	Hypogastrura tullbergi	Adult	5	>336	Hodkinson and Bird (2004)
	Tetracanthella arctica	Adult	5	>336	Hodkinson and Bird (2004)
	Onychiurus arcticus	Adult	5	>336	Hodkinson and Bird (2004)
	Onychiurus groenlandicus	Adult	5	<336	Hodkinson and Bird (2004)
	Folsomia quadrioculata	Adult	5	60	Hodkinson and Bird (2004)
	Isotoma tschernovi	Adult	5	24	Hodkinson and Bird (2004)
	Isotoma anglicana	Adult	5	24	Hodkinson and Bird (2004)
	Folsomia candida	Adult	18	18	Zinkler and Rüssbeck (1986)
	Sminthurides malmgreni	Adult	5	42	Hodkinson and Bird (2004)
Drosophilidae	Drosophila melanogaster	Adult	25	4	Krishnan et al. (1997)
Aphidae	Pemphigus trehernei	Adult	20	38	Foster and Treherne (1976)
Formicidae	Polyrhachis sokolova	Adult	23	11	Nielsen (1997)
Acrididae	Ageneotettix deorum	Adult	20	7	Brust et al. (2007)
Acrididae	Arphia xanthoptera	Adult	20	10	Brust et al. (2007)
Acrididae	Chortophaga viridifasciata	Adult	20	22	Brust et al. (2007)
Acrididae	Melanoplus confusus	Adult	20	7	Brust et al. (2007)
Acrididae	Melanoplus femurrubrum	Adult	20	10	Brust et al. (2007)
Acrididae	Pardalophora haldemani	Adult	20	15	Brust et al. (2007)
Acrididae	Schistocerca gregaria	Adult		8	Hochachka et al. (1993)
Acrididae	Trimerotropis maritima	Adult	20	10	Brust et al. (2007)

few species examined, research on closely related species has revealed substantial differences in anoxia tolerance that appear to be related to ecology (Table 1). For example, individual tiger beetle larvae from two populations of the same species exhibit an approximately 40% difference in survival times (Brust et al. 2005), and adults of the same genus (*Cicindela*) exhibit more than a 400% difference in survival times (Brust and Hoback 2009). Now, let us examine ecological and experimental conditions where insects may be exposed to severe hypoxia and anoxia.

3.1. TERRESTRIAL INSECTS AND FLOODING

A diverse assemblage of insects inhabits the interstitial spaces in soils, and some groups, including crickets, beetles, and wasps, excavate burrows in the substrate. Anderson and Ultsch (1987) reported that oxygen levels near the soil surface are similar to atmospheric concentrations. However, respiration and diffusional exchanges with the surrounding soil in deeper burrows result in minor hypoxia in ant nests and spider burrows and more pronounced hypoxia in burrows of tiger beetles (Anderson and Ultsch 1987). These soil-dwelling organisms face severe hypoxia and anoxia when soils become flooded (Baumgartl et al. 1994).

The frequency, magnitude, and predictability of immersion range from daily exposure for insects dwelling in intertidal substrates to occasional prolonged periods for insects that dwell in flood-prone areas. The most severe flooding periods occur in many tropical areas where soils may be flooded for more than 6 months as is the case in areas of the Amazon rainforest. Terrestrial insects inhabiting these habitats exhibit different survival times and likely utilize different mechanisms to cope with immersion and the resulting severe hypoxia. Experiments to determine survival of different groups in severely hypoxic water have been the research focus of my laboratory.

Initially, experiments were conducted to determine the effects of hypoxia on the larvae of *Cicindela togata*, a species that forms permanent larval burrows in temperate soils that periodically flood for periods of days to weeks. These larvae survive about 100 h of exposure to anoxic water at 25°C and longer at cooler temperatures (Hoback et al. 1998). When flooded, the larvae quickly reduce their metabolic rates by more than 97% and support low rates through anaerobiosis (Hoback et al. 2000). Larvae of *C. togata* do not appear to extract dissolved oxygen from water because survival times did not differ between larvae submerged in anoxic water and those submerged in aerated water. Compared to *C. togata*, a basal lineage of tiger beetle, *Amblycheila cylindriformis*, survives about 34 h and maintains higher metabolic rates (Hoback et al. 2000). Further, these larvae inhabit arid cliffs where they are unlikely to be exposed to immersion.

Additional testing of tiger beetle larvae has been conducted to test the potential of human alteration to waterways to cause the extinction of a floodplain species. We found surprising differences among populations of *Cicindela hirticollis* with those that occupied ocean beach areas having lower tolerance to immersion in anoxic waters compared to those that occupy river shoreline areas which had higher tolerance (Brust et al. 2005). Among tiger beetles, *C. hirticollis* is unusual because the larvae often relocate their burrow. The seashore populations which are exposed to daily immersion appear to maintain higher metabolic rates while immersed and recover more quickly than riverine populations (Brust et al. 2005).

These observations led to the hypothesis that immersion tolerance and ability to withstand anoxia were correlated with flooding risk among tiger beetle species. However, testing by Brust and Hoback (2009) found only partial support for this conclusion. In these experiments, larval tiger beetles survived between 60 and 120 h, and survival differences did not correlate with the likelihood of immersion. Moreover, in these experiments, adult tiger beetles which are highly mobile were tested for survival in similar conditions. We found adults to survive between 10 and 45 h of immersion in anoxic waters (Brust and Hoback 2009). The 45 h survival time by *Cicindela formosa* makes it the current record holder for an adult terrestrial insect immersed in severely hypoxic water. The habitat occupied by this species is dry sand; however, adults seek shelter underground during a 3-month period of inactivity and thus may benefit from immersion tolerance (Brust and Hoback 2009).

Hypoxia as a result of oxygen uptake and CO_2 production may also result from terrestrial invertebrate burrows being sealed during immersion from tides. The rove beetle, *Bledius spectabilis*, lives in the intertidal zone, and females construct and maintain burrows with narrow openings. When these burrows are immersed by tides, the air pressure in the burrow prevents flooding. However, respiration by the beetles and their larvae is likely to create hypoxia within the burrow, and the adult beetles can survive immersion in sea water for up to 36 h (Wyatt 1986). The mangrove ant, *Camponotus andersoni*, employs a similar strategy with adults using their heads to block the nest opening during high tides. Respiration within the nest causes hypoxia, quickly reducing oxygen concentrations to approximately 4% and increases the concentration of CO_2. Under these conditions, the ants utilize anaerobic metabolism and maintain function allowing them to resume activity when the tide recedes (Nielsen and Christian 2007).

3.2. IMMERSION OF ECONOMICALLY IMPORTANT SPECIES

Flooding of agricultural fields to control soil-dwelling pests has been used as a valuable form of cultural control in rice, cranberries, sugarcane, and potato (Teixeria and Averill 2006). Despite the apparent success of these methods, few data exist comparing tolerance of pest species to controlled immersion. Larvae of root-feeding pest species belonging to the genus *Diabrotica* were examined for tolerance to immersion in severely hypoxic water at various temperatures (Hoback et al. 2002). As observed for other species, temperature and stage affected survival times but consistent differences associated with habitat occurred among species. Correlations of immersion tolerance and likely exposure to flooding were also observed for two species of snout beetles that use banana as a host. At 20°C, *Cosmopolites sordidus* survived nine times longer than *Temnoschoita nigroplagiata* (Kölsch 2001). Survival differences correlated with the accumulation of lactate which was slower in *C. sordidus*. Unfortunately, the use of flooding for cultural control in many crops is limited by the vulnerability of these crops to hypoxia associated with immersion (Hoback et al. 2002).

In rangelands, grasshoppers belonging to the family Acrididae are the most significant pests. Periodic outbreaks result in millions of dollars in economic losses. These outbreaks are often attenuated by rainfall, and anecdotal reports of drowning are widespread in the literature. Direct tests of adults and nymphs revealed median survival times in severely hypoxic water ranging from 6 to more than 45 h (Brust et al. 2007) suggesting that flooding of these stages would rarely result in high enough mortality to prevent outbreaks.

3.3. HIGH ALTITUDE HYPOXIA AND ANOXIA ASSOCIATED WITH FREEZING

A number of insects have colonized high altitudes and latitudes adapting to withstand short growing seasons, cold temperatures, hypoxia, and anoxia during frozen periods. Increases in elevation lower the atmospheric pressure, and at 6,000 m, only about 50% of the oxygen is available compared to at sea level (Schmidt-Nielson 1990). Insects in high altitude environments and those found in high latitudes tend to have smaller body sizes and have reduced wings and flight capability, which are potentially responses to hypoxia (Mani 1968). Hypoxia becomes more severe when insects are encased in ice.

Oxygen diffusion through ice is very limited, and the freezing of extracellular fluids, including those at the distal ends of tracheoles, expose freeze-tolerant insect species to anoxia (Storey and Storey 1992). Collembola and high altitude beetles have been extensively researched by Sømme and colleagues. They have demonstrated extensive survival times in anoxia under cold (0°C) conditions. For example, the ground beetle *Pelophila borealis* has been observed to survive over 150 days of anoxia (Sømme 1974). In addition, survival of anoxia at freezing temperatures of more than 120 days has been reported for species of leaf beetles and weevils from high altitudes (Sømme 1974; Meidell 1983). Similar long survival times have been reported for many arctic collembolan species (Sømme and Conradi-Larsen 1977, Leinaas and Sømme 1984; Sømme 1979; Hodkinson and Bird 2004). As these species thaw and resume aerobic respiration, they must also survive reperfusion where damage from reactive oxygen radicals is possible. Insects appear adapted to prevent damage from reactive oxygen radicals by expressing antioxidants in response to both freezing and anoxia (Joanisse and Storey 1998). Moreover, insects appear to be pre-adapted to resist damage during reperfusion by undergoing substantial metabolic reduction upon exposure to hypoxia followed by a slow recovery upon return to normoxic environments (Joanisse and Storey 1998).

3.4. OTHER SEVERELY HYPOXIC ENVIRONMENTS

A number of insects from many orders are adapted to survive and maintain activity at very low levels of atmospheric oxygen. Included among these groups are those that feed as internal parasites of mammals and those that feed on decaying wood, flesh, or dung.

Stomach botfly larvae, *Gastrophilus intestinalis*, inhabit the stomachs of horses and possess adaptations to these low oxygen environments. Larvae possess hemoglobin and apparently store oxygen obtained when the horse swallows air during feeding. Larvae are also able to survive extended periods of severe hypoxia/anoxia (Levenbook 1950) building up succinate during these periods.

Insects that develop on decaying organic material face varying degrees of hypoxia. In wet logs, atmospheres may be reduced to 0.5% oxygen (Paim and Beckel 1964). Despite this challenge, larvae of many species including those of the longhorn beetle, *Orthosoma brunnem*, survive and develop. In this species, larvae behave normally and both feed and grow in atmospheres with as little as 1% oxygen. However, prolonged exposure to slightly lower oxygen levels (0.6%) resulted in death of all larvae (Paim and Beckel 1964).

Members of the dung beetles occupy very hypoxic microhabitats (1–2% oxygen) within cavities in fresh dung pats (Holter 1991) where they feed and rear young. These adult beetles maintain normal rates of respiration and movement under conditions of 1–2% oxygen availability; however, these beetles become quiescent at lower concentrations and eventually expire (Holter and Spangenberg 1997). These findings suggest that the beetles are able to measure small differences in oxygen availability and avoid anoxic conditions where they would perish.

Wingrove and O'Farrell (1999) demonstrated that larval fruit flies, *Drosophila melanogaster*, can tolerate prolonged periods of hypoxia (~1% oxygen) associated with decaying fruits. Larvae exposed to hypoxia exhibited exploratory behaviors beginning with withdrawing from their feeding sites within the media, and increasing mobility, until the larvae left the surface of the media. After extended hypoxic periods, the larvae stopped moving. Wingrove and O'Farrell (1999) demonstrated the role of nitric oxide synthase in the exploratory behavior reaction to hypoxia in the larvae. Using a mutant strain of *Drosophila*, Wingrove and O'Farrell (1999) also showed that the larvae lacking this enzyme failed to leave dangerously hypoxic areas.

3.5. USE OF MODIFIED ATMOSPHERES TO MANAGE INSECT PESTS

Although insects are the most diverse organisms on the planet, relatively few species cause economic loss to humans and are therefore considered pests. Among pest species, many kinds of insects can cause harm to stored products or art, but the majority are in the orders Coleoptera, Blattodea, and Isoptera. Modified atmospheres have been used to manage insect pests of stored foods including grain, nuts, meat, fruits, and vegetables, along with greenhouses and museum collections (Jayas and Jeyamkondan 2002). A variety of gas combinations have been employed and the effectiveness of creating severely hypoxic or anoxic atmospheres has been assessed for management of a number of species (Table 2). In grain storage facilities, atmospheres can be modified by increasing levels of carbon dioxide, decreasing levels of oxygen (usually through the addition of nitrogen), or by hermetically sealing the storage area and allowing the grain's respiration to deplete oxygen and increase carbon dioxide levels (Bailey and Banks 1980).

Table 2. Minimum time reported for 100% mortality of various life stages of common museum pests using modified atmospheres.

Order	Family	Species	O_2 (ppm)	Temperature (°C)	% R.H	Time (h)	Reference
Blattodea	Blattidae	Periplaneta americana	<1,000	25	55	120	Rust and Kennedy (1993)
	Blattellidae	Supella longipalpa	<1,000	25	40	72	Rust and Kennedy (1993)
	Blattellidae	Blattella germanica	<1,000	25	55	24	Rust and Kennedy (1993)
Coleoptera	Anobiidae	Xestobium rufovillosum	300	30	40	72	Valentin (1993)
	Anobiidae	Xestobium rufovillosum	300	30	40	48	Valentin (1993)
	Anobiidae	Lasioderma serricorne	4,200	30	65–70	168	Gilberg (1989)
	Anobiidae	Lasioderma serricorne	<1,000	25	55	192	Rust and Kennedy (1993)
	Anobiidae	Lasioderma serricorne	300	30	40	144	Valentin (1993)
	Anobiidae	Lasioderma serricorne	300	30	40	96	Valentin (1993)
	Anobiidae	Stegobium paniceum	300	30	40	144	Valentin (1993)
	Anobiidae	Stegobium paniceum	300	30	40	96	Valentin (1993)
	Anobiidae	Stegobium paniceum	4,200	30	65–70	168	Gilberg (1989)
	Anobiidae	Anobium punctatum	300	30	40	168	Valentin (1993)
	Anobiidae	Anobium punctatum	300	30	40	120	Valentin (1993)
	Anobiidae	Nicobium sp.	300	30	40	240	Valentin (1993)
	Anobiidae	Nicobium sp.	300	30	40	168	Valentin (1993)

(continued)

Table 2. (continued)

Order	Family	Species	O_2 (ppm)	Temperature (°C)	% R.H	Time (h)	Reference
	Anobiidae	*Lyctus brunneus*	300	30	40	120	Valentin (1993)
	Anobiidae	*Lyctus brunneus*	300	30	40	72	Valentin (1993)
	Anobiidae	*Lyctus brunneus*	<1,000	25	55	120	Rust and Kennedy (1993)
	Anobiidae	*Lyctus brunneus*	4,200	30	65–70	168	Gilberg (1989a)
	Cerambycidae	*Hylotrupes bajulus*	<1,000	25	55	120	Rust and Kennedy (1993)
	Curculionidae	*Sitophilus oryzae*	10	20	12	1,000	Banks and Annis (1977)
	Curculionidae	*Sitophilus oryzae*	10	26	12	500	Banks and Annis (1977)
	Dermestidae	*Attagenus unicolor*	300	30	40	72	Valentin (1993)
	Dermestidae	*Attagenus unicolor*	300	30	40	48	Valentin (1993)
	Dermestidae	*Anthrenus vorax*	<1,000	25	55	72	Rust and Kennedy (1993)
	Dermestidae	*Dermestes lardarius*	<1,000	25	55	96	Rust and Kennedy (1993)
	Dermestidae	*Anthrenus verbasci*	4,200	30	65–70	168	Gilberg (1989)
	Tenebrionidae	*Tribolium confusum*	<1,000	25	55	96	Rust and Kennedy (1993)
Diptera	Drosophilidae	*Drosophila melanogaster*	5,000	30	75	80	Valentin and Preusser (1990)
Isoptera	Kalotermitidae	*Incisitermes minor*	<1,000	25	55	96	Rust and Kennedy (1993)
	Kalotermitidae	*Cryptotermes brevis*	10	22	40	360	Valentin and Preusser (1990)
Lepidoptera	Tineidae	*Tineola bisselliella*	<1,000	25	55	96	Rust and Kennedy (1993)
		Tineola bisselliella	4,200	30	65–70	168	Gilberg (1991)
Thysanura	Lepismatidae	*Thermobia domestica*	<1,000	25	40	48	Rust and Kennedy (1993)

Controlled tests of modified atmospheres have been conducted on a number of species under various conditions ranging from near anoxia (oxygen of 10 ppm) to severe hypoxia (oxygen of 4,200 ppm). All tested species have been observed to survive for at least 24 h (German cockroach), with some species such as the rice weevil surviving more than 40 days of exposure to near anoxia at 20°C (Table 2). As in other situations, increases in temperature reduce survival times. Exposure to severe hypoxia in lower relative humidity appears to reduce insect survival times. For example, survival times of the cigarette beetle, *Lasioderma serricorne*, were reduced from about 190 h to about 96 h when relative humidities were reduced from 60% to 40% (Valentin 1993). These results along with the observations that spiracles open in dung beetles (Chown and Holter 2000) in severe hypoxia suggest that experiments that assess survival times in anoxic gaseous environments should also control humidity.

Insects that feed on stored nuts and grain are adapted to survive periods of severe hypoxia especially at lower temperatures, and using modified atmospheres to control these pests is particularly challenging. Larval beetles of *Hypothenemus obscures* that infest Macadamia nuts by burrowing into the nut meat survived more than 6 days of exposure to pure nitrogen gas to displace oxygen at temperatures above 24°C (Delate et al. 1994). Among stored grain pests, grain feeding Lepidoptera including several species of codling moth have the longest reported survival times. These larvae survive more than 120 days of exposure to atmospheres containing 2.5% oxygen when temperatures are 0°C (Soderstrom et al. 1990). Pest beetles including red flour beetles, *Tribolium castaneum*, and confused flour beetles, *Tribolium confusum*, survived more than 8 days and 3 days exposure to atmospheres containing <0.5% oxygen, respectively (Knipling et al. 1961). However, the grain weevil *Sitophilus granarius* quickly died in atmospheres containing less than 2% oxygen (Donahaye 1990).

Short-term use of modified environments including high CO_2 and exposure to pure nitrogen has also been tested for greenhouse applications (Held et al. 2001). Fungus gnat larvae, *Bradysia* sp., suffered approximately 75% mortality with 18 h exposure to nitrogen gas, while green peach aphids, *Myzus persicae*, western flower thrips, *Frankliniella occidentalis*, and sweetpotato whiteflies, *Bemisia* sp., suffered 100% mortality after 18 h at 22°C (Held et al. 2001).

Museum specimens and organic objects including furniture, paper, and paintings are attacked by a diverse assemblage of insect pests belonging primarily to the orders Blattodea, Isoptera, and Coleoptera. Modified atmospheres have been used for the control of these pests.

3.6. INSECT RESPONSE TO SEVERE HYPOXIA AND ANOXIA

Compared to most vertebrate species, insects exhibit a remarkable ability to recover from lengthy periods of complete anoxia with little apparent damage (Harrison et al. 2006). Upon exposure to hypoxia, insects typically exhibit behavioral changes including increased spiracular opening frequencies (Krafsur and Graham 1970), increased ventilatory movements (Greenlee and Harrison 1998), and increased activity (Heslop et al. 1963). In grasshoppers, exposure to hypoxia triggers escape behaviors (Hochachka et al. 1993; Wegener 1993). As the period of exposure to severe hypoxia increases, terrestrial insects become quiescent, depress metabolism, and switch to anaerobic metabolism (Heslop et al. 1963; Hochachka et al. 1993; Wegener 1993; Hoback et al. 2000).

When insects are deprived of oxygen, the ATP levels decline rapidly, and most insects depress metabolic rates to very low levels. In the desert locust, *Schistocerca gregaria*, metabolic rates are depressed to 6% of basal rates allowing these relatively large insects to survive approximately 8 h of anoxia (Hochachka et al. 1993; Wegener 1993). Brust et al. (2007) found a range of survival among grasshopper species with the most tolerant surviving more than 40 h. The mechanisms allowing different species of Orthoptera to survive different exposure times have not been elucidated. Among congeneric tiger beetles, Hoback et al. (2000) found *Cicindela togata* to depress metabolic rates by more than 95% and to survive anoxia for more than 5 days at moderately warm temperatures (25°C) compared to *Amblycheila* which exhibited less metabolic depression and a shorter survival time in anoxia.

An interesting exception to metabolic depression during anoxia is found in the aquatic midge larva *Chironomus thummi*. This species along with other chironomid larvae possesses hemoglobin which it uses to store oxygen. Once this store of oxygen is depleted, the larva maintains metabolic rates and ATP levels by using alcoholic fermentation for anaerobic energy production (Redecker and Zebe 1988). In aquatic organisms, the production of ethanol as an end product is likely adaptive because alcohol is highly soluble, and secretion from the larvae prevents the buildup of dangerous end products in the organism's tissues. The buildup of lactate or alanine in other insects appears to explain the relationship between recovery times of insects and anoxia exposure times (Kölsch 2001; Zerm et al. 2004b; Brust et al. 2005).

For the other insects studied to date, ATP levels decline within minutes of exposure to anoxia and ADP, AMP, and IMP levels increase. The predominant reported end products of insects using anaerobic metabolism are lactate and alanine (Chefurka 1965; Wegener 1993; Grieshaber et al. 1994; Hoback et al. 2000; Zerm et al. 2004b). In addition to changes in ATP, an increase in the presence of trehalose, a nonreducing disaccharide found in many invertebrates, has been observed (Chen and Haddad 2004). In insects, trehalose provides cell protection against freezing damage, heat stress, dehydration and desiccation, and oxidation. In addition, trehalose provides cell protection against hypoxia and anoxia in fruit flies, *Drosophila* (Chen and Haddad 2004).

As in many other aspects of biology, *Drosophila* has become a model organism for examining the molecular mechanisms of anoxia resistance and recovery. As in other insects, *Drosophila* can survive anoxia for hours and exhibit the typical loss of motor control and anoxic coma. Also as observed in other insects, recovery from exposure to anoxia correlates to exposure time. Research by Ma et al. (1999) revealed the gene "*fau*" that impacts recovery time, and overexpression of this gene by transgenic flies allowed a more rapid recovery from anoxia. Further recent work by Dawson-Scully et al. (2010) revealed that differences in protein kinase result in greater tolerance to hypoxia and greater period prior to the onset of anoxic coma.

4. Prospectus

Observations of insects surviving immersion date back more than a century; however, controlled experiments to examine survival in severely hypoxic or anoxic environments have been conducted for less than 40 years. Surveys of the literature reveal that a limited number of species belonging to few taxonomic groups have been examined. Within these groups, a remarkable ability to withstand severe hypoxia emerges, and as far as I am aware, no insect has been shown to suffer permanent damage after exposure to a few minutes of anoxia as is the case for most vertebrates. The remarkable tolerance to anoxia appears to result from the evolutionary history of the insects, their unique morphology for oxygen delivery and adaptations to prevent exposure to oxidative damage, their ability to depress metabolism, and to tolerate the buildup of anaerobic end products perhaps in association with their use of trehalose. As research progresses, *Drosophila* have become model organisms for examining physiological and genetic mechanisms for anoxia resistance. Perhaps this use is warranted for larvae that are exposed to hypoxia in rotting fruits, but adults would rarely, if ever, encounter hypoxia. Thus, as research advances, the inclusion of insects such as tiger beetle larvae or dung dwelling beetles may greatly advance knowledge of the adaptive significance of hypoxia tolerance mechanisms. The paucity of data for hypoxia tolerance by most insects, the diversity of habitats occupied by insect species, and the emerging genetic techniques to elucidate reasons for differential tolerance offer a wealth of research opportunities. From an applied view, the ability to use modified atmospheres to achieve control of some insect pests provides a relatively inexpensive and environmentally safe alternative to conventional pesticide treatments, though here too more research is needed.

5. References

Anderson JF, Ultsch GR (1987) Respiratory gas concentration in the microhabitats of some Florida arthropods. Comp Biochem and Phys 88:585–588

Bailey SW, Banks FJ (1980) A review of the recent studies of the effect of controlled atmosphere on stored-product pests. In: Shejbal J (ed) Controlled atmosphere storage of grains. Elsevier Scientific Publishing Co, Amsterdam

Banks HJ, Annis PC (1977) Suggested procedures for controlled atmosphere storage of dry grain CSIRO. Aust Div Entomol Tech Pap 13

Baumgartl H, Kritzler K, Zimelka W, Zinkler D (1994) Local Po_2 measurements in the environment of submerged soil microarthropods. Acta Oecologia 15:781–789

Block W, Sømme L (1983) Low temperature adaptations in beetles from the sub-Antarctic island of South Georgia. Polar Biol 2:109–114

Bradley TJ (2000) The discontinuous gas exchange cycle in insects may serve to reduce oxygen supply to the tissues. Am Zool 40:952

Brust ML, Hoback WW (2009) Hypoxia tolerance in adult and larval *Cicindela* tiger beetles varies by life history but not habitat association. Ann Entomol Soc Am 102:462–466

Brust ML, Hoback WW, Skinner KM, Knisley CB (2005) Differential immersion survival by populations of *Cicindela hirticollis* Say (Coleoptera: Cicindelidae). Ann Entomol Soc Am 98:973–979

Brust ML, Hoback WW, Wright RJ (2007) Immersion tolerance in rangeland grasshoppers (Orthoptera: Acrididae). J Orthop Res 16:135–138

Centanin L, Gorr TA, Wappner P (2010) Tracheal remodeling in response to hypoxia. J Insect Physiol 56:447–454

Chapman RF (1982) The insects structure and function. Harvard University Press, Cambridge

Chefurka W (1965) Some comparative aspects of the metabolism of carbohydrates in insects. Annu Rev Entomol 10:345–382

Chen Q, Haddad GG (2004) Role of trehalose phosphate synthase and trehalose during hypoxia: from flies to mammals. J Exp Biol 207:3125–3129

Chown SL, Holter P (2000) Discontinuous gas exchange cycles in *Aphodius fossor* (Scarabaeidae): a test of hypotheses concerning the origins and mechanisms. J Exp Biol 203:397–403

Chown SL, Nicholson SW (2004) Insect physiological ecology. Oxford University Press, New York

Chown SL, Gibbs AG, Hetz SK, Klok CJ, Lighton JRB, Marias E (2006) Discontinuous gas exchange in insects: a clarification of hypotheses and approaches. Physiol Biochem Zool 79:333–343

Conradi-Larsen E-M, Sømme L (1973) Anaerobiosis in the overwintering beetle *Pelophila borealis*. Nature 245:388–390

Contreras HL, Bradley TJ (2009) Metabolic rate controls respiratory pattern in insects. J Exp Biol 212:424–428

Dawson-Scully K, Bukvic D, Chakborty-Chatterjee M, Ferreira R, Milton SL, Sokolowski MB (2010) Controlling anoxic tolerance in adult *Drosophila* via the cGMP-PKG pathway. J Exp Biol 213:2410–2416

Delate KM, Armstrong JW, Jones VP (1994) Postharvest control treatments for *Hypothenemus obscures* (F.) (Coleoptera: Scolytidae) in Macadamia nuts. J Econ Entomol 87:120–126

der Geest V (2007) Behavioral responses of caddisfly larvae (*Hydropsyche angustipennis*) to hypoxia. Contrib Zool 76:255–260

Donahaye E (1990) Laboratory selection of resistance by the red flour beetle, *Tribolium castaneum* (Herbst), to an atmosphere of low oxygen concentration. Phytoparasitica 18:189–202

Foster WA, Treherne JE (1976) The effects of tidal submergence on an intertidal aphid, *Pemphigus trehernei* Foster. J Anim Ecol 45:291–301

Gallon C, Hare L, Tessier A (2008) Surviving in anoxic surroundings: how burrowing aquatic insects create an oxic microenvironment. J N Am Benthol Soc 27:570–580

Gaufin AR (1973) Water quality requirements of aquatic insects. EPA 660/3-73-004. United States Environmental Protection Agency, Corvallis, 86 pp

Gilberg M (1989) Inert atmosphere fumigation of museum objects. Stud Conserv 34:30–34

Gilberg M (1991) The effects of low oxygen atmospheres on museum pests. Stud Conserv 36:93–98

Greenlee KJ, Harrison JF (1998) Acid-base and respiratory responses to hypoxia in the grasshopper *Shistocerca americana*. J Exp Biol 210:2843–2855

Grieshaber MK, Hardewig I, Kreutzer U, Portner H-O (1994) Physiological and metabolic responses to hypoxia in invertebrates. Rev Physiol Biochem Pharmacol 125:43–147

Hamilton J (1885) Hibernation of the Coleoptera. Can Entomol 17:35–38

Harrison J, Frazier MR, Henry JR, Kaiser A, Klok CJ, Rascón B (2006) Responses of terrestrial insects to hypoxia or hyperoxia. Respir Physiol Neurobiol 154:4–17

Held DW, Potter DA, Gates BS, Anderson RG (2001) Modified atmosphere treatments as a potential disinfestations technique for arthropod pests in greenhouses. J Econ Entomol 94:430–438

Heslop JP, Price GM, Ray JW (1963) Anaerobic metabolism in the housefly, *Musca domestica* L. Biochem J 87:35–38

Hetz SK, Bradley TJ (2005) Insects breathe discontinuously to avoid oxygen toxicity. Nature 433:516–519

Hoback WW, Stanley DW (2001) Insects in hypoxia. J Insect Physiol 47:533–542

Hoback WW, Higley LG, Stanley DW, Barnhart MC (1998) Survival of immersion and anoxia by larval tiger beetles, *Cicindela togata*. Am Midl Nat 140:27–33

Hoback WW, Podrabsky JE, Higley LG, Stanley DW, Hand SC (2000) Anoxia tolerance of con-familial tiger beetle larvae is associated with differences in energy flow and anaerobiosis. J Comp Physiol B 170:307–314

Hoback WW, Clark TL, Meinke LJ, Higley LG, Scalzitti JM (2002) Immersion survival differs among three *Diabrotica* species. Entomol Exp Appl 105:29–34

Hochachka PW, Nener JC, Hoar J, Saurez RK (1993) Disconnecting metabolism from adenylate control during extreme oxygen limitation. Can J Zool 71:1267–1270

Hodkinson ID, Bird JB (2004) Anoxia tolerance in high arctic terrestrial microarthropods. Ecol Entomol 29:506–509

Holter P (1991) Concentrations of oxygen, carbon dioxide and methane in the air within dung pats. Pedobiolgica 35:381–386

Holter P, Spangenberg A (1997) Oxygen uptake in coprophilous beetles (*Aphodius*, *Geotrupes*, *Sphaeridium*) at low oxygen and high carbon dioxide concentrations. Physiol Entomol 22:339–343

Jayas DS, Jeyamkondan S (2002) Modified atmosphere storage of grains meats fruits and vegetables. Biosyst Eng 82:235–251

Joanisse DR, Storey KB (1998) Oxidative stress and antioxidants in stress recovery of cold-hardy insects. Insect Biochem Mol Biol 28:23–30

Knipling GD, Sullivan WN, Fulton RA (1961) The survival of several species of insects in a nitrogen atmosphere. J Econ Ent 54:1054–1055

Kölsch G (2001) Anoxia tolerance and anaerobic metabolism in two tropical weevil species (Coleoptera, Curculionidae). J Comp Physiol B 171:595–602

Kölsch G, Jakobi K, Wegener G, Braune HJ (2002) Energy metabolism and metabolic rate of the alder leaf beetle *Agelastica alni* (L.) (Coleoptera, Chrysomelidae) under aerobic and anaerobic conditions: a micorcalorimetric study. J Insect Physiol 48:143–151

Krafsur ES, Graham CL (1970) Spiracular responses of *Aedes* mosquitoes to carbon dioxide and oxygen. Ann Entomol Soc Am 63:691–696

Krishnan SN, Sun YA, Mohsenin A, Wyman RJ, Haddad GG (1997) Behavioral and electrophysiological responses of *Drosophila melanogaster* to prolonged periods of anoxia. J Insect Physiol 43:203–210

Leinaas HP, Sømme L (1984) Adaptations in *Xenylla maritima* and *Anurophorus laricis* (Collembola) to lichen habitats on alpine rocks. Oikos 43:197–206

Levenbook L (1950) The effect of carbon dioxide and certain respiratory inhibitors on the respiration of larvae of the horse bot fly (*Gastrophilus intestinalis* De Geer). J Exp Biol 28:181–202

Lighton JRB (1996) Discontinuous gas exchange in insects. Annu Rev Entomol 41:309–324

Lighton JRB (2007) Respiratory biology: why insects evolved discontinuous gas exchange. Curr Biol 17:645–647

Ma E, Xu T, Haddad GG (1999) Gene regulation by O_2 deprivation: an anoxia-regulated novel gene in *Drosophila melanogaster*. Brain Res Mol Brain Res 63:217–224

Mani MS (1968) Ecology and biogeography of high altitude insects. W.S. Junk N.V. Publishers, Belinfante, 527 pp

Marais E, Klok CJ, Terblanch JS, Chown SL (2005) Insect gas exchange patterns: a phylogenetic perspective. J Exp Biol 14:470–472

Mathews PG, White CR (2011) Discontinuous gas exchange in insects: is it all in their heads? Am Nat 177:130–134

Meidell EM (1983) Diapause, aerobic and anaerobic metabolism in alpine adult *Melasoma collaris* (Coleoptera). Oikos 41:239–244

Meyer SGE (1977) Concentrations of some glycolytic and other intermediates in larvae of *Callitroga macellaria* (F.) (Diptera, Calliphoridae) during anaerobiosis. Comp Biochem Physiol B 58:49–55

Miller MF, Labandeira CC (2002) Slow crawl across the salinity divide: delayed colonization of freshwater ecosystems by invertebrates. GSA Today 12:4–10

Nagell B (1977) Survival of *Cloeon dipterum* (Ephemeroptera) larvae under anoxic conditions in winter. Oikos 29:161–165

Nagell B, Fagerstrom T (1978) Adaptations and resistance to anoxia in *Cloeon dipterum* (Ephemeroptera) and *Nemoura cinera* (Plecoptera). Oikos 30:95–99

Nagell B, Landahl C-C (1978) Resistance to anoxia of *Chironomus plumosus* and *Chironomus anthracinus* (Diptera) larvae. Holarct Ecol 1:333–336

Nielsen MG (1997) Nesting biology of the mangrove mud-nesting ant *Polyrhachis sokolova* Forel (Hymenoptera, Formicidae) in northern Australia. Insectes Sociaux 44:15–21

Nielsen MG, Christian KA (2007) The mangrove ant, *Camponotus andersoni*, switches to anaerobic respiration in response to elevated CO_2 levels. J Insect Physiol 53:505–508

Paim U, Beckel WE (1964) Effects of environmental gases on the motility and survival of larvae and pupae of *Orthosoma brunnem* (Forster) (Col. Cerambycidae). Can J Zool 42:59–69

Quinlan MC, Gibbs AG (2006) Discontinuous gas exchange in terrestrial insects. Respir Physiol Neurobiol 154:18–29

Redecker B, Zebe E (1988) Anaerobic metabolism in aquatic insect larvae: studies on *Chironomus thummi* and *Culex pipiens*. J Comp Physiol B 158:307–315

Rosenberg DM, Resh VH (eds) (1993) Freshwater biomonitoring and benthic macroinvertebrates. Chapman & Hall, New York, 488 pp

Rust M, Kennedy J (1993) The feasibility of using modified atmospheres to control insect pests in museums. Internal report, The Getty Conservation Institute Scientific Program, Marina del Rey

Schmidt-Nielson K (1990) Animal physiology: adaptation and environment, 4th edn. Cambridge University Press, New York, 602 pp

Soderstrom EL, Brandl DG, Mackey B (1990) Responses of codling moth (Lepidoptera: Tortricidae) life stages to high carbon dioxide or low oxygen atmospheres. J Econ Entomol 83:472–475

Sømme L (1974) Anaerobiosis in some alpine Coleoptera. Norsk Entomol Tidskr 21:155–158

Sømme L (1979) Overwintering ecology of alpine Collembola and oribatid mites from the Austrian Alps. Ecol Entomol 4:175–180

Sømme L, Block W (1982) Cold hardiness of Collembola at Signy Island, maritime Antarctic. Oikos 38:168–176

Sømme L, Conradi-Larsen E-M (1977) Anaerobiosis in overwintering collembolans and oribatid mites from windswept mountain ridges. Oikos 29:127–132

Storey KB, Storey JM (1992) Biochemical adaptations for winter survival in insects. In: Steponkus PL (ed) Advances in low-temperature biology, vol 1. JAI Press, London, pp 101–140

Teixeria LAF, Averill AL (2006) Evaluation of flooding for cultural control of *Sparganothis sulfureana* (Lepidoptera: Tortricidae) in cranberry bogs. Environ Entomol 35:670–675

Valentin N (1993) Comparative analysis of insect control by nitrogen, argon, and carbon dioxide in museum, archive, and herbarium collections. Int Biodeterior Biodegrad 32:263–278

Valentin N, Preusser FD (1990) Insect control by inert gases in museums, museum archives, and museum collections. Restaurator 11:22–33

Wang F, Tessier A, Hare L (2001) Oxygen measurements in the burrows of freshwater insects. Freshwater Biol 46:317–327

Ward P, Labandeira C, Laurin M, Berner RA (2006) Confirmation of Romer's Gap as a low oxygen interval constraining the timing of initial arthropod and vertebrate terrestrialization. Proc Natl Acad Sci U S A 103:16818–16822

Wegener G (1993) Hypoxia and post hypoxic recovery in insects: physiological and metabolic aspects. In: Hochachaka PW, Lutz PL, Rosenthal M, Sick T, van den Thillart G (eds) Surviving hypoxia – mechanisms of control and adaptation. CRC Press, Boca Raton, pp 417–432

Westneat MW, Belz O, Blob RW, Fezzaa K, Cooper WJ, Lee W-K (2003) Tracheal respiration in insects visualized with synchrotron x-ray imaging. Science 299:558–560

White CR, Blackburn TM, Terblanche JS, Marias E, Gibernau M, Chown SL (2007) Evolutionary responses of discontinuous gas exchange in insects. Proc Natl Acad Sci U S A 104:8357–8361

Williams AA, Rose MR, Bradley TJ (1997) CO_2 release patters in *Drosophila melanogaster*: the effect of selection for desiccation resistance. J Exp Biol 200:615–624

Wingrove JA, O'Farrell PH (1999) Nitric oxide contributes to behavioral, cellular, and developmental responses to low oxygen in *Drosophila*. Cell 98:105–114

Wyatt TD (1986) How a subsocial intertidal beetle, *Bledius spectabilis*, prevents flooding and anoxia in its burrow. Behav Ecol Sociobiol 19:323–331

Zerm M, Adis J (2003) Exceptional anoxia resistance in larval tiger beetle, *Phaeoxantha klugii* (Coleoptera: Cicindelidae). Physiol Entomol 28:150–153

Zerm M, Walenciak O, Val AL, Adis J (2004a) Evidence for anaerobic metabolism in the larval tiger beetle, *Phaeoxantha klugii* (Col. Cicindelidae) from a central Amazonian floodplain (Brazil). Physiol Entomol 29:483–488

Zerm M, Zinkler D, Adis J (2004b) Oxygen uptake and local PO_2 profiles of submerged larvae of *Phaexantha klugii* (Coleoptera: Cicindelidae), as well as their metabolic rate in air. Physiol Biochem Zool 77:378–389

Zinkler D, Russbeck R (1986) Ecolophysiological adaptations of collembolan to low oxygen concentrations. In: Dallai R (ed) International Seminar on Apterygota. University of Siena, Siena, pp 123–127

Biodata of **James S. Clegg**, author of "*The Unusual Response of Encysted Embryos of the Animal Extremophile, Artemia franciscana, to Prolonged Anoxia.*"

James S. Clegg is Professor Emeritus at Section of Molecular and Cellular Biology, University of California, Davis, and the Bodega Marine Laboratory, Bodega Bay, California. He received Ph.D. in 1961 from The Johns Hopkins University, Baltimore, Maryland. His current research interests include biochemical and biophysical adaptations to extreme stress in invertebrates.

E-mail: **jsclegg@ucdavis.edu**

THE UNUSUAL RESPONSE OF ENCYSTED EMBRYOS OF THE ANIMAL EXTREMOPHILE, *ARTEMIA FRANCISCANA*, TO PROLONGED ANOXIA

JAMES S. CLEGG
Bodega Marine, Laboratory, Section of Molecular and Cellular Biology, University of California, Davis, Bodega Bay, CA 94923, USA

1. Introduction

The general response of ectothermal animals not adapted to anoxia is to enhance pathways of anaerobic energy production, a response that does not result in extended anoxic longevity. In contrast, ectotherms that are well adapted to anoxic conditions (such as rocky intertidal sessile organisms) respond by reducing metabolic rates, a condition known as metabolic rate depression, or MRD (Guppy and Withers 1999; Hochachka and Somero 2002; Withers and Cooper 2010). The anoxic longevity of adapted animals depends heavily on temperature but, in general, extends over periods of weeks to a month or so, based on laboratory studies. Here I will make the case that encysted embryos (sometimes referred to as "cysts") of the primitive crustacean, *Artemia franciscana*, represent the ultimate in MRD, bringing their overall metabolism to a reversible standstill – for years. While interesting in the context of anoxic survival, this response also has important implications for the more general issue of prolonged cellular integrity in the absence of a significant flow of free energy and without macromolecular turnover in a system that is hydrated and at physiological temperature. The teachings of modern biology tell us that survival should not be possible under those conditions.

Early studies by Dutrieu and Chrestia-Blanchine (1966) indicated that encysted embryos survived 5 months in anoxic sea water in the laboratory. That interesting result was not studied further except for a rather preliminary effort (Ewing and Clegg 1969) and a study by Stocco et al. (1972) on the nucleotides of aerobic and anoxic embryos (more on that topic later). Later we confirmed the work of Dutrieu and Chrestia-Blanchine (1966) and found that anoxic longevity extended over even longer periods (Clegg and Jackson 1989; Clegg 1992, 1994). We also found that during prolonged anoxia there was no indication of substrate utilization, end product accumulation, or the presence of an oxygen debt when the embryos were returned to aerobic conditions. I will be describing further results on these matters later in the paper but first the salient features of these embryos will be described in case the reader is not familiar with this system.

2. Encysted Embryos of *Artemia franciscana*

This anostracan crustacean, known as the brine shrimp, inhabits hypersaline environments worldwide (Browne et al. 1991; Abatzopoulos et al. 2002; Muñoz and Pacios 2010). As diagrammed in Fig. 1, it reproduces by one of two modes: either by the direct production of nauplius larvae that are released into the aqueous environment or by the production of gastrula embryos that are encased in a tough chitinuous shell and enter developmental diapause, a state of obligate dormancy (Jackson and Clegg 1996). During encysted embryo production from fertilized eggs important metabolic events take place, including the programmed synthesis of the molecules shown at the top left in Fig. 1. Encysted embryos, now released from the female, are remarkably resistant to a wide variety of external stresses, indicated briefly at the lower left of the figure.

Development resumes only after diapause is terminated by various environmental cues (Drinkwater and Clegg 1991; MacRae 2005; Nambu et al. 2008, 2009; Robbins et al. 2010). In the case of *A. franciscana* from the San Francisco Bay (SFB), the species I will be referring to in this paper, a common terminator is desiccation.

These dry but "activated" embryos are said to be quiescent in the sense that they do not resume metabolism and development only because of lack of cellular water and oxygen.

The activities of quiescent embryos are rapidly brought into play upon adequate hydration under permissive conditions of temperature and molecular oxygen (Clegg and Conte 1980; Clegg and Trotman 2002). Figure 1 illustrates these features and, as mentioned, lists several molecular components to be described

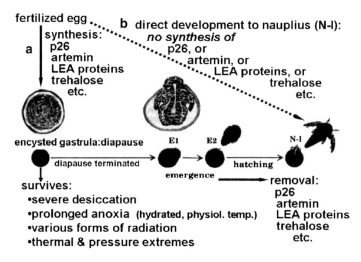

Figure 1. The two reproductive paths: to the diapause embryo (**a**) and directly to the nauplius larva, N-1 (**b**).

later in more detail, that are essential parts of the adaptive repertoire of encysted embryos. Note that these molecules are restricted to the pathway leading to the diapause embryo, are not synthesized during direct development to the nauplius, and are metabolized away as the post-diapause gastrula undergoes further development. In other words, these molecules are encysted embryo specific. I refer to *Artemia* as an animal extremophile not only because of its exceptionally stress-resistant embryos, but also because the motile stages are such remarkable osmoregulators that they thrive in environments so severely hypersaline that the vast majority of metazoans are excluded.

3. The Structure of Encysted Embryos

Figure 2a is a cross section of a hydrated encysted embryo, part of which is shown in part b at a much higher magnification. An important feature is the inner layer of the shell that is impermeable to nonvolatile substances; thus, metabolism is endogenous, requiring only water and molecular oxygen from the environment. To the right of part b are shown some of the organelles of these cells and the percentage of cyst volume that they occupy. An essential point to be made from this brief morphological visit is that, except for the tough and complex shell (Tanguay et al. 2004) the ultrastructure of these embryos does not reveal anything

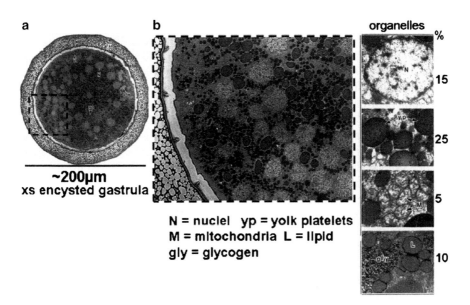

Figure 2. Structural features of encysted embryos: (**a**) diapause gastrula, part of which is shown at higher magnification in (**b**) numbers to the right are percentages of cyst volume occupied by (*from top to bottom*) nuclei, yolk platelets, mitochondria, and lipid droplets.

unusual that can be related to their extraordinary stress-resistance; indeed, their cellular structure seems typical of yolky crustacean embryos that do not exhibit unusual resistance to the stresses that are easily endured by *Artemia*. Therefore, it appears that the basis of such capabilities must be found at the suborganelle/molecular level.

4. The Longevity and Metabolic Status of Anoxic Embryos

Some variability exists in studies aimed at determining anoxic longevity, but we know that these embryos survive for very long periods, in some cases as much as 7 years (Clegg 1997). Other studies confirmed prolonged anoxic survival (Dutrieu and Chrestia-Blanchine 1966; Stocco et al. 1972; Clegg 1994; Clegg and Jackson 1998; Clegg et al. 2000a, b; Warner and Clegg 2001). But I should also note that for reasons not yet known, an occasional batch of anoxic cysts shows reduced hatching after shorter periods of anoxia. The conditions used to establish anoxia are important in that regard. Thus, we learned some time ago that the use of anoxic seawater with non-sterile cysts in sealed glass containers resulted in shorter and variable longevities, and that this result was apparently due to the microbial generation of toxic sulfides from the sulfate in sea water (Clegg and Jackson 1989; Clegg 1992, 1994). Even so, avoiding that problem does not always explain matters.

Despite repeated efforts to detect metabolism in anoxic cysts, no evidence for that was obtained after the first few days of anoxia (see Hontario et al. 1993; Clegg 1997). Microcalorimetry also suggested that anoxic embryos were metabolizing during the first days of anoxia (Hand and Gnaiger 1988; Hand 1998; Hand and Podrabsky 2000). Much longer periods of anoxia were not reported on, presumably because of the very low levels of heat production connected to the technical problem of baseline drift, although I do not know that to be the case. In any event, I believe the calorimetric studies are consistent with other evidence indicating that encysted embryos eventually bring their overall metabolism to a reversible standstill during prolonged anoxia.

In a study dedicated to detecting anoxic metabolism, we followed up an earlier study by Stocco et al. (1972) and found that the nucleotide Gp_4G was slowly broken down during prolonged anoxia, but the rate was extraordinarily slow, sometimes requiring years for detection (Warner and Clegg 2001). Given that, it is noteworthy that we have no idea how (or if) this breakdown is coupled to any free-energy transactions in the cells. Even so, greatly limited enzymatic activity such as this (only three enzymes appear to be involved with Gp_4G during anoxia) is not proof of an ongoing general metabolism, if by that we mean the existence of hundreds of enzymatic reactions under cellular control and integrated into pathways representing, among other things, the coupling of biosynthesis and degradation. I accept the caveat that one cannot prove the *absence* of an anoxic metabolism in these embryos, and no such claim is made. What I do propose is that if there is a "metabolism" in long-term anoxic embryos, it must be so restricted and incredibly slow as to be irrelevant in the general context of the rates of metabolic reactions required to support the existence and integrity of cells at physiological temperatures. Next, I consider some consequences of that contention.

5. The Matter of the Free-Energy Requirement for Living Systems

It is a well-developed principle of biology that living systems are unstable at ordinary temperatures and require a constant and substantial flow of free energy to survive (see Hochachka and Somero 2002), something I will refer to here as the "free energy rule." So what are we to think of the anoxic *Artemia* embryo that violates this rule but, nevertheless, maintains its stability and integrity? In addition, anoxic survival takes place without the turnover of macromolecules, another integral component of cellular integrity, all this over periods of years.

It seems unlikely that *Artemia* is the only animal to have achieved this – are there other exceptions? Copepod cysts (usually referred to as "eggs") taken from cores and other sediments suggest that they seem to qualify. Thus, Marcus et al. (1994) found that viable cysts of marine copepods were recovered from anoxic marine sediments that were dated at about 40 years, and Katajisto (1996) showed that copepod cysts from 10–13-year-old anoxic sediments from the brackish Baltic Sea were still viable. Interestingly, she went on to show that quiescent subitaneous eggs survived over at least a year in the laboratory (Katajisto 2004). Astonishing are the findings of Hairston et al. (1995) who recovered viable copepod cysts from freshwater anoxic pond sediments up to 300 years old. Given that such observations provide no direct evidence that these cysts were ametabolic, the cysts were hydrated and viable; so how else does one account for these results?

Perhaps a more compelling case for complete MRD comes from the laboratory work of Cooper and Van Gundy (1970) showing that 90–95% of the nematodes of *Aphelenchus avenae* recovered after 90 days of continuous anoxia, the longest period they examined. The authors pointed out that the worms had no access to exogenous substrates or other energy sources, and no utilization of endogenous energy sources was detected. I was unaware of this paper, in spite of my interest in the subject, until I came across it by chance in a book chapter by Wharton (2002). I mention this since I believe it is likely that other examples exist but have not attracted attention or simply been overlooked. Another possible factor in "discovery" comes from our work on *Artemia*. In initial studies on anoxia, we evaluated postanoxic recovery by the usual hatching assays, allowing incubation for 48 h under aerobic conditions before determining the percentage of swimming nauplii. In some of those earlier studies, we thought that many of the previously anoxic cysts that did not produce nauplii after 48 h were dead, but subsequently found that their development and hatching were greatly delayed as a function of the duration of previous anoxia (Clegg 1992, 1994), and we documented that relationship (Clegg 1997). The reasons for such a delay are unknown, although one might guess that repair of damage might be part of the reason. This delay effect might also be involved in the study of other systems undergoing anoxia leading to faulty interpretation of survival and longevity results. Finally, it is possible that the question of long-term anoxic survival has not been asked very often since it might be assumed that animals and their life history stages simply don't survive for such long periods in the absence of molecular oxygen – so why should one do an experiment on that?

An interesting paper by Minsky et al. (2002) provides insight into how cells and organisms might survive complete MRD. They suggest that the free energy rule might be escaped through the formation of highly ordered intracellular assemblies that are thermodynamically stable, protected by a physical sequestration that is independent of energy consumption. We have no evidence that such assemblies exist in anoxic embryos, although their ultrastructure reveals some unusual features (see Clegg et al. 2000a).

Returning to the anoxic *Artemia* embryo, I next consider how it endures such prolonged anoxia, without substantial free energy input and the beneficial, seemingly vital, turnover of macromolecules. Although it is obvious that nucleic acid integrity during anoxia is of prime importance, we know almost nothing about how they are protected. Remarkably, it seems that only a single paper has been published on DNA damage and stability during anoxia in *Artemia* encysted embryos (McLennan 2009). His conclusion is that anoxic embryos do not have specific protective molecular mechanisms but instead rely on the active repair of accumulated DNA lesions after anoxia and the resumption of metabolism. Perhaps these make up part of the delay just discussed? But the embryo has other important adaptations in terms of DNA and chromosomal protection. It has long been known that DNA synthesis and cell division are shut down completely during pre-emergence development (Nakanishi et al. 1962; Olson and Clegg 1978; Fig. 1). These shutdowns are a means of minimizing the damage that could result if they were operating during a period when the embryo was being subjected to various stresses. I will deal with the matter of RNA protection later in the paper. The matter of protein stability, notably of globular proteins having tertiary or quaternary structure, has attracted our attention for some time. We know these proteins tend to be inherently unstable in terms of subunit dissociation and unfolding, with the dangerous possibility of aggregation. There is no reason to think that encysted embryo proteins are exceptions because the function of those proteins also requires flexibility, and flexibility confers instability. In other words, I do not think that evolution has produced proteins in these embryos that are not prone to these dangers. In that case, the following question arises:

6. How Are the Proteins of Anoxic Embryos Protected?

Two considerations are involved here: one being the degradation of protein chains by proteolytic attack, and the other the unfolding or denaturation of globular proteins with subsequent aggregation, a very dangerous happening in cells that must be avoided. Although a complete understanding is not at hand, it is quite clear that the anoxic embryo must be very good at handling both of these potentially disastrous possibilities. The question is – how?

6.1. CONTROL OF PROTEASES

Research by Anchordouguy and Hand (1994, 1995) revealed that the ubiquitin-dependent protein breakdown pathway is not operative in anoxic embryos, an important finding in terms of protein integrity. Detailed studies on protease activity and its regulation have come from the laboratory of Al Warner (Warner et al. 1997; Liu and Warner 2006). As might be expected, there is no indication that proteases are active in anoxic embryos. A major protease, a cathepsin-like cysteine protease (CP) is very tightly regulated by a cysteine protease inhibitor (CPI). Interestingly, that inhibition is lost when the embryo emerges as a larva. Clearly, the controlled inhibition of this major protease is a critical adaptation to prolonged anoxia. One can assume that all protease activity is under strict control.

6.2. STRESS PROTEINS/MOLECULAR CHAPERONES

Like virtually all eukaryotic cells and organisms, the cysts contain the important 70-kD family of stress proteins/molecular chaperones (Clegg et al. 1999) as well as hsp90 (Clegg et al. 2000a). It seems unlikely that hsc/hsp70 plays a role in anoxic embryos since the function of these chaperones requires ATP, and it is my contention that this is not available in any appreciable amount. However, this chaperone family is likely to be very important after anoxia ends and active metabolism resumes. Once again, this chaperone activity might be part of the post-anoxic delay. Much more important for anoxic survival is the presence of very large amounts of two small heat shock proteins: an alpha-crystallin named p26, and a ferritin-like protein called artemin. Table 1 gives some of the features of these proteins, both of which we believe are of major importance in protecting the proteins of long-term anoxic embryos. Before discussing these I should point out that their function does not appear to require ATP or any other ribonucleoside di- or triphosphate, and this is interesting in itself.

Table 1. Two major stress proteins in encysted embryos of *Artemia*.

	P parameter	p26	Artemin
1.	Native MW (kD)	560 ± 20	576–600
2.	Number of subunits	28	24
3.	Subunit MW (kD)	20.7	~26
4.	Protein family	α-Crystallin	Ferritin
5.	% of Non-yolk protein	10–15	10–15

6.2.1. Artemin

Studies on this protein date back to the late 1970s (see Warner et al. 2004 for this literature). As seen in Table 1 this protein is very large and extremely abundant in encysted embryos. Although related to the ferritin family it does not contain iron (Chen et al. 2003, 2007). This protein prevents protein aggregation in vitro (Chen et al. 2007), and it is likely that artemin also serves as a molecular chaperone for RNA in general; that is, it is promiscuous for this category of nucleic acid (Warner et al. 2004). In that capacity the RNA of encysted embryos, as well as their proteins, might be protected from degradation or denaturation during prolonged anoxia, and following the return to aerobic conditions and resumption of metabolism. Another reason to believe that artemin is important in the encysted embryo is that it is removed after the nauplius larva is produced, and is not found in any other life history stage, including those embryos that give rise directly to larvae without going through the encysted diapause stage (Fig. 1). That behavior is also shown by the next chaperone to be discussed, p26.

6.2.2. The Small Heat Shock Protein p26

This very large α-crystallin protein was first described in 1994 in encysted embryos (Clegg et al. 1994) and is similar to artemin in several respects (Table 1). A hint about its possible function was its translocation to nuclei in encysted embryos undergoing anoxia and other stresses (Clegg et al. 1994, 1995). That translocation even took place in isolated nuclei in vitro simply by incubating at slightly acid pH (Clegg et al. 1994, 1995), an interesting result because the intracellular pH (pHi) of stressed embryos is known to be acidic (see Busa et al. 1982). The evidence that p26 acts as a molecular chaperone for proteins in general is reasonably strong (Clegg et al. 1995; Liang et al. 1997a, b; Liang and MacRae 1999; Willsie and Clegg 2001, 2002; Viner and Clegg 2001; Crack et al. 2002; MacRae 2003, 2010; Sun and MacRae 2005; Qiu et al. 2006; Clegg 2007), as is the demonstration that p26 in the nuclei of stressed embryos protects lamins, nuclear proteins known to be sensitive to environmental stress (Willsie and Clegg 2002). As mentioned above, a curious feature of the chaperone activity of p26 is that it does not require an input of free energy in vitro. That feature presumably is the case in vivo, in keeping with our contention that the encysted embryo is essentially ametabolic under anoxic conditions.

Other molecular chaperones present in lower amounts are hsp21 and 22 (Qiu and MacRae 2008a, b) and the stress-related transcription cofactor p8 (Qiu and MacRae, 2007). Thus, the anoxic *Artemia* embryo seems well equipped to cope with the dangers of protein unfolding and aggregation.

6.3. LATE EMBRYOGENESIS ABUNDANT (LEA) PROTEINS

These LEA proteins, once thought to be restricted to certain plants, have now been demonstrated in a number of invertebrates that tolerate dehydration and low temperature (see Berjak 2006; Tunnacliffe and Wise 2007; Tunnacliffe et al. 2010).

Not surprisingly LEA proteins are also present in *Artemia* encysted embryos (Hand et al. 2007; Menze et al. 2009) in abundance (Sharon et al. 2009). These proteins seem to play various roles that remain to be described in detail, but there is reason to believe they are important to stress resistance of these embryos, notably in tolerance to desiccation where they may be important in the formation of biological glasses along with other embryo constituents (see above references and articles in Leopold 1986).

6.4. TREHALOSE

This nonreducing disaccharide of glucose is present in *Artemia* encysted embryos in massive amounts, roughly 15% of the dry weight, and there is good evidence that it plays a key role in the exceptional tolerance of these embryos to severe desiccation (see Clegg and Conte 1980; Clegg 2001; Clegg and Trotman 2002). Noteworthy is its synergistic role in the chaperone activity of p26 in vitro, and presumably in vivo (Viner and Clegg 2001). It seems possible that trehalose could play a similar role in the activities of other chaperones, a speculation worth examining. More generally, this sugar might be involved with the creation of an internal environment conducive to the successful participation of a variety of macromolecules.

7. Concluding Comments

The classic cellular stress response is a widespread and intensely studied mechanism permitting survival in the face of a variety of environmental stresses, including anoxia. The response involves the stress-induced upregulation and induction of stress proteins that, after the stress subsides, are returned to pre-stress levels. But that path cannot be followed in encysted embryos during prolonged anoxia because they lack the ability to synthesize RNA and proteins during that time. Instead, the embryo premeditates stress, as it were, by building into its early development a programmed synthesis of stress proteins and other stress-related molecules such as the disaccharide trehalose. In this way the fully formed gastrula is prepared to cope with stresses it might encounter very soon or, in some cases, after many years, without the need to mount a stress response; indeed, it is unable to do so.

The reduction of metabolism to nonmeasurable levels in anoxic embryos indicates that they are exceptions to the free energy rule, an interesting possibility since it suggests that there is something lacking in our understanding about the maintenance of cellular integrity, at least under these conditions. Equally provocative is cellular survival over very long periods without macromolecular turnover. That ability requires incredibly tight inhibition of proteolytic activity as well as mechanisms to prevent unfolding of globular proteins with subsequent dangerous aggregation, or to refold those undergoing denaturation. The evidence shows that these embryos are very well equipped to deal with all these threats.

8. Acknowledgments

Long time collaborations with Professors Tom MacRae and Al Warner, and members of their laboratories, have been very valuable as we all attempt to understand the remarkable animal extremophile, *Artemia*. Partially supported by CRIS project Ca-D*-BML-5207-H, Ag. Exp. Station, University of California.

9. References

Abatzopoulos TJ, Beardmore JA, Clegg JS, Sorgeloos P (eds) (2002) Artemia: basic and applied biology. Kluwer Academic Publishers, Dordrecht
Anchordouguy TJ, Hand SC (1994) Acute blockage of the ubiquitin-mediated proteolytic pathway during invertebrate quiescence. Am J Physiol 267:R895–R900
Anchordouguy TJ, Hand SC (1995) Reactivation of ubiquination in *Artemia franciscana* embryos during recovery from anoxia-induced quiescence. J Exp Biol 198:1299–1305
Berjak P (2006) Unifying perspectives of some mechanisms basic to desiccation tolerance across life forms. Seed Sci Res 16:1–15
Browne RA, Sorgeloos P, Trotman CNA (eds) (1991) *Artemia* biology. CRC Press, Boca Raton
Busa WB, Crowe JH, Matson GB (1982) Intracellular pH and the metabolic status of dormant and developing *Artemia* embryos. Arch Biochem Biophys 216:711–718
Chen T, Amons R, Clegg JS, Warner AH, MacRae TH (2003) Molecular characterization of artemin and ferritin from *Artemia franciscana*. Eur J Biochem 270:137–145
Chen T, Villeneuve TS, Garant KA, Amons R, MacRae TH (2007) Functional characterization of artemin, a ferritin homolog synthesized in *Artemia* embryos during encystment and diapause. FEBS J 274:1093–1101
Clegg JS (1992) Post-anoxic viability and developmental rate of *Artemia franciscana* encysted embryos. J Exp Biol 169:255–260
Clegg JS (1994) The unusual response of *Artemia franciscana* embryos to anoxia. J Exp Zool 270: 332–334
Clegg JS (1997) Embryos of *Artemia franciscana* survive four years of continuous anoxia: the case for complete metabolic rate depression. J Exp Biol 200:467–475
Clegg JS (2001) Cryptobiosis–a peculiar state of biological organization. Comp Biochem Physiol B 128:613–624
Clegg JS (2007) Protein stability in *Artemia* embryos during prolonged anoxia. Biol Bull 212:74–81
Clegg J, Conte FP (1980) A review of the cellular and developmental biology of *Artemia*. In: Persoone G, Sorgeloos P, Roels O, Jaspers E (eds) The brine shrimp artemia, vol 2. Universa Press, Wetteren, pp 11–54
Clegg JS, Jackson SA (1989) Long term anoxia in *Artemia* cysts. J Exp Biol 147:539–543
Clegg JS, Jackson SA (1998) The metabolic status of quiescent and diapause embryos of *Artemia franciscana*. Arch Hydrobiol 52:425–439
Clegg JS, Trotman CAN (2002) Physiological and biochemical aspects of *Artemia* ecology. In: Abatzopoulos TJ, Beardmore JA, Clegg JS, Sorgeloos P (eds) Artemia: basic and applied biology. Kluwer Academic Publishers, Dordrecht, pp 129–170
Clegg JS, Jackson SA, Warner AH (1994) Extensive intracellular translocations of a major protein accompany anoxia in embryos of *Artemia franciscana*. Exp Cell Res 212:77–83

Clegg JS, Jackson SA, Liang P, MacRae TH (1995) Nuclear-cytoplasmic translocations of protein p26 during aerobic-anoxic transitions in embryos of *Artemia franciscana*. Exp Cell Res 219:1–7

Clegg JS, Willsie JK, Jackson SA (1999) Adaptive significance of a small heat shock/alpha-crystallin protein (p26) in encysted embryos of the brine shrimp, *Artemia franciscana*. Am Zool 39:836–847

Clegg JS, Jackson SA, Popov VI (2000a) Long term anoxia in encysted embryos of the crustacean, *Artemia franciscana*: viability, ultrastructure and stress proteins. Cell Tissue Res 301:433–446

Clegg JS, Jackson SA, Hoa NV, Sorgeloos P (2000b) Thermal resistance, developmental rate and heat shock proteins in *Artemia franciscana* from San Francisco Bay and southern Vietnam. J Exp Mar Biol Ecol 252:85–96

Cooper AF, Van Gundy SD (1970) Metabolism of glycogen and neutral lipids by *Aphelenchus avenae* and *Caenorabditis sp.* in aerobic, microaerobic and anaerobic environments. J Nematol 2:305–315

Crack JA, Mansour M, Sun Y, MacRae TH (2002) Functional analysis of a small heat shock/alpha-crystallin protein from *Artemia franciscana*: Oligomerization and thermotolerance. Eur J Biochem 269:1–10

Drinkwater LE, Clegg JS (1991) Experimental biology of cyst diapause. In: Browne RA, Sorgeloos P, Trotman CNA (eds) Artemia biology. CRC Press, Boca Raton, pp 93–118

Dutrieu J, Chrestia-Blanchine D (1966) Résistance des oeufs durables hydratés d'*Artemia salina* à l'anoxie. CR Acad Sci Paris Série D 263:998–1000

Ewing RD, Clegg JS (1969) Lactate dehydrogenase activity and anaerobic metabolism during the development of *Artemia salina*. Comp Biochem Physiol 31:297–307

Guppy MG, Withers PC (1999) Metabolic depression in animals: physiological perspectives and biochemical generalizations. Biol Rev Camb Philos Soc 7:1–40

Hairston NG, Van Brunt RA, Kearns CM, Engstrom DR (1995) Age and survivorship of diapausing eggs in a sediment egg bank. Ecology 76:1706–1711

Hand SC (1998) Quiescence in *Artemia franciscana* embryos: reversible arrest of metabolism and gene expression at low oxygen levels. J Exp Biol 201:1233–1242

Hand SC, Gnaiger E (1988) Anaerobic dormancy quantified in *Artemia* embryos: a test of the control mechanism. Science 239:1425–1427

Hand SC, Podrabsky JE (2000) Bioenergetics of diapause and quiescence in aquatic animals. Thermochim Acta 349:31–42

Hand SC, Jones D, Menze MA, Witt TL (2007) Life without water: expression of plant LEA genes by an anhydrobiotic arthropod. J Exp Zool Part A 307:62–66

Hochachka PW, Somero GN (2002) Biochemical adaptation. Oxford University Press, New York

Hontario R, Crowe JH, Crowe LE, Amat F (1993) Metabolic heat production by *Artemia* embryos. J Exp Biol 178:149–159

Jackson SA, Clegg JS (1996) The ontogeny of low molecular weight stress protein p26 during early development of the brine shrimp, Artemia franciscana. Dev Growth Differ 38:153–160

Katajisto T (1996) Copepod eggs survive a decade in sediments of the Baltic Sea. Hydrobiologia 320:153–159

Katajisto T (2004) Effects of anoxia and hypoxia on the dormancy and survival of subitaneous eggs of *Acartia bifilosa* (Copepoda: Calanoida). Mar Biol 145:751–757

Leopold AC (ed) (1986) *Membranes, metabolism and dry organisms*. Cornell University Press, Ithaca

Liang P, MacRae TH (1999) The synthesis of a small heat shock/alpha-crystallin protein in *Artemia* and its relationship to stress tolerance during development. Dev Biol 207:445–456

Liang P, Amons R, MacRae TH, Clegg JS (1997a) Purification, structure and molecular chaperone activity in vitro of *Artemia* p26, a small heat shock/α-crystallin protein. Eur J Biochem 243:225–232

Liang P, Amons R, Clegg JS, MacRae TH (1997b) Molecular characterization of a small heat-shock/α-crystallin protein from encysted *Artemia* embryos. J Biol Chem 272:19051–19058

Liu L, Warner AH (2006) Further characterization of the cathepsin L-associated protein and its gene in two species of the brine shrimp. Artemia Comp Biochem Physiol A 145:458–467

MacRae TH (2003) Molecular chaperones, stress resistance and development in *Artemia franciscana*. Semin Cell Dev Biol 14:251–258

MacRae TH (2005) Diapause: diverse states of developmental and metabolic arrest. J Biol Res 3:3–14

MacRae TH (2010) Gene expression, metabolic regulation and stress tolerance during diapause. Cell Mol Life Sci 67:2405–2424

Marcus NH, Lutz R, Burnett W (1994) Age, viability and vertical distribution of zooplankton resting eggs from an anoxic basin: evidence of an egg bank. Limnol Oceanogr 39:154–158

McLennan AG (2009) Ametabolic embryos of *Artemia franciscana* accumulate DNA damage during prolonged anoxia. J Exp Biol 212:785–789

Menze MA, Boswell L, Toner M, Hand SC (2009) Occurrence of mitochondria-targeted Late Embryogenesis Abundant (LEA) gene in animals increases organelle resistance to water stress. J Biol Chem 284:10714–10719

Minsky A, Shimoni E, Frenkiel-Krispin D (2002) Stress, order and survival. Nat Rev 3:50–60

Muñoz J, Pacios F (2010) Global biodiversity and geographical distribution of diapausing aquatic invertebrates: the case of the cosmopolitan brine shrimp. Artemia (Branchiopoda, Anostraca) 83:465–480

Nakanishi YH, Iwasaki T, Okigaki T, Kato H (1962) Cytological studies of *Artemia salina*. I. Embryonic development without cell multiplication after the blastula stage in encysted dry eggs. Annot Zool Jpn 35:223–228

Nambu Z, Tanaka S, Nambu F, Nakano M (2008) Influence of temperature and darkness on embryonic diapause termination in dormant *Artemia* cysts that have never been desiccated. J Exp Zool A Ecol Genet Physiol 309:17–24

Nambu Z, Tanaka S, Nambu F, Nakano M (2009) Influence of darkness on embryonic diapause termination in dormant *Artemia* cysts with no experience of desiccation. J Exp Zool A Ecol Genet Physiol 311:182–188

Olson C, Clegg JS (1978) Cell division during the development of *Artemia salina*. Wilhelm Roux's Arch Entwickl Mech Org 184:1–13

Qiu Z, MacRae TH (2007) Developmentally regulated synthesis of p8, a stress-associated transcription cofactor, in diapause-destined embryos of *Artemia franciscana*. Cell Stress Chaperones 12:255–264

Qiu Z, Macrae TH (2008a) ArHsp21, a developmentally regulated small heat-shock protein synthesized in diapausing embryos of *Artemia franciscana*. Biochem J 411:605–611

Qiu Z, MacRae TH (2008b) ArHsp22, a developmentally regulated small heat shock protein produced in diapause-destined *Artemia* embryos, is stress inducible in adults. FEBS J 275:3556–3566

Qiu Z, Bossier P, Wang X, Bojikova-Fournier S, MacRae TH (2006) Diversity, structure, and expression of the gene for p26, a small heat shock protein from *Artemia*. Genomics 88:230–240

Robbins HM, Van Stappen G, Sorgeloos P, Sung YY, MacRae TH, Bossier P (2010) Diapause termination and development of encysted *Artemia* embryos: roles for nitric oxide and hydrogen peroxide. J Exp Biol 213:1464–1470

Sharon MA, Kozarova A, Clegg JS, Vacratsis PO, Warner AH (2009) Characterization of a group 1 late embryogenesis abundant protein in encysted embryos of the brine shrimp *Artemia franciscana*. Biochem Cell Biol 87:415–430

Stocco DM, Beers PC, Warner AH (1972) Effects of anoxia on nucleotide metabolism in encysted embryos of the brine shrimp. Dev Biol 27:479–493

Sun Y, MacRae TH (2005) Small heat shock proteins: molecular structure and chaperone function. Cell Mol Life Sci 62:2460–2476

Tanguay JA, Reyes RC, Clegg JS (2004) Habitat diversity and adaptation to environmental stress in encysted embryos of the crustacean *Artemia*. J Biosci 29:489–501

Tunnacliffe A, Wise MJ (2007) The continuing conundrum of the LEA proteins. Naturwissenschaften 94:791–812

Tunnacliffe A, Hincha DK, Leprince O (2010) LEA proteins: versatility of form and function. In: Lubzens E, Cerda I, Clark M (eds) Dormancy and resistance in harsh environments. Springer, Berlin, pp 91–108

Viner RI, Clegg JS (2001) Influence of trehalose on the molecular chaperone activity of p26, a small heat shock protein. Cell Stress Chaperones 6:126–135

Warner AH, Clegg JS (2001) Diguanosine nucleotide metabolism and the survival of *Artemia* embryos during years of continuous anoxia. Eur J Biochem 268:1569–1576

Warner AH, Jackson SA, Clegg JS (1997) Effect of anaerobiosis on cysteine protease regulation during the embryonic-larval transition in *Artemia franciscana*. J Exp Biol 200:897–908

Warner AH, Brunet RT, MacRae TH, Clegg JS (2004) Artemin is an RNA-binding protein with high thermal stability and potential RNA chaperone activity. Arch Biochem Biophys 424:189–200

Wharton DA (2002) Survival strategies. In: Lee DL (ed) The biology of nematodes. Taylor & Francis, London, pp 389–411

Willsie JK, Clegg JS (2001) Nuclear p26, a small heat shock protein, and its relationship to stress resistance in *Artemia franciscana* embryos. J Exp Biol 204:2339–2350

Willsie JK, Clegg JS (2002) Small heat shock protein p26 associates with nuclear lamins and HSP70 in nuclei and nuclear matrix fractions from stressed cells. J Cell Biochem 84:601–614

Withers PC, Cooper CE (2010) Metabolic depression: a historical perspective. In: Navas CA, Carvalho JE (eds) *Aestivation*. Molecular and physiological aspects. Springer, Berlin, pp 1–24

Biodata of **Daiki D. Horikawa**, author of *"Survival of Tardigrades in Extreme Environments: A Model Animal for Astrobiology."*

Daiki D. Horikawa is currently a Research Fellow at University Paris-Descartes Medical School, Inserm, in France. He obtained his Ph.D. from Hokkaido University in 2007 and continued his studies on tardigrades as a postdoctoral researcher at the University of Tokyo and NASA Ames Research Center. Dr. Horikawa's scientific interests are in the areas of: desiccation tolerance and radiation tolerance of anhydrobiotic organisms, anhydrobiotic engineering, and astrobiology.

E-mail: **horikawadd@gmail.com**

SURVIVAL OF TARDIGRADES IN EXTREME ENVIRONMENTS: A MODEL ANIMAL FOR ASTROBIOLOGY

DAIKI D. HORIKAWA[1,2]
[1] Universety PAris Descartes-site Necker, Inserm U 1001,
75751, Paris Cedex 15, France
[2] Mediterranean Institute for Life Sciences, 21000 Split, Croatia

1. Introduction

Among abiotic environmental factors, desiccation is considered to be the most detrimental stress to organisms. To cope with severe desiccation, some organisms have evolved a survival strategy called anhydrobiosis in which organisms are tolerant of extreme dehydration for an extended period. Organisms showing anhydrobiotic ability are distributed in many taxonomic groups including bacteria, fungi, algae, plants, protozoans, and invertebrate animals such as tardigrades, nematodes, rotifers, and arthropodes (Alpert 2006; Watanabe 2006). Anhydrobiotic organisms have been thought as appropriate model organisms for astrobiological studies (Horneck 2003) because those which fulfill conditions for surviving open space environments must require ability to be tolerant of extreme desiccation induced by space vacuum (Jönsson 2007) and of other extraterrestrial environmental parameters, including extremely high or low temperatures, vacuum, and ultraviolet (UV) and ionizing radiation. Particularly, anhydrobiotic animals can be considered as interesting model organisms for astrobiological research since they could provide an estimate for the probabilities of existence of animal-like extraterrestrial life forms by both ground and flight experiments. Unlike unicellular organisms, anhydrobiotic animals have complicated body structures such as tissues and organs, possibly providing new insight into survival strategies for organisms under extreme environmental conditions.

Among anhydrobiotic animals, tardigrades are considered typical examples of extremotolerant animals because they are known to enter into an anhydrobiotic state and exhibit high resistance to a variety of environmental extreme conditions (Rothschild and Mancinelli 2001; Cavicchioli 2002). Recently, Jönsson et al. (2008) showed that tardigrades in the anhydrobiotic state survived even open space environments. Tardigrades are small invertebrate animals (0.1–1.0 mm in body length) (Fig. 1a), composing a phylum Tardigrada which is phylogenetically close to Nematoda and Onychophora (Dunn et al. 2008). This phylum consists of over 1,000 species, and they are found in various environments from deep sea to high mountains around the world. They are often observed in extremely dry and cold

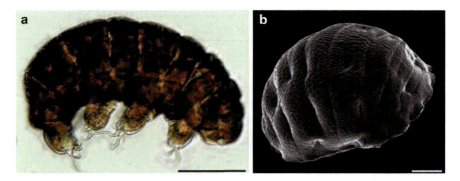

Figure 1. A light micrograph of a hydrated adult (**a**) and a scanning electron micrograph of an anhydrobiotic adult (**b**) of the tardigrade *Ra. varieornatus*. Scale bars = 100 μm (**a**) and 20 μm (**b**) (Reproduced with permission from Horikawa et al. 2008).

areas, such as the Antarctic. Tardigrades can also be found in dry mosses on pavements or lichens on trunks of trees in urban areas. Even terrestrial tardigrades, however, require a film of water surrounding them for foraging, reproduction, and other activities.

2. Anhydrobiosis in Tardigrades

Anhydrobiosis is defined as an ametabolic state induced by dehydration and followed by resurrection when rehydrated (Keilin 1959). When terrestrial tardigrades enter into anhydrobiosis, their body water decreases to between 1% and 3% wt./wt. in association with body contraction (Crowe 1972; Horikawa et al. 2006, 2008). The contracted anhydrobiotic animal, referred to as a tun (Fig. 1b), shows no visible signs of life but can resume activity when placed in a drop of water. Tardigrades have the ability to enter anhydrobiosis at any developmental stages (Horikawa et al. 2008). Pigon and Weglarska (1955) measured oxygen consumption of the tardigrade *Macrobiotus hufelandi* and demonstrated that metabolism in this species became almost completely arrested in the anhydrobiotic state under low humidities (below 48%). Although tardigrades are believed to be ametabolic during anhydrobiosis, tardigrade anhydrobiotes could not survive more than 10 years at room temperature under aerobic conditions (Guidetti and Jönsson 2002). It suggests that the accumulation of oxidative stress causes critical damage to the anhydrobiotic animals under such conditions.

There is a critical desiccation rate for tardigrades to enter anhydrobiosis (Wright 1989; Horikawa and Higashi 2004). The upper critical desiccation rate varies among tardigrade species and reflects moisture conditions in their microhabitat and a degree of their desiccation tolerance. For instance, a species inhabiting a microenvironment in which water evaporates quickly can enter anhydrobiosis at rapid desiccation rate while one inhabiting a relatively moist habitat needs slow desiccation rate to enter into anhydrobiosis successfully.

Extreme water loss causes critical damage on cells, and, therefore, tardigrades and other anhydrobiotic animals must require systems for protecting their cells during anhydrobiotic state. Some types of compatible solutes have been considered to protect cells in anhydrobiotic animals. Nonreducing disaccharide trehalose has been believed to be a protective compound which was found in high concentration (about 10–20% wt./dry wt.) in several anhydrobiotic animals, including nematodes (Madin and Crowe 1975), embryos of the crustacean *Artemia salina* (Clegg 1962), and the insect larvae *Polypedilum vanderplanki* (Watanabe et al. 2003). The role of trehalose is thought to maintain the structures of biomolecules during dehydration by acting as a water-replacement substitute and/or by forming a glassy state (vitrification) (Crowe et al. 1987; Franks et al. 1991). However, tardigrades contain relatively low amounts of trehalose (approximately 0.002–2.3%) in the anhydrobiotic state (Westh and Ramløv 1991; Horikawa et al. 2006; Hengherr et al. 2008). Furthermore, no trehalose was detected in the anhydrobiotic bdelloid rotifers (Lapinski and Tunnacliffe 2003), implying that trehalose is not an essential solute for successful anhydrobiosis. Another candidate molecule responsible for the cell protection against dehydration is late embryogenesis abundant (LEA) proteins which may work as molecular chaperons helping other proteins protected from denaturation (Wise and Tunnacliffe 2004). LEA proteins are expressed in nematodes (Browne et al. 2004), rotifers (Denekamp et al. 2009), tardigrades (Schokraie et al. 2010), Artemia cysts (Hand et al. 2007), and *Po. vanderplanki* (Kikawada et al. 2006), but evidence in vivo for supporting the theory has not been reported yet.

In addition to the compatible solutes, results from recent studies suggest that maintaining DNA integrity is important for successful anhydrobiotic survival in tardigrades (Rebecchi et al. 2009a, b; Neumann et al. 2009). Dehydration causes DNA double strand breaks (DSBs) (Mattimore and Battista 1996), which are critical damage leading to an organism's death. Significant DNA degradation was not detected in anhydrobiotic tardigrade *Paramacrobiotus richtersi* immediately after induction into anhydrobiosis (Rebecchi et al. 2009a, b) but Neumann et al. (2009) observed that there is a positive correlation between DNA damage accumulation and a period of anhydrobiosis in the tardigrade *Milnesium tardigradum*. Survival rate of tardigrades decreases according to a period that they are stored in the anhydrobiotic state (Guidetti and Jönsson 2002), suggesting that DNA damage accumulation presumably caused by oxidation leads to the death of anhydrobiotic tardigrades.

3. Radiation Tolerance

One of the most remarkable characteristics exhibited in tardigrades is the considerable tolerance to ionizing radiation. Hitherto, four tardigrade species have been studied concerning their radiation tolerance. May et al. (1964) reported that LD_{50} of

Table 1. LD50 + standard error at 2, 24, and 48 h after irradiation of gamma-rays (^{60}Co) or ^{4}He ions in *Mi. tardigradum* (Reproduced from Horikawa et al. (2006) with modification).

Radiation	State	LD_{50} (Gy)		
		2 h survival	24 h survival	48 h survival
Gamma	Hydrated	5,500 ± 400	5,200 ± 1,700	5,000 ± 1,900
	Anhydrobiotic	5,500 ± 500	5,400 ± 600	4,400 ± 500
^{4}He ions	Hydrated	7,800 ± 2,500	6,300 ± 600	6,200 ± 1,200
	Anhydrobiotic	7,900 ± 30,000	5,400 ± 2,200	5,200 ± 2,900

X-rays in the species *Macrobiotus areolatus* was around 5 kGy 1 day after irradiation. Jönsson et al. (2005) and Horikawa et al. (2006) examined tolerance of the species *Richtersius coronifer* and *Mi. tardigradum* at adult stage to gamma-radiation, respectively. LD_{50} of gamma-rays (^{137}Cs) 1 day after exposure in *Ri. coronifer* was 4.7 and 3.0 kGy in hydrated (active) and anhydrobiotic states, respectively. *Mi. tardigradum* showed similar results that LD_{50} for gamma-rays using ^{60}Co source 1 day after irradiation was 5.0 and 4.4 kGy in the hydrated (active) and anhydrobiotic states, respectively (Table 1).

As described above, hydrated animals are more radiation tolerant than anhydrobiotic ones in both the *Ri. coronifer* and the *Mi. tardigradum* in terms of post-irradiation short-time survival ability. However, in other anhydrobiotic animals, such as embryos of *A. salina* (Iwasaki 1964) and larvae of *Po. vanderplanki* (Watanabe et al. 2006), anhydrobiotic individuals showed higher tolerance to gamma radiation than hydrated ones. Organisms in an anhydrobiotic state are supposed to be more tolerant of irradiation than in a hydrated state because anhydrobiotes which contain extremely low water contents could avoid damage to biomolecules caused by the indirect radiation action through water molecules compared with hydrated animals. As described above, *A. salina* and *Po. vanderplanki* accumulate a large amount of trehalose in the anhydrobiotic state (Clegg 1962; Watanabe et al. 2002), while anhydrobiotic individuals of *Ri. coronifer* and *Mi. tardigradum* contain relatively low amounts of trehalose (2.3%, <0.2%, wt./dry wt., respectively) (Westh and Ramløv 1991; Horikawa et al. 2006). Yoshinaga et al. (1997) suggest that trehalose plays a role in protecting biomolecules against radiation. Thus, the difference in the radiation tolerance pattern between tardigrades and the other anhydrobiotic animals may be derived from the difference in trehalose concentration in a body during anhydrobiosis. Tardigrades are also known to have less mitotic activity at an adult stage (Bertolani 1970), and it is thought that the absence of somatic cell division enhances radiation tolerance in organisms (Ducoff 1972). These theories can be supported from the results that eggs of the tardigrade *Ramazzottius varieornatus* showed higher tolerance, based on hatchability, to heavy ion irradiation in the anhydrobiotic state than the hydrated state (Horikawa et al. unpublished data). This is because the somatic cell division actively occurs at an embryonic stage in tardigrades (Suzuki 2003; Gabriel et al.

2007), resulting in less tolerance of eggs to radiation in the hydrated state than the anhydrobiotic state.

Long-term survival and reproductive abilities of *Ri. coronifer* and *Mi. tardigradum* after gamma-irradiation were also examined. *Ri. coronifer* was maintained without any food sources (Jönsson et al. 2005) whereas *Mi. tardigradum* was cultured by supplying bdelloid rotifers as food (Horikawa et al. 2006). In both tardigrade species, post-irradiation life span in the hydrated and anhydrobiotic states decreased in a dose-dependent manner (Jönsson et al. 2005; Horikawa et al. 2006). *Ri. coronifer* laid eggs after irradiation with doses up to 5,000 Gy in the hydrated state and up to 2,000 Gy in the anhydrobiotic state. In *Mi. tardigradum*, egg laying was observed in one individual irradiated with 2,000 Gy in the hydrated state. No eggs produced by irradiated animals in both species hatched. Even though the high doses of gamma-irradiation inhibit normal embryonic development in tardigrades, they can be considered as the most radiation tolerant taxonomic group based on these results. The considerable tolerance of tardigrades to radiation suggests that they have effective systems for protecting DNA against radiation or repairing damaged DNA, because such high doses of ionizing radiation cause critical DNA damage, such as DSBs (Mattimore and Battista 1996). There also might be a protective system against radiation-induced oxidation in tardigrades, as protection to oxidative stress may be an essential factor for radiation tolerance in the bacterium *Deinococcus radiodurans* (Daly et al. 2007; Krisko and Radman 2010).

Mi. tardigradum also showed a high short-term survival rate after exposure to high linear energy transfer (high-LET) heavy ions (^{4}He ions), with LD_{50} doses of 6.2 kGy in the hydrated state and 5.2 kGy in the anhydrobiotic one 48 h after irradiation (Table 1) (Horikawa et al. 2006). Further investigations on effects of high-LET heavy ions on tardigrades are necessary for estimating the possibilities for tardigrades to survive in extraterrestrial environments.

4. Tolerance to Low and High Temperatures

Tardigrades have been shown to tolerate extremely low and high temperatures. Several studies have reported that tardigrades survived extreme low temperature in both the hydrated and anhydrobiotic states. Becquerel (1950) demonstrated that the tardigrade *Mi. tardigradum* and *Ra. oberhauseri* survived after exposure to −273°C, near absolute zero temperature, in the anhydrobiotic state. In addition, even in the hydrated state, tardigrade *Ri. coronifer* survived −196°C (Ramløv and Westh 1992). However, remarkable decrease in survival rate was observed in hydrated *Ri. coronifer* when they were exposed to −196°C at rapid cooling rate (approximately 1,500°C min^{-1}) (Ramløv and Westh 1992). Besides, Horikawa et al. (2008) reported that survival rate in *Ra. varieornatus* in the hydrated state was much lower than that in the anhydrobiotic state after direct exposure to −196°C, indicating that tardigrades are more tolerant of low temperature when they are in the anhydrobiotic state than the hydrated state.

Anhydrobiosis confers a high degree of tolerance to high temperatures (70°C and above) on tardigrades. Doyère (1842) reported that the tardigrade *Ma. hufelandi* was resurrected post-heating treatment at 120–125°C. *Mi. tardigradum* and *Ra. oberhauseri* survived 110–151°C for 35 min (Rahm 1921). Ramløv and Westh (2001) studied high temperature tolerance of *Ri. coronifer* and showed that this species could not survive 100°C for 1 h, with LD_{50} temperature of approximately 76°C. In *Ra. varioeornatus*, more than 90% of adult anhydrobiotic specimens survived 1 h exposure to 90°C (Horikawa et al. 2008), and approximately 15% of egg anhydrobiotes survived 80°C for 1 h (Horikawa et al. unpublished data). The cause of death of the anhydrobiotic tardigrades after heat treatment is probably because of loosing the glassy state, a state helping stabilize functional biological structures in an anhydrobiote, by critical upper temperature, as suggested in a study on *Po. vanderplanki* (Sakurai et al. 2008).

Judging from the tolerance ability of tardigrades against extremely low and high temperatures, they could survive temperatures on Mars where the temperature fluctuates between −123°C and 25°C (Diaz and Schulze-Makuch 2006). It is also possible that they survive nearly absolute zero temperature (Horneck 1999) in interplanetary space.

5. Tolerance to Low and High Pressures

Extreme vacuum must cause considerable desiccation and therefore is critical to organisms. In concert with the tardigrade anhydrobiotic ability, animals in the anhydrobiotic state can tolerate open space vacuum. Horikawa et al. (unpublished data) exposed anhydrobiotic eggs of *Ra. varieornatus* to low pressure at 5.3×10^{-4} Pa to 6.2×10^{-5} Pa for 7 days and demonstrated that experiment and demonstrated that 86% of the eggs exposed hatched. Moreover, adult individuals of *Mi. tardigradum* and *Ri. coronifer* tolerated 10-day exposure to vacuum in open space environments (Jönsson et al. 2008). These results strongly suggest that anhydrobiote tardigrades can survive even in anoxic environments in the universe.

Tardigrades have also been shown to exhibit extraordinary tolerance to extremely high pressures in the anhydrobiotic state. Seki and Toyoshima (1998) found tolerance of tardigrades to high pressure for the first time, reporting that two species of tardigrades *Ma. occidentalis* and *Echiniscus japonicus* in the anhydrobiotic state survived after 20 min exposure to high hydrostatic pressure up to 600 MPa when a water-free liquid, perfluoro-hydrocarbon, was used as a pressure medium. The tardigrade *Mi. tardigradum* survived exposure to high hydrostatic pressure at 1.2 GPa (Horikawa et al. 2009) and 7.5 GPa (Ono et al. 2008). Extremely high pressure generally causes physical changes in biomolecules such as DNA and proteins, leading to death of organisms (Abe et al. 1999). It is likely that removal of water from a tardigrade body avoids biomolecule damage by high hydrostatic pressure. Considering the high pressure tolerance in tardigrades, there might be some organisms that exist in the anhydrobiotic state under high pressure environments on extraterrestrial planets.

6. Exposure to Actual and Simulated Extraterrestrial Environments

Hitherto, there have been two flight experiments on tardigrades (Jönsson et al. 2008; Rebecchi et al. 2009a, b). Jönsson et al. (2008) conducted a flight experiment using tardigrades *Ri. coronifer* and *Mi. tardigradum* which were exposed to open space environments at low Earth orbit for 10 days. In this experiment, adult and egg specimens of both species in the anhydrobiotic state were set in the Biopan-6 experimental platform, where temperatures were controlled from 10°C to 39°C, provided by the European Space Agency (ESA) in the FOTON-M3 mission and exposed to three different conditions: space vacuum alone, space vacuum and UV-A and UV-B (UV_{AB}, 280–400 nm spectral range) with a dose of 7,095 kJ m^{-2}, and space vacuum and the full UV range from vacuum UV to UV-A (UV_{ALL}, 116.5–400 nm spectral range) with a dose of 7,577 kJ m^{-2}. In both species, adult and egg specimens exposed to space vacuum alone showed comparable survival rate compared with control specimens which were kept under ground conditions. In addition, adult samples of both species retained normal reproductive activities after space vacuum exposure. On the other hand, space vacuum with UV radiation considerably decreased survival rate of the adult and egg samples in both species. In both species, no eggs exposed to UV radiation hatched. Although UV radiation had high negative effects on survival of anhydrobiotic tardigrades, a few specimens of adult *Mi. tardigradum* survived space vacuum with UV_{ALL}. Rebecchi et al. (2009a, b) examined effects of space environments including microgravity and moderate galactic cosmic radiation on the tardigrade *Pa. richtersi* in hydrated and anhydrobiotic states. *Pa. richtersi* was exposed to space environmental conditions for 12 days in the FOTON M-3 spacecraft. Temperature inside the experimental compartment was between around 18°C and 26°C. The authors found that the space flight did not decrease survival rate of the animals in both states, and hydrated active individuals produced next generation. Interestingly, tardigrades experienced the space flight showed higher antioxidant activity than ground control animals. Although doses of cosmic radiation that the tardigrade samples received in these two flight experiments were quite low (4.5–23.53 mGy in Jönsson et al. (2008) and 1.9 mGy in Rebecchi et al. (2009a, b)), those studies demonstrated that anhydrobiotic animals can survive simultaneous space environmental parameters, and thus can offer the possibilities of interplanetary transfer of anhydrobiotic tardigrades.

Johnson et al. (2011) examined the survival ability of the tardigrade *Ra. varieornatus* in the anhydrobiotic state at a burial depth of 5 mm in regolith after 40-day exposure to simulated Martian environments with temperatures ranging from −40.4°C to 24.0°C, 19.3 Wm^{-2} of UV flux (200–400 nm), 10–22 mbar of atmospheric pressure, and 95.3% CO_2 concentration. Seventy percent of the *Ra. varieornatus* specimens survived after exposure to the simulated Martian environments, implying that tardigrade anhydrobiotes and other anhydrobiotic multicellular organisms can survive on Mars-like planets.

7. Conclusion

The results from the exposure experiments would imply that even animals in the anhydrobiotic state can be transported among planets and suggest a possibility that animal-like organisms thrive on other planets. In order to estimate those possibilities, additional astrobiological studies, such as long-term exposure experiments using tardigrades or other anhydrobiotic animals, seem to be required. A high proportion of the tardigrade anhydrobiotes survived short-term exposure to natural space level vacuum (Jönsson et al. 2008; Horikawa et al. unpublished data), but it is possible that prolonged (e.g., years) extreme vacuum exposure reduces tardigrade survivability because considerable desiccation by space vacuum is thought to cause accumulation of critical DNA damage (Horneck 2003). Although both flight experiments (Jönsson et al. 2008; Rebecchi et al. 2009a, b) did not evaluate effects of natural space temperatures on tardigrades, it seems that tardigrade survivability is not affected by extremely low temperatures based on cold temperature exposure experiments. Since any kinds of chemical reactions in anhydrobiotes can nearly be stopped under extremely low temperatures, it is expected that anhydrobiotic tardigrades may be able to live for prolonged period in interplanetary space environments where the temperature is extremely low. On the other hand, it is likely that tardigrades cannot survive high temperatures more than 150°C according to data from Rahm (1921), meaning that habitable environments for tardigrades can be largely limited by the upper survival temperature for tardigrades. UV radiation and galactic cosmic radiation are thought to be the most detrimental environmental factors for anhydrobiotic tardigrades. Thus, the possibility that tardigrades complete interplanetary travel is considered; in the future, we need to evaluate whether they can be protected against radiation if they are packed inside a meteorite-like material.

Although tardigrade research has not prevailed compared with extremophilic bacteria, such as *D. radiodurans*, largely due to difficulties of culturing tardigrades, recent methodology development of artificial rearing systems for several tardigrade species (Altiero and Rebecchi 2001; Suzuki 2003; Horikawa et al. 2008; Hengherr et al. 2008) makes it possible to conduct molecular and biochemical experiments and thus would accelerate future investigations on mechanisms behind the tardigrade tolerance to extreme environments. Such research might provide new insights into the probabilities of interplanetary transfer and the existence animal-like life forms in extraterrestrial environments.

8. Acknowledgments

I thank Lynn J. Rothschild and John Cumbers from NASA Ames Research Center for providing research advice on my studies. I also thank the NASA Astrobiology Institute Postdoctoral Program for supporting my research project at NASA Ames Research Center.

9. References

Abe F, Kato C, Horikoshi K (1999) Pressure-regulated metabolism in microorganisms. Trends Microbiol 7:447–453

Alpert P (2006) Constraints of tolerance: why are desiccation-tolerant organisms so small or rare? J Exp Biol 209:1575–1584

Altiero T, Rebecchi L (2001) Rearing tardigrades: results and problems. Zool Anz 240:217–221

Becquerel P (1950) La suspension de la vie au dessous de 1/20 K absolu par demagnetization adiabatique de l'alun de fer dans le vide les plus eléve. C R hebd Séances Acad Sci Paris 231:261–263

Bertolani R (1970) Mitosi somatische e constanza cellulare numerica nei Tardigradi. Atti Accad Naz Lincei Rend Ser 8a:739–742

Browne JA, Dolan KM, Tyson T, Goyal K, Tunnacliffe A, Burnell AM (2004) Dehydration-specific induction of hydrophilic protein genes in the anhydrobiotic nematode *Aphelenchus avenae*. Eukaryot Cell 3:966–975

Cavicchioli R (2002) Extremophiles and the search for extraterrestrial life. Astrobiology 2:281–292

Clegg JS (1962) Free glycerol in dormant cysts of the brine shrimp, *Artemia salina*, and its disappearance during development. Biol Bull 122:295–301

Crowe JH (1972) Evaporative water loss by tardigrades under controlled relative humidities. Biol Bull 142:407–416

Crowe JH, Crowe LM, Carpenter JF, Wistrom CA (1987) Stabilization of dry phospholipid bilayers and proteins by sugars. Biochem J 242:1–10

Daly MJ, Gaidamakova EK, Matrosova VY, Vasilenko A, Zhai M, Leapman RD, Lai B, Ravel B, Li SM, Kemner KM, Fredrickson JK (2007) Protein oxidation implicated as the primary determinant of bacterial radioresistance. PLoS Biol 5:769–779

Denekamp NY, Thorne MA, Kube M, Reinhardt R, Lubzens E (2009) Discovering genes associated with dormancy in the monogonont rotifer *Brachionus plicatilis*. BMC Genomics 10:108

Diaz B, Schulze-Makuch D (2006) Microbial survival rates of *Escherichia coli* and *Deinococcus radiodurans* under low temperature, low pressure, and UV-irradiation conditions, and their relevance to possible martian life. Astrobiology 6:332–347

Doyère PLN (1842) Memories sur les tardigrades. Sur le facilité que possedent les tardigrades, les rotifers, les anguillules des toits et quelques autres of animalcules, de revenir à la vie après été completement déssechées. Ann Sci Nat (Ser 2) 18:5

Ducoff HS (1972) Causes of death in irradiated adult insects. Biol Rev 47:211–240

Dunn CW, Hejnol A, Matus DQ, Pang K, Browne WE, Smith SA, Seaver E, Rouse GW, Obst M, Edgecombe GD, Sørensen MV, Haddock SHD, Schmidt-Rhaesa A, Okusu A, Kristensen RM, Wheeler WC, Martindale MQ, Giribet G (2008) Broad phylogenomic sampling improves resolution of the animal tree of life. Nature 452:745–749

Franks F, Hatley RHM, Mathias SF (1991) Materials science and the production of shelf stable biologicals. Pharm Technol Int 3:24–34

Gabriel WN, McNuff R, Patel SK, Gregory TR, Jeck WR, Jones CD, Goldstein B (2007) The tardigrade *Hypsibius dujardini*, a new model for studying the evolution of development. Dev Biol 312:545–559

Guidetti R, Jönsson KI (2002) Long-term anhydrobiotic survival in semi-terrestrial micrometazoans. J Zool 257:181–187

Hand SC, Jones D, Menze MA, Witt TL (2007) Life without water: expression of plant LEA genes by an anhydrobiotic arthropod. J Exp Zool 307A:62–66

Hengherr S, Heyer AG, Köhler HR, Schill RO (2008) Trehalose and anhydrobiosis in tardigrades–evidence for divergence in responses to dehydration. FEBS J 275:281–288

Horikawa DD, Higashi S (2004) Desiccation tolerance of the tardigrade *Milnesium tardigradum* collected in Sapporo, Japan, and Bogor. Indonesia Zool Sci 21:813–816

Horikawa DD, Sakashita T, Katagiri C, Watanabe M, Kikawada T, Nakahara Y, Hamada N, Wada S, Funayama T, Higashi S, Kobayashi Y, Okuda T, Kuwabara M (2006) Radiation tolerance in the tardigrade *Milnesium tardigradum*. Int J Radiat Biol 82:843–848

Horikawa DD, Kunieda T, Abe W, Watanabe M, Nakahara Y, Sakashita T, Hamada N, Wada S, Funayama T, Kobayashi Y, Katagiri C, Higashi S, Okuda T (2008) Establishment of a rearing system of the extremotolerant tardigrade *Ramazzottius varieornatus*: a new model animal for astrobiology. Astrobiology 8:549–556

Horikawa DD, Iwata K, Kawai K, Koseki S, Okuda T, Yamamoto K (2009) High hydrostatic pressure tolerance of four different anhydrobiotic animal species. Zool Sci 26:238–242

Horneck G (1999) Astrobiology studies of microbes in simulated interplanetary space. In: Ehrenfreund P, Krafft C, Kochan H, Pirronello V (eds) Laboratory astrophysics and space research. Springer, Berlin, pp 667–686

Horneck G (2003) Could life travel across interplanetary space? Panspermia revisited. In: Rothschild LJ, Lister AM (eds) Evolution of planet earth. Academic, Amsterdam, pp 109–127

Iwasaki T (1964) Sensitivity of Artemia eggs to the gamma-irradiation. III. The sensitivity and the duration of hydration. J Radiat Res 5:91–96

Johnson AP, Pratt LM, Vishnivetskaya T, Pfiffner S, Bryan RA, Dadachova E, White L, Radtke K, Chan E, Tronnick S, Borgonie G, Mancinelli R, Rotshchild L, Rogoff D, Horikawa DD, Onstott TC (2011) Extended survival of several microorganisms and relevant amino acid and biomarkers under simulated Martian surface conditions as a function of burial depth. Icarus 211:1162–1178

Jönsson KI (2007) Tardigrades as a potential model organism in space research. Astrobiology 7:757–766

Jönsson KI, Harms-Ringdahl M, Torudd J (2005) Radiation tolerance in the eutardigrade *Richtersius coronifer*. Int J Radiat Biol 81:649–656

Jönsson KI, Rabbow E, Schill RO, Harms-Ringdahl M, Rettberg P (2008) Tardigrades survive exposure to space in low Earth orbit. Curr Biol 18:R729–R731

Keilin D (1959) The problem of anabiosis or latent life: history and current concept. Proc R Soc Lond B 150:149–191

Kikawada T, Nakahara Y, Kanamori Y, Iwata K, Watanabe M, McGee B, Tunnacliffe A, Okuda T (2006) Dehydration-induced expression of late-embryogenesis abundant proteins in an anhydrobiotic chironomid. Biochem Biophys Res Commun 348:56–61

Krisko A, Radman M (2010) Protein damage and death by radiation in *Escherichia coli* and *Deinococcus radiodurans*. PNAS 107:14373–14377

Lapinski J, Tunnacliffe A (2003) Anhydrobiosis without trehalose in bdelloid rotifers. FEBS Lett 553:387–390

Madin KAC, Crowe JH (1975) Anhydrobiosis in nematodes: carbohydrate and lipid metabolism during dehydration. J Exp Zool 193:335–342

Mattimore V, Battista JR (1996) Radioresistance of *Deinococcus radiodurans*: functions necessary to survive ionizing radiation are also necessary to survive prolonged desiccation. J Bacteriol 178:633–637

May RM, Maria M, Guimard J (1964) Action différentielle des rayons x et ultraviolets sur le tardigrade *Macrobiotus areolatus*, a l'état actif et desséché. Bull Biol Fr Belg 98:349–367

Neumann S, Reuner A, Brümmer F, Schill RO (2009) DNA damage in storage cells of anhydrobiotic tardigrades. Comp Biochem Physiol A 153:425–429

Ono F, Saigusa M, Uozumi T, Matsushima Y, Ikeda H, Saini NL, Yamashita M (2008) Effect of high hydrostatic pressure on a life of a tiny animal tardigrade. J Phys Chem Solids 69:2297–2300

Pigon A, Weglarska B (1955) Rate of metabolism in tardigrades during active life and anabiosis. Nature 176:121–122

Rahm PG (1921) Biologische und physiologische Beiträge zur Kenntnis de Moosfauna. Z allgem Physiol 20:1–35

Ramløv H, Westh P (1992) Survival of the cryptobiotic eutardigrade *Adorybiotus coronifer* during cooling to $-196°C$: effect of cooling rate, trehalose level, and short-term acclimation. Cryobiology 29:125–130

Ramløv H, Westh P (2001) Cryptobiosis in the eutardigrade *Adorybiotus coronifer*: tolerance to alcohols, temperature and *de novo* protein synthesis. Zool Anz 240:517–523

Rebecchi L, Altiero T, Guidetti R, Cesari M, Bertolani R, Negroni M, Rizzo AM (2009a) Tardigrade resistance to space effects: first results of experiments on the LIFE-TARSE mission on FOTON-M3 (September 2007). Astrobiology 9:581–591

Rebecchi L, Cesari M, Altiero T, Frigieri A, Guidetti R (2009b) Survival and DNA degradation in anhydrobiotic tardigrades. J Exp Biol 212:4033–4039

Rothschild LJ, Mancinelli RL (2001) Life in extreme environments. Nature 409:1092–1101

Sakurai M, Furuki T, Akao K-i, Tanaka D, Nakara Y, Kikawada T, Watanabe M, Okuda T (2008) Vitrification is essential for anhydrobiosis in an African chironomid, *Polypedilum vanderplanki*. PNAS 105:5093–5098

Schokraie E, Hotz-Wagenblatt A, Warnken U, Mail B, Förster F, Dandekar T, Hengherr S, Schill RO, Schnölzer M (2010) Proteomic analysis of tardigrades: towards a better understanding of molecular mechanisms by anhydrobiotic organisms. PLoS One 5:e9502

Seki K, Toyoshima M (1998) Preserving tardigrades under pressure. Nature 395:853–854

Suzuki AC (2003) Life history of *Milnesium tardigradum* Doyère (Tardigrada) under a rearing environment. Zool Sci 20:49–57

Watanabe M (2006) Anhydrobiosis in invertebrates. Appl Entomol Zool 41:15–31

Watanabe M, Kikawada T, Yukuhiro F, Okuda T (2002) Mechanism allowing an insect to survive complete dehydration and extreme temperatures. J Exp Biol 205:2799–2802

Watanabe M, Kikawada T, Okuda T (2003) Increase of internal ion concentration triggers trehalose synthesis associated with cryptobiosis in larvae of *Polypedilum vanderplanki*. J Exp Biol 206:2281–2286

Watanabe M, Sakashita T, Fujita A, Kikawada T, Horikawa DD, Nakahara Y, Wada S, Funayama T, Hamada N, Kobayashi Y, Okuda T (2006) Biological effects of anhydrobiosis in an African chironomid, Polypedilum vanderplanki on radiation tolerance. Int J Radiat Biol 82:587–592

Westh P, Ramløv H (1991) Trehalose accumulation in the tardigrade *Adorybiotus coronifer* during anhydrobiosis. J Exp Zool 258:303–311

Wise MJ, Tunnacliffe A (2004) POPP the question: what do LEA proteins do? Trends Plant Sci 9:13–17

Wright JC (1989) Desiccation tolerance and water-retentive mechanisms in tardigrades. J Exp Biol 142:267–292

Yoshinaga K, Yoshioka H, Kurosaki H, Hirasawa K, Uritani M, Hasegawa M (1997) Protection by trehalose of DNA from radiation damage. Biosci Biotechnol Biochem 61:160–161

Biodata of **Robert M.M. Crawford**, author of "***Long-Term Anoxia Tolerance in Flowering Plants***."

Robert M.M. Crawford is now Professor Emeritus at the University of St. Andrews, Scotland. He obtained his Ph.D. at the University of Liege (Belgium in 1960). Subsequently, he researched in the Univerisities of Freiburg im Breisgau, Oxford and Bern as well as the A.N. Bakh Institute of the Soviet Academy of Sciences in Moscow. He is a Fellow of the Royal Society of Edinburgh and the Linnean Society of London and an associate member of the Academie Royale, Belgium. Until 1999, he was Professor of Plant Ecology in the University of St. Andrews, Scotland, where his research centered on adaptations of plants to extreme environments, with particular attention to problems of flooding tolerance and anoxia, as well as long-term survival in the Arctic. Professor Crawford is the author of ***Plants at the Margin*** (Cambridge University Press, 2008).

E-mail: **rmmc@st-andrews.ac.uk**

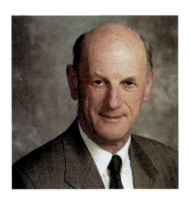

LONG-TERM ANOXIA TOLERANCE IN FLOWERING PLANTS

ROBERT M.M. CRAWFORD
School of Biology, The University of St Andrews, St Andrews KY16 9AL, Scotland, UK

1. Prevalence of Anoxia in Flowering Plants

At first sight, the morphology of flowering plants does not suggest that plant tissues are liable to oxygen deprivation. The aerial expansion of leaves, the ability to aerate roots by diffusion of oxygen generated in the shoot, and the existence of both diffusive and convective mechanisms of oxygen transport (Wegner 2010) argue against the possibility of anoxic conditions existing even in the underground parts of the plant. It might be expected that anoxia would be only a transitory condition in plants, as on a hot day in an inundated soil. However, there are many other more frequent and prolonged situations where oxygen is either deficient (hypoxia) or totally absent (anoxia). Overwintering rhizomes and stolons of marsh and aquatic plants are deprived of oxygen when their shoots die down in autumn. The underground organs gradually become anaerobic, and in spring glycolysis is a major source for the energy required for new shoot growth and emergence from the anaerobic submerged muds of lakes and rivers. Germinating seeds endure a period of enforced hypoxia as metabolism accelerates before the seed coat is ruptured, giving access to air. This is particularly the case for large-seeded species, such as peas and beans. If the germination process is interrupted by cool, wet conditions, the period of enforced anaerobiosis that develops before the rupture of the testa can be fatal. Large fruits also impede oxygen supply to the developing seed, and in these species the seed embryo has to be tolerant of oxygen deprivation. The vascular cambium of trees is separated from the air by the outer bark and secondary phloem, and as a result the cambium can experience sufficient hypoxia to induce anaerobic fermentation (Macdonald and Kimmerer 1991). Xylem sap flow apparently provides an aqueous pathway for facilitating oxygen supply to the wood parenchyma of common birch saplings (*Betula pubescens*), and when this is reduced or interrupted, there can be a 70% reduction in oxygen transport to the sap wood (Gansert 2003). The source of this oxygen however is not clear. It is possible that it is derived from photosynthetic activity in the chlorophyll-containing tissues under the bark and then distributed by the xylem sap flow when it is flowing normally. Hypoxia can be a common condition in tree trunks, and it has been suggested that the formation of heartwood in trees is triggered by low-oxygen conditions and even anoxia (Mugnai and Mancuso 2010).

One somewhat surprising occurrence of anoxia in flowering plants is found in carnivorous plants (Adamec 2007). Measurements of steady-state oxygen levels in excised and intact traps of six species of *Utricularia* always approached zero despite the presence of many microbial commensal species. The oxygen concentration appears to be below that necessary for prey survival and causes the prey to die from suffocation. There are however short periods of higher oxygen levels after firing of the traps to which the internal trap glands and trap commensals are considered to be adapted (see also Adamec 2011).

Maize seedlings, even when growing with free access to air on moist filter paper, have a zone of anaerobic activity in their root tips (Fig. 1) due to the density

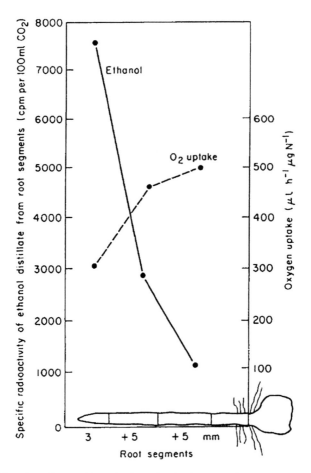

Figure 1. Manifestation of natural hypoxia that exists in the roots of maize seedlings when grown in air. Plotted against the root segment position is the specific activity of the combusted ethanol from the root segments that were incubated with 0.05 M $3^{14}C$ pyruvate for 4 h under air, together with the oxygen consumption of the same segments (Reproduced with permission from Crawford 1992).

of the young meristematic tissue impeding access to oxygen (Crawford 1992). Given that an unstressed root typically has a hypoxic meristem, it is not surprising that any stress that might impair free access to oxygen will increase the degree of anaerobiosis (Armstrong et al. 2009). It has long been known that roots contract when exposed to drought (Faiz and Weatherley 1982). Subsequent research has shown that such contractions can induce anaerobiosis. Measurements of ethanol production in roots of saplings of the tropical pacara earpod tree (*Enterolobium contortisiliquum*) and the lemon-scented gum (*Eucalyptus citriodora*) while growing with free access to air revealed an accumulation of ethanol when exposed to drought. A similar effect has been noted in chickpea (*Cicer arietinum*), carrot (*Daucus carota*), French bean (*Phaseolus vulgaris*), as well as in the cereals barley (*Hordeum vulgare*) and maize (*Zea mays*). Drought-induced anaerobiosis has also been found to have a positive effect in that it can preadapt roots to subsequent severe oxygen deprivation when flooded. Experiments with seedlings of *Eucalyptus citriodora* have shown that drought increases subsequent tolerance of flooding (Scarano 1992). It is therefore evident that oxygen deprivation, in common with many other stresses, can arise in a variety of ways. Flooding, however, is the most obvious cause of oxygen deficiency, even though the opposite stress of drought can also give rise to hypoxic conditions.

It has long been apparent in agriculture that flooding, which reduces oxygen availability, can be injurious to many plant species. The Roman agrarian writer Cato (234–149 BC), in his treatise *De Agri Cultura* written in 160 BC, observed that "although water may be allowed to stand on wheat fields during the cold part of winter it must be removed before the spring" (White 1970). Research into the dangers of oxygen deprivation, as applied to flowering plants, has been stimulated mainly by a desire to improve the flooding tolerance of agricultural crops. Even rice can suffer from anoxic injury when the growing shoots are submerged (Ismail et al. 2009).

Understandably, crop research is primarily concerned with what happens to plants flooded during the growing season, and although seasonal variations in stress tolerance have long been recognized, less attention has been given to the problem of prolonged winter flooding. In particular, agricultural research has neglected the question as to whether the adaptations that allow some crop species to survive short periods of flooding during the growing season are the same as those that enable aquatic and amphibious plants to endure long periods of flooding at any season of the year (Crawford 2003).

1.1. ANOXIA IN THE ARCTIC

Arctic plants which live in habitats with prolonged snow and ice cover can be subjected to lengthy periods of anoxia. Encasement in ice is a barrier to oxygen diffusion which can last from October to June, and sometimes for more than a year when large snow banks fail to melt in summer. As a result, the Arctic, which accounts for 5.5% of the Earth's land surface, is the region where plants are most at risk of having to endure long periods without access to oxygen.

Paradoxically, as a result of climatic warming, this risk is increasing. In polar winters, when the sun fails to rise above the horizon for many weeks or months, the soil remains frozen. During intervals of warm weather arriving from the south, the rain that falls on the frozen soils of the tundra quickly turns to ice. In addition, any water coming from short periods of snowmelt refreezes as an oxygen-impermeable barrier.

It is therefore not surprising that a uniquely high level of anoxia tolerance has been observed when testing plants in situ at high latitudes in Spitsbergen (79°N). Plants, complete with root systems, when placed on moist filter papers in anaerobe jars and kept in the dark at ambient air temperatures for 7 days, are in an environment that is totally devoid of oxygen (Fig. 2). It is therefore possible to observe any deleterious effects either during anoxia or in the subsequent post-anoxic period. From 20 species tested in this manner, 13 were found to be completely tolerant of anoxia and its after effects (see Crawford 2008). It was particularly remarkable that even in summer, at the height of the growing season, so many species did not show any damage to the growing leaves. Even the flowers were undamaged by the anaerobic incubation. Normally, green leaves lose turgor and wither rapidly when deprived of oxygen. Tests on more southern populations of *Saxifraga oppositifolia* from Norway, Iceland, and Scotland failed to find any degree of anoxia tolerance to match that of the Spitsbergen populations as the southern populations were killed by only 4 days of total anoxia (see also Sect. 3.2.1 on foliar tolerance of anoxia).

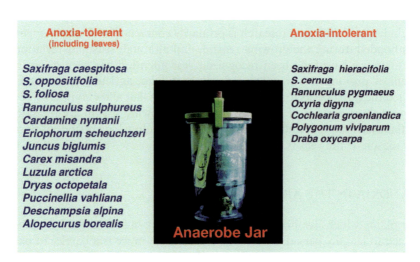

Figure 2. Ability of Arctic species examined in Spitsbergen to survive 7 days of total anoxia in an anaerobe jar in the dark at ambient arctic temperatures. In the anoxia-tolerant species the entire plant including leaves and flowers survived (see also Crawford et al. 1994).

1.2. ANOXIA AND THE AMERICAN CRANBERRY

The importance of carbohydrate reserves in relation to flooding tolerance is a predominant factor in achieving long-term resistance to the imposition of extended periods of anoxia. This is borne out in studies on the American cranberry (*Vaccinium macrocarpon*). Cranberries are sensitive to frost damage, and the vines are grown in specially constructed cranberry bogs, which are flooded in winter (Fig. 3). The water freezes to form an ice blanket, which prevents drying of the evergreen vines by cold wind and minimizes fluctuations in temperature.

Figure 3. Cranberry bog flooded for winter protection of the vines from frost. This method of protection risks exposing the cranberry vines to prolonged periods of oxygen deficiency. Inset: cranberry plant (*Vaccinium macrocarpum*) which has been kept under total anoxia for 21 days, followed by 3 days of post-anoxic recovery. The only visible damage is to the younger leaves which turned pale yellow on return to air (Reproduced from Crawford 2008).

Although frost damage is avoided by this treatment, the ice impairs oxygen diffusion. Consequently, the submerged plants are prone to oxygen deficiency stress, particularly when the ice is covered by snow, which prevents light from reaching the submerged plants. The lack of submerged light stops the evergreen leaves from alleviating their interrupted oxygen supply through photosynthetic activity. Especially dangerous are warmer periods during flooding, when respiration rises and carbohydrate reserves are rapidly depleted, making the plants more sensitive to subsequent ice encasement. The carbohydrate status of the overwintering vines has also a direct influence on fruit set. In spring, a rapid resumption of growth requires adequate carbohydrate supplies, and insufficient carbohydrate levels may be responsible for the low fruit set, which is sometimes observed.

Experimental studies with the American cranberry on the effects of oxygen deprivation have shown that during anoxia (the total withdrawal of available oxygen), the mature leaves exhibit a marked downregulation of metabolism. Carbohydrate consumption and energy metabolism stabilize at low levels soon after the switch from aerobic to anaerobic pathways. Pathways such as TCA cycle and photosynthesis, which are non-operating during the anoxia treatment, are still capable of resuming activity after 28 days of anoxia. Most remarkable however is the rapid recovery (Fig. 4) in photosynthesis, which takes place when the plants are returned to air (Schlüter and Crawford 2003).

1.3. NATURAL ANOXIA IN SEEDS

Most seeds when germinating pass through a period of natural anaerobiosis due to the impermeability of the seed coat (testa). Prompt rupturing of the seed coat is normally essential if anaerobic injury is to be avoided. However, in the seeds of aquatic plants, rupturing of the seed coat for seeds that are lying under water in anaerobic mud does not provide ready access to oxygen. Among aquatic plants, there is a small but significant number of species which can initiate germination in the total absence of oxygen. The list includes rice as well as wild rice (*Zizania* spp.). In these cases, it is only the coleoptile that is extended in order to reach the surface of the water, contact the atmosphere, and transfer oxygen to the seed, allowing subsequent growth of the radicle and leaf. In the anoxic cells of rice coleoptiles, an efficient alcoholic fermentation allows an elevated energy charge to be maintained. Significant RNA and protein syntheses, including phosphorylation and glycosylation, also occur (Bertani et al. 1997). In addition to rice, various species of the genus *Echinochloa,* a graminaceous weed of rice fields, can also germinate under total anoxia.

The ability of *Echinochloa* spp. to match the anoxia tolerance of rice in germination makes the genus a troublesome weed of paddy fields (Kennedy et al. 1980). Texas wild rice (*Zizania texana*) has improved germination at low oxygen concentrations, with significantly higher rates at 0.1–2.0 ppm O_2 and lower germination

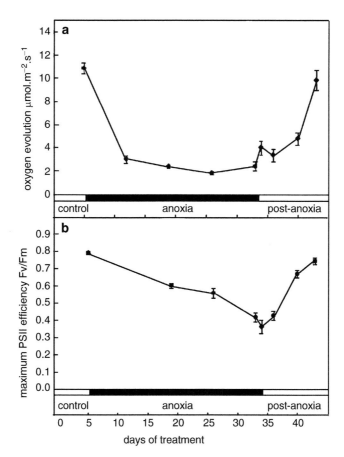

Figure 4. Effects of prolonged anoxia followed by post-anoxic aeration on the recovery of photosynthetic activity of mature cranberry leaves; (**a**) oxygen evolution as measured by an oxygen electrode; (**b**) maximum photosystem II efficiency as measured by fluorescence (Reproduced from Crawford and Schlüter 2003).

at 3.0–5.0 ppm O_2. Anoxia for this species is clearly an important environmental cue for germination. For a species that is sensitive to drought, anaerobic sediments provide the most suitable substrates, as they tend to retain some water even during a dry weather (Power and Fonteyn 1995).

Germination under anaerobic conditions is not restricted to aquatic species. Myxospermous seeds (seeds with a coat that produces mucilage when wet) are found in certain drought-tolerant species. The mucilage not only provides protection against desiccation but also impedes access to oxygen during germination. In the myxospermous Brazilian dry-forest tree *Chorisia speciosa*, germination is capable of proceeding as far as radicle protrusion even in total absence of oxygen

(Joly 1982). The advantages of mucilage in reducing desiccation injury, and possibly also fungal attack, have to be combined with an ability to extend the radicle anaerobically as the mucilage restricts oxygen access to the interior of the seed. If on protrusion the radicle still finds no oxygen available, further development ceases. A similar example of a dryland species being able to germinate under total anoxia is found in the S.E. African leguminous plant, the Cape kafferboom (*Erythrina caffra* – Small et al. 1989). The ecological advantage for this species probably stems from the ability of the seedling to break open the hard testa which, while intact, impedes the entry of oxygen. As with germinating seeds of *C. speciosa,* after the rupture of the testa there is no further development in *E. caffra* unless oxygen becomes available. Other species which are recorded as being able to germinate under very low concentrations of oxygen include cucumber (*Cucumis sativus*) and the tropical Silver cockscomb (*Celosia argentea*) as well as the Indian lotus (*Nelumbo nucifera*).

There are numerous reports of seeds of the Indian lotus (*Nelumbo nucifera*) being able to germinate after centuries of burial in anaerobic mud. In some cases where dates of over a millennium have been claimed for their age, doubts have arisen concerning the carbon samples that were chosen from the neighboring sediments as proxy indicators of the age of the seeds. More recent analyses, in which the carbon dates are taken from samples of the testa, provide more reliable estimates. An examination of 60 lotus fruits excavated from a now dry lakebed in north-eastern China gave ages of 200–500 BP by C^{14} dating. The seeds germinated rapidly but exhibited abnormal phenotypes which resembled those of chronically irradiated plants (Shen-Miller et al. 2002). In cold environments, such as the Arctic tundra, buried seeds can potentially persist in the soil for centuries under peat deposits. A study carried out on a Tussock tundra site in Alaska succeeded on germinating seeds of the small-flowered woodrush (*Luzula parviflora*) and the stiff sedge (*Carex bigelowii*) that had persisted for at least 197 ± 80 years, together with viable seeds associated with wood fragments 295 ± 75 years old (McGraw et al. 1991). For further details of seed longevity, see (Baskin and Baskin 1998).

2. Evolution of Flooding Tolerance

Adapting to anoxia has had a profound influence on plant evolution. Aquatic and marsh plants have many similarities with diving mammals and birds in that they have all evolved from terrestrial ancestors that have left dryland habitats with free access to air for an aquatic or amphibious lifestyle where the supply of oxygen is severely limited. The majority of the plants that can survive with their roots continually submerged in water, or in areas prone to regular flooding (amphibious plants), belong to the monocotyledons. It is over 180 years since the Swiss botanist Augustin Pyramus De Candolle (1778–1841) proposed an aquatic origin for the monocots. It is now considered that the evolution of the monocots, via aquatic or amphibious species, has occurred many times. The actual number of times can be debated but is generally estimated to be in excess of two hundred (Cook 1999). The simplified leaf form, the reduction in secondary thickening, and the reliance on adventitious root systems, in monocots have long been regarded as an ancestral response of certain groups of dryland plants to the wetland conditions (Henslow 1893; Arber 1920). Apart from the morphological changes, there would also have been a number of physiological adaptations, including tolerance of long-term anoxia, especially for plants that have to endure winters in permanently flooded habitats.

The pre-eminent position of monocots in aquatic fringe habitats is exemplified by the dominance of sedges, rushes, and reeds in the wetlands and swamps of the world. Only 4% of the dicotyledonous families can be regarded as aquatic compared with 33% in the monocots (Arber 1920). Given the uncertain divisions between families, this figure can be regarded only as an approximation. However, as there are approximately 4.5 times as many dicot as monocot species in the world's flora, it is all the more remarkable that the monocots hold such dominance over wetland and aquatic habitats. Nevertheless, despite the overwhelming predominance of the monocot families, the dicot family of the water lilies (Nymphaceae) is one of the most ancient families of aquatic flowering plants, having evolved approximately 120 million years ago in the Lower Cretaceous at the dawn of the appearance of the flowering plants (Friis et al. 2001).

For most plants of wetland habitats, the ability to lead an amphibious or aquatic life depends on a combination of mechanisms for *anoxia tolerance* and *anoxia avoidance*, both of which have a long evolutionary history. Survival in such habitats requires a number of adaptive responses not just to the metabolism of overwintering organs but also for combating the other dangers that arise from inhabiting an oxygen-deficient milieu with the presence of potentially toxic reduced soluble ions of iron and manganese, etc. It is therefore the combination of anoxia tolerance with anoxia avoidance mechanisms that makes it possible for many flowering plants to survive prolonged flooding in summer and winter. During the growing season, avoidance mechanisms can predominate with the development of tissues that facilitate aeration from shoot to root. The development of aeration mechanisms has apparently been an early feature in the evolution of large land plants. Morphological and anatomical features which would facilitate gas exchange with the atmosphere can be traced to vascular plants of the Middle and Upper Devonian. The first aeration canals in arboreal plants were found in the lepidodendroids and sigillarians of the Carboniferous era. Strands of aerenchyma-like tissues, connecting the outer and middle cortex of the stem to the leaf mesophyll, were defined as *parichnos*. These areas of aeration tissues were made up of disintegrated cells, which connected the rootlets at the base of the stem to the leaf and lacunar tissue of the mid-cortex. They apparently functioned by taking air in through the leaf and conducting it through the stem to the stigmarian rhizome and associated rootlets. In ancient plant forms, *parichnos* were found in the absence of cauline stomata. Similar structures were initially present in some members of the genus *Pinus*, certain Abietineae, and in the ancient, and still extant, maiden hair tree *Ginkgo biloba* (Hook and Scholtens 1978).

Horsetails (*Equisetum* spp.) are also notable for both their antiquity and early development of internal aerating structures. Investigations on the capacity for convective flow of air from shoot to root in the great horsetail, *Equisetum telmateia*, have brought to light the possibility that convection aerated the massive rhizomes of the genus *Calamites*, the extinct giant horsetails of the Carboniferous. It has therefore been suggested that plant ventilation by a type of "molecular gas pump" may date back more than 350 million years (Armstrong and Armstrong 2009).

3. Survival Strategies for Anoxia Avoidance

Any discussion of anoxia tolerance strategies in plants, if it is to avoid confusion, needs to distinguish between adaptations that allow mainly annual and crop plants to overcome short periods of oxygen deprivation during the growing season and those that permit perennial species to endure prolonged periods of inundation either during the growing season or in the non-growing season as can frequently arise with winter flooding. This distinction reflects the typical dichotomy in relation to potential stress between adaptations that aid avoidance from those that aid tolerance. This distinction is all the more important, given the overwhelming research activity carried out on rice in comparison to other plants. Such is the preponderance of papers on rice that it can give the impression that it is the epitome of anoxia tolerance in plants, when in fact, apart from its ability to germinate under anoxia, for the greater part of its life cycle it is mainly an avoider of anoxia (see Ismail et al. 2009). To understand the nature of long-term adaptations to anoxia, it is necessary also to make comparisons with short-term tolerance of anoxia.

3.1. SHORT-TERM ANOXIA TOLERANCE

Plants that are tolerant of short periods of flooding in summer tend to respond to oxygen deprivation by an acceleration of glycolysis to compensate for the lower yield of ATP per molecule of glucose under anaerobic conditions as compared with normal aerobic respiration. This is basically a short-term strategy as it depends not only on an adequate supply of carbohydrate reserves but also on the ability to disperse the potentially toxic fermentation products. This response to anoxia is commonplace in many organisms and has long been known as the *Pasteur effect* originally discovered in 1861 by Louis Pasteur who described it in the reverse sense as *oxygen conserves carbohydrate* or the *inhibiting effect of oxygen on fermentation* (Pasteur 1861). When flooding takes place during the growing season, this type of response can maintain tissues with an adequate supply of metabolic energy, giving the plant time to improve root aeration through the development of adventitious roots and enhanced *aerenchyma* (air spaces which improve the exchange of gases between the shoot and the root). Anaerobic shoot growth through cell extension is also a short-term response that can be initiated by the action of ethylene. A review of *ethylene-driven underwater growth* has examined the process of signaling in relation to flooding survival in rice (Bailey-Serres and Voesenek 2010).

The onset of anoxia is nearly always accompanied by an increase in carbon dioxide. Carbon dioxide levels can therefore act as reliable indicators of impending oxygen shortages. It is therefore illuminating to see how plants that differ in their short- and long-term tolerance of anoxia react to the presence or absence of imposed high carbon dioxide levels. When such conditions are applied to anoxia-intolerant chickpeas (*Cicer arietinum*), it has been found that anoxia combined with high carbon dioxide levels results in an increase in ethanol production, followed by a severe reduction in survival (Fig. 5).

The reverse situation is found in long-term anoxia-tolerant species (Fig. 6) where high concentrations of carbon dioxide reduce anaerobic respiration and ethanol production (Crawford 2003).

Attempts to improve the flooding tolerance of wheat in Australia and India have demonstrated a further complication between short-term plant survival and the nature of the anaerobic environment. Here the situation becomes complicated as the adverse effects of hypoxia depend on the nature of the soil in relation to the relative concentrations of the elements Mn, Fe, Na, Al, and B. These detrimental ionic toxicities are similar to the well-known interactions of salinity in aggravating the adverse effects of hypoxia (Akhtar et al. 1994). Such diverse element toxicities (or deficiencies) are exacerbated during waterlogging and explain why waterlogging tolerance at one site may not always be replicated at another (Setter et al. 2009).

The above examples illustrate the complexity of improving anoxia tolerance in crops during the growing season, when the plants can endure only brief interruptions to their oxygen supply. So far, it has not been possible to select annual crop varieties that possess the degree of anoxia tolerance that can be found in perennial native species of wetland habitats.

3.2. LONG-TERM ANOXIA TOLERANCE

Like all things in nature, long-term survival under anoxia can show many variations between tolerance of total anoxia and long-term tolerance of hypoxia. In the field, the degree of anaerobiosis can vary, and in studying the effects of oxygen deprivation, both the hypoxic and the anoxic states have to be considered. Experimentally, it is possible in the laboratory to impose conditions of total anoxia using specialized anaerobic incubators, and these can differentiate clearly between species in their innate capacity to withstand anoxia. Table 1 compares the

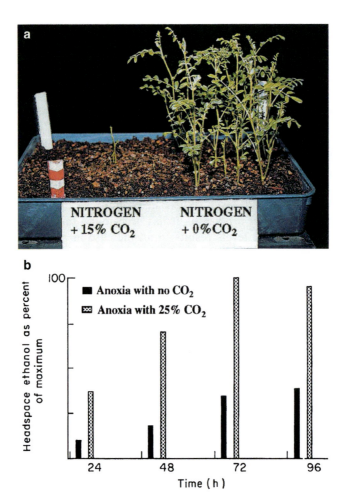

Figure 5. Chickpea (*Cicer arietinum*) survival after varying treatments in an anaerobic incubator. (**a**) The plants on the *left* are failing to recover after incubation in the presence of 15% carbon dioxide. The plants on the *right* had all carbon dioxide removed from the anaerobic atmosphere by potassium hydroxide. (**b**) Ethanol evolution under anoxia as detected in the headspace above chickpea seedlings ($n=3$) (see also Crawford 1992).

survival of detached underground stems and rhizomes of a variety of species subjected to total anoxia in an anaerobic incubator. Root viability is not included as roots are not essential for long-term survival as they can be regenerated after a period of anoxia, provided the rhizome or underground stem survives with viable buds.

Species that are killed by a mere 4 days' total deprivation of oxygen (group 1) can be described safely as intolerant. Group 2 is of interest as in these species, the time of year can have a marked influence on the ability to survive. Reed sweetgrass

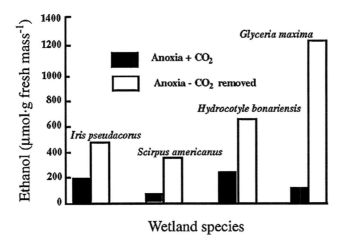

Figure 6. Effect of high carbon dioxide accumulations in a static anaerobic atmosphere as compared with one where carbon dioxide is removed (see text) on the rate of ethanol production by rhizomatous species ($n = 25$) (Reproduced from Crawford 2003).

Table 1. Maximum anoxia tolerance of underground stems and rhizomes as observed in an anaerobic incubator at 20–25°C. Survival was measured in having an ability to grow new shoots after the period of anaerobic incubation (Crawford and Braendle 1996).

Group 1. Species with minimal tolerance of anoxia, surviving only 1–4 days
Carex rostrata
Eriphorum vaginatum
Juncus conglomeratus
Juncus effusus
Oxyria digyna
Saxifraga cernua
Saxifraga hieracifolia
Solanum tuberosum

Group 2. Species with moderate tolerance of anoxia, surviving 4–21 days
Carex papyrus
Carex rostrata
Eleocharis palustris
Eriophorum angustifolium
Festuca vivipara (Iceland)
Filipendula ulmaria
Glyceria maxima
Iris germanica
Mentha aquatica
Phalaris arundinacea
Ranunculus repens
Saxifraga caespitosa
Saxifraga oppositifolia
Tussilago farfara

(continued)

Table 1. (continued)

Group 3. Species with a high tolerance of anoxia, surviving 1–3 months
Acorus calamus
Bolboschoenus maritimus
Deschampsia beringensis (N. Alaska)
Iris pseudacorus
Phragmites australis
Schoenoplectus lacustris
Spartina anglica
Typha latifolia

(*Glyceria maxima*) is such a case. In early summer, when the overwintering carbohydrate reserves have been depleted producing the new season's growth, this species can be killed in as little as 4 days of total experimental anoxia. Earlier in the season before the resumption of spring growth, their anoxia tolerance can be considerably greater. In the third group, there is a high degree of anoxia tolerance at all seasons of the year, with some species surviving more than 3 months. The most notable species so far reported are the sweet flag (*Acorus calamus*) and the yellow flag iris (*Iris pseudacorus*) where the rhizomes can survive for over 2 months in darkness in an anaerobic incubator (see also 1.1.) where the oxygen concentration is less than 0.001% (Schlüter and Crawford 2001).

3.2.1. Foliar Tolerance of Anoxia
In relation to foliar tolerance of anoxia, it should be noted that there are also outstanding examples in some warm-temperate species of long-term anoxia tolerance in whole plants and their leaves. Tests imposing prolonged periods of anoxia on intact plants of the sweet flag (*Acorus calamus*) and flag iris (*Iris pseudacorus*) found the leaves capable of resuming photosynthesis after prolonged periods of total anoxia (Fig. 7). The leaves of the sweet flag have even been able to resume photosynthesis after 96 days of total anoxia (Schlüter and Crawford 2001). Prolonged experiments with seedlings of Arctic and sub-Arctic grasses have also found a remarkable ability to survive anoxia. The limit to anoxia tolerance in these species appears to indicate that long-term survival of anoxia is dependent on having adequate carbohydrate reserves (Crawford and Braendle 1996). Any treatment which reduces carbohydrate levels also diminishes anoxia tolerance. This has also been demonstrated (see below) in various studies on the American cranberry (*Vaccinium macrocarpon*).

3.2.2. Adaptations for Winter Survival Under Anoxia
The range of responses to the imposition of anoxia on plant tissues is summarized in Fig. 8. These are arranged in a successive series of possible responses, some of which are alternative possibilities, while others represent mutually exclusive and opposing reactions. These divisions are never absolute. In the case of avoiding

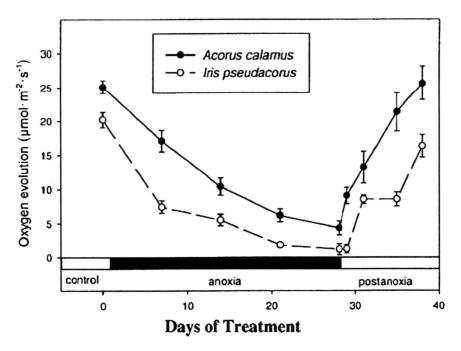

Figure 7. Photosynthetic capacity estimated as oxygen evolution and measured with an oxygen electrode in leaves of *Acorus calamus* and *Iris pseudacorus* immediately on removal from anoxic chamber at intervals of up to 28 days of anoxia (Reproduced with permission from Schlüter and Crawford 2001).

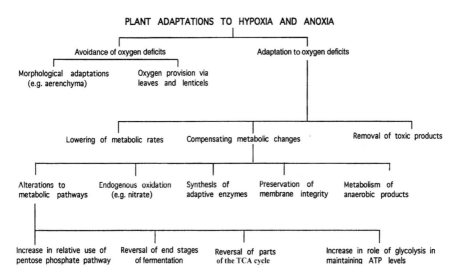

Figure 8. Diagrammatic representation of the diversity of adaptations that can be observed in higher plants which can contribute to their tolerance of flooding (Reproduced with permission from Crawford and Braendle 1996).

or tolerating low oxygen levels, plants can vary by degree in relation to their use of the two opposing strategies. Rhizomes of *Carex rostrata, Juncus effusus, J. conglomeratus,* and *Glyceria maxima* can be killed in as little as 4 days of total anoxia. In these species, the winter foliage can provide access to oxygen either by diffusion or from photosynthetic activity when the base of the shoots remains green. The length of time that the rhizomes can survive under experimentally imposed anoxia appears to be a function of the level of carbohydrate supply and respiratory activity.

For northern agriculture, winter survival under anaerobic conditions is an important consideration in the cultivation of many winter cereals as well as perennial pasture and meadow grasses. Anoxia tolerance has therefore been a subject for extensive study in relation to ice-encasement injury in forage crops in eastern Canada and Iceland (Gudleifsson 1997). In a Canadian experiment (Bertrand et al. 2001), anoxic stress was imposed on a number of forage crop species by enclosing potted plants in gas-tight winter conditions in an unheated greenhouse. It was found that near-anaerobic conditions were reached after 60 days of enclosure for orchard grass (*Dactylis glomerata*), alfalfa (*Medicago sativa*), and red clover (*Trifolium pratense*) and after 80 days for timothy (*Phleum pratense*). Red clover and orchard grass were the most anoxia-sensitive species, followed by alfalfa and timothy being the most resistant.

The concentration of ethanol increased in response to oxygen deprivation and reached the highest value in the sensitive red clover, whereas its concentration was the lowest in the anoxia-tolerant timothy.

The expression of the ADH gene (*Adh*) was also markedly lower in timothy than in the other three species. It was concluded that the greater resistance of timothy to anaerobic conditions at low temperatures is associated with a slower glycolytic metabolism (Bertrand et al. 2001). It appears therefore that timothy, in common with native overwintering amphibious plants, downregulates its metabolism under anoxia and thus maintains higher carbohydrate reserves during a period of oxygen deprivation when deprived of oxygen. This strategy is one that will favour winter survival and spring regrowth.

Ice encasement is normally a long-term stress, lasting weeks or even months. It is therefore relevant to note that in more resistant plants, it is associated with a downregulation of anaerobic metabolism. This is in contrast with the acceleration of metabolism that is used by plants that are limited to tolerating only a short-term stress. These variations in anoxia tolerance can also be related to differing ecological distributions of the species (Braendle and Crawford 1987). Not all wetland species possessing rhizomes are, however, tolerant of anoxia (e.g., *Juncus* spp., see above).

3.2.3. End Products of Glycolysis and the Accumulation of the Oxygen Debt
An ever-present source of injury to plants is the excessive accumulation of glycolytic end products, such as ethanol and carbon dioxide, which can arise during prolonged periods of anaerobiosis. These reduced products represent an *oxygen debt*,

which is eventually repaid when the tissues are restored to air or the debt is translocated from submerged organs to aerial shoots. Whether or not such accumulations are harmful to the plant tissue in question depends partly on the rate of production and partly on the ease with which they can be removed. Removal mechanisms include a general loss by diffusion or by specific transport mechanisms, such as the transpiration stream or the mass movement of gases. For this reason, any discussion of the susceptibility of plant tissues to the potential hazards of accumulating glycolytic end products needs to be related to the physiology of the whole plant. The flag iris (*Iris pseudacorus*), which is very tolerant of prolonged anoxia, together with other aquatic species, also has a high degree of porosity in its rhizomes, which aids aeration and facilitates the dispersal of ethanol. Estimations of ethanol accumulation in roots, rhizomes, and leaves of the anoxia-tolerant species *Acorus calamus* and *Iris pseudacorus* kept under prolonged anoxia found that ethanol levels reached a plateau after 20–30 days of anaerobic incubation (Schlüter and Crawford 2001). It is also notable that the levels in the roots and rhizomes remained relatively low compared with those in the leaves. The leaves have a greater facility to remove ethanol by dispersion, which enables the plants as a whole to balance ethanol production by dispersion (Fig. 9).

Doubt has been expressed as to whether or not ethanol is toxic to plant tissues in the concentrations that are commonly found after periods of anaerobic incubation (Jackson et al. 1982). However, if anoxia-sensitive organs, such as rhizomes or germinating seedlings, are kept under anoxic environments where the anaerobic atmosphere is kept moving so that ethanol and carbon dioxide do not accumulate, then the anaerobic life of the tissues is commonly prolonged. This has been observed for several species of germinating seeds. Factorial experiments in which static and moving environments are combined with the addition or removal of carbon dioxide have shown that the greatest mortality is found when static environments cause the tissues to be exposed to high carbon dioxide concentrations without ethanol removal (Crawford et al. 1987). The presence of carbon dioxide induced an increase in glycolytic activity, and it was concluded that the deleterious interaction between carbon dioxide and ethanol was due to the increase in ethanol accumulated during the anoxia treatment (see Figs. 5–6).

The moment when ethanol becomes most dangerous to plant tissues appears to be on readmission to air, as the oxidation of ethanol by catalase causes a surge in acetaldehyde concentrations.

Catalase can oxidize a range of different toxins, such as *formaldehyde, formic acid, phenols*, and *alcohols*. In doing so, it uses hydrogen peroxide according to the following reaction:

$$H_2R + H_2O_2 \xrightarrow{Catalase} R + 2H_2O$$

By oxidizing ethanol, catalase increases the damage to plant membranes as acetaldehyde is considerably more phytotoxic for these tissues than ethanol (Studer and Braendle 1988). Studies such as this, which follow the course of ethanol

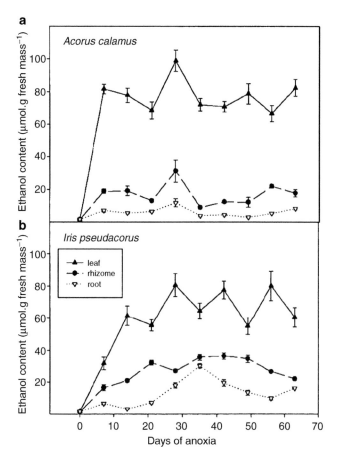

Figure 9. Ethanol content in roots, rhizomes, and leaves of *Acorus calamus* (a) and *Iris pseudacorus* (b) kept under anoxia for 63 days ($n=6$). Note that in these experiments carried out with intact plants, after 20 days of anoxia, a plateau is reached in the concentration of ethanol, with considerable quantities being found in the leaves from where it is readily dispersed (Reproduced with permission from Schlüter and Crawford 2001).

accumulation and its subsequent post-anoxic fate, explain the earlier controversy as to whether or not ethanol is toxic to plant tissues, as those that doubted its toxicity had not looked at the reactions that take place as anaerobic plants return to air after a period of anoxia.

Ethanol is not the only substance used by plants to accumulate and disperse the oxygen debt. There are other compounds not directly involved in glycolysis and in the tricarboxylic acid cycle that serve to accumulate or translocate hydrogen away from the hypoxic and anoxic tissues. The process of oxygen debt removal cannot be examined in detached tissues in anaerobic incubation vessels, as the process is dependent on the functioning of the intact plant. Compounds that can serve to disperse the oxygen debt, apart from ethanol, include organic and amino acids and amines. These hydrogen-rich compounds are less toxic and serve as efficient transporters of the *oxygen debt* without causing tissue damage when access to oxygen is restored. In the temperate marine flowering plant eelgrass (*Zostera marina*), there is an active diurnal cycling of amino acids and amines which accumulates the oxygen debt by night and disperses it by day (Pregnall 2004). As a result of these alternative hydrogen transporters, during the night anoxic conditions prevail but with little production of ethanol. There is instead an accumulation of alanine and γ-aminobutyric acid, which can account for 70% of the total amino acid pool. Upon resumption of shoot photosynthesis and oxygen transport to the roots, the accumulated alanine and γ-aminobutyric acid levels decline rapidly.

Anoxic stress also induces a strong change in sugar, protein, and amino acid metabolism in many higher plants. Sugars are rapidly consumed through the anaerobic glycolysis to sustain energy production. Protein degradation under anoxia is a mechanism to release free amino acids contributing in this way to maintaining the osmotic potential of the tissue under stress. Among free amino acids, a particular role is played by glutamic acid, being a precursor of some characteristic compounds of anaerobic metabolism, e.g., alanine, γ-aminobutyric acid, and putrescine (Reggiani and Bertani 2003).

In birches (*Betula* spp.), the active upward movement of xylem sap in spring carries with it substantial quantities of soluble carbohydrates, ethanol, and organic and amino acids. This flow has been suggested as the means whereby roots of some woody plants compensate for the lack of oxygen in the atmosphere of flooded soils by exporting hydrogen to the shoot and thus maintaining the redox balance of the inundated root system (Crawford 2008). The upward movement of hydrogen in reduced compounds as a constituent of the xylem flow is a more efficient means of repaying the oxygen debt of submerged organs than the slow diffusion of gaseous oxygen through more than a meter of woody tissue. This replacement of oxygen import by hydrogen export does however have a metabolic cost, namely the provision of a high carbohydrate store in the overwintering root.

The flow of sap in birch trees should not be confused with that of the sugar maple (*Acer saccharum*). In the latter, the sap flow is dependent on diurnal freeze-thaw cycles affecting the trunk and stem and is not a function of root pressure. In birch, however, root pressure drives the spring ascent of sap (Kozlowski and Pallardy 1997). It is relevant to note, however, that in the bottomland trees of the USA, namely *Fraxinus pennsylvanica* (green ash) and the equally flood-tolerant *Nyssa aquatica* (water tupelo), high pre-flood root-starch concentrations are important in allowing these species to survive inundation (Gravatt and Kirby 1998).

4. Post-anoxic Injury and the Dangers of Un-flooding

Re-exposure to oxygen after a period of deprivation can cause serious injury both to plant and animal tissues. In animals, this type of damage is usually referred to as post-ischemic injury and is a potential source of tissue damage after heart attacks and organ transplants. In plant tissues, re-exposure of underground perennating organs and roots to air after a period of inundation can cause post-anoxic injury. Thus, in the anoxia-intolerant *Iris germanica*, there is little sign of injury to the rhizomes when they are first taken from an anaerobic incubator after 15 days of oxygen deprivation. However, after 6 h exposure to air, there is extensive peroxidative damage, and the condition of the rhizomes rapidly deteriorates (Hunter et al. 1983).

The return to air generates reactive oxygen species (ROS) to which the tissues of many plant species become vulnerable after a prolonged absence of oxygen and cessation of aerobic metabolism. While flooded, the lack of oxygen causes the transition metals involved in aerobic metabolism (e.g., the iron in cytochromes etc.) to become reduced. The turnover rate of the enzymes needed for aerobic metabolism is also much diminished. Consequently, a sudden re-emergence into air as water tables subside presents an aerobic shock to unprepared tissues. Tissues prone to this type of injury become soft and spongy on return to air and rapidly lose their cell constituents. In these cases, oxygen exerts a definite toxic effect, and membranes are destroyed irreversibly. This type of injury arises from the post-anoxic generation of *reactive oxygen species* (ROS), typically oxygen (O_2^-) and hydroxy radicals ($*HO_2$). Both O_2^- and $*HO_2$ can undergo spontaneous dismutation to produce H_2O_2:

$$H^+ + O_2^- \rightarrow *HO_2$$
$$*HO_2 + *HO_2 \rightarrow H_2O_2 + O_2$$

In the post-anoxic state when cytochromes have been reduced by a previous period of anoxia, the Fenton reaction can be particularly active. This is the iron-salt-dependent decomposition of dihydrogen peroxide, generating highly reactive hydroxyl radicals.

$$Fe^{2+} \text{complex} + H_2O_2 \rightarrow Fe^{3+} \text{complex} + *OH + OH^-$$

As mentioned above, ethanol that has accumulated under anoxia is also rapidly oxidized to acetaldehyde by the presence of H_2O_2 and catalase (N.B. not alcohol dehydrogenase). As mentioned above, the oxidation product, namely acetaldehyde, is highly damaging to membranes and therefore probably contributes more to membrane damage at the post-anoxic stage than ethanol during anoxia. Other enzymatic sources of damaging active radicals include the action of xanthine oxidase and superoxide dismutase on dioxygen.

$$O_2 \xrightarrow{\text{Xanthine oxidase}} O_2^-$$
$$O_2^- + O_2^- + 2H \xrightarrow{\text{Superoxide dismutase}} H_2O_2$$

Most amphibious plant species, and also those Arctic species that face the risk of oxygen deprivation due to prolonged encasement in ice, show a remarkable capacity to survive post-anoxic injury. This is achieved by having an efficient antioxidant defence system to protect cell membranes, cellular proteins, and genetic material from oxidative damage. In plant tissues, the antioxidant defence system consists of a wide range of molecules, both hydrophilic and hydrophobic, which can protect both aqueous and lipid phases in live tissues. The molecular antioxidant arsenal in plants consists of ascorbate, glutathione, tocopherols, tocotrienols, carotenoids, and xanthophylls. In addition, many of the often neglected phenolic compounds in plants also have considerable antioxidant activity (see Blokhina and Fagerstedt 2010). To maintain the redox status and to regenerate antioxidants in their active form, an array of enzymes is also required to support these antioxidative defences. These include dehydroascorbate reductase, thioredoxin reductase, glutathione reductase, lipoamide dehydrogenase, and thiol transferase (Blokhina et al. 2003).

The highly anoxia-tolerant sweet flag (*Acorus calamus*) is an example of a rhizomatous species that maintains high concentrations of antioxidants, including ascorbate, phenolics, and glutathione, throughout the period in which the plant is under anoxia as well as the enzymes ascorbate reductase, peroxidase, and catalase. The yellow flag iris (*Iris pseudacorus*) has a tolerance of anoxia that is only slightly inferior to that of *A. calamus* and is also notable for the ability while under anoxia to synthesize superoxide dismutase, a key enzyme for active radical detoxification on return to air (Monk et al. 1987).

5. Anoxia Sensing in Plants

The low K_m for cytochrome oxidase (cytochrome a_3, between K_m 0.1 and 1.0 mM) with respect to oxygen hinders any plant tissue from sensing directly the concentration of oxygen that it is available until the *Pasteur point* is reached at about 0.3% by volume, which is less than 1% of present atmospheric level. However, various aspects of cell metabolism can sense oxygen deficits above this level. One example is the effect of hypoxia-accumulated carbon dioxide on cytoplasmic pH, commonly referred to as cytoplasmic acidosis (Felle 2010). Other changes can be brought by diminished or altered nutrient uptake and induction of oxygen-sensing hemoglobins and improved aeration by the creation of lysigenous aerenchyma through programmed cell death (Mancuso and Shabala 2010).

Exposure to high concentration of carbon dioxide, which is commonly associated with low oxygen concentrations, is sufficient to trigger marked responses in metabolism that can eventually prove detrimental to plant survival. High concentrations of carbon dioxide (10–15%) act differently in plants, depending on whether they are adapted to short- or long-term periods of anoxia. Germinating chickpeas, (*Cicer arietinum*) as discussed above (see Fig. 5), are more intolerant of anoxia under high concentrations of carbon dioxide due to its action in accelerating glycolysis and ethanol production which are detrimental to survival (Crawford and Zochowski 1984). By contrast, the anoxia-tolerant rhizomes of aquatic plants exhibit a downregulation of glycolysis and are not influenced by high carbon dioxide levels (see Crawford 2008).

Recently, it has been observed that certain stress-induced hemoglobins which are now known to be widespread in the plant kingdom appear to have a function in relation to hypoxia tolerance (Dordas 2009). There are different types of hemoglobins in plants. The first to be discovered were those that are associated with nitrogen-fixing nodules. These are referred to as symbiotic hemoglobins (sHb). There is also a group of non-symbiotic hemoglobins (nsHb), which fall into two cases. Class 1 hemoglobins (nsHb-1 s) have a very high affinity for oxygen, while class 2 hemoglobins (nsHb-2 s) have a lower affinity for oxygen. The class 1 non-symbiotic hemoglobins are similar to the symbiotic hemoglobins in that they are expressed as a result of various stresses, including hypoxia.

Class 1 hemoglobins (Hbs) which are induced in plant cells under hypoxic conditions have a high affinity for oxygen, with Km values two orders of magnitude lower than that of cytochrome oxidase, permitting the utilization of oxygen by the molecule at extremely low oxygen concentrations (Igamberdiev et al. 2005). Sometimes this function is referred to as "signaling." Strictly speaking, as the hemoglobin system can function at extremely low concentrations of oxygen, it is not a signal alerting the plant to the near exhaustion of its oxygen supply, neither is it a store. The amount of oxygen sequestered by the plant hemoglobins is too small in this respect. It should instead be regarded as a scavenger of the last remnant oxygen which facilitates responses which reduce the effects of hypoxia. The presence of these plant hemoglobins reduces the levels of nitric oxide (NO) that are produced from nitrate ion during hypoxia and improves the redox and energy status of the hypoxic cell.

It has been suggested that a metabolic pathway involving NO and Hb provides an alternative type of respiration to mitochondrial electron transport under conditions of limited oxygen. Hb in hypoxic plants acts as part of a soluble, terminal, NO dioxygenase system, yielding nitrate ion from the reaction of oxyHb with NO. Nitric oxide is mainly formed due to anaerobic accumulation of nitrite. The overall reaction sequence, referred to as the Hb/NO cycle, consumes NADH and maintains ATP levels via an as yet unknown mechanism (Dordas 2009). There is still much uncertainty about the extent of the effects of hemoglobins in relation to stress. However, it would appear that in relation to hypoxia their main function is to alleviate the effects of oxygen deprivation for short periods and thus provide a respite that enables the plant to develop aerenchyma and possibly new adventitious roots so as to improve their tissue aeration and escape from the dangers of prolonged hypoxia. Other signals include increases in the cytosolic free calcium, the release of programmed cell-death signals from mitochondria, and an increase in the gaseous plant hormone ethylene, which eventually lead to programmed cell death and the development of lysigenous aerenchyma in roots. This will then aid oxygen diffusion and reduce the effects of hypoxia. It is not apparent as yet that they have any role in relation to long-term tolerance of anoxia (Fagerstedt 2010).

6. Ecological Advantages of Anoxia Tolerance

The fact that all eukaryote species have a need for certain common resources at some time in their life cycle creates a situation where an evolutionary advantage can be gained by those individuals which can survive longer without the resource than their competitors. Oxygen is consumed by all flowering plants and must therefore be considered as an essential resource. All species must have access to oxygen at some time in their life cycle. However, a species that is able to survive without oxygen longer than its competitors has access to habitats that are denied to its more oxygen-demanding competitors. This phenomenon has been termed *deprivation indifference* (Crawford et al. 1989). This advantage has evolutionary parallels in animals where diving reptiles, mammals, and birds have access to food resources that are unavailable to species confined to terrestrial habitats. As well as an access to resources, the anaerobic environment provides a refuge from predation and disturbance both for plants and animals. The monotypic stands of *Phragmites australis* and *Spartina anglica* in anaerobic muds are but two examples of species which have access to habitats where they suffer little interspecific competition or grazing pressure. The marine angiosperms *(Zostera* spp.) are further extreme examples of higher plants with a high tolerance of anoxia and access to sites free from competition.

Some of the species which inhabit wetlands and marshes and are anoxia-tolerant live in highly productive habitats, namely the swamps and salt marshes of flooded river basins and deltas. In these habitats, the ability to "do without" oxygen for a while does not diminish the ability to survive in these potentially productive sites. For flowering plants, their relationship with oxygen has evolved with many subtle differences. Not least is the ability to survive for a period in an anaerobic environment. Variations in the availability of oxygen are therefore powerful factors in creating biodiversity. Drought and flooding can both be overcome through anoxia tolerance. In seeds, an oxygen-impermeable testa protects anoxia-tolerant embryos and germinating seedlings from drought. Flood-tolerant rhizomes and tubers that overwinter in anaerobic muds can dominate productive wetland habitats. These numerous variations in anoxia tolerance provide striking examples of the unending capacity of evolution to overcome the potential vicissitudes of fluctuating environments.

7. Acknowledgments

I am much indebted to Dr. L Adamec (Trebon), Professor W. Armstrong (Hull), and Professor K.V. Fagerstedt (Helsinki) for careful readings and useful suggestions in the preparation of this text.

8. References

Adamec L (2007) Oxygen concentrations inside the traps of the carnivorous plants *Utricularia* and *Genlisea* (Lentibulariaceae). Ann Bot 100:849–856
Adamec L (2011) Ecophysiological look at plant carnivory: why are plants carnivorous? In: Seckbach J, Dubinski Z (eds) All flesh is grass. Plant-animal interrelationships, vol 16, Cellular origin, life in extreme habitats and astrobiology. Springer, Heidelberg, pp 455–489
Akhtar J, Gorham J, Qureshi RH (1994) Combined effect of salinity and hypoxia in wheat (*Triticum aestivum*) and wheat Thinopyrum amphiploids. Plant Soil 166:47–54
Arber AR (1920) Water plants: a study of aquatic angiosperms. Cambridge University Press, Cambridge
Armstrong J, Armstrong W (2009) Record rates of pressurized gas-flow in the great horsetail, Equisetum telmateia. Were Carboniferous Calamites similarly aerated? New Phytol 184:202–215
Armstrong W, Webb T, Darwent M, Beckett PM (2009) Measuring and interpreting respiratory critical oxygen pressures in roots. Ann Bot 103:281–293
Bailey-Serres J, Voesenek L (2010) Life in the balance: a signaling network controlling survival of flooding. Curr Opin Plant Biol 13:489–494
Baskin CC, Baskin JM (1998) Seeds, ecology, biogeography, and evolution of dormancy and germination. Academic, San Diego
Bertani A, Brambilla I, Mapelli S, Reggiani R (1997) Elongation growth in the absence of oxygen: the rice coleoptile. Russ J Plant Physiol 44:543–547
Bertrand A, Castonguay Y, Nadeau P, Laberge S, Rochette P, Michaud R et al (2001) Molecular and biochemical responses of perennial forage crops to oxygen deprivation at low temperature. Plant Cell Environ 23:1085–1093
Blokhina O, Fagerstedt KV (2010) Reactive oxygen species and nitric oxide in plant mitochondria: origin and redundant regulatory systems. Physiol Plant 138:447–462
Blokhina O, Virolainen E, Fagerstedt KV (2003) Antioxidants, oxidative damage and oxygen deprivation stress: a review. Ann Bot 91:179–194
Braendle R, Crawford RMM (1987) Rhizome anoxia tolerance and habitat specialization in wetland plants. In: Crawford RMM (ed) Plant life in aquatic and amphibious habitats. Blackwell Scientific Publications, Oxford, pp 397–410
Cook CDK (1999) The number and kinds of embryo-bearing plants which have become aquatic: a survey. Evol Systematics 2:79–102
Crawford RMM (1992) Oxygen availability as an ecological limit to plant distribution. Adv Ecol Res 23:93–185
Crawford RMM (2003) Seasonal differences in plant responses to flooding and anoxia. Can J Bot 81:1224–1246
Crawford RMM (2008) Plants at the Margin – ecological limits and climate change. Cambridge University Press, Cambridge
Crawford RMM, Braendle R (1996) Oxygen deprivation stress in a changing climate. J Exp Bot 47:145–159
Crawford RMM, Zochowski ZM (1984) Tolerance of anoxia and ethanol toxicity in chickpea seedlings (*Cicer arietinum* L.). J Exp Bot 35:1472–1480
Crawford RMM, Monk LS, Zochowski ZM (1987) Enhancement of anoxia tolerance by removal of volatile products of anaerobiosis. In: Crawford RMM (ed) Plant life in aquatic and amphibious habitats. Blackwell Scientific Publications, Oxford, pp 375–384
Crawford RMM, Studer C, Studer K (1989) Deprivation indifference as a survival strategy in competition: advantages and disadvantages of anoxia tolerance in wetland vegetation. Flora 182:189–201
Crawford RMM, Chapman HM, Hodge H (1994) Anoxia tolerance in high Arctic vegetation. Arct Alp Res 26:308–312
Dordas C (2009) Nonsymbiotic hemoglobins and stress tolerance in plants. Plant Sci 176:433–440
Fagerstedt KV (2010) Programmed cell death and aerenchyma formation under hypoxia. In: Mancuso S, Shabala S (eds) Waterlogging signalling and tolerance in plants. Springer Science, Heidelberg, pp 99–118

Faiz SMA, Weatherley PE (1982) Root contraction in transpiring plants. New Phytologist 92: 333–343

Felle HH (2010) pH signalling during anoxia. In: Mancuso S, Shabala S (eds) Waterlogging signalling and tolerance in plants. Springer, Heidelberg, pp 79–98

Friis E, Pedersen MKR, Crane PR (2001) Fossil evidence of water lillies in the Early Cretaceous. Nature 410:357–360

Gansert D (2003) Xylem sap flow as a major pathway for oxygen supply to the sapwood of birch (*Betula pubescens* Ehr.). Plant Cell Environ 26:1803–1814

Gravatt DA, Kirby CJ (1998) Patterns of photosynthesis and starch allocation in seedlings of four bottomland hardwood tree species subjected to flooding. Tree Physio 18:411–417

Gudleifsson BE (1997) Survival and metabolite accumulation by seedlings and mature plants of timothy grass during ice encasement. Ann Bot 79:93–96

Henslow G (1893) A theoretical origin of endogens from exogens, through self-adaptation to an aquatic habitat. J Linnean Society - Bot 29:485–528

Hook DD, Scholtens JR (1978) Adaptation and flood tolerance of tree species. In: Hook DD, Crawford RMM (eds) Plant life in anaerobic environments. Ann Arbor Science, Ann Arbor, pp 299–331

Hunter MI, Hetherington AM, Crawford RMM (1983) Lipid peroxidation – a factor in anoxia intolerance in *Iris* species. Phytochemistry 22:1145–1147

Igamberdiev AU, Baron K, Manac'h-Little N, Stoimenova M, Hill RD (2005) The haemoglobin/nitric oxide cycle: Involvement in flooding stress and effects on hormone signalling. Ann Bot 96:557–564

Ismail AM, Ella ES, Vergara GV, Mackill DJ (2009) Mechanisms associated with tolerance to flooding during germination and early seedling growth in rice (*Oryza sativa*). Ann Bot 103:197–209

Jackson MB, Herman B, Goodenough A (1982) An examination of the importance of ethanol in causing injury to flooded plants. Plant Cell Environ 5:163–172

Joly CA (1982) Variation in tolerance and metabolic responses to flooding in some tropical trees. J Exp Bot 33:799–809

Kennedy RA, Barett SCH, Van der Zee D, Rumpho ME (1980) Germination and seedling growth under anaerobic conditions in *Echinochloa crus-galli* (barnyard grass). Plant Cell Environ 3: 243–248

Kozlowski TT, Pallardy SG (1997) Physiology of woody plants. Academic, San Diego

Macdonald RC, Kimmerer TW (1991) Ethanol in the stems of trees. Physiol Plant 82:582–588

Mancuso S, Shabala S (2010) Waterlogging signalling and tolerance in plants. Springer Science, Heidelberg

McGraw JB, Vavrek MC, Bennington CC (1991) Ecological genetic variation in seed banks 1. Establishment of a time transect. J Ecol 79:617–625

Monk LS, Braendle R, Crawford RMM (1987) Catalase activity and post-anoxic injury in monocotyledonous species. J Exp Bot 38:233–246

Mugnai S, Mancuso S (2010) Oxygen transport in the sapwood of trees. In: Mancuso S, Shabala S (eds) Waterlogging signalling and tolerance in plants. Springer, Heidelberg, pp 61–75

Pasteur L (1861) Expériences et vues nouvelles sur la nature des fermentations. Comptes Rendus de l'Académie des Sciences (Paris) 52:1260–1264

Power P, Fonteyn PJ (1995) Effects of oxygen concentration and substrate on seed germination and seedling growth of Texas wildrice (*Zizania texana*). Southwest Nat 4:1–4

Pregnall AM (2004) Effects of aerobic versus anoxic conditions on glutamine synthetase activity in eelgrass (*Zostera marina* L.) roots: regulation of ammonium assimilation potential. J Exp Mar Biol Ecol 311:11–24

Reggiani R, Bertani A (2003) Anaerobic amino acid metabolism. Russ J Plant Physiol 50:733–736

Scarano FR (1992) The effects of ontogeny and environmental oscillations on plant responses to oxygen deprivation. PhD thesis, University of St Andrews, pp 208

Schlüter U, Crawford RMM (2001) Long-term anoxia tolerance in leaves of *Acorus calamus* L. and *Iris pseudacorus* L. J Exp Bot 52:2213–2225

Schlüter U, Crawford RMM (2003) Metabolic adaptation to prolonged anoxia in leaves of American cranberry (*Vaccinium macrocarpon*). Physiol Plant 117:492–499

Setter TL, Waters I, Sharma SK, Singh KN, Kulshreshtha N, Yaduvanshi NPS et al (2009) Review of wheat improvement for waterlogging tolerance in Australia and India: the importance of anaerobiosis and element toxicities associated with different soils. Ann Bot 103:221–235

Shen-Miller J, Schopf JW, Harbottle G, Cao RJ, Ouyang S, Zhou KS et al (2002) Long-living lotus: germination and soil gamma-irradiation of centuries-old fruits, and cultivation, growth, and phenotypic abnormalities of offspring. Am J Bot 89:236–247

Small JGC, Potgieter GP, Botha, FC (1989). Anoxic seed-germination of erythrina-caffra - ethanol fermentation and response to metabolic-inhibitors. Journal of Experimental Botany 40:375–381

Studer C, Braendle R (1988) Postanoxische Effekte von Äthanol in Rhizomen von *Glyceria maxima* (Hartm.) Holmberg, *Iris germanica* (Cav.) Trin. Botanica Helv 98:111–121

Wegner LH (2010) Oxygen transport in waterlogged plants. In: Mancuso S, Shabala S (eds) Waterlogging signalling and tolerance in plants. Springer, Heidelberg, pp 3–22

White KD (1970) Roman farming. Thames & Hudson, London

PART IV: FORAMINIFERA

Koho
Piña-Ochoa
Heinz
Geslin
Mallon
Glock
Schönfeld
Cardich
Morales
Quipúzcoa
Sifeddine
Gutiérrez

Biodata of **Karoliina A. Koho** and **Elisa Piña-Ochoa**, authors of chapter *"Benthic Foraminifera: Inhabitants of Low-Oxygen Environments."*

Dr. Karoliina A. Koho is currently postdoctoral research fellow at Utrecht University in the Netherlands. She received her Ph.D. from Utrecht University in 2008. Her thesis focused on ecology of deep-sea benthic foraminifera from submarine canyons of Portuguese continental margin. At the moment, she works on benthic foraminifera from low oxygen settings, including Arabian Sea Oxygen Minimum Zone. Her other research interests include among others benthic foraminiferal ecology and proxy development, foraminiferal survival mechanisms under low oxygen condition, and the role of foraminifera in the marine nitrogen cycle.

E-mail: **K.A.Koho@uu.nl**

Dr. Elisa Piña-Ochoa is currently a Marie Curie postdoctoral researcher at the Center for Geomicrobiology (Aarhus University, Denmark). She obtained her Ph.D. from Complutense University of Madrid (Spain) in 2007. Her dissertation work focused on the biogeochemistry of the nitrogen cycle in aquatic environments at the Institute of Natural Resources-CSIC. Her research interests are in the field of marine N biogeochemistry, identifying and exploring the factors and microorganisms that mediate and regulate N-cycling processes. She is currently working on the new discovery of widespread nitrate-storing and denitrifying foraminifera, studying their survival mechanisms and contribution to the removal of fixed nitrogen to the world's oceans.

E-mail: **elisa.ochoa@biology.au.dk**

Karoliina A. Koho   **Elisa Piña-Ochoa**

BENTHIC FORAMINIFERA: INHABITANTS OF LOW-OXYGEN ENVIRONMENTS

KAROLIINA A. KOHO[1] AND ELISA PIÑA-OCHOA[2]

[1]*Department of Earth Sciences, Faculty of Geosciences, Utrecht University, Budapestlaan 4, 3584 CD Utrecht, The Netherlands*
[2]*Center for Geomicrobiology, Institute of Biological Sciences, Aarhus University, DK-8000 Aarhus, Denmark*

1. Introduction

Benthic foraminifera are single-celled eukaryotic organisms, which are abundant throughout the marine realm. Due to their good fossilization potential, foraminifera are commonly used tools by paleoceanographers to reconstruct changes in the past oceanic conditions; such studies can be based on both foraminiferal assemblage composition and/or calcium carbonate test chemistry. In addition, in more recent years the importance of benthic foraminifera in carbon cycling and their role in the benthic food web has become an important study direction. Several carbon-labeling and fatty acid studies have now shown that benthic foraminifera are involved in the initial rapid breakdown of fresh phytodetritus on the seafloor (e.g., Moodley et al. 2000, 2002; Witte et al. 2003; Suhr and Pond 2006; Nomaki et al. 2009); the importance of their role may equal to that of bacteria (Moodley et al. 2002). Especially in oxygen-deficient settings, where the occurrence of other meiofauna and macrofauna becomes limited due to lack of oxygen, foraminifera play a primary role in the initial breakdown and cycling of carbon (Woulds et al. 2007). Foraminifera themselves are consumed by a wide variety of organisms, e.g., selective and nonselective deposit feeders and specialized predators, thus forming an important link between lower and higher levels of the deep-sea food web (Gooday et al. 1992).

Advances in modern techniques, including both laboratory and field equipment, have greatly enhanced our ability to study foraminifera. In the past, deep-sea studies have been constrained by the remoteness of the environment, and thus have been dominated by the sediment coring approach to recover samples. Modern equipment, such as remotely operated vehicles (ROV) armed with cameras, and long-term deep sea observation sites (e.g., the deep sea observatory in Sagami Bay, Japan, and EuroSITES) have already allowed us to collect more precise and detailed observations of changes in the deep seafloor environment through space and time (e.g., Kitazato et al. 2000, 2003; Gooday et al. 2010). Furthermore, the development of new and more precise instrumentation and the greater involvement of biologists have led to new techniques (e.g., molecular genetics, biomarker analyses, nonlethal fluorogenic probes for accurate identification of living individuals, microsensors and stable isotopes, and trace element geochemistry). These types of studies will continue to enhance our understanding of heterogeneity and seasonal to interannual dynamics in these ecosystems, providing a better context for understanding foraminiferal ecology. Finally, advances in ecosystem modeling (both mathematical and conceptual) also play a role in our understanding, or considering, of how observations may be explained.

In the following sections, the low-oxygen tolerance of benthic foraminifera will be outlined, first, in settings where the oxygen stress mainly occurs seasonally (e.g., in fjords), and second, in regions of permanent oxygen stress (e.g., oxygen minimum zones (OMZs) and isolated sedimentary basins). Oxygen is also gradually depleted in the pore waters of sediments underlying oxic bottom waters, the rate of depletion dependent on local parameters, such as primary productivity and circulation. The effect of pore water redox chemistry on the vertical distribution of foraminifera in sediments is discussed in Sect. 3. In addition to field observations, laboratory evidence from foraminiferal cultures is presented in Sect. 4. Finally, foraminiferal survival mechanisms, including intracellular nitrate storage and denitrification, chloroplast sequestration, and potential bacterial symbionts are discussed, and potential future study directions are suggested.

2. Spatial Distribution: Foraminiferal Communities Living in Low-Oxygen Settings

Before proceeding further with the discussion of foraminiferal distribution in oxygen-depleted environments, it is necessary to define the thresholds, which are used to identify such environments on the basis of oxygen depletion. In this chapter, for simplicity and continuity with current literature (e.g., Levin et al. 2009), we will use the following three definitions, both in the context of bottom water and sediment pore water oxygen concentrations: "oxic" if the dissolved oxygen concentration in the environment is higher than 62.5 μM (=1.42 ml l^{-1}, 2 mg l^{-1}), "hypoxic" if oxygen falls below 62.5 μM, and anoxic if oxygen is 0 μM (with or without the presence of hydrogen sulfide). Several other terms, e.g., dysoxic and suboxic, are also commonly used in literature to refer to low-oxygen settings. In this chapter, they will not be considered further. It should be noted, however, when referring to permanent oxygen minimum zones (OMZs), that the commonly quoted threshold of dissolved oxygen concentration of <22.3 μM (=0.5 ml l^{-1}, 0.7 mg l^{-1}) is adopted (Helly and Levin 2004; Paulmier and Ruiz-Pino 2009).

Unfortunately, many field studies regarding the distribution of modern foraminifera do not report detailed bottom water and/or pore water oxygen measurements, and even fewer studies include measurements of other electron acceptors like nitrate and/or sulfate. In this chapter, we concentrate on literature which contains at least some of these measurements; thus, studies that do not meet these criteria are not considered here.

2.1. SEASONALLY LOW-OXYGEN ENVIRONMENTS

Seasonal hypoxia ($O_2 < 62.5$ µM) or even anoxia ($O_2 = 0$) develops in many estuarine, fjord, and coastal environments (Helly and Levin 2004). In many cases, these conditions occur naturally due to freshwater input or restricted circulation of water masses, related to density stratification or the presence of physical barriers such as sills. However, anthropogenic impacts can also trigger or exacerbate oxygen stress. For example, changes in land use may lead to enhanced freshwater runoff and nutrient loading, reinforcing stratification and increasing the oxygen demand by eutrophication (e.g., Diaz and Rosenberg 2008; Rabalais et al. 2010). Some benthic foraminifera appear very tolerant to (seasonal) hypoxic/anoxic conditions, forming an abundant part of the benthic ecosystem at such sites. Here, we outline foraminiferal communities found in such settings. Examples include Scandinavian fjords, Louisiana Coast (Gulf of Mexico), and Adriatic Sea.

In Havstens Fjord, located on the Swedish west coast, seasonal anoxia prevails for about 5 months a year, leading to the formation of a low-diversity benthic faunal assemblage, dominated by two foraminiferal taxa: *Elphidium magellanicum* and *Stainforthia fusiformis* (Gustafsson and Nordberg 2000). *S. fusiformis* is common in many Scandinavian fjords, which are also suffering from oxygen depletion. In the deep basin of Gullmar Fjord, *S. fusiformis* became the dominant fauna, replacing the traditional, more diverse Skagerrak-Kattegat community following a severe low-oxygen event (<0.5 ml O_2 l^{-1}, or 22.3 µM) in the winter of 1979–1980 (Filipsson and Nordberg 2004; Nordberg et al. 2009). The annual low-oxygen events now occurring in Gullmar Fjord have prevented the original foraminiferal fauna from recovering. *S. fusiformis* was also recorded in the Norwegian Drammesfjord where it successfully pioneered colonization of a formally anoxic basin following a reduction of organic matter discharge from the catchment (Alve 1995a; Bernhard and Alve 1996). However, it should be noted that the distribution of *S. fusiformis* is not confined to low-oxygen settings, and it is also found in some well-ventilated sites throughout the NW European shelf seas (Alve 1994, 2003), in hydrographic frontal regions (Scott et al. 2003), and physically disturbed sediments (Alve and Murray 1997). Rather, it seems that this species is able to take advantage of rapidly changing environmental conditions and is highly adaptable to various sources of environmental stress (oxygen, physical disturbance, hydrographic fronts, salinity; Alve 2003), thus bearing a very opportunistic character.

Seasonal hypoxia is also a common environmental problem on the Louisiana shelf in the northern Gulf of Mexico. The hypoxia is thought to have expanded since the 1950s, and the development has been related to the expansive use of artificial nitrogen fertilizers on the adjacent continent draining into the Gulf via the Mississippi river (e.g., Turner and Rabalais 1994; Platon et al. 2005; Osterman et al. 2005). Distinct, low-diversity, high-dominance foraminiferal populations are found in the affected areas. Blackwelder et al. (1996) noted a high abundance of *Buliminella morgani* and *Epistominella vitrea* in the area impacted by the Mississippi plume. Furthermore, two foraminiferal hypoxia indicators have been developed for the region of the Gulf of Mexico: the "A-E" (*Ammonia-Elphidium*) index for shallow continental waters (<30-m water depth; Sen Gupta et al. 1996; Platon and Sen Gupta 2001) and the "PEB" (*Pseudononion, Epistominella, Buliminella*) index for shelf and slope environments (13–70-m water depth; Osterman 2003). The A-E index is based on field observations of a relative increase in the abundance of *Ammonia parkinsoniana* in comparison to *Elphidium excavatum* in areas subject to hypoxia. However, the significance of the index has been recently questioned by Brunner et al. (2006) who suggest that the index responds to oxygen depletion only indirectly as a response to another environmental variable (e.g., change in benthic trophic conditions) that also correlates with oxygen. The PEB index, on the other hand, is based on cumulative percent of three foraminiferal species, *Pseudononion atlanticum*, *Epistominella vitrea*, and *Bulimina morgani*, in the assemblage, and shows highest values (±20%) in regions affected by hypoxia (Osterman 2003).

Seasonal hypoxia is also encountered in the northern Adriatic Sea (e.g., Jorissen et al. 1992; Degobbis et al. 2000). The hypoxia is partially attributed to natural causes (Tomasino 1996); however, increase in the frequency and duration of hypoxia has been related to anthropogenic nutrient and sediment input from the Po River drainage basin (Molin et al. 1992, Barmawidjaja et al. 1995). The foraminiferal standing stocks in the regions affected by the hypoxia have been reported to be significantly lower than in the adjacent areas with higher bottom water concentrations (Jorissen et al. 1992). A temporal study of Duijnstee et al. (2004), covering a period of two years (August 1996 to July 1998), also reported a drop in the foraminiferal standings stock, coinciding with the onset of hypoxia at a study site close to Po River drainage basin. Another site located somewhat further from the Po River also experienced decline in bottom water oxygenation; however, the concentrations did not drop below the hypoxic threshold. At this site, foraminifera were able to benefit from the eutrophication and seemed to increase in numbers. At the hypoxic site, the abundance drop was seen for all species of foraminifera; however, some appeared to show a higher tolerance than others, including *Nonionella turgida*, *Bolivina* spp., *Eggerella* spp., and *S. fusiformis*.

2.2. PERMANENT LOW-OXYGEN ENVIRONMENTS

Permanent (stable over decadal timescales) low-oxygen conditions occur naturally at midwater depth (approx. 100–1,200 m) along the continental margins of the eastern Pacific, off western Africa, and in the northern Indian Ocean (Arabian Sea). Some controversy has surrounded the definition of these low-oxygen zones, known as oxygen minimum zones (OMZs), in recent years (Paulmier and Ruiz-Pino 2009). Here, we constrain our definition of the OMZ to that of Helly and Levin (2004) and Paulmier and Ruiz-Pino (2009), in which the dissolved oxygen threshold of <0.5 ml l^{-1} (22 μM) was applied. The OMZs commonly occur below upwelling regions where nutrient-rich waters are brought to the surface, fueling high primary productivity, which in turn leads to elevated oxygen demand at depth. The vertical structure of the OMZ (Fig. 1) is typically divided into three sections: the upper boundary where oxygen first decreases, the core where the minimum oxygen content is recorded, and the lower boundary where oxygen begins to increase again. Generally, the core of the OMZ is very stable, whereas the position of the upper and lower boundaries in the water column can fluctuate seasonally (Helly and Levin 2004). It should also be noted that the OMZs are the largest low-oxygen areas in the world's oceans, and recent estimates suggest that their total extent is ±30.4 million km^2, equal to 8% of the global seafloor area (Paulmier and Ruiz-Pino 2009).

The OMZs are also important in respect to global biogeochemical cycling (Burdige 2007). Most crucially, they form major sites of organic carbon storage (e.g., van der Weijden et al. 1999; Muller-Karger et al. 2005), and are responsible for 30–50% of global reactive nitrogen loss by denitrification (Codispoti et al. 2001). As the extent of the OMZs is likely to increase in the future due to human impact (Keeling et al. 2010), changes can also be anticipated in respect to carbon and nutrient cycles in these settings.

A limited number of studies have been undertaken to date on the foraminiferal communities living in the OMZs, most of which were conducted in the Arabian Sea. These suggest that abundant communities are present in the OMZs, although the species diversity and richness are generally low. The living communities are also mainly made of perforated calcareous taxa, although some agglutinated species are also present. A clear species zonation is observed across the Arabian Sea OMZ, with a distribution expected to reflect changes in bottom water oxygenation and food availability (e.g., Jannink et al. 1998; Schumacher et al. 2007; Fig. 1). The upper OMZ (±150–250 m), where bottom water concentrations can fluctuate seasonally, is strongly dominated by *Uvigerina* ex. gr. *semionata* (Larkin and Gooday 2009), although this species is present also in the upper part of the core of the OMZ (down to ±500 m; Schumacher et al. 2007; Larkin and Gooday 2009). In addition, Erbacher and Nelskamp (2006) reported dominance of *Globobulimina* spp., *Bolivina pacifica*, and *B. pseudopunctata* in and around a sulfur-oxidizing bacterial mat in the OMZ off Pakistan, although the abundances were four times greater outside the mat than within. The true core of the OMZ (>250–800 m) is characterized by *Bulimina exilis*, *Bolivina dilatata*, and *Ammodiscus* spp. (Jannink et al. 1998; Schumacher et al. 2007). Around 1,000-m water depth in the lower OMZ, *Uvigerina peregrina*, *Rotaliatinopsis semiinvoluta*, and *Ehrenbergina trigona* become dominant (Hermelin and Shimmield 1990; Jannink et al. 1998; Schumacher et al. 2007). Below the OMZ (>1,300 m), where oxygen concentrations increase over 45 μM (±0.5 ml/l), *Bulimina aculeata* and *Epistominella exigua* are typically found (Jannink et al. 1998). In addition, the abundance of large agglutinated and soft-walled foraminifera increases with greater water depth (Gooday et al. 2000).

The foraminiferal communities in the OMZ of the Eastern Tropical North and South Pacific have been less well documented (Brandy and Rodolfo 1964;

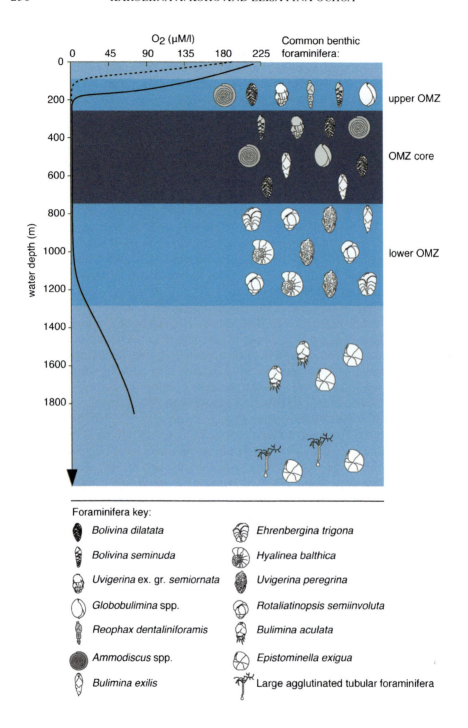

Smith 1964; Ingle et al. 1980; Halfar and Ingle 2003). Studies of Levin et al. (2002) and Tapia et al. (2008) note that calcareous foraminifera constitute the majority of the living benthic community along the depth transect through the OMZ, and that these thrive in the low-oxygen/high organic matter load conditions. The communities in the OMZ have low diversity and are dominated by few specialized species, including *Nonionella auris, N.* cf. *stella, Bolivina seminuda, Stainforthia* sp., and *Bulimina costata* (Tapia et al. 2008; Høgslund et al. 2008).

High primary productivity at some continental margin settings, like Sagami Bay in Japan, may lead to permanently hypoxic bottom waters, although the hypoxia is not severe enough for these regions to be considered fully developed OMZs (dissolved bottom water O_2 content between 22 and 62 µM). In Sagami Bay, a stable oxygen minimum of 55–60 µmol is observed between ±1,200- and 1,400-m water depth (Glud et al. 2009). Time-series studies, examining seasonal changes in benthic foraminiferal communities in Sagami Bay at 1,450-m water depth, were carried out by Kitazato and Ohga (1995), Ohga and Kitazato (1997), and Kitazato et al. (2000). Their results highlight the importance of seasonal phytodetritus input, even under prevailing low-oxygen conditions, on the deep-sea foraminiferal communities. Following a spring phytodetritus deposition, opportunistic species, like *Bolivina pacifica* and *Textularia kattegatensis* living close to the sediment water interface, responded rapidly and reproduced. The period of reproduction was followed by a period of growth during autumn and winter, as was noted in the increase in the size of the specimens. In contrast, other abundant species, *Globobulimina affinis* and *Chilostomella ovoidea*, which live deeper in the sediment (see more on microhabitat in Sect. 3), showed less clear seasonal behavior and were estimated to have a longer life cycle of 2 years. Similar foraminiferal communities were observed to dominate the site in a recent study of Glud et al. (2009), in which *G. affinis* and *G. pacifica* made up over 45% of total assemblage. In addition, *B. spissa, C. ovoidea, Cyclammina cancellata*, and *Uvigerina akitaensis* were found in relatively high abundances.

Permanent low-oxygen conditions can also develop in isolated marine basins, such as Santa Barbara Basin off California and Cariaco Basin off the coast of Venezuela, where bottom water exchange is restricted and surface primary productivity

◄─────────────────────────────

Figure 1. Schematic representation of selected, common benthic foraminifera found in and around the Arabian Sea Oxygen Minimum Zone (*OMZ*). Distribution of species from the upper OMZ is based on Schumacher et al. (2007) and Larkin and Gooday (2009); from the OMZ core on Hermelin and Shimmield (1990), Jannink et al. (1998), Gooday et al (2000), Erbacher and Nelskamp (2006), Schumacher et al. (2007), and Larkin and Gooday (2009); from the lower OMZ on Hermelin and Shimmield (1990), Jannink et al. (1998), and Schumacher et al. (2007); and from below OMZ on Hermelin and Shimmield (1990), Jannink et al. (1998), Gooday et al. (2000), and Schumacher et al. (2007). The dissolved bottom water oxygen profile is modified from Cowie and Levin (2009). The upper boundary of the OMZ fluctuates seasonally; the *dashed line* represents values during the inter-monsoon period (spring, April–May), while the *solid line* shows the profile measured during the late monsoon period (autumn, late August–September). Dissolved O_2 in the OMZ core is 0 µM and in the upper and lower boundary <22 µM (or <0.5 ml/l).

is relatively high. The sediments from Santa Barbara Basin are typically partially or continually laminated. The dissolved oxygen concentrations in the deep basin (>500 m) may fluctuate seasonally between ±12 and 0 µM (Bernhard and Reimers 1991) but remain poorly oxygenated for the whole year. The sediments are inhabited by a small number of low-oxygen-tolerant taxa (Bernhard et al. 1997, 2003; Bernhard and Reimers 1991). *Nonionella stella* dominates the assemblage, making up to 83% of the total assemblage (Bernhard et al. 1997). Other common calcareous species include *Chilostomella ovoidea* and *Bolivina seminuda*. The only common agglutinated species is *Textularia earlandi* (Bernhard and Reimers 1991). For details on foraminiferal studies regarding other low-oxygen California borderland basins (i.e., San Pedro Basin, Santa Monica Basin, San Nicolas and San Diego Basins), see Bandy 1964; Silva et al. 1996 and Kaminski et al. (1995); Reiter 1959 and Mackensen and Douglas (1989); and Douglas and Heitman (1979), respectively.

To our knowledge, no extensive ecological studies of benthic foraminifera or other (macro)/meiofauna have been carried out in the Cariaco Basin, where permanently anoxic, occasionally sulfidic conditions (Jacobs et al. 1987) have prevailed for the last ±15,000 years (Dean et al. 1999). However, despite the hostile nature of the environment, living benthic foraminifera, *Virgulinella fragilis*, have been recovered from the sediments of this bathyal basin (Bernhard 2003).

The world's largest anoxic basin is the Black Sea (Caspers 1957; Brewer 1971, Glen & Arthur 1985) where, due to a poor ventilation mechanism, oxygen is only present in the upper 100 m of water column, the deeper parts being completely anoxic and measuring very high H_2S and NH_4^+ concentrations (Gregoire and Soetaert 2010). In addition, methane gas vents are a well-known feature of this basin. Relatively few ecological studies have been carried out across the oxic to anoxic transition zone in the Black Sea, and therefore little is known about the ecosystem functioning in this unique setting. However, a recent study by Sergeeva et al. (2010) draws attention to the role of benthic foraminifera in the Black Sea. Numerous soft-shelled foraminifera and some calcareous taxa were found to inhabit the depth transect across the oxygenated lower boundary to hydrogen sulfide–containing zone (120–240 m). The highest numbers were recorded between 150- and 160-m water depth where bottom water oxygen content was virtually zero (less than <0.07 µM measured using Winkler titration). Furthermore, the diversity of soft-shelled foraminifera was high with over 40 morphotypes recognized, including both organic-walled "allogromiids" and agglutinated "saccaminids." The taxonomy of soft-shelled foraminifer is in its infancy, and hence, many of the morphotypes remain undescribed.

3. Foraminiferal Depth Zonation in Sediment: Oxic, Hypoxic, and Anoxic Microhabitats

Benthic foraminifera are typically found living in the top 10 cm of sediment, with different species inhabiting specific sediment depths, or microhabitats (Corliss 1985; Gooday 1986). Generally, four microhabitats are described (Fig. 2): epifaunal,

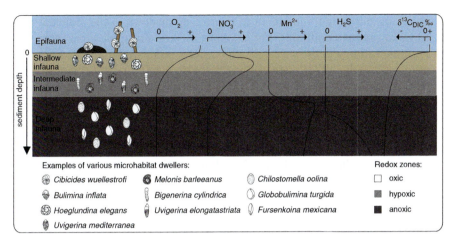

Figure 2. Idealized representation of foraminiferal microhabitats in marine sediments. Species are selected from literature; see text for individual species references. Epifauna is strictly found above the sediment–water interface in direct contact with bottom waters. Shallow infauna is present in the top centimeter(s) of sediment where pore water oxygen is generally not limiting. Intermediate infauna is found deeper, in the hypoxic sediments, where pore water nitrate also declines. Deep infauna is present at several centimeters depth in sediment in the anoxic sediment units. The schematized pore water profiles of dissolved oxygen, nitrate, manganese, and hydrogen sulfide are based on data presented by Froelich (1979), where the order of oxidants consumed (O_2, NO_3^-) or reduced compounds generated (Mn^{2+}, H_2S) is shown to be related to decreasing energy yield (The isotopic composition of dissolved inorganic carbon ($\delta^{13}C_{DIC}$) is schematized from Grossman (1984)).

referring strictly to species that live above the sediment–water interface, e.g., encrusting on worm burrows or shells, and hence being in direct contact with bottom waters; shallow infauna, referring to species that live in the oxic upper centimeters of sediment; intermediate infauna, living at a few centimeters depth, often in hypoxic conditions where pore water oxygen starts to decline but nitrate is plentiful; and finally, deep infaunal species, which occupy habitats at several centimeters depth in sediment and are often found below the pore water oxygen and nitrate penetration depth. However, it should be noted that due to the mobile nature of foraminifera, this scheme is not fixed and exceptions are always possible.

The microhabitat distribution is generally well developed in (outer) continental shelf (e.g., de Stigter et al. 1998; Langezaal et al. 2006; Duchemin et al. 2005) and continental slope settings (e.g., Corliss and Emerson 1990; Jorissen et al. 1998; de Stigter et al. 1998; Fontanier et al. 2002; Koho et al. 2008a; Mojtahid et al. 2010), where bottom waters are well oxygenated but the fresh organic matter flux is high enough to support a diverse benthic community. Therefore, it appears that the in-sediment distribution is mainly controlled by the sediment redox conditions and organic matter supply. A theoretical model known as the TRophic OXygen model (TROX) was developed by Jorissen et al. (1995) to describe the foraminiferal distribution in sediment. This model also demonstrates the dynamic nature of microhabitats, allowing foraminifera to migrate up or down in the sediment with the prevailing redox conditions. For example, in very eutrophic areas where the oxygen penetration depth is very shallow, deep infaunal foraminifera can be encountered already in the surface sediments, shallow infauna and epifauna being absent altogether. In contrast, in areas with deeper oxygen penetration depth, deep infauna is found deeper in the sediment, tracking the redox boundary, while shallow infauna is present at the surface.

However, recent advances in the foraminiferal research note that not only oxygen but also other electron acceptors (namely nitrate) may play an important role in the foraminiferal distribution in sediment (Risgaard-Petersen et al. 2006; Piña-Ochoa et al. 2010a, b; Koho et al. 2011). As this chapter focuses on foraminifera inhabiting low-oxygen environments, in the following section some detail will be given on common intermediate and deep infaunal foraminifera.

3.1. FORAMINIFERAL SPECIES LIVING IN HYPOXIC AND ANOXIC MICROHABITATS

3.1.1. Intermediate Infauna

Many of the intermediate infaunal foraminifera appear cosmopolitan and are found to occupy similar niches in various ocean basins. One of the most typical intermediate infaunal foraminifera, inhabiting upper to middle continental slope sediments, is *Melonis barleeanus*. This species is commonly found close to or just above the oxygen penetration depth, where pore water nitrate is still present (e.g., Jorissen et al. 1995, 1998; Schönfeld 1997; Fontanier et al. 2005; Koho et al. 2008a; Mojtahid et al. 2010), although exceptions do occur. For example, Schönfeld (2001) found the abundance maximum of *M. barleeanus* in the continental slope of Gulf of Cadiz (western Iberian margin) to coincide with a well-oxygenated nepheloid layer (a fluffy sediment surface horizon). Other good examples of intermediate infaunal species, also found in the upper and middle continental slope environment, include *Uvigerina elongatastriata* and *Bigenerina cylindrica*, also known as *Clavulina cylindrica* (Fontanier et al. 2003; Koho et al. 2008a; Mojtahid et al. 2010). These taxa are also commonly encountered in sediments close to the oxygen penetration where pore water nitrate is still abundant. In addition, their in-sediment distribution has been reported to be very similar to *M. barleeanus* (Fontanier et al. 2003; Koho et al. 2008a) or in some cases, somewhat shallower, especially for *C. cylindirica* (Mojtahid et al. 2010).

In continental shelf settings, the above-mentioned species are usually absent or present in low numbers only. Instead, the intermediate infaunal habitat can be occupied by species, such as *Nonion scaphum* (Langezaal et al. 2006), *Bolivina striatula*, *Nonion fabum*, and *Nonionella turgida* (Duchemin et al. 2005). However, their in-sediment depth distribution appears less clear in relation to pore water chemistry than for intermediate infauna encountered on the continental slope. For example, no clear correlation was observed for the habitat depth of *N. scaphum* and pore water redox chemistry in the seasonal survey of Langezaal et al. (2006), spanning from October 1997 to April 2001. The average living depth of *N. scaphum* fluctuated between ±1- and 4.5-cm depth, while oxygen and nitrate penetration remained relatively low through the study, below ±1 cm and ±2.5 cm, respectively. The foraminiferal population dynamics in shelf settings appear to be tightly related to trophic conditions and food availability. For example, many of the Bolivinids encountered in continental shelf settings appear very opportunistic in nature, and their abundance seems to be related to availability of labile organic matter (Langezaal et al. 2006). Thus, food availability may also play a role in the foraminiferal microhabitat in these settings. For example, the abundance of *N. scaphum* was seen to peak at spring-early summer and was generally lower during winter, thus responding to trophic conditions. In addition, the in-sediment depth distribution was more evenly spread during winter months, while in spring and/or summer, a maximum was typically recorded within the top 1.5 cm. For notes on other (intermediate) infaunal species, see Table 1.

3.1.2. Deep Infauna

The best-documented cosmopolitan deep infaunal foraminifera are *Globobulimina* spp. and *Chilostomella* spp. Several authors have described encountering either one or both of these taxa in anoxic sediments deep below the pore water oxygen and/or nitrate penetration depth (e.g., Corliss and Emerson 1990; Rathburn and Corliss 1994; Jorissen et al. 1998; Schönfeld 2001; Fontanier et al. 2003, 2005; Koho et al. 2007, 2008a; Glud et al. 2009). Furthermore, they are commonly found both in continental shelf and slope settings, although their average living depth may be shallower on the shelf due to the more eutrophic nature and compressed redox zonation of these sites (e.g., Schönfeld 2001; Fontanier et al. 2002; Langezaal et al. 2006; Koho et al. 2008a).

The in-sediment distribution of *Globobulimina* typically peaks below or close to the zero oxygen boundary (Schönfeld 2001; Fontanier et al. 2003; Licari et al. 2003; Koho et al. 2008a), where nitrate can also be absent from the pore waters (Corliss and Emerson 1990; Jorissen et al. 1998; Risgaard-Petersen et al. 2006). However, the complete in-sediment depth range that *Globobulimina* occupies can also include well-oxygenated sediments; thus, a broad in-sediment habitat is not uncommon (e.g., Jorissen et al. 1998; Kitazato et al. 2000; Mojtahid et al. 2010). The in-sediment distribution of *Chilostomella* has been described to be similar to that of *Globobulimina*, typically peaking close to the zero-oxygen boundary where nitrate may or may not be present (e.g., Langezaal et al. 2006; Koho et al. 2008a, b). However, some studies have also noted a shallower in-sediment distribution, suggesting a more intermediate infaunal microhabitat (Fontanier et al. 2002).

Other deep infauna include *Fursenkoina* spp., for which Jorissen et al. (1998) reported *Fursenkoina mexicana* living in the sulfate reduction zone below oxygen penetration depth on the continental slope of northwest Africa. Furthermore, Koho et al. (2007) found *F. bradyi* and *Fursenkoina* sp. living together with *Globobulimina*

Table 1. List of benthic foraminiferal species encountered in some oxygen-stressed environments (no discrimination between pore water and bottom water oxygen stress)

Species	Oxygenation preference	NO_3^- storage	Marine environment (bathymetric range)	Microhabitat	Other information	References (examples)
Ammonia beccarii	Oxic to hypoxic	No[a]	Coastal–continental shelf	Shallow infauna	Cosmopolitan, polluted areas	Zaninetti (1984), Moodley and Hess (1992), Alve and Murray (1999), and Mojtahid et al. (2008)
Ammonia tepida	Oxic to hypoxic	No[a]	Intertidal–mudflat	Shallow infauna	Cosmopolitan	Zaninetti (1984), Moodley and Hess (1992), and Debenay et al. (1998)
Ammonia parkinsoniana	Oxic to hypoxic	?	Coastal–continental shelf	Shallow infauna	Cosmopolitan	Moodley and Hess (1992) and Pawlowski et al. (1995)
Bigenerina cylindrica	Oxic to anoxic	Yes[a]	Continental shelf and slope; canyon	Intermediate infauna	Eutrophic taxa, potential denitrifier (rate note measured)	Fontanier et al. (2003), Koho et al. (2008a), and Hess and Jorissen (2009)
Bolivina alata	Oxic to hypoxic	Yes[a]	Continental shelf, bathyal canyon	Shallow–intermediate infauna	Potential denitrifier (rate not measured)	Langezaal et al. (2006), Fontanier et al. (2002, 2008), and Hess and Jorissen 2009
Bolivina dilatata	Hypoxic to anoxic	?	Continental shelf–upper slope	Shallow–intermediate infauna	Opportunistic, cosmopolitan, including Arabian Sea OMZ	Jannink et al. (1998), Langezaal et al. (2006), and Larkin and Gooday (2009)
Bolivina skagerrakensis	Oxic to hypoxic	Yes[b–cf]	Continental shelf	Deep infauna–epifauna	Potential denitrifier (rate not measured)	Qvale and Nigam (1985)
Bolivina plicata	Oxic to hypoxic	Yes[a]	Continental shelf	Deep infauna–epifauna	Able to denitrify	Bandy (1964) and Piña-Ochoa et al. (2010a, b)

Species	Oxygen	Denitrify	Location	Microhabitat	Notes	References
Bolivina seminuda	Oxic to hypoxic	Yes[a]	Continental shelf	Shallow infauna	Stress-tolerant, able to denitrify	Douglas (1981), Langezaal et al. (2006), and Piña-Ochoa et al. (2010a, b)
Bolivina spissa	Hypoxic	Yes[c]	Continental slope–bathyal	Shallow–intermediate infauna	Potential denitrifier (rate not measured)	Douglas (1981) and Glud et al. (2009)
Bolivina subaenariensis	Oxic to hypoxic	Yes[a]	Coastal-canyon	Shallow infauna	Able to denitrify	Fontanier et al. (2002), Hess and Jorissen (2009), and Piña-Ochoa et al. (2010a, b)
Bolivina pacifica	Oxic to hypoxic	?	Continental shelf to slope	Shallow–intermediate infauna	Found in Arabian Sea OMZ in and around sulfur-oxidizing bacterial mat	Ingle et al. (1980), Douglas (1981), Ohga and Kitazato (1997), and Erbacher and Nelskamp (2006)
Bulimina aculeata	Oxic to hypoxic	Yes[a]	Continental shelf–slope	Shallow–intermediate infauna	Cosmopolitan, double microhabitat, eutrophic taxa, potential denitrifier (rate not measured)	Fontanier et al. (2002) and Mojtahid et al. (2008)
Bulimina elongata	Anoxic	Yes[b,cf]	Continental shelf	Deep infauna	Potential denitrifier (rate not measured)	Ingle et al. (1980), Moodley (1990), and Murray (1992)
Bulimina exilis	Hypoxic to anoxic	?	Continental slope	Deep infauna	Dominant species in the Arabian Sea OMZ core	Jannink et al. (1998) and Schumacher et al. (2007)
Bulimina marginata	Oxic, hypoxic, anoxic	Yes[a]	Continental shelf–slope	Shallow–intermediate–deep infauna	Cosmopolitan, double microhabitat, eutrophic taxa, potential denitrifier (rate not measured)	Barmawidjaja et al. (1992), Nordberg et al. (2000), Fontanier et al. (2002), and Murray (2003)

(continued)

Table 1. (continued)

Species	Oxygenation preference	NO_3^- storage	Marine environment (bathymetric range)	Microhabitat	Other information	References (examples)
Buliminella morgani	Hypoxic to anoxic	?	Continental shelf–upper slope	Shallow infauna	Muddy sediment	Blackwelder et al. (1996) and Platon and Sen Gupta (2001)
Chilostomella oolina	Anoxic	Yes[a]	Continental shelf–slope	Deep infauna	Potential denitrifier (rate not measured)	Bernhard (1992), Rathburn et al. (1996), Jorissen et al. (1998), and Koho et al. (2007)
Chilostomella ovoidea	Anoxic	Yes[c]	Lower continental shelf–bathyal	Deep infauna	Potential denitrifier (rate not measured)	Bernhard et al. (1997), Ohga and Kitazato (1997), Kitazato et al. (2000), Schönfeld (2001), and Glud et al. (2009)
Cyclammina cancellata	Oxic to hypoxic	Yes[a]; no[c]	Continental shelf–slope	Shallow infauna and/or epifauna	Potential denitrifier (rate not measured)	Schönfeld (1997)
Elphidium magellanicum	Oxic to hypoxic	?	Coastal–continental shelf	Shallow infauna	Swedish fjords	Gustafsson and Nordberg (2000)
Fursenkoina mexicana	Hypoxic/anoxic	?	Continental shelf–upper bathyal	Deep infauna	Opportunistic	Ohga and Kitazato (1997) and Jorissen et al. (1998)
Globobulimina affinis	Anoxic	Yes[a,c]	Coastal–continental shelf and slope	Deep infauna	Potential denitrifier (rate not measured)	Schönfeld (2001), Fontanier et al. (2002), and Glud et al. (2009)
Globobulimina auriculata f. arctica	Anoxic	Yes[a]	Coastal–continental shelf	Deep infauna	Potential denitrifier (rate not measured); Greenland	Sen Gupta and Hayes (1979) and Jennings et al. (2004)
Globobulimina pacifica	Anoxic	Yes[c]	Continental slope–bathyal	Deep infauna	Potential denitrifier (rate not measured)	Mackensen and Douglas (1989), Bernhard et al. (2001), and Glud et al. (2009)

Globobulimina turgida	Anoxic	Yes[a]	Coastal–continental shelf and slope	Deep infauna	Able to denitrify	Alve (1995b), Gustafsson and Nordberg (1999), Risgaard-Petersen et al. (2006), and Piña-Ochoa et al. (2010a, b)
Goesella flintii	Oxic to hypoxic	Yes[a]	Continental shelf	Epifauna	Potential denitrifier (rate not measured)	Walton (1955), Ikeya (1970), and McGann (2002)
Haynesina germanica	Oxic and anoxic	No[a]	Intertidal–mud flat	Shallow infauna	Cosmopolitan, husbands chloroplasts	Knight and Mantoura (1985) and Alve and Murray (1999)
Hyalinea balthica	Oxic to anoxic	Yes[a]	Continental shelf–slope	Epifauna–infauna	Potential denitrifier (not measured); cosmopolitan	Hermelin and Shimmield (1990), Evans et al. (2002), and Filipsson and Nordberg (2004)
Melonis barleeanus	Hypoxic to anoxic	Yes[a]	Lower continental shelf to bathyal	Intermediate infauna	Potential denitrifier (rate not measured)	Jorissen et al. (1998), Fontanier et al. (2002), Koho et al. (2008a), and Mojtahid et al. (2010)
Nonion scaphum	Hypoxic to anoxic	No[a]	Coastal–continental shelf	Intermediate infauna	Opportunistic, tolerant to strong environmental stress	Langezaal et al. (2006) and Mojtahid et al. (2008, 2010)
Nonionella stella	Hypoxic to anoxic	Yes[b, cf]	Continental shelf–slope	Intermediate to deep infauna	Able to denitrify	Silva et al. (1996), Bernhard et al. (1997), Gustafsson and Nordberg (2001), Diz et al. (2004), Risgaard-Petersen et al. (2006), and Høgslund et al. (2008)
Pseudononion atlanticum	Oxic to hypoxic	?	Coastal–continental shelf	Shallow infauna	PEP index species Louisiana coast	Osterman (2003)

(continued)

Table 1. (continued)

Species	Oxygenation preference	NO_3^- storage	Marine environment (bathymetric range)	Microhabitat	Other information	References (examples)
Rectuvigerina phlegeri	Hypoxic to anoxic	Yes[a]	Coastal–continental shelf	Shallow–intermediate infauna	Tolerant to strong environmental stress, potential denitrifier (rate not measured)	Bartels-Jónsdóttir et al. (2006b) and Mojtahid et al. (2008, 2010)
Reophax dentaliniformis	Oxic to hypoxic	?	Continental shelf–slope	Shallow to intermediate infauna	Present in the upper OMZ Arabian Sea	Kaminski et al. (1995) and Larkin and Gooday (2009)
Stainforthia sp var I	Oxic to hypoxic	Yes[a]	Intertidal–continental shelf and slope	Shallow–intermediate–deep infauna	Opportunistic species, able to denitrify	Risgaard-Petersen et al. (2006) and Høgslund et al. (2008)
Stainforthia fusiformis	Oxic to anoxic	?	Intertidal–continental shelf and slope	Shallow–intermediate–deep infauna	Opportunistic species	Jorissen et al. (1992), Alve (1993, 1994), Bernhard and Alve (1996), and Scott et al. (2003)
Textularia tenuissima	Oxic to hypoxic	Yes[b,cf]	Continental shelf	Shallow–intermediate–deep infauna	Potential denitrifier (rate not measured)	Alve and Murray (1997) and Gustafsson et al. (2000)
Uvigerina akitaensis	Hypoxic	Yes[c]	Continental slope	Infauna	Potential denitrifier (rate not measured)	Ikeya (1970), Ohga and Kitazato (1997), Kitazato et al. (2000), and Glud et al. (2009)
Uvigerina ex. gr. *semionata*	Hypoxic to anoxic	?	Lower continental shelf–upper slope	Shallow infauna	Endemic to Arabian Sea. Present in the OMZ core	Schumacher et al. (2007) and Larkin and Gooday (2009)
Uvigerina elongatastriata	Hypoxic to anoxic	Yes[a]	Continental shelf–upper slope	Intermediate infauna	Able to denitrify	Schönfeld (1997), Fontanier et al. (2003), Koho et al. (2008a), and Piña-Ochoa et al. (2010a, b)

Species	Environment	NO$_3^-$ storage	Microhabitat	Notes	References
Uvigerina mediterranea	Oxic to hypoxic	Yes[a]	Continental shelf–upper slope	Shallow to intermediate infauna	Potential denitrifier (rate not measured), usually confined to oxic sediment	Altenbach et al. (1999), Fontanier et al. (2002), and Koho et al. (2008a)
Uvigerina peregrina	Oxic to hypoxic	Yes[a]	Coastal–continental shelf	Intermediate infauna	Potential denitrifier (rate not measured), cosmopolitan	Ohga and Kitazato (1997), Bernhard (1992), Altenbach et al. (1999), Fontanier et al. (2002), and Koho et al. (2008a)
Valvulineria bradyana	Hypoxic to anoxic	Yes[a]	Coastal–continental shelf–bathyal canyon	Intermediate infauna	Tolerant to strong environmental stress, high productivity marker species, able to denitrify	Jorissen (1987), Verhallen (1991), Fontanier et al. (2002), Koho et al. (2007), Piña-Ochoa et al. (2010a, b), and Mojtahid et al. (2010)
Valvulineria laevigata	Anoxic	Yes[b, cf]	Coastal–continental shelf	Intermediate infauna	Able to denitrify	Corliss and Fois (1990), Kurbjeweit et al. (2000), and Piña-Ochoa et al. (2010a, b)
Virgulinella fragilis	Anoxic	?	Coastal–continental shelf to slope	Deep infauna	Coexists with sulfur-oxidizing bacteria, sequester chloroplasts, found in Cariaco basin	Bernhard (2003), Takata et al. (2003), Erbacher and Nelskamp (2006), and Tsuchiya et al. (2009)

Environmental preferences (incl. marine environment, typical microhabitat, and other related information) are given as well as presence or absence of intracellular nitrate (NO$_3^-$) pool. Presence of nitrate pool unknown "?". If NO$_3^-$ pool has been measured in a species but the rate of denitrification is unknown, the species has been labeled as a potential denitrifier

Nitrate storage measured by [a]Piña-Ochoa et al. (2010a, b), [b, cf]also by Piña-Ochoa et al. (2010a, b), but species name with "*cf.*" notation, [c]Glud et al. (2009)

spp. below the oxygen penetration depth in the Nazaré canyon on the Portuguese continental margin. The depth distribution of deep infauna has been linked to specialized feeding habits like digestion of refractory organic matter (Schmiedl et al. 2000, Fontanier et al. 2005) or feeding on specific bacterial colonies living around the zero oxygen redox boundaries (Schönfeld 2001). In addition, the highly specialized deep infaunal niche would provide the deep infaunal foraminifera a rather stable habitat with low risk of predation and low competition with other (metazoan) fauna (Fontanier et al. 2002). The recent discovery of denitrification among Foraminifera (Risgaard-Petersen et al. 2006; Piña-Ochoa et al. 2010a, b), however, suggests that the microhabitat distribution of deep infauna may be related to respiration pathway and migration activity in respect to refueling of their nitrate storage (Koho et al. 2011; see Sect. 5 for more details).

3.1.3 Infauna with Variable Microhabitat

The preferred microhabitats of some taxa, such as *Bulimina marginata* and *Bulimina aculeata*, appear less clear, and they have been described to live both deep in hypoxic and anoxic microhabitats (e.g., Alve 1995a; Ohga and Kitazato 1997; Bernhard and Alve 1996; Fontanier et al. 2002) and in shallower oxic (infaunal) habitats (e.g., Corliss 1991; Barmawidjaja et al. 1992; Rathburn and Corliss 1994; Schönfeld 2001). In some studies, a bimodal (surface and subsurface maxima) distribution has also been observed (Jorissen et al. 1998; Mackensen et al. 2000). Furthermore, these species, especially *B. marginata*, have a wide environmental distribution, ranging from stratified fjords (Gustafsson and Nordberg 2000) to continental shelf (e.g., Bartels-Jónsdóttir et al. 2006a; Duchemin et al. 2005) and slope settings (e.g., Jorissen et al. 1992, 1998; Fontanier et al. 2002). The erratic microhabitat distribution has been linked to sediment bioturbation or bioirrigation, i.e., polychaete tubes creating oxygenated pockets and specialized niches for foraminifera to inhabit deep sediment zones (Ohga and Kitazato 1997). Alternatively, the microhabitat of these taxa may be linked to food availability and related migration activity in sediment, with foraminifera following optimum trophic and redox conditions (Ohga and Kitazato 1997; Mackensen et al. 2000).

3.2. TEST CHEMISTRY: RELATION TO FORAMINIFERAL MICROHABITAT

The microhabitat of benthic foraminifera is also recorded in the carbonate chemistry of their test, and this signature can give an indication of their calcification environment or habitation depth in sediment. With increasing sediment depth and changing redox chemistry, the pore water signature of $\delta^{13}C$ becomes increasingly lighter (more negative) as organic matter is remineralized and light carbon-12 is released into the pore waters (e.g., McCorkle et al. 1985; McCorkle and Emerson 1988; Fig. 2). A corresponding $\delta^{13}C$ shift is observed in foraminiferal test carbonate (e.g., McCorkle et al. 1990, 1997; Schmiedl et al. 2004; Fontanier et al. 2006, 2008), although due to fractionation and possible other vital effects during calcification, the foraminiferal $\delta^{13}C$ signature is often offset from that of inorganically precipitated carbonate (Grossman 1984). Nevertheless, light $\delta^{13}C$ values have been recorded in deep infaunal taxa, such as *Globobulimina* spp. and *Chilostomella* spp. (e.g., McCorkle et al. 1997; Schmiedl et al. 2004; Fontanier et al. 2006, 2008), while heavy values have been found in the epifaunal species *Cibicides wuellestrofi* (e.g., Mackensen et al. 1993; McCorkle et al. 1997; Schmiedl et al. 2004), reflecting their preferred microhabitat in sediment.

The trace metal signature of foraminiferal test carbonate can also provide information on their microhabitat. Many trace metals are redox sensitive, and thus are either released to pore water or precipitated out of solution at certain thresholds of reduction potential (see a recent review by Tribovillard et al. 2006). For example, under low-oxygen conditions, solid-phase manganese oxides are reduced to Mn^{2+} and released into the pore water (Fig. 2). Foraminifera living and calcifying in this environment incorporate trace concentrations of manganese in their carbonate shell. Reichart et al. (2003) measured Mn/Ca ratios in foraminifera collected from Bay of Biscay using laser ablation mass spectrometry (LA-ICP-MS), a modern technique that allows high-resolution elemental analysis of individual foraminiferal tests or even individual chambers. Furthermore, the technique allows a distinction to be made between the surface coating (causing potential contamination in conventional mass spectrometry measurements) and authentic carbonate signature of foraminifera. The authors found a clear enrichment of Mn in specimens of *Hoeglundina elegans* collected from stations with shallow oxygen penetration depth (15 mm) and high pore water Mn content, while specimens collected from stations with deeper oxygen penetration had Mn content below the detection limit. Furthermore, Ní Fhlaithearta et al. (2010) measured relatively higher Mn/Ca ratios in *H. elegans* within sapropel S1, an organic rich sediment sequence deposited under very low-oxygen conditions some 10 000–6,000 years ago in the Mediterranean. Examination of redox-sensitive trace metal chemistry in foraminiferal test calcite is in its infancy, and in the future may provide far more detailed insight into foraminiferal ecology and microhabitat distribution.

4. Experimental Evidence

In addition to field studies, laboratory experiments are an essential part of ecological studies, allowing the effects of specific environmental parameters on the foraminiferal community to be investigated. To date, several laboratory studies have been conducted to investigate the survival and behavior of benthic foraminiferal species in hypoxic, anoxic, and sulfidic conditions. The exposure time to low-oxygen conditions has ranged from days (Moodley and Hess 1992) to ±3–4 weeks (Bernhard 1993; Alve and Bernhard 1995; Bernhard and Alve 1996; Ernst and van der Zwaan 2004; Koho et al. 2011) and longer (Moodley et al. 1997, 1998a; Pucci et al. 2009; Koho et al. 2011). All of the experiments highlighted the generally high survival rates of foraminifera in low-oxygen conditions. However, the responses were often species specific, with some taxa able to withstand the experimental conditions better than others. For example, the study of Bernhard and Alve (1996) showed that the numbers of living *Adercotryma glomeratum, Psammosphaera bouwmanni,* and *Stainfortia fusiformis* were not significantly different following >3 week incubation in nitrogen-purged (anoxic) seawater than in the aerated control conditions. In contrast, the survival of *Bulimina marginata* was considerably reduced, thus indicating lower tolerance to anoxia. Furthermore, intracellular adenosine triphosphate levels (ATP; indicative of cellular energy) varied under the anoxic and control settings. While the ATP levels of *A. glomeratum* and *S. fusiformis* were reduced in comparison to the controls in the anoxic incubations, the ATP levels of *P. bouwanni* were not significantly affected. The lower ATP levels in *A. glomeratum* and *S. fusiformis* may suggest that these species lower their metabolism to overcome the low-oxygen conditions. Low ATP levels were also detected in *B. marginata* from the nitrogen-purged setting. A decrease in the number of living foraminifera and diversity of the assemblage were observed in the experiment of Moodley et al. (1997) following a long-term incubation in anoxic conditions. Furthermore, exposure to hydrogen sulfide negatively affects the long-term survival in anoxia (Moodley et al. 1998a). Generally, hard-shelled foraminifera were more resistant to anoxic conditions than soft-walled species, especially the numbers of perforate calcareous species, such as *Nonionella* and *Stainforthia,* which did not appear to be significantly influenced by the low-oxygen conditions (Moodley et al. 1997). *Stainforthia* also appeared to be the most tolerant species to hydrogen sulfide, increasing in relative abundance from ±25% to ±60% following the 66-day incubation in anoxic, sulfidic conditions (Moodley et al. 1998a).

The influence of pore water redox chemistry on the in-sediment distribution and mobility of benthic foraminifera has also been studied by laboratory experiments (Alve and Bernhard 1995; Moodley et al. 1998b; Gross 2000; Duijnstee et al. 2003; Geslin et al. 2004), and attempts have been made to tackle the paradigm of food availability versus redox chemistry as the main controller of the foraminiferal microhabitat (Ernst and van der Zwaan 2004). All such experiments showed that foraminifera generally respond to low-oxygen conditions by migrating toward better-ventilated sediments (generally the core top; except in study of Geslin et al. 2004 where deeper sediments were better oxygenated). For instance, in the experiment of Alve and Bernhard (1995), the foraminiferal population (indicated by the cumulative 85% depth level) shifted from deeper than 2 cm to less than 0.5-cm depth in sediment when bottom water oxygen content fell from >134 to 9 µM (or >3– 0.2 ml/l). Similarly, the average living depth of foraminiferal species was observed to decrease up to a centimeter following an upward shift in the oxygen penetration depth in the studies of Duijnstee et al. (2003) and Ernst and van der Zwaan (2004). However, once again some taxa showed a clearer reaction than others to changes in bottom water or pore water redox conditions.

In the study of Bernhard and Alve (1995), *Nonionella labradorica, Bolivinellina pseudopunctata,* and *Epistominella vitrea* all migrated to shallower microhabitats, and some specimens were also found encrusting on polychaete tubes following the induction of hypoxic conditions. *S. fusiformis* was still present at 2-cm depth in sediment, although generally the population had shifted upward.

General inertia of some intermediate and deep infaunal taxa (e.g., *Globobulimina* spp. and *Melonis barleeanus*) to changes in bottom water oxygenation has been reported by several authors (Geslin et al. 2004; Ernst and van der Zwaan 2004). The in-sediment distribution of these taxa appears to be controlled by the sediment redox chemistry, as they were consistently reported in anoxic and hypoxic microhabitats. Furthermore, addition of labile organic matter on the sediment surface did not trigger these infaunal species to migrate; instead they remained at depth (Ernst and van der Zwaan 2004; Koho et al. 2008b), suggesting that on time scales of months, sediment redox chemistry is more important than food supply in controlling the distribution of foraminifera in the sediment.

5. Survival Strategies in Low-Oxygen Environments

The majority of organisms require oxygen to survive; for many benthic eukaryotes, the critical limiting oxygen concentration is approximately 0.5 ml/l (Levin et al. 2000). Below this value, the meio- and macrofaunal community structure and overall taxonomic composition change. However, there are some metabolic strategies that presumably allow benthic foraminifers to survive in such apparently adverse conditions. These strategies are discussed below.

5.1. INTERNAL NITRATE POOL AND DENITRIFICATION

It has been argued that there are no true anaerobic eukaryotes and that the anaerobic world is confined to Bacteria and Archaea (Fenchel and Finlay 1995; Tielens et al. 2002). However in 2006, a collaboration between foraminifer geologists and microbiologists led to the discovery of two benthic species, *Globobulimina turgida* and *Nonionella stella,* that were able to accumulate intracellular nitrate stores in concentrations of several mM, and that these could be respired, in the absence of oxygen, to dinitrogen gas (Risgaard-Petersen et al. 2006). The amount of nitrate detected in an individual *G. turgida* – above 10 mM or more than 500-fold higher than the maximum concentration in the sediment pore water – was estimated to be sufficient to support respiration for over a month (Risgaard-Petersen et al. 2006). This ability was later confirmed in a study of the OMZ benthic community on the continental shelf off Chile (Høgslund et al. 2008), where the two dominant taxa *Nonionella* cf. *stella* and *Stainforthia sp.* were found to contain nitrate at even higher concentrations than *G. turgida* (35–180 mM, respectively). Their intracellular nitrate concentrations were up to 15,000 times higher than the ambient maximum concentration. In Japanese deep ocean margin sediments, *Globobulimina affinis, Globobulimina pacifica, Uvigerina akitaensis, Bolivina spissa*, and *Textularia* sp. (Glud et al. 2009) were also shown to accumulate nitrate.

The widespread ability of benthic foraminifera and Gromia to accumulate (>100 mM) and respire on nitrate was demonstrated by Piña-Ochoa et al. (2010a, b). More than half of the 67 tested species sampled from different marginal marine environments, including sediments below OMZs, continental slopes, shelves, and coastal sediments, showed the trait. Furthermore, the investigated species represented all major foraminiferal groups (miliolids, rotaliids, textulariids; see Table 1 in Piña-Ochoa et al. 2010a, b and in this chapter). In addition, the ability was also shown in Gromia, a eukaryotic organism related to foraminifera. The widespread occurrence among distantly related organisms suggests an ancient origin of the trait. All tested denitrifying species could also respire with oxygen (Piña-Ochoa et al. 2010a, b), even those collected in oxygen-free environments, such as the Peruvian OMZ. The denitrifying foraminifers should, therefore, be regarded as facultative anaerobes.

A specific advantage of using nitrate for respiration is the possibility of accumulating this nontoxic ion to high concentrations in the cell, and thereby obtaining the ability to sustain respiratory activity and exploit anoxic sediments in periods when oxygen and/or nitrate are absent from the environment. This behavior of foraminifera is very similar to the strategy used by other nitrate-storing prokaryotes, such as large white sulfur bacteria in the Beggiatoaceae family, that frequently dominate nitrate-rich and oxygen-depleted sediments of the seafloor (Fossing et al. 1995; McHatton et al. 1996; Schulz et al. 1999). The differences are mainly related to e-donor preferences and NO_3^- reduction pathways (Beggiatoaceae-affiliated bacteria use NO_3^- for H_2S oxidation and reduce NO_3^- to NH_4^+, while *G. turgida* reduces NO_3^- to N_2, Huettel et al. 1996; Otte et al. 1999; Risgaard-Petersen et al. 2006). *Thioploca* and *Beggiatoa* are able to store NO_3^- in vacuoles and to transport it deep into oxygen- and nitrate-free environments where, likewise with foraminifera, it is used for respiration (Jørgensen and Gallardo 1999; Schulz et al. 1999; Høgslund et al. 2009). Therefore, the mobile nature of many infaunal foraminifera and their commonly observed migration activity may be related to refueling and subsequent depletion of their nitrate storage (Koho et al. 2011).

Although nitrate respiration seems common among foraminifera, little is known about the metabolic activity of foraminifera in anoxic environments (Bernhard 1993; Bernhard and Alve 1996), and it remains unknown if energy gained from nitrate respiration is enough to sustain growth, reproduction, and completion of their lifecycle. Data from Bernhard and Reimers (1991) and Moodley et al. (1997, 1998a) suggest that foraminifera, typical of oxygen-depleted environments, cannot survive prolonged anoxia. Nevertheless, a recent experimental study showed that *G. turgida* are capable of long-term survival (up to 86 days) under simulated oxygen-free conditions if nitrate is readily available to sustain cellular respiration (Piña-Ochoa et al. 2010a, b). In the absence of desired electron acceptors (oxygen and nitrate), the survival rates of *G. turgida* are reduced (average of 25 days), but foraminifera can still survive up to 2 months if utilizing their internal nitrate pool only. Therefore, the facultative anaerobe capacity of *G. turgida* and the relatively long-term survival through the intracellular nitrate pool allow foraminifera to reside wide sediment depth ranges and inhabit various sediment redox zones.

The foraminiferal nitrate respiration has also implications for geochemical cycling of major nutrients. It has been estimated that the removal of fixed nitrogen by foraminiferal respiration may in some areas equal to the importance of bacterial denitrification (Piña-Ochoa et al. 2010a, b). A first estimation of foraminiferal denitrification in various environments was reported in Piña-Ochoa et al. (2010a, b) by combining the abundance of living nitrate-storing foraminifers and denitrification rates measured on isolated specimens. For example, in the shelf sediments from the Skagerrak and a canyon of the Bay of Biscay, where denitrifying foraminifera are abundant, it was estimated that foraminifers were responsible for 60–70% of the measured benthic denitrification activity. Likewise, in the OMZ off Chile, foraminifers may account for almost 70% of the nitrogen loss from the sediment. In the Arabian Sea OMZ, however, the foraminifers are likely to contribute less (between 9% and 15%) to denitrification due to their lower abundances. In estuaries (i.e., Tagus prodelta, Portugal), foraminiferal denitrification may also play a significant role in the loss (8–50%) of combined nitrogen, and it may thereby mitigate coastal eutrophication. However, there are also sites where the abundance of denitrifying foraminifers is too low to contribute significantly to nitrogen loss via denitrification. In the Sagami Bay (Japan), the small population of foraminifers apparently contributes only 4% to benthic N_2 production. Therefore, the quantitative importance of foraminiferal denitrification should therefore call for further studies on their role on the marine global nitrogen cycling.

5.2. FORAMINIFERAL CELL ULTRASTRUCTURE, INCLUDING CHLOROPLAST SEQUESTRATION

The survival and possible physiological and morphological adaptations of foraminiferans exposed to anoxia and possibly sulfidic habitats have been previously reviewed by Bernhard 1996 and later by Bernhard and Sen Gupta (1999). These reviews dealt mainly with experimental and ultrastructure evidence suggesting that certain benthic foraminiferans are microaerophiles, requiring only trace amounts of oxygen.

Some physiological adaptations include sequestration of chloroplasts, assimilative nitrate reductase from preserved diatom chloroplasts, microaerophiles characteristics, encystment, metabolic depression (i.e., dormancy), and aggregation of endoplasmic reticulum and peroxisomes, providing as yet undefined metabolic advantage permitting foraminifera to survive in these environments (Leutenegger

1984; Travis and Bowser 1986; Cedhagen 1991; Bernhard and Alve 1996; Moodley et al. 1998a, b; Grzymski et al. 2002; Bernhard and Bowser 2008).

For example, benthic foraminifera inhabiting anoxic sediment zones from Santa Barbara Basin, and co-occurring with sulfur-oxidizing bacteria, have been reported to possess numerous peroxisome complexes with endoplasmic reticulum (Bernhard and Bowser 2008). This cellular adaptation (not found in all foraminiferal species) may enhance the tolerance of some foraminiferal species to otherwise hostile conditions. Among others, hydrogen peroxide (H_2O_2) is produced during spontaneous chemical oxidation of sulfide. Prolonged exposure to this strong oxidizer is potentially toxic for many organisms and detrimental to biological membranes (Lesser 2006; Genestra 2007). As peroxisomes serve several metabolic functions, including conversion of H_2O_2 to H_2O, their high abundance in the foraminifera in question must play an important role. The authors suggest that foraminifera are able to utilize the oxygen derived from the breakdown of environmentally and metabolically produced H_2O_2.

Another physiological adaptation that foraminiferans have to thrive in anoxic (or in very low oxygen) environments may be related to occurrence of numerous mitochondria (Bernhard and Bowser 2008). Leutenegger and Hansen (1979) already noted that a number of foraminiferal species from low-oxygen (<0.3 ml/l, ±14 µM) environments had a high density of mitochondria concentrated near pore openings of the cell. In contrast, specimens from higher-oxygen environments had fewer mitochondria, which were more evenly distributed throughout the cell. Thus, pores might be important in gas exchange in foraminiferans (Corliss 1985), allowing a faster rate of molecular transport in low-oxygen environments and pointing to a function related to respiration. Mitochondria are also the most likely place for denitrification to occur (Risgaard-Petersen et al. 2006), although traditionally mitochondria are associated with aerobic respiration only. Nevertheless, evidence of a flexible biochemistry of the mitochondrion is now growing (Tielens et al. 2002), and it appears that many mitochondria are able to generate ATP without oxygen. Therefore, further attention should be made with the evolutionary origin of a possible nitrate respiring mitochondrion.

5.3. BACTERIAL SYMBIONTS

First, it should be noted that in the study of Risgaard-Petersen et al. (2006), no bacterial symbionts were observed to be involved in the denitrification within *G. turgida*, and that the denitrifying enzymes in the foraminifera's own organelles and not endosymbiotic nitrate-reducing bacteria were responsible for nitrate storage and subsequent denitrification. However, the role of potential bacterial symbiont has been previously investigated and proposed as survival mechanisms for foraminifera living in anoxic settings. For example, endosymbiotic bacteria (e.g., methanogens and sulfate-reducing bacteria similar to those seen in certain ciliates) have been noted within vacuoles of foraminiferal cytoplasm in four different species from Santa Barbara Basin (*Buliminella tenuata, Globocassidulina cf. biora, Nonionella stella,* and *Spiculodendron corallicolun*) (Bernhard and Reimers 1991; Bernhard 1996; Bernhard et al. 2000). Furthermore, Bernhard and Sen Gupta (1999) suggested that symbiotic bacteria benefit foraminiferans by oxidizing sulfide and consequently detoxifying the pore water. However, microbial association within foraminiferal phylogeny has yet not been addressed to be a widespread phenomenon. It also remains to be explained why only specific specimens of selected species may contain bacterial endosymbionts and not all, if there is a body of evidence (Lopez 1979) indicating that endosymbionts are acquired by hosts over their whole lifetime and they are not passed from one generation to another. Further studies are required to address these issues.

6. Outlook into Future Research Directions

As well as further studies of foraminifera in OMZs, more detailed investigations are needed into Gromiids, whose ecology is largely unknown. Recently published measurements of intracellular nitrate accumulation show that these large (up to centimeters in size) single-celled eukaryotes are new and promising candidates for an anaerobic eukaryote outside the group of foraminifera. It has also been speculated that they may also play an important role in carbon cycling, as they tend to occur in high abundances in organically enriched environments.

Future field studies of benthic foraminifera should aim to use consistent methodologies. Especially to ensure a comprehensive understanding of foraminiferal microhabitat in sediment and its relation to redox chemistry, detailed oxygen, nitrate, and sulfate pore water profiles should always be measured.

Widespread occurrence of nitrate accumulation and denitrification upends the classical view on mechanisms responsible for nitrogen loss from the world's oceans, which forces us to consider foraminiferal activity as a part of the global N-cycle. The contribution of eukaryotes in ocean sediments to the removal of fixed N by respiration may equal the importance of bacterial denitrification. Some species from the Peruvian OMZ were also found to be able to produce N_2O (Piña-Ochoa et al. 2010a, b), which may require further studies to quantify their possible contribution as greenhouse gas sources.

Future studies on active microbial associations within foraminiferal phylogeny, as well as on functional genes involved in foraminiferal denitrification, may help to enhance our understanding of the microbial effects, biochemistry, and genetics of denitrification in Eukarya.

More information on ultrastructure of foraminifera is needed. Comprehensive study of ultrastructure of strictly aerobic versus facultative anaerobe foraminifera may shed more light on the differences in cellular structure.

Current global warming is spreading of the oceanic (low oxygen) dead zones (Diaz and Rosenberg 2008), causing changes in benthic communities. Currently, hypoxia and anoxia are among the most widespread deleterious anthropogenic influences on estuarine and marine environments (Diaz and Rosenberg 2008). Future field studies of foraminifera can help us to not only evaluate these impacts in the modern day environments but also allow the establishment of the natural baseline conditions through paleoenvironmental reconstructions.

7. Acknowledgments

We are very grateful to Nils Risgaard-Petersen for his always inspiring, fruitful, and enthusiastic support in our joint collaboration in the role of benthic foraminifera in the N-cycle. We also acknowledge our colleagues Emmanuelle Geslin, Frans Jorissen, and Tomas Cedhagen for their help with an earlier version of Table 1. Tom Jilbert is thanked for proofreading and general comments regarding the structure of the chapter.

8. References

Altenbach AV, Pflaumann U, Schiebel R, Thies A, Timm S, Trauth M (1999) Scaling percentages and distributional patterns of benthic foraminifera with flux rates or organic carbon. J Foraminifer Res 29:173–185

Alve E (1994) Opportunistic features of the foraminifer Stainforthia fusiformis (Williamson): evidence from Frierfjord, Norway. J Micropaleontol 13:24

Alve E (1995a) Benthic foraminiferal distribution and recolonization of formerly anoxic environments in Drammensfjord, southern Norway. Mar Micropaleontol 25:169–186

Alve E (1995b) Benthic foraminiferal response to estuarine pollution: a review. J Foraminifer Res 25:190–203

Alve E (2003) A common opportunistic foraminiferal species as an indicator of rapidly changing conditions in a range of environments. Estuar Coast Shelf Sci 57:501–514

Alve E, Bernhard JM (1995) Vertical migratory response of benthic foraminifera to controlled oxygen concentrations in an experimental mesocosm. Mar Ecol Prog Ser 116:137–152

Alve E, Murray JW (1997) High benthic fertility and taphonomy of foraminifera: a case study of the Skagerrak, North Sea. Mar Micropaleontol 31:157–175

Alve E, Murray JW (1999) Marginal marine environments of the Skagerrak and Kattegat, a baseline study of living (stained) benthic foraminiferal ecology. Palaeogeogr Palaeoclimatol Palaeoecol 146:171–193

Bandy OL (1964) Foraminiferal trends associated with deep-water sands, San-Pedro and Santa-Monica basins, California. J Paleontol 38:138–148

Barmawidjaja DM, Jorissen FJ, Puskaric S, Van Der Zwaan GJ (1992) Microhabitat selection by benthic foraminifera in the northern Adriatic Sea. J Foraminifer Res 22:297–317

Barmawidjaja DM, Van der Zwaan GJ, Jorissen FJ, Puskaric S (1995) 150 years of eutrophication in the northern Adriatic Sea: evidence from a benthic foraminiferal record. Mar Ecol 122:367–384

Bartels-Jónsdóttir HB, Knudsen KL, Schönfeld J, Lebreiro SM, Abrantes FG (2006a) Recent benthic foraminifera from the Tagus Prodelta and Estuary, Portugal: microhabitats, assemblage composition and stable isotopes. Zitteliana Reihe A: Mitteilungen der Bayerischen Staatssammlung fur Palaontologie und Geologie 46:91–104

Bartels-Jónsdóttir HB, Knudsen KL, Abrantes F, Lebreiro S, Eiríksson J (2006b) Climate variability during the last 2000 years in the Tagus Prodelta, western Iberian Margin: Benthic foraminifera and stable isotopes. Mar Micropaleontol 59:83–103

Bernhard JM (1992) Benthic foraminiferal distribution and biomass related to pore-water oxygen content – Central California continental slope and rise. Deep-Sea Res A 39:585–605

Bernhard JM (1993) Experimental and field evidence of Antarctic foraminiferal tolerance to anoxia and hydrogen sulfide. Mar Micropaleontol 20:203–213

Bernhard JM (1996) Microaerophilic and facultative anaerobic benthic foraminifera: a review of experimental and ultrastructural evidence. Revue de Paleobiologie 15:261–275

Bernhard JM (2003) Potential symbionts in bathyal foraminifera. Science 299:861

Bernhard JM, Alve E (1996) Survival, ATP pool, and ultrastructural characterization of benthic foraminifera from Drammensfjord (Norway): response to anoxia. Mar Micropaleontol 28:5–17

Bernhard JM, Bowser SS (2008) Peroxisome proliferation in foraminifera inhabiting the chemocline: an adaptation to reactive oxygen species exposure? J Eukaryot Microbiol 55:135–144

Bernhard JM, Reimers CE (1991) Benthic foraminiferal population fluctuations related to anoxia: Santa Barbara Basin. Biogeochemistry 15:127–149

Bernhard JM, Sen Gupta BK (1999) Foraminifera of oxygen-depleted environments. In: Sen Gupta BK (ed) Modern foraminifera. Kluwer Academic Publishers, Dordrecht, pp 201–216

Bernhard JM, Sen Gupta BK, Borne PF (1997) Benthic foraminiferal proxy to estimate dysoxic bottom-water oxygen concentrations: Santa Barbara Basin, U.S. Pacific continental margin. J Foraminifer Res 27:301–310

Bernhard JM, Buck KR, Farmer MA, Bowser SS (2000) The Santa Barbara Basin is a symbiosis oasis. Nature 403:77–80

Bernhard JM, Buck KR, Barry JP (2001) Monterey cold-seep biota: assemblages, abundance and ultrastructure of living foraminifera. Deep-Sea Res I 48:2233–2249

Bernhard JM, Visscher PT, Bowser SS (2003) Submillimeter life positions of bacteria, protists, and metazoans in laminated sediments of the Santa Barbara Basin. Limnol Oceanogr 48:813–828

Blackwelder P, Hood T, Alvarez-Zarikian C, Nelsen TA, McKee B (1996) Benthic foraminifera from the NECOP study area impacted by the Mississippi River plume and seasonal hypoxia. Quat Int 31:19–36

Brandy O, Rodolfo KS (1964) Distribution of foraminifera and sediments, Peru-Chile Trench area. Deep-Sea Res 11:817–837

Brewer PG (1971) Hydrographic and chemical data from the Black Sea. Woods Hole Oceanographic Institution. Technical Report Reference No. 71–65

Brunner CA, Beall JM, Bentley SJ, Furukawa Y (2006) Hypoxia hotspots in the Mississippi Bight. J Foraminifer Res 36:95–107

Burdige D (2007) Preservation of organic matter in marine sediments: controls, mechanisms, and an imbalance in sediment organic carbon budgets? Chem Rev 107:467–485

Caspers H (1957) Black Sea and Sea of Azov. Geol Soc Am Mem 67:803–890

Cedhagen T (1991) Retention of chloroplasts and bathymetric distribution in the sublittoral foraminiferan Nonionellina labradorica. Ophelia 33:17–30

Codispoti LA, Brandes JA, Christensen JP, Devol AH, Naqvi SWA, Paerl HW, Yoshinari T (2001) The oceanic fixed nitrogen and nitrous oxide budgets: moving targets as we enter the anthropocene? Sci Mar 65:85–105

Corliss BH (1985) Microhabitats of benthic foraminifera within deep-sea sediments. Nature 314:435–438

Corliss BH (1991) Morphology and microhabitat preferences of benthic foraminifera from the northwest Atlantic Ocean. Mar Micropaleontol 17:195–236

Corliss BH, Emerson S (1990) Distribution of rose bengal stained deep-sea benthic foraminifera from the Nova Scotian continental margin and Gulf of Maine. Deep Sea Res A Oceanogr Res Pap 37:381–400

Corliss BH, Fois E (1990) Morphotype analysis of deep-sea benthic foraminifera from the northwest Gulf of Mexico. Palaios 5:589–605

Cowie GL, Levin LA (2009) Benthic biological and biogeochemical patterns and processes across an oxygen minimum zone (Pakistan margin, NE Arabian Sea). Deep-Sea Res II 56:261–270

De Stigter HC, Jorissen FJ, Van Der Zwaan GJ (1998) Bathymetric distribution and microhabitat partitioning of live (Rose Bengal stained) benthic foraminifera along a shelf to bathyal transect in the southern Adriatic Sea. J Foraminifer Res 28:40–65

Dean WE, Piper DZ, Peterson LC (1999) Molybdenum accumulation in Cariaco basin sediment over the past 24 k.y.: a record of water-column anoxia and climate. Geology 27:507–510

Debenay JP, Bénéteau E, Zhang J, Stouff V, Geslin E, Redois F, Fernandez-Gonzalez M (1998) Ammonia beccarii and Ammonia tepida (Foraminifera): morphofunctional arguments for their distinction. Mar Micropaleontol 34:235–244

Degobbis D, Precali R, Ivancic I, Smodlaka N, Fuks D, Kveder S (2000) Long-term changes in the northern Adriatic ecosystem related to anthropogenic eutrophication. Int J Environ Pollut 13:495–533

Diaz RJ, Rosenberg R (2008) Spreading dead zones and consequences for marine ecosystems. Science 321:926–929

Diz P, Francés G, Costas S, Souto C, Alejo I (2004) Distribution of benthic foraminifera in coarse sediments, Ría de Vigo, NW Iberian margin. J Foraminifer Res 34:258–275

Douglas RG (1981) Paleoecology of continental margin basins: A modern case history from the borderland of southern California. In: Douglas RG, Colburn IP, Gorsline DS (eds) Depositional systems of active continental margin basins. Short course notes, society of economic and petroleum mineralogist, Pacific Section. San Franciso, pp 121–156

Douglas RG, Heitman HL (1979) Slope and basin benthic foraminifera of the California borderland. Geol of cont slop 27:231–246

Duchemin G, Jorissen FJ, Andrieux-Loyer F, Le Loch F, Hily C, Philippon X (2005) Living benthic foraminifera from "la grande vasière", French atlantic continental shelf: faunal composition and microhabitats. J Foraminifer Res 35:198–218

Duijnstee IAP, Ernst SR, Van Der Zwaan GJ (2003) Effect of anoxia on the vertical migration of benthic foraminifera. Mar Ecol Prog Ser 246:85–94

Duijnstee I, de Lugt I, Noordegraaf HV, Van Der Zwaan GJ (2004) Temporal variability of the northern foraminiferal densities in Adriatic Sea. Mar Micropaleontol 50:125–148

Erbacher J, Nelskamp S (2006) Comparison of benthic foraminifera inside and outside a sulphur-oxidizing bacterial mat from the present oxygen-minimum zone off Pakistan (NE Arabian Sea). Deep-Sea Res 1 Oceanogr Res Pap 53:751–775

Ernst S, Van Der Zwaan B (2004) Effects of experimentally induced raised levels of organic flux and oxygen depletion on a continental slope benthic foraminiferal community. Deep-Sea Res I Oceanogr Res Pap 51:1709–1739

Evans JR, Austin WEN, Brew DS, Wilkinson IP, Kennedy HA (2002) Holocene shelf sea evolution offshore northeast England. Mar Geol 191:147–164

Fenchel T, Finlay BJ (1995) Ecology and evolution in anoxic worlds, Oxford series in Ecology and Evolution. Oxford University Press, Oxford

Filipsson HL, Nordberg K (2004) A 200-year environmental record of a low-oxygen fjord, Sweden, elucidated by benthic foraminifera, sediment characteristics and hydrographic data. J Foraminifer Res 34:277–293

Fontanier C, Jorissen FJ, Licari L, Alexandre A, Anschutz P, Carbonel P (2002) Live benthic foraminiferal faunas from the Bay of Biscay: faunal density, composition, and microhabitats. Deep-Sea Res I 49:751–785

Fontanier C, Jorissen FJ, Chailloua G, David C, Anschutz P, Lafon V (2003) Seasonal and interannual variability of benthic foraminiferal faunas at 550 m depth in the Bay of Biscay. Deep-Sea Res (1 Oceanogr Res Pap) 50:457–494

Fontanier C, Jorissen FJ, Chaillou G, Anschutz P, Grémare A, Griveaud C (2005) Live foraminiferal faunas from a 2800 m deep lower canyon station from the Bay of Biscay: faunal response to focusing of refractory organic matter. Deep-Sea Res I 52:1189–1227

Fontanier C, MacKensen A, Jorissen FJ, Anschutz P, Licari L, Griveaud C (2006) Stable oxygen and carbon isotopes of live benthic foraminifera from the Bay of Biscay: microhabitat impact and seasonal variability. Mar Micropaleontol 58:159–183

Fontanier C, Jorissen FJ, Lansard B, Mouret A, Buscail R, Schmidt S, Kervé P, Buron F, Zaragosi S, Hunault G, Ernoult E, Artero C, Anschutz P, Rabouille C (2008) Live foraminifera from the open slope between Grand Rhône and Petit Rhône Canyons (Gulf of Lions, NW Mediterranean). Deep-Sea Res I 55:1532–1553

Fossing H, Gallardo VA, Jorgensen BB, Huttel M, Nielsen LP, Schulz H, Canfield DE, Forster S, Glud RN, Gundersen JK, Kuver J, Ramsing NB, Teske A, Thamdrup B, Ulloa O (1995) Concentration and transport of nitrate by the mat forming sulphur bacterium Thioploca. Nature 374:713–715

Froelich PN (1979) Early oxidation of organic matter in pelagic sediments of the eastern equatorial Atlantic: suboxic diagenesis. Geochim Cosmochim Acta 43:1075–1090

Genestra M (2007) Oxyl radicals, redox-sensitive signalling cascades and antioxidants. Cell Signal 19:1807–1819

Geslin E, Heinz P, Jorissen F, Hemleben C (2004) Migratory responses of deep-sea benthic foraminifera to variable oxygen conditions: laboratory investigations. Mar Micropaleontol 53:227–243

Glen CR, Arthur MA (1985) Sedimentary and geochemical indicators of productivity and oxygen contents in modern and ancient basins. The Holocene Black Sea as the type anoxic basin. Chemical Geology 48:325–354

Glud RN, Thamdrup B, Stahl H, Wenzhoefer F, Glud A, Nomaki H, Oguri K, Revsbech NP, Kitazatoe H (2009) Nitrogen cycling in a deep ocean margin sediment (Sagami Bay, Japan). Limnol Oceanogr 54:723–734

Gooday AJ (1986) Meiofaunal foraminiferans from the bathyal Porcupine Seabight (north-east Atlantic): size structure, standing stock, taxonomic composition, species diversity and vertical distribution in the sediment. Deep-Sea Res 33:1345–1373

Gooday AJ, Levin LA, Linke P, Heeger T (1992) The role of benthic foraminifera in deep-sea food webs and carbon cycling. In: Rowe GT, Pariente V (eds) Deep-sea food chains and the global carbon cycle. Kluwer Academic Publishers, Dordrecht, pp 63–91

Gooday AJ, Bernhard JM, Levin LA, Suhr SB (2000) Foraminifera in the Arabian Sea oxygen minimum zone and other oxygen-deficient settings: taxonomic composition, diversity, and relation to metazoan faunas. Deep-Sea Res II 47:25–54

Gooday AJ, Malzone MG, Bett BJ, Lamont PA (2010) Decadal-scale changes in shallow-infaunal foraminiferal assemblages at the Porcupine Abyssal Plain, NE Atlantic. Deep-Sea Res II 57:1362–1382

Gregoire M, Soetaert K (2010) Carbon, nitrogen, oxygen and sulfide budgets in the Black Sea: a biogeochemical model of the whole water column coupling the oxic and anoxic parts. Ecol Model 221:2287–2301

Gross O (2000) Influence of temperature, oxygen and food availability on the migrational activity of bathyal benthic foraminifera: evidence by microcosm experiments. Hydrobiologia 426:123–137

Grossman EL (1984) Carbon isotopic fractionation in live benthic foraminifera-comparison with inorganic precipitate studies. Geochim Cosmochim Acta 48:1505–1512

Grzymski J, Schofield OM, Falkowski PG, Bernhard JM (2002) The function of plastids in the deep-sea benthic foraminifer, Nonionella stella. Limnol Oceanogr 47:1569–1580

Gustafsson M, Nordberg K (1999) Benthic foraminifera and their response to hydrography, periodic hypoxic conditions and primary production in the Koljö Fjord on the Swedish west coast. J Sea Res 41:163–178

Gustafsson M, Nordberg K (2000) Living (Stained) benthic Foraminifera and their response to the seasonal hydrographic cycle, periodic hypoxia and to primary production in Havstens Fjord on the Swedish west coast. Estuar Coast Shelf Sci 51:743–761

Gustafsson M, Nordberg K (2001) Living (stained) benthic foraminiferal response to primary production and hydrography in the deepest part of the Gullmar Fjord, Swedish West Coast, with comparison to Höglund's 1927 material. J Foraminifer Res 31:2–11

Gustafsson M, Dahloff I, Blanck H, Hall P, Molander S, Nordberg K (2000) Benthic foraminiferal tolerance to tri-n-butylin (TBT) pollution in an experimental mesocosm. Mar Pollut Bull 40:1072–1075

Halfar J, Ingle JC (2003) Modern warm-temperate and subtropical shallow-water benthic foraminifera of the southern Gulf of California, Mexico. J Foraminifer Res 33:309–329

Helly JJ, Levin LA (2004) Global distribution of naturally occurring marine hypoxia on continental margins. Deep-Sea Res I 51:1159–1168

Hermelin JOR, Shimmield GB (1990) The importance of the oxygen minimum zone and sediment geochemistry in the distribution of Recent benthic foraminifera in the northwest Indian Ocean. Mar Geol 91:1–29

Hess S, Jorissen FJ (2009) Distribution patterns of living benthic foraminifera from Cap Breton canyon, Bay of Biscay: faunal response to sediment instability. Deep-Sea Res I 56:1555–1578

Høgslund S, Revsbech NP, Cedhagen T, Nielsen LP, Gallardo VA (2008) Denitrification, nitrate turnover, and aerobic respiration by benthic foraminiferans in the oxygen minimum zone off Chile. J Exp Mar Biol Ecol 359:85–91

Høgslund S, Revsbech NP, Kuenen JG, Jørgensen BB, Gallardo VA, Van De Vossenberg J, Nielsen JL, Holmkvist L, Arning ET, Nielsen LP (2009) Physiology and behaviour of marine Thioploca. ISME J 3:647–657

Huettel M, Forster S, Klöser S, Fossing H (1996) Vertical migration in the sediment-dwelling sulfur bacteria Thioploca spp. in overcoming diffusion limitations. Appl Environ Microbiol 62:1863–1872

Ikeya N (1970) Population ecology of benthonic foraminifera in Ishukari Bay, Hokkaido, Japan. Rec Ocenogr Works Jpn 10:173–191

Ingle JC Jr, Keller G, Kolpack RL (1980) Benthic foraminiferal biofacies, sediments and water masses of the southern Peru- Chile Trench area, southeastern Pacific Ocean. Micropaleontology 26:113–150

Jacobs L, Emerson S, Huested SS (1987) Trace metal geochemistry in the Cariaco Trench. Deep Sea Res A 34:965–981

Jannink NT, Zachariasse WJ, Van Der Zwaan GJ (1998) Living (Rose Bengal stained) benthic foraminifera from the Pakistan continental margin (northern Arabian Sea). Deep-Sea Res I: Oceanogr Res Pap 45:1483–1513

Jennings AE, Weiner NJ, Helgadottir G (2004) Modern foraminiferal faunas of the southwestern to northern Iceland shelf: oceanographic and environmental controls. J Foraminifer Res 24:180–207

Jørgensen BB, Gallardo VA (1999) Thioploca spp.: filamentous sulfur bacteria with nitrate vacuoles. FEMS Microbiol Ecol 28:301–313

Jorissen FJ (1987) The distribution of benthic foraminifera in the Adriatic Sea. Mar Micropaleontol 12:21–48

Jorissen FJ, Barmawidjaja DM, Puskaric S, Van Der Zwaan GJ (1992) Vertical distribution of benthic foraminifera in the northern Adriatic Sea: the relation with the organic flux. Mar Micropaleontol 19:131–146

Jorissen FJ, De Stigter HC, Widmark JGV (1995) A conceptual model explaining benthic foraminiferal microhabitats. Mar Micropaleontol 26:3–15

Jorissen FJ, Wittling I, Peypouquet JP, Rabouille C, Relexans JC (1998) Live benthic foraminiferal faunas off Cape Blanc, NW-Africa: community structure and microhabitats. Deep-Sea Res I 45:2157–2188

Kaminski MA, Boersma A, Tyszka J, Holbourn AEL (1995) Response of deep-water agglutinated foraminifera to dysoxic conditions in the California Borderland basins. In: Kaminski MA, Geroch S, Gasinski MA (eds) Proceedings of the fourth international workshop on agglutinated foraminifera. Grzybowski Foundation Special Publication, Krakow, pp 131–140

Keeling RF, Körtzinger A, Gruber N (2010) Ocean deoxygenation in a warming world. Ann Rev Mar Sci 2:199–299

Kitazato H, Ohga T (1995) Seasonal changes in deep-sea benthic foraminiferal populations: results of long-term observations at Sagami Bay, Japan. In: Sakai H, Nozaki Y (eds) Biogeochemical processes and ocean flux in the Western Pacific. Terra Scientifi Publishing Company, Tokyo, pp 331–342

Kitazato H, Shirayama Y, Nakatsuka T, Fujiwara S, Shimanaga M, Kato Y, Okada Y, Kanda J, Yamaoka A, Masuzawa T, Suzuki K (2000) Seasonal phytodetritus deposition and responses of bathyal benthic foraminiferal populations in Sagami Bay, Japan: preliminary results from "Project Sagami 1996-1999". Mar Micropaleontol 40:135–149

Kitazato H, Nakatsuka T, Shimanaga M, Kanda J, Soh W, Kato Y, Okada Y, Yamaoka A, Masuzawa T, Suzuki K, Shirayama Y (2003) Long-term monitoring of the sedimentary processes in the central part of Sagami Bay, Japan: rationale, logistics and overview of results. Prog Oceanogr 57:3–16

Knight R, Mantoura RFC (1985) Chlorophyll and carotenoid pigments in foraminifera and their symbiotic algae: analysis by high performance liquid chromatography. Mar Ecol Prog Ser 23:241–249

Koho KA, Kouwenhoven T, de Stigter H, van der Zwaan G (2007) Benthic foraminifera in the Nazaré Canyon, Portuguese continental margin: sedimentary environments and disturbance. Mar Micropaleontol 66:27–51

Koho KA, García R, de Stigter HC, Epping E, Koning E, Kouwenhoven TJ, van der Zwaan GJ (2008a) Sedimentary labile organic carbon and pore water redox control on species distribution of benthic foraminifera: a case study from Lisbon-Setúbal Canyon (southern Portugal). Prog Oceanogr 79:55–82

Koho KA, Langezaal AM, van Lith YA, van der Duijnstee IAP, Zwaan GJ (2008b) The influence of a simulated diatom bloom on deep-sea benthic foraminifera and the activity of bacteria: a mesocosm study. Deep-Sea Res I Oceanogr Res Pap 55:696–719

Koho KA, Piña-Ochoa E, Geslin E, Risgaard-Petersen N (2011) Survival and nitrate uptake mechanisms of foraminifers (*Globobulimina turgida*): laboratory experiments. FEMS Microbiol Ecol 75:273–283

Kurbjeweit F, Schmiedl G, Schiebel R, Hemleben C, Pfannkuche O, Wallmann K, Schäfer P (2000) Distribution, biomass and diversity of benthic foraminifera in relation to sediment geochemistry in the Arabian Sea. Deep-Sea Research Part II: Topical Studies in Oceanography 47:2913–2955

Langezaal AM, Jorissen FJ, Braun B, Chaillou G, Fontanier C, Anschutz P, van der Zwaan GJ (2006) The influence of seasonal processes on geochemical profiles and foraminiferal assemblages on the outer shelf of the Bay of Biscay. Cont Shelf Res 26:1730–1755

Larkin KE, Gooday AJ (2009) Foraminiferal faunal responses to monsoon-driven changes in organic matter and oxygen availability at 140 and 300 m water depth in the NE Arabian Sea. Deep-Sea Res (2 Top Stud Oceanogr) 56:403–421

Lesser MP (2006) Oxidative stress in marine environments: biochemistry and physiological ecology. Annu Rev Physiol 68:253–278

Leutenegger S (1984) Symbiosis in benthic foraminifera: specificity and host adaptations. J Foraminifer Res 14:16–25

Leutenegger S, Hansen HJ (1979) Ultrastructural and radiotracer studies of pore function in foraminifera. Mar Biol 54:11–16

Levin LA, Gage JD, Martin C, Lamont PA (2000) Macrobenthic community structure within and beneath the oxygen minimum zone, NW Arabian Sea. Deep-Sea Res II 47:189–226

Levin L, Gutiérrez D, Rathburn A, Neira C, Sellanes J, Muñoz P, Gallardo V, Salamanca M (2002) Benthic processes on the Peru margin: a transect across the oxygen minimum zone during the 1997–98 El Niño. Prog Oceanogr 53:1–27

Levin LA, Ekau W, Gooday AJ, Jorissen F, Middelburg JJ, Naqvi SWA, Neira C, Rabalais NN, Zhang J (2009) Effects of natural and human-induced hypoxia on coastal benthos. Biogeosciences 6:2063–2098

Licari LN, Schumacher S, Wenzhofer F, Zabel M, Mackensen A (2003) Communities and microhabitats of living benthic foraminifera from the Tropical East Atlantic: impact of different productivity regimes. J Foraminifer Res 33:10–31

Lopez E (1979) Algal chloroplasts in the protoplasm of three species of benthic foraminifera: taxonomic affinity, viability and persistence. Mar Biol 53:201–211

Mackensen A, Douglas RG (1989) Down-core distribution of live and dead deep-water benthic foraminifera in box cores from the Weddell Sea and the California continental borderland. Deep Sea Res A 36:879–900

Mackensen A, Hubberten H-W, Bickert T, Fischer G, Futterer DK (1993) The δ13C in benthic foraminiferal tests of Fontbotia wuellerstorfi (Schwager) relative to the δ13C of dissolved inorganic carbon in Southern Ocean deep water; implications for glacial ocean circulation models. Paleoceanography 8:587–610

Mackensen A, Schumacher S, Radke J, Schmidt D (2000) Microhabitat preferences and stable carbon isotopes of endobenthic foraminifera: clue to quantitative reconstruction of oceanic new production? Mar Micropaleontol 40:233–258

McCorkle DC, Emerson SR (1988) The relationship between pore water carbon isotopic composition and bottom water oxygen concentration. Geochim Cosmochim Acta 52:1169–1178

McCorkle DC, Emerson SR, Quay PD (1985) Stable carbon isotopes in marine porewaters. Earth Planet Sci Lett 74:13–26

McCorkle DC, Keigwin LD, Corliss BH, Emerson SR (1990) The influence of microhabitats on the carbon isotopic composition of deep-sea benthic foraminifera. Paleoceanography 5:161–185

McCorkle DC, Corliss BH, Farnham CA (1997) Vertical distributions and stable isotopic compositions of live (stained) benthic foraminifera from the North Carolina and California continental margins. Deep-Sea Res 1 Oceanogr Res Pap 44:983–1024

McGann M (2002) Historical and modern distributions of benthic foraminifers on the continental shelf of Monterey Bay, California. Mar Geol 181:115–156

McHatton SC, Barry JP, Jannasch HW, Nelson DC (1996) High nitrate concentrations in vacuolate, autotrophic marine Beggiatoa spp. Appl Environ Microbiol 62:954–958

Mojtahid M, Jorissen F, Pearson TH (2008) Comparison of benthic foraminiferal and macrofaunal responses to organic pollution in the Firth of Clyde (Scotland). Mar Pollut Bull 56:42–76

Mojtahid M, Griveaud C, Fontanier C, Anschutz P, Jorissen FJ (2010) Live benthic foraminiferal faunas along a bathymetrical transect (140–4800 m) in the Bay of Biscay (NE Atlantic). Rev Micropaleontol 53:139–162

Molin D, Guidoboni E, Lodovisi A (1992) Muscilage and the phenomena of algae in the history of the Adriatic: Periodization and the anthropic context (17th-20th centuries). Sci Total Environ 511–524

Moodley L (1990) Southern North Sea seafloor and subsurface distribution of living benthic foraminifera. Neth J of Sea Res 27:57–71

Moodley L, Hess C (1992) Tolerance of infaunal benthic foraminifera for low and high oxygen concentrations. Biol Bull 183:94–98

Moodley L, Van Der Zwaan GJ, Herman PMJ, Kempers L, Van Breugel P (1997) Differential response of benthic meiofauna to anoxia with special reference to Foraminifera (Protista: Sarcodina). Mar Ecol Prog Ser 158:151–163

Moodley L, Schaub BEM, Van Der Zwaan GJ, Hermanl PMJ (1998a) Tolerance of benthic foraminifera (Protista: Sarcodina) to hydrogen sulphide. Mar Ecol Prog Ser 169:77–86

Moodley L, van der Zwaan GJ, Rutten GMW, Boom RCE, Kempers AJ (1998b) Subsurface activity of benthic foraminifera in relation to porewater oxygen content: laboratory experiments. Mar Micropaleontol 34:91–106

Moodley L, Boschker HTS, Middelburg JJ, Pel R, Herman PMJ, De Deckere E, Heip CHR (2000) Ecological significance of benthic foraminifera: ^{13}C Labelling experiments. Mar Ecol Prog Ser 202:289–295

Moodley L, Middelburg JJ, Boschker HTS, Duineveld GCA, Pel R, Herman PMJ, Heip CHR (2002) Bacteria and foraminifera: key players in a short-term deep-sea benthic response to phytodetritus. Mar Ecol Prog Ser 236:23–29

Muller-Karger FE, Varela R, Thunell R, Luerssen R, Hu CM, Walsh JJ (2005) The importance of continental margins in the global carbon cycle. Geophys Res Lett 32. doi:10.1029/2004GL021346

Murray JW (1992) Distribution and population dynamics of benthic foraminifera from the southern North Sea. J Foraminifer Res 22:114–128

Murray JW (2003) An illustrated guide to the benthic foraminifera of the Hebridean shelf, west of Scotland, with notes on their mode of life. Palaeontol Electron 5(1), 31 pp

Ní Fhlaithearta S, Reichart G-J, Jorissen FJ, Fontanier C, Rohling EJ, Thomson J, de Lange GJ (2010) Reconstructing the sea floor environment during sapropel formation using trace metals and sediment composition. Paleoceanography 25(4). doi:10.1029/2009PA001869

Nomaki H, Ohkouchi N, Heinz P, Suga H, Chikaraishi Y, Ogawa NO, Matsumoto K, Kitazato H (2009) Degradation of algal lipids by deep-sea benthic foraminifera: an in situ tracer experiment. Deep-Sea Res I 56:1488–1503

Nordberg K, Gustafsson M, Krantz AL (2000) Decreasing oxygen concentrations in the Gullmar Fjord, Sweden, as confirmed by benthic foraminifera, and possible association with NAO. J Mar Syst 23:303–316

Nordberg K, Filipsson HL, Linné P, Gustafsson M (2009) Stable oxygen and carbon isotope information on the establishment of a new, opportunistic foraminiferal fauna in a Swedish Skagerrak fjord basin, in 1979/1980. Mar Micropaleontol 73:117–128

Ohga T, Kitazato H (1997) Seasonal changes in bathyal foraminiferal populations in response to the flux of organic matter (Sagami Bay, Japan). Terra Nova 9:33–37

Osterman LE (2003) Benthic foraminifers from the continental shelf and slope of the Gulf of Mexico: an indicator of shelf hypoxia. Estuar Coast Shelf Sci 58:17–35

Osterman LE, Poore RZ, Swarzenski PW, Turner RE (2005) Reconstructing a 180 yr record of natural and anthropogenic induced low-oxygen conditions from Louisiana continental shelf sediments. Geology 33:329–332

Otte S, Kuenen JG, Nielsen LP, Paerl HW, Zopfi J, Schulz HN, Teske A, Strotmann B, Gallardo VA, Jørgensen BB (1999) Nitrogen, carbon, and sulfur metabolism in natural Thioploca samples. Appl Environ Microbiol 65:3148–3157

Paulmier A, Ruiz-Pino D (2009) Oxygen minimum zones (OMZs) in the modern ocean. Prog Oceanogr 80:113–128

Pawlowski J, Bolivar I, Farhni J, Zaninetti L (1995) DNA analysis of "Ammonia beccarii" morphotypes: one or more species? Mar Micropaleontol 26:171–178

Piña-Ochoa E, Høgslund S, Geslin E, Cedhagen T, Revsbech NP, Nielsen LP, Schweizer M, Jorissen F, Rysgaard S, Risgaard-Petersen N (2010a) Widespread occurrence of nitrate storage and denitrification among Foraminifera and Gromiida. PNAS 107:1148–1153

Piña-Ochoa E, Koho KA, Geslin E, Risgaard-Petersen N (2010b) Survival and life strategy of foraminifer, Globobulimina turgida, through nitrate storage and denitrification: laboratory experiments. Mar Ecol Prog Ser 417:39–49

Platon E, Sen Gupta BK (2001) Benthic foraminiferal communities in oxygen-depleted environments of the Louisiana continental shelf. Coast Hyp: Conseq for Living Resour Ecosyst 58:147–163

Platon E, Sen Gupta BK, Rabalais NN, Turner RE (2005) Effect of seasonal hypoxia on the benthic foraminiferal community of the Louisiana inner continental shelf: the 20th century record. Mar Micropaleontol 54:263–283

Pucci F, Geslin E, Barras C, Morigi C, Sabbatini A, Negri A, Jorissen FJ (2009) Survival of benthic foraminifera under hypoxic conditions: results of an experimental study using the Cell Tracker Green method. Mar Pollut Bull 59:336–351

Qvale G, Nigam R (1985) Bolivina skagerrakensis, a new name for Bolivina cf. B. robusta, with notes on its ecology and distribution. J Foraminifer Res 15:6–12

Rabalais NN, Diaz RJ, Levin LA, Turner RE, Gilbert D, Zhang J (2010) Dynamics and distribution of natural and human-caused hypoxia. Biogeosciences 7:585–619

Rathburn AE, Corliss BH (1994) The ecology of living (stained) deep-sea benthic foraminifera from the Sulu Sea. Paleoceanography 9:87–150

Rathburn AE, Corliss BH, Tappa KD, Lohmann KC (1996) Comparison of ecology and stable isotopic compositions of living (stained) benthic foraminifera from the Sulu and South China Seas. Deep-Sea Res I 43:1617–1646

Reichart G-J, Jorissen F, Anschutz P, Mason PRD (2003) Single foraminiferal test chemistry records the marine environment. Geology 31:355–358

Reiter M (1959) Seasonal variations in intertidal foraminifera of Santa-Monica Bay, California. J paleontol 33:606–631

Risgaard-Petersen N, Langezaal AM, Ingvardsen S, Schmid MC, Jetten MSM, Op Den Camp HJM, Derksen JWM, Piña-Ochoa E, Eriksson SP, Nielsen LP, Revsbech NP, Cedhagen T, Van Der Zwaan GJ (2006) Evidence for complete denitrification in a benthic foraminifer. Nature 443:93–96

Schmiedl G, De Bovée F, Buscail R, Charrière B, Hemleben C, Medernach L, Picon P (2000) Trophic control of benthic foraminiferal abundance and microhabitat in the bathyal Gulf of Lions, western Mediterranean Sea. Mar Micropaleontol 40:167–188

Schmiedl G, Pfeilsticker M, Hemleben C, Mackensen A (2004) Environmental and biological effects on the stable isotope composition of recent deep-sea benthic foraminifera from the western Mediterranean Sea. Mar Micropaleontol 51:129–152

Schönfeld J (1997) The impact of the Mediterranean Outflow Water (MOW) on benthic foraminiferal assemblages and surface sediments at the southern Portuguese continental margin. Mar Micropaleontol 29:211–236

Schönfeld J (2001) Benthic foraminifera and pore-water oxygen profiles: a re-assessment of species boundary conditions at the western Iberian Margin. J Foraminifer Res 31:86–107

Schulz HN, Brinkhoff T, Ferdelman TG, Hernández Mariné M, Teske A, Jørgensen BB (1999) Dense populations of a giant sulfur bacterium in namibian shelf sediments. Science 284:493–495

Schumacher S, Jorissen FJ, Dissard D, Larkin KE, Gooday AJ (2007) Live (Rose Bengal stained) and dead benthic foraminifera from the oxygen minimum zone of the Pakistan continental margin (Arabian Sea). Mar Micropaleontol 62:45–73

Scott GA, Scourse JD, Austin WEN (2003) The distribution of benthic foraminifera in the Celtic Sea: the significance of seasonal stratification. J Foraminifer Res 33:32–61

Sen Gupta BK (1999) Introduction to modern foraminifera. In: Sen Gupta BK (ed) Modern foraminifera. Kluwer Academic Publishers, Dordrecht

Sen Gupta BK, Hayes WB (1979) Recognition of Holocenic benthic foraminiferal facies by recurrent group analysis. J Foraminifer Res 9:233–245

Sen Gupta BK, Turner RE, Rabalais NN (1996) Seasonal oxygen depletion in continental-shelf waters of Louisiana: historical record of benthic foraminifers. Geology 24:227–230

Sergeeva NG, Anikeeva OV, Gooday AJ (2010) Soft-shelled, monothalamous foraminifera from the oxic/anoxic interface (NW Black Sea). Micropaleontology 56:393–407

Silva KA, Corliss B, Rathburn AE, Thunell R (1996) Seasonality of living benthic foraminifera from San Pedro basin, California borderland. J Foraminifer Res 26:71–93

Smith PB (1964) Ecology of benthonic species: recent foraminifera off Central America. US Geol Surv Prof Pap 429B:1–51

Suhr SB, Pond DW (2006) Antarctic benthic foraminifera facilitate rapid cycling of phytoplankton-derived organic carbon. Deep-Sea Res II 53:895–902

Takata H, Murakami S, Seto K, Sakai S, Tanaka S, Takayasu K (2003) Foraminiferal assemblages in Aso-Kai Lagoon, central Japan. Laguna 10:113–118 (in Japanese with English abstract)

Tapia R, Lange CB, Marchant M (2008) Living (stained) calcareous benthic foraminifera from recent sediments off Concepcion, central-southern Chile (similar to 36 degrees S). Rev Chil Hist Nat 81:403–416

Tielens AGM, Rotte C, Van Hellemond JJ, Martin W (2002) Mitochondria as we don't know them. Trends Biochem Sci 27:564–572

Tomasino MG (1996) Is it feasible to predict 'slime blooms' or 'mucilage' in the northern Adriatic Sea? Ecol Model 84:189–198

Travis JL, Bowser SS (1986) Microtubule-dependent reticulopodial motility: is there a role for actin? Cell Motil. Cytoskeleton 6:146–152

Tribovillard N, Algeo TJ, Lyons T, Riboulleau A (2006) Trace metals as paleoredox and paleoproductivity proxies: an update. Chem Geol 232:12–32

Tsuchiya M, Grimm GW, Heinz P, Stögerer K, Ertan KT, Collen J, Brüchert V, Hemleben C, Hemleben V, Kitazato H (2009) Ribosomal DNA shows extremely low genetic divergence in a world-wide distributed, but disjunct and highly adapted marine protozoan (Virgulinella fragilis, Foraminiferida). Mar Micropaleontol 70:8–19

Turner RE, Rabalais NN (1994) Coastal eutrophication near the Mississippi river delta. Nature 368:619–621

Van Der Weijden CH, Reichart GJ, Visser HJ (1999) Enhanced preservation of organic matter in sediments deposited within the oxygen minimum zone in the northeastern Arabian Sea. Deep-Sea Res I 46:807–830

Verhallen PJ (1991) Late Pliocene to Early Pleistocene Mediterranean mud-dwelling foraminifera; influence of a changing environment of community structure and evolution. Utrecht Micropaleontol Bull 40:1–219

Walton WR (1955) Ecology of living benthonic foraminifera, Todos Santos Bay area, Baja California. J Paleontol 29:952–1018

Witte U, Wenzhöfer F, Sommer S, Boetius A, Heinz P, Aberle N, Sand M, Cremer A, Abraham W-R, Jørgensen BB, Pfannkuche O (2003) In situ experimental evidence of the fate of a phytodetritus pulse at the abyssal sea floor. Nature 424:763–766

Woulds C, Cowie GL, Levin LA, Andersson JH, Middelburg JJ, Vandewiele S, Lamont PA, Larkin KE, Gooday AJ, Schumacher S, Whitcraft C, Jeffreys RM, Schwartz M (2007) Oxygen as a control on seafloor biological communities and their roles in sedimentary carbon cycling. Limnol Oceanogr 52:1698–1709

Zaninetti L (1984) Les foraminifers du salin de Bras del Port (Santa Pola, Espagne) avec remarques sur le distribution des ostracodes. Revista d'investigacions geologiques 38(39):123–138

Biodata of **Petra Heinz** and **Emmanuelle Geslin**, authors of *"Ecological and Biological Response of Benthic Foraminifera Under Oxygen-Depleted Conditions: Evidence from Laboratory Approaches."*

Dr. Petra Heinz is working in the Micropaleontology group of the Department for Geosciences at the University of Tübingen, Germany. She obtained her Ph.D. at the University of Tübingen in 1999 and continued her studies and research there. In 2001, she got the International Brönnimann Award. Dr. Heinz's scientific interests concentrate on the ecology and biology of foraminifera, with main focus on recent benthic foraminifera of the deep sea. She is exploring the ecological demands of foraminifera by field investigations, complemented by laboratory studies and in situ experiments. She is responsible for the culture laboratory of benthic foraminifera at the Department for Geosciences in Tübingen.

E-mail: **petra.heinz@uni-tuebingen.de**

Dr. Emmanuelle Geslin is a Research Scientist at the Laboratory of Actual and Fossil Bio-indicators (BIAF) at the University of Angers (France). She holds a Ph.D. degree in Actuo-micropaleontology in 1999. In 2000 she was rewarded a "Deutsche Forschungsgemeinschaft" fellowship for 2 years and started her work on ecological and experimental aspects of benthic foraminifera in the laboratory of Prof. Christoph Hemleben in Tübingen (Germany). She worked with deep-sea foraminifera under controlled conditions in laboratory in order to understand the impact of various oxygen conditions on the vertical distribution of deep-sea foraminifera. She obtained a permanent staff status at the University of Angers in 2002 and developed a new culture laboratory for deep-sea foraminifera. She collaborated with the Geomicrobiological group of Aarhus University (Dk) in order to study the anaerobic metabolism of foraminifera.

E-mail: **emmanuelle.geslin@univ-angers.fr**

ECOLOGICAL AND BIOLOGICAL RESPONSE OF BENTHIC FORAMINIFERA UNDER OXYGEN-DEPLETED CONDITIONS: EVIDENCE FROM LABORATORY APPROACHES

PETRA HEINZ[1] AND EMMANUELLE GESLIN[2]
[1]*Department for Geosciences, University of Tübingen, Hölderlinstr. 12, 72074 Tübingen, Germany*
[2]*Laboratoire d'Etude des Bio-indicateurs Actuels et Fossiles (BIAF) and LEBIM, University of Angers, 2 Boulevard Lavoisier, Angers Cedex 49045, France*

1. Introduction

Foraminifera are unicellular organisms with characteristic net-like pseudopodia (granuloreticulopodia). These amoeboid protists, ranging in size between a few micrometers and several centimeters, are an enormously successful group, abundant in all marine ecosystems, and their fossils are known since the late Precambrian and early Cambrian (Vachard et al. 2010). The diversity of foraminifera with respect to shape, size, habitat, trophic mechanisms, and feeding behaviors is impressive compared to other protists. The ability of most species to produce stable agglutinated or secreted outer protective tests, with the high potential to fossilize, makes this group very attractive to paleontologists. At the same time, biologists were not interested in these protists with their shells hampering the direct observation of most cellular processes. In the past, some experiments were involved to understand how foraminifera can be cultured under laboratory conditions in order to understand their ecological and food requirements (e.g., Bradshaw 1955; Murray 1963; Arnold 1954; Muller and Lee 1969) or their physiological characteristics such as growth (e.g., Bradshaw 1957). Experiments testing the response of foraminifera to low-oxygen conditions started at the beginning of the 1990ths years (e.g., Moodley and Hess 1992; Bernhard 1993; Alve and Bernhard 1995; Hannah and Rogerson 1997; Moodley et al. 1997) and continues intensively until now (e.g., Duijnstee et al. 2003; Geslin et al. 2004; Ernst and Van der Zwaan 2004; Risgaard-Petersen et al. 2006; Piña-Ochoa et al. 2010b; Koho et al. 2011).

Individual species are adapted to different habitats on the seafloor ranging from suboxic to well-oxygenated sediments (Murray 2006). Foraminifera are aerobes, but many species are facultative anaerobes (Bernhard 1989, 1993; Moodley and Hess 1992; Sen Gupta and Machain-Castillo 1993; Risgaard-Petersen et al. 2006; Høgslund et al. 2008; Piña-Ochoa et al. 2010a, b). In fact certain species can survive and inhabit oxygen-depleted and even anoxic and/or sulfidic sediments, at least during short time periods (several weeks to months) (e.g., Corliss and Emerson 1990; Bernhard 1992; Jannink et al. 1998; Jorissen et al. 1998; Szarek et al. 2007; Piña-Ochoa et al. 2010b). But when anoxia extended over long time periods (several months to years), like in silled basins or during sapropel-formation, foraminifera disappear, like all benthos in such laminated sediments (Bernhard 1996; Schmiedl et al. 2003). Benthic foraminiferal taxa that can occur in oxygen-depleted habitats form characteristic assemblages, named low-oxygen foraminiferal assemblages (LOFAS) (Bernhard and Sen Gupta 1999). Mechanisms responsible for the anaerobic respiration of foraminifera were unknown until 2006. Capacity of nitrate storage and denitrification have been proved in 2006 for two species (Risgaard-Petersen et al. 2006) and were confirmed for many foraminiferal species in various marine environments (Høgslund et al. 2008; Piña-Ochoa et al. 2010a).

In order to gain further insights into the foraminiferal response to hypoxic/anoxic events and to physiological mechanisms and strategies to get anaerobic respiration, field investigations have to be complemented by laboratory studies. In laboratory investigations, it is possible to control specific conditions and keep them stable while single parameters can be varied during the experimental period. This allows recording the response of the foraminifera to this factor. Additionally, life observations can be made about important ecological processes, and physiological characters can be measured and help to understand the biological mechanisms. It offers complementary insights on important aspects of the life of benthic foraminifera. The outcome provides crucial constraints on the ecological demands and limits of benthic foraminiferal faunas and single species. In 1996, a review about experimental work on "microaerophilic and facultative anaerobic benthic foraminifera…" was published by Bernhard. A considerable amount of new experiments and new insights concerning this topic were made since then. Here we now want to focus on laboratory approaches and experiments to analyze the ecological and biological responses of benthic foraminifera under oxygen-depleted conditions. In situ studies in oxygen-depleted environments (e.g., Bernhard and Alve 1996; Jannink et al. 1998; Schumacher et al. 2007; Szarek et al. 2007; Høgslund et al. 2008) bring very interesting observations which will be introduced in this book in the chapter of Koho and Piña-Ochoa ("In situ studies about foraminiferal communities living in low oxygen setting").

2. Laboratory Methodologies

Different laboratory studies have been performed in the recent years to examine the influence of oxygen on benthic foraminifera. These investigations include observations of behavior in response to oxygen changes, analyses of foraminiferal influence on oxygen distribution, and determination of survival times under oxygen-depleted or anoxic conditions and physiological adaptations.

Experimental working with foraminifera in oxygen-depleted environments can be problematic because of the determination of living foraminifera during the experiment. Observation of unstained or stained (e.g., with Rose Bengal) cytoplasm inside the shell is not sufficient to judge vitality when incubation times are performed under oxygen-depleted conditions because cytoplasm and color can be retained in dead specimens for weeks to months (e.g., Bernhard 1988; Hannah and Rogerson 1997). Even the presence of debris around tests or apertures, interpreted as indirect evidence of pseudopodial activity, can lead to wrong identifications, because foraminifera can secrete adhesive substances (Langer and Gehring 1993) that passively bind particles, and these bindings may be stable some short time period after death. Observation of pseudopodia, cytoplasmic streaming, or moving of individuals is a time-consuming but reliable determination of living foraminifera. Different biochemical techniques have been developed during the last years to simplify the identification. The Adenosine Triphosphate Assay is a biochemical method to assess foraminiferal vitality by Adenosine Triphosphate (ATP) analysis (Delaca 1986; Bernhard and Reimers 1991). It is a terminal method, causing death of treated organisms (Bernhard et al. 1995). Advantages of ATP analysis include its accuracy and the ability to analyze large populations in about the same time as required by Rose Bengal processing (Bernhard 2000). Disadvantages include the fact that alive specimens need to be analyzed soon (e.g., 4–6 h) after sampling, which is normally not a problem when experiments are performed in laboratory. Another biochemical method is the spectrofluorimetric analysis of individuals with fluorogenic probes such as fluorescein diacetate (FDA) (Bernhard et al. 1995) or Cell Tracker Green (CTG) (Bernhard et al. 2006; Morigi and Geslin 2009). Fluorogenic probes are nonfluorescent compounds that produce a fluorescent product after modification by intracellular esterases that are only active in living individuals (Fig. 1). FDA is a fluorogenic probe which cannot be fixed by formalin (Bernhard et al. 1995). It can be used in two different ways: by fluorescence intensity quantification (Bernhard et al. 1995; Geslin et al. 2004) or by fluorescence observation (Piña-Ochoa et al. 2010b). Foraminifera are incubated with FDA for several hours. The former procedure then measures the fluorescence intensity using a spectrofluorimeter.

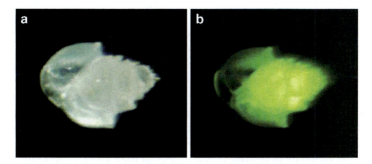

Figure 1. Alive *Bulimina marginata* observed using (**a**) a normal light binocular, (**b**) an epifluorescent binocular.

The fluorescence intensity is related to the foraminiferal vitality. For the second way, the fluorescence intensity will be observed by eye using an epifluorescence binocular or stereomicroscope. Both methods are just practicable if only few foraminifera have to be analyzed. Cell Tracker Green is often more practicable than FDA because the fluorescence can be fixed with formalin (Bernhard et al. 2006, 2009; Pucci et al. 2009). The fluorescence will be observed with an epifluorescence binocular. Therefore, the Cell Tracker Green can be used comparable to Rose Bengal, but it needs longer incubation times and it is more expensive. A new method was proposed by Borrelli et al. (2010) which is based on the determination of the metabolically active fraction by means of Fluorescent in situ Hybridization (FISH). This method was not used for experimental studies under low-oxygen conditions yet.

Oxygen-depleted conditions are often reached in the experiment by flushing seawater or sediment with nitrogen. Depending on the experimental set-up, nitrogen can be inserted in a water reservoir, which is connected with the culture vessels or aquaria (e.g., Moodley et al. 1997; Piña-Ochoa et al. 2010b; Koho et al. 2011), or directly in the overlaying water of the culture vessels (e.g., Panchang et al. 2006; Pucci et al. 2009). Other studies used gas or air flushed tubes positioned in the sediment to reach desirable oxygen distribution patterns (e.g., Moodley et al. 1998b).

In some cases, oxygen-depleted conditions were reached by closing the mesocosm containing organic-matter-rich sediment (Ernst and Van der Zwaan 2004). Working with oxygen-depleted or even anoxic experimental set-ups needs necessarily the use of oxygen sensors that control oxygen concentration in the overlying water and more specifically in the sediment. In order to obtain oxygen profiles in the sediment, electrochemical sensors like steel-reinforced electrodes or glass electrodes are easy to use and allow obtaining one-dimensional oxygen distribution. Various sizes of tips are available according to the wanted precision and according to the type of sediment. In muddy sediment, very fine tip glass electrodes (few decades of μm) are often used, but such electrodes are quite fragile (Fig. 2a). However, a one-dimensional oxygen concentration profile cannot show the occurrence of oxic micro-niches created by bioturbation. Another type of oxygen sensors are specific optodes (optical sensors) (Fig. 2b). The tip of the optode is glowing in blue light. Therefore, it is possible to orientate in the sediment, which can be very useful when taking profiles directly in a burrow or next to foraminifera, in a trail or in a cyst (Figs. 2b and 3). Additionally, optodes are not as fragile as glass electrodes and can easily be repaired if they break. In order to get more details of the spatial oxygen concentration, a two-dimensional (2D) oxygen distribution in the water and sediment can be recorded using planar optodes defined by Holst et al. (1998) (Fig. 2c). Planar optodes are excellent tools for laboratory-based investigations. They allow recording the two-dimensional O_2 concentration dynamic in benthic communities at high spatial and temporal resolution. Holst and Grunwald (2001) improved this technique by developing a new modular luminescence lifetime imaging system (MOLLI) which enables the use

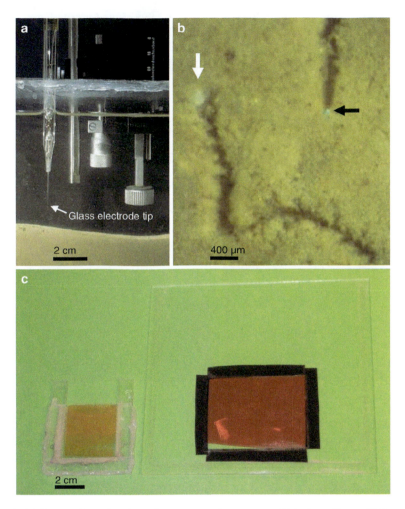

Figure 2. (a) Oxygen glass microelectrode to measure the oxygen profile in the sediment. (b) Burrow of a foraminifer in the sediment and optode measurement system to quantify the oxygen concentration in the sediment and burrow. *Black arrow* indicates the micro-optode (light is sent through the fiber-optic sensor during the measurement; the tip at the end of the optode is shining blue). *White arrow* shows the end of a trail built by *Globocassidulina subglobosa*; part of the test is still visible. (c) Two transparent planar optodes fixed on the glass walls of various aquaria.

of transparent planar optodes allowing the observation of samples while taking 2D oxygen images. 2D oxygen images around benthic foraminifera were made by Geslin et al. (2002) (Fig. 4).

Several technical problems can arise when using these sensors: analytic detection limit of oxygen under trace concentrations of oxygen, oxygen consumption of the sensor during measurement, aging of the sensors, sensor damage in

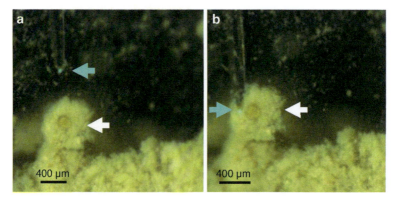

Figure 3. Micro-optode measures oxygen concentration inside a cyst of *Ammodiscus catinus*. (**a**) Optode is positioned above the cyst. (**b**) Optode carefully penetrated the cyst. *Blue arrows* indicate the tip of the optode. *White arrows* show the cyst of *Ammodiscus catinus*; the brown test is visible inside.

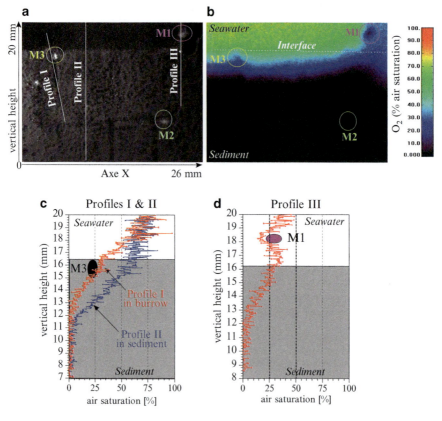

Figure 4. Black and white image (**a**), its 2D oxygen distribution image (**b**), and the extracted oxygen profiles (**c, d**). (**a**) Three specimens of *Masselina secans* are visible (M1 on the glass, M2 and M3 in the sediment). The *white lines* correspond to the extracted oxygen profiles (I–III) introduced on (c, d). (**b**) Corresponding 2D oxygen image shows oxygen depletion around the specimens M1 and M3 which are located on oxygenated layer. M2 is located in anoxic sediment. (**c, d**) Oxygen profiles extracted from 2D oxygen distribution images. The comparison between profiles I and II shows the oxygen content in the burrow which presents the foraminifer is lower than in the sediment. Profile III shows an oxygen depletion around the specimen M1 corresponding to its own oxygen consumption by respiration.

coarse sediments, etc., which have to be overcome during laboratory work and need some skills and experience.

3. Survival Experiments

Among meiofauna, foraminifera seem to be the group which is least affected by in situ hypoxia (Josefson and Widbom 1988; Moodley et al. 1998a). However, until 1997, some authors are still suspicious that foraminifera living deep in the sediment are able to be active (Hannah and Rogerson 1997). Now it is accepted that foraminiferal species differ in their tolerance to anoxia (Bernhard and Sen Gupta 1999; Piña-Ochoa et al. 2010a). Some species appear to be more resistant than others such as *Stainforthia, Nonionella, Nonion* whereas *Reophax* and *Bulimina marginata* can be considered less-tolerant species (Moodley et al. 1997; Bernhard and Alve 1996). Hard-shelled foraminifera seem to be better resistant to prolonged anoxia than soft-shelled or agglutinated (Moodley et al. 1997). Another interesting observation is that chambers of specimens formed under dysaerobic conditions had a significantly higher porosity, mainly due to larger pores (Moodley and Hess 1992; Kitazato and Tsuchiya 1999). According to the data of in situ observations and laboratory experiments, it is proved that some species of benthic foraminifera are facultative anaerobes such as *Globobulimina turgida* (e.g., Risgaard-Petersen et al. 2006; Piña-Ochoa et al. 2010b). In a number of studies, experimental settings were performed to test foraminiferal survival under anoxic or oxygen-depleted conditions for a specific period of time (e.g., Moodley and Hess 1992; Cedhagen 1993; Bernhard 1993; Bernhard and Alve 1996; Hannah and Rogerson 1997; Moodley et al. 1997; Pucci et al. 2009). Experiments differ between incubation times, oxygen concentrations, and investigated species. Laboratory experiments conducted under anoxic conditions allow giving survival times which are only the minimum time range of tolerance under tested conditions. The highest survival time recorded (proved by H_2S smell or by no oxygen detection using microelectrodes) is more than roughly 80 days for the species *Stainforthia* and *Globobulimina turgida* (Moodley et al. 1997; Piña-Ochoa et al. 2010b). However, differences in survival rates of single experiments, ATP concentrations, and ultrastructural observations show that foraminiferal species differ in responses of oxygen depletion (Bernhard and Alve 1996; Moodley et al. 1997). In nature, anoxic conditions often appear together with sulfidic conditions. Only very few laboratory experiments tested survival under anoxic and sulfidic conditions (Bernhard 1993; Moodley et al. 1998a), with different H_2S concentrations and foraminiferal faunas. Low H_2S concentrations (500 nM H_2S) did not significantly influence the foraminifera within 30 days (with ATP levels in living foraminifera under H_2S conditions comparable to control foraminifera (Bernhard 1993)). Higher H_2S concentrations (5 µM H_2S) were also tolerated by foraminifera for some time, showing cytoplasmic movement inside foraminiferal cells, but between 21 and 42 days, they died (Moodley et al. 1998a).

4. Orientation in the Sediment: Foraminiferal Aerotaxis?

In nature, the bulk of benthic foraminifera colonize the sediment–water interface and the upper sediment layers, where high-quality organic matter is available. But some species can inhabit deeper levels in the sediment where oxygen is absent and high-quality organic matter is less abundant and mostly refractory, with a low nutritional value. Such assemblages are referred to as the deep infauna (Jorissen 1999). The oxygen concentration in the pore water decreases with increasing sediment depths. Chemical differences between the sediment surface and deeper horizons are thus pronounced, and foraminiferal species can orientate and settle down in the preferred horizon.

Field investigations showed that seasonal variations in the depth of the redox front and the thickness of the oxygenated layer affect the faunal structure and the distribution of benthic foraminifera (Barmawidijaja et al. 1992; Kitazato and Ohga 1995; Ohga and Kitazato 1997); more details are given in the chapter of Koho and Piña-Ochoa. Similar observations were made in the laboratory. Nomaki et al. (2005) reported a laboratory feeding experiment using sediments and living benthic foraminifera from central Sagami Bay, an oxygen-depleted area. The response of the organisms to simulated seasonal nutrition events were tested and recorded. Different food levels were examined and the behavior and microhabitat distribution were compared. After food supply, many foraminifera left their microhabitat, migrated up to more shallow sediment layers, and ingested it. Shallow- and intermediate-living species respond faster than deep-living foraminifera. Dissolved oxygen concentration at the sediment surface and the oxygen penetration depth in the fed aquaria changed to micro-oxic conditions during the experiment. Do the foraminifera migrate to reach the food particles, or do they want to escape oxygen conditions which get from bad to worse? In nature, the most important environmental factors, both food and oxygen supply, are often very closely related to each other as it is shown by the TROX model (Jorissen et al. 1995). Higher organic fluxes lead to higher oxygen consumption during oxidation and degradation of more organic matter, which regularly initiate lower oxygen supply. Laboratory investigations can elucidate the separate impact of these two factors. The influence of food during well-oxygenated laboratory conditions was described, e.g., in Heinz (2001) and Heinz et al. (2001, 2002). Increased nutrition supply leads to elevated foraminiferal standing stocks, while the vertical distribution in the sediment and the microhabitat preferences did not change. This was confirmed by experiments of Ernst and van der Zwaan (2004), who showed that adding organic matter affects only some shallow-living, opportunistic taxa, while oxygen changes can induce strong responses, especially concerning the vertical distribution of foraminifera.

Oxygen is essential for efficient generation of cellular energy in aerobically grown cells, and it is not surprising that motile cells have developed strategies for seeking environments with optimal concentrations of oxygen. This phenomenon is called *aerotaxis*. Dysoxic to anoxic conditions can be one important stimulus for foraminiferal migration and microhabitat selection. To study this topic in more detail, it is important to keep trophic conditions stable while available oxygen varies. However, even if food is not added in the experiment, keeping control cores under well-oxygenated conditions accelerates the degradation of labile organic matter whereas this degradation is slower under oxygen-depleted conditions. Consequently, the oxic cores became oligotrophic earlier. This phenomenon was observed by Pucci et al. (2009), who showed a higher density of the total foraminiferal fauna under low-oxygen condition after 15 days of experiment. Species reacted differently. *Eggerella scabra*, *Nonionella turgida*, and *Nouria polymorphinoides* possessed the same trend than the total fauna whereas *Bulimina* spp. and *Pseudoeponides falsobeccarii* demonstrated similar densities under both oxic conditions. After longer incubation (until 69 days), the total density of foraminifera was equal under both conditions.

Foraminifera can move in the sediment to avoid uncomfortable oxygen conditions or to reach the favorable living zone. Experimental oxygen changes appear to induce strong changes in the vertical distribution of foraminifera within short time periods. Laboratory experiments demonstrated that foraminifera vertically migrate upward in the sediment, when seawater oxygen conditions decrease gradually (Alve and Bernhard 1995; Ernst and van der Zwaan 2004; Geslin et al. 2004). Migration in the sediments started when oxygen diminished below 2 mL $O_2 L^{-1}$, but different species showed different time responses (Alve and Bernhard 1995). Reoxygenation of the aquaria led to the migration of most foraminifera back to the original depth distribution. Ernst and van der Zwaan (2004) showed that most foraminiferal species respond to experimental anoxic conditions by climbing up in the sediment toward better oxic levels, except for some deep infaunal species like *Globobulimina* species, which are adapted to anoxia and prefer these conditions. Such observations with *G. turgida* were also made by Koho et al. (2011). In Geslin et al. (2004), the response of five dominant species to oxygen variations regarding microhabitat selection and migration directions was given. Changes in oxygen gradients triggered foraminiferal movement and change of the microhabitat. An inversed oxygen profile with the highest oxygen concentrations in the deepest sediment layer and anoxic conditions at the sediment–water interface causes an inversed vertical distribution of the five tested taxa. Shallow infaunal species were recorded on the bottom of the aquarium, whereas the deep-infaunal-living *Globobulimina affinis* moved upward to the sediment surface. For these five species, and probably for many other benthic foraminifera, oxygen gradients and maybe other chemical gradients seem to be very important for microhabitat selection, and even may be the main factor to orientate inside the sediment. Interestingly, migration speed measurements of different foraminiferal species recorded that the moving speed of *Quinqueloculina lamarckiana* was not greatly affected by oxygen depletion (Gross 2000). Foraminifera in laboratory experiments can migrate through anoxic sediments (Moodley et al. 1998b). Koho et al. (2011) observed migration of *G. turgida* under anoxic condition. Some specimens were found 3 cm deeper after 30 days of anoxia. This phenomenon was also directly observed using planar optodes (see section below, Fig. 4). When foraminifera were suddenly buried in anoxic sediment layers, some (but not all) migrated upward to aerated horizons (Moodley et al. 1998b). But significant parts of buried individuals stayed in the anoxic sediment.

5. Sediment Oxidation and the Influence of Bioturbation

Foraminifera may play an important role in the benthic system, often forming a considerable part of the meiofauna. The abundance of living foraminifers can be very high, reaching 1000–2000 specimens per 10 cm^2 (Gooday 1986) and even higher. Foraminiferal rhizopodial activities in the sediment comprise locomotion, burrow formation, cyst formation, and deposit feeding (Gross 2000; Heinz et al. 2005). Bioturbation by benthic macrofauna is known to have a major influence on sediment–water fluxes of oxygen (Aller 1982; Forster and Graf 1992, 1995; Heip et al. 2001). Macrofauna burrows allow penetration of oxygen to depths of several centimeters, altering the spatial distribution of biogeochemical zones within the pore water (Mortimer et al. 1999; Meyers et al. 1988). Hemleben and Kitazato (1995) suggested that foraminifera may play an important role on the sediment ventilation by bioturbation. High mixing rates ($Db \approx 0.2$ cm^2 day^{-1}) were measured by foraminiferal assemblages in laboratory experiments (Gross 2002). A first estimate of sediment displacement by traces of allogromiids produced in culture experiments suggests that there is a fourfold mixing per year of the first sediment layer by foraminifera alone (Gross 2002).

Laboratory experiments were also investigated to measure the microdistribution of oxygen concentration in sediment containing living foraminifera in order to evaluate their impact on the total amount of oxygen in the sediment (Geslin et al. 2002; Geslin unpublished data). The two-dimensional O_2 concentration distribution around *Masselina secans* living above and in the sediment was measured using planar optodes. One specimen of *Masselina secans* was constructing a vertical burrow (Fig. 4a, specimen M3), and when it has reached the more oxygenated sediment–water interface, foraminiferal oxygen consumption was detected (Fig. 4b). The comparison between the oxygen profiles measured at the level of the burrow containing the specimen M3 and of the sediment without any foraminifer has shown lower oxygen concentrations in the burrow compared to the sediment (Fig. 4c). Foraminiferal occurrence in burrows did not induce its reoxygenation. Oxygen consumption by respiration can also be observed around one specimen attached to the aquarium wall above the sediment (Fig. 4a, b, d, specimen M1, profile III). According to these observations, we can conclude that foraminifers create oxygen-depleted microenvironments in their burrows and in the micro-surrounding environment because of their own respiration. Finally, the foraminiferal burrows did not induce oxygen enrichment in the sediment as observed for macrofauna burrows.

An important next step of experimental studies in future would now be the further development of oxygen-depletion in situ experiments, which could excellently complement and continue laboratory work directly at the seafloor (Bernhard et al. 2009).

6. Foraminiferal Metabolism Under Oxic and Anoxic Conditions

Benthic foraminifera are facultative anaerobes, meaning they respire oxygen under oxic conditions, and they are able to live using other metabolism pathways under anoxia. Aerobic respiration rates were measured on benthic foraminifera since 1961 when Bradshaw was showing a relation between respiration rates and body size for *Ammonia tepida*. Later authors reported that respiration rates measured for only few species (Lee and Muller 1973; Hannah et al. 1994; Nomaki et al. 2007; Moodley et al. 2008). Recently, a work based on 17 benthic species with various test sizes has been published (Geslin et al. 2011). The authors proposed correlation between respiration rate and foraminiferal biovolume: $R = 7.9 \times 10^{-7} * \text{BioVol}^{0.85}$, where respiration rate is expressed in nl $O_2 h^{-1}$ ($n = 67$; $R^2 = 0.74$; $F = 193$; $p < 0.0001$). A comparison between foraminiferal aerobic respiration and other meiofaunal groups suggests that benthic foraminifera have lower respiration rates than the others, probably because of their lower metabolism rates. The measured rates allow estimating the foraminiferal input to aerobic C-mineralization. On the marine shelf environment, this contribution appears to be low (Geslin et al. 2011).

Laboratory observations demonstrated that foraminifera tolerate oxygen-depleted conditions and anoxia (e.g., Pucci et al. 2009; Piña-Ochoa et al. 2010b). Before 2006, various hypotheses based on laboratory observations were suggested to explain the physiological adaptations to this tolerance, most of them summarized in Bernhard (1996). One possibility is that foraminifera consume oxygen by mitochondria located in their pseudopods which they can extend to at least ten times the shell diameter (Travis and Bowser 1991) to reach better oxygenated sediment layers. Some species use mitochondria activity in extended pseudopods to maintain oxidative phosphorylation (Bernhard and Sen Gupta 1999). But an experimental study showed that cyanide, an electron-transport inhibitor, did not stop intracellular motility in two allogromiid species (Travis and Bowser 1986). This suggests that at least the tested species can survive without oxidative phosphorylation. *Virgulinella fragilis*, living under microaerophilic, sulphidic conditions, and some other calcareous species showed potential bacterial endosymbionts (Bernhard and Reimers 1991; Bernhard 1993, 1996, 2003). But no prokaryotic symbionts were described in investigated foraminifera from, e.g., cold-seep sites (Bernhard et al. 2001). Other unusual ultrastructural features like sequestered chloroplasts (see Table 1 in Bernhard and Bowser 1999; Leutenegger 1984; Cedhagen 1991; Bernhard and Alve 1996) or peroxisome-endoplasmic reticulum complexes (P-ER complexes) (see Table 1 in Bernhard and Bowser 2008; Nyholm and Nyholm 1975; Bernhard and Alve 1996; Bernhard 1996) were observed at different foraminifera from oxygen-depleted environments. In 2006, the first evidence for a complete denitrification process in a benthic foraminifer was observed, which proves at least one anaerobic metabolism pathway (Risgaard-Petersen et al. 2006). Various recent papers focus on this ability (Høgslund et al. 2008; Piña-Ochoa et al. 2010a, 2010b; Koho et al. 2011). *Globobulimina turgida* is facultative anaerobe using denitrification respiration to live under anoxia (Risgaard-Petersen et al. 2006). Many species of benthic foraminifera are able to accumulate nitrate in the cytoplasm (Piña-Ochoa et al. 2010a). The ability to store intracellular nitrate is found to be associated with the ability of these organisms to respire nitrate into N_2 through complete denitrification (Risgaard-Petersen et al. 2006; Høgslund et al. 2008; Piña-Ochoa et al. 2010a). This capacity may be one reason for their successful colonization of diverse marine sediment environment. The widespread occurrence among distantly related organisms suggests an ancient origin of the trait (Piña-Ochoa et al. 2010a). Total foraminifer-mediated denitrification in various environments can be estimated, showing that benthic foraminifera may contribute 8–80% of the measured denitrification activity.

7. References

Aller RC (1982) The effects of macrobenthos on chemical properties of marine sediment and overlying water. In: McCall PL, Tevesz MJS (eds) Animal-sediment relations. Plenum Press, New York, pp 53–102

Alve E, Bernhard JM (1995) Vertical migratory response of benthic foraminifera to controlled oxygen concentrations in an experimental mesocosm. Mar Ecol Prog Ser 116:137–151

Arnold ZM (1954) Culture methods in the study of living foraminifera. J Paleontol 28:404–416

Barmawidjaja DM, Jorissen FJ, Puskaric S, Van der Zwaan GJ (1992) Microhabitat selection by benthic foraminifera in the northern Adriatic Sea. J Foramin Res 22:297–317

Bernhard JM (1988) Postmortem vital staining in benthic foraminifera: duration and importance in population and distributional studies. J Foramin Res 18:143–146

Bernhard JM (1989) The distribution of benthic foraminifera with respect to oxygen concentrations and organic carbon levels in shallow-water Antarctic sediments. Limnol Oceanogr 34:1131–1141

Bernhard JM (1992) Benthic foraminiferal distribution and biomass related to pore-water oxygen content: central California continental slope and rise. Deep-Sea Res 39:585–605

Bernhard JM (1993) Experimental and field evidence of Antarctic foraminiferal tolerance to anoxia and hydrogen sulfide. Mar Micropaleontol 20:203–213

Bernhard JM (1996) Microaerophilic and facultative anaerobic benthic foraminifera: a review of experimental and ultrastructural evidence. Rev Paléobiol 15:261–275

Bernhard JM (2000) Distinguishing live from dead foraminifera: methods review and proper applications. Micropaleontology 46:38–46

Bernhard JM (2003) Potential symbionts in bathyal foraminifera. Science 299:861

Bernhard JM, Alve E (1996) Survival, ATP pool, and ultrastructural characterization of foraminifera from Drammensfjord (Norway): response to anoxia. Mar Micropaleontol 28:5–17

Bernhard JM, Bowser S (1999) Benthic foraminifera of dysoxic sediments: chloroplast sequestration and functional morphology. Earth-Sci Rev 46:149–165

Bernhard JM, Bowser S (2008) Peroxisome proliferation in foraminifera inhabiting the chemocline: an adaptation to reactive oxygen species exposure? J Eukaryot Microbiol 55:135–144

Bernhard JM, Reimers CE (1991) Benthic foraminiferal population fluctuations related to anoxia. Santa Barbara Basin, U.S. Pacific continental margin. J Foramin Res 27:301–310

Bernhard JM, Sen Gupta BK (1999) Foraminifera of oxygen-depleted environments. In: Sen Gupta BK (ed) Modern foraminifera. Kluwer Academic Publishers, Dordrecht, pp 201–216

Bernhard JM, Newkirk SG, Bowser SS (1995) Towards a nonterminal viability assay for foraminiferan protists. J Eukaryot Microbiol 42:357–367

Bernhard JM, Buck KR, Barry JP (2001) Monterey Bay cold-seep biota: assemblages, abundance, and ultrastructure of living foraminifera. Deep Sea Res Pt I 48:2233–2249

Bernhard JM, Habura A, Bowser SS (2006) An endobiont-bearing allogromiid from the Santa Barbara Basin: implications for the early diversification of foraminifera. J Geophys Res 111:G03002

Bernhard JM, Barry JP, Buck KR, Starczak VR (2009) Impact of intentionally injected carbon dioxide hydrate on deep-sea benthic foraminiferal survival. Glob Change Biol 15:2078–2088

Borrelli C, Sabbatini A, Luna GM, Morigi C, Danovaro R, Negri A (2010) Determination of the metabolically active fraction of benthic foraminifera by means of fluorescent in situ hybridization (FISH). Biogeosciences Discuss 7:7475–7503

Bradshaw JS (1955) Preliminary laboratory experiments on ecology of foraminiferal populations. Micropaleontology 1:351–358

Bradshaw JS (1957) Laboratory studies on the rate of growth of the foraminifer, "*Streblus beccarii* (Linné) *var. tepida* (Cushman)". J Paleontol 31:1138–1147

Cedhagen T (1991) Retention of chloroplasts and bathymetric distribution in the sublittoral foraminiferan *Nonionella labradorica*. Ophelia 37:17–30

Cedhagen T (1993) Taxonomy and biology of *Pelosina arborescens* with comparative notes on *Astrorhiza limicola* (Foraminiferida). Ophelia 37:143–162

Corliss BH, Emerson S (1990) Distribution of Rose Bengal stained deep-sea benthic foraminifera from the Nova Scotian continental margin and Gulf of Maine. Deep-Sea Res 37:381–400

Delaca TE (1986) Determination of benthic rhizopod biomass using ATP analyses. J Foramin Res 16:285–292

Duijnstee IAP, Ernst SR, van der Zwaan GJ (2003) Effect of anoxia on the vertical migration of benthic foraminifera. Mar Ecol Prog Ser 246:85–94

Ernst S, Van der Zwaan B (2004) Effects of experimentally induced raised levels of organic carbon flux and oxygen depletion on a continental slope benthic foraminiferal community. Deep-Sea Res Pt I 51:1709–1739

Forster S, Graf G (1992) Continuously measured changes in redox potential influenced by oxygen penetrating from burrows of *Callianassy subterranea*. Hydrobiologia 235(236):527–532

Forster S, Graf G (1995) Impact of irrigation on oxygen flux into the sediment: intermittent pumping by *Callianassa subterranea* and "piston-pumping" by *Lanice conchilega*. Mar Biol 123:335–346

Geslin E, Köhler-Rink S, Franke U, Heinz P, Holst G, Hemleben Ch (2002) Two dimensional oxygen distribution in sediment containing benthic foraminifers: laboratory study using planar optodes. In: Revets SA (ed) Forams 2002, volume of abstracts. University of Western Australia, Perth

Geslin E, Heinz P, Jorissen F, Hemleben Ch (2004) Migratory responses of deep-sea benthic foraminifera to variable oxygen conditions: laboratory investigations. Mar Micropaleontol 53:227–243

Geslin E, Risgaard-Petersen N, Lombard F, Metzger E, Langlet D, Jorissen F (2011) Respiration rates of benthic foraminifera as measured with oxygen microsensors. J Exp Mar Biol Ecol 396:108–114

Gooday AJ (1986) Meiofaunal foraminiferans from the bathyal Porcupine Seabight (northeast Atlantic): size structure, standing stock, taxonomic composition, species diversity and vertical distribution in the sediment. Deep-Sea Res 33:1345–1373

Gross O (2000) Influence of temperature, oxygen and food availability on the migrational activity of bathyal benthic foraminifera: evidence by microcosm experiments. Hydrobiologica 426:123–137

Gross O (2002) Sediment interactions of foraminifera: implications for food degradation and bioturbation processes. J Foramin Res 32:414–424

Hannah F, Rogerson A (1997) The temporal and spatial distribution of foraminiferans in marine benthic sediments of the Clyde Sea area, Scotland. Estuar Coast Shelf Sci 44:377–383

Hannah F, Rogerson A, Laybourn-Parry J (1994) Respiration rates and biovolumes of common benthic foraminifera (protozoa). J Mar Biol Assoc U K 74:301–312

Heinz P (2001) Laboratory feeding experiments: response of deep-sea benthic foraminifera to simulated phytoplankton pulses. Rev Paléobiol 20:643–646

Heinz P, Kitazato H, Schmiedl G, Hemleben Ch (2001) Response of deep-sea benthic foraminifera from the Mediterranean Sea to simulated phytoplankton pulses under laboratory conditions. J Foramin Res 31:210–227

Heinz P, Hemleben Ch, Kitazato H (2002) Time-response of cultured deep-sea benthic foraminifera to different algal diets. Deep-Sea Res Pt I 49:517–737

Heinz P, Geslin E, Hemleben Ch (2005) Laboratory observations of benthic foraminiferal cysts. Mar Biol Res 1:149–159

Heip CHR, Duineveld G, Flach E, Graf G, Helder W, Herman PMJ, Lavaleye M, Middelburg JJ, Pfannkuche O, Soetaert K, Soltwedel T, de Stigter H, Thomsen L, Vanaverbeke J, de Wilde P (2001) The role of the benthic biota in sedimentary metabolism and sediment-water exchange processes in the Goban Spur area (NE Atlantic). Deep-Sea Res Pt II 48:3223–3243

Hemleben C, Kitazato H (1995) Deep-sea foraminifera under long time observation in the laboratory. Deep-Sea Res 42:827–832

Høgslund S, Revsbech NP, Cedhagen T, Nielsen LP, Gallardo VA (2008) Denitrification, nitrate turnover, and aerobic respiration of benthic foraminiferans in the oxygen minimum zone off Chile. J Exp Mar Biol Ecol 359:85–91

Holst G, Grunwald B (2001) Luminescence lifetime imaging with transparent oxygen optodes. Sensor Actuator B 74:78–90

Holst G, Kohls O, Klimant I, König B, Kühl M, Richter T (1998) A modular luminescence lifetime imaging system for mapping oxygen distribution in biological samples. Sensor Actuator B 51:163–170

Jannink NT, Zachariasse WJ, Van der Zwaan GJ (1998) Living (Rose Bengal stained) benthic foraminifera from the Pakistan continental margin (northern Arabian Sea). Deep-Sea Res Pt I 45:1483–1513

Jorissen F (1999) Benthic foraminiferal microhabitats below the sediment-water interface. In: Sen Gupta BK (ed) Modern foraminifera. Kluwer Academic Publishers, Dordrecht, pp 161–179

Jorissen FJ, De Stigter HC, Widmark JGV (1995) A conceptual model explaining benthic foraminiferal microhabitats. Mar Micropaleontol 26:3–15

Jorissen FJ, Wittling I, Peypouquet JP, Rabouille C, Relexans JC (1998) Live benthic foraminiferal faunas off Cape Blanc, NW-Africa: community structure and microhabitats. Deep-Sea Res Pt I 45:2157–2188

Josefson AB, Widbom B (1988) Differential response of benthic macrofauna and meiofaunal to hypoxia in the Gullmar fjord basin. Mar Biol 100:31–40

Kitazato H, Ohga T (1995) Seasonal changes in deep-sea benthic foraminiferal populations: results of long-term observations at Sagami Bay, Japan. In: Sakai H, Nozaki Y (eds) Biogeochemical processes and ocean flux in the Western Pacific. Terra Scientific Publishing Company, Tokyo, pp 331–342

Kitazato H, Tsuchiya M (1999) Why are foraminifera useful proxies for modern and ancient marine environments? An example using *Ammonia beccarii* (LINNE) from brackish bay environments. Kagoshima University Research Center for the Pacific Islands, Occasional Papers 32:3–17

Koho K, Piña-Ochoa E, Geslin E, Risgaard-Petersen N (2011) Survival and nitrate uptake mechanisms of foraminifers (*Globobulimina turgida*): laboratory experiments. FEMS Microbiol Ecol 75:273–283

Langer MR, Gehring CA (1993) Bacterial farming: a possible feeding strategy of some smaller, motile formainifera. J Foramin Res 23:40–46

Lee JJ, Muller WA (1973) Trophic dynamics and niches of salt marsh foraminifera. Am Zool 13:215–223

Leutenegger S (1984) Symbiosis in benthic foraminifera: specificity and host adaptations. J Foramin Res 14:16–35

Meyers MB, Powell EN, Fossing H (1988) Movement of oxybiotic and thiobiotic meiofauna in response to changes in pore-water oxygen and sulfide gradients around macro-infaunal tubes. Mar Biol 98:395–414

Moodley L, Hess C (1992) Tolerance of infaunal benthic foraminifera for low and high oxygen concentrations. Biol Bull 183:94–98

Moodley L, Van der Zwaan GJ, Herman PMJ, Kempers L, van Breugel P (1997) Differential response of benthic meiofaunal to anoxia with special reference to foraminifera (Protista: Sarcodina). Mar Ecol Prog Ser 158:151–163

Moodley L, Schaub BEM, Van der Zwaan GJ, Herman PMJ (1998a) Tolerance of benthic foraminifera (Protista: Sarcodina) to hydrogen sulphide. Mar Ecol Prog Ser 169:77–86

Moodley L, Van der Zwaan GJ, Rutten GMW, Boom RCE, Kempers AJ (1998b) Subsurface activity of benthic foraminifera in relation to porewater oxygen content: laboratory experiments. Mar Micropaleontol 34:91–106

Moodley L, Steyaert M, Epping E, Middelburg JJ, Vincx M, van Avesaath P, Moens T, Soetaert K (2008) Biomass-specific respiration rates of benthic meiofauna: demonstrating a novel oxygen micro-respiration system. J Exp Mar Biol Ecol 357:41–47

Morigi C, Geslin E (2009) Quantification of benthic foraminiferal abundance. In: Danovaro R (ed) Methods for the study of deep-sea sediments, their functioning and biodiversity (from viruses to megafauna). CRC Press, Boca Raton

Mortimer RJG, Davey JT, Krom MD, Watson PG, Frickers PE, Clifton RJ (1999) The effect of macrofauna on porewater profiles and nutrient fluxes in the intertidal zones of the Humber Estuary. Estuar Coastal Shelf Sci 48:683–699

Muller WA, Lee JJ (1969) Apparent indispensability of bacteria in foraminiferan nutrition. J Protozool 16:471–478

Murray JW (1963) Ecological experiments on foraminiferida. J Mar Biol Assoc UK 43:621–642

Murray JW (ed) (2006) Ecology and applications of benthic foraminifera. Cambridge University Press, Cambridge

Nomaki H, Heinz P, Hemleben Ch, Kitazato H (2005) Behavior and response of deep-sea benthic foraminifera to freshly supplied organic matter: a laboratory feeding experiment in microcosm environments. J Foramin Res 35:103–113

Nomaki H, Yamaoka A, Shirayama Y, Kitazato H (2007) Deep-sea benthic foraminiferal respiration rates measured under laboratory conditions. J Foramin Res 37:281–286

Nyholm K-G, Nyholm P-G (1975) Ultrastructure of monothalamous foraminifera. Zoon 2:117–122

Ohga T, Kitazato H (1997) Seasonal changes in bathyal foraminiferal populations in response to the flux of organic matter (Sagami Bay, Japan). Terra Nova 9:33–37

Panchang R, Nigam R, Linshy V, Rana SS, Ingole B (2006) Effect of oxygen manipulations on benthic foraminifera from Central West Coast of India: a preliminary laboratory experiment. Indian J Mar Sci 35:235–239

Piña-Ochoa E, Høgslund S, Geslin E, Cedhagen T, Revsbech NP, Nielsen LP, Schweizer M, Jorissen F, Rysgaard S, Risgaard-Petersen N (2010a) Widespread occurrence of nitrate storage and denitrification among foraminifera and gromiids. Proc Natl Acad Sci 107:1148–1153

Piña-Ochoa E, Koho K, Geslin E, Risgaard-Petersen N (2010b) Survival and life strategy of foraminifer, Globobulimina turgida, through nitrate storage and denitrification: laboratory experiments. Mar Ecol Prog Ser 417:39–49

Pucci F, Geslin E, Barras C, Morigi C, Sabbatini A, Negri A, Jorissen F (2009) Survival of benthic foraminifera under hypoxic conditions: results of an experimental study using the cell tracker green method. Mar Pollut Bull 59:336–351

Risgaard-Petersen N, Langezaal AM, Ingvardsen S, Schmid MC, Jeten MSM, Ob den Camp HJM, Derksen JWM, Piña-Ochoa E, Eriksson SP, Nielsen LP, Revsbech NP, Cedhagen T, Van der Zwaan GJ (2006) Evidence for complete denitrification in a benthic foraminifer. Nature 443:93–96

Schmiedl G, Mitschele A, Beck S, Emeis K-C, Hemleben C, Schulz H, Sperling M, Weldeab S (2003) Benthic foraminiferal record of ecosystem variability in the eastern Mediterranean Sea during times of sapropel S5 and S6 deposition. Paleogr Paleoclim Paleoecol 190:139–164

Schumacher S, Jorissen FJ, Dissard D, Larkin KE, Gooday AJ (2007) Live (Rose Bengal stained) and dead benthic foraminifera from the oxygen minimum zone of the Pakistan continental margin (Arabian Sea). Mar Micropaleontol 62:45–73

Sen Gupta BK, Machain-Castillo ML (1993) Benthic foraminifera in oxygen-poor habitats. Mar Micropaleontol 20:183–201

Szarek R, Nomaki H, Kitazato H (2007) Living deep-sea benthic foraminifera from the warm and oxygen-depleted environment of the Sulu Sea. Deep-Sea Res II 54:145–176

Travis JL, Bowser SS (1991) The mobility of Foraminifera. In: Lee JJ, Anderson OR (eds) Biology of the foraminifera. Academic, London, pp 91–155

Travis JL, Browser SS (1986) Microtubule-dependent reticulopodial motility: is there a role for actin? Cell Motil Cytoskeleton 6:2–14

Vachard D, Pille L, Gaillot J (2010) Palaeozoic foraminifera: systematics, palaeoecology and responses to global changes. Rev Micropal 53:209–254

Biodata of **Jürgen Mallon**, **Nicolaas Glock**, and **Joachim Schönfeld**, authors of "*The Response of Benthic Foraminifera to Low-Oxygen Conditions of the Peruvian Oxygen Minimum Zone.*"

Jürgen Mallon is currently a Ph.D. student in the Faculty of Paleoceanography at the IFM-GEOMAR at the University of Kiel, Germany. He studied Geology and Paleontology at the University of Hamburg, Germany, where he obtained his diploma. Now he works on benthic foraminifera of the Peruvian oxygen minimum zone regarding to the collaboration project SFB754 "Climate – Biogeochemistry Interactions in the Tropical Ocean," funded by the DFG. The scientific interests are the ecology and distribution of low-oxygen benthic foraminifera and their applications in paleoceanographical and biogeochemical research.

E-mail: **jmallon@ifm-geomar.de**

Nicolaas Glock is currently working on his Ph.D. thesis at the IFM-GEOMAR Leibniz Institute for Marine Research. He obtained his diploma in chemistry at the Philipps-University in Marburg in 2008. The topic of his Ph.D. thesis is the development of a set of geochemical and micropaleontological proxies for the reconstruction of past redox conditions in the Peruvian oxygen minimum zone.

E-mail: **nglock@ifm-geomar.de**

Jürgen Mallon   **Nicolaas Glock**

Dr. Joachim Schönfeld is currently a staff scientist from the Research Unit Paleoceanography of Leibniz Institute for Marine Sciences IFM-GEOMAR, Kiel, Germany. He obtained his Dr. rer. nat. from the University of Kiel in 1989 and continued research at the Federal Bureau of Geosciences and Natural Resources, Hannover, GEOMAR Research Center for Marine Geosciences, Kiel, at Department 5 - Geowissenschaften, Bremen University and with Tethys Geoconsulting Ltd., Kiel, Germany. Doctor Schönfeld's scientific interests are in benthic foraminifera, their taxonomy, ecology, Mesozoic and Cenozoic paleoceanography, stratigraphy and climate history.

E-mail: **jschoenfeld@ifm-geomar.de**

THE RESPONSE OF BENTHIC FORAMINIFERA TO LOW-OXYGEN CONDITIONS OF THE PERUVIAN OXYGEN MINIMUM ZONE

JÜRGEN MALLON[1,2], NICOLAAS GLOCK[1,2], AND JOACHIM SCHÖNFELD[2]
[1]*Christian-Albrechts-University Kiel, Sonderforschungsbereich 754, Kiel, Germany*
[2]*Leibniz Institute of Marine Sciences, Ifm-Geomar, Wischhofstrasse 1-3, D-24148 Kiel, Germany*

1. Introduction

The upwelling system off the NW coast of South America constitutes one of the most widespread oxygen minimum zones. Severe oxygen depletion extends over large volumes of subsurface water masses between 80 and about 550 m water depth and effects a horizontal zonation of specially adapted faunal assemblages. The vertical distribution of population density and diversity of benthic foraminifera changes dramatically following the bottom water oxygen gradient. Many studies in areas where dissolved oxygen levels are low or even anoxic revealed that foraminiferal life flourishes under oxygen depleted conditions, resulting in conspicuously high foraminiferal population density but low species diversity (e.g., Phleger and Soutar 1973; Bernhard et al. 1997; Gooday et al. 2000; Levin et al. 2002). The inhabiting fauna of such kind of oxygen depleted environment like the Peruvian OMZ requires special adaptations. Some benthic foraminiferal species we found in this study are known to be able to withstand anoxic or even sulfidic conditions for several weeks (Bernhard and Reimers 1991; Bernhard and Sen Gupta 1999; Geslin et al. 2004; Moodley et al. 1997).

Some of these adaptations, which have been observed in benthic foraminifera at low oxygen levels, were the use of chloroplasts (Bernhard and Alve 1996; Bernhard and Bowser 1999; Grzymski et al. 2002), bacterial endo- and ectosymbionts (Bernhard and Reimers 1991; Bernhard et al. 2001, 2009), intracellular nitrate storage (Risgaard-Petersen et al. 2006; Glud et al. 2009; Piña-Ochoa et al. 2010), peroxisomes (Bernhard and Bowser 2008), clustering of mitochondria behind pore plugs (Leutenegger and Hansen 1979; Bernhard et al. 2001) and high pore-density values of the tests (Glock et al. 2011).

On the sea floor bathed by OMZs, high amounts of organic matter accumulate due to enhanced preservation and provide a high supply of food for deposit feeders like benthic foraminifera. A concept which describes the effect of these two interacting parameters on the vertical distribution of living benthic foraminifera is the TROX-model of Jorissen et al. (1995). The maximum living depth of foraminifera in an oligotrophic environment depends on the available organic matter in the sediment. Oxygen availability is, on the other hand, the controlling factor which confines the maximum living depth of benthic foraminifera in eutrophic environments. The distribution patterns of living (Rose Bengal stained) species in this study mostly follow the TROX-model.

In this study, we describe the faunal composition and distribution patterns of living benthic foraminifera from surface sediments of 13 multicorer stations along a composite depth section off Peru, consisting of a main transect at 11°S, three stations at 12°S and one at 10°S. The data were compared with bottom water [O_2] in order to depict the relationship between species distribution and [O_2]. A further aim of this study is the development of a benthic foraminiferal proxy for reconstruction of ancient bottom water [O_2] for the Peruvian OMZ.

2. Study Area

The study area is located in the central part of the Peruvian OMZ between 10°S and 12°S off Peru (Fig. 1). Coastal water masses are subjected to wind-driven upwelling and carry cold nutrient-rich deep water to the surface. High primary production rates in the euphotic zone cause oxygen depletion resulting in an OMZ extending from northern Chile to near equatorial latitudes off the west coast of South America. Dissolved oxygen values, derived from conductivity-temperature-depth-profiler (CTD) measurements between 10°S and 12°S, show oxic (>45 μmol/kg) values down to 31 m water depth. Beneath these surface waters, dissolved oxygen levels decrease to dysoxic conditions (5–45 μmol/kg), and even reach microxic to anoxic levels with values <5 μmol/kg between 88 and 522 m. At least as deep as 944 m, levels begin to increase again to dysoxic conditions. For the classification of microxic, dysoxic and oxic conditions, we refer to limits of <5, 5–45 and >45 μmol/kg (Bernhard and Sen Gupta 1999), respectively. The classification and ordering of different oxygen levels into several groups is not defined in a common sense. Therefore, different classifications are in use, for a further example, Kaiho (1994), who subdivided four levels of oxygen conditions – anoxic (<0.1 ml/l), dysoxic (0.1–0.3 ml/l), suboxic (0.3–1.5 ml/l) and oxic (>1.5 ml/l). It is important to note that the actual oxygen concentration in the OMZ core certainly reaches anoxic levels (<2 nmol/l, Revsbech et al. 2009), but the resolution of the CTD sensors used on R/V Meteor cruise M77 did not resolve extremely low dissolved oxygen concentrations (<2 μmol/kg). Consequently, we should assume that the CTD derived [O_2] values for the levels between 214 and 375 m water depth and values slightly lower than 2.5 μmol/kg were even lower. It is likely that they were in fact anoxic (Revsbech et al. 2009).

The sediments in this area are predominantly olive-green, organic-rich muds. High amounts of filamentous sulfide-oxidizing, nitrate-reducing bacteria (mainly *Thioploca* spp. and *Beggiatoa* spp.) form dense mats in some areas. Nematodes, annelids and small oligochaetes, together with benthic calcareous

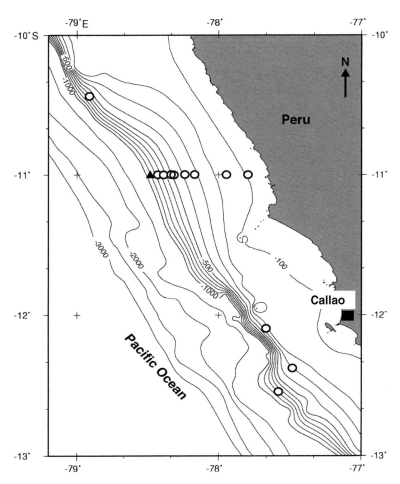

Figure 1. Study area with sampling locations (*black dots*) and CTD station (*black triangle*) offshore Peru.

foraminifera, are common organisms in samples of the central part of the OMZ. Recent bioturbation by macroinvertebrates is nearly absent. The first bioturbating macroinvertebrates, e.g., small gastropods, occur at stations deeper than ca. 460 m water depth (at 11°S transect). Further characteristic features found in few sediment samples at the lower OMZ boundary are phosphorite nodules. They occurred at 465 and 516 m.

The degradation of sinking organic material is strongly attenuated in oxygen-depleted environments. Thus, the organic matter accumulates and leads to high organic carbon values. The organic carbon content (C_{org}) at 11°S varies between 3% and 11% in the upper 0.5 cm sediment interval and shows an inverse relationship to $[O_2]$ within the OMZ, but from the lower OMZ boundary to the deeper stations follows the increasing trend of [O_2] (Fig. 3). Highest C_{org} contents with 10.6% and 10.9% were observed at the 319 and 376 m stations.

3. Materials and Methods

3.1. SAMPLE PROCESSING

Thirteen sediment cores from the west coast off Peru between 10 and 12°S were analyzed for this study (Table 1). They were recovered during October and November 2008 with using a multicorer with 10-cm tube diameter. Immediately after recovery, one tube of the array with the most even sediment surface was selected and brought to a laboratory container with ambient temperature of 4°C, where the sediment core was sliced with a spatula in 2, 3 or 5 mm intervals from 0 to 10 mm, in 5 mm intervals from 10 to 40 mm and in 10 mm intervals from 40 to 50 mm sediment depth. Subsamples were filled in Kautex™ bottles and stained with ethanol + Rose Bengal (2 g/l) for distinguishing living from dead individuals. Under cool and dark conditions, they were transported and stored for further procedures.

Stained subsamples were gently wet sieved with tap water through 2000 and 63 μm screens. To beware of contamination, the sieves were thoroughly cleaned and treated with Metyhlene Blue before every use. The >2,000 μm fractions (pebbles, gastropod shells etc.) were dried and archived in polyethylene bags. The 63–2,000 μm fractions were dried at 40°C and weighed before they were picked. If >63 μm fractions contained large amounts of sand, they were soaked again with tap water and wet picked. At least 100 stained specimens were picked per

Table 1. Sampling locations.

Site	Longitude (W)	Latitude (S)	Water depth (m)	Bottom water oxygen concentration (μmol/kg)
M77/1-540/MUC-49	77°47.40'	11°00.01'	79	5.28
M77/1-470/MUC-29	77°56.60'	11°00.02'	145	3.33
M77/2-635/MUC-5-4	77°40.07'	12°05.66'	214	2.37
M77/1-616/MUC-81	77°29.05'	12°22.69'	302	2.20
M77/1-473/MUC-32	78°09.94'	11°00.01'	317	2.25
M77/1-449/MUC-19	78°09.97'	11°00.01'	319	2.25
M77/1-482/MUC-34	78°14.17'	11°00.01'	375	2.26
M77/1-456/MUC-22	78°19.23'	11°00.01'	465	3.79
M77/1-516/MUC-40	78°20.00'	11°00.00'	511	4.83
M77/1-553/MUC-54	78°54.07'	10°26.38'	521	4.83
M77/1-487/MUC-38	78°23.17'	11°00.00'	579	8.59
M77/1-459/MUC-25	78°25.60'	11°00.03'	697	15.72
M77/1-622/MUC-85	77°34.74'	12°32.75'	823	26.67

subsample in order to attain statistically reliable values. Some subsamples contained less than 100 stained specimens. In some cases, it was necessary to split the subsamples into manageable aliquots by using an ASC™ sample microsplitter. Only individuals containing more than 50% of all chambers brightly stained were considered as individuals that were living at the time of sampling. After they were picked, the specimens were sorted, fixed in Plummer cells and counted. In our study, we concentrated on the topmost multicorer subsamples, whereas the faunal census data of subsurface samples are provided online at: http://doi.pangaea.de/10.1594/PANGAEA.757092.

3.2. ENVIRONMENTAL DATA

For a comparison of foraminiferal distribution patterns with the bottom water dissolved oxygen concentrations, we used only the top-slices of every multicorer. The [O_2] was measured with an optode sensor mounted to a conductivity-temperature-depth profiler (CTD) at station M77/1-424-RO9 from 928 m water depth further off shore. The concentrations of dissolved oxygen as displayed by the sensor were calibrated by Dr. S. Sommer and S. Kriwanek using Winkler-titration method. The [O_2] from the CTD was projected to the respective water depths of stations where the multicorer for foraminiferal studies has been deployed. The underlying assumption of near-horizontal oxygen isolines was corroborated by in situ measurements from benthic chamber lander deployments (S. Sommer, pers. comm.). Sedimentary organic carbon concentrations (C_{org}) were obtained with a flush combustion element analyzer on freeze-dried surface sediment samples by Dr. C. Hensen.

4. Species Distribution Patterns of Benthic Foraminifera in the OMZ Off Peru

The sampling transect of our study comprises a depth range from 79 to 823 m and intersects the OMZ. Sixty-nine living benthic foraminiferal species were identified from the multicorer samples, but in the five samples subjected to [O_2] lower than 2,5 µmol/kg, just eight to fourteen living species were found. Calcareous taxa were dominating the assemblages throughout the OMZ. Agglutinated taxa were negligible within the OMZ (<3% of the total to a water depth of 579 m). However, at stations deeper than 579 m with [O_2] exceeds 8.5 µmol/kg, they comprised more than 11% of the fauna and reached their maximum with 19.7% at 697 m water depth.

The comparatively low diversities in the OMZ are displayed by the averaged *Fisher's* α index. The stations within the OMZ core (214–375 m water depth, [O_2] <2.5 µmol/kg) yielded values around 1.9, whereas the α index was 3.8 at stations with higher oxygen values.

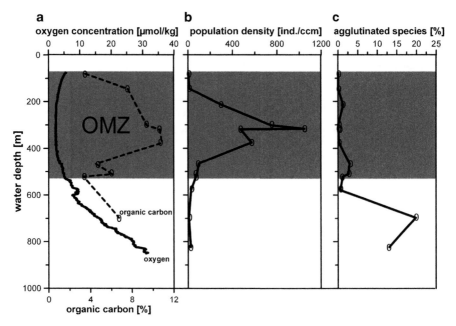

Figure 2. Oxygen concentration and organic carbon (**a, b**), population density (**b**) and percentage of agglutinated species (**c**) plotted against water depth. Density data were derived from samples taken from 10–12°S and sediment depth intervals from 0 to 6 mm. Organic carbon content is represented by the top 10 mm of each sediment core and obtained from nine stations between 11 and 12°S. Winkler-calibrated dissolved oxygen concentration measured at 11°S.

Highest values of foraminiferal population densities occurred in the surface sediments. All of the examined samples that were considered in this study showed the highest population densities between 2 and 3 mm sediment depth. The population densities showed a marked, inverse correlation with ambient bottom water $[O_2]$ (Fig. 2). The stations of the OMZ core showed population densities between 473 and 1,045 ind./ccm, reaching the maximum at 317 m water depth.

Although most of the species abundances as observed in our samples varied among the stations, several species showed conspicuous density patterns which compare to the dissolved bottom water oxygen concentrations. Systematic variations of species percentages with respect to water depth dependent $[O_2]$ showed the most common species of the Peruvian OMZ, in particular *Angulogerina angulosa*, *Bolivina costata*, *Bolivina interjuncta*, *Bolivina plicata*, *Bolivina seminuda*, *Bolivinita minuta*, *Cancris carmenensis*, *Fursenkoina fusiformis*, *Nonionella stella*, *Uvigerina peregrina* and *Valvulineria glabra*. The majority of species within the immediate OMZ belong to the family *Bolivinitidae*. These species comprise 76–95% of the living assemblages in the samples between 214 and 375 m water depth (Fig. 3).

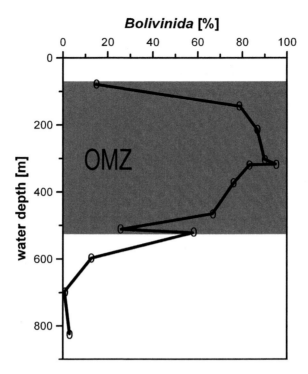

Figure 3. Percentages of stained *Bolivinida* (*B. alata*, *B. costata*, *B. interjuncta*, *B. pacifica*, *B. plicata*, *B. seminuda*, *B. serrata* and *B. spissa*) as fraction of total stained calcareous species in relation to oxygen concentration of a CTD profile along 11°S off Peru.

At OMZ stations between 302 and 375 m water depth, *Bolivina seminuda* was the dominant species and accounted for 42–76% of the living fauna. This demonstrated that maximum values of *B. seminuda* in samples where the diversity was very low mutually indicate very low [O_2] as previously demonstrated by Oberhänsli et al. (1990). *Bolivina plicata* was more restricted to very low oxygen concentrations and occurred with 5–25% in 302–375 m water depth. It showed values of <1% or even disappears at higher oxygen concentrations.

Another dominating bolivinid which reached its maximum abundance at slightly higher oxygen concentrations was *B. interjuncta*. This species showed two maxima with 52% at 465 m water depth and 20% at 521 m water depth. *Bolivina interjuncta* is restricted to the lower part of the OMZ. Other characteristic bolivinids were *Bolivina alata* and *Bolivina spissa*. They occurred also at the lower part of the OMZ. *B. alata* was found between 511 and 576 m water depth. It reached its maximum at 521 m water depth with 15.9% and < 2.2% at 511 and 576 m water depth. *Bolivina* spissa occurred at 521 m water depth and below but reached not more than 3.3%. Living specimens of *Bolivina costata*, which is a characteristic species of the Peru Margin (Resig 1981 and references therein; Resig 1990), were found only in a depth range from 79 to 521 m water depth.

Its relative abundance varied significantly throughout the transect and reached two maxima with 11 and 14.7–14.9% at 79 and 317–319 m water depth, respectively. The upper OMZ boundary was dominated by *Nonionella stella* and *Fursenkoina fusiformis*. Interestingly, *N. stella* was found only at the shallowest station with 44% and with 1% at 375 m water depth.

Fursenkoina fusiformis accounts for 29.6% at 79 m water depth but then decreases rapidly with decreasing $[O_2]$ and becomes absent at 302 m water depth. This species re-occurred in a second interval from 375 to 579 m water depth. However, it did not account for more than 7.5% there. A further common species of the OMZ is *C. carmenensis*. It occurred between 214 and 521 m water depth, whereas it reached its maximum abundance of 37.4% at 511 m water depth. At the shallower stations with lower $[O_2]$, *C. carmenensis* just comprised 0.3-8% of the living assemblages. Stations below the lower OMZ boundary were, in particular, dominated by four species which were absent or very rare within the OMZ core and at shallower stations. *Uvigerina peregrina* and *Bolivinita minuta* appeared at 521 m water depth and 511 m water depth ($[O_2] > 4.8$ µmol/kg), respectively. They reached their maximum relative abundances at 823 and 579 m water depth, with corresponding $[O_2]$ of 27.7 and 8.6 µmol/kg. Maximum values were 45.8% and 31% for *U. peregrina* and *B. minuta*, respectively. The relative abundance of *Angulogerina angulosa*, absent in all stations at shallower than 697 m depth, increased from 2.5% at 697 m water depth to 7.7% at 823 m. *Cassidulina crassa* first occurred at 465 m water depth and reached abundances between 0.5% and 1.9% down to 579 m water depth. At 697 m, it reached 19.1% and 6.6% at 823 m. Together with *B. minuta, U. peregrina* and *A. angulosa* dominated the 697 and 823 m stations. Living specimens of the deep infaunal species *Globobulimina pacifica*, which was usually common under nearly anoxic conditions, were found in our study only at sediment depths greater than 4 mm and at 697 m water depth. The microhabitat depth increased to between 5 and 20 mm at 823 m water depth.

The lower OMZ boundary, where $[O_2]$ exceeds 3.8 µmol/kg, seems to provide optimum conditions for many species which are absent within the immediate OMZ core. A good preservation of available organic matter prevailed at this depth, and together with a higher oxygen contents, this may lead to the conspicuously higher species richness.

5. Discussion

5.1. SPECIES PATTERNS

The foraminiferal distribution patterns from the Peruvian OMZ seemingly followed the ambient bottom water oxygen concentrations. Species diversity and population density were inversely correlated. The abundance of agglutinated taxa showed a positive correlation with [O_2]. It seems that most agglutinated species require [O_2] >8.5 µmol/kg. This confirms the results of previous studies (Gooday et al. 2000; Levin et al. 2002) which showed that agglutinated taxa are not well adapted to extremely low-oxygen conditions. Probably, these taxa are disadvantaged by the absence of pores or low pore-density of their tests.

We found a predominance of *Bolivinitidae* at stations with lowest oxygen concentrations. Obviously, some species of this family are well adapted to low-oxygen conditions. But it is important to consider that not bolivinids in general are typical inhabitants of oxygen-depleted environments. Instead, each species has its characteristic oxygen requirement. Figure 4 shows that *Bolivina seminuda* has been found in all stations, from 79 to 823 m water depth. Thus, it is able not

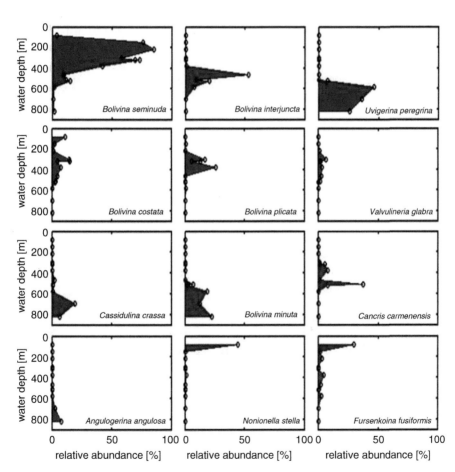

Figure 4. The relative abundance of some "living" calcareous species vs. water depth dependent [O_2] distribution (relative abundance obtained from surface sediment samples: 0–2 mm, 0–5 mm, 2–4 mm or 3–6 mm).

only to live under normal oxygen conditions but also under extreme low oxygen concentrations where it reaches its highest abundance.

Bolivina costata, which is a characteristic species of the Peru Margin (Resig 1981 and references therein; Resig 1990), was found in a depth range from 79 to 521 m. This is in agreement with previous studies where this species has been found at even shallower stations but also down to a water depth of >2,000 m (Resig 1981 and references therein). However, these studies did not use any method for distinguishing live from dead fauna. Based on our study, we assume that *B. costata* is actually restricted to the upper bathyal of the OMZ-affected Peruvian Margin.

Living specimens of *C. carmenensis* were found between 214 and 521 m water depth and showed highest abundance at 511 m water depth (37.4%). This is consistent with the studies of Resig (1981, 1990) who characterized *C. carmenensis* as an upper bathyal species ranging from 151 to 500 m. The highest abundances correlate with oxygen concentrations of 4.8 µmol/kg. The distribution of *Nonionella stella* we found in the Peruvian OMZ corresponded to the observed depth range of *N. stella* (denominated as *N. miocenica*) in the study of Resig (1990). She found specimens only to a depth of about 80 m. We can not confirm that *N. stella* is restricted to extremely low $[O_2]$ like it has been observed by Bernhard et al. (1997), where *N. stella* dominated the assemblages in all the samples with $[O_2]$ <2 µM.

The distributions of most infaunal and epifaunal species from the surface sediments are in agreement with the TROX-model by Jorissen et al. (1995). At stations with lowest $[O_2]$ between 79 and 317 m water depth, the fauna of the topmost subsamples was composed of infaunal taxa (i. e. *Bolivina seminuda, B. plicata, B. interjuncta, B. costata, Fursenkoina fusiformis, Nonionella stella*), whereas species of the genera *Planulina* and *Cibicidoides*, which prefer an epibenthic lifestyle or, at least, dwell on the sediment surface, were very scarce or absent in these samples.

The deep infaunal species *G. pacifica*, which is adapted to extremely low or even anoxic conditions, has not been found in any of the OMZ samples but just in deeper sediment intervals of the deepest stations where $[O_2]$ again was relatively high. This observation was in disagreement with the conceptual model of Jorissen et al. (1995). They assumed an upward migration of this species resulting in a subsurface maximum under low-oxygen conditions. Probably, *G. pacifica* lived deeper in the sediment or might be restricted by redox fronts in the upper few millimeters or centimeters. A similar process has been invoked by Jannink et al. (1998) for *G. pacifica* in the northern Arabian Sea.

The lower OMZ boundary, where $[O_2]$ exceeds 3.8 µmol/kg, seems to provide optimum conditions for few opportunistic species, which were absent within the OMZ core like *A. angulosa, U. peregrina* and *B. minuta*. Good preservation of available organic matter together with higher oxygen contents may have lead to the enhanced species richness at stations deeper than 597 m where oxygen concentrations exceeded 8.5 µmol/kg.

When comparing the species composition from the Peruvian OMZ with the faunas from other oceanic low-oxygen habitats, it emerges that just a few species dominate these assemblages. This results in low diversities but high population densities which were generally recognized also in other OMZs (Phleger and Soutar 1973, Baja California, Mexico; Gooday et al. 2000; Schumacher et al. 2007; den Dulk et al. 1998 and Jannink et al. 1998 for the Arabian Sea OMZ). The extremely high population densities of a very few species within the OMZ core were certainly an interplay of different factors. Low predation pressure coupled with high food availability and low competition over a long time period may lead to development of specially adapted species which can inhabit and successfully reproduce under such extreme oxygen deficiency (Gooday 1986). Another concept for a successful life under low-oxygen conditions may be a radiative reproduction strategy (r-strategy) which results in a very high reproduction rate and thus in high standing stocks (Sjoerdsma and van der Zwaan 1992; Jannink et al. 1998). Further, high diversity patterns of the lower OMZ boundary and below are explainable with a predominance of "k-strategic" species. They need much energy for competition against many other species.

The distinct faunal change of benthic species across the Peruvian OMZ is comparable with the observations from the Arabian Sea. As described in Schumacher et al. (2007) for the OMZ off Pakistan, we also found a specific fauna within the OMZ (e.g., *B. seminuda, B. costata, B. interjuncta*) and a more cosmopolitan fauna in the lower part and below the OMZ (*U. peregrina, B. minuta, A. angulosa*). An oxygen dependent, horizontal zonation of cosmopolitan and specific species was considered to result from stable conditions in an upwelling system, in which a very few species have adapted to extreme oxygen-depleted conditions (Schumacher et al. 2007).

Many of the above mentioned species, especially those from the low-oxygen habitats, are known to be able to store nitrate in cell-bound vacuoles and use it for respiration. This has been proven for *Nonionella* cf. *stella* by Risgaard-Petersen et al. (2006). But also *Bolivina alata, Bolivina plicata, Bolivina seminuda, Cyclammina cancellata* and *Uvigerina peregrina*, which were common in our samples, are able to store and respire nitrate (Piña-Ochoa et al. 2010). The ability to use nitrate as an alternative energy source may certainly be an advantage for these species to live in this extreme environment.

5.2. A NEW PROXY FOR ESTIMATION OF BOTTOM WATER OXYGEN CONCENTRATIONS

Some of the dominating Peruvian OMZ species showed a conspicuous trend of their abundance and population density patterns which corresponded to the ambient bottom water oxygen concentrations. Due to their $[O_2]$ requirements, each species showed its maximum abundance at a distinct water depth (Fig. 4). This resulted in a distinct species succession with oxygen levels. The following main species were involved: *Angulogerina angulosa, Bolivina alata, Bolivina costata, Bolivina interjuncta, Bolivina plicata, Bolivina seminuda, Bolivinita minuta, Cancris carmenensis, Cassidulina crassa, Fursenkoina fusiformis, Uvigerina peregrine* and *Valvulineria glabra*. Their abundances exhibited an overlapping pattern which allowed estimations of the prevailing [O_2] on a relatively fine scale (Fig. 4).

We observed six distinct assemblages indicating different levels of [O_2]. The first assemblage indicating [O_2] <2.5 µmol/kg was dominated by *B. seminuda* (42-85%) with co-occurrence of *B. plicata* (5-25%), *B. costata* (1-14%) and *V. glabra* (1-6-6%). At slightly higher [O_2], between 3.3 and 4.8 µmol/kg, *B. seminuda* (9.5-76%) dominated this assemblage together with *B. interjuncta* (up to 52%). *B. alata* appeared with 2–15%., whereas *B. plicata* disappeared. The third assemblage at [O_2] 8.6 µmol/kg was characterized by *U. peregrina* and *B. minuta* with 45.7% and >10%, respectively. *B. seminuda* and *B. alata* became very rare (<2%). *B. interjuncta* decreased to 7.5%; *B. costata* and *Valvulineria glabra* were absent. At [O_2] of 15.7 µmol/kg, we found a fourth assemblage with dominance of *U. peregrina* (35.7%) together with *C. crassa* (19%) and *B. minuta* (12.7%). *B. seminuda* further decreased (<1%) and *B. interjuncta* disappeared. *Angulogerina angulosa* first occurred (2.5%) at this level. Agglutinated species, which were barely found at lower [O_2], strongly increased from <3% to 19%. The fifth assemblage indicated [O_2] 27.7 µmol/kg. Here we found still a dominance of *U. peregrina* with *B. minuta* but with a higher proportion of *B. minuta* (22.7%) than *U. peregrina* (25.6%). *Angulogerina angulosa* increased to 7.7% and *C. crassa* decreased to 6.6%. Agglutinated species were making up 13% in this assemblage.

Due to their recent occurrence, it is important to mention that the assemblages which are dominated by *C. carmenensis, B. minuta* and *A. angulosa* may serve as indicator for reconstructing oxygen values prevailing at the lower OMZ boundary.

The upper OMZ boundary was characterized by high abundances of *N. stella* and *F. fusiformis*. At the two shallowest station (79 m water depth [O_2] 5.3 µmol/kg), *N. stella* and *F. fusiformis* reached relative abundances of 44% and 29.6%. At deeper stations, *N. stella* was absent in the surface sediment samples except for 375 m water depth where it occurred with 1.1%. *Fursenkoina fusiformis* decreased except for 375 m water depth where it contributed 1.1%. *Fursenkoina fusiformis* decreased at deeper stations but occasionally occurred with low abundances (<5%) throughout the OMZ to a water depth of 579 m. In a former study by Bernhard et al. (1997), *N. stella* from the Santa Barbara Basin was referred as the dominating species of all stations with [O_2] <2 µM, thus leading to the conclusion that *N. stella* may serve as a proxy for oxygen concentrations <2 µM. Our observation cannot support this conclusion for the Peruvian OMZ. It appears that the distribution pattern is more likely to be explained by the requirement of the availability of labile organic matter which was reported to be highest at the upper OMZ boundary off Peru (Neira et al. 2001 and Levin et al. 2002).Our observations on the distribution of *U. peregina* suggest that this species may be used as an indicator species for lower OMZ boundary conditions where organic carbon concentrations are relatively high but [O_2] rather low(Altenbach and Sarnthein 1989; and Rathburn and Corliss 1994).

The distribution patterns of the described species can together be used as valuable tool for reconstructing past bottom water oxygen levels on a relative scale. However, it has to be proven that the sediments are not seriously affected by dissolution, and that fossil assemblages are not seriously biased by the taphonomic loss of arenaceous or delicate calcareous species.

6. Acknowledgements

We thank Dr. S. Sommer and S. Kriwanek for providing us with the calibrated oxygen data and Dr. C. Hensen for providing the C_{org} and [NO_3] data. A. Bleyer, B. Domeyer and R. Ebbinghaus are gratefully acknowledged for the analytical work onboard. Very special thanks go to Dr. M. E. Perez for supporting us with taxonomy and also for fruitful discussions and for providing us access to the foraminiferal collection at the National History Museum, London. This study was financially supported by the DFG through the Sonderforschungsbereich 754: "Climate – Biogeochemistry Interactions in the Tropical Ocean."

Appendix 1: Faunal Reference List

Species	References
Angulogerina angulosa (Williamson, 1858)	Loeblich and Tappan (1994): p. 487, pl. 250, figs. 13–20.
Bolivina alata (Seguenza, 1862)	van Marle (1991): p. 305, pl. 17, figs. 1–2.
Bolivina costata (d'Orbigny, 1839)	Resig (1981): p. 647, pl. 1, fig. 1.
Bolivina interjuncta (Cushman, 1926)	Boltovskoy and Theyer (1970): p. 304, pl. 1, figs. 8, 9.
Bolivina plicata (d'Orbigny, 1839)	Resig (1981): p. 647, pl. 1, figs. 3, 4.
Bolivina seminuda (Cushman, 1911)	Resig (1981): p. 655, pl. 5, fig. 14.
Bolivina serrata (Natland, 1938)	Whittaker (1988): p. 100, pl. 13, figs. 1–3.
Bolivina spissa (Cushman, 1926)	Resig (1981): p. 647, pl. 1, fig. 7.
Bolivina minuta (Natland, 1938)	Resig (1981): p. 647, pl. 1, fig. 9.
Cancris carmenensis (Natland, 1950)	Resig (1981): p. 649, pl. 2, fig. 9–11.
Cassidulina crassa (d'Orbigny, 1839)	van Marle (1991): p. 289, pl. 9, figs. 13–15.
Cyclammina cancellata (Brady, 1879)	Zheng (2001): pl. 52, figs. 3–6.
Fursenkoina fusiformis (Williamson, 1858)	Murray (1971): p. 184, pl. 77, figs. 1–5.
Globobulimina pacifica (Cushman, 1927)	Loeblich and Tappan (1994): p. 480, pl. 243, figs. 13–16.
Nonionella stella (Cushman & Moyer, 1930)	Narayan et al. (2005): p. 147, pl. 4, fig. 23.
Uvigerina peregrina (Cushman, 1923)	Frezza and Carboni (2009): p. 56, pl. 2, fig. 15.
Valvulineria glabra (Cushman, 1927)	Loeblich and Tappan (1994): p. 505, pl. 268, figs. 1–3.

Appendix 2: Supplementary Data

Supplementary data to this article can be found online at: http://doi.pangaea.de/10.1594/PANGAEA.757092

7. References

Altenbach AV, Sarnthein M (1989) Productivity record in benthic foraminifera. In: Berger WH, Smetacek VS, Wefer G (eds) Production in the ocean: present and past. Wiley, New York, pp 255–269

Bernhard JM, Alve E (1996) Survival, ATP pool, and ultrastructural characterization of benthic foraminifera from Drammensfjord (Norway): response to anoxia. Mar Micropaleontol 28:5–17

Bernhard JM, Bowser SS (1999) Benthic foraminifera of dysoxic sediments: chloroplast sequestration and functional morphology. Earth-Sci Revs 46:149–165

Bernhard JM, Bowser SS (2008) Peroxisome Proliferation in Foraminifera Inhabiting the Chemocline: An Adaptation to Reactive Oxygen Species Exposure? J Eukaryot Microbiol 55(3):135–144

Bernhard JM, Reimers CE (1991) Benthic foraminiferal population fluctuations related to anoxia: Santa Barbara Basin. Biogeochemistry 15:127–149

Bernhard JM, Sen Gupta BK (1999) Foraminifera of oxygen-depleted environments. In: SenGupta BK (ed) Modern foraminifera. Kluwer Academic Publishers, Great Britain, pp 201–216

Bernhard JM, Sen Gupta BS, Borne PF (1997) Benthic foraminiferal proxy to estimate dysoxic bottom-water oxygen concentrations. Santa Barbara Basin, U.S. Pacific continental margin. J Foraminife Res 27:301–310

Bernhard JM, Buck KR, Barry JP (2001) Monterey Bay cold-seep biota: Assemblages, abundance, and ultrastructure of living foraminifera. Deep-Sea Res I 48:2233–2249

Bernhard JM, Bowser SS, Goldstein S (2010) An ectobiont-bearing foraminiferan, *Bolivina pacifica*, that inhabits microxic pore waters: cell-biological and paleoceanographic insights: Environ Microbiol 12:2107–2119

Boltovskoy E, Theyer F (1970) Foraminiferos recientes de Chile Central. Revista del Museo Argentino de Ciencias Natururales. Bernard Rivadavia 2(9):280–397

den Dulk M, Reichart GJ, Memon GM, Roelofs EMP, Zachariasse WJ, van der Zwaan GJ (1998) Benthic foraminiferal response to variations in surface water productivity and oxygenation in the northern Arabian Sea. Mar Micropaleontol 35:43–66

Frezza V, Carboni MG (2009) Distribution of recent foraminiferal assemblages near the Ombrone River mouth (Northern Tyrrhenian Sea, Italy). Rev de micropaleontol 52:43–66

Geslin E, Heinz P, Jorissen F, Hemleben Ch (2004) Migratory responses of deep-sea benthic foraminifera to variable oxygen conditions: laboratory investigations. Mar Micropaleontol 53:227–243

Glock N, Eisenhauer A, Milker Y, Liebetrau V, Schönfeld J, Mallon J, Sommer S, Hensen C (2011) Environmental influences on the pore-density of *Bolivina spissa* (Cushman). J Foraminifer Res 41(1):22–32

Glud RN, Thamdrup B, Stahl H, Wenzhoefer F, Glud A, Nomaki H, Oguri K, Revsbech NP, Kitazato H (2009) Nitrogen cycling in a deep ocean margin sediment (Sagami Bay, Japan). Limnol Oceanogr 54(3):723–734

Gooday AJ (1986) Meiofaunal foraminiferans from the bathyal Porcupine Seabight (northeast Atlantic): size structure, standing stock, taxonomic composition, species diversity and vertical distribution in the sediment. Deep Sea-Res 33(10):1345–1373

Gooday AJ, Bernhard JM, Levin LA, Suhr SB (2000) Foraminifera in the Arabian Sea oxygen minimum zone and other oxygen deficient settings: taxonomic composition, diversity, and relation to metazoan faunas. Deep-Sea Res II 47:25–54

Grzymski J, Schofield OM, Falkowski PG, Bernhard JM (2002) The function of plastids in the deep-sea benthic foraminifer, *Nonionella stella*. Limnol Oceanogr 47(6):1569–1580

Jannink NT, Zachariasse WJ, van der Zwaan GJ (1998) Living (Rose bengal stained) benthic foraminifera from the Pakistan continental margin (northern Arabian Sea). Deep-Sea Res I 45:1483–1513

Jorissen FJ, de Stigter HC, Widmark JGV (1995) A conceptual model explaining benthic foraminiferal microhabitats. Marine Micropaleontol 26:3–15

Kaiho K (1994) Benthic foraminiferal dissolved-oxygen index and dissolved-oxygen levels in the modern ocean. Geology 22:719–722

Leutenegger S, Hansen HJ (1979) Ultrastructural and radiotracer studies of pore function in Foraminifera. Mar Biol 54:11–16

Levin LA, Gutiérrez D, Rathburn A, Neira C, Sellanes J, Muñoz P, Gallardo V, Salamanca M (2002) Benthic processes on the Peru margin: a transect across the oxygen minimum zone during the 1997–98 El Niño. Progr Oceanogr 53:1–27

Loeblich AR, Tappan H (1994) Foraminifera of the Sahul Shelf and Timor Sea, Cushman Foundation for Foraminiferal Research. Spec Publ 31:1–661

Moodley L, Van der Zwaan GJ, Herman PMJ, Kempers L, Van Breugel P (1997) Differential response of benthic meiofauna to anoxia with special reference to Foraminifera (Protista: Sarcodina). Mar Ecol Prog Series 158:151–163

Murray J (1971) An atlas of British recent foraminiferids. Heinemann Education Books, London

Narayan YR, Barnes CR, Johns MJ (2005) Taxonomy and biostratigraphy of Cenozoic foraminifers from Shell Canada wells, Tofino Basin, offshore Vancouver Island, British Columbia. Micropaleontology 51(2):101–167

Neira C, Sellanes J, Levin LA, Arntz WE (2001) Meiofaunal distributions on the Peru margin: relationship to oxygen and organic matter availability. Deep-Sea Res I 48:2453–2472

Oberhänsli H, Heinze P, Diester-Haass L, Wefer G (1990) Upwelling off Peru during the last 430,000 yr and its relationship to the bottom-water environment, as deduced from coarse grain-size distributions and analyses of benthic foraminifers at holes 679D, 680B, and 681B, Leg 112. Proc Ocean Drilling Prog Scientific Results 112:369–390

Phleger FB, Soutar A (1973) Production of benthic foraminifera in three east Pacific oxygen minima. Micropaleontology 19:110–115

Piña-Ochoa E, Høgslund S, Geslin E, Cedhagen T, Revsbech NP, Nielsen LP, Schweizer M, Jorissen F, Rysgaard S, Risgaard-Petersen N (2010) Widespread occurrence of nitrate storage and denitrification among Foraminifera and *Gromiida*. Proc Natl Acad Sci 19:1148–1153

Rathburn AE, Corliss BH (1994) The ecology of living (stained) deep-sea benthic foraminifera from the Sulu Sea. Paleoceanography 9(1):87–150

Resig JM (1981) Biogeography of benthic foraminifera of the northern Nazca plate and adjacent continental margin. Geol Soc of Am Memoir 154:619–666

Resig JM (1990) Benthic foraminiferal stratigraphy and Paleoenvironments off Peru, Leg 112. Proc Ocean Drilling Prog Scientific Results 112:263–296

Revsbech NP, Larsen LH, Gundersen J, Dalsgaard T, Ulloa O, Thamdrup B (2009) Determination of ultra-low oxygen concentrations in oxygen minimum zones by the STOX sensor. Limnol Oceanogr Methods 7:371–381

Risgaard-Petersen N, Langezaal AM, Ingvardsen S, Schmid MC, Jetten MSM, Op den Camp HJM, Derksen JWM, Piña-Ochoa E, Eriksson SP, Nielsen LP, Revsbech NP, Cedhagen T, van der Zwaan GJ (2006) Evidence for complete denitrification in a benthic foraminifer. Nature 443:93–96

Schumacher S, Jorissen FJ, Dissard D, Larkin KE, Gooday AJ (2007) Live (Rose Bengal stained) and dead benthic foraminifera from the oxygen minimum zone of the Pakistan continental margin (Arabian Sea). Mar Micropaleontol 62:45–73

Sjoerdsma PG, van der Zwaan GJ (1992) Simulating the effect of changing organic flux and oxygen content on the distribution of benthic foraminifera. Mar Micropaleontol 19:163–180

van Marle LJ (1991) Eastern Indonesian, Late Cenozoic smaller benthic Foraminifera. Verh. K. Nederl. Akad. Wetensch. North Holland, Amsterdam

Whittaker J (1988) Benthic Cenozoic foraminifera from Ecuador, British Museum (Natural History). Great Britain, London

Zheng S-Y, Zhaoxian F (2001) Granuloreticulosa: foraminifera, agglutinated foraminifera, Fauna Sinica: Invertebrata: 26. Science Press, Beijing

Biodata of **Jorge Cardich, María Morales, Luis Quipúzcoa, Abdelfettah Sifeddine**, and **Dimitri Gutiérrez**, authors of "*Benthic Foraminiferal Communities and Microhabitat Selection on the Continental Shelf Off Central Peru.*"

Jorge Cardich is a Peruvian marine biologist. He obtained his bachelor degree in 2009 in the Universidad Nacional Mayor de San Marcos (Peru). He is a graduate student in the Master's Program in Marine Sciences of the Universidad Peruana Cayetano Heredia (Peru) since 2010, and is associated to the Marine Benthos Laboratory of the Peruvian Institute of Marine Research (IMARPE). His scientific interests are: biological oceanography, biogeochemistry, benthic ecology, and biology of benthic foraminifera.

E-mail: **jorge.cardich.s@upch.pe**

Geologist **María Morales** is in charge of the Area of Micropaleontology from the Peruvian Geological Survey (Instituto Geológico Minero Metalúrgico – INGEMMET) and is Professor of Micropaleontology and Palynology at Universidad Nacional Mayor de San Marcos (Peru). She obtained her B.Sc. degree in Geology in 1991 in the Universidad Nacional Mayor de San Marcos. Her main scientific interests are: marine geology, Cenozoic micropaleontology, foraminiferal ecology and paleoenvironmental reconstruction.

E-mail: **mmorales@ingemmet.gob.pe**

Jorge Cardich

María Morales

Luis Quipúzcoa is a marine biologist from the Universidad Ricardo Palma (Peru) and is currently a research scientist in the Marine Benthos Laboratory of IMARPE. He obtained his B.Sc. degree in Biology in 1996. His scientific interests are biological oceanography and ecology of benthic macrofauna.

E-mail: **lquipuzcoa@imarpe.gob.pe**

Abdelfettah Sifeddine is currently a Research Director of the Institut de Recherche pour le Développement (IRD-France). He obtained his Ph.D. in Geosciences at the Université d'Orléans (France) in 1991. His scientific interests are: aquatic and marine organic geochemistry, paleo-climatology and paleoceanography in tropical ecosystems during the Late Cuaternary.

E-mail: **abdel.sifeddine@ird.fr**

Luis Quipúzcoa   **Abdelfettah Sifeddine**

Dr. Dimitri Gutiérrez is a biological oceanographer of IMARPE and Director of the Master's Program in Marine Sciences of the Universidad Peruana Cayetano Heredia. He obtained a Ph.D. degree in Oceanography at the Universidad de Concepción (Chile) in 2000. Since then he has been the leader of the Marine Benthos Laboratory of IMARPE. His scientific interests are: benthic responses to natural and human-induced anoxia, paleo-reconstruction of the Peruvian upwelling ecosystem history, and global warming impacts on the upwelling ecosystem.

E-mail: **dgutierrez@imarpe.gob.pe**

BENTHIC FORAMINIFERAL COMMUNITIES AND MICROHABITAT SELECTION ON THE CONTINENTAL SHELF OFF CENTRAL PERU

JORGE CARDICH[1,2,3], MARÍA MORALES[4]
LUIS QUIPÚZCOA[2], ABDELFETTAH SIFEDDINE[5,6],
AND DIMITRI GUTIÉRREZ[1,2,3]

[1] *Facultad de Ciencias y Filosofía, Programa Maestría en Ciencias del Mar, Universidad Peruana Cayetano Heredia, Av. Honorio Delgado 430, Lima 31, Peru*
[2] *Dirección de Investigaciones Oceanográficas, Instituto del Mar del Perú (IMARPE), Av. Gamarra y Gral. Valle, s/n, Chucuito, Callao, Peru*
[3] *Joint International Laboratory 'Dynamics of the Humboldt Current System' (LMI DISCOH), Lima, Peru*
[4] *Laboratorio de Micropaleontología, Instituto Geológico Minero Metalúrgico (INGEMMET), Av. Canadá 1470, Lima 41, Peru*
[5] *Centre IRD France-Nord, LOCEAN, UMR 7159, 32 Avenue Henri Varagnat, 93143 Bondy cedex, France*
[6] *Departamento de Geoquimica, Universidade Federal Fluminense, LMI PALEOTRACES, Niteroi, RJ, Brazil*

1. Introduction

Oxygen and food availability are critical factors in the ecology of benthic foraminifera, influencing on the composition of the communities and on their vertical distribution (Jorissen et al. 1995). Several groups of benthic foraminifera are known to tolerate oxygen deficiency (e.g., Gooday et al. 2000; Bernhard et al. 2000), and some of them are able to survive under anoxic conditions (Moodley et al. 1997, 1998) by means of physiological adaptations (Risgaard-Petersen et al. 2006; Piña-Ochoa et al. 2010) or symbiotic associations with bacteria (Bernhard 2003; Bernhard et al. 2006).

The upper Central Peruvian margin is subjected to extreme dysoxic bottom waters and to strong phytodetrital carbon fluxes (Levin et al. 2001; Gutiérrez et al. 2009). Here, the Tropical South Eastern Pacific (TSEP) oxygen minimum zone (OMZ; O_2 <22 µmol.l^{-1}) is intensified because of the enhanced consumption driven by the upwelling-supported high productivity (Helly and Levin 2004; Graco et al. 2007). The oxycline can be very shallow, reaching up to 20–40 m depth (Gutiérrez et al. 2008). The continental shelf sediments display suboxic to anoxic conditions, driven by anaerobic carbon recycling and by oxygenation episodes reaching the bottom water (Suits and Arthur 2000; Gutiérrez et al. 2008).

Benthic foraminifera composes one of the major groups that dominate the biota of the upper Peruvian margin (Levin et al. 2002). The calcareous species *Bolivina seminuda* has been determined as the most characteristic one of the TSEP-OMZ sediments off the Peruvian and Chilean coasts (Páez et al. 2001; Levin et al. 2002; Tapia et al. 2008). There have been reported metabolic adaptations of several species from the TSEP-OMZ to thrive under anoxic conditions and profit from the large amount of available food (Risgaard-Petersen et al. 2006; Høgslund et al. 2008; Piña-Ochoa et al. 2010). Nevertheless, stilllittle is known about the composition and spatial change of living benthic foraminifera in the continental shelf sediments (Khusid 1974), which are subjected to varying levels of oxygenation and duration of anoxia (Gutiérrez et al. 2008).

Our study gives new insights into benthic foraminiferal ecology off Callao (12ºS), focusing on the differences between foraminiferal communities from the inner-shelf, prone to sulfidic conditions, and the dysoxic outer-shelf off Callao. Here we assess spatial (topmost sediment) and vertical distribution (0–5 cm) of benthic foraminifera in relationship with environmental biogeochemical conditions. Our results provide elements for the calibration of benthic foraminiferal assemblages as paleoenvironmental proxies in the Peruvian margin.

2. Material and Methods

The samples were collected in April 2009 aboard the RV José Olaya Balandra, from five stations along the cross-shelf transect off Callao (12ºS) (Fig. 1a). Two of them (E1 and E2) are located in the inner-shelf (<10 nm off the coast), one in the mid-shelf (E3) and the last two ones on the outer-shelf (E4 and E5, 20–30 nm off the coast). Water column and bottom water dissolved oxygen concentrations, temperatures and salinities were determined, employing a CTD Seabird SBE 19+ instrument. Characteristics of the stations are shown in Table 1.

A mini-multicorer (MUC) was employed to sample benthic foraminifera and sediment/pore water properties. Samples of benthic foraminifera were collected directly from the MUC tubes (=9.6 cm). Two core tubes were collected at each station from independent multicorer casts. The cores were sliced at one centimeter intervals for the upper 5 cm. The slices of sediment were immediately preserved and stained following Rathburn and Corliss (1994) methodology. Samples were allowed to stain with Rose Bengal for at least 1 week before being analyzed. Only specimens with all chambers stained, except for the last one (the youngest), were counted (Tapia et al. 2008). After staining, samples were wet-sieved using 63, 150 and 500 μm mesh. In order to describe the vertical distribution, we use the Average Living Depth index (ALD_x; Jorissen et al. 1995).

The total stock of living foraminifera was determined from the integration of the data at all levels from 0 to 5 cm depth. Relative abundance of taxa was

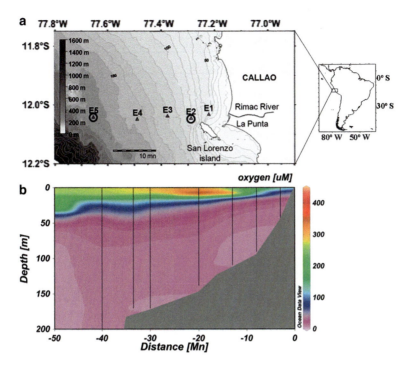

Figure 1. (a) Map showing sampling-site locations in the shelf off Callao; (b) Dissolved oxygen (μM) section in the water column.

calculated from the raw density data. Finally, for vertical distributions, densities were normalized to 50 cm^3 sediment for each sediment level. Here we present results from the top 1 cm of the entire transect and from the 0–5 cm of stations E2 and E5.

Sedimentary organic matter properties were determined in the top centimeter. Total carbon and total nitrogen were determined using a *Thermo electron* CNS elemental analyzer. The ratio between microbe-bacterial carbon and total carbon (Bact C:TC) was used as a proxy of the content of labile organic matter. Microbe-bacterial carbon was estimated according to the procedure by Bratbak and Dundas (1984). For this, bacterial cells were double-stained with Acridine Orange and DAPI (Kuwae and Hosokawa 1999) to be counted and measured under epifluorescence microscopy. Chlorophyll-*a* and phaeopigments were also measured, following the fluorometric method described in Gutiérrez et al. (2000). At stations E2 and E5, pore water was collected with rhizon samplers from the top 5 cm, with a 1 cm resolution. Dissolved sulfide in the pore water was determined on $ZnCl_2$-preserved samples by the colorimetric methylene blue method of Cline (1969).

Table 1. Location, bottom water physical factors, topmost sediment properties (0–1 cm) and community parameters of the sample stations.

	Sample stations				
	E1	E2	E3	E4	E5
Latitude (S)	12° 01.95′	12° 02.75′	12° 01.46′	12° 02.98′	12° 02.43′
Longitude (W)	77° 13.24′	77° 17.14′	77° 22.45′	77° 29.12′	77° 39.00′
Physical factors	–	–	–	–	–
Water column depth (m)	48	94	117	143	178
Bottom-water temperature (°C)	14.8	14.6	14.4	14.0	13.6
Salinity	34.98	34.98	34.97	34.96	34.94
Bottom-water oxygen (μM)	6.70	5.02	4.95	8.37	5.69
Sediment properties	–	–	–	–	–
% C	4.67	6.76	8.71	11.77	12.75
% N	0.55	0.85	1.19	1.43	1.8
Chlorophyll-a (μg.g^{-1})	90.09 ± 37.05	83.83 ± 57.77	228.67 ± 23.77	160.71 ± 25.24	76.03 ± 3.32
Chl-a:Pha (g.g^{-1})	0.23 ± 0.05	0.17 ± 0.05	0.18 ± 0.02	0.13 ± 0.02	0.12 ± 0.01
H$_2$S (nmol.cm^{-3})	n.d.	111.69	n.d.	n.d.	0.00
H$_2$S (μmol.cm^{-2}) (0–5 cm)	n.d.	1802.07	n.d.	n.d.	2.87
Bact C:TC (mg.g^{-1})	29.23 ± 1.67	40.90 ± 0.10	25.20 ± 0.89	15.04 ± 4.45	15.53 ± 5.77
Community parameters	–	–	–	–	–
0–1 cm	–	–	–	–	–
Density (ind. 50 cm^{-3})	4156 ± 1522	3343 ± 109	4904 ± 343	3832 ± 673	3533 ± 588
Main groups (%)	–	–	–	–	–
Calcareous	90.1	94.4	85.3	84.8	84.7
Agglutinated	0.5	0.0	6.0	8.7	10.8
Tectinous	9.4	5.5	8.8	6.5	4.5
Size fraction contribution (%)	–	–	–	–	–
63–150 μm	87.6	83.1	68.5	70.2	65.7
>150 μm	12.4	16.9	31.5	29.8	34.3
0–5 cm	–	–	–	–	–
Standing stock (ind. 50 cm^{-2})	n.d.	4457 ± 394	n.d.	n.d.	4396 ± 666
Size fraction contribution (%)	–	–	–	–	–
63–150 μm	n.d.	84.7	n.d.	n.d.	65.0
>150 μm	n.d.	15.3	n.d.	n.d.	35.0
Dominant species (%)	*B. costata* (73.3)	*B. costata* (78.1; *69.2*)	*B. humilis* (38.9)	*B. humilis* (38.5)	*B. humilis* (39.4; *38.2*)

Values in italics of the relative abundance of dominant species correspond to total stock when available

3. Results

3.1. OCEANOGRAPHIC SETTING AND SEDIMENT PROPERTIES

The OMZ was well developed over the shelf (Fig. 1b), as expected from the enhanced thermal stratification and development of algal blooms towards the end of the productive season (Gutiérrez et al. 2008). BWDO ranged from 5.0 to 8.4 $\mu mol.l^{-1}$ (Table 1), corresponding to extreme dysoxic conditions (Bernhard and Sen Gupta 1999).

Surface sediments are diatomaceous-rich silty-clay muds. In the inner-shelf stations, a black, sulfidic flocculent layer, resulting from settling aggregates of phytodetritus, characterized the sediment–water interface. In the mid- and outer-shelf stations, no sulfide odor was detected, and superficial *Thioploca* mats were developed, especially at stations E3 and E4.

Total carbon (4.67–12.75%) and total nitrogen (0.55–1.80%) tended to increase with water depth (Spearman's $r = 1.00$, $p < 0.05$) (Table 1). Previous determinations of carbonate contents in the surface sediments have resulted in values below 10% (Dimitri Gutiérrez, unpubl. observations), so that the organic fraction should explain the spatial variation and over 90% of the carbon values. Bact C:TC ratios decreased with water depth (Spearman's $r = 0.80$, $p < 0.10$), being almost three times higher at the inner-shelf stations than at stations E4 and E5 (Table 1). Surface Chl-*a* content was not significantly correlated with water depth (Spearman's $r = -0.20$, $p > 0.70$). Instead, the Chlorophyll-*a* to Phaeopigments ratio (Chl-*a*:Pha) decreased with water depth (Spearman's $r = 0.90$, $p < 0.05$), and was negatively correlated to total carbon and total nitrogen (Spearman's $r = 0.99$, $p < 0.05$). Porewater H_2S contents from the 0–1 and 0–5 cm intervals recorded in the shallower station E2 were respectively two and three orders of magnitude higher than those in the outer-shelf station E5 (Table 1), confirming field observations.

3.2. BENTHIC FORAMINIFERAL ASSEMBLAGES AND VERTICAL DISTRIBUTION

Topmost sediment (0–1 cm) densities (>63 μm) ranged between $3,343 \pm 109$ ind.50 cm^{-3} (E2) and $4,904 \pm 343$ ind.50 cm^{-3} (E3) (Table 1); there were no significant differences among sample sites (Kruskal-Wallis test, $H_{(4,10)}$, $p = 0.567$). These densities were over 75% of the total standing stock in the top 5 cm at stations E2 and E5. The larger (>150 μm) fraction contribution to topmost sediment density was greater offshore, ranging from 12.4% (E1) to 34.3% (E5). Individuals larger than 500 μm were observed only at station E5 with very low abundances (~2 ind.50 cm^{-2}), and all of them corresponded to the calcareous foraminifer *Cancris auriculus* (Fichtel and Moll). Therefore, they were included in the larger fraction.

The communities at all stations were mainly comprised by calcareous foraminifera. Tectinous (soft-shelled) foraminifera tended to decrease with water

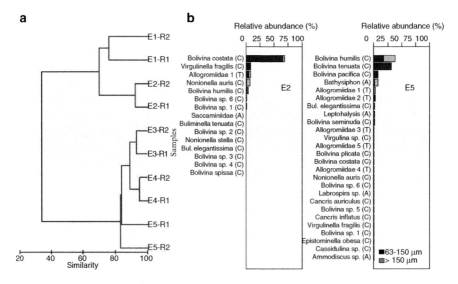

Figure 2. (a) Cluster analysis for all the topmost sediment samples (0–1 cm); (b) Relative abundance of species of benthic foraminifera by size fractions from stations E2 and E5. *C* Calcareous, *A* Agglutinated, *T* Tectinous.

depth; yet a relative minimum in abundance was observed at station E2. In contrast, agglutinated foraminifera increased in abundance with water depth (Table 1).

Cluster analysis based on the benthic foraminiferal assemblage composition makes clear the existence of two communities (Fig. 2). The inner-shelf group (E1 and E2) was dominated by *Bolivina costata* d'Orbigny (73–78%). Other important species were *Nonionella auris* (d'Orbigny) (7–13%), the spherical allogromiid morphotype 1 (5.6–9.2%), and *Virgulinella fragilis* Grindel and Collen (1.8–2.3%). The more diverse outer-shelf group (E3, E4 and E5) presented *Bolivina seminuda* Cushman var. *humilis* Cushman and McCulloch (= *Bolivina humilis*; 38.5–39.4%) and *Buliminella subfusiformis* Cushman var. *tenuata* Cushman (= *Buliminella tenuata* 33.5–36.1%) as dominant species, followed by *Bolivina pacifica* Cushman and McCulloch (6.0–6.7%), *Bathysiphon* sp., which increased with water depth from 3.1% to 9.1%, and five allogromiid morphotypes, which represented 4.5–8.8% of the total densities.

At stations E2 and E5, some of the species exhibited preferences in their vertical distribution (Fig. 3). The dominant species at both stations were concentrated at the top sediment layer: *B. costata* ($ALD_5 = 0.8 \pm 0.0$ cm), *B. humilis* ($ALD_5 = 0.7 \pm 0.0$ cm) and *Buliminella tenuata* ($ALD_5 = 0.7 \pm 0.0$ cm). The agglutinated species *Bathysiphon* sp. was also concentrated almost entirely in the topmost sediment ($ALD_5 = 0.6 \pm 0.0$ cm). *N. auris* and *B. pacifica* appeared slightly deeper ($ALD_5 = 0.9 \pm 0.1$ cm and $ALD_5 = 0.9 \pm 0.1$ cm, respectively). The allogromiid morphotype 1 was distributed primarily down to the second centimeter ($ALD_5 = 1.2 \pm 0.1$ cm) at station E2, but it exhibited a subsurface peak at station

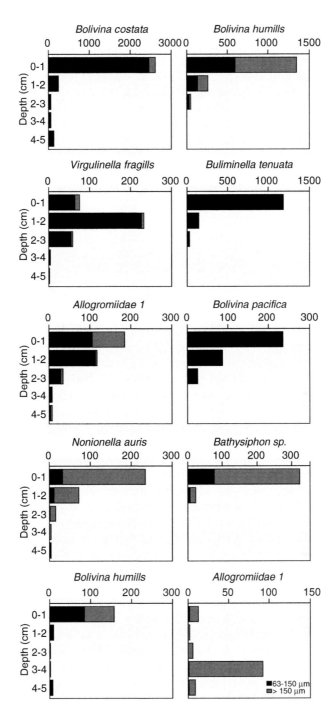

Figure 3. Vertical distribution of the main five benthic foraminifera by size fractions at stations E2 (*left column*) and E5 (*right column*). Species appear by order of relative abundance.

E5 (3.2±0.0 cm). The deepest distribution of all calcareous species was exhibited by *V. fragilis*, characterized by a subsurface maximum peak at the 1–2-cm interval ($ALD_5 = 1.5 \pm 0.3$ cm) (Fig. 3). Most of the dominant species that attained a large "adult" size at both stations presented a shallower ALD_5 for the "adult" individuals (> 150 μm) than for the "young" individuals (63–150 μm). However, *N. auris*, from station E2, exhibited similar vertical distributions for the "adult" population ($ALD_5 = 0.9 \pm 0.1$ cm) and for the "young" ones ($ALD_5 = 0.9 \pm 0.3$ cm). It is remarkable that at both stations *B. humilis* "adult" individuals presented a shallower ALD_5 than the one of the "young" individuals. Large individuals from the morphotype 1 at station E5 were distributed much deeper than the large individuals at station E2 ($ALD_5 = 3.2 \pm 0.0$ cm versus $ALD_5 = 0.8 \pm 0.2$ cm, respectively).

4. Discussion

4.1. BENTHIC FORAMINIFERAL ASSEMBLAGES AND BIOGEOCHEMICAL CONDITIONS

The total densities of foraminifera obtained in our study (~3,000–5,000 ind.50 cm^{-2}) are similar to the estimates by Phleger and Soutar (1973) for the Callao shelf (180 m) and by Páez et al. (2001) for Mejillones del Sur Bay, off Northern Chile (23°S). In addition, the observed densities are comparable to those recorded at similar depths in the OMZ of the Arabian Sea (Schumacher et al. 2007). Data from the larger fraction are comparable to the values observed in the shelf stations off Concepción (Central Chile, 36°S), where the >180 μm fraction was used (Tapia et al. 2008). However, our results were lower than those reported at bathyal depths, as in the OMZ core off Callao (Levin et al. 2002), the Santa Barbara Basin (Bernhard et al. 1997), and the Arabian Sea by an order of magnitude (Gooday et al. 2000).

The foraminiferal community in the Callao shelf was largely composed (>80%) by small specimens (63–150 μm) in the inner-shelf. The dominant species of the inner-shelf benthic foraminiferal community was *B. costata*, a coastal species which is also dominant near the coastline in this region (0–5 m; Verano 1974). Two of the subdominant species of this community, *N. auris* and *V. fragilis*, are considered as characteristic species in other sedimentary settings with pronounced chemoclines and strong sulfide production (Risgaard-Petersen et al. 2006; Leiter and Altenbach 2010). On the other hand, both *B. humilis* (= *B. seminuda*) and *Buliminella tenuata* dominated the outer-shelf community. *B. seminuda* Cushman has been reported as dominant under severe oxygen depletion, such as the Peruvian upper slope within the OMZ core (Pérez et al. 2002) and other dysoxic environments (Sen Gupta and Machain-Castillo 1993; Páez et al. 2001).

Lower carbon and nitrogen contents and higher ratios of microbe bacterial carbon to total carbon in the inner-shelf surface sediments suggest that the labile fraction is greater and benthic carbon recycling is more intense here than offshore. This is supported by the greater sulfide release in the inner-shelf, which given the similar range of low bottom water oxygen content, is further indicative of greater (anaerobic) respiration rates of the organic matter at the study period. In addition, the lower Chl-a:phaeopigment ratios towards the outer shelf show that the arriving phytodetritus is more degraded with water depth.

The two communities determined for the shelf off Callao are in agreement with two benthic foraminiferal biofacies described by Resig (1990): the shelf biofacies (<150 m), which was characterized by *B. costata* and *N. auris* as dominant species and the oxygen-minimum biofacies (150–300 m), characterized by *Bolivinellina humilis* (= *Bolivina seminuda* var. *humilis*), as dominant species. We interpret that the "shelf biofacies" is indicative of the dominance of high fluxes of labile organic matter and sedimentary sulfidic conditions near the coast. Also, from our observations, the "oxygen-minimum biofacies" is likely indicative of past average sedimentary "suboxic" or "postoxic" conditions (Tyson and Pearson 1991; Bernhard and Sen Gupta 1999).

4.2. VERTICAL DISTRIBUTION AND MICROHABITAT SELECTION

A shallow distribution of the total stock is indicative of a benthic foraminiferal community under high sedimentary organic matter contents, oxygen depletion, and a redox front positioned close to the sediment surface (Jorissen et al. 1995; Murray 2001). Some species appear to present adaptations to inhabit deeper into the sediment in order to take advantage from food availability in the subsurface layers. In our study, the "young" individuals of most dominant species appear to present deeper distributions than the "adult" individuals. This might be explained by the smaller individuals' greater surface-to-volume ratios, which are advantageous in oxygen-depleted environments for inhabiting deeper into the sediment (Shepherd et al. 2007).

A particular case was *Nonionella auris*, which presented a deeper distribution of its larger specimens than that of the "young" individuals. This pattern has been explained for cases when reproduction or early growth takes place near the water–sediment interface, whereas the "adult" individuals can thrive deeper into the sediment (Shepherd et al. 2007). Risgaard-Petersen et al. (2006) reported that *N.* cf. *stella* from the Central Chile shelf is able to perform denitrification, by using its stored intracellular nitrate, which is 3,000 times as concentrated as the maximum pore-water nitrate. We note that the description of this species by Risgaard-Petersen et al. (2006) in fact appears to correspond to *N. auris* (Resig 1990). We postulate that the abundance of *N. auris* in the sulfidic inner-shelf stations might be explained by the combination of (1) its nitrate storage capacity that allows the continuation of the denitrification activity when nitrate is depleted in the pore water (Risgaard-Petersen et al. 2006), (2) the oxygenation episodes that bring incoming nitrate-rich waters to the bottom (Gutiérrez et al. 2008), and (3) the high porosity of the flocculent surface sediment in the inner-shelf that permits nitrate replenishment in the pore waters. In turn, the pattern of vertical distribution among young/adult individuals might result from the larger storage and motility capacities that "adult" individuals present over the "young" individuals, so that the former ones can stay longer deep into the sediment.

The particular deep vertical distribution of allogromiids in the Callao shelf sediments provides more insight on this group which was thought to be less tolerant, in a general sense, to oxygen depletion than calcareous taxa (Gooday et al. 2000). However, there is evidence that some allogromiids present sulfur-oxidizing prokaryotic endobionts in anoxic sediments, supporting the idea that symbiogenesis evolved in this foraminiferal group during early foraminiferal diversification (Bernhard et al. 2006).

As indicated by the higher ratios of Bact C:TC and higher sulfide concentrations, inner-shelf stations are closer to the eutrophic, anoxic end-member of the TROX model (Jorissen et al. 1995) than the outer-shelf stations, which lay on the "dysoxic" range of the model. Despite the differences in the content of labile organic matter and the redox conditions, benthic foraminifera show nearly equal shallow vertical distributions for the inner-shelf station E2 and for the outer-shelf station E5. For the inner-shelf, the overall vertical distribution is influenced by the deeper distribution of some species (*V. fragilis*, *N. auris*) which are known to present metabolic/symbiotic adaptations. For the outer-shelf station E5, the superficial distribution of benthic foraminifera cannot be explained by redox stress. The importance of competitive interactions for food is thought to increase with weaker redox stress, according to the "TROX-2" model (Van der Zwaan et al. 1999). We observed that nematodes exhibit lower relative densities at the topmost layer here than at shallower sites (Jorge Cardich, unpubl. observations), which is consistent with competitive advantage of benthic foraminifera for exploiting fresh phytodetritus (Nomaki et al. 2005).

4.3. *VIRGULINELLA FRAGILIS* AND H_2S CONCENTRATION

The abundance of *V. fragilis* in the inner-shelf off Callao is characteristic of settings subjected to sulfide release. *V. fragilis* is present in sulfide-rich environments, such as the shelf off Namibia (Leiter and Altenbach 2010), the Cariaco Basin (Bernhard 2003), sulfur-oxidizing bacterial mats in bathyal sediments in the Arabian Sea (Erbacher and Nelskamp 2006), and the Aso-kai lagoon in Japan (Takata et al. 2005), among others (see Revets 1991 and Erbacher and Nelskamp 2006 for a review). Erbacher and Nelskamp (2006) remarked the occurrence of *V. fragilis* at stations with *Beggiatoa/Thioploca* mats in the Arabian Sea (>100 m depth), but in our study *Thioploca* mats were not developed in the inner-shelf inhabited by *V. fragilis* (Table 1).

V. fragilis presents kleptoplasts and sulfur-oxidizing prokaryotic endobionts, allowing thriving under anoxic conditions (Bernhard 2003). Moreover, *V. fragilis* presents peroxisome-endoplasmic reticulum complexes that enable oxygen uptake from H_2O_2 (Bernhard and Bowser 2008). Given these features *V. fragilis* has been

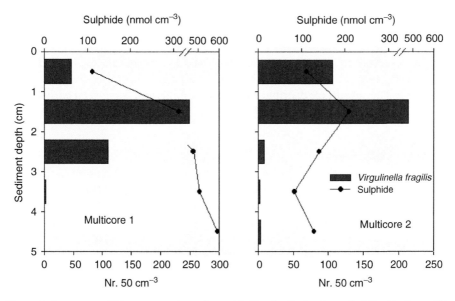

Figure 4. Pore-water sulfide concentrations along with *Virgulinella fragilis* vertical distribution at E2.

proposed as a potential proxy of sulfate-reducing environments in sediment records (Bernhard 2003).

The singular profile of *V. fragilis* into the sediment in this study along with pore water sulfide might be explained by these adaptations. Here at the sulfide-rich station E2, *V. fragilis* exhibited its abundance peak at the second centimeter into the sediment (Fig. 4), where sulfide concentration increased compared to the 0–1 cm section.

On the first multicore cast, from the topmost centimeter to the second centimeter, sulfide content in the pore water increased from 111.7 to 310.9 nmol.cm^{-3}, and then sulfide concentrations were rather homogeneous at deeper levels (Fig. 4a). On the second multicore cast, sulfide content increased from 111.5 to 210.1 nmol.cm^{-3}, before decreasing at deeper levels (Fig. 4b).

In both casts, the highest density of *V. fragilis* was found in the base of the upper cline of sulfide concentrations. These observations are consistent with the findings of Erbacher and Nelskamp (2006), who suggested that *V. fragilis* is concentrated near the boundary at which sulfide is produced.

5. Conclusions

1. Two benthic foraminiferal assemblages were determined in the shelf off Callao: the inner-shelf community, dominated by *Bolivina costata*, and the outer-shelf, dominated by *Bolivina humilis* and *Buliminella tenuata*.

2. Labile organic matter and redox conditions appeared to be the main factors governing the diversity and vertical distribution of benthic foraminifera in the Callao continental shelf.
3. The tolerance to sulfidic conditions exhibited by *Virgulinella fragilis* adds more evidence that this species could be used as a paleoindicator of anoxic environments. The subsurface distribution of allogromiids and *Nonionella auris* are also consistent with symbiotic and physiological adaptations to anoxic conditions.

6. Acknowledgements

This study was supported by the Instituto del Mar del Perú (IMARPE) and its project MINIOX and by the project JEAI MIXPALEO (IRD, IMARPE and INGEMMET). We would like to thank our colleagues C. Machado, E. Enríquez, R. Marquina, B. Cuevas, A. Pérez, V. Aramayo, and J. Solís. As well, we acknowledge the crew of the RV José Olaya, without their help this study would not have been possible. We also are grateful to Dr. Elena Pérez and to Sergio Mayor, who helped us to confirm species identifications and to Dr. Anthony Rathburn, who provided us with valuable suggestions since the early stage of this research.

7. References

Bernhard JM (2003) Potential symbionts in bathyal foraminifera. Science 299:861

Bernhard JM, Bowser SS (2008) Peroxisome proliferation in foraminifera inhabiting the chemocline: an adaptation to reactive oxygen species exposure? J Eukaryot Microbiol 55:135–144

Bernhard JM, Sen Gupta BK (1999) Foraminifera of oxygen-depleted environments. In: Sen Gupta BK (ed.) Modern Foraminifera. Kluwer Academic, Great Britain, pp 201–216

Bernhard JM, Sen Gupta BK, Borne PF (1997) Benthic foraminiferal proxy to estimate dysoxic bottom-water oxygen concentrations: Santa Barbara Basin, U.S. Pacific Continental Margin. J Foraminifer Res 27:301–310

Bernhard JM, Buck KR, Farmer MA, Bowser SS (2000) The Santa Barbara Basin is a symbiosis oasis. Nature 403:77–80

Bernhard JM, Habura A, Bowser SS (2006) An endobiont-bearing allogromiid from the Santa Barbara Basin: implications for the early diversification of foraminifera. J Geophys Res 111:G03002. doi: 10.1029/2005JG000158

Bratbak G, Dundas I (1984) Bacterial dry matter content and biomass estimation. Appl Environ Microbiol 48:755–757

Cline JD (1969) Spectrophotometric determination of hydrogen sulfide in natural waters. Limnol Oceanogr 14:454–458

Erbacher J, Nelskamp S (2006) Comparison of benthic foraminifera inside and outside a sulphur-oxidizing bacterial mat from the present oxygen-minimum zone off Pakistan (NE Arabian Sea). Deep-Sea Res Pt I 53:751–775

Gooday AJ, Bernhard JM, Levin LA, Suhr SB (2000) Foraminifera in the Arabian Sea oxygen minimum zone and other oxygen-deficient settings: taxonomic composition, diversity, and relation to metazoan faunas. Deep-Sea Res Pt II 47:25–54

Graco M, Ledesma J, Flores G, Girón M (2007) Nutrientes, oxígeno y procesos biogeoquímicos en el sistema de surgencias de la corriente de Humboldt frente a Perú. Rev Peru Biol 14:117–128

Gutiérrez D, Gallardo VA, Mayor S, Neira C, Vásquez C, Sellanes J, Rivas M, Soto A, Carrasco F, Baltazar M (2000) Effects of disolved oxygen and fresh organic matter on the bioturbation potential of macrofauna in sublittoral bottoms off central Chile, during the 1997-98 El Niño. Mar Ecol Progr Ser 202:208–210

Gutiérrez D, Enríquez E, Purca S, Quipúzcoa L, Marquina R, Graco M (2008) Oxygenation episodes on the continental shelf of central Peru: remote forcing and benthic ecosystem response. Progr Oceanogr 79:177–189

Gutiérrez D, Sifeddine A, Field D, Ortlieb L, Vargas G, Chávez F, Velazco F, Ferreira V, Tapia P, Salvatteci R, Boucher H, Morales MC, Valdés J, Reyss JL, Campusano A, Boussafir M, Mandeng-Yogo M, García M, Baumgartner T (2009) Rapid reorganization in ocean biogeochemistry off Peru towards the end of the Little Ice Age. Biogeosciences 6:835–848

Helly JJ, Levin LA (2004) Global distribution of naturally occurring marine hypoxia on continental margins. Deep-Sea Res Pt I 51:1159–1168

Høgslund S, Revsbech NP, Cedhagen T, Nielsen LP, Gallardo VA (2008) Denitrification, nitrate turnover, and aerobic respiration by benthic foraminiferans in the oxygen minimum zone off Chile. J Exp Mar Biol Ecol 359:85–91

Jorissen FJ, de Stigter HC, Widmark JGV (1995) A conceptual model explaining benthic foraminiferal microhabitats. Mar Micropaleontol 26:3–15

Khusid TA (1974) Distribution of benthic foraminifers off the west coast of South America. Oceanology 14:900–904

Kuwae T, Hosokawa Y (1999) Determination of abundance and biovolume of bacteria in sediments by dual staining with 49,6-Diamidino-2-Phenylindole and Acridine Orange: relationship to dispersion treatment and sediment characteristics. Appl Environ Microbiol 65:3407–3412

Leiter C, Altenbach AV (2010) Benthic foraminifera from the diatomaceous mud belt off Namibia: characteristic species for severe Anoxia. Palaeontol Electron 13:11A:19p

Levin LA, Etter RJ, Rex MA, Gooday AJ, Smith CR, Pineda J, Stuart CT, Hessler RR, Pawson D (2001) Environmental influences on regional deep-sea species diversity. Annu Rev Ecol Syst 132:51–93

Levin L, Gutiérrez D, Rathburn AE, Neira C, Sellanes J, Muñoz P, Gallardo V, Salamanca M (2002) Benthic processes on the Peru margin: a transect across the oxygen minimum zone during the 1997-98 El Niño. Progr Oceanogr 53:1–27

Moodley L, van der Zwaan GJ, Herman PMJ, Kempers L, van Breugel P (1997) Differential response of benthic meiofauna to anoxia with special reference to Foraminifera (Protista: Sarcodina). Mar Ecol Progr Ser 158:151–163

Moodley L, Schaub BEM, van der Zwaan GJ, Herman PMJ (1998) Tolerance of benthic foraminifera (Protista: Sarcodina) to hydrogen sulphide. Mar Ecol Progr Ser 169:77–86

Murray JW (2001) The niche of benthic foraminifera, critical thresholds and proxies. Mar Micropaleontol 41:1–7

Nomaki H, Heinz P, Nakatsuka T, Shimanaga M, Kitazato H (2005) Species-specific ingestion of organic carbon by deep-sea benthic foraminifera and meiobenthos: in situ tracer experiments. Limnol Oceanogr 50:134–146

Páez M, Zúñiga O, Valdés J, Ortlieb L (2001) Foraminíferos bentónicos recientes en sedimentos micróxicos de la bahía Mejillones del Sur (23°S). Chile Rev Biol Mar Oceanogr 36:129–139

Pérez ME, Rathburn AE, Levin LA, Deng WB (2002) The ecology of benthic foraminifera of the Peru oxygen minimum zone. Dissertation, GSA Annual Meeting in Colorado Convention Center

Phleger FB, Soutar A (1973) Production of benthic foraminifers in three east Pacific oxygen minima. Micropaleontology 19:110–115

Piña-Ochoa E, Høgslund S, Geslin E, Cedhagen T, Revsbech NP, Nielsen LP, Schweizerf M, Jorissen F, Rysgaard S, Risgaard-Petersen N (2010) Widespread occurrence of nitrate storage and denitrification among Foraminifera and Gromiida. Proc Natl Acad Sci USA 107:1148–1153

Rathburn AE, Corliss BH (1994) The ecology of living (stained) deep-sea benthic foraminifera from the Sulu Sea. Paleoceanography 9:87–150

Resig JM (1990) Benthic foraminiferal stratigraphy and paleoenvironments off Perú, Leg 112. Proc Ocean Drill Progr Sci Results 112:263–296

Revets SA (1991) The nature of Virgulinella Cushman, 1932 and the implications for its classification. J Foraminifer Res 21:293–298

Risgaard-Petersen N, Langezaal AM, Ingvardsen S, Schmid MC, Jetten MSM, Op den Camp HJM, Derksen JWM, Piña-Ochoa E, Eriksson SP, Nielsen LP, Revsbech NP, Cedhagen T, van der Zwaan GJ (2006) Evidence for complete denitrification in a benthic foraminifer. Nature 443:93–96

Schumacher S, Jorissen FJ, Dissard D, Larkin KE, Gooday AJ (2007) Live (Rose Bengal stained) and dead benthic foraminifera from the oxygen minimum zone of the Pakistan continental margin (Arabian Sea). Mar Micropaleontol 62:45–73

Sen Gupta BK, Machain-Castillo ML (1993) Benthic foraminifera in oxygen-poor habitats. Mar Micropaleontol 20:183–201

Shepherd AS, Rathburn AE, Pérez ME (2007) Living foraminiferal assemblages from the Southern California margin: a comparison of the >150, 63–150, and >63 μm fractions. Mar Micropaleontol 65:54–77

Suits NS, Arthur MA (2000) Bacterial production of anomalously high dissolved sulfate concentrations in Peru slope sediments: steady-state sulfur oxidation, or transient response to end of El Niño? Deep-Sea Res Pt I 47:1829–1853

Takata H, Seto K, Sakai S, Tanaka S, Takayasu K (2005) Correlation of *Virgulinella fragilis* Grindell and Collen (benthic foraminiferid) with near-anoxia in Aso-kai Lagoon, central Japan. J Micropalaeontol 24:159–167

Tapia R, Lange CB, Marchant M (2008) Living (stained) calcareous benthic foraminifera from recent sediments off Concepción, central-southern Chile (36°S). Rev. Chil Hist Nat 81:403–416

Tyson R, Pearson T (1991) Modern and ancient continental shelf anoxia: an overview. In: Tyson R, Pearson T (eds.) Modern and ancient continental shelf anoxia, vol 58. Geological Society, London, pp 1–24

Van der Zwaan GJ, Duijnstee IAP, den Dulk M, Ernst SR, Jannink NT, Kouwenhoven TJ (1999) Benthic foraminifers: proxies or problems? A review of paleoecological concepts. Earth Sci Rev 46:213–236

Verano R (1974) Foraminíferos del litoral del departamento de Lima. Rev Peru Biol 1:63–80

PART V:
ZONES AND REGIONS

Buck
Rabalais
Bernhard
Barry
Stachowitsch
Riedel
Zuschin
Sergeeva
Gooday
Mazlumyan
Kolesnikova
Lichtschlag
Kosheleva

Anikeeva
Saccà
Stoeck
Behnke
Fritz
Pfannkuchen
Struck
Hengherr
Strohmeier
Brümmer
Por
Edgcomb
Biddle

Biodata of **Kurt R. Buck**, **Nancy N. Rabalais**, **Joan M. Bernhard**, and **James P. Barry**, authors of *"Living Assemblages from the "Dead Zone" and Naturally Occurring Hypoxic Zones."*

Kurt R. Buck, who is a Senior Research Specialist at Monterey Bay Aquarium Research Institute, is an oceanographer specializing in quantitative enumeration, ecology, and imaging of marine protists and bacteria. Upper water column communities from Antarctic and Arctic sea ice to equatorial regions were his initial focus. Currently, he is working with deep-sea sediment communities, including those from hypoxic zones. Buck has a degree in Oceanography (M.S., Texas A&M University, 1979).

E-mail: **buku@mbari.org**

Dr. Nancy N. Rabalais is a professor at the Louisiana Universities Marine Consortium and is its executive director. Her research focuses on the dynamics of hypoxic environments, interactions of large rivers with the coastal ocean, estuarine and coastal eutrophication, benthic ecology, and science policy. She is well known for her research and outreach concerning the linkages of the Mississippi River and the "dead zone" in the northern Gulf of Mexico. Rabalais earned her Ph.D. in Zoology from The University of Texas at Austin in 1983.

E-mail: **nrabalais@lumcon.edu**

Kurt R. Buck

Nancy N. Rabalais

Dr. Joan M. Bernhard, who is a Senior Scientist at Woods Hole Oceanographic Institution, is a biogeochemist with a major focus on the adaptations and ecology of protists living in the chemocline. Her work is largely in the bathyal to abyssal deep sea but also in high latitudes. Bernhard has degrees in Geology (M.S. 1984, University of California Davis) and Biological Oceanography (Ph.D. 1990, Scripps Institution of Oceanography, University of California San Diego), and did postdoctoral work in cell biology. Her multidisciplinary training gives her a unique perspective into anoxic habitats. For more information, see Biodata of the Editors.

E-mail: **jbernhard@whoi.edu**

Dr. James P. Barry is a Senior Scientist at the Monterey Bay Aquarium Research Institute (MBARI) whose research program focuses on the effects of climate change on ocean ecosystems. After training at Scripps Institution of Oceanography, Barry joined the science staff at the MBARI in the early 1990s. In addition to climate change, his research interests are broad, spanning topics such as (1) the biology and ecology of chemosynthetic biological communities in the deep-sea, (2) coupling between upper ocean and seafloor ecosystems in polar and temperate environments, (3) the biology of deep-sea communities, and (4) the biology of submarine canyon communities. Dr. Barry has helped inform Congress on ocean acidification, ocean carbon sequestration, and climate change by speaking at congressional hearings, briefings, and meetings with congressional members.

E-mail: **barry@mbari.org**

Joan M. Bernhard

James P. Barry

LIVING ASSEMBLAGES FROM THE "DEAD ZONE" AND NATURALLY OCCURRING HYPOXIC ZONES

KURT R. BUCK[1], NANCY N. RABALAIS[2],
JOAN M. BERNHARD[3], AND JAMES P. BARRY[1]
[1]*Monterey Bay Aquarium Research Institute, 7700 Sandholdt Road, Moss Landing, CA 95039, USA*
[2]*Louisiana Universities Marine Consortium, 8124 Hwy. 56, Chauvin, LA 70344, USA*
[3]*Geology and Geophysics Department, Woods Hole Oceanographic Institution, MS #52, Woods Hole, MA 02543, USA*

1. Introduction

"Dead" or hypoxic zones are defined as the seafloor and the water column immediately above the seafloor, with oxygen concentrations less than 2 mg l^{-1}. Persistent and seasonal "dead" zones have been recognized for decades to centuries (Diaz and Rosenberg 1995). The low oxygen concentrations in dead zones result from the aerobic decay of locally derived primary productivity that is fueled by anthropogenic nutrient increases (e.g., fertilizer runoff). The term "dead" in this context refers to the fact that macro/megafauna (e.g., metazoans living in or on the surface of the sediment >300 μm in size) that cannot move away from these low oxygen concentrations succumb to prolonged hypoxia as well as to fish and shrimp that will avoid hypoxic regions (Rabalais et al. 2002). The resultant downturn of coastal and continental shelf fisheries in some of the roughly 400 hypoxic zones or episodes is well documented (Diaz and Rosenberg 2008). Some portion of the infaunal assemblage (micro-and meiofauna which are protists and metazoans in the <300-μm size class) has been documented as surviving the persistence of hypoxia (Murrell and Fleeger 1989; Sawyer et al. 1997).

Other benthic habitats with persistent hypoxia, not attributable to anthropogenic sources, have been known for decades. The continental shelf off eastern boundary current upwelling regimes (e.g., continental shelf of Peru and Chile), silled basins (Santa Barbara Basin), and cold seeps (Monterey Bay) are examples of natural hypoxic regions. The presence of a well-established oxygen minimum zone in the water column between 300- and 1,000-m water depth is important to the existence of these natural hypoxic zones. Even with the lately noted spread of "dead" zones, naturally occurring hypoxic zones are probably greater in a real extent (e.g., the Peruvian and Chilean continental shelf) (Gallardo and Espinoza 2007). There are also shallow natural hypoxic zones that have had their infaunal assemblages quantitatively documented (Bernard and Fenchel 1995; Levin et al. 2009).

Both "dead" zones and natural hypoxic zones are on the low end of the spectrum of oxygen concentrations from fully saturated to anoxia found in marine benthic environments. The diversity and abundance of mega- and macrofaunal organisms scale to this spectrum of such a critical factor for aerobic life. Smaller organisms (e.g., micro- and meiofauna) thrive at the lower end of this spectrum (Bernhard et al. 2000). Analyzing and comparing the smaller components of infauna from these environments, near the lower end of the oxygen spectrum allows extrapolation to possible community structure in anoxic/hypoxic environments, in the past, when they were more widespread (Canfield et al. 2006). We have had an ongoing program looking at the infauna from some of these natural hypoxic zones on the west coast of North America (Buck and Barry 1998; Bernhard et al. 2000; Bernhard and Buck 2004; Buck and Bernhard 2002). Here we compare and contrast results regarding natural hypoxic zones we have been working on with the so-called "dead" zone of the Gulf of Mexico.

2. Methods

The three areas (Table 1) sampled are the "dead" zone in the Gulf of Mexico (Rabalais and Turner 2001), the silled basin off Santa Barbara (Bernhard et al. 2000), and a cold seep community of Monterey Bay (Buck and Barry 1998). We took sediment with various samplers and subsampled a quantitative amount of sediment to a depth of 1 cm, preserved this with cacodylate-buffered glutaraldehyde, and prepared the samples using a protocol described in Starink et al. (1994). Briefly, a density gradient of colloidal silica (Percoll) and filtered seawater is layered with sediment/preservative and centrifuged. The sediment sinks while the organisms remain suspended. The supernatant is filtered, stained with a nuclear stain, and enumerated with an epifluorescence microscope. Organisms are placed into functional groups, measured and converted to biovolume and subsequently carbon using standard algorithms. Oxygen concentrations were determined using a Hydrolab CTD (Gulf of Mexico), an ROV-mounted Sea-Bird Electronics CTD (Monterey Bay), or the Micro-Winkler method of Broenkow and Cline (1969) for Santa Barbara Basin.

Table 1. Samples taken to compare "dead" and "natural" hypoxic zones.

Location	Depth (m)	Oxygen (μM)
"Dead" zone GOM	11–18	5.3–5.9
Monterey Bay seep	911	29
Santa Barbara Basin	600	1

Oxygen concentrations are in the water overlying the sediment
GOM Gulf of Mexico

Samples destined for electron microscopy emanated from aliquots of quantitative samples. Following sediment grain separation, organisms were rinsed of Percoll, concentrated by centrifugation, osmicated, dehydrated, and either embedded for transmission electron microscopy (TEM) sectioning or osmicated, dehydrated, and critical point dried for scanning electron microscopy (SEM).

3. Results and Discussion

The quantitative analysis of 0–1-cm sediment samples from these three disparate hypoxic regions reveals three different meiofaunal/protistan assemblages (Fig. 1). The "dead zone" is dominated by flagellates and ciliates. While epi- or endobionts were associated with a substantial portion (23–41%) of flagellates, mostly euglenozoa (Buck et al. 2000), they were associated with the vast preponderance (>90%) of the ciliates (Figs. 1–3). The Monterey Bay cold seeps, while hosting biomass roughly equivalent to that from the "dead zone," had assemblages dominated by nematodes. Flagellates and ciliates, both with and without symbionts, are present but in low biomass (Figs. 1–3).

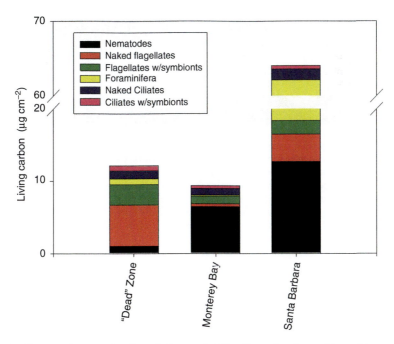

Figure 1. Biomass of nematodes, foraminifera, naked flagellates, flagellates with symbionts, naked ciliates, and ciliates with symbionts from the top 1 cm of three hypoxic zones. Note the scale break for the Living Carbon axis (The data from the cold seep and SBB emanate from samples that comprise part of the datasets presented in Buck and Barry (1998) and Bernhard et al. (2000)).

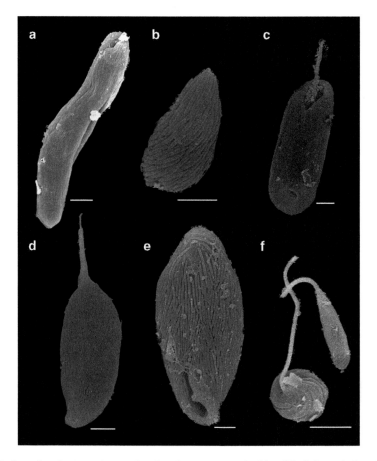

Figure 2. Scanning electron micrographs of euglenozoa covered with epibiotic bacteria from hypoxic zones. (**a, b**) Euglenozoa from Monterey Bay seep site. (**a**) Ventral view with the insertion point for the flagella visible at the anterior end of the cell. (**b**) Dorsal view of a cell. (**c**) Ventral view of a cell from the "dead" zone with interruptions in the bacterial coat at the flagella insertion and at the posterior end. (**d–f**) Euglenozoa from Santa Barbara Basin. (**d**) Ventral view of a *Calkinsia*-like cell. (**e**) Dorsal view of a cell showing the deep flagellar insertion channel. (**f**) Two euglenozoa with bacteria found only at the posterior ends of the cell. Scale bar = 10 μm except for b where it is = 1 μm.

In contrast to the "dead zone" and the Monterey Bay cold seep, the assemblage biomass from Santa Barbara Basin was approximately an order of magnitude higher and dominated by foraminifera (Fig. 1). Many of the foraminifera from Santa Barbara Basin possess prokaryotic endosymbionts (Bernhard et al. 2000), while those foraminifera from Monterey Bay cold seeps are usually without symbionts (Bernhard et al. 2001). Flagellates and ciliates from Santa Barbara Basin were usually associated with symbionts (Figs. 1–3; Bernhard et al. 2000; Edgcomb et al. 2010). One other study of a shallow hypoxic site that quantifies the protistan and nematode ele-

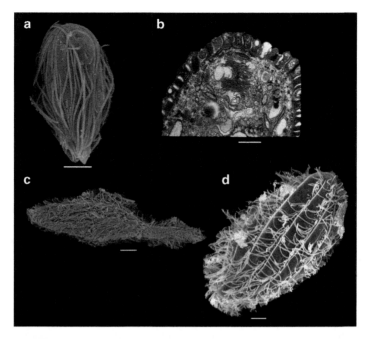

Figure 3. Electron micrographs of euglenozoa and ciliates covered with epibiotic bacteria from hypoxic zones. ns. (**a, c, d**) Scanning electron micrographs, (**b**) transmission electron micrograph. (**a**) Euglenozoa from Santa Barbara Basin with elongated bacteria loosely attached to the cell. (**b**) Longitudinal section through the posterior end of an euglenozoa from Monterey Bay seep showing the closely associated bacteria. (**c, d**) Ciliates from Santa Barbara Basin. (**c**) Ciliate with numerous large epibionts. (**d**) Ciliate with epibionts along the longitudinal furrows close to the flagellar insertion points. Scale bars = 10 μm, except for b where it = 1 μm.

ments is that of Bernard and Fenchel (1995) from Niva Bay, Denmark (Table 2). They report the lowest concentrations of flagellates and ciliates but the highest concentration of nematodes (Table 2).

Both the Monterey cold seep and the Santa Barbara Basin sample were substantially different in biomass and/or assemblage composition from nearby control sites (data not presented), suggesting that the natural hypoxic sites' assemblages are responding to environmental cues unique to them (Buck and Barry 1998; Bernhard et al. 2000). Due to the widespread nature of the "dead" zone, the comparison with temporally appropriate unaffected control sites is outside the realm of this study.

The suggestion that "dead zones" are anything but dead is not a new one. The realization that prokaryotes would thrive in oxygen levels as low as 2 mg l^{-1} is not particularly profound either since they have been documented as thriving at much lower O_2 concentrations (e.g., Jørgensen and Gallardo 1999). Metazoan meiofauna (e.g., nematodes and harpacticoid copepods) have been documented

Table 2. Abundance (individuals cm^{-2}, 1 cm deep) of major infaunal constituents from the "dead" zone, the two natural hypoxic zones, and a shallow hypoxic site in Niva Bay (Bernard and Fenchel 1995).

	Hypoxic zone (GOM)	Cold seep (Monterey Bay)	Santa Barbara Basin	Niva Bay
Flagellates	157,084	35,209	118,000	8,700
Ciliates	1,324	1,288	943	851
Nematodes	165	196	76	338

The data from the cold seep and SBB emanate from samples that comprise part of the datasets presented in Buck and Barry (1998) and Bernhard et al. (2000)

to decrease in abundance in "dead zones" (Murrell and Fleeger 1989), and our findings support their conclusion (Fig. 1). However, the biomass and diversity of the micro- and nanofauna from the "dead zone" have not previously been documented. Silled basins, such as Santa Barbara Basin, with their high biomass and high protistan diversity (Bernhard et al. 2000) are characterized by even lower oxygen concentrations than the "dead zone"; hence, the presence of abundant and diverse assemblages in the "dead zone" is not unexpected. The continental shelf of the eastern South Pacific is another naturally occurring hypoxic site that hosts prokaryotes as well as micro- and nanofauna (Gallardo and Espinosa 2007). Although there are no direct quantitative data from these regions, one might reasonably expect rich micro- and nanofauna to be present.

There are a number of factors that potentially promote the assemblages of small-sized (e.g., micro- and meiofauna) organisms associated with "dead" zones and natural hypoxic zones. Principally, the small body size of the micro- and nanofauna allows them to thrive in low oxygen concentrations because molecular diffusion is the operative mechanism for obtaining oxygen at small sizes (Fenchel and Findlay 2008). Combine this advantage with hypometabolism and perhaps an elevated abundance of bacteria in the dead zones, and there would be an environment well suited for small-sized organisms. The dearth of larger infauna/predators may make it less likely for smaller organisms to be subjected to predation.

We can only conjecture about the impact of this micromeiofaunal assemblage upon the "dead" zone. This assemblage may consume oxygen, further decreasing its concentration at the sediment/water interface. The "microbial loop" that the presence of micro- and meiofauna implies may be important to the onset and duration of the "dead" zone. Results from water column heterotrophic nanoflagellate experiments in hypoxic conditions indicate that this size class of organism is unaffected by lower O_2 concentrations and maintains predatory pressure upon bacteria (Park and Cho 2002). The robust biomass and functional group diversity common to these hypoxic site assemblages indicate that these types of environments are oases (Bernhard et al. 2000) rather than dead zones for protistan life and that symbiotic relationships are common (Edgcomb et al. 2010). The highest biomass and greatest percentage of organisms with symbionts were found at the lowest oxygen concentration (e.g., Santa Barbara Basin; Tables 1 and 2, Fig. 1).

While our findings generally support the treatise that foraminifera and nematodes are dominants in hypoxic benthos (Levin et al. 2002), the presence of other small organisms (e.g., microfauna) are also substantial biomass contributors and may display a disproportionate metabolic effect due to their small size. In addition, the "dead" zone fauna were virtually bereft of nematodes and foraminifera, perhaps indicating an intrinsic difference between seasonal and persistent hypoxic zones.

We have not attempted to make this a comprehensive treatise on the variability of the types of hypoxic environments found at the sediment/water interface nor do we suggest that these three disparate environments are a comprehensive treatment of all hypoxic zones. Rather, we have presented examples from three well-known, geographically distinct hypoxic zones that suggest an inherent diversity in the assemblages associated with hypoxic zones. The enhanced presence of paleontologically significant components such as foraminifera in some extant hypoxic environments we present (e.g., Santa Barbara Basin) may be important for studying ancient communities; however, substantial parts of these assemblages are not preserved.

Finally, preliminary reports indicate decreased water column oxygen concentrations associated with bacteria metabolizing the oil from the Deepwater Horizon oil spill (Biello 2010). It is likely that these hypoxic environments in the Gulf of Mexico will only expand, making the understanding of the organisms that comprise them more critical than ever.

4. Acknowledgments

We thank the captain, crew, and scientific staff of the research vessel *Pelican* for assistance in collection of specimens and the National Oceanographic and Atmospheric Administration for funding under grant number NA06NPS4780197. This is contribution number 135 of the NOAA, Center for Sponsored Coastal Ocean Research program.

5. References

Bernard C, Fenchel T (1995) Mats of colorless sulphur bacteria. II. Structure, composition of biota and successional patterns. Mar Ecol Prog Ser 128:171–179

Bernhard JM, Buck KR (2004) Eukaryotes of the Cariaco, Soledad, and Santa Barbara Basins: protists and metazoans associated with deep-water marine sulfide-oxidizing microbial mats and their possible effects on the geologic record. In: Amend JP, Edwards KJ, Lyons TW (eds) Sulfur biogeochemistry – past and present, Geological Society of America Special Paper 379. Geological Society of America, Boulder, pp 35–47

Bernhard JM, Buck KR, Farmer MA, Bowser BB (2000) The Santa Barbara Basin is a symbiosis oasis. Nature 403:77–80

Bernhard JM, Barry JP, Buck KR (2001) Monterey Bay cold-seep biota: assemblages, abundance, and ultrastructure of living foraminifera. Deep-Sea Res I 48:2233–2249

Biello D (2010) Biological breakdown. Sci Am 303:14–17

Buck KR, Barry JP (1998) Monterey Bay cold seep infauna: quantitative comparison of bacterial mat meiofauna with non-seep control sites. Cah Biol Mar 39:333–335

Buck KR, Bernhard JM (2002) Protistan-prokaryotic symbioses in deep-sea sulfidic sediments. In: Seckbach J (ed) Symbioses. Kluwer, Dordrecht, pp 507–517

Buck KR, Barry JP, Simpson AGB (2000) Monterey Bay cold seep biota: Euglenozoa with chemoautotrophic bacterial epibionts. Eur J Protistol 36:117–126

Canfield DE, Rosing MT, Bjerrum C (2006) Early anaerobic metabolisms. Philos Trans R Soc B Biol Sci 361:1819–1834

Diaz RJ, Rosenberg R (1995) Marine benthic hypoxia: a review of ecological effects and the behavioural responses of benthic macrofauna. Oceanogr Mar Biol Annu Rev 33:245–303

Diaz RJ, Rosenberg R (2008) Spreading dead zones and consequences for marine ecosystems. Science 321:926–929

Edgcomb VP, Breglia SA, Yubuki N, Beaudoin D, Patterson DJ, Leander BS, Bernhard JM (2010) Identity of epibiotic bacteria on symbiontid euglenozoans in O_2-depleted marine sediments: evidence for symbiont and host co-evolution. Int Soc Microb Ecol XX:1–13

Fenchel T, Finlay B (2008) Oxygen and the spatial structure of microbial communities. Biol Rev 83:553–569

Gallardo VA, Espinoza C (2007) New communities of large filamentous sulfur bacteria in the eastern South Pacific. Int Microbiol 10:97–102

Jørgensen BB, Gallardo VA (1999) *Thioploca* spp.: filamentous sulfur bacteria with nitrate vacuoles. FEMS Microbiol Ecol 28:513–518

Levin LA, Gutiérrez D, Rathburn A, Neira C, Sellanes J, Muñoz P, Gallardo V, Salamanca M (2002) Benthic processes on the Peru margin: a transect across the oxygen minimum zone during the 1997–98 El Niño. Prog Oceanogr 53:1–27

Levin LA, Ekau W, Gooday AJ, Jorissen F, Middelburg JJ, Naqvi SWA, Neira C, Rabalais NN, Zhang J (2009) Effects of natural and human-induced hypoxia on coastal benthos. Biogeoscience 6:2063–2098

Murrell MC, Fleeger JW (1989) Meiofauna abundance on the Gulf of Mexico continental shelf affected by hypoxia. Cont Shelf Res 9:1049–1062

Park JS, Cho BC (2002) Active heterotrophic nanoflagellates in the hypoxic water-column of the eutrophic Masan Bay. Korean Mar Ecol Prog Ser 230:35–45

Rabalais NN, Turner RE (2001) Hypoxia in the northern Gulf of Mexico: description, causes, and changes. In: Rabalais NN, Turner RE (eds) Coastal and estuarine studies: coastal hypoxia, consequences for living resources and ecosystems. AGU, Washington, DC

Rabalais NN, Turner RE, Wiseman WJ (2002) Gulf of Mexico hypoxia, aka "The dead zone". Annu Rev Ecol Syst 33:235–263

Sawyer TK, Nerad TA, Rabalais NN, McLaughlin SM (1997) Protozoans isolated from Louisiana shelf sediments subject to hypoxia/anoxia with emphasis on freshwater amoebae and marine flagellates. Bull Mar Sci 61:859–867

Starink M, Bar-Gilissen MJ, Bak RP, Cappenberg TE (1994) Quantitative centrifugation to extract benthic protozoa from freshwater sediments. Appl Environ Microbiol 60:167–173

Biodata of **Michael Stachowitsch**, **Bettina Riedel**, and **Martin Zuschin**, authors of *"The Return of Shallow Shelf Seas as Extreme Environments: Anoxia and Macrofauna Reactions in the Northern Adriatic Sea."*

Dr. Michael Stachowitsch received his Ph.D. (1980) from the University of Vienna (B.Sc., University of Pittsburgh) and has been working in the Northern Adriatic Sea since 1974, initially on hermit crab symbioses, then on marine snow events, anoxia, and other disturbances to sublittoral benthic communities.

E-mail: **stachom5@univie.ac.at**

Dr. Bettina Riedel obtained her Ph.D. in 2009 and is currently PostDoc at the Department of Marine Biology (University of Vienna). Her research interests include coastal and estuarine benthic ecology and anthropogenic threats to biodiversity (eutrophication, anoxia, fishing activities, climate change) and bioindication.

E-mail: **bettina.riedel@univie.ac.at**

Professor Martin Zuschin is Associate Professor at the Department of Palaeontology of the University of Vienna. His research interests are in benthic ecology and diversity through the Cenozoic and the evolutionary ecology and paleoecology of the marine biosphere.

E-mail: **martin.zuschin@univie.ac.at**

Michael Stachowitsch   **Bettina Riedel**   **Martin Zuschin**

THE RETURN OF SHALLOW SHELF SEAS AS EXTREME ENVIRONMENTS: ANOXIA AND MACROFAUNA REACTIONS IN THE NORTHERN ADRIATIC SEA

MICHAEL STACHOWITSCH[1], BETTINA RIEDEL[1], AND MARTIN ZUSCHIN[2]

[1]Department of Marine Biology, University of Vienna, Althanstrasse 14, 1090 Vienna, Austria
[2]Department of Paleontology, University of Vienna, Althanstrasse 14, 1090 Vienna, Austria

1. The Problem

It is a fundamental dogma of ecology that ecosystems change. The long-standing efforts center around detecting, quantifying (rates and directions), and determining the drivers of that change. Today, the focus is often on distinguishing which changes are "natural" and which are "unnatural," whereby the former are presumably acceptable, and the latter, anthropogenically driven changes, may require remedial action.

Shallow marine environments can serve as a model for this package of issues. Why? Small-scale, gradual change is an omnipresent feature of both marine and terrestrial ecosystems today. Complete, sudden and larger-scale collapses, however, are rare in the marine environment. Such worst-case ecosystem degradation scenarios are tangible, irrefutable case studies of major ecosystem shifts. Here, there can be no doubt about whether change is taking place and no debate about whether the endpoint is a matter of concern. Precisely this scenario is increasingly being encountered in connection with hypoxia/anoxia. It involves coastal eutrophication driven by nutrient loading from point and nonpoint sources, coupled with losses of filtering and buffering capacity of vegetation and suspension feeders (Zhang et al. 2010). Worldwide, anthropogenic eutrophication promotes this phenomenon (Diaz and Rosenberg 2008; Breitburg et al. 2009; Gooday et al. 2009; Rabalais et al. 2010). No other human impacts or pollution sources – oil, heavy metals, organohalogens, radioactivity, marine debris – have this impact, namely, complete and widespread ecosystem collapse characterized by extensive mass mortalities of all key species (Gray et al. 2002; Wu 2002; Sala and Knowlton 2006; Vaquer-Sunyer and Duarte 2008; Levin et al. 2009). Accordingly, shallow coastal seas that suffer hypoxic and anoxic conditions represent the most stressful habitat conceivable, much more so than any "stably extreme" habitat, i.e., characterized by extreme yet constant environmental conditions.

Mass mortalities are a prime example – on a large scale and at many levels – of how benthic communities and the overall system respond to perturbation. So-called dead zones are the ultimate manifestation and symptom of this phenomenon. The term dead zone was coined in the late 1980s to describe large-scale midsummer hypoxic/anoxic areas, known since the early 1970s and mapped since 1985 in the Gulf of Mexico (Rabalais et al. 2002; Rabalais et al. 2007). It now refers, in general, to areas suffering intermittent, periodic, or permanent hypoxia and anoxia leading to depauperated benthic communities. Today, nearly 415 eutrophic and hypoxic coastal systems have been identified worldwide: of these, 169 are hypoxic, 233 are areas of concern, and 13 are systems in recovery (Diaz and Rosenberg 2008; World Resources Institute, www.wri.org/map/world-hypoxic-and-eutrophic-coastal-areas). Recognized dead zones range from Scandinavian and Baltic waters in Europe to the Chesapeake Bay and the Gulf of Mexico in the USA to Japanese waters. The frequency and extension of hypoxic and anoxic zones are increasing worldwide (Diaz 2001), and dead zones in the world's oceans are at the top of the list of emerging environmental challenges (UNEP 2004). Moreover, the phenomenon appears to be spreading beyond shallow, semi-enclosed seas (Gilbert et al. 2009). The recent intensification of severe inner-shelf hypoxia and the novel rise of water-column anoxia in larger ecosystems (California Current Large Marine Ecosystem: CCLME) reflect this trend (Levin 2003; Chan et al. 2008).

2. The Model System

The Northern Adriatic Sea is a model for this phenomenon. It combines many features known to be associated with low dissolved oxygen (DO) events (Stachowitsch and Avcin 1988): it is semi-enclosed, shallow (<50 m), and is characterized by soft bottoms, a high riverine input (mainly from the Po River), high productivity, and long water residence times (Ott 1992). As elsewhere in the northern hemisphere, this constellation can be associated with seasonal hypoxia and anoxia in late summer/early fall.

In general, hypoxia and anoxia can develop in the Northern Adriatic during late summer through a combination of certain meteorological and hydrological conditions such as calm weather and water stratification (Franco and Michelato 1992; Malej and Malačič 1995). Low DO events and benthic mortalities have been noted here periodically for centuries (Crema et al. 1991), but their frequency and severity have markedly increased during recent decades. High anthropogenic input of nutrients into the Northern Adriatic (Justić et al. 1995; Danovaro 2003; Druon et al. 2004) has led to a higher production and deposition of organic matter to the bottom than the available oxygen supply to allow its decomposition (Rabalais and Turner 2001). The average long-term decrease in water body transparency here over the twentieth century, accompanied by a historical decrease in bottom DO since the early twentieth century, has been outlined by Justić (Justic et al. 1987; Justic 1987). Since the 1980s, severe oxygen deficiencies have been reported here regularly (Hrs-Brenko et al. 1994; Stachowitsch and Fuchs 1995; Penna et al. 2004). The impacted areas range from restricted zones (several square kilometers; Stachowitsch 1992) to approx. 250 km^2 (Faganeli et al. 1985) to 4,000 km^2 (Stefanon and Boldrin 1982; D. Degobbis, personal communication 1995), ultimately affecting every region of the Northern Adriatic (Fig. 13 in Ott 1992). Such mortalities were often accompanied and promoted by severe marine snow events (Stachowitsch et al. 1990).

The soft bottoms in the Northern Adriatic feature conspicuous, widely distributed macroepifaunas (Fedra 1978; Zuschin et al. 1999). These consist largely of decimeter-scale, interspecific, high-biomass aggregations termed multi-species clumps (Fedra et al. 1976), or bioherms: one or more shelly hard substrates provide the base for sessile, suspension-feeding colonizers (mostly sponges, ascidians, anemones, or bivalves), which in turn serve as an elevated substrate for additional vagile and hemi-sessile organisms (e.g., brittle stars, crabs, and sea cucumbers) (Zuschin and Pervesler 1996). The presence of a well-developed infauna is expressed in the early designations of the benthic communities here (*Schizaster chiajei*-community) based on infauna-oriented sampling methods (grabs) by Vatova (1949) and later authors (Gamulin-Brida 1967; Orel and Mennea 1969; Orel et al. 1987; Occhipinti-Ambrogi et al. 2002). An underwater TV-camera sled (Machan and Fedra 1975; Fedra 1978), combined with SCUBA diver taken samples and in situ experimentation, led to the distinction of a predominant, wide-ranging epifauna-based benthic community named the ORM community based on the biomass dominants, the brittle star *Ophiothrix quinquemaculata*, the sponge *Reniera* sp., and the ascidians *Microcosmus* spp. The mean biomass, measured as wet weight, amounted to 370 (±73) g/m^2 (Fedra et al. 1976).

The early TV-camera sled work encountered a large zone of decaying benthic organisms in the central Gulf of Trieste, Northern Adriatic Sea, in 1974 (Fig. 1 in Fedra et al. 1976). Oxygen deficiency was postulated for this "graveyard" phenomenon. Over a 2-week period in September 1983, the course of a mass mortality event in the Gulf of Trieste was more comprehensively documented in situ (Stachowitsch 1984). Within 2–3 days, all sponges and the brittle star *Ophiothrix quinquemaculata*, which together make up over 60% of the community biomass, were dead. The death of sponges, key components of multi-species clumps, killed sponge-dwellers and other bioherm-associated species. This was accompanied by the emergence of the complete spectrum of macroinfauna organisms, including (in approximate order of emergence) holothurians, burrowing shrimp, sea urchins, polychaetes, sipunculids, and bivalves. A spectrum of stress behaviors – involving avoidance patterns and directed toward reaching more oxygenated water layers – was recorded from arm-tipping in brittle stars and swimming in *Squilla mantis* to hermit crabs emerging from their gastropods shells. Many motile organisms aggregated on higher substrates (sediment mounds) and sessile organisms extended certain body parts higher up into the water column (e.g., the tentacle crowns of sea anemones). Mortality was very rapid. Within 2 days, an estimated 90% of the biomass had died. Certain elements of the fauna apparently survived, including the gastropod *Hexaplex trunculus* and the sea anemone *Ragactis pulchra*.

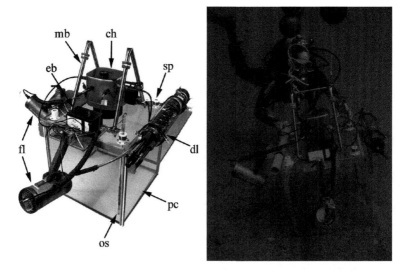

Figure 1. The Experimental Anoxia Generating Unit (*EAGU*) consisting (clockwise from the top) of camera housing (*ch*), sensor ports (*sp*), datalogger (*dl*), plexiglass chamber (*pc*), oxygen sensor (*os*), flashes (*fl*), external batteries (*eb*), and protective metal brackets (*mb*). Right photo: the chamber is positioned over selected macrofauna organisms, and anoxia is generated by respiration within the sealed-off water body.

3. The Experimental Solution

One might conclude that the severity of the problem means extensive available documentation and full insight into the processes. This is not the case because the precise onset of hypoxia/anoxia is related to local weather and hydrological conditions and continues to be difficult to predict. Moreover, the course of mortality is often rapid, within a few days. The result is a patchy narrative of low DO events. Laboratory experiments on responses to decreasing oxygen concentrations also have drawbacks: while conditions can be precisely controlled, they cannot reveal natural behavioral responses, intra- and interspecies interactions, mortality sequences, and community-level processes. One strategy to overcome these methodological shortcomings and to help close the gap in our knowledge is in situ experimentation (Fig. 1).

The approach involves inducing and fully documenting the course of small-scale experimental anoxia on the seafloor. The University of Vienna's Experimental Anoxia Generating Unit (EAGU) combines photodocumentation with detailed chemo-physical analyses and allows the behaviors and mortalities of benthic organisms to be analyzed from the onset (Stachowitsch et al. 2007). This 50 × 50 × 50 cm plexiglass unit is designed as a chamber with a detachable instrument lid that bears a digital camera, flashes, a sensor array (oxygen, H_2S, temperature, and pH), and a datalogger.

EAGU can be deployed in an open configuration (aluminum frame that permits water exchange) to document normal behavior or in a closed configura-

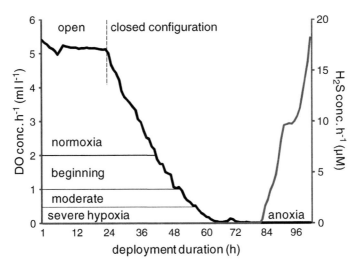

Figure 2. Dropping oxygen values (*dark line*) after deployment of closed configuration, and increasing H_2S values after generation of anoxia. Five dissolved oxygen thresholds were distinguished.

tion (plexiglass chamber) that seals the enclosed area off and induces hypoxia and anoxia. A typical deployment consists of a 24-h open configuration followed by ca. 3 days in the closed configuration (until anoxia is generated, mortality of most species is reached, and the initial behaviors of the most resistant species documented).

The EAGU concept has yielded a total of more than a dozen experiments. The system induced anoxia within 2 d in every case, followed by increasing H_2S values (Fig. 2). Photographs in 6-min intervals and sensor values in 1-min intervals provided a detailed picture of responses and a correlation with oxygen and H_2S. The images are processed into time-lapse sequences and also analyzed image by image. The system triggered responses in both the macroepifauna and macroinfauna. On one level, the observed responses and their sequence corresponded to those of earlier reports here and elsewhere, e.g., arm-tipping in brittle stars (Figs. 3 and 4), the upward movement of mobile species to higher levels on available substrates (Haselmair et al. 2010), and the emergence of infauna (Riedel et al. 2008b).

Thus, the EAGU recreated conditions experienced during typical hypoxia and anoxia events and provided previously unknown detail. The emergence patterns and relative emergence times of infauna individuals became visible, as did speed and distance of post-emergence movement. The system also captured species (e.g., infaunal shrimp) whose in situ behavior has never been documented before, either under normoxic or anoxic conditions. On the community level, a complete range of parallel and successive responses were combined in the individual experiments (Fig. 5).

One unexpected phenomenon was predation of brittle stars by sea anemones (Riedel et al. 2008a). In three deployments, five predatory events were recorded

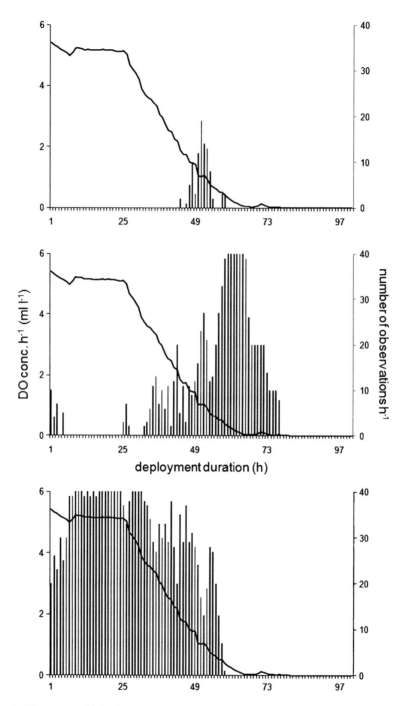

Figure 3. Three stages of behavioral responses of the brittle star *Ophiothrix quinquemaculata* to oxygen depletion: suspension-feeding posture (*top*), arm-tipping (*middle*), and clinging to substrate.

Figure 4. Three stages of behavioral responses of the brittle star *Ophiothrix quinquemaculata* to oxygen depletion. Top: normal suspension-feeding posture with arms outstretched. Middle: arm-tipping with central disc held elevated above the substrate. Bottom: clinging flat on the substrate. In a final stage (see Fig. 5b), the brittle stars often fall onto the sediment surface and may overturn.

involving four anemones (three *Cereus pedunculatus* and one *Calliactis parasitica*). Under near anoxic conditions, the anemones made contact with, pulled in, and consumed the brittle stars. The duration of each predatory event lasted from 1 to 7.5 h, and regurgitation was recorded in three of the five events. This phenomenon

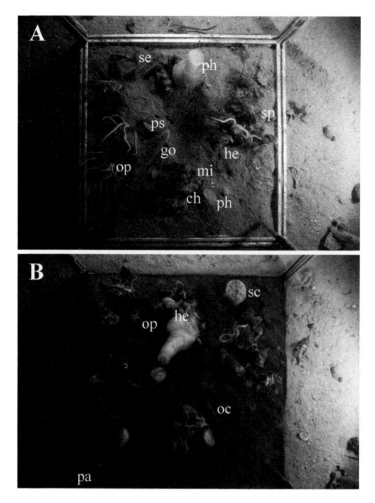

Figure 5. (a) Initial deployment phase – open configuration. (b) Final image of closed configuration after anoxia. Organism abbreviations: ch, the bivalve *Chlamys varia*; go, the fish *Gobius niger*; he, the gastropod *Hexaplex trunculus*; mi, the ascidian *Microcosmus* sp.; oc, the sea cucumber *Ocnus planci*; op, the brittle star *Ophiothrix quinquemaculata*; ph, the ascidian *Phallusia mammilata*; ps, the sea urchin *Psammechinus microtuberculatus*; sc, the infaunal sea urchin S*chizaster canaliferus*; se, serpulid tubeworm. (Note darkened sediment, brittle stars dead on sediment, and emerged sea urchins sc: three individuals from top right to bottom left). See www.marine-hypoxia.com for video version.

was interpreted to reflect the different tolerances and behaviors of the species experiencing oxygen deficiency: the anemones are not only highly resistant to anoxia but apparently also benefit by taking advantage of more vulnerable prey. Thus, in a narrow window of opportunity, the combined increased activity of anemones, coupled with moribund brittle stars, increased predatory efficiency. This may help explain why sea anemones are often a dominant element in the composition of post-mass mortality benthic communities in the Northern Adriatic Sea.

4. The Outlook

Within the lifetime of a single generation of marine biologists, the status of the marine environment has changed dramatically, yielding new "extreme" habitats; recent mapping efforts show that nearly half of all marine habitats are strongly affected by multiple threats (Halpern et al. 2008). Overall, the greatest cumulative impact is on coastal ecosystems (Lotze et al. 2006) – the very systems we rely on for most of our marine goods and services (Worm et al. 2006). They will experience the largest change in biodiversity should present trends in human activity continue (Jenkins 2003). Conservation efforts typically focus on higher trophic levels, and restoring ecosystem structure and function remains an unattainable goal (Lotze et al. 2006).

The rise in new "extreme" habitats is reflected in the current status of overall research – a paradigm shift from basic biological/ecological to conservation-related issues. The traditional research landscape has been fundamentally re-aligned in an attempt to identify and understand perturbed systems, prevent further losses, restore impacted habitats, as well as preserve what remains intact. The four eras of marine research identified by Rupert Riedl (1980) – namely, that of seafarers, of oceanographic expeditions, of marine stations, and of field research – have been amended to include a fifth era (Stachowitsch 2003) devoted to studying deteriorated ecosystems. It differs from earlier eras in its underlying conservation aim, narrower scientific focus, and increasing societal component, ever more centralized and directed funding, greater urgency, and in the reduced validity/interpretability of studying damaged objects. Correspondingly, an ever larger percentage of published papers has a conservation bent (for a literature analysis regarding cetaceans, for example, see Rose et al. 2011), and numerous new journals have a clear conservation focus both in title and content.

The rise of new "extreme" environments has also affected benthic research. One trend has been to bring scientists ever closer to their object of study, i.e., to the bottom as divers or with remotely operated or autonomous instruments. The EAGU experimental approach circumvents both the time limitations of direct diver observations and the unlikelihood of being on-site and fully equipped at critical phases of hypoxia/anoxia events. It successfully mimicked the biological processes observed during previous anoxia in the Northern Adriatic Sea. Moreover, many new organisms could be studied at a new level of detail, including the correlation of all behaviors and mortalities with specific oxygen thresholds. The experimental results are valid both from a broader temporal and spatial perspective. The short temporal scale of the deployments, i.e., the generation of hypoxia, anoxia, behavioral modifications, and mortalities within a few days, agrees well with the rapid course of events observed in a previous mortality here ("rapid death, slow recovery": Stachowitsch 1991, 1984). This approach therefore captures almost the full range of community responses, albeit with potential deficits in documenting the final responses of the most tolerant species. The small spatial scale also yielded more broadly applicable results. There are no larger-scale factors here (other adjoining habitats, greater structural heterogeneity, or relief features) that would enable alternative behaviors and escape reactions or directions. The relative habitat homogeneity, coupled with the sessile and hemi-sessile life habits of most filter- and suspension-feeding species – and the restricted movements of common mobile forms such as hermit crabs – means that the phenomena observed in the 50 × 50 cm chambers reflect those at much larger scales.

At the same time, the EAGU has recently been deployed to document immediate post-anoxia developments. The first time-lapse results show the quick arrival of a distinct succession of scavenging organisms: initially fish, followed by hermit crabs, and finally gastropods. The issue of scale will play a greater role in interpreting the EAGU-based recovery results than the respective behavior-generating results. In larger-scale anoxia in the Adriatic and elsewhere, the greater distances involved will considerably hinder the immediate immigration of all but the largest and most mobile scavengers and predators. Nonetheless, the sequence of arrival would probably reflect the one we observed above in the films, with fish arriving first. A first look at the EAGU-related longer temporal aspects of recovery (revisiting deployment "plots" at intervals of months or a year) confirms the "slow recovery" aspect after community destruction: no obvious recolonization has taken place. This aspect is still being evaluated but, clearly, in larger dead zones, the relative contribution of immigration to repopulation will shift more toward other strategies (larval settlement).

Ecosystem collapses and slow recoveries will severely impede the role benthic communities play in sublittoral soft bottoms. In the shallow Northern Adriatic, the benthos is not merely a receiving compartment. Complex feedback processes enable the benthic subsystem to control and help dampen oscillations in the pelagic subsystem (Ott 1992). Ott and Fedra (1977) estimated that the suspension feeders here can remove all the suspended material in the water column every 20 days. This is on the same order of magnitude as calculated for the Oosterschelde (Herman and Scholten 1990), Swedish waters (Loo and Rosenberg 1989), the USA (Cloern 1982), and France (Hily 1991). Such communities have therefore been termed as a "natural eutrophication control" (Officer et al. 1982). They help stabilize the entire ecosystem and their loss makes the system more sensitive to perturbations. Other key functional processes for the overall system, such as bioturbation and related sedimentary activity, may also be altered by hypoxia/anoxia and the corresponding loss of biodiversity (Rosenberg 2001; Levin 2002; Solan et al. 2004; Sala and Knowlton 2006).

Shelf anoxias have a long geological history and were particularly abundant during sea level highstands and during periods of warm temperatures. They are considered as a prime reason for mass extinctions in the fossil record (Hallam and Wignall 1999; Hallam 2004). The recognition that anoxia has played a major role in structuring benthic communities in continental shelf environments, in both the past and present, has led early efforts to find a common terminology and to describe parallel phenomena (Tyson and Pearson 1991). In the case of the

Northern Adriatic shelf, for example, recent efforts have been devoted to drawing parallels between Paleozoic stationary suspension feeders on the sediment surface and the above-described modern, epifauna-dominated ORM community (McKinney and Hageman 2006; McKinney 2007). One argument is that this striking epifauna depends on the presence of stable hard substrata on the seafloor, is very sensitive to sediment input by flood events and storm-induced sediment resuspension, and is related to seasonally high productivity (Zuschin and Stachowitsch 2009). Moreover, the elevation of organisms above the sediment-water interface in multi-species clumps or bioherms helps to survive hypoxia – a typical seasonal feature of the Adriatic shelf and of many ancient epeiric seas. Accordingly, the gradual disappearance of large epicontinental seas along with their low sedimentation rates and frequent bottom-water hypoxia during the Mesozoic may have supported the more global replacement of this archaic epifauna by modern, bivalve-dominated infaunas.

The current status of the ORM community makes it unlikely that it fulfills its pre-mortality regulatory capacity. Such impacted communities, however, can provide crucial information. They serve as a long-term memory of disturbance events because they indicate the status on three levels: behavioral responses over the short-term, mortality or survival over the midterm, and recolonization over the long-term. In the case of the Northern Adriatic, benthic communities were designated as "eutrophication's memory mode" (Stachowitsch 1992). Depending on community responses on the behavioral and composition levels, the benthos stores information on prior disturbances over years and even decades. The specific configuration of this memory will depend on the severity of the event, the area affected, and the nature of the recolonization process (e.g., immigration and larval settlement).

The experimental approach will lead to a catalog of species-specific behaviors that can be related to distinct oxygen thresholds. Combined with mortality sequences, this will allow us to better define sensitive and more tolerant (indicator) species. The goal is to synthesize the results into a community-based scenario, providing a generally valid framework for determining the status of benthic communities exposed to hypoxia based on species compositions/functional groups. The reactions of benthic organisms to hypoxia/anoxia are expected to be the same whether natural or pollution-related, ancient or modern. Therefore, the behavioral reactions, mortality sequences, and post-anoxia conditions and developments will provide valuable information for a wide range of relevant disciplines.

5. References

Breitburg DL, Craig JK, Fulford RS, Rose KA, Boynton WR, Brady DC, Ciotti BJ, Diaz RJ, Friedland KD, Hagy JD III, Hart DR, Hines AH, Houde ED, Kolesar SE, Nixon SW, Rice JA, Secor DH, Targett TE (2009) Nutrient enrichment and fisheries exploitation: interactive effects on estuarine living resources and their management. Hydrobiologia 629:31–47

Chan F, Barth JA, Luchenco J, Kirincich A, Weeks H, Peterson WT, Menge BA (2008) Emergence of anoxia in the California large marine ecosystem. Science 319:920

Cloern JE (1982) Does benthos control phytoplankton biomass in South San Francisco Bay? Mar Ecol Prog Ser 9:191–202

Crema R, Castelli A, Prevedelli D (1991) Long term eutrophication effects on macrofaunal communities in Northern Adriatic Sea. Mar Pollut Bull 22(1):503–508

Danovaro R (2003) Pollution threats in the Mediterranean Sea: An overview. Chem Ecol 19:15–32

Diaz RJ (2001) Overview of hypoxia around the world. J Environ Qual 30:275–281

Diaz RJ, Rosenberg R (1995) Marine benthic hypoxia: a review of its ecological effects and the behavioral responses of benthic macrofauna. Oceanogr Mar Biol Annu Rev 33:245–303

Diaz RJ, Rosenberg R (2008) Spreading dead zones and consequences for marine ecosystems. Science 321:926–929

Druon JN, Schrimpf W, Dobricic S, Stips A (2004) Comparative assessment of large-scale marine eutrophication: North Sea area and Adriatic Sea as case studies. Mar Ecol Prog Ser 272:1–23

Faganeli J, Avcin A, Fanuko N, Malej A, Turk V, Tusnik P, Vriser B, Vukovic A (1985) Bottom layer anoxia in the central part of the Gulf of Trieste in the late summer of 1983. Mar Pollut Bull 16:75–78

Fedra K (1978) On the ecology of the North Adriatic Sea. Wide-range investigations on the benthos: the Gulf of Trieste. Memorie di Biogeografia Adriat 9:69–87

Fedra K, Ölscher EM, Scherübel C, Stachowitsch M, Wurzian RS (1976) On the ecology of a North Adriatic benthic community: distribution, standing crop and composition of the macrobenthos. Mar Biol 38:129–145

Franco P, Michelato A (1992) Northern Adriatic Sea: oceanography of the basin proper and the western coastal zone. Sci Total Environ, pp 35–62 (Suppl.)

Gamulin-Brida H (1967) The benthic fauna of the Adriatic Sea. Oceanogr Mar Biol Annu Rev 5:535–568

Gilbert D, Rabalais NN, Diaz RJ, Zhang J (2009) Evidence for greater oxygen decline rates in the coastal ocean than in the open ocean. Biogeosciences Discuss 6:9127–9160

Gooday AJ, Jorissen F, Levin LA, Middelburg JJ, Naqvi SWA, Rabalais NN, Scranton M, Zhang J (2009) Historical records of coastal eutrophication-induced hypoxia. Biogeosciences Discuss 6:2567–2658

Gray JS, Wu RS, Or YY (2002) Effects of hypoxia and organic enrichment on the coastal marine environment. Mar Ecol Prog Ser 238:249–279

Hallam A (2004) Catastrophes and lesser calamities. The causes of mass extinctions. Oxford University Press, Oxford

Hallam A, Wignall PB (1999) Mass extinctions and sea-level changes. Earth Sci Rev 48:217–250

Halpern BS, Walbridge S, Selkoe KA, Kappel CV, Micheli F, D'Agrosa C, Bruno JF, Casey KS, Ebert C, Fox HE, Fujita R, Heinemann D, Lenihan HS, Madin EMP, Perry MT, Selig ER, Spalding M, Steneck R, Watson R (2008) A global map of human impact on marine ecosystems. Science 319:948–952

Haselmair A, Stachowitsch M, Zuschin M, Riedel B (2010) Behaviour and mortality of benthic crustaceans in response to experimentally induced hypoxia and anoxia in situ. Mar Ecol Prog Ser 414:195–208. doi:10.3354/meps08657

Herman PMJ, Scholten H (1990) Can suspension-feeders stabilize estuarine ecosystems? In: Barnes M, Gibson RN (eds.) Trophic relationships in the marine environment. Proceeding of the 24th European Marine Biology Symposium. Aberdeen University Press, pp 104–116

Hily C (1991) Is the activity of benthic suspension feeders a factor controlling water quality in the Bay of Brest? Mar Ecol Prog Ser 69:179–188

Hrs-Brenko M, Medakovic D, Labura Z, Zahtila E (1994) Bivalve recovery after a mass mortality in the autumn of 1989 in the northern Adriatic Sea. Period Biol 96(4):455–458

Jenkins M (2003) Prospects for biodiversity. Science 302:1175–1177

Justic D (1987) Long-term eutrophication of the Northern Adriatic Sea. Mar Pollut Bull 18:281–284

Justic D, Legovic T, Rottini-Sandrini L (1987) Trend in the oxygen content 1911–1984 and occurrence of benthic mortality in the northern Adriatic Sea. Estuar Coast Shelf Sci 25:435–445

Justic D, Rabalais NN, Turner RE (1995) Stoichiometric nutrient balance and origin of coastal eutrophication. Mar Pollut Bull 30:41–46

Levin LA (2002) Oxygen minimum zone influence on the community structure of deep-sea benthos. In: Thurston RV (ed.) Proceedings of the sixth international symposium on fish physiology, toxicology, and water quality, La Paz, Mexico. U.S. Environmental Protection Agency, Ecosystems Research Division, Athens, pp 121–133

Levin LA (2003) Oxygen minimum zone benthos: adaptation and community response to hypoxia. Oceanogr Mar Biol Annu Rev 41:1–45

Levin LA, Ekau W, Gooday AJ, Jorissen F, Middelburg JJ, Naqvi W, Neira C, Rabalais NN, Zhang J (2009) Effects of natural and human-induced hypoxia on coastal benthos. Biogeosciences Discuss 6:3563–3654

Loo LO, Rosenberg R (1989) Bivalve suspension-feeding dynamics and benthic-pelagic coupling in an eutrophicated marine bay. J Exp Mar Biol Ecol 130:253–276

Lotze HK, Lenihan HS, Bourque BJ, Bradbury RH, Cooke RG, Kay MC, Kidwell SM, Kirby MX, Peterson CH, Jackson JBC (2006) Depletion, degradation, and recovery potential of estuaries and coastal seas. Science 312:1806–1809

Machan R, Fedra K (1975) A new towed underwater camera system for wide-range benthic surveys. Mar Biol 33:75–84

Malej A, Malačič V (1995) Factors affecting bottom layer oxygen depletion in the Gulf of Trieste (Adriatic Sea). Annales 7:33–42

McKinney F (2007) The northern Adriatic ecosystem: deep time in a shallow sea. Columbia University Press, New York

McKinney F, Hageman SJ (2006) Paleozoic to modern marine ecological shift displayed in the northern Adriatic Sea: Sedimentary Record 5:4–8

Occhipinti-Ambrogi A, Favruzzo M, Savini D (2002) Multi-annual variations of macrobenthos along the Emiglia-Romagna Coast (Northern Adriatic). PSZN Mar Ecol 23:307–319

Officer CB, Smayda TJ, Mann R (1982) Benthic filter feeding: a natural eutrophication control. Mar Ecol Prog Ser 9:203–210

Orel G, Mennea B (1969) I popolamenti betonici di alcuni tipi di fondo mobile del Golfo di Trieste. Pubbl Staz Zool Napoli 37:261–276

Orel G, Marocco R, Vio E, Del Piero D, Della Seta G (1987) Sedimenti e biocenosi bentoniche tra la foce del Po ed il Golfo di Trieste (Alto Adriatico). Bull Ecol 18:229–241

Ott JA (1992) The Adriatic benthos: problems and perspectives. In: Columbo G, Ferrari I, Ceccherelli VU, Rossi R (eds.) Marine eutrophication and population dynamics, 25th EMBS Olsen & Olsen, Fredensborg, Denmark, pp 367–378

Ott J, Fedra K (1977) Stabilizing properties of a high-biomass benthic community in a fluctuating ecosystem. Helgoländer Meeresunters 30:485–494

Penna N, Capelacci S, Ricci F (2004) The influence of the Po River discharge on phytoplankton Bloom dynamics along the coastline of Pesaro (Italy) in the Adriatic Sea. Mar Pollut Bull 48(3–4):321–326

Rabalais NN, Turner RE (2001) Hypoxia in the Northern Gulf of Mexico: description, causes and change. Coastal and Estuarine Studies 58:1–36

Rabalais NN, Turner RE, Scavia D (2002) Beyond science into policy: Gulf of Mexico hypoxia and the Mississippi River. BioScience 52:129–142

Rabalais NN, Turner RE, Sen Gupta BK, Boesch DF, Chapman P, Murrell MC (2007) Hypoxia in the northern Gulf of Mexico: does the science support the plan to reduce, mitigate, and control hypoxia? Estuaries Coasts 30(5):753–772

Rabalais NN, Diaz RJ, Levin LA, Turner RE, Gilbert D, Zhang J (2010) Dynamics and distribution of natural and human-caused coastal hypoxia. Biogeosciences 7:585–619

Riedel B, Stachowitsch M, Zuschin M (2008a) Sea anemones and brittle stars: unexpected predatory interactions during induced in situ oxygen crises. Mar Biol 153:1075–1085

Riedel B, Zuschin M, Haselmair A, Stachowitsch M (2008b) Oxygen depletion under glass: behavioural responses of benthic macrofauna to induced anoxia in the Northern Adriatic. J Exp Mar Biol Ecol 367:17–27

Riedl R (1980) Marine ecology – a century of changes. PSZNI Mar Ecol 1:3–46

Rose NA, Janiger D, Parsons ECM, Stachowitsch M (2011) Shifting baselines in scientific publications: a case study using cetacean research. Mar Policy 35:477–482

Rosenberg R (2001) Marine benthic faunal succession stages and related sedimentary activity. Sciencia Mar 65(suppl 2):107–109

Sala E, Knowlton N (2006) Global marine biodiversity trends. Annu Rev Environ Resour 31:93–122

Solan M, Cardinale BJ, Downing AL, Engelhardt KAM, Ruesink JL, Srivastava DS (2004) Extinction and ecosystem function in the marine benthos. Science 306:1177–1180

Stachowitsch M (1984) Mass mortality in the Gulf of Trieste: the course of community destruction. PSZNI Mar Ecol 5(3):243–264

Stachowitsch M (1991) Anoxia in the Northern Adriatic Sea: rapid death, slow recovery. In: Tyson RV, Pearson TH (eds.) Modern and ancient continental shelf anoxia. Geological Society Special Publication No. 58, London, pp 119–129

Stachowitsch M (1992) Benthic communities: eutrophication's "memory mode". In: Vollenweider RA, Marchetti R, Viviani R (eds.) Marine coastal eutrophication. Science of the Total Environment, suppl 1992, pp 1017–1028

Stachowitsch M (2003) Viewpoint: research on natural marine ecosystems: a lost era. Mar Pollut Bull 46:801–805

Stachowitsch M, Avcin A (1988) Eutrophication-induced modifications of benthic communities. In: Eutrophication of the Mediterranean Sea: receiving capacity and monitoring of long-term effects, vol 49. Unesco Technical Reports in Marine Science, Bologna, pp 67–80

Stachowitsch M, Fuchs A (1995) Long-term changes in the benthos of the Northern Adriatic. Annales 7:7–16

Stachowitsch M, Fanuko N, Richter M (1990) Mucus aggregates in the Adriatic Sea: an overview of stages and occurrences. PSZNI Mar Ecol 11(4):327–350

Stachowitsch M, Riedel B, Zuschin M, Machan R (2007) Oxygen depletion and benthic mortalities: the first in situ experimental approach to documenting an elusive phenomenon. Limnol Oceanogr:Methods 5:344–352

Stefanon A, Boldrin A (1982) The oxygen crisis of the Northern Adriatic Sea waters in late fall 1977 and its effects on benthic communities. In: Blanchard J, Mair J, Morrison I (eds.) Diving science symposium – Proceedings 6th symposium of the Confederation Mondiale des Activites Subaquatique. Natural Environmental Research Council, London, pp 167–175

Tyson RV, Pearson TH (1991) Modern and ancient continental shelf anoxia: an overview. In: Tyson RV, Pearson TH (eds.) Modern and ancient continental Shelf Anoxia, vol 58. Geological Society Special Publication, London, pp 1–24

UNEP United Nations Environment Programme (2004) Geo year book 2003. GEO Section/UNEP, Nairobi

Vaquer-Sunyer R, Duarte CM (2008) Thresholds of hypoxia for marine biodiversity. PNAS 105:15452–15457

Vatova A (1949) La fauna bentonica dell'alto e medio Adriatico. Nova Thalassia 1(3):1–110

Worm B, Barbier EB, Beaumont N, Duffy JE, Folke C, Halpern BS, Jackson JBC, Lotze HK, Micheli F, Palumbi SR, Sala E, Selkoe KA, Stachowicz JJ, Watson R (2006) Impacts of biodiversity loss on ocean ecosystem services. Science 314:787–790

Wu RSS (2002) Hypoxia: from molecular responses to ecosystem responses. Mar Pollut Bull 45:35–45

Zhang J, Gilbert D, Gooday AJ, Levin L, Naqvi SWA, Middelburg JJ, Scranton M, Ekau W, Pena A, Dewitte B, Oguz T, Monteiro PMS, Urban E, Rabalais NN, Ittekkot V, Kemp WM, Ulloa O, Elmgren R, Escobar-Briones E, Van der Plas AK (2010) Natural and human-induced hypoxia and consequences for coastal areas: synthesis and future development. Biogeosciences 7:1443–1467

Zuschin M, Pervesler P (1996) Secondary hardground-communities in the Northern Gulf of Trieste, Adriatic Sea. Senckenbergiana maritima 28:53–63

Zuschin M, Stachowitsch M (2009) Epifauna-dominated benthic shelf assemblages: lessons from the modern Adriatic Sea. Palaios 24:211–221

Zuschin M, Stachowitsch M, Pervesler P, Kollmann H (1999) Structural features and taphonomic pathways of a high-biomass epifauna in the northern Gulf of Trieste, Adriatic Sea. Lethaia 32(2):299–317

Biodata of **Nelli G. Sergeeva, Andrew J. Gooday, Sofia A. Mazlumyan, Elena A. Kolesnikova, Anna Lichtschlag, Tetiana N. Kosheleva**, and **Oksana V. Anikeeva** authors of the chapter "*Meiobenthos of the Oxic/Anoxic Interface in the Southwestern Region of the Black Sea: Abundance and Taxonomic Composition.*"

Professor Nelli G. Sergeeva is currently the head of the Benthos Ecology Department in Institute of Biology of the Southern Seas (IBSS) of Ukrainian Academy of Science. She obtained her PhD from the IBSS in 1974 continued her studies and research of ecology of free-living nematodes at the IBSS (Sevastopol). Prof. Sergeeva scientific interests are in the areas: benthic ecosystem; biodiversity of meiobenthos; systematic and ecology of free-living Nematodes; deep-sea benthos; methane gas seeps associated fauna.

E-mail: **nserg05@gmail.com**

Professor Andrew J. Gooday is a deep-sea biologist at the UK's National Oceanography Centre, Southampton, and before that at the Institute of Oceanographic Sciences. He has broad interests in the ecology, biodiversity and biogeography of modern benthic foraminifera, xenophyophores and gromiids. Dr. Gooday obtained his PhD in 1973, in Geology, at University of Exeter, UK. Then he joined an oceanography institute and evolved into a biologist.

E-mail: **ang@noc.soton.ac.uk**

Nelli G. Sergeeva   **Andrew J. Gooday**

Dr. Sofia A. Mazlumyan is currently the senior research associate of Benthos Ecology Department (IBSS). She obtained her PhD from the IBSS in 1989. Dr. Mazlumyan scientific interests are in the areas: modeling of socio-economic importance of sea ecosystem; mathematical methods, describing changes in structure of bottom communities; biodiversity of benthos.

E-mail: **mazlmeister@gmail.com**

Dr. Elena A. Kolesnikova is currently the leader scientist of Benthos Ecology Department (IBSS). She obtained her PhD from the IBSS in 1981. Dr. Kolesnikova scientific interests are in the areas of: ecology, biodiversity, composition of meiobenthos communities; taxonomic and ecology of the harpacticoides.

E-mail: **elan.kolesnikova@gmail.com**

Dr. Anna Lichtschlag is currently working as a postdoctoral research associate in the Max-Planck-Institute for Marine Microbiology (MPI) in Bremen, Germany. She received her PhD in 2009 from the University of Bremen, working in the microsensor group of the MPI. Her dissertation work focused on the biogeochemistry and microbiology of marine cold seep systems. Dr. Lichtschlag scientific interests are in the in situ quantification of marine benthic biogeochemical processes, biogeochemistry and primary production at cold seeps, marine methane cycle, and the iron and sulfur geochemistry of marine sediments.

E-mail: **alichtsc@mpi-bremen.de**

Sofia A. Mazlumyan   **Elena A. Kolesnikova**   **Anna Lichtschlag**

Tetiana N. Kosheleva is Ph.D. student of Benthos Ecology Department. She graduated from the National Technical University (Sevastopol), Dep. Ecology, 2006. Tatiana Kosheleva scientific interests are in the areas of: taxonomy and ecology of coastal and deep-water free-living marine nematodes.

E-mail: **Alinka8314@gmail.com**

Oksana V. Anikeeva is currently the junior scientist of Benthos Ecology Department (IBSS). Oksana Anikeeva scientific interests are in the areas of: soft-shelled monothalamous foraminifers, diversity soft-shelled monothalamous foraminifers.

E-mail: **oksana.anikeeva@gmail.com**

Tetiana N. Kosheleva   **Oksana V. Anikeeva**

MEIOBENTHOS OF THE OXIC/ANOXIC INTERFACE IN THE SOUTHWESTERN REGION OF THE BLACK SEA: ABUNDANCE AND TAXONOMIC COMPOSITION

NELLI G. SERGEEVA[1], ANDREW J. GOODAY[2], SOFIA A. MAZLUMYAN[1], ELENA A. KOLESNIKOVA[1], ANNA LICHTSCHLAG[3], TETIANA N. KOSHELEVA[1], AND OKSANA V. ANIKEEVA[1],

[1]*Institute of Biology of the Southern Seas NASU, Sevastopol, Ukraine*
[2]*National Oceanography Centre, Southampton SO14 3ZH, UK*
[3]*Max-Planck-Institute for Marine Microbiology, Bremen, Germany*

1. Introduction

Among the seas of Mediterranean basin, the Black Sea exhibits a number of striking features. The combined effect of great depth, considerable desalination of surface waters caused by river discharges, and the influx of saline deep water from the Sea of Marmara, creates a distinct stratification of the water column into an upper, relatively thin (150–250 m deep) oxic zone with plentiful flora and fauna and a huge anoxic zone (from 150 to 250 m depth down to the deepest point) with high hydrogen sulfide concentrations.

The interactions between these oxic and anoxic water masses are of great interest to researchers in different scientific fields. Study of the boundary region between them has shown that the redox zone is a good indicator of oxygen concentrations and also corresponds to the distribution of other parameters, namely temperature, pH, ammonium, manganese, phosphates, and the transparency of the water (Stunzhas and Yakushev 2006). The anoxic Black Sea water mass is separated from the oxic zone by a layer, several tens of meters thick, where oxygen concentrations decrease to 0.1–0.3 ml·L^{-1} (= 4.5–13.5 µmol L^{-1}) and hydrogen sulfide levels increase correspondingly (Vinogradov and Flint 1987). Long-term monitoring of the distribution of oxygen and sulfide across this boundary indicates that it is situated at a depth of 80–100 m in the center and 160–250 m at the periphery (Eremeev and Konovalov 2006).

Where the oxic/anoxic interface in the water column impinges on the seafloor, it creates a strong benthic gradient in oxygen and hydrogen sulfide concentrations. This O_2/H_2S-transition zone is highly dynamic, characterized by varying concentrations of oxygen and hydrogen sulfide in the bottom water, and oscillates above and below its average depth. The taxonomic composition and distribution of benthic fauna inhabiting the depth zone where the oxic/anoxic interface zone meets the sea floor is of special interest.

Specific communities of benthic organisms adapted to reduced oxygen concentrations occur within this zone in the NW Black Sea (Kiseleva 1998; Sergeeva and Zaika 2000; Zaika and Sergeeva 2008). This benthic region is termed the "periazoic zone" (Bacesco 1963). It has been studied in parts of the Black Sea off the coast of Romania (Gomoiu et al. 2008; Surugiu 2005), Bulgaria (Marinov 1978) and Ukraine (Sergeeva and Zaika 2000; Zaika and Sergeeva 2008). In recent years, attention has focused on the meio and microbenthos living in the deeper, sulfidic part of the Black Sea; there have been several reports of live and active eukaryotes from this hostile environment (Zaika 2008; Zaika and Sergeeva 2009).

Faunal studies in hypoxic/anoxic and sulfidic environments involve certain problems, particularly when Rose-Bengal staining is used to identify "live" specimens. Although easy to use, this protein stain has the disadvantage that it can color dead as well as live cytoplasm (Bernhard 2000), giving rise to "false positives" (Bernhard et al. 2006). Organic matter decomposes at a slower rate in hypoxic than in fully oxic sediments (van der Weijden et al. 1999; Cowie 2005). Under hypoxic conditions, the decomposition of animal tissues and protozoan cytoplasm may be retarded (Jorissen et al. 1994), while in anoxic sediments, the carcasses of animals and protozoans are likely to be preserved (Danovaro et al. 2010). Thus, unless they are seen alive, the unequivocal recognition of living organisms can only be achieved by observing tracer uptake, intact ultrastructural features, or by using a vital fluorogenic probe such as CellTracker Green, which fluoresces inside live cells after the original molecule has been modified by enzymatic activity (Bernhard et al. 2006; Danovaro et al. 2010).

Here, we review information on the composition and structure of Rose-Bengal-stained meiobenthic communities (protozoans and metazoans) along depth gradients that span the oxic/anoxic transition in the NW part of the Black Sea. Our review is based on a combination of published data and results obtained from new samples collected in 2007 during RV *Meteor* cruise 72/2. We also provide arguments to support our contention that the faunal patterns observed are real rather than artifacts of preservation under hypoxic/anoxic and sulfidic conditions.

2. Previous Research

Since the discovery of the anoxic zone, which occupies 77% of the floor of the Black Sea, it has been assumed that, except for bacteria, this vast region is a lifeless desert (Nikitin 1938; Kiseleva 1979; Klenov 1948; Kriss 1959; Sorokin 1962; Zhizhchenko 1974). Kiseleva (1979) recognized three zones in the Black Sea, reflecting the dominance of different size classes of organism. The region from the tide mark to 120–150 m depth is inhabited by macro (>1 mm), meio (0.1–1.0 mm), and microbenthic (<0.1 mm) organisms. The second zone, between 120–150 and 250–300 m depth, is characterized by meio and microbenthos, while the third, from 250 to 300 m to the deepest point, is occupied only by bacteria (Kiseleva 1979). According to this scheme, the lower border of eukaryotic life in the Black

Sea is located at a depth of ~250–300 m. In the second half of the twentieth century, an additional "periazoic" zone was recognized at 120–200 m depth in the lower part of the *Modiolula phaseolina* community off the Romanian coast (Bacesco 1963; Gomoiu et al. 2008)

Because the deeper part of the Black Sea basin is anoxic, it was long assumed that the aerobic zoobenthos did not penetrate into this region (Zaika et al. 1999). As a result, relatively little research has been conducted on the biota in the anoxic layers, limiting our understanding of the faunal changes across the oxic/anoxic interface and the occurrence of organisms at depths where conditions are permanently sulfidic (Luth and Luth 1997, 1998; Revkov and Sergeeva 2004). Moreover, although the occurrence, spatial distribution, and environmental role of methane seeps in the Black Sea are well studied, less is known about the bottom fauna associated with these features (Sergeeva and Gulin 2007, 2009), particularly those located across the oxic/anoxic interface.

Comparative studies of the structure of Black Sea benthic communities in areas with and without methane seepages were conducted for the first time in 1993–1994. The biomass and biological activity of bottom communities was similar at all the sites studied. Areas of seepage were characterized by larger body sizes among the macrobenthos (Luth and Luth 1998). The diversity and abundance of the meiobenthos were both relatively high within the oxic/anoxic transition zone across the same field of methane seeps (Sergeeva 2003a, b, 2004a; Sergeeva and Gulin 2007). Analyses of benthic eukaryotes (protozoa and metazoa), based on the collections of six cruises over the period 1993–2007 across the oxic/anoxic interface in the region of the Dnieper paleo-delta, revealed the presence of 23 higher taxa (Sergeeva and Gulin 2009). The species identified included both stenobiont and eurybiont forms adapted to hypoxia and the presence of hydrogen sulfide and methane (Sergeeva and Zaika 2000; Zaika et al. 1999; Zaika and Sergeeva 2008). The density of the meiobenthos varied from 2,400 to 53,000 ind.$\cdot m^{-2}$ at depths of 182–252 m in the submarine Dnieper Canyon, where CH_4 concentrations in bottom sediments ranged from 2.40 to 5.75 nmol$\cdot m^{-3}$. The meiobenthos was represented by Ciliata, Foraminifera, Nematoda, Polychaeta, Bivalvia, Gastropoda, Amphipoda, and Acarina. Nematoda and Foraminifera were the dominant groups (Sergeeva and Gulin 2007). Over a slightly shallower depth range (70–235 m), Revkov and Sergeeva (2004) reported 12 higher meiobenthic taxa in areas with methane seepage in the Dnieper Canyon region; Nematoda, Turbellaria, Ostracoda, and Acarina occurred at the deepest sites (230–235 m).

3. Methods and Materials

New samples were collected during RV *Meteor* cruise 72/2 (February–March 2007) on the open slope northwest of the Crimea Peninsula (Boetius 2007). Ten stations were chosen along a transect from the oxic into the anoxic zone (Fig. 1, Table 1), crossing the shelf and the NW Crimea slope in an area characterized

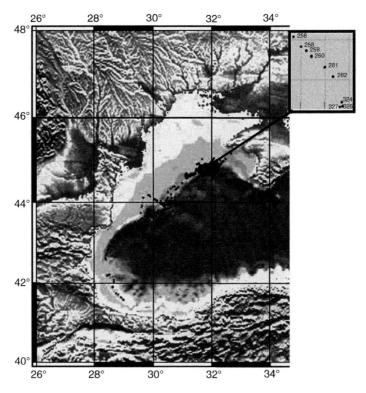

Figure 1. Stations were chosen to span the lower boundary oxygen layer, and the upper anoxic zone, crossing the shelf and the NW Crimean slope (based on Egorov et al. 2003; Boetius 2007; Sergeeva et al. 2010).

Table 1. Change of concentration of O_2 and sulfide with water depth in the NW part of the Black Sea.

Year	2007	2007	1994[a]	1994[a]
Depth (m)	O_2 mmol· L^{-1}	H_2S µmol· L^{-1}	O_2 mmol· L^{-1}	H_2S µmol· L^{-1}
120	0.14	0	–	–
123	–	–	0.006	–
130	–	–	0.006	0
132	0.12	0	–	–
140	0.09	0	–	–
150	0.07	0	0.004	6.3
160	0.06	<1	0.004	6.8
170	0.07	<1	–	–
188	–	–	0	14

Oxygen and sulfide concentrations in 2007 (M72/2) were measured in the overlying water of retrieved MUC cores. Oxygen was measured according to the Winkler method (Winkler 1888) and samples for sulfide were immediately fixed in ZnAc and measured with the photometric methylene blue method (Cline 1969) in the home laboratory. Pangaea event label: M72/2_256, M72/2_258 to 262 (www.pangaea.de)
[a]Data from Pimenov et al. (1998), – No data

by methane seeps (Egorov et al. 2003). The main field of methane seeps in the northwestern part of the Black Sea is located near the submarine Dnieper Canyon (Egorov et al. 1998), where they occupy an area 105 km in length and 43 km in width, covering 1,623 km^2 of seafloor and a depth range of 35–785 m. Bottom-water oxygen concentrations range from 0.14 mmol L^{-1} at 120 m water depth to 0.06–0.07 mmol L^{-1} at 160–170 m and zero at 188 m (Table 1) (Pimenov et al. 1998). There is a corresponding rise in sulfide concentrations from zero at 130 m to 14 µmol L^{-1} at 188 m. Bottom sediments are sandy silt with large quantities of bivalve, ostracod, and foraminiferan shells at 120–140 m depth, and silt with molluscan shell debris and bacterial aggregations at 150–240 m depth. The sediment surface is covered by various kinds of organic detritus, including the remains of macroalgae and microalgae, as well as bacterial mats of different sizes.

Samples for biological studies were obtained at eight sites located at 10 m intervals between 120 and 190 m, and from two additional sites located at 210 and 240 m water depth. They were collected using a modified version of the Barnett multiple corer (MUC), a device that takes virtually undisturbed sediment samples (Barnett et al. 1984). At each station, three replicate sediment cores (from a single multicorer deployment) were sectioned into the following horizontal layers: 0–1, 1–2, 2–3, and 3–5 cm. All sediment sections were preserved in 75% alcohol, which we know from previous experience, preserves morphological structures without distortion. We avoided prior fixation in formalin in order not to damage calcareous taxa. The sediments were washed through sieves with a mesh size of 1 mm and 63 µm and stained with Rose Bengal solution before being sorted in water under a microscope for "live" (stained) organisms. We extracted only those specimens that stained intensely with Rose Bengal and showed no sign of morphological damage. All of the organisms isolated were counted and identified to higher taxa, or in the case of nematodes and harpacticoids to species level. Protists were identified as gromiids based on test characteristics, e.g., test shape and the presence of an oral capsule (Rothe et al. 2010). However, because gromiids can be confused with allogromiid foraminiferans, confirmation of these identifications must await molecular analyses and an examination of the test-wall ultrastructure. Details of methods for estimating bacterioplankton production and bacterial chemosynthesis (Sect. 4.1.2) are given by Sorokin and Sorokina (2008) and Gulin (1991), respectively.

4. Meiobenthos of the Oxic/Anoxic Interface

We focus here on the meiobenthos, regarded as protozoans and metazoans retained on the 63 µm mesh (i.e., the 63 µm to 1 mm size fraction). We also included individuals of meiobenthic taxa retained on the 1 mm mesh, mainly large nematodes. The meiobenthos comprises the permanent meiofauna (eumeiobenthos) and the temporary meiofauna (pseudomeiobenthos), the latter represented by the juvenile stages of macrobenthos (Bougis 1950; Chislenko 1961).

The eumeiobenthos constitutes 81% to almost 100% of the bottom community (Table 2), while the pseudomeiobenthos constitutes only 0.3–19% (Table 3).

Table 2. Abundance (indiv m^{-2}) of eumeiobenthos and constituent taxa along the depth gradient near the Dnieper Canyon.

Taxon	Depth				
	120 m	130 m	140 m	150 m	160 m
Gromia	21,945±5,315	20,461±4,975	1,409±178	856±375	352±100
Ciliophora	8,154±1,967	1,510±396	705±125	7,097±3,110	12,231±3,830
Foraminifera	15,352±3,042	11,929±2,368	5,335±1,224	35,586±10,059	66,893±22,210
Nematoda	1,347,675±251,752	245,073±21,690	202,289±41,461	422,548±95,913	727,719±171,078
Kinorhyncha	302±52	0	201±53	0	0
Harpacticoida	91,758±8,427	54,381±7,570	28,514±5,504	36,441±6,487	10,268±2,097
Ostracoda	705±170	4,681±761	201±52	252±39	0
Acarina	2,215±443	755±173	176±27	0	151±66
Tardigrada	34,227±7,237	14,798±3,531	604±158	151±39	101±44
other	0	0	2,215±580	0	201±88
Total	1,511,158±263,978	345,358±37,428	236,391±49,178	492,461±113,451	814,293±192,525

Taxon	170 m	180 m	190 m	210 m	240 m
Gromia	289±61	828±419	221±153	265±129	276±170
Ciliophora	5,965±3,018	10,754±5,442	11,144±7,722	1,324±648	7,172±4,444
Foraminifera	64,439±21,599	717±362	607±420	66±33	441±273
Nematoda	0	0	0	0	0
Kinorhyncha	0	0	0	0	0
Harpacticoida	3,196±710	0	276±115	0	0
Ostracoda	0	0	0	0	0
Acarina	0	0	0	0	0
Tardigrada	0	0	0	0	717±297
other	55,266±27,967	0	110±58	0	0
Total	191,556±60,055	17,516±5,468	21,074±10,000	11,651±5,708	10,758±3,795

Table 3. Abundance (indiv m^{-2}) of pseudomeiobenthos and constituent taxa along the depth gradient near the Dnieper Canyon.

Taxon	Depth				
	120 m	130 m	140 m	150 m	160 m
Coelenterata	5,536 ± 1201	32,616 ± 8,291	0	10,973 ± 2,874	14,194 ± 4,643
Oligochaeta	302 ± 82	0	0	2,919 ± 765	101 ± 44
Polychaeta	2,970 ± 488	2,190 ± 626	9,261 ± 2,425	6,896 ± 1,579	21,341 ± 5,605
Turbellaria	151 ± 41	453 ± 124	0	0	0
Nemertini	0	0	0	151 ± 39	0
Bivalvia	5,285 ± 1,210	8,154 ± 1,531	1,233 ± 246	0	0
Gastropoda	0	0	0	50 ± 13	0
other	0	151 ± 66	201 ± 52	0	0
Total	14,244 ± 2,825	43,564 ± 9,484	56,172 ± 14,395	20,989 ± 5,269	35,636 ± 9,594
	170 m	180 m	190 m	210 m	240 m
Coelenterata	4,492 ± 2,273	0	0	0	0
Oligochaeta	302 ± 51	0	0	0	0
Polychaeta	2,076 ± 741	55 ± 28	55 ± 29	0	0
Turbellaria	38 ± 19	110 ± 56	0	0	0
Nemertini	76 ± 38	0	0	0	0
Bivalvia	76 ± 38	0	0	0	0
Gastropoda	0	0	0	0	0
other	151 ± 76	110 ± 55	0	0	0
Total	7210 ± 2705	276 ± 139	55 ± 29	0	0

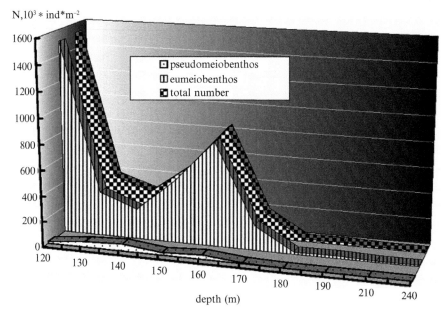

Figure 2. Distribution of meiobenthos abundance (N) along depth gradient.

Eumeiobenthos was found at all studied depths with two well-defined peaks of abundance at 120 m ($O_2 = 0.14$ mmol L^{-1}, sulfide = zero) and 160 m ($O_2 = 0.06$ mmol L^{-1}, sulfide = <1 µmol L^{-1}) (Table 1). The maximum abundance of eumeiobenthos corresponds to the 120 m peak; the minimum abundance is recorded at 240 m depth. Pseudomeiobenthos was most abundant at 140 m depth ($O_2 = 0.09$ mmol L^{-1}, sulfide = zero) and disappeared at 190 m (Fig. 2, Tables 2 and 3). These two categories are combined hereafter and referred to as the meiobenthos.

The meiobenthos present between 120 and 240 m included the following 16 taxa: Ciliophora, Gromia, Foraminifera (soft-shell and hard- shell forms), Nematoda, Kinorhyncha, Harpacticoida, Acarina, Ostracoda, and Tardigrada, and among the pseudomeiobenthos, Coelenterata, Oligochaeta, Polychaeta, Turbellaria, Nemertini, Bivalvia, and Gastropoda. Among these, the Nematoda were dominant and the Foraminifera and Ciliophora subdominant.

Biodiversity at the major taxon level declines with increasing water depth, although the decrease is not linear and there are minor peaks at 170 and 190 m (Fig. 3). Many of the species recognized along this transect are either completely new to science or new records for the Black Sea.

Maximum concentrations of meiofauna are usually found in the surface layers of bottom sediments. The vertical distribution of meiobenthos in the upper 5 cm of sediment has been analyzed at different water depths in the study area (Fig. 4).

Overall, the 0–1 cm layer contained 70% of total meiobenthos abundance, compared to 16% in the 1–2 cm layer, 8% in the 2–3 cm layer, and 6% in the

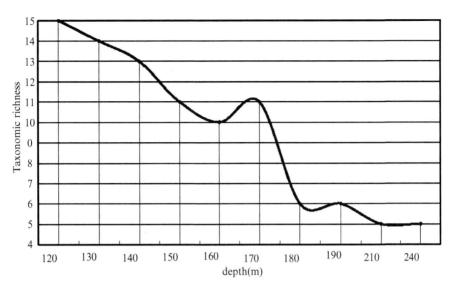

Figure 3. Higher taxon richness of meiobenthos along depth gradient near Dnieper Canyon.

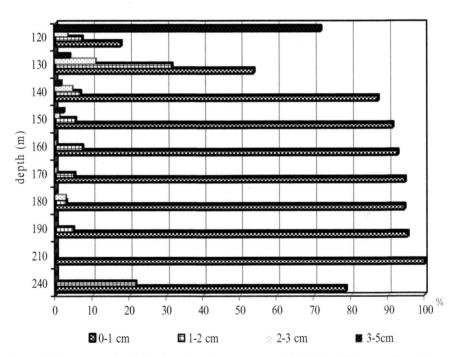

Figure 4. Proportion of meiobenthos (including protozoans) inhabiting different sediment layers along depth transect near Dnieper Canyon.

3–5 cm layer. The 0–1 cm layer yielded the largest number of specimens at all points along the transect, except at 120 m depth, where the meiobenthos was concentrated in the deepest (3–5 cm) sediment layer. The highest concentration of meiobenthos in the 1–2 cm layer was at 210–240 m in the anoxia zone.

4.1. PROTOZOA

Protozoans from the deep-water Dnieper Canyon area comprised Gromiida, Ciliophora, and Foraminifera, notably soft-shelled, single-chambered (monothalamous) taxa. Foraminifera were most abundant at depths of 150–170 m, gromiids at depths of 120–130 m, while ciliates were fairly evenly distributed across the transect (Fig. 5). We did not find any testate amoebae of the kind reported by Golemansky (2007 and papers cited therein) from supralittoral sediments on the Bulgarian Black Sea coast.

4.1.1. Gromiida

Our knowledge of gromiids from sublittoral sediments across the oxic/anoxic interface is limited (Sergeeva and Gulin 2009). Although organisms resembling gromiids were found at all studied depths, they were rare except at 120 and 130 m (Figs. 6 and 7). Preliminary examination of this material revealed three species, all of which are believed to be new to science. Specimens were concentrated in the 3–5 cm layer at the 120 m site and, to a lesser extend, at 140 m. At other depths, they were more or less confined to the top 1 cm layer.

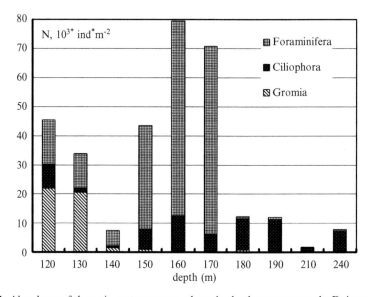

Figure 5. Abundance of the main protozoan taxa along the depth transect near the Dnieper Canyon.

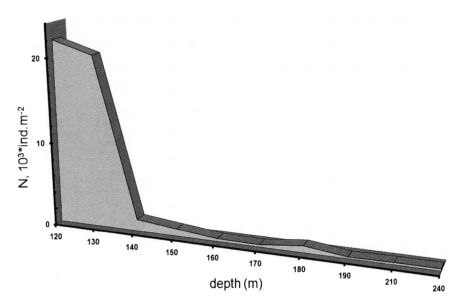

Figure 6. Trends in the abundance of gromiids along the depth transect near Dnieper Canyon.

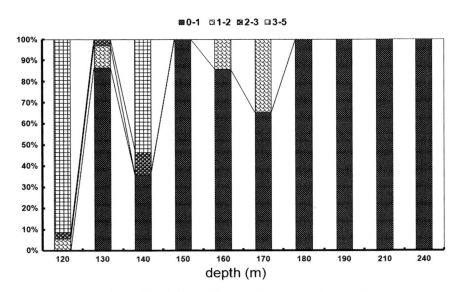

Figure 7. Proportion of gromiids inhabiting different sediment layers along depth transect.

4.1.2. Ciliophora

The following account is based on Zaika and Sergeeva (2008, 2009), who described the bathymetric distribution of the benthic ciliates at depths from 120 to 2,075 m near the Dnieper Canyon and the Sorokin Trough (eastern part of the Black Sea)

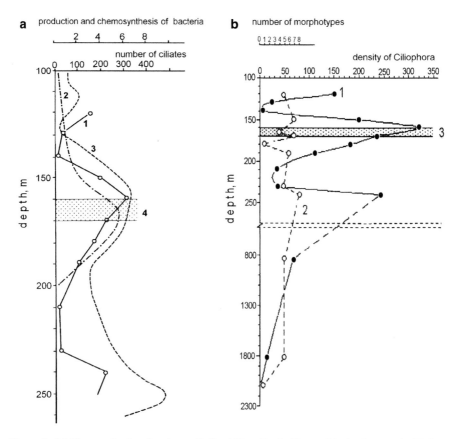

Figure 8. (a) Changes in the abundance (indiv. 100 cm^2) of ciliates with depth near the Dnieper Canyon. *1* number of ciliates, *2* production of bacteria, *3* intensity of bacterial chemosynthesis, and *4* border of hydrogen sulfide. (b) Vertical profiles of total abundance and number of ciliate morphotypes. *1* total abundance, *2* numbers of morphotypes, and *3* upper boundary of hydrogen sulfide. After Zaika and Sergeeva (2009).

(Fig. 8). They occurred in all studied samples of near bottom water, sediment surface detritus, and in the upper layer (0–1 cm) of sediment (Fig. 9). Different forms were present, including free living and attached forms, among which more than 30 morphospecies were recognized. Ciliates exhibited abundance peaks at 120, 160–190, and 240 m (Fig. 8).

4.1.3. Foraminifera

Previous studies. A total of 104 species of foraminifera has been recognized in the Black Sea (Yanko and Troitskaya 1987; Yanko and Vorobjeva 1990, 1991; Temelkov et al. 2006), in addition to the monothalamous supralittoral species reviewed by Golemansky (2007). The "hard-shelled" species are fairly well studied, but information about the monothalamous, soft-shelled taxa ("allogromiids"

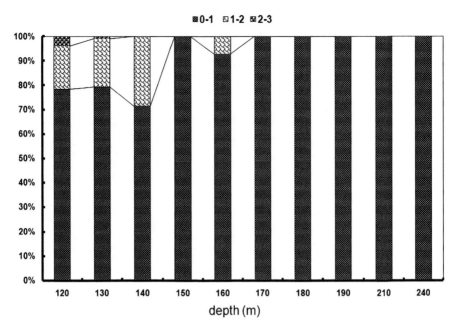

Figure 9. Proportion of ciliates inhabiting different sediment layers along the depth gradient near the Dnieper Canyon.

and "saccamminids") is rather limited. These protists were first reported in the Black Sea towards the end of the twentieth century (Golemansky 1974, 1999; Sergeeva and Kolesnikova 1996). It is now clear that they are a conspicuous and diverse element of the Black Sea coastal and deep-water benthic communities (Anikeeva 2003; Anikeeva and Sergeeva 2001; Golemansky 2007; Gooday et al. 2006; Revkov and Sergeeva 2004; Sergeeva and Anikeeva 2004).

The abundance of allogromiids and saccamminids reaches 116,000 indiv.m^{-2} at 260 m depth west of the Crimea (Sergeeva 2003b). Sergeeva et al. (2010) recognized a total of 40 informal groupings (morphospecies or morphotypes), and one described species, of monothalamous foraminifera, either organic-walled allogromiids or agglutinated saccamminids at sites along the 2007 *Meteor* transect in the region of the submarine Dnieper Canyon. Within the lower oxygen to upper hydrogen sulfide border zone, monothalamous foraminifera were more numerous than multichambered calcareous taxa. Both groups reached their highest abundance between 150 and 170 m, with a sharp peak at 160 m.

New data. In addition to the information reported by Sergeeva et al. (2010), new data about the vertical distribution of deep-water soft-shelled foraminiferans within the sediment column have been obtained across the oxic/anoxic interface in the Dnieper Canyon study area (Fig. 10). Most specimens were found in the top layer (0–1 cm) or in the overlying surface detritus. *Tinogullmia* sp., which occurred between 120 and 180 m, inhabited the 0–1 and 1–2 cm sediment layers.

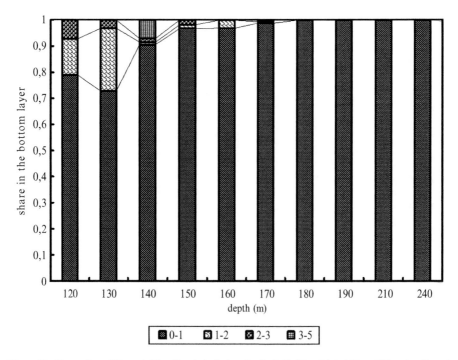

Figure 10. Proportion of foraminifera (hard-shelled and soft-shelled taxa) inhabiting different sediment layers along the depth gradient near the Dnieper Canyon.

Tinogullmia cf. *riemanni* and *Goodayia rostellatum* was also found in these upper two horizons, although they were concentrated in the 0–1 cm layer. *Bathyallogromia* sp. 2 was the only species that occurred in all three studied horizons and increased in abundance with depth into the sediment. Allogromiid sp. J was found only at 120 m water depth, where it was confined to the 1–2 and 2–3 cm layers, being equally distributed between these horizons. Saccamminid sp. 4 occurred only in the 2–3 cm horizon at 120 m. Thus, our new data suggest that ~70–75% of the studied soft-walled foraminiferal species are confined to the top layer (0–1 cm) of the sediments, presumably in order to gain access to the limited oxygen available in the bottom water, as well as to food.

4.2. METAZOA

A distinctive metazoan meiobenthic community is associated with the oxic/anoxic interface of the NW of the Black Sea (Sergeeva 2003b; Sergeeva and Zaika 2000; Sergeeva et al. 2008; Zaika 1999). In the area that we studied, it included Hydrozoa (two unknown species), Nematoda (including some unusual species), Polychaeta (two species), several species of Harpacticoida, and two undescribed species of Tardigrada. Here, we focus on the three most abundant groups, the nematodes, polychaetes, and harpacticoids. Of these, the nematodes were always the dominant (70–100%) metazoan meiobenthic organisms and the only taxon present at all investigated depths (Fig. 11).

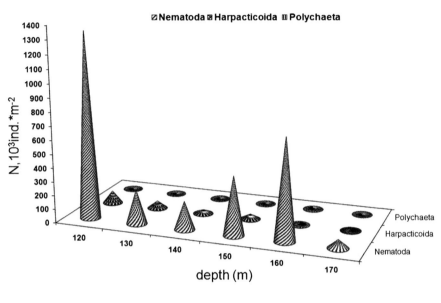

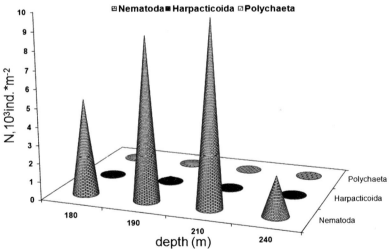

Figure 11. Abundance (10^3 indiv. m^{-2}) of the three main metazoan meiofaunal taxa along the depth transect near the Dnieper Canyon.

4.2.1. Free-Living Nematoda

Previous studies. In samples collected during 1994 in the NW Black Sea, the maximum abundance of nematodes was 490,000 ind. m^{-2} at a depth of 134 m and 478,800 indiv.m^{-2} at 77 m. Densities declined to 271,000, 185,600, and 10,200 indiv. m^{-2} at depths of 146, 151, and 172 m, respectively (Revkov and Sergeeva 2004), and a small number (1,000 indiv.m^{-2}) were recorded at the 232 m site. Nematodes were represented by 143 species at sites along this transect, with 69 species being recognized at 150 m, and 33–63 at depths between 120 and 140 m. The taxonomic composition was similar to that found on the Black Sea shelf, suggesting that some nematode species are eurybiotic. However, 38 species and 6 genera have been found only in the study area, where oxygen is either absent or present in minimal concentrations, and are previously unreported from the Black Sea.

New data. In samples collected during 2007, nematode densities reached 1,348,000 indiv.m^{-2} at 120 m depth and 727,700 indiv.m^{-2} at 160 m (Table 2, Fig. 12). They persisted to depths of 210 and 240 m at densities of 10,030 and 1,520 indiv.m^{-2}, respectively. Preliminary and ongoing analyses of species diversity at depths between 120 and 240 m includes 23 families and 6 orders. Our data show that some nematodes can live in permanently anoxic environments due to a high tolerance to hydrogen sulfide conditions.

The nematode fauna is unique at the studied depths. It includes 90 species and 9 genera unknown from the Black Sea, and 19 species, 9 genera, and 1 family recognized for the first time in this basin. The fauna includes stenobiontic and eurybiontic forms adapted to living in the redox zone. The species richness of

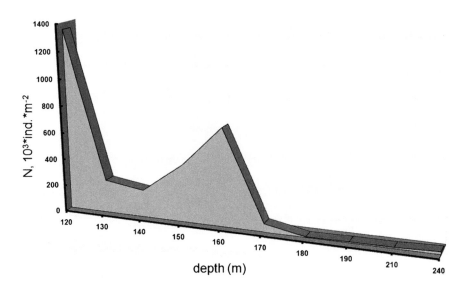

Figure 12. Abundance (10^3 indiv. m^{-2}) of nematodes along the depth transect near the Dnieper Canyon.

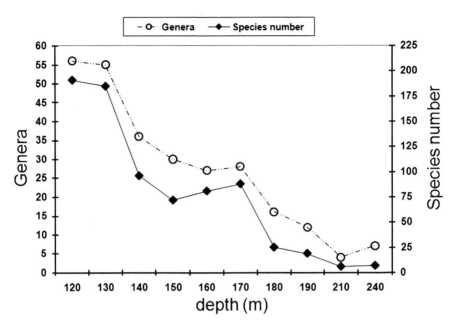

Figure 13. Distribution of nematode species and genera along the depth gradient near the Dnieper Canyon.

free-living nematodes decreased with increasing depth, highest values being found at depths of 120 and 130 m with a second peak in number of species and genera located at a depth of 170 m ($O_2 = 0.07$ mmol L^{-1}, sulfide = <1 μmol L^{-1}) (Fig. 13). Species adapted to anaerobic conditions occurred at depths ranging from 190 to 240 m. They included species (e.g., *Sabatieria pulchra*) usually found at shallower depths in the Black Sea. Among the nematodes at 150–240 m were species confined to these depths; these included *Quadricoma* sp. *U, Cobbionema* sp., *Sabatieria* sp.1, *Linhomoeus* sp. *X, Paralinhomoeus* sp 1, *Theristus* sp.1, *Theristus sp. A, Theristus* sp. 4, *Monhystera* sp. *A, Campylaimus* sp. *4, Spirinia* sp 1, *Metalinhomoeus* sp. *O, Linhomoeus* sp. 1, *Aponema* sp. 1, *Microlaimus* sp. 2, and *Campylaimus* sp. 8.

The vertical distribution of nematodes in the sediment varied with water depth (Fig. 14). The 0–1 cm layer was occupied at all depths, and the 1–2 cm layer was also consistently inhabited, except at 210 m where the nematodes were confined to the upper layer. In most samples, nematodes were concentrated (81–96%) in the 0–1 cm layer. At 120 m, however, 77% were found in the 3–5 cm layer and at 240 m, where densities were much lower, 77% occurred in the 1–2 cm horizon. The 2–3 cm layer yielded a small proportion (0–16%) of nematodes at all water depths (Fig. 14). Studies are presently ongoing to determine the vertical distribution of individual species within the sediments.

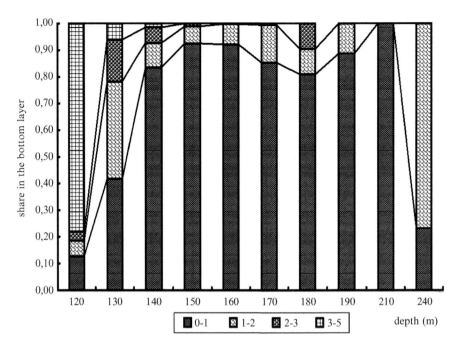

Figure 14. Proportion of nematodes inhabiting different sediment layers along the depth gradient near the Dnieper Canyon.

4.2.2. Polychaeta

Previous studies. Larval polychaetes belonging to two species are abundant in the deepest oxygenated waters across the basin throughout the year. One of them, which occurs in relatively low numbers, has been identified as *Protodrilus* sp. The second larval species is much more common. The growth of these larvae to the adult stage in the laboratory at IBSS led to their description as *Vigtorniella zaikai*, the type species of a new genus (Kiseleva 1992). Detailed information on the vertical distribution of zooplankton showed that they are pioneer metazoans inhabiting the narrow transition layer between oxic and anoxic water (Kiseleva 1959, 1990, 1998). These polychaetes have been found in all parts of the Black Sea (Zaika 1999; Sergeeva and Zaika 2000; Murina et al. 2006). This reflects the specific habitat requirements of the larvae, which are associated with a water layer characterized by constant low temperatures (8°C) and low oxygen concentrations (0.2–0.3 mL L^{-1} = 9–13.5 μmol L^{-1}). They are representatives of the metazoan plankton living in severely hypoxic water near the boundary of the hydrogen sulfide zone (Sergeeva and Zaika 2000). Where this layer impinges on the seafloor dwelling specimens of *Protodrilus* sp. and *Vigtorniella zaikai* were discovered in 1994 in the Black Sea around the shelf break to the west of the Crimean peninsula (Sergeeva et al. 1996). They dominated the meiobenthic community on at depths

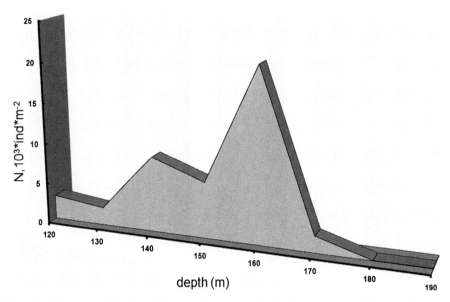

Figure 15. Abundance (10^3 indiv. m^{-2}) of polychaetes along the depth transect near the Dnieper Canyon.

of ~100–150 m, the larvae settle and adopt a benthic mode of life. Bottom the seafloor immediately above the hydrogen sulfide zone (Sergeeva and Zaika 2000; Zaika et al. 1999).

New data. In our new core samples, *Protodrilus* sp. and *Vigtorniella zaikai* were the only species present at depths between 140 and 160 m, where they were responsible for peaks in polychaete abundance (Fig. 15). At shallower sites (120–130 m), polychaetes were represented by species that are common in the middle part of the sublittoral zone. The polychaetes were concentrated (85–100%) in the upper layer of sediment across the transect, except at 120 m, where specimens were more evenly distributed through the sediment column (50% in the 0–1 cm layer, 31% in the 1–2 cm layer, and 9.5% in each of the two deeper layers) (Fig. 16).

4.2.3. Harpacticoida

Harpacticoid copepods (including nauplia) are a consistent component of benthic communities at depths of 120–190 m in the study area, with peaks in abundance at 120 m and 150 m (Fig. 17).

Harpacticoids were distributed unevenly in different sediment layers (Fig. 18). The upper layer (0–1 cm) was inhabited at all sites, except for the 180 m sample, which was devoid of harpacticoids. These crustaceans penetrated deeper into the sediment at the two shallowest sites. At 120 m, 51% were found in the 0–1 cm, 19% in the 1–2 cm, 5% in the 2–3 cm, and 25% in the 3–5 cm layer.

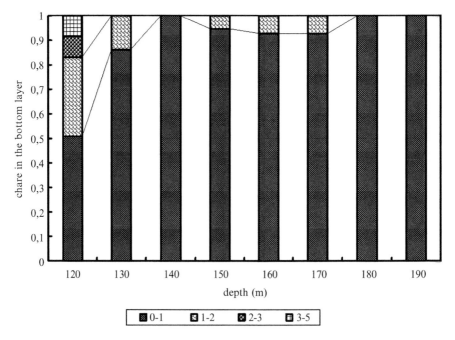

Figure 16. Proportion of polychaetes inhabiting different sediment layers along the depth gradient near the Dnieper Canyon.

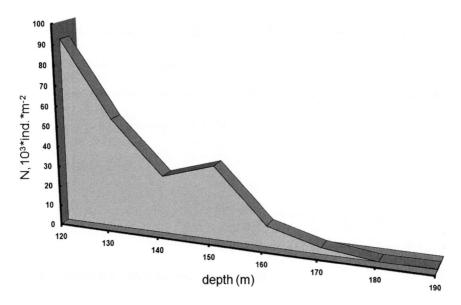

Figure 17. Abundance (10^3 indiv.m^{-2}) of harpacticoids (including nauplii) along the depth transect near the Dnieper Canyon.

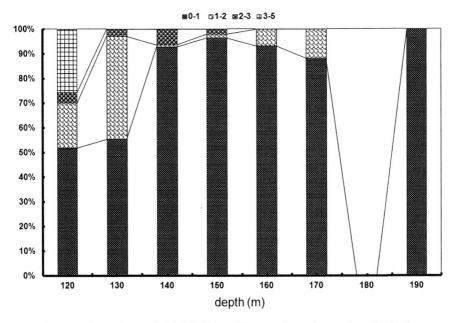

Figure 18. Proportion of harpacticoids inhabiting different sediment layers along the depth transect near the Dnieper Canyon.

Table 4. Harpacticoid species abundance (ind.m^{-2}) without nauplii along the depth transect near the Dnieper Canyon.

Species	Depth					
	120 m	130 m	140 m	150 m	160 m	170 m
Paramphiascopsis longirostris	38,175	2,068	8,638	9,690	1,510	361
Haloshizopera pontarchis	0	0	0	0	252	0
Amphascella subdebilis	10,411	0	0	0	0	0
Ameira parvula	7,982	0	0	0	252	0
Ameira sp.	0	414	0	0	0	0
Mesochra sp.	3,470	414	9,357	4,845	1,007	180
Enhydrosoma sp.	0	0	0	0	0	180
Laophonte sp.	0	0	0	0	0	0
Laophonte setosa	0	0	0	0	0	180
Normanella serrata	3470	0	0	0	0	0
Archesola typhlops	13882	827	0	969	0	180
Total without Nauplii	**77,392**	**3,723**	**17,995**	**15,504**	**3,021**	**1,081**

At 130 m, 55% were found in the 0–1 cm, 41% in the 1–2 cm, and 4% in the 2–3 cm layer. In contrast, between 87% and 95% of harpacticoids were concentrated in the top 1 cm at 140–170 m, while at the deepest site 190 m they present only in the upper layer. The harpacticoids unevenly on including only 11 species at 120–170 m (Table 4).

5. Are the Faunal Patterns Real or Artifacts of Preservation?

In this study, we identified "live" protists and other meiofaunal organisms based on Rose-Bengal staining. Because of the known limitations of this method (Bernhard 2000; Bernhard et al. 2006; Danovaro et al. 2010), particularly in hypoxic/anoxic and sulfidic settings, it is possible that some of the stained organisms found in Black Sea sediments were dead when collected. On the other hand, certain ciliate taxa have long been known to occur in anoxic sediments (e.g., Fenchel and Finlay 1995), while foraminifera possess various morphological and physiological adaptations to hypoxia (Bernhard and Sen Gupta 1999) and apparently can survive without oxygen by accumulating and respiring nitrate (Risgaard-Petersen et al. 2006; Piña-Ochoa et al. 2010). Recently, the existence of metazoans (loriciferans) that can live permanently in anoxic habitats (hypersaline basins in the deep Mediterranean) has been convincingly demonstrated (Danovaro et al. 2010).

Such unequivocal evidence is not yet available for the benthos from the deeper parts of the Black Sea. Nevertheless, a number of considerations suggest that at least some of the organisms were alive, rather than having been transported from adjacent oxygenated areas.

1. Specimens collected within the oxic/anoxic transition zone (120–240 m water depth) stained intensely with Rose Bengal and showed no sign of morphological damage (Fig. 19). Danovaro et al. (2010) also observed that the live loriciferans they studied were very strongly stained and morphologically intact, in contrast to other meiofaunal organisms that stained weakly and were therefore assumed to be transported carcasses.
2. Meiobenthic densities within the transition zone are similar to, and even exceed, those in sublittoral areas.
3. Our samples yielded a characteristic fauna of benthic organisms, adapted to limited oxygen concentrations. Among the main components were soft-shelled foraminiferans, large numbers of undescribed nematode species, the polychaetes *Vigtorniella zaikai* and *Protodrilus* sp., and two unknown tardigrade species, all of which appear to be confined to the transition zone.
4. The nematode and harpacticoid populations included specimens of all sizes (life stages), as well as gravid females containing eggs.
5. High ciliate densities appear to be associated with concentrations of bacteria, suggesting that these may be live populations.

Additional evidence that eukaryotes can live in the anoxic Black Sea comes from observations of live organisms. During a study of bottom sediments associated with methane gas hydrates in the Sorokin Trough (NE Black Sea), an endemic species of Cladocera was discovered in the hydrogen sulfide zone at depths of 1990 and 2,140 m (Sergeeva 2004b, c). This was described as *Pseudopenilia bathyalis,* the type species of a new genus (Sergeeva 2004b, c) and family Pseudopenilidae (Korovchinsky and Sergeeva 2008). One specimen of the new cladoceran was observed alive in a sample from the deeper site. In April–May

2010, during a cruise on the RV *Maria S. Merian,* we studied bottom sediments from 250 to 300 m depth in the Bosporus Strait region in order to search for live fauna in the permanent hydrogen sulfide zone of the Black Sea. Using light microscope, we observed actively moving protozoans (large ciliates) and metazoans (free-living nematodes). These observations were recorded on video. Assuming that they were indigenous, then these organisms provide good evidence that eukaryotes can live under anoxic/sulfidic conditions in the Black Sea. However, this question requires further investigation.

6. Discussion

The Black Sea is the largest and best-studied permanently anoxic body of deep water in the world. Since the discovery of sulfidic conditions in the 1880s, it has been widely assumed that the deeper parts are inhabited only by prokaryotes. However, the IBSS NAS of Ukraine (Sevastopol) has studied the possible occurrence of benthic protozoans and metazoans in the oxic/anoxic transition and permanently anoxic zones of the Black Sea for more than 30 years (Sergeeva 2000a, b; Sergeeva 2001; Zaika 1999; Zaika et al. 1999). The previous and new observations on meiobenthos summarized in the present paper were made at depths (120–240 m) corresponding to the transition between increasingly hypoxic but non-sulfidic bottom water and the anoxic/sulfidic zone. The poorly defined boundary between these two domains is located at approximately 150–180 m.

Our new data suggest that the oxic/anoxic transition zone supports a rich protozoan and metazoan biota. Although organisms are found in samples from deeper, anoxic/sulfidic areas, high faunal densities are typically located in depths where oxygen disappears. Both calcareous and monothalamous (soft-shelled) foraminifera exhibit a sharp peak at ~160 m (Sergeeva et al. 2010). Among the ciliates, abundances maxima are located at 120, 160–190, and at 240 m, where they are possibly associated with concentrations of bacterial cells in the area of transition between oxic and anoxic/sulfidic conditions. These observations echo those of Zubkov et al. (1992) who described a unique community of ciliates from the oxic/anoxic interface in the central Black Sea and off the coasts of Bulgaria and Georgia. *Pleuronema marinum* and members of the ciliate families Tracheliidae, Holophryidae, and Amphileptidae were concentrated in hypoxic waters ($O_2 = 15$ µmol L^{-1}). *Askenasia* sp. (family Mesodiniidae), many individuals of which were covered in epizoic bacteria, inhabited the lower part of the hypoxic layer and the upper part of the sulfidic zone ($H_2S = 6$ µmol L^{-1}). Zubkov et al. (1992) suggested that the ciliates were feeding on large sulfur bacteria. The distribution of ciliates in the Black Sea has parallels in other meromictic water bodies. In particular, molecular analyses have revealed the presence of protists, including ciliates, around and below the oxic/anoxic boundary in the Cariaco Basin (Caribbean Sea) (Stoeck et al. 2003) and in anoxic waters in the Framvaren Fjord (Norway), where bottom-water sulfide levels are 25 times higher than in the Black

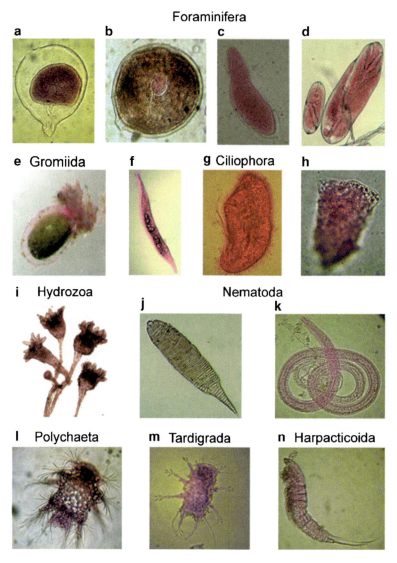

Figure 19. Examples of meiobenthos found in hypoxic/anoxic sediments at 120–2,075 m depth in the Black Sea. (**a–d**) Foraminifera, 140–170 m. (**e**) Gromida, 120–150 m. (**f–h**) Ciliophora, 190–240 m. (**f**) 2,075 m. (**g**) 170 m. (**h**) (**j–k**) Nematoda, 120–240 m. (**l**) Polychaeta, 120–150 m. (**m**) Tardigrada, 120–160 m. (**n**) Harpacticoida, 120–170 m.

Sea (Behnke et al. 2006). Earlier, distinct peaks of ciliates around oxic/anoxic boundaries were revealed using classical microscopic techniques in Danish fjords (Fenchel et al. 1990) and in Lake Cisó (Spain) (Massana and Pedros-Alio 1994).

Very large numbers of ciliated protozoa, contained symbiotic algae (*Chlorella* spp.), were reported to live beneath the oxic/anoxic boundary in a stratified freshwater pond (Finlay et al. 1996). The three most abundant metazoan taxa in our samples (nematodes, harpacticoids, and polychaetes) likewise display abundance maxima around the level where oxygen disappears. In particular, the polychaetes *Protodrilus* sp. and *Vigtorniella zaikai* are concentrated in this narrow zone in our new material, as well as elsewhere on the Black Sea margin (Bacesco 1963), suggesting that they can be considered as indicators of severe hypoxia in this basin.

The high abundance of different meiobenthos taxa at the depths with a low concentration of dissolved bottom-water oxygen is undoubtedly not accidental. These concentrations are reminiscent of the "edge effects" observed at the upper and/or lower boundaries of many continental margin oxygen minimum zones (Levin 2003; Gooday et al. 2009). OMZ edge effects generally coincide with rising oxygen levels across the upper or lower boundaries, whereas in the Black Sea, and perhaps other anoxic basins, abundance maxima appear where oxygen has almost disappeared. In both cases, however, the abundance of benthic organisms appears to be a response to an enhanced food supply.

7. Future Perspectives

Our data suggest that some benthic eukaryotes can tolerate anoxic and sulfidic conditions. Further comparative studies of shallow- and deep-water meiobenthic communities in the Black Sea are necessary in order to establish which species are characteristic and indicative of hypoxic/anoxic conditions as well as the proportion of Rose-Bengal-stained specimens that were alive when captured. The recent paper of Danovaro et al. (2010) provides a model for the kinds of analyses that could be conducted. The application of CellTracker Green to freshly collected samples is one obvious way forward. In the case of ciliates, a taxonomic study would reveal whether they include genera typical of anoxic sediments.

Further studies of the Black Sea seep fauna will yield more information about the taxonomic composition of benthos in the transitional oxic/anoxic water masses. They should reveal, among other things, the relationship between the diversity and abundance of meiofauna and concentrations of hydrogen sulfide and methane. The specific physiological and biochemical processes that facilitate the survival of eukaryotes in such "extreme" environments are important questions for future studies too.

8. Acknowledgements

A part of this work was supported from EC 7th FP project HYPOX 226213 and 6th FP project HERMES GOCE-CT-2005-511324. AJG was supported by the Oceans 2025 project funded by the Natural Environment Research Council, UK, and the

Collaborative Project HERMIONE (contract number 226354) funded under the European Commission's Framework seven program. We thank an anonymous reviewer for detailed comments that helped to improve the paper considerably. We are grateful to Professor Antje Boetius for providing an opportunity to participate in RV *Meteor* cruise 72/2 and RV *M.S. Merian* cruise 15/1, and Katerina Ivanova for sample collection, Felix Janssen for help with the O_2 and H_2S data, and Evgenia Babich for support in preparing the figures.

9. References

Anikeeva OV (2003) Monothalamous soft-shelled foraminifera in Sevastopol bay (Crimea, the Black Sea): biodiversity of coastal marine ecosystems functional aspects. In: High-level scientific conference, Renesse, 11–15 May 2003. Renesse Abstract, pp 69–70

Anikeeva OV, Sergeeva NG (2001) Distribution of the foraminifera on the Crimean shelf (Black Sea). In: Book of abstracts, conference on ecology problems of Azov–Black Sea basin: modern state and prediction, Sevastopol, Sept 2001, pp 42–44 (In Russian)

Bacesco M (1963) Contribution a la biocenologie de la mer Noir. L'etage periazoique et la facies dreissenifere leurs caracteristiques. Rapp Commun Int Mer Mediterr 17:107–122

Barnett PR, Watson J, Conelly J (1984) A multiple corer for taking virtually undisturbed samples from shelf, bathyal and abyssal sediments. Oceanol Acta 7:399–408

Behnke A, Bunge J, Barger K, Breiner H-W, Alla V, Stoeck T (2006) Microeukaryote community pattern along O_2/H_2S gradient in supersulphuridic anoxic fjord (Framvaren, Norwey). Appl Environ Microbiol 72:3626–3636

Bernhard JM (2000) Distinguishing live from dead foraminifera: methods review and proper applications. Micropaleontology 46(suppl 1):38–46

Bernhard JM, Sen Gupta BK (1999) Foraminifera in oxygen-depleted environments. In: Sen Gupta BK (ed) Modern foraminifera. Kluwer Academic, Dordrecht/Boston/London, pp 201–216

Bernhard JM, Ostermann DR, Williams DS, Blanks JK (2006) Comparison of two methods to identify live benthic foraminifera: a test between rose Bengal and CellTracker Green with implications for stable isotope paleoreconstructions. Paleoceanography 21:PA4210. doi:10.1029/2006PA001290

Boetius A (2007) R/V *Meteor* Cruise 72–2 MICROHAB. Istanbul (Turkey) – Istanbul (Turkey) Short cruise report, Max Planck Institute for Marine Microbiology, 23 Feb–13 Mar 2007, 11 pp

Bougis B (1950) Mèthode pour l'etude quantitative de la microfaune de fonde marine (meiobenthos). Vie Milieu 1:23–38

Chislenko LL (1961) On the existence of dimensional gap in the marine fauna of the littoral and sublittoral. In: Proceedings of academy of sciences of the USSR, New Series 137:431–435 (In Russian)

Cline JD (1969) Spectrophotometric determination of hydrogen sulfide in natural waters. Limnol Oceanogr 14:454–458

Cowie G (2005) The biogeochemistry of Arabian Sea surficial sediments: a review of recent studies. Prog Oceanogr 65:260–289

Danovaro R, Dell'Anno A, Pusceddu A, Gambi C, Heiner I, Kristensen RM (2010) The first metazoa living in permanently anoxic conditions. BMC Biol 2010(8):30

Egorov VN, Luth U, Luth C, Gulin MB (1998) Gas seeps in the submarine Dnieper Canyon Black Sea: acoustics, video and trawl data. In: Luth U, Luth C, Thiel H (eds) MEGASEEPS-methane gas seeps exploration in the Black Sea, vol 14. Berichte aus dem Zentrum für Meeres- und Klimatoforsch, Hamburg, pp 11–21

Egorov VN, Polykarpov GG, Gulin SB, Artemov J, Stokozov MA, Kostova SK (2003) Modern conception about forming-casting and ecological role of methane gas seeps from bottom of the Black Sea. Mar Ecol J 7:5–26 (In Russian)

Eremeev VN, Konovalov SK (2006) On the budget and the distribution of oxygen and sulfide in the Black Sea water. Mar Ecol J 5:5–30

Fenchel T, Finlay BJ (1995) Ecology and evolution in anoxic worlds. Oxford University Press, Oxford

Fenchel T, Kristensen LD, Rasmussen L (1990) Water column anoxia: vertical zonation of planktonic protozoa. Mar Ecol Prog Serv 62:1–10

Finlay BJ, Maberly SC, Esteban GF (1996) Spectacular abundance of ciliates in anoxic pond water: contribution of symbiont photosynthesis to host respiratory oxygen requirements. FEMS Microbiol Ecol 20:229–235

Golemansky VG (1974) *Lagenidopsis valkanovi* gen. n., sp. n – un nouveau thécamoebien (Rhizopoda: Testacea) du psammal supralittoral des mers. Acta Protozoologica 13:1–4

Golemansky VG (1999) *Lagynis pontica* n. sp., a new monothalamous rhizopod (Granuloreticulosea: Lagynidae) from the Black Sea littoral. Acta Zool Bulg 51:3–13

Golemansky VG (2007) Testate amoebas and monothalamous foraminifera (Protozoa) from the Bulgarian Black Sea coast. In: Fet V, Popov A (eds) Biogeography and ecology of Bulgaria. Springer, Dodrecht, pp 555–570

Gomoiu M-T, Begun T, Teaca A (2008) Macrobenthos distribution along the depth gradient on the North-western Black Sea. *IGCP 521* "Black Sea-Mediterranean Corridor during last 30 ky: Sea level change and human adaptation". Extended Abstracts, pp 63–65

Gooday AJ, Anikeeva OV, Sergeeva NG (2006) *Tinogullmia lukyanovae* sp. nov. – a monothalamous, organic-walled foraminiferan from the coastal Black Sea. J Mar Biol Assoc UK 86:43–49

Gooday AJ, Levin LA, Aranda da Silva A, Bett BJ, Cowie GL, Dissard D, Gage JD, Hughes DJ, Jeffreys R, Lamont PA, Larkin KE, Murty SJ, Schumacher S, Whitcraft C, Woulds C (2009) Faunal responses to oxygen gradients on the Pakistan margin: a comparison of foraminiferans, macrofauna and megafauna. Deep-Sea Res II 56:488–502

Gulin MB (1991) The study of microbial processes of sulfate reduction and chemosynthesis in the water environment of the Black Sea. PhD. Sevastopol, 20 p (In Russian)

Jorissen FJ, Buzas MA, Culver SJ, Kuehl SA (1994) Vertical distribution of living benthic foraminifera in submarine canyons off New Jersey. J Foramin Res 24:28–36

Kiseleva MI (1959) Distribution of polychaete larvae in the plankton of the Black Sea. Proc Sevastopol Biol Station 12:160–167 (In Russian)

Kiseleva MI (1979) *Zoobenthos*. The principles of biological productivity of the Black Sea. Naukova Dumka Publication House, Kiev, pp 208–241 (In Russian)

Kiseleva MI (1990) On the absence of larvae of pelagic polychaete in the Black Sea. Zool J 69:132–133 (In Russian)

Kiseleva MI (1992) New genus and species of the family Chrysopetalidae Polychaeta from the Black sea. Zool J 71:128–132 (In Russian)

Kiseleva MI (1998) Peculiarity of vertical distribution of polychaetes family Protodrilidae and Nerilidae in the Black Sea. Zool J 77:533–539 (In Russian)

Klenov MV (1948) The geology of sea. Uchpedgiz, Moscow, 495p (In Russian)

Korovchinsky NM, Sergeeva NG (2008) A new family of the order Ctenopoda (Crustacea: Cladocera) from the depths of the Black Sea. Zootaxa 1795:57–66

Kriss AE (1959) Marine microbiology (deepwater). M. Publishing house USSR Academy Science, Moscow, 455p

Levin LA (2003) Oxygen minimum zone benthos: adaptation and community response to hypoxia. Oceanogr Mar Biol 41:1–45

Luth C, Luth U (1997) A benthic approach to determine long-term changes of the oxic/anoxic interface in the water column of the Black Sea. In: Hawkins LE, Hutchinson S (eds) The responses of marine organisms to their environments. Proceedings of the 30th European marine biology symposium, Southampton, United Kingdom, 18–22 Sept 1995. Southampton Oceanography Centre, Southampton, pp 231–242

Luth U, Luth C (1998) Benthic meiofauna and macrofauna of a methane seep area south-west of the Crimean Peninsula, Black Sea. In: Luth U, Luth C, Thiel H (eds) MEGASEEPS-methane gas seeps exploration in the Black Sea, vol 14. Berichte aus dem Zentrum für Meeres- und Klimatoforsch, Hamburg, 113–126

Marinov T (1978) Qualitative composition and quantitative distribution of the meiobenthos of the Bulgarian Black sea waters, 1. Proc Inst of Fish, Varna 16:35–49 (In Bulgarian)

Massana R, Pedros-Alio C (1994) Role of anaerobic ciliates in planktonic food webs: abundance, feeding, and impact on bacteria in the field. Appl Environ Microbiol 60:1325–1334

Murina VV, Kideis AE, Ustin F, Toklu B (2006) Occurrence of bathypelagic larvae of the polychaete, Vigtorniella zaikai in the southern part of Black Sea. Mar Ecol J 5:57–62

Nikitin VN (1938) The lower boundary of benthic fauna and its distribution in the Black Sea. Rep Ac Sci USSR 21:341–345 (In Russian)

Pimenov NV, Rusanov II, Poglazova MN, Mityushina LL, Sorokin DYu, Khmelenina VN, Trotsenko YuA (1998) Bacterial mats on coral-shaped carbonate structures in methane areas of the Black Sea. In: Luth U, Luth C, Thiel H (eds) MEGASEEPS-methane gas seeps exploration in the Black Sea, vol 14. Berichte aus dem Zentrum für Meeres- und klimatoforsch, Hamburg, pp 37–50

Piña-Ochoa E, Høgslund S, Geslin E, Cedhagen T, Revsbech NP, Nielsen LP, Schweizer M, Jorissen F, Rysgaard S, Risgaard-Petersen N (2010) Widespread occurrence of nitrate storage and denitrification among Foraminifera and Gromiida. Proc Natl Acad Sci 107:1148–1153

Revkov NK, Sergeeva NG (2004) Current state of the zoobenthos at the Crimean shores of the Black Sea. In: Ozturk B, Mokievsky VO, Topalog˘lu B (eds) International workshop of the Black Sea Benthos, Istanbul, 18–23 Apr 2004, pp 189–217

Risgaard-Petersen N, Langezaal AM, Ingvardsen S, Schmid MC, Jetten MS, Op den Camp HJ, Derksen JW, Pina-Ochoa E, Eriksson SP, Nielsen LP, Revsbech NP, Cedhagen T, van der Zwaan GJ (2006) Evidence for complete denitrification in a benthic foraminifer. Nature 443:93–96

Rothe N, Gooday AJ, Cedhagen T, Hughes JA (2010) Biodiversity and distribution of the genus *Gromia* (Protista, Rhizaria) in the deep Weddell Sea (Southern Ocean). Polar Biol 33. doi:10.1007/s00300-010-0859-z

Sergeeva NG (2000a) Biological diversity of bottom sediments in hydrogen-sulphide zone of the Black Sea: the distribution by depths, the stratification in sediment depth. In: The geology of the Black and the Azov Seas, Kiev, pp 314–330 (In Russian)

Sergeeva NG (2000b) About the problem of biological diversity in the Black Sea deep-water benthos. Ekologiya Morya 50:57–62 (In Russian)

Sergeeva NG (2001) Meiobenthos of deep-water hydrogen sulphide zone of the Black Sea. Hydrobiology J 37:3–9 (In Russian)

Sergeeva NG (2003a) Meiobenthos in the region with the methane gas seeps. In: Eremeev VN, Gaevskaya AV (eds) Modern conditions of biological diversity in the nearshore zone of Crimea Peninsula (the Black Sea sector). Ekosi-Gidrophizika, Sevastopol, pp 258–267 (In Russian)

Sergeeva NG (2003b) Meiobenthos of the Crimean fluffy bottom bed. In: Eremeev VN, Gaevskaya AV (eds) Modern condition of biological diversity in near-shore zone of the Crimea (the Black Sea sector). Ekosi-Gidrophizika, Sevastopol, pp 248–251 (In Russian)

Sergeeva NG (2004a) Structure and distribution of meiobenthos in the region of methane gas seeps from the Black Sea bottom. Hydrobiological J 40:45–56

Sergeeva NG (2004b) *Pseudopenilia bathyalis* gen. n., sp. n. (Crustacea, Branchiopoda, Ctenopoda) – the inhabitant of hydrogen-sulphide zone of the Black Sea. In: Proceedings of the 30th international conference pacem in Maribus, Kiev/Sevastopol, 27–30 Oct 2003, pp 556–560

Sergeeva NG (2004c) *Pseudopenilia bathyalis* gen. n., sp. n. (Crustacea, Branchiopoda, Ctenopoda), an inhabitant of the hydrogen-sulphide zone of the Black Sea. Vestnik Zoologii 38:37–42

Sergeeva NG, Anikeeva OV (2004) New Black Sea foraminifera from the Allogromiidae family. In: 4th international congress "Environmental Micropaleontology, Microbiology and Meiobenthology", Isparta, 13–18 Sept 2004, pp 179–180

Sergeeva NG, Anikeeva OV (2008) *Goodayia rostellatum* gen. nov. sp. n. (PROTOZOA) – a monothalamous foraminiferan from the Black Sea. Vestnik Zoologii 42:467–471

Sergeeva NG, Anikeeva OV, Gooday AJ (2010) Soft-shelled (monothalamous) foraminifera from the oxic/anoxic interface (NW Black Sea). Micropaleontology 56:393–407

Sergeeva NG, Gulin MB (2007) Meiobenthos from an active methane seepage area in the NW Black Sea. Mar Ecol-Evol Persp 28:152–159

Sergeeva N, Gulin S (2009) Benthic fauna of the methane seeps in the Dnieper Paleo-Delta: comparative analysis IGCP 521-INQUA 0501 Fifth Plenary Meeting and Field Trip, Turkey, 22–31 Aug 2009, pp 151–152

Sergeeva NG, Kolesnikova EA (1996) The results of meiobenthic studies in the Black Sea. Ekologiya Morya 45:54–62 (In Russian)

Sergeeva NG, Zaika VE (2000) Ecology of Polychaeta from bordering pelagic and benthic communities of the Black Sea. Rep Acad Sci Ukraine 1:197–201

Sergeeva NG, Zaika VE (2008) Ciliophora in hydrogen sulfide zone of the Black Sea. Mar Ecol J 7:80–85

Sergeeva NG, Zaika VE, Kisseleva MI (1996) Life cycle and ecological demands of larval and adult *Vigtorniella zaikai* Kisseleva 1992 (Chrysopetalidae) in the Black Sea. Bull Mar Sci 60:622–623

Sergeeva NG, Zaika VE, Lichtschlag A (2008) Preliminary data on the presence of diverse benthic ciliate species in deep anoxic Black Sea. In: 5th international conference "Environmental Micropaleontology, Microbiology and Meiobenthology", Chennai, India, 17–25 Feb 2008. Chennai, pp 279–282

Sorokin Yu (1962) Microflora of the Black Sea bottom. Microbiologia 31:899–903

Sorokin YuI, Sorokina OV (2008) Primary production and bacterioplankton dynamics in the Black Sea during the cold season. Mar Ecol J 7:65–75

Stoeck T, Taylor G, Epstein S (2003) Novel eukaryotes from the permanently anoxic Cariaco basin (Caribbean Sea). Appl Environ Microbiol 2003:5656–5663

Stunzhas PA, Yakushev EV (2006) On fine hydrochemical structure of redox-zone in the Black Sea by the measurements results of the open oxygen sensor and by the bathometric data. Oceanology 46:650–672 (In Russian)

Surugiu V. (2005) The use of polychaetes as indicators of eutrophication and organic enrichment of coastal waters: a study case – Romanian Black Sea Coast. Analele Stiintifice ale Universitatii "Al.I. Cuza" Iasi, s. Biologie animala, LI

Temelkov BK, Golemansky VG, Todorov MT (2006) Updated checklist of the recent foraminifera from the Bulgarian Black Sea coast. Acta Zool Bulg 58:17–36

van der Weijden CH, Reichart GH, Visser HJ (1999) Enhanced preservation of organic matter in sediments deposited within the oxygen minimum zone in the northeastern Arabian Sea. Deep-Sea Res I 46:807–830

Vinogradov AA, Flint MV (1987) Current state of the Black Sea ecosystem. Nauka, Moscow, 232 pp (In Russian)

Winkler LW (1888) The determination of dissolved oxygen in water. Ber Dtsch Chem Ges 21:2843–2857

Yanko VV, Troitskaya TS (1987) Late Quaternary foraminifera of the Black Sea. Nauka, Moscow (In Russian)

Yanko VV, Vorobjeva LV (1990) Modern foraminifera of Sea of Azov and Kerch strait. Ekologiya Morya 35:29–34 (In Russian)

Yanko VV, Vorobyova LV (1991) Foraminifera of the Bosphorus and the Black Sea. Ekologiya Morya 39:47–51 (In Russian)

Zaika VE (1999) Specific pelagic and benthic communities of the Black Sea in the hydrogen sulfide zone. Biology of the Sea. Vladivostok 25:480–482 (In Russian)

Zaika VE (2008) Is there animal life at the Black sea great depths? Mar Ecol J 7:5–11

Zaika VE, Sergeeva NG (2008) The boundary change of benthic settlement of polychaetes *Protodrilus* sp. and *Vigtorniella zaikai* in the Black Sea. Mar Ecol J 7:49–53 (In Russian)

Zaika VE, Sergeeva NG (2009) The vertical distribution of the deep-water ciliates in the Black sea. Mar Ecol J 8:30–34 (In Russian)

Zaika VE, Sergeeva NG, Kisseleva MI (1999) Two polychaete species bordering deep anoxic waters in the Black Sea. Tavricheskiy Medico-Biologicheskiy Vestnik, no 1–2: 56–60

Zhizhchenko BP (1974) Methods of palaeogeographical investigations at oil and gas fields. Nedra Publishing House, Moscow, 375 pp (In Russian)

Zubkov MV, Sazhin AF, Flint MV (1992) The microplankton organisms at the oxic-anoxic interface in the pelagial of the Black Sea. FEMS Microbiol Lett 101:245–250

Biodata of **Alessandro Saccà**, author of *"The Role of Eukaryotes in the Anaerobic Food Web of Stratified Lakes."*

Dr. Alessandro Saccà is a research fellow in the field of aquatic microbial ecology at the University of Messina (Messina, Italy). He got a diploma from the Faculty of Mathematical, Physical and Natural Sciences of the University of Messina and earned his Ph.D. from the same university with a dissertation on the role of predation in the microbial food web of anoxic aquatic environments (2004). He carried out most of his postdoctoral activities at the Institute for the Coastal Marine Environment of the National Research Council in Messina and at the Department of Animal Biology and Marine Ecology of the University of Messina. He also collaborated with the Group of Molecular Microbial Ecology, University of Girona (Catalunya, Spain), and with the Department of Integrative Biology, University of Guelph (Ontario, Canada).

His scientific interests focus on the carbon and energy flux through the planktonic microbial food web, with particular regard to chemically stratified aquatic environments characterized by bottom anoxic waters. He is also concerned about the biology and ecology of free-living ciliated protozoa and of anoxygenic photosynthetic bacteria.

E-mail: **asacca@unime.it**

THE ROLE OF EUKARYOTES IN THE ANAEROBIC FOOD WEB OF STRATIFIED LAKES

ALESSANDRO SACCÀ
Department of Animal Biology and Marine Ecology,
University of Messina, Viale Ferdinando Stagno d'Alcontres 33,
98166 Messina, Italy

1. Chemically Stratified Aquatic Environments

Molecular oxygen (O_2) has become widespread throughout the biosphere thanks to its massive production by early prokaryote organisms performing oxygenic photosynthesis. The water column of present aquatic environments is generally oxygenated from the surface down to the bottom, along with the interstitial water of the uppermost layer of sediments. On the contrary, anoxia occurs in all natural waters where the rate of oxygen consumption exceeds the rate at which this molecule is supplied. Although molecular oxygen is spontaneously consumed by oxidative reactions and, through enzymatic catalysis, by any living organism respiring aerobically, there are efficient mechanisms that replenish the depleted stock. In fact, the supply of oxygen in the water is dependent upon diffusion, turbulence, advection, and also, in the photic zone, oxygenic photosynthesis. Since the most effective transport mechanism by far is advection, anoxia is typical of enclosed basins where physical barriers and density stratifications limit the advection of O_2 to deep waters (Grasshoff 1975), causing the water column to become both physically and chemically stratified. Stratified aquatic environments with anoxic bottom waters are found in the sea as well as on continents.

There are several types of stratified basins. The most common one in coastal areas is associated with a strong halocline (salinity gradient), which is the result of a net outflow of low-salinity water from a positive estuary: The halocline prevents low-salinity shallow oxygenated waters from mixing with high-salinity deep anoxic layers. Examples of this type of basin are the Black Sea, the Baltic Sea, and many fjords, as well as coastal lakes and estuaries. A second type of basin arises because of a strong thermocline preventing the mixing of surface and deep waters. The Cariaco Trench, located on the northern continental shelf of Venezuela, is the world's largest truly marine anoxic body of water and is an outstanding example of this type of basin. A third type is due to submerged mineralized springs bringing a continuous supply of denser water into the lower layer of the basin. Stratification, at high latitudes and/or altitudes, may also develop due to deep-water accumulation of salts precipitated by freezing out from a surface ice layer. A final type is originated due to the interactions between tectonic processes, fluid migration, and dissolution of evaporitic rocks. In the sea it is represented by the deep hypersaline anoxic basins (DHABs), such as the ones present in the Eastern Mediterranean at depths varying between 3,300 m and 3,700 m along the Mediterranean ridge (Jongsma et al. 1983; Westbrook and Reston 2002). This chapter mainly focuses on freshwater and saline lakes, although most notions apply also to marine environments.

In ecology, lakes are most commonly discriminated based on their mixing behavior, since this directly influences their ecological processes. The first hierarchical discrimination is generally made between holomictic and meromictic lakes. In holomictic lakes (the most common type), the water circulates from the surface down to the bottom at least once a year, resulting in homogenization of dissolved oxygen and nutrient concentrations throughout the water mass. Meromictic lakes, on the contrary, are characterized by a persistent physical and chemical stratification of the water column and are distinguished by two segregated zones: an upper layer that undergoes mixing, defined as the mixolimnion (Hutchinson 1937), and a stagnant lower layer, defined as the monimolimnion (Findenegg 1935). Typically, the monimolimnion is denser than the mixolimnion, and this maintains stability since water circulation can occur only to a depth where the mixing forces are greater than the stabilizing ones. While the mixolimnion is well oxygenated, chiefly due to advection processes, the monimolimnion is anoxic and usually characterized by a vertical gradient of hydrogen sulfide (H_2S) concentrations, reaching a maximum at the water/sediment interface. A transition zone exists between these two layers, named the chemocline (Hutchinson 1937), where oxygen concentration sharply decreases with depth.

Meromictic lakes are both freshwater and saline, and are scattered all around the world. Meromixis, however, is a complex and still debated phenomenon, and discrimination between holomictic and meromictic basins is not always straightforward (compare Lewis 1983 and Hakala 2004 and the references therein). At any rate, from most ecological aspects, seasonally stratified lakes are comparable to meromictic lakes during the period of water column stratification.

2. Microbial Zonation in Stratified Lakes

Stratified lakes, and particularly meromictic ones, have long been objects of scientific interest because of their importance in biogeochemical processes (Sorokin and Donato 1975; Cloern et al. 1983; McDonough et al. 1986; Zehr et al. 1987; Bloem and Bär-Gilissen 1989; Macek et al. 1994; Oremland et al. 2000; Lehours et al. 2005). In fact, these lakes are inhabited by communities of free-living prokaryotes that are able to exploit virtually every ecological niche within the chemical gradients of the water column. Anoxic waters, along with sediments, host populations of fermenting, sulfate reducing, denitrifying, and methanogenic representatives of both Bacteria and Archaea, which thrive according to the local redox potential and the specific substrate availability. In the lower part of the chemocline, each of the of microaerophilic chemoautotroph and photoheterotroph populations settles at its preferred depth according to its microhabitats, while anoxygenic photoautotrophs bloom in the upper layer of the anoxic and sulfidic zone, provided that enough light is available. Finally, cyanobacteria may flourish throughout the photic zone, both in oxygenated and in anoxic waters.

Whenever light is not limiting and high concentrations of H_2S provide reducing power in great supply, anaerobic phototrophs are often the dominant organisms and can form dense populations within narrow layers, the so-called "plates." The development of a plate of phototrophic sulfur bacteria in permanently or periodically stratified lakes with an anoxic bottom layer is a widespread phenomenon (Takahashi and Ichimura 1968; Trüper and Genovese 1968; Biebl and Pfennig 1979; Parkin and Brock 1980; Guerrero et al. 1985; Overmann and Tilzer 1989; Mas et al. 1990; Overmann et al. 1991; Vicente et al. 1991). However, in temperate stratified lakes, chemoclines occur at such a depth that light is the limiting factor for the growth of photosynthetic sulfur bacteria during most of the year. In summer, the establishment of a thermocline strengthens the water column stability, and the oxic–anoxic boundary moves upward, along with the photosynthetic sulfur bacteria assemblage, to a depth where light intensity is no longer limiting. This is the reason why in some meromictic lakes, during summer, photosynthetic sulfur bacteria account for a considerable fraction of the primary production: 28.7% on average (according to Overmann 1997) and up to 83–85% (after Culver and Brunskill 1969 and van Gemerden and Mas 1995, respectively).

Regarding eukaryotes, since the last decade, it has been considered that only several protists were able to complete their whole life cycle under anoxic conditions. Fenchel et al. (1990) noticed that in stratified lakes with an anoxic bottom layer, a vertical zonation of anaerobic and microaerophilic ciliates, reflecting the chemical stratification of the environment exists, similar to the typical zonation of interstitial microfauna in marine sandy sediments (Fenchel 1969). The authors distinguished three main groups of ciliated protozoa: those living exclusively in the oxygenated surface layer, those dwelling in the redoxcline (redox discontinuity layer) and, finally, the strictly anaerobic forms which are found only in anoxic and chemically reduced waters (Fenchel et al. 1990). Furthermore, it has been observed that, when stratification is seasonal, the development of an anoxic bottom layer is associated with an exodus of otherwise benthic ciliates to the water column: microaerophilic ciliates migrate upwards until they find their ideal oxygen tension (Fenchel and Finlay 1984), while obligatory anaerobes live as plankton in the anoxic zone (Fenchel et al. 1990; Hayward et al. 2003).

3. The Hidden Biodiversity

Since recently, most of our notions of protistan diversity have been based on cultivation or manual single cell collection and subsequent microscopy analyses. Studies relying on these techniques have led to the conclusion that the anoxic protistan community below the chemocline of stratified aquatic environments is relatively poor in species (Fenchel et al. 1990, 1995).

It is now known, however, that many, if not most, protists are rather small sized (<5 μm) and have only few morphological characters useful for taxonomic identification. Moreover, as evolution is not necessarily accompanied by morphological alteration, many different evolutionary lineages or sexually isolated taxon groups are morphologically indistinguishable or are even "cryptic species" (Bickford et al. 2006; Hausmann et al. 2006; Stoeck et al. 2008). Protists from extreme environments, that are highly adapted to their natural habitat, hardly survive manual collection and microscopy preparations, while as for traditional culturing techniques, they are highly selective and detect only a very small proportion of organisms in an environmental sample.

In the past decade, microbial diversity research has been revolutionized by phylogenetic analyses of eukaryotic 18 S rRNA genes from environmental samples which are amplified and cloned into bacterial vectors for subsequent Sanger sequencing (for a review see Epstein and López-García 2008). Molecular diversity studies of microbial eukaryotes in anoxic aquatic environments revealed that protist groups are far more genetically heterogeneous than their morphological diversity suggests, and unveiled an exceptionally high ecological complexity in anaerobic protist communities (Dawson and Pace 2002; Stoeck and Epstein 2003; Stoeck et al. 2003; Lefranc et al. 2005; Richards et al. 2005; Slapeta et al. 2005; Takishita et al. 2007; Zuendorf et al. 2006; Lefèvre et al. 2007; Alexander et al. 2009; Stock et al. 2009).

However, even the largest SSU rDNA clone library has not approached sampling saturation and fails to accurately characterize the diversity of eukaryotic microbes in an environmental sample (e.g. Edgcomb et al. 2002; Stoeck et al. 2006, 2007). These shortcomings are in part due to the massive genetic diversity of protists in most environments and also to a number of methodological issues with the clone library approach.

The setting up of the first massively parallel pyrosequencing platform (Margulies et al. 2005) has inaugurated the next-generation sequencing (NGS) era (see Voelkerding et al. 2009 for a review). A promising NGS technique, defined as high-throughput parallel tag 454 sequencing, offers an unprecedented scale of sampling for the molecular detection of microbial diversity (Stoeck et al. 2010) and allows for more comprehensive insights into the diversity of protistan communities. The use of NGS for surveys of molecular diversity of eukaryotic microorganisms is not exempt of limitations, though, (Medinger et al. 2010) and should be combined with appropriate statistical tools (Stoeck et al. 2009) and well-designed sampling strategies (Nolte et al. 2010).

4. Anaerobic Metabolism in Free-Living Protozoa

The importance of free-living phagotrophic protists in the carbon flux of aquatic systems is well known following the introduction of the "microbial food loop" model by Azam et al. (1983), according to which prokaryotes are exploited (and controlled) by small protozoa; in turn, these are fed upon by larger protozoa, and the overall grazing activity is accompanied by the excretion of substances that are taken up by the bacteria in an almost closed circuit. This paradigm was integrated later by Rassoulzadegan (1993), who introduced the notion of the "microbial food web" for the more open system that also includes the autotrophic pico- and nanoplankton. There is evidence, however, that the impact of phagotrophic protozoa on prey populations in anoxic water compartments is lower than in oxygenated ones, possibly due to their less-efficient metabolism (Fenchel and Finlay 1990).

In fact, among the eukaryotes that live in the absence of oxygen, some base their energetic metabolism upon simple cytosolic fermentations in which the electrons resulting from glycolysis are transferred to pyruvate, with lactate or ethanol excreted as end products. Many anaerobic protozoa can ferment pyruvate into acetate and hydrogen; this occurs in special organelles referred to as hydrogenosomes that contain a hydrogenase enzyme. Hydrogenosomes occur in a wide spectrum of anaerobic protists, including ciliates, amoeboflagellates, chytridiomycete fungi, and parabasalids (Müller 1993; Embley et al. 1995; Hackstein et al. 1999, 2001; Kulda 1999; Van der Giezen et al. 2002; Voncken et al. 2002). Both morphologically and enzymatically, hydrogenosomes resemble mitochondria (Biagini et al. 1997; Finlay and Fenchel 1989) and have evolved from these organelles independently in the various taxa of anaerobic protozoa (Boxma et al. 2005; Martin 2005). Various hydrogenosome-bearing ciliates harbor methanogenic endosymbionts, which depend upon the H_2 production by hydrogenosomes for their growth and survival. The intracellular methanogens probably gain energy by reducing CO_2 with the waste H_2 from host hydrogenosomes (Jones et al. 1987). The benefits to the host are less obvious, and in most ciliates, they are unknown. However, some large ciliates including *Metopus contortus* and *Plagiopyla frontata* do benefit, since their growth rate and yield are reduced if the methanogens are specifically inhibited (Fenchel and Finlay 1991b). In some anaerobic ciliates, the methanogens even undergo dramatic morphological changes during physical association with hydrogenosomes (Embley and Finlay 1994). This kind of physiological interaction, called anaerobic syntrophy, is widespread in anaerobic habitats (Fenchel and Finlay 1995; Finlay and Fenchel 1992).

In eukaryotes living in oxygen-poor (hypoxic) or oxygen-free (anoxic) conditions, a mitochondrial ATP synthesis can also occur. Such protists bear mitochondria which produce ATP with the help of proton-pumping electron transport, using terminal electron acceptors other than O_2 so that their excreted end product of electron transport is not H_2O but nitrite (NO_2^-), nitric oxide (NO), succinate, and so forth, depending on the electron acceptor (nitrate, nitrite, fumarate, etc.). Anaerobically functioning mitochondria need an alternative terminal oxidase to be able to use substrates other than oxygen as the final electron acceptor. Organisms with anaerobic mitochondria can be broadly divided into two different classes: those that use an electron acceptor present in the environment, such as NO_3^-, and those that use an endogenously produced organic electron acceptor, such as fumarate (Tielens et al. 2002). Examples of the first class are the most common among anaerobic protists, and nitrate respiration has been reported from several ciliates (Finlay et al. 1983).

In anoxic waters, it is also possible to find protozoa respiring aerobically: Some ciliates have a typical aerobic metabolism but harbor endosymbiotic algae that, by producing oxygen, enable their hosts to survive in oxygen-free environmental conditions. In fact, at a depth where the photon irradiance is enough for a positive net photosynthesis (and hence net oxygen evolution) by its symbionts, each host ciliate may live aerobically even if surrounded by anoxic water (Finlay et al. 1996). The ciliate also manipulates its symbionts in order to release a photosynthate (usually maltose) for its own metabolism requirements (Finlay 1990).

5. Protozoa in Anaerobic Food Webs

Studies on the trophic role of planktonic protozoan populations in anoxic aquatic ecosystems have seldom been carried out (Finlay et al. 1991; Guhl and Finlay 1993; Massana and Pedrós-Alió 1994; Saccà et al. 2009). Furthermore, only a few of them have directly addressed the in situ impact of predation by phagotrophic protozoa both on prey populations and on predator populations (Massana and Pedrós-Alió 1994; Saccà et al. 2009). As a consequence, our knowledge of the carbon and energy flow in anaerobic food webs is less accurate than of that which takes place in aerobic ones.

One question that may arise regards what differentiates anaerobic food webs from aerobic ones under the trophic aspect. First of all, anaerobic metabolism, as pointed out above, is less efficient than aerobic metabolism in terms of ATP yield, and secondly, the gross growth efficiency [(assimilated C)/(consumed C)] of anaerobic eukaryotes is lower (about 10% vs. 40%) than that of aerobic ones. This is in accordance with the appreciably lower ratio usually found between predator and prey biomass in anaerobic communities (about one fourth, on average, according to Fenchel and Finlay 1990). However, in stratified lakes, there is evidence that the prokaryotes tend to be larger (Fig. 1) and more abundant in anoxic waters than in oxygenated ones (Fenchel et al. 1990; Cole et al. 1993).

Larger prey cell size results in a higher amount of energy and carbon obtained by bacterivorous protozoa from each single food item ingested, and also, given a similar prey capture efficiency in both aerobic and anaerobic bacterivorous protozoa, the grazing rate would be enhanced in anoxic habitats due to the higher encounter frequency between predator and prey.

Since potential growth rate can be estimated as

$$\mu = \ln [1+(0.1 \times G \times C_b/C_c)]$$

where G is the predator-specific grazing rate; C_b and C_c are the average carbon content of bacterial and protozoan cells, respectively; and 0.1 is the gross growth efficiency for anaerobic protozoa, the presence of more and larger prey cells in

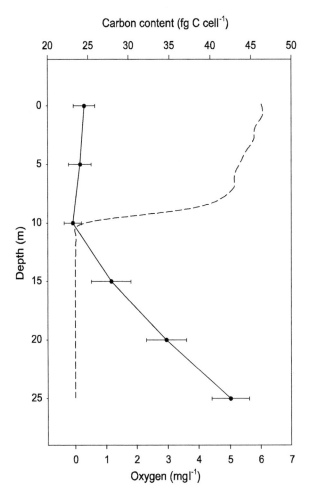

Figure 1. Vertical profile showing picoplankton mean cell carbon content (± 1 SE) in femtograms (fg) at each sampled depth (0, 5, 10, 15, 20, 25 m) in the coastal meromictic Lake Faro (Sicily, Italy) during an annual cycle ($n=11$). *Dashed line* represents the dissolved oxygen concentration (mg/l) in July (the oxic/anoxic interface was close to 10 m throughout the year). A pronounced and steady increase in carbon content with depth is apparent starting from 15 down to 25 m. (Saccà et al. 2008, unpublished data).

anoxic habitats would result in higher potential growth rates (and, accordingly, a higher biomass) than those predicted based only on the gross growth efficiency of anaerobic metabolism.

Anoxic habitats are also remarkable for the high proportion of eukaryotic microbial species that live in symbiotic association with prokaryotes. *Strombidium purpureum*, as an example, harbors endosymbiotic purple non-sulfur photosynthetic bacteria, while, as discussed in the previous section, many phagotrophic ciliates host methanogen endosymbionts. Moreover, numerous ciliated and flagellated protists, as well as some amoeboid protists, living in anoxic or hypoxic environments are regularly associated with ectobiotic prokaryotes (epixenosomes), the identity and functions of which are often unknown. In most cases, a metabolic cooperation including an exchange of substances seems likely, and sometimes, portions of the ectobionts are phagocytosed and digested (see Radek 2010 for a review). A variety of anaerobic ciliates (e.g., representatives of the genera *Parablepharisma*, *Metopus*, and *Sonderia*) are known for supporting ectosymbiotic prokaryotes, chiefly sulfate reducers (Fenchel et al. 1977; Fenchel and Finlay 1991a). Symbiotic associations in phagotrophic protozoa may have a significant effect on biogeochemical cycles: For example, the digestion of sulfide-oxidizing bacteria by ciliates that harbor methanogenic bacteria provides a short bridge between the anaerobic sulfur and carbon cycles (Finlay et al. 1991).

Another important difference between aerobic and anaerobic food webs is that anoxic assemblages apparently consist exclusively of prokaryotes and unicellular eukaryotes (Goulder 1972, 1974; Finlay 1980; Bark 1981), while metazoan predators, the most effective in terms of clearance, are virtually absent. It is generally assumed that the shortness of the anaerobic food chains with respect to aerobic ones (2–3 vs. 5–6 trophic levels, respectively) and the lack of metazoans completing their whole life cycle in anoxic conditions are due to the low growth efficiency of anaerobic microbial phagotrophs and thus to the small amount of energy available for higher trophic levels (Fenchel and Finlay 1990). This matter is probably more complex, however, as implied by the unexpectedly high biodiversity recently found in many anoxic habitats (see Sect. 3) and by the discovery of benthic anaerobic metazoans living in extreme and resource-poor anoxic habitats (Danovaro et al. 2010). Moreover, the above hypothesis does not give an exhaustive explanation for the lack of anaerobic bacterivorous metazoans, which would feed at the same trophic level as many phagotrophic protists do, in the same way as rotifers, cladocerans, and copepod nauplii, among others, commonly do in oxygenated habitats. Finally, the occurrence of considerably high biomasses of phagotrophic protozoa during blooms of anaerobic phototrophic prokaryotes (Fig. 2) apparently conflicts with this assumption.

In fact, during blooms of photosynthetic sulfur bacteria, the proportion between predator and prey biomass can be higher than is usually found in anoxic compartments. Saccà et al. (2008) found a mean proportion of 21.36% between the biomass of bacterivorous protozoa (ciliates specialized on bacterial food, plus phagotrophic nanoflagellates) and the biomass of total picoplankton in correspondence with a bloom of the green sulfur bacterium *Chlorobium phaeovibrioides*. At the same depth, but not in the bloom period, the mean proportion was only 6.8%. This discrepancy was possibly due to the high gross growth rate of green sulfur bacteria (GSB) during their blooms, allowing a considerable biomass of phagotrophic protozoa to build up.

Some species of phagotrophic protozoa indeed appear to depend on photosynthetic sulfur bacteria for food (Finlay et al. 1991). Certain anaerobic protozoa actively hunt for photosynthetic sulfur bacteria and even track their seasonal

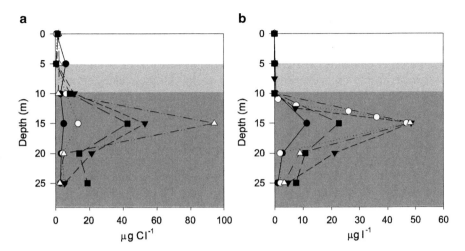

Figure 2. Vertical profiles of (**a**) phagotrophic ciliate biomass and (**b**) picoplankton bacteriochlorophyll *e* concentration in Lake Faro, at monthly intervals, from May to September 2004. *Black circles*, May; *white circles*, June; *black triangles*, July; *white triangles*, August; *black squares*, September. *Background color* indicates the water column stratification: mixolimnion, clear; chemocline, *grey*; anoxic monimolimnion, *dark grey* (Redrawn from Saccà et al. 2009).

migrations regardless of the distribution of the total bacterial abundance (Guhl and Finlay 1993). As a result, the biomass of phagotrophic protozoa can be strictly correlated with the bacteriochlorophyll concentration – a well-known biomarker for specific groups of photosynthetic sulfur bacteria (see Borrego et al. 1997) – independent of the biomass of total prokaryotes (Saccà et al. 2009). The considerably high biomass of anaerobic protozoa that often accompanies photosynthetic sulfur bacterial blooms may potentially support an additional trophic level of phagotrophs.

6. The Role of Metazoa

A high biomass of protists and/or bacteria may also attract motile aerobic metazoans somewhat adapted to low oxygen tensions, which are able to swim downward into anoxic waters to feed. Those metazoans are often in a reproduction/hatching stage of their life cycle during the bloom period of photosynthetic sulfur bacteria (late spring–summer) and can take advantage of such an additional food source. Vertical migrations between the oxygenated and anoxic zones, besides enhancing trophic resources, can also serve as a defense mechanism against fish predation for small metazoans (De Meester and Vyverman 1997; Kršinić et al. 2000).

Sorokin (1970) and Takahashi and Ichimura (1968) demonstrated the assimilation of radiocarbon-labeled photosynthetic sulfur bacteria by daphnids, while Overmann et al. (1999a) proved a direct carbon transfer from anaerobic phototrophic bacteria to copepods. Culver and Brunskill (1969) found that the vertical distribution of zooplankton in a meromictic lake was clearly skewed toward the chemocline, where O_2 concentration was low or equal to zero and a photosynthetic sulfur bacteria plate was present. All these clues imply that photosynthetic sulfur bacteria plates often constitute sites of high primary productivity in chemically stratified aquatic ecosystems. Such high productivity is exploited by both anaerobic phagotrophic protists and by those aerobic metazoans that are tolerant to low oxygen tensions and to the presence of sulfides. The latter, by migrating between the oxygenated and anoxic zones, export part of the anaerobic primary production to the mixolimnion, thus making it available to the aerobic food web of metazoans.

Anoxygenic photosynthetic production can be also channeled into the aerobic food web independent of metazoan vertical migrations. The phenomenon of massive upwellings of photosynthetic sulfur bacteria from the chemocline into the mixolimnion can occur at the end of the bloom period (usually early autumn), contributing significantly to the carbon and energy flux through the aerobic food web in stratified lakes (Overmann et al. 1999b).

The scenario depicted above appears limited to a small proportion of aquatic ecosystems on Earth (<0.5% according to Bertine and Turekian 1973; Emerson and Huested 1991; Morford and Emerson 1999), but the presence of anoxic and H_2S-containing (euxinic) water on a global scale was more common in past eons relative to the present one (Phanerozoic). Molecular oxygen began to accumulate in the atmosphere and in the surface of the ocean ca. 2,400 million years ago (Mya), but the persistent oxygenation of water masses throughout the oceans developed much later, perhaps beginning as recently as 580–550 Mya (Canfield 1998). For much of the transitional interval, moderately oxic surface waters overlaid an oxygen minimum zone (OMZ) that tended toward euxinia. A global oxygen deficiency in the deep ocean, and perhaps euxinia during most, if not all, of the Proterozoic (and likely extending into the early Phanerozoic), is thought to have occurred (Arnold et al. 2004; Lyons et al. 2009). In mid-Proterozoic oceans, a reduced partial pressure of oxygen and warmer temperatures (causing lower O_2 solubility) would have increased the sulfide availability for anoxygenic photoautotrophs, enhancing their potential contribution to the overall primary production (Johnston et al. 2009). Anoxygenic photosynthesis was widespread in the ancient oceans then (as is demonstrated by molecular fossils of pigments produced by anoxygenic phototrophs, Brocks et al. 2005), and what now happens only in stratified basins with shallow chemoclines was perhaps common for an interval at least 1,000 Mya in duration. Subsequent expansions of the OMZ in the oceans occurred repeatedly during the Cambrian period, characterized by the appearance of most of the major groups of complex animals. Therefore, the existence of adaptations to anoxic and sulfidic conditions in metazoans whose lineage originated that far in the past is not surprising.

7. References

Alexander E, Stock A, Breiner H-W, Behnke A, Bunge J, Yakimov MM, Stoeck T (2009) Microbial eukaryotes in the hypersaline anoxic L'Atalante deep-sea basin. Environ Microbiol 11:360–381

Arnold GL, Anbar AD, Barling J, Lyons TW (2004) Molybdenum isotope evidence for widespread anoxia in mid-proterozoic oceans. Science 304:87–90

Azam F, Fenchel T, Field JG, Gray JS, Meyer-Reil LA, Thingstad F (1983) The ecological role of water-column microbes in the sea. Mar Ecol Prog Ser 10:257–263

Bark AW (1981) The temporal and spatial distribution of planktonic and benthic protozoan communities in a small productive lake. Hydrobiologia 85:239–255

Bertine KK, Turekian KK (1973) Molybdenum in marine deposits. Geochim Cosmochim Acta 37:1415–1434

Biagini GA, Hayes AL, Suller MTE, Winters C, Finlay BJ, Lloyd D (1997) Hydrogenosomes of Metopus contortus physiologically resemble mitochondria. Microbiology 143:1623–1629

Bickford D, Lohman DJ, Sohdi NS, Ng PKL, Meier R, Winker K, Ingram K, Das I (2006) Cryptic species as a window on diversity and conservation. Trends Ecol Evol 22:148–155

Biebl H, Pfennig N (1979) Anaerobic CO_2 uptake by phototrophic bacteria. A review. Arch Hydrobiol Beih Ergeb Limnol 12:48–58

Bloem J, Bär-Gilissen M-JB (1989) Bacterial activity and protozoan grazing potential in a stratified lake. Limnol Oceanogr 34:297–309

Borrego CM, Garcia-Gil LJ, Vila XP, Figueras JB, Abella CA (1997) Distribution of bacteriochlorophyll homologues in natural populations of brown-colored phototrophic sulfur bacteria. FEMS Microbiol Ecol 24:301–309

Boxma B, de Graaf RM, van der Staay GWM, van Alen TA, Ricard G, Gabaldón T, van Hoek AHAM, Moon-van der Staay SY, Koopman WJH, van Hellemond JJ, Tielens AGM, Friedrich T, Veenhuis M, Huynen MA, Hackstein JHP (2005) An anaerobic mitochondrion that produces hydrogen. Nature 434:74–79

Brocks JJ et al (2005) Biomarker evidence for green and purple sulphur bacteria in a stratified Palaeoproterozoic sea. Nature 437:866–870

Canfield DE (1998) A new model for Proterozoic ocean chemistry. Nature 396:450–453

Cloern JE, Cole BE, Oremland RS (1983) Autotrophic processed in meromictic Big Soda Lake, Nevada. Liminol Oceanogr 28:1049–1061

Cole JJ, Pace ML, Caraco NF, Steinhart GS (1993) Bacterial biomass and cell size distributions in lakes: more and larger cells in anoxic waters. Limnol Oceanogr 38(8):1627–1632

Culver DA, Brunskill GJ (1969) Fayetteville Green Lake, New York. 5. Studies of primary production and zooplankton in a meromictic marl lake. Limnol Oceanogr 14:862–873

Danovaro R, Dell'Anno A, Pusceddu A, Gambi C, Heiner I, Kristensen RM (2010) The first metazoa living in permanently anoxic conditions. BMC Biol 8:30

Dawson SC, Pace NR (2002) Novel kingdom-level eukaryotic diversity in anoxic environments. Proc Natl Acad Sci USA 99:8324–8329

De Meester L, Vyverman W (1997) Diurnal residence of the larger stages of the Calanoid Copepod *Acartia tonsa* in the Anoxic Monimolimnion of a tropical Meromictic Lake in New Guinea. J Plankton Res 19:425–434

Edgcomb VP, Kysela DT, Teske A, de Vera GA, Sogin ML (2002) Benthic eukaryotic diversity in the Guaymas Basin hydrothermal vent environment. Proc Natl Acad Sci USA 99:7658–7662

Embley TM, Finlay BJ (1994) The use of small subunit rRNA sequences to unravel the relationships between anaerobic ciliates and their methanogen endosymbionts. Microbiology 140:225–235

Embley TM, Finlay BJ, Dyal PL, Hirt RP, Wilkinson M, Williams AG (1995) Multiple origins of anaerobic ciliates with hydrogenosomes within the radiation of aerobic ciliates. Proc R Soc Lond Ser B Biol Sci 262:87–935

Emerson SR, Huested SS (1991) Ocean anoxia and the concentrations of molybdenum and vanadium in seawater. Mar Chem 34:177–196

Epstein SS, López-García P (2008) "Missing" protists: a molecular prospective. Biodivers Conserv 17:261–276

Fenchel T (1969) The ecology of marine microbenthos IV Structure and function of the benthic ecosystem, its chemical and physical factors and the microfauna communities with special reference to the ciliated protozoa. Ophelia 6:1–182

Fenchel T, Finlay BJ (1984) Geotaxis in the ciliated protozoon *Loxodes*. J Exp Biol 110:17–33

Fenchel T, Finlay BJ (1990) Anaerobic free-living protozoa: growth efficiencies and the structure of anaerobic communities. FEMS Microbiol Ecol 74:269–276

Fenchel T, Finlay BJ (1991a) The biology of free-living anaerobic ciliates. Eur J Protistol 26:201–215

Fenchel T, Finlay BJ (1991b) Endosymbiotic methanogenic bacteria in anaerobic ciliates: significance for the growth efficiency of the host. J Protozool 38:18–22

Fenchel T, Finlay BJ (1995) Ecology and evolution in anoxic worlds. Oxford University Press, Oxford

Fenchel T, Perry T, Thane A (1977) Anaerobiosis and symbiosis with bacteria in free-living ciliates. J Protozool 24:154–163

Fenchel T, Kristensen LD, Rasmussen L (1990) Water column anoxia: vertical zonation of planktonic protozoa. Mar Ecol Prog Ser 62:1–10

Fenchel T, Bernard C, Esteban G, Finlay BJ, Hansen PJ, Iversen N (1995) Microbial diversity and activity in a Danish fjord with anoxic deep water. Ophelia 43:45–100

Findenegg I (1935) Limnologische Untersuchungen im Kärntner Seengebiete. Ein Beitrag zur Kenntnis des Stoffhaushaltes in Alpenseen. Int Rev ges Hydrobiol 32:369–423

Finlay BJ (1980) Temporal and vertical distribution of ciliophoran communities in the benthos of a small eutrophic loch with particular reference to the redox profile. Freshw Biol 10:15–34

Finlay BJ (1990) Physiological ecology of free-living protozoa. Adv Microb Ecol II:1–35

Finlay BJ, Fenchel T (1989) Hydrogenosomes in some anaerobic protozoa resemble mitochondria. FEMS Microbiol Lett 65:311–314

Finlay BJ, Fenchel T (1992) Methanogens and other bacteria as symbionts of free-living anaerobic ciliates. Symbiosis 14:375–390

Finlay BJ, Span ASW, Harman JMP (1983) Nitrate respiration in primitive eukaryotes. Nature 303:333–336

Finlay BJ, Clarke KJ, Vicente E, Miracle MR (1991) Anaerobic ciliates from a sulphide-rich solution lake in Spain. Eur J Protistol 27:148–159

Finlay BJ, Maberly SC, Esteban GF (1996) Spectacular abundance of ciliates in anoxic pond water: contribution of symbiont photosynthesis to host respiratory oxygen requirements. FEMS Microbiol Ecol 20:229–235

Goulder R (1972) The vertical distribution of some ciliated protozoa in the plankton of an eutrophic pond during the summer stratification. Freshw Biol 2:163–176

Goulder R (1974) The seasonal and spatial distribution of some benthic ciliated protozoa in Esthwaite Water. Freshw Biol 4:127–147

Grasshoff K (1975) The hydrochemistry of landlocked basins and fjords. In: Riley JP, Skirrow G (eds.) Chemical oceanography, 2nd edn. Academic, London

Guerrero R, Montesinos E, Pedrós-Alió C, Esteve I, Mas J, van Gemerden H, Hofman PAG, Bakker JF (1985) Phototrophic sulfur bacteria in two Spanish lakes: vertical distribution and limiting factors. Limnol Oceanogr 30:919–931

Guhl BE, Finlay BJ (1993) Anaerobic predatory ciliates track seasonal migration of planktonic photosynthetic bacteria. FEMS Microbiol Lett 107:313–316

Hackstein JHP, Akhmanova A, Boxma B, Harhangi HR, Voncken FGJ (1999) Hydrogenosomes: eukaryotic adaptations to anaerobic environments. Trends Microbiol 7:441–447

Hackstein JHP, Akhmanova A, Voncken F, van Hoek AHAM, van Alen T, Boxma B, Moon-van der Staay SY, van der Staay G, Leunissen J, Huynen M, Rosenberg J, Veenhuis M (2001) Hydrogenosomes: convergent adaptations of mitochondria to anaerobic environments. Zoology 104:290–302

Hakala A (2004) Meromixis as a part of lake evolution – observations and a revised classification of true meromictic lakes in Finland. Boreal Environ Res 9:37–53

Hausmann K, Selchow P, Scheckenbach F, Weitere M, Arndt H (2006) Cryptic species in a morphospecies complex of heterotrophic flagellates: the case study of *Caecitellus* spp. Acta Protozool 45:415–431

Hayward BH, Droste R, Epstein SS (2003) Interstitial ciliates: benthic microaerophiles or planktonic anaerobes? J Eukaryot Microbiol 50:356–359

Hutchinson EG (1937) A contribution to the limnology of arid regions. Trans Conn Acad Arts Sci 33:47–132

Johnston DT, Wolfe-Simon F, Pearson A, Knoll AH (2009) Anoxygenic photosynthesis modulated Proterozoic oxygen and sustained Earth's middle age. Proc Natl Acad Sci 106:16925–16929

Jones WJ, Nagle DP, Whitman WB (1987) Methanogens and the diversity of archaebacteria. Microbiol Rev 51:135–177

Jongsma D, Fortuin AR, Huson W, Troelstra SR, Klaver GT, Peters JM, Van Harten D, De Lange GJ, Ten Haven L (1983) Discovery of an anoxic basin within the Strabo Trench, Eastern Mediterranean. Nature 305:795–797

Kršinić F, Carić M, Viličić D, Cigleneǎki I (2000) The Calanoid Copepod *Acartia italica* Steuer, phenomenon in the small saline lake Rogoznica (Eastern Adriatic Coast). J Plankton Res 22:1441–1464

Kulda J (1999) Trichomonads, hydrogenosomes and drug resistance. Int J Parasitol 29:199–212

Lefèvre E, Bardot C, Noël C, Carrias J-F, Viscogliosi E, Amblard C, Sime-Ngando T (2007) Unveiling fungal zooflagellates as members of freshwater picoeukaryotes: evidence from a molecular diversity study in a deep meromictic lake. Environ Microbiol 9:61–71

Lefranc M, Thenot A, Lepere C, Debroas D (2005) Genetic diversity of small eukaryotes in lakes differing by their trophic status. Appl Environ Microbiol 71:5935–5942

Lehours A-C, Bardot C, Thenot A, Debroas D, Fonty G (2005) Anaerobic microbial communities in Lake Pavin, a unique meromictic lake in France. Appl Environ Microbiol 71:7389–7400

Lewis WM Jr (1983) A revised classification of lakes based on mixing. Can J Fish Aquat Sci 40:1779–1787

Lyons TW, Anbar AD, Severmann S, Scott C, Gill BC (2009) Tracking Euxinia in the ancient ocean: a multiproxy perspective and Proterozoic case study. Annu Rev Earth Planet Sci 37:507–534

Macek M, Vilaclara G, Lugo A (1994) Changes in protozoan assemblage structure and activity in a stratified tropical lake. Mar Microb Food Webs 8:235–149

Margulies M, Egholm M, Altman WE et al (2005) Genome sequencing in microfabricated high-density picolitre reactors. Nature 437:376–380

Martin W (2005) The missing link between hydrogenosomes and mitochondria. Trends Microbiol 13:457–459

Mas JC, Pedrós-Alió C, Guerrero R (1990) In situ specific loss and growth rates of purple sulfur bacteria in Lake Cisó. FEMS Microbiol Ecol 73:271–281

Massana R, Pedrós-Alió C (1994) Role of anaerobic ciliates in planktonic food webs: abundance, feeding, and impact on bacteria in the field. Appl Environ Microbiol 60:1325–1334

McDonough RJ, Sanders RW, Porter KG, Kirchman DL (1986) Depth distribution of bacterial production in a stratified lake with an anoxic hypolimnion. Appl Environ Microbiol 11:1199–1204

Medinger R, Nolte V, Pandey RV, Jost S, Ottenwälder B, Schlötterer C, Boenigk J (2010) Diversity in a hidden world: potential and limitation of next-generation sequencing for surveys of molecular diversity of eukaryotic microorganisms. Mol Ecol 19(s1):32–40

Morford JL, Emerson S (1999) The geochemistry of redox sensitive trace metals in sediments. Geochim Cosmochim Acta 63:1735–1750

Müller M (1993) The hydrogenosome. J Gen Microbiol 139:2879–2889

Nolte V, Pandey RV, Jost S, Medinger R, Ottenwälder B, Boenigk J, Schlötterer C (2010) Contrasting seasonal niche separation between rare and abundant taxa conceals the extent of protist diversity. Mol Ecol 19:2908–2915

Oremland RS, Dowdle PR, Hoeft S, Sharp JO, Schaefer JK, Miller LG, Switzer Blum J, Smith RL, Bloom NS, Wallschlaeger D (2000) Bacterial dissimilatory reduction of arsenate and sulfate in meromictic Mono lake, California. Geochim Cosmochim Acta 64:3073–3084

Overmann J (1997) Mahoney Lake: a case study of the ecological significance of phototrophic sulfur bacteria. In: Jones JG (ed.) Advances in microbial ecology, vol 15. Plenum Press, New York, pp 251–288

Overmann J, Tilzer MM (1989) Control of primary productivity and the significance of phototrophic bacteria in a meromictic kettle lake, Mittlerer Buchensee, West-Germany. Aquat Sci 51:261–278

Overmann J, Beatty JT, Hall KJ, Pfennig N, Northcote TG (1991) Characterization of a dense, purple sulfur bacterial layer in a meromictic salt lake. Limnol Oceanogr 36:846–859

Overmann J, Hall KJ, Northcote TG, Beatty JT (1999a) Grazing of the copepod *Diaptomus connexus* on purple sulphur bacteria in a meromictic salt lake. Environ Microbiol 1:213–21

Overmann J, Hall KJ, Northcote TG, Ebenhöh W, Chapman MA, Beatty T (1999b) Structure of the aerobic food chain in a meromictic lake dominated by purple sulfur bacteria. Arch Hydrobiol 144:127–156

Parkin TB, Brock TD (1980) Photosynthetic bacterial production in lakes: the effects of light intensity. Limnol Oceanogr 25:711–718

Radek R et al (2010) Adhesion of bacteria to protists. In: König H (ed.) Prokaryotic cell wall compounds. Springer, Berlin/Heidelberg, pp 429–456

Rassoulzadegan F (1993) Protozoan patterns in the Azam-Ammerman's bacteria–phytoplankton mutualism. In: Guerrero R, Pedrós-Alió C (eds.) Trends in microbial ecology. Spanish Society for Microbiology, Barcelona, pp 435–439

Richards TA, Vepritskiy AA, Gouliamova DE, Nierzwicki-Bauer SA (2005) The molecular diversity of freshwater picoeukaryotes from an oligotrophic lake reveals diverse, distinctive and globally dispersed lineages. Environ Microbiol 7:1413–1425

Saccà A, Guglielmo L, Bruni V (2008) Vertical and temporal microbial community patterns in a meromictic coastal lake influenced by the Straits of Messina upwelling system. Hydrobiologia 600:89–104

Saccà A, Borrego C, Renda R, Triadò X, Bruni V, Guglielmo L (2009) Predation impact of ciliated and flagellated protozoa on a summer bloom of purple sulfur bacteria in a meromictic coastal lake. FEMS Microbiol Ecol 70:42–53

Slapeta J, Moreira D, López-García P (2005) The extent of protist diversity: insights from molecular ecology of freshwater eukaryotes. Proc Biol Sci 272:2073–2081

Sorokin YI (1970) Interrelations between sulphur and carbon turnover in meromictic lakes. Arch Hydrobiol 66:391–446

Sorokin YI, Donato N (1975) On the carbon and sulphur metabolism in the meromictic Lake Faro (Sicily). Hydrobiologia 47:241–252

Stock A, Jürgens K, Bunge J, Stoeck T (2009) Protistan diversity in suboxic and anoxic waters of the Gotland Deep (Baltic Sea) as revealed by 18S rRNA clone libraries. Aquat Microb Ecol 55:267–284

Stoeck T, Epstein SS (2003) Novel eukaryotic lineages inferred from SSU rRNA analyses in oxygen-depleted marine environments. Appl Environ Microbiol 69:2657–2663

Stoeck T, Taylor GT, Epstein SS (2003) Novel Eukaryotes from the permanently Anoxic Cariaco Basin (Caribbean Sea). Appl Environ Microbiol 69:5656–5663

Stoeck T, Hayward B, Taylor GT, Varela R, Epstein SS (2006) A multiple PCR-primer approach to access the microeukaryotic diversity in environmental samples. Protist 157:31–43

Stoeck T, Kasper J, Bunge J, Leslin C, Ilyin V, Epstein SS (2007) Protistan diversity in the arctic: a case of paleoclimate shaping modern biodiversity? PLoS One 2:e728

Stoeck T, Jost S, Boenigk J (2008) Multigene phylogeny of clonal *Spumella*-like strains, a cryptic heterotrophic nanoflagellate isolated from different geographic regions. Int J Syst Evol Microbiol 58:716–724

Stoeck T, Behnke A, Christen R, Amaral-Zettler L, Rodriguez-Mora MJ, Chistoserdov A, Orsi W, Edgcomb VP (2009) Massively parallel tag sequencing reveals the complexity of anaerobic marine protistan communities. BMC Biol 7:72

Stoeck T, Bass D, Nebel M, Christen R, Jones MDM, Breiner H-W, Richards TA (2010) Multiple marker parallel tag environmental DNA sequencing reveals a highly complex eukaryotic community in marine anoxic water. Mol Ecol 19:21–31

Takahashi M, Ichimura S (1968) Vertical distribution of organic matter production of photosynthetic sulfur bacteria in Japanese lakes. Limnol Oceanogr 13:644–655

Takishita K, Tsuchiya M, Kawato M, Oguri K, Kitazato H, Maruyama T (2007) Genetic diversity of microbial eukaryotes in anoxic sediment of the saline meromictic lake Namako-Ike (Japan): on the detection of anaerobic or anoxic-tolerant lineages of eukaryotes. Protist 158:51–64

Tielens AGM, Rotte C, van Hellemond J, Martin W (2002) Mitochondria as we don't know them. Trends Biochem Sci 27:564–572

Trüper HG, Genovese S (1968) Characterization of photosynthetic sulfur bacteria causing red water in Lake Faro (Messina, Sicily). Limnol Oceanogr 13:225–232

van der Giezen M et al (2002) Conserved properties of hydrogenosomal and mitochondrial ADP/ATP carriers: a common origin for both organelles. EMBO J 21:572–579

van Gemerden H, Mas J (1995) Ecology of phototrophic sulphur bacteria. In: Blankenship RE, Madigan MT, Bauer CE (eds.) Anoxygenic photosynthetic bacteria, vol 11, Advances in photosynthesis. Kluwer Academic Publishers, Dordrecht, pp 49–85

Vicente E, Rodrigo MA, Camacho A, Miracle MR (1991) Phototrophic prokaryotes in a karstic sulphate lake. Verh Intern Ver Limnol 24:998–1004

Voelkerding KV, Dames SA, Durtschi JD (2009) Next-generation sequencing: from basic research to diagnostics. Clin Chem 55:641–658

Voncken FGH et al (2002) Multiple origins of hydrogenosomes: functional and phylogenetic evidence from the ADP/ATP carrier of the anaerobic chytrid *Neocallimastix* sp. Mol Microbiol 44:1441–1454

Westbrook GK, Reston TJ (2002) The accretionary complex of the Mediterranean ridge: tectonics, fluid flow and the formation of brine lakes. Mar Geol 186:1–8

Zehr JP, Harvey RW, Oremland RS, Cloern JE, George LH, Lane JL (1987) Big Soda Lake (Nevada). 1. Pelagic bacterial heterotrophy and biomass. Limnol Oceanogr 32:781–793

Zuendorf A, Bunge J, Behnke A, Barger KJ-A, Stoeck T (2006) Diversity estimates of microeukaryotes below the chemocline of the anoxic Mariager Fjord, Denmark. FEMS Microbiol Ecol 58:476–491

Biodata of **Thorsten Stoeck** and **Anke Behnke**, authors of "*The Anoxic Framvaren Fjord as a Model System to Study Protistan Diversity and Evolution.*"

Professor Dr. Thorsten Stoeck currently holds the position as Head of the Ecology Department, Faculty of Biology, at the University of Kaiserslautern. He obtained his Ph.D. from the University of Kaiserslautern in 1999; he has spend a two-year postdoctoral period at the Senckenberg Research Institute, Marine Biology, Wilhelmshaven, followed by a two-year postdoctoral period at the Marine Science Center of Northeastern University, Boston, MA, USA. In 2004, he returned to the University of Kaiserslautern with an Emmy Noether Independent Junior Research Group funded by the Deutsche Forschungsgemeinschaft. In 2008, he was awarded a Heisenberg Professorship by the Deutsche Forschungsgemeinschaft and obtained the position as Head of the Ecology Department from the University of Kaiserslautern. Professor Stoeck's main scientific interests focus on microbial (eukaryotic) life in (extreme) marine environments, with emphasis on (1) microeukaryote diversity and biogeography using molecular and culturing approaches, (2) the ecological significance of microeukaryote diversity, (3) the early evolution of eukaryotes and the reconstruction of evolutionary relationships between eukaryote species using molecular markers, and (4) mechanisms of adaptation to extreme environments.

E-mail: **stoeck@rhrk.uni-kl.de**

Dr. Anke Behnke is currently working as a postdoctoral fellow in the Faculty of Biology, Department of Ecology, at the University of Kaiserslautern, Germany. She obtained her Ph.D. from the University of Kaiserslautern in 2007 and continued her studies and research there in the areas of: diversity of protists in anoxic marine systems, spatial and temporal patterns of protists, and causes and consequences of changes of protistan communities.

E-mail: **behnke@rhrk.uni-kl.de**

THE ANOXIC FRAMVAREN FJORD AS A MODEL SYSTEM TO STUDY PROTISTAN DIVERSITY AND EVOLUTION

THORSTEN STOECK AND ANKE BEHNKE
Department of Ecology, University of Kaiserslautern,
Erwin Schroedinger Str. 14, Kaiserslautern D-67663,
Germany

1. An Early Evolution of Protists in an Anoxic World?

The pioneering pictures of the universal tree of eukaryote life, which were based on the phylogenetic analyses of small subunit ribosomal rRNA genes, provided a first overview of the evolutionary relationships between the organisms of this kingdom (Woese et al. 1990; Pace 1997). In these pictures, which were confirmed by phylogenetic tress constructed from protein genes (e.g., translation elongation factor (EF); Hashimoto et al. 1997), a paraphyletic group of protists, including parabasalids, diplomonads, and microsporidia, consistently branched at the base of the eukaryotic tree. These organisms are united by a simple ultrastructure and morphology, as well as a lack of microbodies, mitochondria, peroxisomes, and a strictly anaerobe lifestyle. This led to the conclusion that these amitochondrial eukaryotes emerged prior to the endosymbiotic event, which eventually resulted in the acquisition of mitochondria in the evolutionary history of eukaryotes. As a consequence, these assumed ancient and primitive eukaryote lineages were united in the eukaryote kingdom Archezoa proposed by Cavalier-Smith in 1989 (Cavalier-Smith 1989; Fig. 1).

The Archezoa were thought to be contemporary descendants of a nucleated, phagotrophic, and amitochondriate lineage that had evolved under anaerobic conditions, including the host for the endosymbiont, giving rise to mitochondria. However, this idea was shaken to the core for the first time only few years after the erection of this kingdom. The Archezoa, as defined by Cavalier-Smith, also included the Archamoebae based on the morphological and ultrastructural simplicity shared with other archezoan lineages. They were even assumed to be the most ancient Archezoa (Cavalier-Smith 1991). However, in a search of the *Entamoeba histolytica* genome, a representative of the Archezoa, Clark and Roger (1995) discovered molecular relics of a mitochondrial symbiont, namely, pyridine nucleotide dehydrogenase and a 60-kDa chaperonin (cpn69). In a phylogenetic analysis, both genes emerged as homologues of nuclear-encoded mitochondrial proteins of other eukaryotes. Most mitochondrial proteins are encoded by the nuclear genome, which again is closely related to the cpn60 in alpha-proteobacteria, the mitochondrial symbiont. This leads to the conclusion that Archamoebae

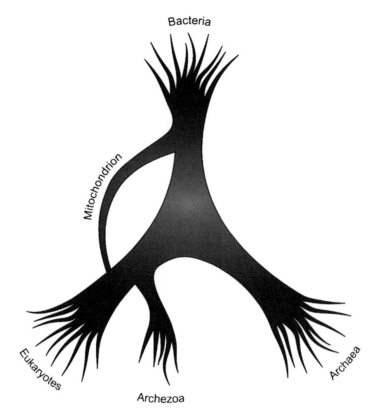

Figure 1. Schematic drawing of the Archezoa hypothesis. Based on phylogenetic SSU rRNA and translation factor 1 (EF), the amitochondrial Archezoa (including parabasalids, diplomonads, microsporidia, and Archamoebae) branched "early" in the eukaryotic tree of life, preceding the radiation of the crown eukaryotes. This led to the hypothesis that Archezoa are contemporary descendants of a nucleated, phagotrophic, and amitochondriate lineage that had evolved under anaerobic conditions, including the host for the endosymbiont, giving rise to mitochondria (Figure drawn by Sabine Filker).

evolved from mitochondrial-bearing ancestors (Clark and Roger 1995). A number of follow-up studies reported mitochondrial proteins from all archezoan key groups (Roger 1999), giving rise to the idea that the common ancestor of all archezoa was mitochondriate. This was supported by the detection of derived double-membraned mitochondrial organelles like hydrogenosomes in parabasalids (Lindmark and Muller 1973), releasing energy not from oxidative but from substrate phosphorylation and the conversion of pyruvate or malate into acetate, carbon dioxide, and hydrogen gas. Likewise, Mai et al. (Mai et al. 1999), as well as Tovar and colleagues (Tovar et al. 1999), discovered mitosomes in *Entamoeba histolytica*, a previously overlooked, highly reduced form of mitochondria, which appear to have no direct involvement in ATP synthesis. Bit by bit, evidence emerged for mitochondrial remnants or derived organelles for a number of Archezoa, including microsporidia, parabasalids, and diplomonads (Peyretaillade et al. 1998; Williams et al. 2002; Tovar et al. 2003). A thorough reexamination of phylogenetic trees based on SSU rRNA genes subsequently revealed that the basal position of the Archezoa is an artifact based on fast-evolving species that are attracted to the base of the eukaryotic tree of life (Philippe et al. 2000). The kingdom Archezoa has therefore been abandoned, and accordingly, the general idea of an evolution of eukaryotes in an anoxic world as well.

However, these two subjects (falsification of the Archezoa and an evolution of early eukaryotes in an anoxic world) are not necessarily mutually exclusive! Even though the common view is that rising oxygen levels at the time of eukaryote evolution prompted the need for an oxidative-metabolizing organelle, it is not unlikely that mitochondria are traces of an anaerobic past. Evidence for this assumption comes from the expanded metabolic repertoire of the "modified" mitochondria, apart from the typical mitochondrial function like iron-sulfur (FeS) cluster assembly and amino acid degradation. Because excellent reviews that summarize the metabolic repertoire of anaerobically functioning forms of mitochondria are available (Embley and Martin 2006; Mentel and Martin 2008), here we only give a very brief overview of the various mitochondrial derivates with their anaerobic biochemistry (Table 1). Interestingly, anaerobe energy

Table 1. Representative organisms with anaerobic biochemical metabolism and organelles. Anaerobic mitochondrial derivates are distributed throughout all eukaryotic, suggesting that anaerobic biochemistry in eukaryotes might be a very ancient trait conserved from earlier phases of Earth's history, which supports survival in anoxic environments (Information in this table is largely assembled from Mentel and Martin (2010)).

Supergroup	Group	Example	Organelle
Opisthokonta	Metazoa	*Spinoloricus sp.*	HLO
		Arenicola marina	FAM
		Neocallimastix frontalis	Hydrogenosomes
	Fungi	*Fusarium oxysporum*	FAM
	Microsporidia	*Encephalitozoon cuniculli*	Mitosomes
Amoebozoa	Archamoebae	*Entamoeba histloytica*	Mitosomes
		Mastigamoeba balamuthi	Hydrogenosomes
Rhizaria	Foraminifer	*Valvulineria bradyana*	FAM
	Gromiida	*Gromia sp.*	FAM
Chromalveolates	Ciliophora	*Nyctotherus ovalis*	MLO with genome
		Loxodes sp.	FAM
		Cyclidium porcatum	Hydrogenosomes
	Apicomplexa	*Cryptosporidium parvum*	Mitosomes
	Stramenopiles	*Blastocystis hominis*	MLO with genome
Archaeplastida	Chlorophyta	*Chlamydomonas reinhardtii*	FAM
Excavata	Diplomonadida	*Giardia lamblia*	Mitososmes
	Parabasalia	*Trichomonas vaginalis*	Hydrogenosomes
	Euglenida	*Euglena gracilis*	FAM

FAM facultative anaerobic mitochondria, *HLO* hydrogenosome-like organelles, *MLO* mitochondria-like organelle metabolism and mitochondria-like organelles are distributed among all major eukaryote lineages. A recent key discovery even reported the first strictly anaerobe animal, a Loricifera, from the sediment of a deep-sea hypersaline and sulfidic Mediterranean basin, spending its entire life cycle in the absence of oxygen and possessing hydrogenosome-like organelles (Danovaro et al. 2010). Similar to anaerobic metabolism ubiquitously distributed throughout the eukaryotic tree of life, the capability to cope with sulfide is a (probably very ancient) acquisition of eukaryotes in a variety of evolutionary lineages. Why is this important when considering the possibility of an anoxic past in eukaryote evolution?

Anoxic environments often times are characterized by sulfidic conditions. Sulfide (as H_2S or HS^-) inhibits the cytochrome c oxidase in the mitochondrial respiratory chain, thus being highly toxic at low levels. Therefore, eukaryotes thriving in anoxic, sulfidic environments need to detoxify sulfide. The mechanisms are diverse and include (but are not restricted to) ectosymbiotic bacteria of priapulid worms, which oxidize sulfide as their energy source (Oeschger and Janssen 1991), endosymbiotic sulfide-oxidizing bacteria of oligochaete worms (Dubilier et al. 2001), and the direct detoxification of sulfide into thiosulphate by sulfide-oxidizing mitochondria in mussels (Doeller et al. 2001). Even though not much is known about sulfide detoxification in protists, direct hydrogen sulfide consumption (Searcy 2006) and sulfide metabolism through symbiotic bacteria (Bernhard and Buck 2004) have been reported from different lineages.

To sum up, anaerobic forms of mitochondria, anaerobic energy metabolism, and means to cope with sulfide are no rare exceptions among eukaryotes, but might have existed long before the major eukaryote radiation. It seems unlikely that eukaryotes only invaded anoxic and sulfidic environments relatively recently and have acquired this anaerobic repertoire by adaptive evolution *de novo* and *in situ*. Therefore, contemporary eukaryotes thriving in oxygen-depleted environments can be seen as "evidence for evolution in the Darwinian sense of descent with modification, with the traits that support survival in anaerobic environments having been conserved from earlier phases of Earth history" (Mentel and Martin 2008).

This view finds support when we now add some recent knowledge collected from geologists on Earth's history. Almost certainly, the eukaryote origin and diversification took place in the Proterozoic ocean about 1.45–1.2 billion years (Brocks et al. 1999) ago. While molecular fossils of biological lipid biomarkers (C28–C30 sterans) detected in shales from the Pilbara Craton, Australia, suggested that eukaryotes are at least 2.7 Ga old (Brocks et al. 1999), the oldest eukaryote fossil (Grypania) has been reported from 1.8 Ga old rocks (Han and Runnegar 1992). However, both are heavily disputed. A recent study, applying a novel technology, the NanoSIMS ion microprobe (secondary ion mass spectrometry with 50-nm resolution), is inconsistent with an indigenous origin for the lipid biomarkers in rocks from the Archaean eon (Rasmussen et al. 2008), and the authors conclude that the biomarkers must have entered the rock after peak metamorphism and that they do not provide evidence for the existence of eukaryotes 2.7 Ga ago (see also Fischer 2008). Also, the identification of Grypania as a eukaryote is uncertain because the morphology of these fossils may point to chain-forming bacteria rather than to an early eukaryote. The evidence for the eukaryote origin that is hardly under debate is acritarchs defining the lower bound for early eukaryotes in general (1.45 Ga; Javaux et al. 2001) and *Bangiomorpha*, which morphologically is hardly distinguishable from extant red algae (Butterfield 2000), as the lower bound for photosynthetic eukaryotes. Of course, the respective organisms must have evolved quite a number of years before they became fossilized.

While the partial pressure of atmospheric oxygen rose substantially about 2.4–2.0 Ga ago (Kump 2008), ocean anoxia might have persisted well into the late Mesoproterozoic Era about 1.2–1.0 Ga ago (Canfield 1998). This so-called Canfield-ocean scenario is nicely reviewed by Anbar and Knoll (2002). In short, principally based on the disappearance of banded iron formations (BIFs), the classical view on ocean oxygenation is that deep oceans became oxygenized already 1.8 Ga ago (Holland 1984) because they require anoxic deep waters (oxygenation would result in insoluble Fe-oxyhydroxides, removing Fe and precluding BIF). However, this argument ignores the fact that the Fe-sulfides also are hardly soluble. The differences in marine sulfides and sulfates before and after the Mesoproterozoic (indicating, that bacterial reduction strongly depleted sulfates) and an increase of sulfates in the Neoproterozoic was only possible after major oxygenation events. Until then, deepwater oxygenation seems insufficient to support microbial sulfur disproportionation (Canfield 1998). This, together with simple modeling of ocean redox, provides strong support for the prevalence of highly sulfidic deep ocean waters at the time of eukaryote origin and early diversification (see Anbar and Knoll (2002) and references within). Due to exceptional nitrogen-stress (sulfidic conditions!) and phosphate limitations, the diversification of photosynthetic eukaryotes was probably very limited well into the Neoproterozoic eon. Other chemical tracers like molybdenum provide further support for the sulfidic Canfield ocean (Scott et al. 2008).

This plausible scenario of anoxic and sulfidic Proterozoic ocean, in concerto with the evolutionary history of eukaryotes as well as their anaerobic metabolism and mitochondrial derivates scattered throughout the eukaryote tree of life, accentuates the significance of anoxic waters in the quest for the origin of eukaryote life.

2. The Norwegian Framvaren Fjord: An Ideal Model for Anoxic and Sulfidic Systems

While studies of eukaryotes in natural anaerobic environments are still in their infancies, substantial progress has been made in biogeochemical processes, specifically in marine anoxic habitats. In the past few years, the same permanently anoxic study sites have been targeted repeatedly, including the Black Sea, the Caribbean Cariaco basin off Venezuela, deep hypersaline anoxic basins in the Eastern Mediterranean deep sea, and the Framvaren Fjord, all of which have gained a "model character" when it comes to the study of anaerobic organisms or processes and geological history. Other frequently studied anoxic sites, which, however, may be subjected to periodic oxygenation events and therefore are not further considered in this chapter, include the Gotland Deep in the Central Baltic Sea, the Canadian Saanich Inlet, the Danish Mariager Fjord, the oxygen minimum zone (OMZ) in the Arabian Sea, or coastal upwelling systems like the Benguela upwelling system off Namibia.

What now makes the Norwegian Framvaren Fjord a model for anoxic systems? Generally, in science, a model *is a representation containing the essential structure of some object, system, or event in the real world* (Stockburger 1996). The knowledge and understanding that scientists have about the world are often represented in the form of models. This includes that a model system for anoxic natural habitats is well studied and represents the general characteristics and "behavior" of such systems (certain specific details ignored). To fulfill these criteria, such a system should meet the criteria of stability and ease of access, as is the case for Framvaren Fjord. The environmental setting of the fjord is reviewed in Skei (1988). Geomorphologically the result of glaciation and deglaciation, the 184 m deep, 1.5 km long, and maximal 1 km wide (5.8-km^2 surface) fjord is located in South Norway (58°10′N, 6°45′W, Fig. 2).

After retreat of the inland ice sheet, some 8,000 year ago, the fjord changed from an oxic fjord, which was connected to the open sea, to a meromictic lake. Isostatic uplift resulted in a physical barrier that eventually separated the basin from the sea. Only in the 1850s, a man-made breakthrough of this barrier opened the basin to the sea again. This created a shallow (2 m deep) and narrow (10 m wide) sill (Fig. 2b), allowing oxygenated seawater to penetrate into the basin. Because of the higher density of this water, it sank to the fjord's bottom, where it quickly turned anoxic (see below and Fig. 5). Vertical cliffs that result in a V-shaped cross-section of the fjord are the general outline of the fjord's shores. This geological structure, together with a very restricted water inflow through the sill and constrained freshwater supply (catchment area 31 km^2 with no major rivers pouring out into the fjord), creates a very stable water column that can principally be divided in four major layers, including the low-salinity surface layer (0–2 m, above sill depth), the oxygenated subsurface (−18 m), separated by a large clear pycnocline from the deep water (−100 m) where steep gradients in water chemistry are characteristic, and the bottom water down to 184 m where changes in water chemistry are only small. Comparisons of more recent studies with the initial studies in Framvaren in the 1930s (Ström 1936) showed that this structuring is extremely stable, making the fjord ideal for biological and (biogeo)chemical long-term studies. For example, studies performed on the vertical distribution of microbial consortia along the redox-gradient (Behnke et al. 2006) showed that oxic–anoxic interphases should be preferably investigated at a spatial resolution of centimeters in order to understand the complexity of such ecosystems and processes involved in the function (McDonald et al. 2007) of these ecosystems. Framvaren Fjord is an enclosed and stable marine system offering and enabling

Figure 2. South–north view of the Framvaren Fjord in southern Norway with its two basins. (**a**) Vertical cliffs rise from the fjord's bottom, locally forming very steep and deep shorelines. (**b**) The narrow and shallow sill that connects the 180 m deep Framvaren Fjord with the Helvik Fjord. Together with the V-shaped geomorphology of the fjord, this sill results in very restricted water exchange with the adjacent water body, preventing mixing of water below 18 m depth in the Framvaren Fjord. The foreground shows the Helvik Fjord, which is directly connected to the sea; the connection between Helvik and Framvaren starts behind the bridge visible in the picture (Figure 2a by T. Stoeck, Fig. 2b by S. Filker).

an ideal stage for such studies. Further characteristics that make Framvaren Fjord a highly interesting study site include, for example, its extremely high concentrations of sulfide. With 8 mM sulfide in the bottom waters, Framvaren Fjord is ca. 25 times more sulfidic than the deep water in the Black Sea below 1,000 m (0.3–0.36 mM) and even 133 times more sulfidic than bottom waters (1,200 m) in the Cariaco basin (0.06 mM), which is the world's largest truly marine permanently anoxic deep-sea basin.

Such high sulfide concentrations are hardly matched by any other waters known to date, making Framvaren a "supersulfidic" environment ideally suited to study sulfur chemistry in a natural laboratory and to study biological mechanisms to cope with highly toxic sulfide (see above). Furthermore, the redox interphase (pycnocline) is located in the photic zone. This distinguishes the stratification in Framvaren from most other major (and frequently targeted) anoxic water bodies, like the Black Sea or Cariaco. This allows a number of studies and experiments like, for example, anoxygenic photosynthesis (Cohen et al. 1975) in bacterioplankton not only to reveal the function (McDonald et al. 2007) of this microbial ecosystem but also to better reconstruct Proterozoic oxygenation of the Earth (Johnston et al. 2009).

For reasons outlined above, the Framvaren Fjord, indeed, deserves its model character, and Framvaren (together with the Black Sea) belongs to the most investigated anoxic water in the world (Dryssen 1999). Since the 1930s, more than 80 studies have been published that report research results from the Framvaren Fjord (a list of references is available from the authors upon request). Most of these studies, however, focus on the biogeochemical processes, the chemistry, and the geology of the fjord. Only few studies have targeted microbes, including protists, and the ecology of this supersulfidic fjord.

3. Diversity of Protists

Protists are referred to as the kingdom Protista (Haeckel 1878) or Protoctista (Hogg 1861). However, phylogenetically, this group does not represent a natural assemblage, i. e., it is not monophyletic. Basically, the organisms are grouped according to their small size (most of them are about 1 to 100 µm) and the relatively simple grade of cellular organization (unicellular or multicellular without specialized tissues) (Hausmann et al. 2003). Protists occur in all of the six major phylogenetic eukaryotic phyla described by Adl et al. (2005, 2007). Actually, four of these phyla are exclusively composed of protists, namely, excavates, chromalveolates, rhizaria, and amoebozoa. Concerning their morphology, ecology, and metabolism, protists are very diverse (Hausmann et al. 2003), and it is therefore not surprising that they occur in nearly all kinds of habitats and have a globally widespread distribution. Very likely, they are the most abundant eukaryotes on earth (Patterson 1999). They are an essential component of microbial food webs (Sherr and Sherr 2000) and play an important role in biogeochemical cycles (Summons 1993).

Until the mid-1990s, the study of protistan diversity was essentially based on investigations of morphology and ultrastructure and on culturing approaches. However, because of their small size, lots of protists do not show a sufficient number of structural characters for their certain distinction and identification.

Furthermore, the predominant proportion of protists has withstood previous culturing efforts. Therefore, it is very likely that a high extent of protistan diversity is still undiscovered, and there are contradictory ideas about the total number of protistan species on Earth. Estimates range from 12,000 (Finlay 2001) or 19,000 (Corliss 1999) species for all protozoa to 300,000 (Foissner 2008). Gaining insights into the diversity of anaerobic protists and the structure and ecology of anaerobic communities is especially hampered due to additional methodological difficulties (Fenchel and Finlay 1995). Already trace amounts of oxygen inhibit or kill many of these organisms. Therefore, investigation by light microscopy and cultivation could only reveal a fractional amount of diversity, and anaerobic environments are virtually unexplored (Cavalier-Smith 2004).

However, in the past 15 years, a molecular approach has been established to assess microbial communities even in the most extreme environments (Barns et al. 1994; López-Garcia et al. 2001; Amaral Zettler et al. 2002). This procedure is based on the sequencing and phylogenetic analysis of small subunit ribosomal RNA (SSU rRNA) genes that have been amplified and cloned from the environment. It proved to be a useful tool to analyze the composition of environmental samples (Pace 1997; Caron et al. 2004) by placing the sequence of this marker gene in a phylogenetic tree (Olsen et al. 1986; Woese 1987). Instead of considering species, one deals with phylotypes or operational taxonomic units (OTUs). These are groups of sequences, which are defined either on the basis of band patterns after digestion with restriction enzymes (restriction fragment length polymorphism [RFLP]) or according to sequence similarities, i.e., phylotypes comprise sequences showing a predefined minimum sequence similarity, for example, 98%.

To date, several environmental SSU rRNA gene surveys have focused on protistan diversity in aquatic anoxic habitats (e.g., Dawson and Pace 2002; Edgcomb et al. 2002; López-Garcia et al. 2003; Stoeck and Epstein 2003; Stoeck et al. 2003a). Most of these detected unexpectedly high species richness (Dawson and Pace 2002; Edgcomb et al. 2002; Stoeck et al. 2003a), with several highly divergent SSU rRNA gene sequences possibly representing valid novel phylotypes at high taxonomic ranks (Berney et al. 2004). Groups of sequences that exclusively comprised uncultured organisms, for example, the marine stramenopiles (MAST) (Massana et al. 2004b) or the uncultured marine alveolates (López-Garcia et al. 2001), were recovered. In addition to this, a meta-analysis of environmental SSU rRNA gene surveys identified site-specific anoxic sequence groups (Richards and Bass 2005). However, other SSU rRNA gene surveys of anoxic aquatic environments revealed relatively low phylogenetic richness of protistan communities (Luo et al. 2005; Takishita et al. 2005) and no indication of novel high-level eukaryotic lineages (Luo et al. 2005).

This latter finding supports the hypothesis that there may be few, if any, previously unknown protist phyla or "new kingdoms" left to discover (Finlay 2002; Cavalier-Smith 2004). Nevertheless, it is undeniable that at least at lower phylogenetic level, for example, on class level and below, there is an impressive hitherto unknown diversity, as demonstrated by molecular environmental surveys (Moreira and López-Garcia 2002; Berney et al. 2004; Richards and Bass 2005). Moreover, investigations of isolates and cultured strains demonstrated a high genetic diversity within established protistan morphospecies according to analyses based on SSU rRNA gene analyses (Boenigk et al. 2005), on internal transcribed spacer sequences (Behnke et al. 2004), on combined data of the two (Katz et al. 2005; Rodriguez et al. 2005), and on multilocus sequence analysis (Katz et al. 2006; Slapeta et al. 2006). Some authors doubt the biological relevance of this genetic variety (Whitfield 2005) if there is no obvious phenotypic counterpart. However, others call into question the validity of a species concept based on phenotype only (Epstein and López-Garcia 2008) and emphasize the importance of applying molecular methods as additional tools for exploring protist speciation, ecological specialization, and biogeography (Foissner 2006).

4. Protists in the Framvaren Fjord

The first (published) investigations of microorganisms in the Framvaren Fjord started in 1980 (Ormerod 1988; Sørensen 1988). These studies focused on phototrophic and nonphototrophic bacteria and their distribution in the upper 25 m of the water column, while microeukaryotes were only taken into account by measuring the content of algal chlorophyll (Sørensen 1988). The taxonomic identity of the latter organisms remained more or less unresolved, and only the most dominant species (i.e., *Skeletonema costatum*, Bacillariophyta, and *Gyrodinium aureolum*, Dinophyta) were determined. Heterotrophic protists were not investigated at all even though total bacterial counts (Ormerod 1988) suggested a sufficient food source at least for bacterivores. The diversity of whole microeukaryote communities for the first time was surveyed in 2004 (Behnke et al. 2006), applying the culture-independent approach based on the phylogenetic analysis of eukaryotic SSU rRNA gene sequences that were amplified and cloned from environmental samples (Caron et al. 2004). The overall protistan diversity initially detected in the Framvaren Fjord was relatively high, with 92 different phylotypes (i.e., groups of sequences sharing at least 98% sequence similarity) (Behnke et al. 2006). Sequences were distributed across all major eukaryotic taxa (Fig. 3); however, the major proportion (phylotypes as well as sequences) could be assigned to alveolates and stramenopiles.

Figure 3. Phylogenetic tree showing the position of protistan SSU phylotypes in the anoxic waters of the Framvaren Fjord. This tree includes 236 distinct environmental phylotypes. One phylotype includes sequences that share at least 98% sequence similarity in their SSU rRNA primary structure. Additionally, the sequence of one isolate (FV18-8TS; Stoeck et al. 2005) was included in the analysis. Sequences were affiliated to a reference parsimony tree (ca. 50,000 sequences) by using the Quick-add-Parsimony utility of ARB (Ludwig et al. 2004). Lineages comprising clones derived from the Framvaren Fjord are indicated in *bold*; *numbers* refer to phylotypes in the respective taxon.

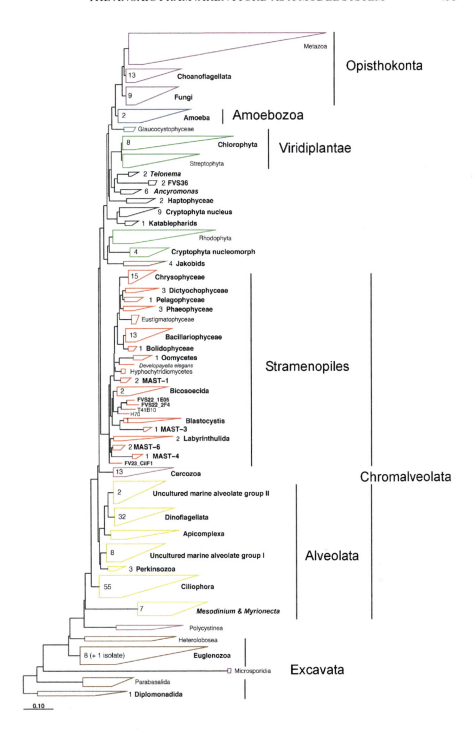

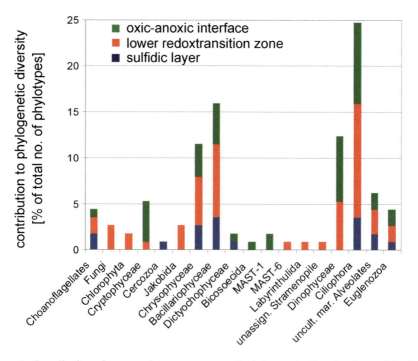

Figure 4. Contribution of taxonomic groups to overall phylogenetic diversity detected in three protistan communities along the vertical O_2/H_2S gradient in Framvaren Fjord sampled in May 2004. Phylotypes were defined to encompass clones that exhibited at least 98.0% sequence similarity based on a pairwise comparison of the SSU rRNA gene sequences. *Colors* refer to the three different protistan communities.

Within the alveolates, ciliates were the most abundant and diverse taxonomic group identified (67% of all alveolate phylotypes, 24% of total number of phylotypes, Fig. 4). This finding was not unexpected, taking into account that these organisms are major consumers of bacteria in planktonic food webs (Porter et al. 1985). Moreover, many of them have independently adapted to an anoxic lifestyle either as obligate anaerobes, like, for example representatives of the order Entodiniomorphida or Odontostomatida, or as facultative anaerobes, as they can be found within the order Karyorelictida (Fenchel and Finlay 1995).

In anoxic waters of the Framvaren Fjord among others, members of the families Plagiopylidae, Strombiidae, Nyctotheridae, Cycliidae, and Prorodontidae were detected, all of which are known to include anaerobic or micro-oxic representatives.

Stramenopiles, heterotrophic representatives as well as autotrophs, were recovered with numerous phylotypes from all three sampling depths (Behnke et al. 2006). The occurrence of the latter ones may seem unusual, as algal photosynthesis at depths below 18 m is restrained by light availability (Sørensen 1988; Fig. 5).

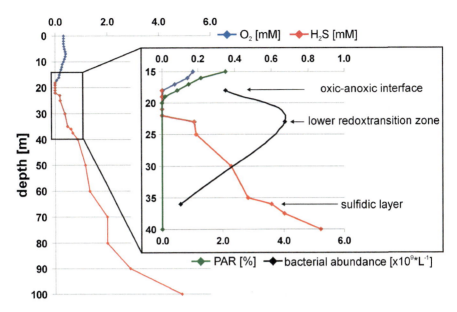

Figure 5. Sampling site characteristics of the Framvaren Fjord in May 2004. Overview shows O_2 and H_2S profile of the upper 100 m of the water column. Detailed section displays part of water column sampled for diversity studies (oxic–anoxic interface, lower redox transition zone, sulfidic layer), including information about photosynthetic active radiation (PAR; Data from (Sørensen 1988) and bacterial abundances. *Upper axes*: *blue* O_2 [mM], *red* H_2S [mM]; *lower axis*: *green* PAR [%], *black* bacterial abundances [cells * 10^9 * L^{-1}].

However, some of these organisms, e.g., chrysophytes and diatoms, in fact have heterotrophic capability and are able to prosper in oxygen-depleted environments without light (Lylis and Trainor 1973; Hellebust and Lewin 1977; Davidson and John 2001).

An important aspect of the Framvaren study was that instead of sampling a single spot within the anoxic system, different samples along the vertical O_2/H_2S gradient were investigated. Because evidence from microscopic (Fenchel et al. 1995) and molecular (Stoeck et al. 2003a) analyses accumulated that protistan community structure changes significantly along such gradients, in the Framvaren study, three different communities were sampled, namely, the oxic–anoxic interface, the lower redox transition zone, and a deeper sulfidic layer (Behnke et al. 2006; Fig. 5). As these layers differ in a variety of abiotic and abiotic factors, like, e.g., O_2 and H_2S content, photosynthetic active radiation, and bacterial abundances (Fig. 5), it was assumed that they provide ground for different species adapted to the different environmental conditions. Indeed, the proposed communal division was confirmed by three results: First of all, the extent of protistan diversity (in terms of the detected number of phylotypes) was very unequal, with the lower redox transition zone community showing the highest number of unique phylotypes (60 phylotypes (98% cutoff); Fig. 6).

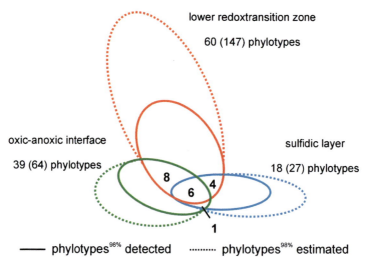

Figure 6. Phylotypes shared among protistan communities in different depths of the Framvaren Fjord. Venn diagram showing phylotypes (98% sequence similarity cutoff) shared among SSU rRNA gene sequence libraries derived from three sampling depths (oxic–anoxic interface, lower redox transition zone, sulfidic layer) along the vertical O_2/H_2S gradient in the Framvaren Fjord in May 2004. *Continuous lines* refer to number of phylotypes detected; the area of an oval is proportional to the size of the corresponding clone library. *Numbers* in *overlapping areas* are the numbers of phylotypes shared by the relevant libraries (2- and 3-set phylotypes); the overlap area is proportional to amount. *Numbers* in *nonoverlapping areas* correspond to the number of sample-specific phylotypes (1-set phylotypes). *Dash lines* refer to diversity estimates (parametric model: inverse Gaussian-mixed Poisson). *Colors* refer to the three different protistan communities.

Because all three libraries turned out to be more or less far from saturation (Behnke et al. 2006), true richness of communities was estimated by applying parametric abundance models (Hong et al. 2006) as well as nonparametric estimators (Chao 2005). While for the clone library derived from the sulfidic layer, only 27 different phylotypes were estimated (inverse Gaussian-mixed Poisson model), this value was 147 for the clone library of the lower redox transition zone (Fig. 6); these differences were also statistically confirmed by a conservative, asymptotically valid, Bonferroni-corrected test at a level of $\alpha = 0.05$ (Behnke et al. 2006). The second evidence for the communal division was the extremely low number of phylotypes shared between the different communities: Only 21% of all phylotypes were detected in at least two of the three libraries (Fig. 6). Even for taxonomic groups that were abundant in all three communities, like, e.g., ciliates, on the phylotype level, it became obvious that only a small proportion of phylotypes occurred in more than one library (6 of 21 unique 98% phylotypes; Behnke et al. 2006). Finally, phylotypes derived from the different communities were very unequally distributed across the major eukaryotic phyla. Cryptophytes, for example, displayed the highest diversity at the oxic–anoxic interface (5 unique

98% phylotypes, Fig. 4), while only one was detected in the lower redox transition zone and none in the sulfidic layer. Cryptophytes are phototrophs in the first place, and only one aplastidic genus (*Goniomonas*) is known. However, many of them can switch to a heterotrophic mode (e.g., Lewitus et al. 1991). Among other algae, cryptophytes are known to constitute populations of so-called deep-living algae (Gasol et al. 1993) that are well adapted to low-light conditions (Overmann and Tilzer 1989) and may contribute to a major proportion of algal biomass (Gasol et al. 1992) and primary production (Gasol et al. 1993). Studies in other stratified systems already revealed that cryptophytes could form dense algal populations at chemoclines in stratified waters, usually in those layers where oxygen is still present, at least in very small amounts (Vyverman and Tyler 1995; Camacho and Vicente 1998). However, they were also detected in layers, where oxygen was absent and in the presence of sulfide (Gasol et al. 1993), where they probably switched to a heterotrophic, fermentative lifestyle (Morgan and Kalff 1975). Therefore, their detection in the Framvaren Fjord in the anoxic water column with a preferential occurrence at the oxic–anoxic boundary layer (i.e., highest phylotype diversity at the upper border of the chemocline) is not surprising. Other groups detected exclusively in one of the communities are, for example, fungi, jakobids, and labyrinthulida, all of which seemed to be restricted to the lower redox transition zone or are representatives of the MAST-1 cluster, detected at the oxic–anoxic interface only (Fig. 4). This cluster was characterized as a group of truly planktonic marine aerobic organisms (Massana et al. 2004b), which may explain the restriction of MAST-1 clones to the suboxic chemocline.

Such deductions of ecological properties of communities and community members, i.e., the analysis and interpretation of biodiversity patterns, have to be made carefully and benefit from a sound knowledge of the stability of recorded spatial patterns over time (Ramette and Tiedje 2007). From microscopy-based studies assaying spatial and/or temporal patterns in protistan plankton, it is not only known that community structures change along physicochemical stratification gradients (Bark and Goodfellow 1985; Fenchel et al. 1995) but also that there is a variety of biotic and abiotic factors resulting in distinctive spatio-temporal patterns of plankton communities (Kahru et al. 1990; Kahru and Nommann 1990; Reul et al. 2005; Rolland et al. 2009). The knowledge gained from studies like these has long been ignored in molecular ecology assays of protistan plankton, and several investigations addressed the spatial variation of genetic microbial diversity patterns from distant geographic locations sampled just once. Just few molecular studies sampled the same site repeatedly (Diez et al. 2001b; Massana et al. 2004a; Romari and Vaulot 2004; Medlin et al. 2006; McDonald et al. 2007).

As the vertical O_2/H_2S gradient in the Framvaren Fjord is relatively stable (Skei 1988) and only the exact location and thickness of the chemocline might vary (2–5 m; Behnke et al. 2010a), it was assumed that communities derived from the very same layer (e.g., oxic–anoxic boundary layer) sampled in different seasons will also be relatively similar, at least when compared to communities derived from other depths. Indeed, the reinvestigation of the three protistan communities

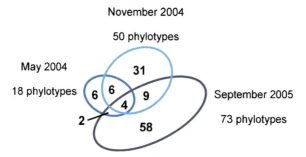

Figure 7. Phylotypes shared among protistan communities in the Framvaren Fjord sampled in different seasons. Venn diagram showing phylotypes (98% sequence similarity cutoff) shared among SSU rRNA gene sequence libraries derived from the sulfidic layer sampled in three seasons (May 2004, November 2004, and September 2005). The area of an oval is proportional to the size of the corresponding clone library. *Numbers* in *overlapping areas* are the numbers of phylotypes shared by the relevant libraries (2- and 3-set phylotypes); the overlap area is proportional to amount. *Numbers* in *nonoverlapping areas* correspond to the number of sample-specific phylotypes (1-set phylotypes). *Colors* refer to the three sampling seasons.

in the Framvaren Fjord (Behnke et al. 2010a) revealed that the communal division of the three investigated anoxic layers can be observed regardless of the sampling season: Few protistan phylotypes were shared between the different sampling depths, and some phylogenetic groups still appeared to be restricted to one or two of the depths investigated, like, e.g., fungi and jakobids. Unexpected, however, was the observation that the temporal variation within each habitat was as pronounced as between habitats. Only 18–28% of all phylotypes detected within one layer were found in more than one season, and communities from the same layer appeared not necessarily to be more similar to each other than to communities from another depth (Behnke et al. 2010a). Such seasonal variability is not without consequences for the deduction of biodiversity patterns in an ecological context, what becomes, for example, evident when considering the sulfidic layer investigated. The clone library in the first sampling season (May 2004; Behnke et al. 2006) was characterized by a low number of phylotypes. However, the same community sampled in September showed the highest diversity compared to all other Framvaren libraries (Fig. 7), and this difference was confirmed by statistical analyses. This observation challenged previous assumptions that due to the levels of sulfide (0.47–0.67 mM) and relatively low bacterial counts (as a food source for bacterivores), this depth might support only a limited number of organisms compared to the oxic–anoxic interface of the lower redox transition zone (Behnke et al. 2006). The observed temporal variations demonstrate the importance of considering the time dimension when investigating spatial patterns of biodiversity. When comparisons between habitats are based on single samplings, observed differences either may reflect true spatial differences or may just be due to temporal variation within the local community structure. In the latter case, communities may be more similar when sampled at corresponding points in time (e.g., same season), whereas in the first case, community patterns will never be similar regardless of the sampling season (Behnke et al. 2010a).

However, the investigation of diversity, the interpretation of its patterns, and the deduction of factors, biotic or abiotic, responsible for spatial and/or temporal variability of microbial communities do not only benefit from multiple samplings and knowledge about stability of recorded patterns but also from having most comprehensive data sets possible. Even though the molecular clone library approach provided deeper insight into microbial communities compared to traditional microscopy- and culturing-based studies (Pace 1997; Caron et al. 2004), still this method allows only for taking snapshots from communities under study, and most clone libraries published to date are heavily undersampled (Stoeck et al. 2006). It became obvious that microbial diversity in more or less all investigated habitats is larger than expected, but it is hard to tell exactly how large "larger" is (Epstein and López-Garcia 2008), and it is impossible to determine "who" is part of the missed proportion of diversity.

The shortcomings of the clone library approach are in part due to the tremendous genetic diversity of protists in most environments and also due to a number of methodological issues, like, e.g., the relatively high expense and labor intensity, and limit of power of subsequent analyses, as, for example, the comparison of different samples (Stoeck et al. 2010). The recent development of a high-throughput sequencing technology (454 sequencing; Margulies et al. 2005) compensated for at least some limitations, e.g., by eliminating the ligation and transformation steps of the clone library approach (Sogin et al. 2006). Therefore, 454 sequencing allow for a much deeper and more comprehensive investigation of microbial samples (Sogin et al. 2006; Amaral-Zettler et al. 2009). The application of this new technology for surveying protistan diversity in the Framvaren Fjord revealed that indeed the diversity is higher than reported (and estimated) before (Stoeck et al. 2009, 2010). For example, for the oxic–anoxic interface, the clone library approach revealed 41 to 52 different phylotypes (98% cutoff) depending on the sampling season, and 59 to 89 were estimated (nonparametric estimator) (Behnke et al. 2006, 2010a). The 454 sequencing, however, recovered 723 and 772 different phylotypes (97% cutoff) (no. of phylotype depends on the hypervariable SSU rRNA gene region investigated: V4 and V9, respectively), even when handling the data conservatively in order to compensate for a possible overestimation of diversity due to methodological issues (Behnke et al. 2010b). More than that, it seemed like these more comprehensive data sets allow for a more reliable comparison of samples, at least when compared to clone library data. While analyses of the latter data did not uncover insightful patterns of communities (Behnke et al. 2010a), the 454 pyrosequencing data revealed that communities derived from the same site and the same layer sampled in different seasons harbor more identical phylotypes compared to other sites and/or layers (Stoeck et al. 2009). Thus, it appears that the application of deep sequencing technologies may have the potential to reveal the complexity and patterns of protistan communities and demonstrated that diversity of protists in the Framvaren Fjord is at least one order of magnitude greater than reported previously (Stoeck et al. 2009, 2010; Behnke et al. 2010b).

5. Novel Protistan Lineages from the Anoxic Framvaren Fjord

The application of the SSU rRNA gene-based surveys for the investigation of overall protist diversity in extreme environments revealed the existence of both sequences closely related to already described taxa as well as a wealth of sequences more or less divergent from what was known before (e.g., Diez et al. 2001a; López-Garcia et al. 2001, 2003; Dawson and Pace 2002; Edgcomb et al. 2002; Stoeck and Epstein 2003; Stoeck et al. 2003a). Even if some of the latter sequences may represent known organisms that were just not yet investigated using molecular methods, the mere extent of divergent sequences supports the idea that at least part of them, indeed, is derived from novel organisms that were missed using traditional methods like light microscopy and culturing approaches (Epstein and López-Garcia 2008).

In the Framvaren Fjord, even within the ciliates, one of the best described protistan groups, novel phylogenetic lineages at different taxonomic levels were detected. Some sequences formed clades within well-described classes (e.g., within the Oligohymenophorea), while for others, it was not even possible to place them in a particular class (FV18_2A12, acc. no. DQ310196.1). Together with two other environmental clones (Lefranc et al. 2005), FV18_2A12 formed a well-supported, deeply branching clade, which appeared to be a sister group of the established ciliate class Plagiopylea and may deserve designation at the same taxonomic level (Behnke et al. 2006). This is remarkable because after two centuries of research on their systematics (Ehrenberg 1838; Kahl 1930–1935; Foissner et al. 1991, 1992, 1994, 1995), some investigators contend that most, if not all, ciliate species have already been described (Finlay et al. 1996; Finlay 2002). Whatever the correct taxonomic status of the clade that was discovered, its very existence suggests that it is too early to establish limits for protistan diversity, even for the best studied taxa.

The most divergent sequences from the anoxic waters of the Framvaren Fjord, however, were detected within the Euglenozoa (Behnke et al. 2006, 2010a). One example for this was sequences branching in an uncultured euglenozoan sequence cluster and that were repeatedly retrieved from the lower redox transition zone and the sulfidic layer. Originally, this cluster consisted exclusively of sequences from anoxic marine environments in Northern Europe (Behnke et al. 2006, 2010a; Zuendorf et al. 2006;) and South America (Stoeck et al. 2003a, 2006), and only recently, the taxonomic identity of these sequences was discovered (Yubuki et al. 2009). Those environmental sequences are affiliated with *Calkinsia aureus*, an isolate that was originally collected from anoxic sediments near Woods Hole (USA) and described as a member of the euglenid family Petalomonidae (Lackey 1960). However, a reinvestigation based on ultrastructure and molecular phylogenies revealed that this organism, together with the abovementioned environmental clones, constitutes a fourth novel euglenozoan subclade referred to as the "Symbiontida" (Yubuki et al. 2009). Another example for highly divergent sequences within the Euglenozoa was a sequence retrieved from the oxic–anoxic interface (FV18_1E2, acc. no. DQ310342.1; Behnke et al. 2006) that was related to an environmental clone from anoxic hydrothermal sediment (AT4-56, acc. no. AF530520.1; López-Garcia et al. 2003). These two sequences are of special importance for the phylogeny of kinetoplasids, one of the euglenozoan subclades. Based on SSU rRNA gene analyses, trypanosomatids were assumed to have emerged from the bodonids (Callahan et al. 2002; Simpson et al. 2002). But at the same time, these analyses showed that there is a huge evolutionary distance between kinetoplastids and their closest relatives, the diplonemids and euglenids, making correct inference of the phylogenetic relationships within the kinetoplastids very difficult. However, AT4-56, together with FV18_1E2, emerged robustly at the base of the kinetoplastids, breaking the long branch which leads to diplonemids and euglenids. Using this sequence as a close out-group in phylogenetic analyses resulted in a much more stable and resolved kinetoplastid phylogeny (Moreira et al. 2004). Until the recovery of FV18_1E2, AT4-56 was the only representative of this basal kinetoplastid "group"; only the Framvaren clone confirmed that this lineage is a real biological entity. This example demonstrates that anoxic systems contain sequences that are very important for reconstruction of eukaryote phylogeny.

One problem with the SSU rRNA gene approach, however, is that it provides little information beyond the fact of an organism's existence, its molecular phylogeny, and maybe its distribution in nature. On the basis of the phylogenetic affiliation of environmental SSU rRNA gene sequences, one has the possibility to deduce characteristics of the organisms behind the sequences, e.g., regarding the question if they are able to prosper in oxygen-depleted environments. Nevertheless, these deductions remain speculative, and the approach yields no definite information about the organism's cellular identity, its physiological capacities, or *in situ* abundances. But sequence information can be used to gain additional information. Based on nucleotide sequences, oligonucleotide probes specific for certain groups of protists can be designed for groups of uncultured organisms as well as for cultured ones. The application of these probes on environmental samples and the use of epifluorescence microscopy (fluorescence *in situ* hybridization [FISH]) yield information about occurrence and abundance of the targeted organisms in the environment under study (DeLong et al. 1989; Amann et al. 1990, 1995). It allows for the *in situ* identification of morphological traits and quantification of organisms that exhibit no sufficient morphological data for distinction, like, e.g., picoplanktonic microbes (Massana et al. 2002, 2004b; Beardsley et al. 2005). Additionally, the combination of FISH with scanning electron microscopy (FISH-SEM) can reveal morphology and ultrastructure of organisms (Stoeck et al. 2003b) and enables the visualization of "novel," previously undescribed microbes. This technique was successfully tested on a wide range of protists (alveolates, stramenopiles, kinetoplastids, and cryptomonads; Stoeck et al. 2003b) and was only previously applied for the identification of a clade of uncultured marine stramenopiles (MAST-12; Kolodziej and Stoeck 2007). Nevertheless, to get detailed information about physiology and ecological habits of the organisms under study, it is necessary to carry on culturing efforts to isolate specific organisms and to enable laboratory experiments.

In the Framvaren Fjord, within the framework of a diversity survey using molecular and culturing approaches, an organism was isolated from the oxic–anoxic interface according to its phylogenetic position branched within an undescribed neobodonid sequence clade (Kinetoplastida, Euglenozoa) (Stoeck et al. 2005). Its isolation allowed for the detailed investigation of its ultrastructure, its taxonomic state and molecular phylogeny, and some ecophysiological capacities, like tolerance limits for different oxygen concentrations, pH, salinity, and temperature. Based on morphological and molecular data, the organism was proposed to represent a new genus within the order Neobodonida (Kinetoplastea) and was referred to as its type species, *Actuariola framvarensis* (Stoeck et al. 2005). The ecophysiological experiments demonstrated that the species could survive under suboxic conditions, even though they are not optimal and growth rates are lower compared to normoxic conditions. Strictly anoxic conditions were not tolerated. These findings allow for the assumption that the oxic–anoxic interface may represent the borderline of the habitat of *A. framvarensis*. However, this hypothesis has to be proven by the application of a FISH probe that is designed for *A. framvarensis* and that represents an *in situ* tool to access presence and abundance of this organism in nature (Stoeck et al. 2005).

6. Conclusions

The ultimate proof (if it exists) that the origin of eukaryotes was in oxygen-free environments is still missing. However, by no means, this implies that such evidence does not exist. Locations like the Framvaren Fjord in Norway and also other anoxic environments might be ideal to further screen for evidence. Systems like the Framvaren are ideal sources to study (ancient?) mechanisms of sulfide detoxification. Further studies of mitochondria in organisms thriving in such systems as the Framvaren Fjord might elucidate whether first eukaryotes evolved in an anoxic world. Unfortunately, working on organisms from such environments is still a major technical challenge due to the difficult accessibility and cultivability of free-living anaerobes. However, new strategies developed in the past few years are about to open new windows and doors to such extreme environments, and future efforts will be in order to meet this challenge by specific sampling and cultivation strategies.

7. References

Adl SM, Simpson AGB, Farmer MA, Andersen RA, Anderson OR, Barta JR et al (2005) The new higher level classification of eukaryotes with emphasis on the taxonomy of protists. J Eukaryot Microbiol 52:399–451

Adl SM, Leander BS, Simpson AGB, Archibald JM, Anderson OR, Bass D et al (2007) Diversity, nomenclature, and taxonomy of protists. Syst Biol 56:684–689

Amann RI, Krumholz L, Stahl DA (1990) Fluorescent-oligonucleotide probing of whole cells for determinative, phylogenetic, and environmental studies in microbiology. J Bacteriol 172(2):762–770

Amann RI, Ludwig W, Schleifer KH (1995) Phylogenetic identification and in situ detection of individual microbial cells without cultivation. Microbiol Rev 59:143–169

Amaral Zettler LA, Gomez F, Zettler E, Keenan BG, Amils R, Sogin ML (2002) Eukaryotic diversity in Spain's river of fire. Nature 417:137

Amaral-Zettler LA, McCliment EA, Ducklow HW, Huse SM (2009) A method for studying protistan diversity using massively parallel sequencing of V9 hypervariable regions of small-subunit ribosomal RNA genes. PLoS ONE 4:e6372

Anbar AD, Knoll AH (2002) Proterozoic ocean chemistry and evolution: a bioinorganic bridge? Science 297:1137–1142

Bark AW, Goodfellow JG (1985) Studies on ciliated protozoa in Eutrophic Lakes 2. Field and laboratory studies on the effects of oxygen and other chemical gradients on ciliate distribution. Hydrobiologia 124:177–188

Barns SM, Fundyga RE, Jeffries MW, Pace NR (1994) Remarkable archaeal diversity detected in a Yellowstone National Park hot spring environment. Proc Natl Acad Sci USA 91:1609–1613

Beardsley C, Knittell K, Amann R, Pernthaler J (2005) Quantification and distinction of aplastidic and plastidic marine nanoplankton by fluorescence in situ hybridization. Aquat Microb Ecol 41:163–169

Behnke A, Friedl T, Chepurnov VA, Mann DG (2004) Reproductive compatibility and rDNA sequence analyses in the *Sellaphora pupula* species complex (Bacillariophyta). J Phycol 40:193–208

Behnke A, Bunge J, Barger K, Breiner H-W, Alla V, Stoeck T (2006) Microeukaryote community patterns along an O_2/H_2S gradient in a supersulfidic anoxic Fjord (Framvaren, Norway). Appl Environ Microbiol 72:3626–3636

Behnke A, Barger KJ, Bunge J, Stoeck T (2010a) Spatio-temporal variations in protistan communities along an O-2/H2S gradient in the anoxic Framvaren Fjord (Norway). FEMS Microbiol Ecol 72:89–102

Behnke A, Engel M, Christen R, Nebel M, Klein R, Stoeck T (2010b) Depicting more accurate pictures of protistan community complexity using pyrosequencing of hypervariable SSU rRNA gene regions. Environ Microbiol 13(2):340–349

Berney C, Fahrni J, Pawlowski J (2004) How many novel eukaryotic 'kingdoms'? Pitfalls and limitations of environmental DNA surveys. BMC Biol 2:13

Bernhard JM, Buck KR (2004) Eukaryotes of the Cariaco, Soledad, and Santa Barbara Basins: Protists and metazoans associated with deep-water marine sulfide oxidizing microbial mats and their possible effects on the geologic record. In: Amend JP, Edwards KJ, Lyons TW (eds) Sulfur biogeochemistry – past and present, Geological Society of America Special Paper. Geological Society of America, Boulder 379:35–38

Boenigk J, Pfandl K, Stadler P, Chatzinotas A (2005) High diversity of the '*Spumella*-like' flagellates: an investigation based on the SSU rRNA gene sequences of isolates from habitats located in six different geographic regions. Environ Microbiol 7:685–697

Brocks JJ, Logan GA, Buick R, Summons RE (1999) Archean molecular fossils and the early rise of eukaryotes. Science 285:1033–1036

Butterfield NJ (2000) *Bangiomorpha pubescens* n. gen., n. sp.: implications for the evolution of sex, multicellularity, and the Mesoproterozoic/Neoproterozoic radiation of eukaryotes. Paleobiology 26:386–404

Callahan HA, Litaker RW, Noga EJ (2002) Molecular taxonomy of the suborder Bodonina (Order Kinetoplastida), including the important fish parasite, *Ichthyobodo necator*. J Eukaryot Microbiol 49:119–128

Camacho A, Vicente E (1998) Carbon photoassimilation by sharply stratified phototrophic communities at the chemocline of Lake Arcas (Spain). FEMS Microbiol Ecol 25:11–22

Canfield DE (1998) A new model for Proterozoic ocean chemistry. Nature 396:450–453

Caron DA, Countway P, Brown MV (2004) The growing contributions of molecular biology and immunology to protistan ecology: molecular signatures as ecological tools. J Eukaryot Microbiol 51:38–48

Cavalier-Smith T (1989) Molecular phylogeny. Archaebacteria and Archezoa. Nature 339:l00–01

Cavalier-Smith T (1991) Archamoebae: the ancestral eukaryotes? Biosystems 25:25–38

Cavalier-Smith T (2004) Only six kingdoms of life. Proc Biol Sci 271:1251–1262

Chao A (2005) Species estimation and applications. In: Balakrishnan N, Read CB, Vidakovic B (eds) Encyclopedia of statistical sciences. Wiley, New York, pp 7907–7916

Clark CG, Roger AJ (1995) Direct evidence for secondary loss of mitochondria in Entamoeba histolytica. Proc Natl Acad Sci USA 92:6518–6521

Cohen Y, Jorgensen BB, Padan E, Shilo M (1975) Sulfide-dependent anoxygenic photosynthesis in cyanobacterium Oscillatoria-limnetica. Nature 257:489–492

Corliss JO (1999) Biodiversity, classification, and numbers of species of protists. In: Raven P, Williams T (eds) Nature and human society: the quest for a sustainable world. Proceedings of 2nd National Forum on Biodiversity. National Academic Press, Washington, DC, 28–31 Oct 1997

Danovaro R, Dell'Anno A, Pusceddu A, Gambi C, Heiner I, Kristensen RM (2010) The first metazoa living in permanently anoxic conditions. BMC Biol 8:30

Davidson K, John EH (2001) The grazing response of the heterotrophic microflagellate Paraphysomonas vestita when ingesting phytoplankton prey. Protistology 2:22–23

Dawson SC, Pace NR (2002) Novel kingdom-level eukaryotic diversity in anoxic environments. Proc Natl Acad Sci USA 99:8324–8329

DeLong EF, Wickham GS, Pace N (1989) Phylogenetic stains: ribosomal RNA-based probes for the identification of single cells. Science 243:1360–1363

Diez B, Pedrós-Alio C, Massana R (2001a) Study of genetic diversity of eukaryotic picoplankton in different oceanic regions by small-subunit rRNA gene cloning and sequencing. Appl Environ Microbiol 67:2932–2941

Diez B, Pedrós-Alio C, Marsh TL, Massana R (2001b) Application of denaturing gradient gel electrophoresis (DGGE) to study the diversity of marine picoeukaryotic assemblages and comparison of DGGE with other molecular techniques. Appl Environ Microbiol 67:2942–2951

Doeller JE, Grieshaber MK, Kraus DW (2001) Chemolithoheterotrophy in a metazoan tissue: thiosulfate production matches ATP demand in ciliated mussel gills. J Exp Biol 204:3755–3764

Dryssen DW (1999) Framvaren and the Black Sea – similarities and differences. Aquat Geochem 5:59–73

Dubilier N, Mulders C, Ferdelman T, de Beer D, Pernthaler A, Klein M et al (2001) Endosymbiotic sulphate-reducing and sulphide-oxidizing bacteria in an oligochaete worm. Nature 411:298–302

Edgcomb VP, Kysela DT, Teske A, de Vera Gomez A, Sogin ML (2002) Benthic eukaryotic diversity in the Guaymas Basin hydrothermal vent environment. Proc Natl Acad Sci USA 99:7658–7662

Ehrenberg CC (1838) Die Infusionsthierchen als vollkommene Organismen. Voss, Leopold, Leipzig

Embley TM, Martin W (2006) Eukaryotic evolution, changes and challenges. Nature 440:623–630

Epstein SS, López-Garcia P (2008) "Missing" protists: a molecular perspective. Biodivers Conserv 17:261–276

Fenchel T, Finlay BJ (1995) Ecology and evolution in anoxic worlds. Oxford University Press, Oxford

Fenchel T, Bernard C, Esteban G, Finlay BJ, Hansen PJ, Iversen N (1995) Microbial diversity and activity in a Danish Fjord with anoxic deep-water. Ophelia 43:45–100

Finlay BJ (2001) Protozoa. In: Levin SA (ed) Encyclopedia of biodiversity. Academic, San Diego

Finlay BJ (2002) Global dispersal of free-living microbial eukaryote species. Science 296:1061–1063

Finlay BJ, Corliss JO, Esteban GF, Fenchel T (1996) Biodiversity at the microbial level: the number of free-living ciliates in the biosphere. Q Rev Biol 71:221–237

Fischer WW (2008) Biogeochemistry: life before the rise of oxygen. Nature 455:1051–1052

Foissner W (2006) Biogeography and dispersal of micro-organisms: a review emphasizing protists. Acta Protozoologica 45:111–136

Foissner W (2008) Protist diversity and distribution: some basic considerations. Biodivers Conserv 17:235–242

Foissner W, Blatterer H, Berger H, Kohmann F (1991) Taxonomische und ökologische Revision des Saprobiensystems – Band I: Cyrtophorida, Oligotrichida, Hypotrichia, Colpodea. Bartels und Wernitzs Druck, Munich

Foissner W, Berger H, Kohmann F (1992) Taxonomische und ökologische Revision des Saprobiensystems – Band II: Peritrichia, Heterotrichida, Odontostomastida. Bartels und Wernitzs Druck, Munich

Foissner W, Berger H, Kohmann F (1994) Taxonomische und ökologische Revision des Saprobiensystems – Band III: Hymenostomata, Prostomatida, Nassulida. Bartels und Wernitzs Druck, Munich

Foissner W, Berger H, Blatterer H, Kohmann F (1995) Taxonomische und ökologische Revision des Saprobiensystems – Band IV: Gymnostomatea, Loxodes, Suctoria. Bartels und Wernitzs Druck, Munich

Gasol JM, Peters F, Guerrero R, Pedrosalio C (1992) Community Structure in Lake Ciso – biomass allocation to trophic groups and differing patterns of seasonal succession in the metalimnion and epilimnion. Archiv Fur Hydrobiologie 123:275–303

Gasol JM, Garciacantizano J, Massana R, Guerrero R, Pedrosalio C (1993) Physiological ecology of a metalimnetic cryptomonas population – relationships to light, sulfide and nutrients. J Plankton Res 15:255–275

Haeckel E (1878) Das Protistenreich. Eine populäre Übersicht über das Formengebiet der niedersten Lebewesen. Mit einem wissenschaftlichen Anhange: System der Protisten. Ernst Günther's Verlag, Leipzig

Han TM, Runnegar B (1992) Megascopic eukaryotic algae from the 2.1-billion-year-old negaunee iron-formation, Michigan. Science 257:232–235

Hashimoto T, Nakamura Y, Kamaishi T, Hasegawa M (1997) Early evolution of eukaryotes inferred from the amino acid sequences of elongation factors 1alpha and 2. Arch Protistenk 148:287–295

Hausmann K, Hülsmann N, Radek R (2003) Protistology. Schweizerbart'sche Verlagsbuchhandlung, Science Publishers, Stuttgart

Hellebust JA, Lewin J (1977) Heterotrophic nutrition. In: Werner D (ed) The biology of diatoms. University of California Press, Berkeley, pp 169–197

Hogg J (1861) On the distinction of a plant and an animal, and on a fourth kingdom of nature. Edingb New Phil J 12:216–225

Holland HD (1984) The chemical evolution of the atmosphere and oceans. Princeton University Press, Princeton

Hong SH, Bunge J, Jeon SO, Epstein SS (2006) Predicting microbial species richness. Proc Natl Acad Sci USA 103:117–122

Javaux EJ, Knoll AH, Walter MR (2001) Morphological and ecological complexity in early eukaryotic ecosystems. Nature 412:66–69

Johnston DT, Wolfe-Simon F, Pearson A, Knoll AH (2009) Anoxygenic photosynthesis modulated Proterozoic oxygen and sustained Earth's middle age. Proc Natl Acad Sci USA 106:16925–16929

Kahl DM (1930–1935) Urtiere oder Protozoa. I. Wipertiere oder Ciliata (Infusoria), eine Barbeitung der freilebenden und ectocomensalen Infusorien der Erde, unter Ausschluss der marinen Tintinnidae. Gustav Fischer, Jena

Kahru M, Nommann S (1990) The phytoplankton spring bloom in the Baltic Sea in 1985, 1986 – Multitude of spatiotemporal scales. Continental Shelf Research 10:329–354

Kahru M, Leppanen JM, Nommann S, Passow U, Postel L, Schulz S (1990) Spatiotemporal mosaic of the phytoplankton spring bloom in the Open Baltic Sea in 1986. Mar Ecol Prog Ser 66:301–309

Katz LA, McManus GB, Snoeyenbos-West OLO, Griffin A, Pirog K, Costas B, Foissner W (2005) Reframing the 'Everything is everywhere' debate: evidence for high gene flow and diversity in ciliate morphospecies. Aquat Microb Ecol 41:55–65

Katz LA, Snoeyenbos-West O, Doerder FP (2006) Patterns of protein evolution in *Tetrahymena thermophila*: implications for estimates of effective population sizekl. Mol Biol Evol 23:608–614

Kolodziej K, Stoeck T (2007) Cellular identification of a novel uncultured marine stramenopile (MAST-12 Clade) small-subunit rRNA gene sequence from a Norwegian Estuary by use of fluorescence in situ hybridization-scanning electron microscopy. Appl Environ Microbiol 73:2718–2726

Kump LR (2008) The rise of atmospheric oxygen. Nature 451:277–278

Lackey JB (1960) *Calkinsia aureus* gen. et sp. nov., a new marine euglenid. Trans Am Microsc Soc 79:105–107

Lefranc M, Thenot A, Lepere C, Debroas D (2005) Genetic diversity of small eukaryotes in lakes differing by their trophic status. Appl Environ Microbiol 71:5935–5942

Lewitus AJ, Caron DA, Miller KR (1991) Effects of light and glycerol on the organization of the photosynthetic apparatus in the facultative heterotroph *Pyrenomonas salina* (Cryptophyceae). J Phycol 27:578–587

Lindmark DG, Muller M (1973) Hydrogenosome, a cytoplasmic organelle of the anaerobic flagellate Tritrichomonas foetus, and its role in pyruvate metabolism. J Biol Chem 248:7724–7728

López-Garcia P, Rodriguez-Valera F, Pedrós-Alio C, Moreira D (2001) Unexpected diversity of small eukaryotes in deep-sea Antarctic plankton. Nature 409:603–607

López-Garcia P, Philippe H, Gail F, Moreira D (2003) Autochthonous eukaryotic diversity in hydrothermal sediment and experimental microcolonizers at the Mid-Atlantic Ridge. Proc Natl Acad Sci USA 100:697–702

Ludwig W, Strunk O, Westram R, Richter L, Meier H, Yadhukumar et al (2004) ARB: a software environment for sequence data. Nucleic Acids Res 32:1363–1371

Luo Q, Krumholz LR, Najar FZ, Peacock AD, Roe BA, White DC, Elshahed MS (2005) Diversity of the microeukaryotic community in sulfide-rich Zodletone Spring (Oklahoma). Appl Environ Microbiol 71:6175–6184

Lylis JC, Trainor FR (1973) Heterotrophic capabilities of *Cyclotella meneghiniana*. J Phycol 9:365–369

Mai Z, Ghosh S, Frisardi M, Rosenthal B, Rogers R, Samuelson J (1999) Hsp60 is targeted to a cryptic mitochondrion-derived organelle ("crypton") in the microaerophilic protozoan parasite Entamoeba histolytica. Mol Cell Biol 19:2198–2205

Margulies M, Egholm M, Altman WE, Attiya S, Bader JS, Bemben LA et al (2005) Genome sequencing in microfabricated high-density picolitre reactors. Nature 437:376–380

Massana R, Guillou L, Diez B, Pedrós-Alio C (2002) Unveiling the organisms behind novel eukaryotic ribosomal DNA sequences from the ocean. Appl Environ Microbiol 68:4554–4558

Massana R, Balague V, Guillou L, Pedrós-Alio C (2004a) Picoeukaryotic diversity in an oligotrophic coastal site studied by molecular and culturing approaches. FEMS Microbiol Ecol 50:231–243

Massana R, Castresana J, Balague V, Guillou L, Romari K, Groisillier A et al (2004b) Phylogenetic and ecological analysis of novel marine stramenopiles. Appl Environ Microbiol 70:3528–3534

McDonald SM, Sarno D, Scanlan DJ, Zingone A (2007) Genetic diversity of eukaryotic ultraphytoplankton in the Gulf of Naples during an annual cycle. Aquat Microb Ecol 50:75–89

Medlin LK, Metfies K, Mehl H, Wiltshire K, Valentin K (2006) Picoeukaryotic plankton diversity at the Helgoland time series site as assessed by three molecular methods. Microb Ecol 52:53–71

Mentel M, Martin W (2008) Energy metabolism among eukaryotic anaerobes in light of Proterozoic ocean chemistry. Philos Trans R Soc Lond B Biol Sci 363:2717–2729

Mentel M, Martin W (2010) Anaerobic animals from an ancient, anoxic ecological niche. BMC Biol. http://www.biomedcentral.com/1741-7007/8/32

Moreira D, López-Garcia P (2002) The molecular ecology of microbial eukaryotes unveils a hidden world. Trends Microbiol 10:31–38

Moreira D, López-Garcia P, Vickerman K (2004) An updated view of kinetoplastid phylogeny using environmental sequences and a closer outgroup: proposal for a new classification of the class Kinetoplastea. Int J Syst Evol Microbiol 54:1861–1875

Morgan K, Kalff J (1975) The winter dark survival of an Algal Flagellate – *Cryptomonas erosa* (Skuja). Verh Int Ver Limnol 19:2734–2740

Oeschger R, Janssen HH (1991) Histological studies on Halicryptus spinulosus (Priapulida) with regard to environmental hydrogen sulfide resistance. Hydrobiologia 222(1):1–12

Olsen GJ, Lane DJ, Giovannoni SJ, Pace NR, Stahl DA (1986) Microbial ecology and evolution – a ribosomal-RNA approach. Annu Rev Microbiol 40:337–365

Ormerod KS (1988) Distribution of some non-phototrophic bacteria and active biomass (Atp) in the permanently anoxic Fjord Framvaren. Mar Chem 23:243–256

Overmann J, Tilzer MM (1989) Control of primary productivity and the significance of photosynthetic bacteria in a Meromictic Kettle Lake – Mittlerer-Buchensee, West-Germany. Aquat Sci 51:261–278

Pace NR (1997) A molecular view of microbial diversity and the biosphere. Science 276:734–740

Patterson DJ (1999) The diversity of eukaryotes. Am Nat 154:S96–S124

Peyretaillade E, Broussolle V, Peyret P, Metenier G, Gouy M, Vivares CP (1998) Microsporidia, amitochondrial protists, possess a 70-kDa heat shock protein gene of mitochondrial evolutionary origin. Mol Biol Evol 15:683–689

Philippe H, Lopez P, Brinkmann H, Budin K, Germot A, Laurent J et al (2000) Early-branching or fast-evolving eukaryotes? An answer based on slowly evolving positions. Proc Biol Sci 267:1213–1221

Porter KG, Sherr EB, Sherr BF, Pace M, Sanders RW (1985) Protozoa in Planktonic food webs. J Protozool 32:409–415

Ramette A, Tiedje JM (2007) Biogeography: an emerging cornerstone for understanding Prokaryotic diversity, ecology, and evolution. Microb Ecol 53:197–207

Rasmussen B, Fletcher IR, Brocks JJ, Kilburn MR (2008) Reassessing the first appearance of eukaryotes and cyanobacteria. Nature 455:1101–1104

Reul A, Rodriguez V, Jimenez-Gomez F, Blanco JM, Bautista B, Sarhan T et al (2005) Variability in the spatio-temporal distribution and size-structure of phytoplankton across an upwelling area in the NW-Alboran Sea, (W-Mediterranean). Continental Shelf Research 25:589–608

Richards TA, Bass D (2005) Molecular screening of free-living microbial eukaryotes: diversity and distribution using a meta-analysis. Curr Opin Microbiol 8:240–252

Rodriguez F, Derelle E, Guillou L, Le Gall F, Vaulot D, Moreau H (2005) Ecotype diversity in the marine picoeukaryote *Ostreococcus* (Chlorophyta, Prasinophyceae). Environ Microbiol 7:853–859

Roger AJ (1999) Reconstructing early events in Eukaryotic evolution. Am Nat 154:S146–S163

Rolland A, Bertrand F, Maumy M, Jacquet S (2009) Assessing phytoplankton structure and spatio-temporal dynamics in a freshwater ecosystem using a powerful multiway statistical analysis. Water Res 43:3155–3168

Romari K, Vaulot D (2004) Composition and temporal variability of picoeukaryote communities at a coastal site of the English Channel from 18S rDNA sequences. Limnol Oceanogr 49:784–798

Scott C, Lyons TW, Bekker A, Shen Y, Poulton SW, Chu X, Anbar AD (2008) Tracing the stepwise oxygenation of the Proterozoic ocean. Nature 452:456–U455

Searcy DG (2006) Rapid hydrogen sulfide consumption by Tetrahymena pyriformis and its implications for the origin of mitochondria. Eur J Protistol 42:221–231

Sherr EB, Sherr BF (2000) Marine microbes. An overview. In: Kirchman DL (ed) Microbial ecology of the oceans. Wiley-Liss, New York, pp 13–46

Simpson AG, Lukes J, Roger AJ (2002) The evolutionary history of kinetoplastids and their kinetoplasts. Mol Biol Evol 19:2071–2083

Skei JM (1988) Framvaren – Environmental Setting. Mar Chem 23:209–218

Slapeta J, Lopez-Garcia P, Moreira D (2006) Global dispersal and ancient cryptic species in the smallest marine eukaryotes. Mol Biol Evol 23:23–29

Sogin ML, Morrison HG, Huber JA, Welch DM, Huse SM, Neal PR et al (2006) Microbial diversity in the deep sea and the underexplored "rare biosphere". Proc Natl Acad Sci USA 103:12115–12120

Sørensen K (1988) The distribution and biomass of phytoplankton and phototrophic bacteria in Framvaren, a permanently anoxic Fjord in Norway. Mar Chem 23:229–241

Stockburger DW (1996) Introductory Statistics. Missouri State University, Springfield

Stoeck T, Epstein S (2003) Novel eukaryotic lineages inferred from small-subunit rRNA analyses of oxygen-depleted marine environments. Appl Environ Microbiol 69:2657–2663

Stoeck T, Taylor GT, Epstein SS (2003a) Novel eukaryotes from a permanently anoxic Cariaco Basin (Caribbean Sea). Appl Environ Microbiol 69:5656–5663

Stoeck T, Fowle WH, Epstein SS (2003b) Methodology of protistan discovery: from rRNA detection to quality scanning electron microscope images. Appl Environ Microbiol 69:6856–6863

Stoeck T, Schwarz MVJ, Boenigk J, Schweikert M, von der Heyden S, Behnke A (2005) Cellular identity of an 18S rRNA gene sequence clade within the class Kinetoplastea: the novel genus Actuariola gen. nov (Neobodonida) with description of the type species Actuariola framvarensis sp nov. Int J Syst Evol Microbiol 55:2623–2635

Stoeck T, Hayward GT, Taylor RV, Epstein S (2006) A multiple PCR-primer approach to access the microeukaryotic diversity in environmental samples. Protist 157:31–43

Stoeck T, Behnke A, Christen R, Amaral-Zettler LA, Rodriguez-Mora M, Chistoserdov A et al (2009) Massively parallel tag sequencing reveals the complexity of anaerobic marine protistan communities. BMC Biol 7:72

Stoeck T, Bass D, Nebel M, Christen R, Jones MDM, Breiner HW, Richards TA (2010) Multiple marker parallel tag environmental DNA sequencing reveals a highly complex eukaryotic community in marine anoxic water. Mol Ecol 19:21–31

Ström KM (1936) Land-lockes waters. Hydrography and bottom deposits in badly-ventilated Norwegian fjords with remarks upon sedimentation under anaerobic conditions. Norske Videnskaps Akademi, Oslo

Summons RE (1993) Biogeochemical cycles: a review of fundamental aspects of organic matter formation, preservation and composition. In: Engel MH, Macko SA (eds) Organic geochemistry. Plenum Press, New York, pp 3–21

Takishita K, Miyake H, Kawato M, Maruyama T (2005) Genetic diversity of microbial eukaryotes in anoxic sediment around fumaroles on a submarine caldera floor based on the small-subunit rDNA phylogeny. Extremophiles 9:185–196

Tovar J, Fischer A, Clark CG (1999) The mitosome, a novel organelle related to mitochondria in the amitochondrial parasite Entamoeba histolytica. Mol Microbiol 32:1013–1021

Tovar J, Leon-Avila G, Sanchez LB, Sutak R, Tachezy J, van der Giezen M et al (2003) Mitochondrial remnant organelles of Giardia function in iron-sulphur protein maturation. Nature 426:172–176

Vyverman W, Tyler P (1995) Fine-Layer Zonation and Short-Term Changes of Microbial Communitites in 2 Coastal Meromictic Lakes (Madang Province, Papua-New-Guinea). Archiv Fur Hydrobiologie 132:385–406

Whitfield J (2005) Biogeography: is everything everywhere? Science 310:960–961

Williams BA, Hirt RP, Lucocq JM, Embley TM (2002) A mitochondrial remnant in the microsporidian Trachipleistophora hominis. Nature 418:865–869

Woese CR (1987) Bacterial evolution. Microbiol Rev 51:221–271

Woese CR, Kandler O, Wheelis ML (1990) Towards a natural system of organisms: proposal for the domains Archaea, Bacteria, and Eucarya. Proc Natl Acad Sci USA 87:4576–4579

Yubuki N, Edgcomb VP, Bernhard JM, Leander BS (2009) Ultrastructure and molecular phylogeny of *Calkinsia aureus*: cellular identity of a novel clade of deep-sea euglenozoans with epibiotic bacteria. BMC Microbiol. http://www.biomedcentral.com/1471-2180/9/16

Zuendorf A, Bunge J, Behnke A, Barger KJ, Stoeck T (2006) Diversity estimates of microeukaryotes below the chemocline of the anoxic Mariager Fjord, Denmark. FEMS Microbiol Ecol 58:476–491

Biodata of **Gisela B. Fritz, Martin Pfannkuchen, Ulrich Struck, Steffen Hengherr, Stephan Strohmeier**, and **Franz Brümmer**, authors of *"Characterizing an Anoxic Habitat: Sulfur Bacteria in a Meromictic Alpine Lake."*

Dr. Gisela B. Fritz is currently working as manager and advisor for the degree program Technical Biology at the University of Stuttgart, Germany. After completing her Master thesis research on the restoration ecology of a salt marsh in Oregon, USA, she obtained her Ph.D. from the University of Stuttgart in the field of limnology in 2009. Her scientific interests are in the areas of limnic and marine restoration ecology and aquatic neobiota.

E-mail: **gisela.fritz@bio.uni-stuttgart.de**

Dr. Martin Pfannkuchen finished the study path of technical biology of the University of Stuttgart with a diploma work on intracellular bacteria in free-living Amoeba. He was later awarded a Ph.D. in marine biology from the University of Stuttgart in 2007. His research interests are cell biology and developmental biology of invertebrates. He currently works at the Center for Marine Research of the Ruder Boskovic Institute in Rovinj, Croatia, where the main research topics are the integration of evolutionary genetics and ecological research in the Adriatic Sea.

E-mail: **Martin.Pfannkuchen@cim.irb.hr**

Dr. Ulrich Struck is head of the stable isotope facilities of the Museum für Naturkunde, Leibniz Institute for Research on Evolution and Biodiversity at the Humboldt University Berlin since 2006. His Diploma as well as his Ph.D. at Kiel University (obtained in 1992), occupied the sedimentology and paleo-ecology of benthic foraminifera in the European Arctic. More than 2 years were spent on research vessels and boats. Present focus in research and teaching is given to global organic matter cycles, trophic pathways, food webs, and induced isotopic fractionations.

E-mail: **Ulrich.struck@mfn-berlin.de**

Gisela B. Fritz

Martin Pfannkuchen

Ulrich Struck

Dr. Steffen Hengherr is currently working as a lecturer for chemistry and biology at the JFS School in Kirchheim u. Teck, Germany. He obtained his Ph.D. from the University of Stuttgart in 2009 in the field of physiological ecology and biochemistry of cryptobiotic tardigrades. His scientific interests are in the areas of physiology, biochemistry, and biophysics of desiccation and cold tolerance in invertebrates.

E-mail: **steffen.hengherr@bio.uni-stuttgart.de**

Dipl. biol (t.o.) Stephan Strohmeier is currently freelancing as consultant in various metagenomic projects applying Next Generation Sequencing techniques. In his diploma thesis appointed at University of Stuttgart, he developed a new assembly algorithm (WiMSeEx) for simultaneous sequenced DNA components. His scientific interests are in the areas of bioinformatics and metagenomics of marine and soil biospheres.

E-mail: **strohmeier@bio.uni-stuttgart.de**

Prof. Dr. Franz Brümmer is currently the head of the Department of Zoology at the Biological Institute of the University of Stuttgart. After his Diploma Thesis in Biology on three dimensionally growing cell cultures (multicellular spheroids), he got his Ph.D. in 1987 from the University of Stuttgart (doctoral adviser: Prof Dr. Dieter Hülser) in the field of biophysics (intercellular communication). After his postdoctoral lecturer qualification (Venia legendi in Zoology & Biophysics), he was appointed as Professor extraordinarius from the University of Stuttgart in 2001. His scientific interests are in the fields of bioactive metabolites, biomaterials as well as ecological adaptations studying aquatic invertebrates with a focus on sponges and echinoderms, and scientific diving. He is teaching students from the degree programs Technical Biology and Environmental Engineering.

E-mail: **franz.bruemmer@bio.uni-stuttgart.de**

Steffen Hengherr **Stephan Strohmeier** **Franz Brümmer**

CHARACTERIZING AN ANOXIC HABITAT: SULFUR BACTERIA IN A MEROMICTIC ALPINE LAKE

GISELA B. FRITZ[1], MARTIN PFANNKUCHEN[2], ULRICH STRUCK[3], STEFFEN HENGHERR[1], STEPHAN STROHMEIER[1], AND FRANZ BRÜMMER[1]

[1]*Department of Zoology, Biological Institute, University of Stuttgart, 70569 Stuttgart, Germany*
[2]*Ruđer Bošković Institute, Rovinj, Croatia*
[3]*Museum für Naturkunde, Leibniz-Institut für Evolutions- und Biodiversitätsforschung, Humboldt Universität Berlin, Invalidenstraße 43, 10115 Berlin, Germany*

1. Meromixis in Lake Alat

Anoxic layers, mainly characterized by the diverse bacterial communities, in meromictic lakes can provide short- to long-term habitats for eukaryotic organisms. Some using these extreme habitats for foraging purposes (Sacca et al. 2009) or are even part of the very unique biocoenosis themselves (Esteban et al. 1993; Stöck et al. 2007). Without detailed knowledge on the prokaryotic communities, in-depth description of eukaryotic inhabitants is fragmentary. Meromictic lakes with their distinct water layers and stable physical conditions offer the possibility to study bacterial communities in the distinct zone of the chemocline. Although a productive habitat, the deep water layers are only pleasurable to extremophiles, such as anaerobic sulfur bacteria or endemic ciliates (Stöck et al. 2007). The alpine Lake Alat in Bavaria, Germany provides such an opportunity (Fig. 1.). Almost during the entire year, three distinct layers can be observed in the water column: an oxygen-rich trophogenic mixolimnion, a usually narrow chemocline, and an anoxic, tropholytic monimolimnion. The permanent chemocline is stabilized by density differences of sulfate-rich water supplied to the deeper water layers (Struck et al. 2009) and mostly rain water with low ion content fed to the mixolimnion (Del Don et al. 2001; Peduzzi et al. 2003a). Lake Alat, similar to Lake Cadagno in Switzerland (Del Don et al. 2001; Tonolla et al. 2004), is surrounded by dolomite and gypsum bedrocks, in particular the Raibl formation (Schöner et al. 2003; Thuro et al. 2000), which are the constant source for sulfate.

Unlike in dimictic lakes or other freshwater systems, sulfur bacteria in meromictic lakes are not only forming layers on sediment or small clouds mainly

Figure 1. The alpine Lake Alat is located at 825 m in the European Alps (*left*). Dr. Martin Pfannkuchen presenting a water sample obtained within the dense layer of purple sulfur bacteria (*right*) (Photos by A.V. Altenbach).

during summer; a thick, red bloom of phototrophic sulfur bacteria can occur within a well-defined layer during the entire year (Imhoff 2006a).

Besides stable physicochemical conditions, light in certain quantity and quality is a limiting factor for growth of photosynthetic bacteria, such as the Chromatiaceae (Overmann and Tilzer 1989). Although extremely low-light-adapted phototrophic sulfur bacteria containing bacteriochlorophyll *e* can be found at very low light intensities (Overmann et al. 1992), absorption maxima for bacteriochlorophyll α range from approximately 800 to 900 nm wavelength (Imhoff 2006a). Due to the water clarity of alpine lakes – Lake Alat is located at 868 m – such deeper layer of sulfur bacteria can develop. A certain amount of oxygen can be tolerated by some species, and they are even flexible in their metabolism, using oxygen for respiratory energy transformations. Thus, sulfur bacteria in Lake Alat can be found still in slightly oxygenated waters.

The stable chemocline provides ideal conditions to analyze the bacterial diversity within this layer. Additionally, due to the high density, high numbers of bacteria can be obtained. Earlier studies showed that up to 45% of the bacterial content such a layer contributed to γ-proteobacteria and less than 10% to members of such groups as the δ-proteobacteria (Tonolla et al. 1999, 2000). Our study focuses on the description of the diversity of the red bacterial bloom of Lake Alat.

2. Analyzing Bacterial Compositions in Lake Alat

Physical parameters such as temperature, conductivity, and pH were measured in May 2005 using a WTW Multiprobe (WTW multiprobe 350i with a WTW MPP 350 Multielectrode; WTW Weilheim, Germany), while oxygen was measured with a Multiline P4 probe (WTW).

Water samples out of the sulfur bacterial layer were taken using double bottles without bottom after Brümmer (DOBB) (Brümmer et al. in preparation) by SCUBA divers. Approximately 150–200 mL samples were centrifuged with a hand centrifuge. Pellets of bacteria were redissolved in ethanol or proteinase K and stored at 4°C or −20°C, respectively.

Samples for the measurement of particulate organic C (POC) and N concentrations (PN) were prepared by filtering variable amounts of lake water (0.18–1.0 L) onto precombusted (500°C/2 h) and preweighed GF/F filters (Whatman, Maidstone, UK; 47 mm in diameter). The filters were then dried for 48 h at 40°C and kept dry in a desiccator until analysis at the University of Munich. Particulate organic carbon and nitrogen concentration measurements were performed with a THERMO NA 2500 elemental analyzer coupled with a THERMO/Finnigan Conflo II interface to a THERMO/Finnigan MAT Delta plus isotope ratio mass spectrometer. Standard deviations of C and N concentration measurements of replicates of the lab standard (peptone) were <3%. The amount of suspended matter (seston) was calculated from the weight difference of the dry empty filter and the filtered sample plus the sample after drying at 40°C overnight per liter of filtered water.

DNA was obtained from the abovementioned, redissolved bacterial pellets by a phenol–chloroform extraction or using a standard DNA preparation kit (Macherey-Nagel, Düren, Germany). DNA concentrations were measured using a NanoDrop®; ND-100 spectrophotometer (Peqlab, Erlangen, Germany). Bacterial SSU (small sub unit) rDNA was amplified applying the 610 forward and 630 reverse primers universal for all bacteria. Due to previous studies, it was expected that a diverse group of members of the Chromatiales might be present. Hence, in addition, a primer set specific for the 16S rDNA of Chromatiaceae (Bosshard et al. 2000) was used in a nested PCR using the PCR products obtained with the universal primers. Amplification was performed in 20 µl total reaction volume with 200 ng template DNA, 1 U Taq polymerase (TaKaRa BIO INC., Shiga, Japan), 0.2 mM l-1 dNTP, and 2 mM l-1 of each oligonucleotide primer. The sequence was amplified with an initial denaturation step of 10 min at 94°C, followed by 30 cycles of 30 s at 94°C, 30 s at 52°C, and 1 min at 72°C, and a final extension of 10 min at 72°C. PCR products were purified using a NucleoSpin® Extract II kit (Macherey-Nagel). Subsequently, they were ligated into pGEM-T Easy Vector (Promega, Madison, USA) and introduced into chemical competent *Escherichia coli* (DH5α) cells following the manufacturer's instructions. Plasmid DNA was subsequently isolated from cultures with the NucleoSpin® Plasmid Kit (Macherey-Nagel). Sequencing reactions were performed with M13forward as well as M13reverse primers by Macrogen (Seoul, Korea).

Sequences were edited and aligned manually as well as subsequently analyzed using the ARB (version 06.11.28) software package (Ludwig et al. 2004; Peplies et al. 2008; Pruesse et al. 2007). Blastn nucleotide search was performed with the megablast option searching NCBIs nucleotide collection (Altschul et al. 1997).

3. Biotic and Abiotic Factors in Lake Alat

The thermocline was located somewhat below 5 m water depth and was about 6 m wide (Fig. 2). The oxygen maximum was near the beginning of the thermocline

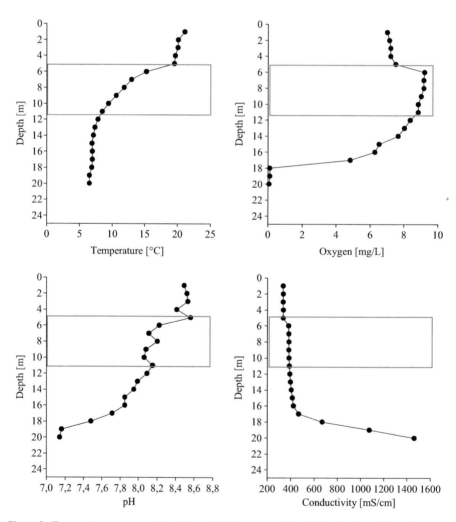

Figure 2. Temperature, oxygen, pH, and conductivity measured in Lake Alat from the surface to a depth of 20 m. The thermocline reaching from 5 to 11 m in depth, as indicated with the *grey box*.

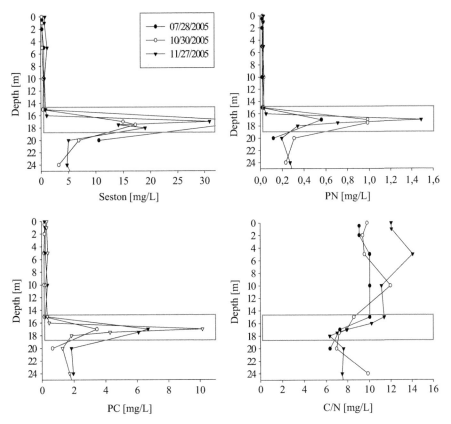

Figure 3. Seston, nitrogen concentration, carbon concentrations, as well as the carbon, nitrogen ration of Lake Alat measured from the surface to a depth of 24 m (Data from Falk 2011). The approximate location of the chemocline is indicated by the *grey box*.

(Fig. 2). At about 17 m of depth, conductivity increased abruptly, while pH dropped significantly to 7 (Fig. 2).

A strong increase in POC, PC, and PN was observed at about 17 m (Fig. 3). Visual observations showed light to reach the chemocline at 16 m but did not reach further than to the dense bacterial layer shown in Fig. 2.

Light and electron microscopic evaluation of the pellet sample showed bacteria with inclusions typical for Chromatiaceae like *Chromatium okenii*. In addition, a few other bacteria with a different morphology could also be found in the water samples.

After DNA extraction and amplifications of bacterial 16S rDNA, we retrieved sequences of 30 different clones. A blasting search in the NCBI nucleotide collection (nr) retrieved sequences with maximal similarities between 89% and 100%. Table 1 shows the corresponding sequences from this investigation with their most similar matches from the NCBI nucleotide collection.

Table 1. Lake Alat sequences blastn (megablast), optimized for highly similar sequences, against the nucleotide collection number. In bold the sequences used in for the strict consensus analysis. Uncult. = uncultured. Identity = percentage of similarity of sampled sequences to closest relative in the database.

Sequence ID	Best matches NCBI Database (blastn)		Identity
	Accession number	Description	
Alat_1-2-M13F	AJ389623.1	Uncult. sulfate-reducing bacterium	94%
Alat_1-4-M13F	AJ389623.1	Uncult. sulfate-reducing bacterium	93%
Alat_1-7-M13F	EU722352.1	Uncult. bacterium	98%
Alat_2-1-M13F	DQ676365.1	Uncult. Gram-positive bacterium	91%
Alat_2-2-M13F	AB478667.1	Uncult. Chromatiaceae bacterium	97%
Alat_2-5-M13F	FJ437772.1	Uncult. bacterium	100%
Alat_2-6-M13F	EU722352.1	Uncult. bacterium clone	98%
Alat_3-2-M13F	AB479054.1	Uncult. Chromatiaceae bacterium	96%
Alat_3-5-M13F	AB478667.1	Uncult. Chromatiaceae bacterium	93%
Alat_4-1-M13F	AB478667.1	Uncult. Chromatiaceae bacterium	97%
Alat_4-2-M13F	AB478667.1	Uncult. Chromatiaceae bacterium	99%
Alat_6a-1-M13F	AB478667.1	Uncult. Chromatiaceae bacterium	94%
Alat_6a-3-M13	FAY327188.1	Uncult. bacterium	93%
Alat_6a-6-M13F	AJ223235.1	*Lamprocystis purpureus*	96%
Alat_6a-7-M13F	AB479054.1	Uncult. Chromatiaceae bacterium	97%
Alat_2-1-M13F	DQ676365.1	Uncult. Gram-positive bacterium	91%
Alat_2-1-M13R	DQ676365.1	Uncult. Gram-positive bacterium	91%
Alat_2-6-M13F	EF648155.1	Uncult. bacterium	96%
Alat_2-6-M13R	EF648155.1	Uncult. bacterium	92%
Alat_1_11-M13F	FJ437772.1	Uncult. bacterium	92%
Alat_1_8-M13F	AM690350.1	*Allochromatium vinosum*	89%
Alat_1_9-M13F	FJ437772.1	Uncult. bacterium	97%
Alat_2_8-M13F	AJ389623.1	Uncult.sulfate-reducing bacterium	98%
Alat_2_9-M13F	AJ389623.1	Uncult.sulfate-reducing bacterium	98%
Alat_3_10-M13F	AM086644.1	*Lamprocystis purpurea*	95%
Alat_3_6-M13F	AB478667.1	Uncult. Chromatiaceae bacterium	98%
Alat_4_5-M13F	AB478667.1	Uncult. Chromatiaceae bacterium	99%
Alat_4_6-M13F	AB479054.1	Uncult. Chromatiaceae bacterium	97%
Alat_4_7-M13F	AB478667.1	Uncult. Chromatiaceae bacterium	98%
Alat_4_8-M13F	AB479054.1	Uncult. Chromatiaceae bacterium	98%
Alat_1-1	AJ389623.1	Uncult. sulfate-reducing bacterium	95%
Alat_2-1	DQ676365.1	Uncult. Gram-positive bacterium	91%
Alat_2-6	EF648155.1	Uncult. bacterium	94%
Alat_3-1	AB479054.1	Uncult. Chromatiaceae bacterium	96%
Alat_4-4	AJ223235.1	*Lamprocystis purpureus*	99%
Alat_6a-4	AJ223235.1	*Lamprocystis purpureus*	95%

Uncult. uncultured, *Identity* percentage of similarity of sampled sequences to closest relative in the database

Very high similarities within the range of the species limit (97%) were only found once with *Amoebobacter* (*Lamprocystis*) *purpureus* and several times with uncultured bacteria. Generally, the retrieved sequences from Lake Alat show slight similarities to sequences from mainly undetermined or not culturable species from the Chromatiales (γ-proteobacteria) and the δ-proteobacteria. Both bacterial groups are known to inhabit environments like the one described here (Imhoff 2006a, b). The

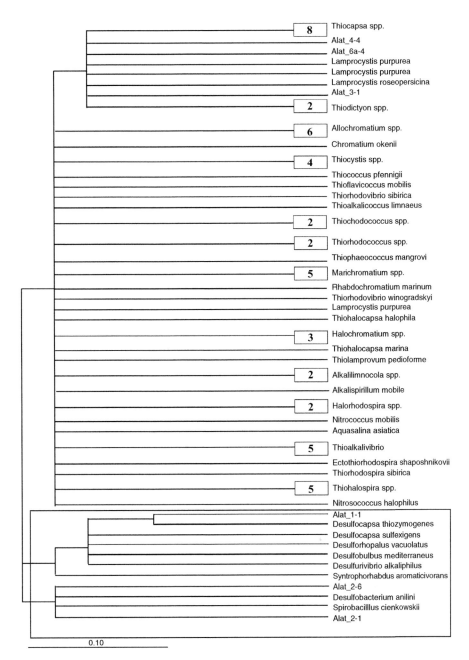

Figure 4. Strict consensus with assembled sequences. All sequences within the grey box are δ-proteobacteria, all others belong to the γ-proteobacteria. Numbers within the small boxes indicated the number of sequences combined within this branch.

sequence Alat_4-4 can be positively identified as *Lamprocystis purpureus*, with about 99% sequence identity, while all other retrieved sequences show too high dissimilarities to all available sequences, implying that they represent unknown species so far.

The corresponding strict consensus tree is shown in Fig. 4. It demonstrates the assembled sequences Alat_4-4 and Alat 6a-4 to cluster with *Lamprocystis purpurea* and Alat_3-1 to cluster with *Allochromatium*, *Chromatium*, and *Thiocystis* species, while Alat_1-1, Alat_2-6, and Alat_2-1 are grouped with δ-proteobacteria but showing not enough similarity to be assigned to a specific group.

4 Sulfuretum in Lake Alat

Extreme environments such as anaerobic, sulfur-rich lakes can provide habitat to a diverse community of eu- and prokaryotes, especially when these habitats are stable and permanent. Lake Alat is such a habitat, showing a typical, seasonally thermal and constant chemical stratification in the water column for a meromictic lake. The oxygen maximum is located near the thermocline which is common on sunny, summer days in clear waters, as the photosynthetic plankton is following their optimal light intensity/quality and inorganic nutrients. With light being crucial in the development of a phototrophic bacteria layer, light attenuation is important in the location not only of the thermo/chemocline but also in the formation of a phototrophic bacterial layer (Guerrero et al. 1985). This necessity of light for photosynthesis can also lead to a much faster rate of reproduction at the top of the chemocline and a slow growth in the middle of the bacterial layer (Overmann et al. 1991; Van Gemerden et al. 1985). The increase of conductivity is in correspondence with the increased dissolved ion concentration, which also stabilizes the chemocline in meromictic lakes, separating the surface waters from the deeper water (Del Don et al. 2001; Peduzzi et al. 2003a; Tonolla et al. 2004). Particulate organic carbon and nitrogen concentrations increase drastically at 17 m indicating high biomasses of bacteria. In correspondence with reducing oxygen contents and visual observation, this is about the location of the chemocline.

In other lakes, 50% of 129 clones belonged to the γ-proteobacteria with below 10% belonging to each of the β-proteobacteria, the green sulfur bacteria within the δ-proteobacteria, the Firmicutes, and others (Tonolla et al. 2004), which was confirmed in these studies by an in situ hybridization.

In the metalimnion of stratified lakes, three distinct communities of phototrophic microorganisms can be distinguished: a layer with cyanobacteria and eukaryotic phytoplankton, one with Chromatiaceae and filamentous green bacteria, and one with green sulfur bacteria (Vila et al. 1996, 1998). These communities are mixed before the onset of the stratification. *Chromatium okenii* is mainly found near the top of the thermocline (Tonolla et al. 2004)], where light conditions and available sulfur are optimal. Members of the Desulfobacteriaceae, δ-proteobacteria, were found dominantly in the monimolimnion (Overmann and Tilzer 1989; Tonolla et al. 2005a, b).

Furthermore, there seem to be seasonal changes in the dominance of a bacterial species within the chemocline. *Chromatium okenii* dominates in the summer months, whereas *Lamprocystis purpurea* is the dominant species in fall and winter (Bosshard et al. 2000).

Populations have also been observed being more evenly distributed during the months of low light intensities but showed microstratifications during the summer months (Tonolla et al. 2003). Smaller cell populations are more evenly distributed throughout the chemocline (Tonolla et al. 2004).

Our results show the narrow layer on top of the chemocline still reached by light to be densely populated by only a few bacterial species belonging to the Chromatiaceae and the sulfate-reducing bacteria within the δ-proteobacteria. Both bacterial groups are well known to inhabit this kind of environments (Peduzzi et al. 2003a, b; Tonolla et al. 2000, 2004).

Nevertheless, the sequences we retrieved from the samples of this chemocline show only in two cases significant similarities to already published sequences. This indicates the extreme but stable kind of the environmental conditions in the chemocline of the alpine Lake Alat, presumably inhabited by groups of bacteria so far not known from other habitats. The fact that 29 of 30 retrieved sequences belong to undescribed or unsequenced bacterial species highlights the habitat of the chemocline of Lake Alat to be a very specific, so far unmatched, and stable enough to possibly allow speciation of bacteria.

Such a stable environment also provides, however, an extreme but sheltered habitat for eukaryotes, whether only for foraging or completing their whole life cycle (Esteban et al. 1993, 2009).

5. Acknowledgments

We thank the city of Füssen and the Hotel Alatsee for their great support of our research. We also thank Andy Reuner and Inge Polle of the Department of Zoology, Universität Stuttgart for their support in the laboratory.

6. References

Altschul SF, Madden TL, Schäffer AA, Zhang J, Zhang Z, Miller W, Lipman DJ (1997) Gapped BLAST and PSI-BLAST: a new generation of protein database search programs. Nucleic Acids Res 25:3389–3402

Bosshard PP, Stettler R, Bachofen R (2000) Seasonal and spatial community dynamics in the meromictic Lake Cadagno. Arch Microbiol 174:168–174

Brümmer F, Fritz GB, Pfannkuchen M (in prep) DOBB: Water sampling device for SCUBA divers to obtain in situ data

Del Don C, Hanselmann KW, Peduzzi R, Bachofen R (2001) The meromictic alpine Lake Cadagno: orographical and biogeochemical description. Aquat Sci 63:70–90

Esteban GF, Finlay BJ, Embley TM (1993) New species double the diversity of anaerobic ciliates in a Spanish lake. FEMS Microbiol Lett 109:93–100

Esteban GF, Finlay BJ, Clarke KJ (2009) Sequestered organelles sustain aerobic microbial life in anoxic environments. Environ Microbiol 11:544–550

Falk M (2011) Physico-chemical stratification and stability of the meromictic lake Alat in Bavaria, Fuessen. BSc. Thesis Geosciences, Freie Universität. Berlin, pp 36

Guerrero R, Montesinos E, Pedros A, Esteve I, Mas J, Van Gemerden H, Hofman PAG, Bakker JF (1985) Phototrophic sulfur bacteria in two Spanish lakes vertical distribution and limiting factors. Limnol Oceanogr 30:919–931

Imhoff J (2006a) The phototrophic alpha-proteobacteria. In: Dworkin M, Falkow S, Rosenberg E, Schleifer K-H, Stackebrandt E (eds.) The prokaryotes: a handbook on the biology of bacteria. Springer, New York

Imhoff J (2006b) The Chromatiaceae. In: Dworkin M, Falkow S, Rosenberg E, Schleifer K-H, Stackebrandt E (eds.) The prokaryotes: a handbook on the biology of bacteria. Springer, New York

Ludwig W, Strunk O, Westram R, Richter L, Meier H, Yadhukumar, Buchner A, Lai T, Steppi S, Jobb G, Förster W, Brettske I, Gerber S, Ginhart AW, Gross O, Grumann S, Hermann S, Jost R, König A, Liss T, Lüßmann R, May M, Nonhoff B, Reichel B, Strehlow R, Stamatakis A, Stuckmann N, Vilbig A, Lenke M, Ludwig T, Bode A, Schleifer K-H (2004) ARB: a software environment for sequence data. Nucleic Acids Res 32:1363–1371

Overmann J, Tilzer MM (1989) Control of primary productivity and the significance of photosynthetic bacteria in a meromictic kettle lake. Mittlerer Buchensee, West-Germany. Aquat Sci 51:261–278

Overmann J, Beatty JT, Hall KJ, Pfennig N, Northcote TG (1991) Characterization of a dense, purple sulfur bacterial layer in a meromictic salt lake. Limnol Oceanogr 36:846–859

Overmann J, Cypionka H, Pfennig N (1992) An extremely low-light-adapted phototrophic sulfur bacterium from the Black Sea. Limnol Oceanogr 37:150–155

Peduzzi S, Tonolla M, Hahn D (2003a) Vertical distribution of sulfate-reducing bacteria in the chemocline of Lake Cadagno, Switzerland, over an annual cycle. Aquat Microb Ecol 30:295–302

Peduzzi S, Tonolla M, Hahn D (2003b) Isolation and characterization of aggregate-forming sulfate-reducing and purple sulfur bacteria from the chemocline of meromictic Lake Cadagno, Switzerland. FEMS Microbiol Ecol 45:29–37

Peplies J, Kottmann R, Ludwig W, Glockner FO (2008) A standard operating procedure for phylogenetic inference (SOPPI) using (rRNA) marker genes. Syst Appl Microbiol 31:251–257

Pruesse E, Quast C, Knittel K, Fuchs BM, Ludwig W, Peplies J, Glockner FO (2007) SILVA: a comprehensive online resource for quality checked and aligned ribosomal RNA sequence data compatible with ARB. Nucleic Acids Res 35:1–9

Sacca A, Borrego CM, Renda R, Triado-Margarit X, Bruni V, Guglielmo L (2009) Predation impact of ciliated and flagellated protozoa during a summer bloom of brown sulfur bacteria in a meromictic coastal lake. FEMS Microbiol Ecol 70:42–53

Schöner R, Scholz H, Krumm H (2003) Die mittelalterliche Eisengewinnung im Füssener Land (Ostallgäu und Außerfern): neue Ergebnisse zum Abbau und zur Verhüttung der Eisenerze aus dem Wettersteinkalk. Arch Lagerst forsch Geol B 24:193–218

Stöck T, Brümmer F, Foissner W (2007) Evidence for local ciliate endemism in an Alpine Anoxic lake. Microb Ecol 54:478–486

Struck U, Falk M, Altenbach AV, Pollehne F, Schneider M (2009) Nitrogen and carbon isotope ratios in suspended matter and dissolved inorganic carbon in a meromictic lake of the northern Alps (Bavaria, Germany). In: ASLO aquatic sciences meeting 2009, Nice, 25–30 Jan 2009

Thuro K, Baumgärtner W, Esslinger C (2000) Gypsum karst problems along an alpine motorway tunnel. In: International conference on geotechnical and geological engineering, Melbourne, 19–24 Nov 2000

Tonolla M, Demarta A, Peduzzi R, Hahn D (1999) In situ analysis of phototrophic sulfur bacteria in the chemocline of meromictic Lake Cadagno (Switzerland). Appl Environ Microbiol 65:325–330

Tonolla M, Demarta A, Peduzzi S, Hahn D, Peduzzi R (2000) In situ analysis of sulfate-reducing bacteria related to *Desulfocapsa thiozymogenes* in the chemocline of meromictic Lake Cadagno (Switzerland). Appl Environ Microbiol 66:820–824

Tonolla M, Peduzzi S, Hahn D, Peduzzi R (2003) Spatio-temporal distribution of phototrophic sulfur bacteria in the chemocline of meromictic Lake Cadagno (Switzerland). FEMS Microbiol Ecol 43:89–98

Tonolla M, Peduzzi S, Demarta A, Peduzzi R, Hahn D (2004) Phototropic sulfur and sulfate-reducing bacteria in the chemocline of meromictic Lake Cadagno, Switzerland. J Limnol 63:161–170

Tonolla M, Bottinelli M, Demarta A, Peduzzi R, Hahn D (2005a) Molecular identification of an uncultured bacterium ("morphotype R") in meromictic Lake Cadagno, Switzerland. FEMS Microbiol Ecol 53:235–244

Tonolla M, Peduzzi R, Hahn D (2005b) Long-term population dynamics of phototrophic sulfur bacteria in the chemocline of Lake Cadagno, Switzerland. Appl Environ Microbiol 71:3544–3550

Van Gemerden H, Montesinos E, Mas J, Guerrero R (1985) Diel cycle of metabolism of phototrophic purple sulfur bacteria in Lake Ciso, Spain. Limnol Oceanogr 30:932–943

Vila X, Dokulil M, Garcia-Gil LJ, Abella CA, Borrego CM, Baneras L (1996) Composition and distribution of phototrophic bacterioplankton in the deep communities of several central European lakes: the role of light quality. Arch Hydrobiol Spec Issue Advanc Limnol 48:183–196

Vila X, Abella CA, Figueras JB, Hurley JP (1998) Vertical models of phototrophic bacterial distribution in the metalimnetic microbial communities of several freshwater North-American kettle lakes. FEMS Microbiol Ecol 25:287–299

Biodata of **Francis Dov Por**, author of *"**Ophel, the Newly Discovered Hypoxic Chemolithotrophic Groundwater Biome: A Window to Ancient Animal Life**."*

Professor Francis Dov Por is Emeritus Professor of Zoology of the Institute of Life Sciences of the Hebrew University in Jerusalem. Obtained his Ph.D. at this university in 1962. Headed the Department of Zoology, co-Founder of the Eilat Marine Biology Laboratory, was Secretary of the Fauna et Flora Palaestina Committee of the Israel Academy and past Head of the Israel National Collections of Natural History. Founder Chairman of the International Society of Zoological Sciences. He studies copepod and crustacean taxonomy, hydrobiology, zoogeography and aspects of animal evolution.

E-mail: **fdpor@netvision.net.il**

OPHEL, THE NEWLY DISCOVERED HYPOXIC CHEMOLITHOTROPHIC GROUNDWATER BIOME: A WINDOW TO ANCIENT ANIMAL LIFE

FRANCIS DOV POR
Department of Evolution, Ecology and Behaviour, National Collections of Natural History, The Hebrew University of Jerusalem, Givat Ram, 91904 Jerusalem, Israel

1. Introduction

One day in April 2006, Israel Naaman, a master's student in speleology of Prof. Amos Frumkin of the Hebrew University, put his hands in a pool he discovered in a pitch-dark karstic cavity of a shaft at the bottom of a 100-m deep limestone quarry near Ramle in Israel, he suddenly felt them nibbled by scores of inch-long blind prawns. A few days later, his torch discovered around the pool many spectral bodies of dead scorpions that disintegrated at touch. It was like landing on an unknown planet (Fig. 1).

Indeed, the discovery of the living world in what came to be designated as the Ayyalon Cave revealed the existence of an unknown biome of life, which I named Ophel after the Hebrew word for darkness (אופל) (Por 2007).

At Ayyalon, a deep groundwater environment became, by accident, directly accessible through the collapsing bottom of the quarry. The local fresh aquifer was known already for some time to be injected by deep artesian thermo-mineral waters (30°C, 528 mg/l chloride, 5.6 mg/l H_2S) (Frumkin and Gvirtzman 2006). In the pool discovered and around it, where the fresh water and the deeper mineral water meet and mix, abundant bacterial mats of the *Beggiatoa* type were found as well as a diverse fauna. The density of the aquatic crustacean population encountered was impressive (Por 2007).

2. A Speleological Background

The history of speleological research has been accompanied by an unresolved paradox. Thousands of mainly crustacean species were discovered in the subterranean waters, belonging to tens of exclusively subterranean orders and families. However, the researchers collected only the springs, the pumped water and the accessible caves, obviously only the marginal populations of these groundwater biota. According to the reigning paradigm, the subterranean fauna feeds on photosynthesized organic matter from aboveground, which is in insufficient and episodic in supply and therefore is adapted to a chronic inanition (Hüppop 2000).

Figure 1. The sulfidic pool at Ayyalon (Photo by Israel Naaman).

Although the terrestrial component of the subterranean fauna may have easier access to epigean resources, the problem seeks an answer for the aquatic component. It is difficult to imagine that many tens of high-order aquatic taxa could have evolved in environments that were dependent on accidental and insecure food supply. Danielopol et al. (2000) surmised that healthy and normal populations of subterranean biota ought to exist under viable life conditions in the unexplored and as yet inaccessible underworld.

In the late 1980s, a dry cave was discovered at Movile in Romania, which contained a sulfidic anaerobic pond. On the rich sulfur bacteria mats of the cave, tens of new species of arthropod stygofauna were found, and many are still being described (Sarbu et al. 1996; Sarbu 2000). Only two species of aquatic stygobionts, both belonging to the Niphargidae amphipods, were found, whereas the terrestrial fauna, amply supplied by the bacterial food, is unusually and uniquely diverse and luxuriant. Movile became a trail-opening example of a subterranean ecosystem which is exclusively based on chemoautotrophy.

Summers-Engel (2007) and Porter et al. (2009) reviewed the additional, relatively scarce bacteria-based faunistic data from other dry sulfidic karst caves. The blind prawn *Typhlocaris galilea*, the largest subterranean invertebrate, was known since a century ago, from the outflow of En Nur, a strong exurgent hot spring of the Tabgha ("Heptapegon") group of springs on the shores of Lake Kinneret (the Sea of Galilee) in Israel. In 1966, Dr. Moshe Tsurnamal, who studied the prawn, dived into the spring despite the incipient speleological diving equipment and reached a void in the aquifer where the prawn was found feeding on sulfur bacteria. Three other subterranean crustaceans were added in the report which resulted; among them was the thermosbaenacean *Tethysbaena relicta* (Tsurnamal and Por 1968) already reported from a hot spring near the Dead Sea (Por 1962). Under the circumstances of those times, we did not give enough priority to those findings.

When the Ayyalon find occurred, more than 40 years later and after the resounding case of the chemoautotrophic Movile system, the half-forgotten experience from En Nur helped me to apply the idea of chemoautotrophy to groundwater ecosystems. In both En Nur and Ayyalon, there are sibling species of blind prawns and of thermosbaenacean sulfide shrimp. In both, temperatures are about 30°C, and the bacterial mats develop in the redox interface between two aquifers of different signatures. Just like in Movile, the abundant bacterial food allows for the development of extremely massive aquatic populations (Por 2007).

For the interest of the readers of the present volume, it is important to point out that the minimal amount of oxygen dissolved in the groundwater, needed to oxidize sulfide to sulfate or to oxidize methane, was enough to build and to maintain an autarchic food chain in the complex aquascape of the groundwaters.

3. The Biota of Ayyalon

Life as we know at Ayyalon is concentrated in the pool and on the wetland around its shores. A few bore holes reaching down to the groundwater at some distance from the pool-containing void were also collected by Dr. Ch. Dimentman, from the Hebrew University, who is in charge of the zoological collecting in the Ayyalon project.

The bacterioflora consists of filamentous *Beggiatoa*-like sulfur bacteria. There exists to date no more accurate identification to the species level, especially of the encountered nonfilamentous types of bacteria. The mats present in abundance the usual sulfur inclusions. According to Dimentman (personal communication), there are also some types of ciliate protozoans moving among the bacteria.

Absolutely dominant in the water of the pool at the second level of dimension are the cyclopoid copepods *Metacyclops longimaxillis* (Defaye and Por 2010). The small crustaceans, measuring up to 0.70 mm, appear in massive populations at all their reproductive stages. Another slightly larger cyclopoid, *Metacyclops subdolus* Kiefer, widespread in circum-Mediterranean subterranean waters and springs, appears in the pool with isolated specimens but seems to be frequent in the surrounding bore holes (Defaye and Por 2010).

Swarming in the water in immense numbers are the 3–4-mm-large, upside-down swimming sulfide shrimps of the subterranean order Thermosbaenacea. The new species is *Tethysbaena ophelicola* Wagner (Fig. 2). A related species, *T. relicta*, was already reported many years ago from a sulfurous spring of Hamei Zohar on the shores of the Dead Sea (Por 1962). While the *T. relicta* reports were based on a few specimens damaged and obviously swept out from their

Figure 2. An in vivo photograph of a *Tethysbaena ophelicola* Wagner population, by the Ayyalon Cave, Israel (Photo by N. Ben Eliahu).

groundwater habitat, the Ayyalon *T. ophelica* was observed, collected, and brought alive to the laboratory in countless numbers. One could observe the females carrying the eggs in their dorsal pouch, the unique feature of this order, as well as all the young stages. Likewise, one could easily observe the individuals gorged with bacterial food, even at the earliest free life stages.

The major animal in the pool, the blind prawn *Typhlocaris ayyaloni*, a new species (Tsurnamal 2008) measuring 5–6 cm belongs to the family Typhlocarididae, the giants of the subterranean invertebrates, which are endemic to the Eastern Mediterranean (Fig. 3). The abundant population of *Typhlocaris ayyaloni* in the pool of Ayyalon consists only of adult specimens. The blind prawn of the Galilee *Typhlocaris galilea* (see above) too is found only in adult stages. In both sites, the water temperature is 30°C. Tsurnamal (1978) has kept *T. galilea* in aquaria at 25–26°C, which is normal groundwater temperature, and obtained reproductive populations. Therefore, the Ayyalon species presumably feeds in the sulfidic pool too, but reproduces elsewhere where hydrographic conditions are less extreme. In fact, Dimentman (personal communication) has caught specimens of *T. ayyaloni* in baited bore holes in more remote sites.

The terrestrial faunal component at Ayyalon is dominated by springtails preliminarily identified by Louis Deharveng, Paris, as belonging eventually to genus *Troglopedetes* (personal communication, 2011). The Collembola have an important role in the food chain of Ayyalon, living on the wet rock and

Figure 3. *Typhlocaris ayyaloni* Tsurnamal, from the cave of Ayyalon, Israel (Photo by D. Darom).

pool surface, probably feeding on bacteria. The micro-predator false scorpion *Ayyalonia dimentmani* represents a new tribe of the Pseudoscorpiones, the Ayyaloniini Čurčič (2008). According to its author, the false scorpion would represent a relic of an ancient tropical terrestrial fauna.

Even more remarkable were the several exoskeleton remains of 5-cm long scorpions which were found on higher dry ground in the void around the pond. They were described as *Akrav israchanani*, representing a new family, Akravidae (Levy 2007). It is assumed that the scorpions lived when the water level in the aquifer was higher. One can only speculate that their prey could have been perhaps the juvenile prawns. As a parenthesis, the Akravidae are only the 13th family of the Scorpiones!

The Ophel ecosystem, as discovered at Ayyalon, has therefore an aquatic component, or a mainly crustacean food chain, and a wetland component with arthropod taxa. It answers to long-asked questions about the biology and evolution of the crustacean stygofauna but reveals also a strange and unexpected arachnid and possibly also apterygote fauna in the air pockets and on the wet substrates surrounding the chemosynthetic sites.

4. The Functioning of the Ophel Biome

Some 15% of the continents consist of karst formations (Culver and Pipan 2009). In Israel too, where karst covers much of the country, the meaning of the Ophel hypothesis, as shown by the sites of Ayyalon and of the Dead Sea–Lake

Kinneret graben, is that the underground below our very feet is inhabited. Ophel is a tridimensional aquatic biome, hundreds of meters thick, which exists wherever fresh, circulating hypoxic groundwater is injected by confined fossil aquifers. Since the confined aquifers are anaerobic and, as a rule, warmer, sulfidic and richer in electrolytes, a chemocline with the fresh aquifer forms in which oxidative bacterial chemoautotrophy takes place.

A set of biota live in the chemocline and feed on the bacterial biomass and are the primary or secondary consumers of the Ophel biome. Characteristically, they are encountered in huge reproductive populations. At Ayyalon, *Tethysbaena*, *Metacyclops longimaxillis*, and probably the collembolan *Troglopedetes* belong to this category, jointly with the yet unidentified protozoans. The first of them the sulfide shrimp, are bulk consumer of bacteria (Fig. 4), whereas the much smaller *Metacyclops* are swimming filter feeders (Defaye and Por 2010). The springtail is a surface-bound bacterial grazer. Judging by its way of life, this first category of Ophel biota are lentic species, i.e., a species bound to still water conditions, which in our case would be hydrogen-sulfide-infested subterranean pools where two groundwaters of different hydrochemical signatures are superposed.

The very rich bacterial biomass which is chemolithotrophically produced in these quiet microenvironments attracts visitors, such as *Typhlocaris ayyaloni* and *Metacyclops subdolus*, in the case of Ayyalon. These visitors live and reproduce in fresh aquifers with normal temperature, commuting between the feeding places. Like the prawn, they are probably omnivorous active swimmers and crawlers, able to move around in the network of the karstic system. They distribute the chemolithotrophic bioproduction over all the catchment of the respective aquifer. They are the ones that appear also in the caves and at the spring heads. They are the lotic species among the Ophel biota, probably the more diverse among subterranean biota. While the lentic species are strictly specialized bacterial feeders, the lotic ones are probably all food opportunists.

Much of the chemoautotrophic bioproduction is probably also passively spread by turbulence and flow as bits of flocculent bacterial mats, with their microfauna also carried along the subterranean ducts, as witnessed also by the damaged and fragmented specimens of *Tethysbaena* encountered in the springs. All this is food for scavengers.

Looked through the eyes of a speleologist, a cavern is inhabited by marginal biota that belong to centrifugal biomes: that of the subterranean chemoautotrophic Ophel and that of the epigean photoautotrophic ones.

To better understand the functioning of Ophel, one could imagine it to be like the network of a series of oases and the surrounding desert or a series of vital watering holes and the surrounding dry savanna. Of course, the comparison is not much more than a figurative one.

A schematic representation of an artesian lentic environment, probably the most typical variety of the ophelic biome, is presented in the figure below (Fig. 5).

A WINDOW TO ANCIENT ANIMAL LIFE 471

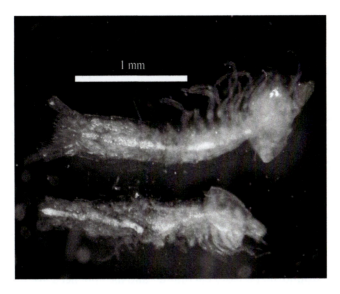

Figure 4. Juvenile *Tethysbaena ophelicola* Wagner from Ayyalon Cave, with intestines gorged with bacteria.

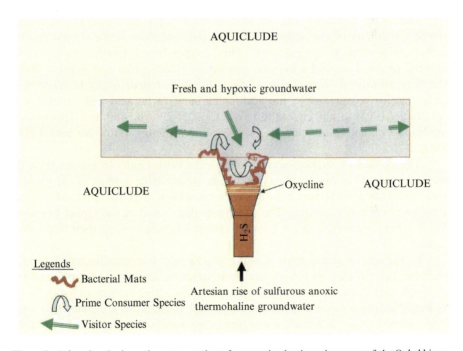

Figure 5. A functional schematic representation of an artesian lentic environment of the Ophel biome.

5. Permanent Hypoxia in the Dark

The definition of hypoxic conditions is lax. Authors use any value below the normoxic DO concentrations in fresh water of 8–10 ppm. In the groundwaters, the values are usually around 1 ppm, whereas the vadose zone and in the interstitial water in the river beds, are richer in O_2. At such low hypoxic DO values, the smallest metabolic activity, either by bacteria or by animals associated with stagnant conditions, leads to anoxia. DO values below 1 ppm are often encountered in groundwaters. DO values of ≤0.2 ppm are considered to be anoxic. Confined fossil aquifers are, as a rule, anoxic or anaerobic. There are, however, considerable technical difficulties to measure very low oxygen values in the field.

Working on the postulate that a metazoan organism needs the presence of a certain amount of molecular oxygen at least temporarily for several basic biochemical and cellular processes, the subterranean aquatic biota are facing some crucial problems. These have been summed up recently and discussed by Summers-Engel (2007).

Aboveground metazoans encounter anoxia and need to survive it only temporarily. In the subterranean waters, the near-limit hypoxic–anoxic conditions are a basic environmental parameter, and therefore, the aquatic subterranean fauna had to develop mechanisms of survival under the permanent menace of oxygen scarcity. This circumstance has not been sufficiently emphasized in the speleological literature.

The main cause of the curious, almost total evolutionary exclusivity of the Crustacea in the subterranean aquatic medium is their arthropodan hemocyanin respiratory pigment. This copper-based protein is especially adapted to function under hypoxic conditions (Hochachka and Somero 2002). The high affinity of hemocyanin to O_2 is fine-tuned structurally to respond to variations in oxygen supply, the accumulation of lactic acid and variations of pH (Reiber 1995). It has high affinity to O_2 under hypoxic and high-temperature conditions, changing adequately its polymeric structure (Hourdez and Lallier 2007).

The possession of hemocyanin allows the crustaceans to extract even the smallest amounts of dissolved oxygen which are available in most of the groundwater spaces and still to lead a life of active movement, even though food is scarce or often temporarily lacking. The arachnids discovered in the Ophel biome of Ayyalon belong also to a taxon known to possess hemocyanin as their respiratory pigment.

In the chemoautotrophic feeding places where ample bacterial food is available, the crustaceans have to confront anaerobic conditions for more or less prolonged periods without losing their capacity of active movement for intensive feeding. This poses a special problem. In general, crustaceans are unable to reduce their metabolic rates, like mollusks or free-living worms. The lactic acid-producing type of anaerobic respiration, which yields little ATP per molecule glucose, would not allow for active swimming and feeding under microaerobic or anaerobic conditions. Therefore, an additional capacity is employed to make sustained long-term anaerobic activity possible. This is the malate pathway (Shapiro and Bobkova 1975), a modified Krebs cycle, known in parasitic nematodes, which uses malate dismutase. In short, under conditions of prolonged anaerobiosis that PEP is converted directly to oxaloacetate, then to malate and finally into in propionate. As much as 15% of the aerobic ATP production per glucose molecule can be obtained in this way, and there is no accumulation of toxic lactic acid. This pathway has been found as functioning also in the thermosbaenacean *Tulumella* sp. and other crustaceans living in the microaerobic chemocline of an anchialine cave by measuring the elevated values of malate dehydrogenase (Bishop et al. 2004). I presume this is the case also for the ophelic thermosbaenaceans, expecting concrete proof.

When food is abundant, like in the case of the bacterial mats of the Ophel biome, even at the low energy yield of this improved anaerobic respiration rate, massive populations of active crustaceans can flourish.

Anaerobic conditions go hand-in-hand with accumulation of free H_2S. In the case of the ophelic biome, hydrogen sulfide is a permanent presence in the oxycline and below it. This gas is toxic to all animals, and its detoxification to thiosulfate requires oxygen (Hochachka and Somero 2002). Moving in and out of the strictly anoxic layer is not a problem for swift swimmers like *Typhlocaris*, but is certainly one for the small and less active *Tethysbaena* and *Metacyclops* of Ayyalon.

The process of sulfide detoxification remains still unexplained for the mid-oceanic hot vent giant annelid worms. It is also the case for a geographically much nearer phenomenon. In Lake Kinneret, which has an anaerobic hypolimnion during 8–9 months every year, the nematode *Eudorylaimus andrassyi* and the oligochete *Euilyodrilus heuscheri* live in the deep muddy bottom of the lake the year round. Kept under controlled anaerobic laboratory conditions, both species survived for 4 months, although the experimental bottles with the specimens contained already high amounts of H_2S 2 weeks after sealing them. The nematodes even reproduced in the bottles after 2 months of anaerobism (Por and Masry 1968). This is the longest reported anaerobic survival time of free-living metazoans (Schmidt-Rhaesa 2007) and perhaps also of maximal survival time to sulfide toxicity.

6. The Thermosbaenacean Extremophily and Its Heuristic Value

All the 35 known species of Thermosbaenacea are specific first consumers on sulfur bacteria, either in stratified anchialine cave waters or groundwaters. This entails also specific adaptations to micro-anaerobic conditions and, without doubt, also sulfide toxicity, as discussed above (Wagner 1994; Wagner in press).

The mass appearance of the bacteria-feeding thermosbaenacean *Tethysbaena ophelicola* in the sulfidic warm pond of Ayyalon was of a heuristic importance for the idea of the existence of the worldwide Ophel biome. Without the need to penetrate and access further groundwater sites of the chemolithotrophic Ophel biome, its existence can be safely assumed by the mere existence of the thermosbaenacean sulfide shrimp in the groundwaters of the different continents Por (2011). The thermosbaenacean sulfide shrimp had to adapt also to the dangerous presence or vicinity of high temperatures, which often accompany the chemolithotrophic systems which are their specific feeding grounds.

Thermosbaena mirabilis, the species discovered in 1924, which lives in thermal waters at a scorching temperature of 45°C in Tunisia (Barker 1959) and gave its name to the whole subsequently discovered order, is the most heat-resistant known metazoan. *Tethysbaena relicta* near the Dead Sea was found in a thermal spring at 31°C (Por 1962) while the temperature inside the feeding groundwater is certainly higher. *T. somala* was reported from a hot spring in Somalia at 31°C and *T. halophila* in the warm sulfuric cave lake of Asclepius in Dalmatia. The Ayyalon species lives at a temperature reaching 30°C.

As a rule, temperatures of a few degrees above 30° are lethal to metazoans, the tube worms and decapod crustaceans of the black smokers of the mid-oceanic ridges included. The famous Pompeii worm *Alvinella* of the mid-oceanic black smokers was thought to withstand 80°C within its tubes. It turned out that the tubes were damaged when the temperature was measured in situ, and that under laboratory condition, the worm survived for some hours only for up to 40°C. Similar are the resistance limits of the vent shrimps *Mirocaris* and *Rimicaris* that are well adapted to warm temperatures of up to 25°C. For all these, see a review by Shillito et al. (2006). And one does not have to forget that the shrimps move in and out of the hot plume into the ice-cold abyssal ocean, and the worm completes there its larval development, whereas the thermosbaenacean lives permanently and develops in the hot water.

The thermosbaenaceans are, in my view, exceptional in being a taxon at the order level, which is specifically adapted to life at high temperature.

Such an adaptation is necessary in the ophelic biome, since the water of the confined aquifer below the bacteria-rich chemocline will always have high temperatures, often reaching 30° and more, because of the geothermal gradient, but mainly because of its deep tectonic hydrological connections.

The thermosbaenaceans are an order which evolved to adapt to the extremophilic conditions of the ophelic biome. In this, they resemble the vestimentiferan worms which accompany characteristically the mid-oceanic hot vents. Like them, their worldwide presence is an indication of the worldwide existence of the ophelic biome.

7. The Antiquity of Ophel and a Crustacean Look at Evolution

The discovery of chemoautotrophy-based ecosystems in the groundwaters and of the entire animal taxa, like the Thermosbaenacea that are evolutionarily adapted to them, naturally raises the question of the antiquity of this type of ecosystems. Much has already been speculated about the primordial role of the oceanic hot vents in the evolution of life and of animal life in particular. Since its discovery by Fenchel and Riedl (1970), the so-called thiobios of the fauna of the sulfide-infested sands served also to such speculations about early anoxic evolution, until finally defined by Platt (1981) as being microaerobic and not anaerobic. The crustacean-like class Cephalocarida is typical for this microaerobic environment, as well as the order Mystacocarida.

If we add the class Remipedia, which is typical for the hypoxic layer in the anchialine cave waters, we have tens of exclusive subterranean high-order crustacean-type taxa adapted to hypoxic waters. This endemism speaks of old age. Some of such taxa, like the above-mentioned ones, or also others like the Speleogriphacea, are real living fossils.

On the contrary, the very diverse and specific crustacean fauna of the mid-oceanic hot vents, despite its ecological importance, is represented by caridean shrimp families probably of relatively young Miocene origin and by opportunistic species of galatheid crabs. But then also, the deep sea was most probably colonized only after the last Cretaceous oceanic anoxic event.

Within the groundwaters, the ophelic biome was probably much more protected from major global perturbations. It could have been in existence since the time when there was enough free oxygen to sufficiently aerate the vadose water reaching the aquifers. Then, through oxidation, a new and abundant bacterial food resource was created, available to a set of subterranean, mainly aquatic animals. These have to be able though to withstand the specific environmental conditions of hypoxia and often of sulfide infestation and increased temperatures which characterize these environments.

The crustaceans or allied arthropods must have been adapted already to the late Proterozoic hypoxic environments reigning well before the fossiliferous and more oxygen-rich Cambrian. Molecular phylogeny calculations indicate a very old age to arthropodan hemocyanin, and recent experience shows that this kind of calculations is, in general, confirmed afterward by hard geological facts. Schmidt-Rhaesa (2007), based on Burmester (2001), considers that the biochemical phenoloxydase superfamily of hemocyanin might have generated the "sudden" appearance of the hard cuticles and skeletons at the Precambrian–Cambrian border by producing the oxygen-consuming molecules important for sclerotization. The extremely diverse fossil fauna of crustacean-like fossils in the Cambrian Lagerstätten and especially the small specimens of the Örsten fauna of Waloszek (2003) gives us all the reason to consider that aquatic arthropod diversity was already at least as great by that time as it is today.

The biota of Ophel could have colonized the groundwaters underlying the continents much earlier than the late Ordovician–Silurian landfall of the photoautotrophic ecosystems. In a protected underground, using small amounts of oxygen and, first of all, powered by a bacterial chemolithotrophic source of energy which was in existence for millions of years, the biota of Ophel did not have to wait for the establishment of a protective ozone screen and an established upland vegetal cover upland in order to function.

The sunlit aboveground photosynthetic biomes and the dark chemosynthetic biomes are parallel and have a parallel historical evolution, albeit not in isolation one from another. Opportunities for adaptation to the two life resources and types of food chains continued to act through natural selection, as new species chose darkness or vice versa. In this sense, not all the species of Ophel are primary, old inhabitants of the chemoautotrophic world, like the Thermosbaenacea. The prawns Typhlocarididae although belonging to an ancestral family, are caridean decapods that went underground probably at a later time. The two species of *Metacyclops* of Ayyalon, which belong to the great and most probably artificial genus of cyclopoid copepods (Defaye and Por 2010), represent two strategies and, *ipso facto*, two stages of subterranean colonization.

Finally, the finding of the new arachnids at Ayyalon, the scorpions of the family Akravidae, the pseudoscorpions of tribe Ayyaloniini (Levy 2007; Čurčič 2008), and the rich collembolan population raises an interesting possibility. Little (1983) amply discussed the way in which the arthropods, chilopods, arachnids and apterygotes colonized land, either directly from the sea or through wetlands. The first tracks and fossils of small forms are already from the late Ordovician and early Silurian, and the problem is the level and amount of terrestrial vegetation-based food encountered by these pioneer arthropods. If, like the Ayyalon arthropods, they could have inhabited at an early date the wet bacterial mats in the ophelic biomes, there would have been an additional way of colonizing the land. Through the intermediary of sulfidic wet caves, they could have emerged to the surface once there was sufficient detritic food and prey to compare with that underground.

8. Codicil

The chemolithotrophic bacterial mats of the ophelic groundwater biome appear to have been an attractive medium for animal life throughout evolutionary history. The unseen prokaryote majority of the globe (Whitman et al. 1998) to which they belong seems to present, however, an overall low lithotrophic productivity rate as compared to the phototrophic one. The food chains of Ophel are short, when compared to the much longer ones of the plant-based world. This chapter, one of the very many ultimate consequences of these long food chains, will not be, I hope, too long for the patient reader.

9. References

Barker D (1959) The distribution and systematic position of the Thermosbaenacea. Hydrobiologia 13:209–235
Bishop RE, Kakuk B, Torres JJ (2004) Life in the hypoxic and anoxic zones: metabolism and proximate composition of Caribbean troglobitic crustaceans, with observations on the water chemistry of anchialine caves. J Crustac Biol 24(3):379–393
Burmester T (2001) Molecular evolution of the of the arthropod hemocyanin superfamily. Mol Biol Evol 18:184–195
Culver DC, Pipan T (2009) The biology of caves and other subterranean habitats. Oxford University Press, New York
Čurčič BPM (2008) *Ayyalonia dimentmani* n. g., n. sp. (Ayyaloniini,n. trib.,Chthoniidae, Pseudoscorpiones) from a cave in Israel. Arch Biol Sci Belgrade 60(1):331–339
Danielopol DL, Pospisil P, Rouch R (2000) Biodiversity in groundwater: a large-scale view. Tree 15(6):223–225
Defaye D, Por FD (2010) *Metacyclops* (Copepoda, Cyclopidae) from Ayyalon Cave, Israel. Crustaceana 83(4):399–423
Fenchel T, Riedl R (1970) The sulfide system: a new biotic system underneath the oxidized layer of marine sand bottoms. Mar Biol 7:225–268
Frumkin A, Gvirtzman H (2006) Cross-formational rising groundwater at an artesian karstic basin: the Ayyalon Saline Anomaly, Israel. J Hydrol 318:316–333
Hochachka PW, Somero GN (2002) Biochemical adaptation. Mechanism and process in physiological evolution. Oxford University Press, New York
Hourdez S, Lallier FH (2007) Adaptations to hypoxia in hydrothermal vent and cold-seep invertebrates. Rev Environ Sci Biotechnol 6:143–159
Hüppop K (2000) How do cave animals cope with food scarcity in caves? In: Culver DC, Humphreys WF (eds) Ecosystems of the world, vol 30, Subterranean ecosystems. Elsevier, Amsterdam, pp 159–188
Levy G (2007) The first troglobite scorpion from Israel and a new chaetoid family (Arachnida: Scorpiones). Zool Middle East 40:91–96
Little C (1983) The colonisation of land. Cambridge University Press, New York
Platt HM (1981) Meiofaunal dynamics and the origin of the Metazoa. In: Forey PL (ed) The evolving biosphere. Cambridge University Press, London, pp 207–216
Por FD (1962) Un nouveau thermosbenacé, *Monodella relicta* n.sp. dans la depression de la Mer Morte. Crustaceana 3(4):304–310
Por FD (2007) Ophel: a groundwater biome based on chemoautotrophic resources. The global significance of the Ayyalon cave finds. Hydrobiologia 592:1–10
Por FD (2011) Groundwater life: New biospeleological views resulting from the Ophel paradigm. Travaux de l' Institut de Speleologie "Emile Racovitza" t.L. pp 61–76
Por FD, Masry D (1968) Survival of a nematode and an oligochaeta species in the anaerobic benthal of Lake Tiberias. Oikos 19:388–391
Porter ML, Engel AS, Kane TC, Kane BK (2009) Productivity-diversity relationship from chemolithoautotrophically based sulfidic karst systems. Int J Speleol 38(1):27–40
Reiber CL (1995) Physiological adaptations of crayfish in the hypoxic environment. Am Zool 35:1–11
Sarbu SM (2000) Movile cave a chemoautotrophically based groundwater ecosystem. In: Wilkens H, Culver DC, Humphreys WF (eds) Ecosystems of the world, vol 30, Subterranean ecosystems. Elsevier, Amsterdam, pp 319–343
Sarbu SM, Kane TC, Kinkle BK (1996) A chemoautotrophically based groundwater. Science 272:1953–1955
Schmidt-Rhaesa A (2007) The evolution of organ systems. Oxford University Press, New York
Shapiro AZ, Bobkova AN (1975) The role of malate dehydrogenase in adaptation to hypoxia in invertebrates. J Evol Biochem Physiol 11:478–479

Shillito B, Le Bris N, Hourdez S, Ravaux J, Clottin D, Caprais J-C, Jollivet D, Gaill F (2006) Temperature resistance studies on the deep-sea vent shrimp *Mirocaris fortunata*. J Exp Biol 209:945–955

Summers-Engel A (2007) Observations on the biodiversity of sulfidic karst habitats. J Cave and Karst Stud 69(1):187–206

Tsurnamal M (1978) The biology and ecology of the blind prawn Typhlocaris galilea Calman (Decapoda, Caridea). Crustaceana 34(3):195–213

Tsurnamal M (2008) A new species of the stygobitic blind prawn Typhlocaris Calman 1906 (Decapoda, Palaemonidae, Typhlocaridinae) from Israel. Crustaceana 81(4):487–501

Tsurnamal M, Por FD (1968) The subterranean fauna associated with the blind palaemonid prawn *Typhlocaris galilea* Calman. Int J Speleol 3:219–223

Wagner HP (1994) A monographic review of the Thermosbaenacea (Crustacea: Peracarida). Zoologische Verhandelingen 291:1–338

Wagner HP (in press) *Tethysbaena ophelicola* (Crustacea, Thermosbaenacea), a new prime consumer of the Ophel biota, Ayyalon Cave, Israel.Crustaceana

Waloszek D (2003) Cambrian "Örsten"-type preserved arthropods and the phylogeny of crustacea. In: Legakis A, Sfenthourakis S, Polymeni R, Thessalou-Legaki M (eds) The new panorama of animal evolution. Proceedings of the XVIII international congress of zoology, Athens, pp 67–89

Whitman WB, Coleman DC, Wiebe WJ (1998) Prokaryotes: The unseen majority. Proc Natl Acad Sci USA 95:6578–6583

Biodata of **Virginia P. Edgcomb** and **Jennifer F. Biddle**, authors of *"Microbial Eukaryotes in the Marine Subsurface?"*

Dr. Virginia P. Edgcomb is a research specialist in the Department of Geology and Geophysics at the Woods Hole Oceanographic Institution (WHOI), Woods Hole, MA. She received her Ph.D. from the University of Delaware (Department of Biology) in 1997. As postdoctoral researcher, she spent 3 years at the Marine Biological Laboratory where she was involved in studies of early eukaryotic evolution and of microbial diversity at hydrothermal vents, followed by 2 years at WHOI where she studied the tolerance of several marine prokaryotes to extreme conditions found at hydrothermal vents. Her current research interests include the diversity and evolution of protists, the microbial ecology of microoxic and anoxic marine environments, and symbioses between protists and prokaryotes in extreme environments, including hypersaline anoxic basins, anoxic and sulfidic marine water column and sedimentary environments, and subsurface marine sediments.

E-mail: **vedgcomb@whoi.edu**

Dr. Jennifer F. Biddle is an assistant professor in the School of Marine Science and Policy at the University of Delaware, Lewes, Delaware. She received her Ph.D. from the Pennsylvania State University in 2007. As postdoctoral researcher, she spent 3 years at the University of North Carolina at Chapel Hill, working on microbial populations of the subsurface and hydrothermal systems of Guaymas Basin. Her current research interests include the diversity of archaea, biogeochemical cycling in estuarine systems, geobiology, microbialite biology, and subsurface marine sediments.

E-mail: **jfbiddle@udel.edu**

Virginia P. Edgcomb   **Jennifer F. Biddle**

MICROBIAL EUKARYOTES IN THE MARINE SUBSURFACE?

VIRGINIA P. EDGCOMB[1] AND JENNIFER F. BIDDLE[2]
[1]*Geology and Geophysics Department, Woods Hole Oceanographic Institution, Woods Hole, MA 02543, USA*
[2]*College of Earth, Ocean and the Environment, University of Delaware, Lewes, DE 19958, USA*

1. Introduction

Marine sediments cover more than two-thirds of the Earth's surface and have been estimated to contain as much as one-third of Earth's prokaryotic biomass (Whitman et al. 1998). Despite this, relatively little is known about this deep biosphere, and essentially nothing is known about the presence of microbial eukaryotes (protists) in sediments deeper than a few centimeters. Through consumption of dissolved organic matter and by selective grazing in subsurface horizons where bacterial and/or archaeal numbers are high, protists may significantly impact carbon cycling in the marine subsurface. An understanding of the biogeochemical activities, composition, and temporal and spatial dynamics of marine subsurface communities is essential for accurate modeling of nutrient cycling in this vast subsurface biosphere.

The first comprehensive survey of microbial life in deeply buried marine sediments, Leg 201 of the Ocean Drilling Program (in 2002), collected samples from the equatorial Pacific Ocean and the Peru Margin (D'Hondt et al. 2004). An analysis of eukaryotic small-subunit ribosomal RNA gene (18S rRNA) sequences obtained from some of these deep marine subsurface sediments, the first metagenome, and a cultivation study have been completed (Biddle et al. 2005, 2008; Edgcomb et al. 2011). Results show that fungi dominate eukaryotic life in the buried marine subsurface. Aside from these studies, the marine subsurface remains largely unexplored regarding the eukaryotes. Gathering some additional information in this chapter from the literature on other deep marine studies of eukaryotes helps to inform the results of those few deep subsurface studies and to direct future investigations of deep subsurface eukaryotic communities.

Marine fungi are predominantly comprised of an ecologically defined group of filamentous ascomycetes, their anamorphs, and yeasts (Kohlmeyer and Kohlmeyer 1979). Studies of marine fungi have primarily focused on tropical mangroves, salt marshes, and open ocean regimes where they are known to be involved in the degradation of organic matter (Hyde et al. 1998; Jobard et al. 2010;

Kohlmeyer 1979). Fungi are known to play key roles in ecologically important relationships as pathogens and symbionts in algae and higher plants and animals, and to represent a significant portion of the biomass in terrestrial systems, and to play important roles in biogeochemical cycles and food webs (Bass et al. 2007; Burgaud et al. 2010; Gadd 2007). Despite the high number of fungal species described to date (Adl et al. 2005; Richards and Bass 2005), we are only now gaining an understanding of their presence in deep marine environments (e.g., Bass et al. 2007; Burgaud et al. 2009, 2010; Edgcomb et al. 2002; Jobard et al. 2010; López-García et al. 2001a, 2003, 2007; Stoeck and Epstein 2003).

The study by Edgcomb et al. (2011) indicates that fungi increasingly dominate the subsurface eukaryotic microbial population with sediment depth. The recovery of ribosomal RNA of fungi at these depths suggests that these sequences come from living cells. Many of the sequences detected to date are close relatives of terrestrial fungi. If these fungal populations are truly active, this may have global implications for ocean carbon and nitrogen cycling, given the extent of the deep subsurface biosphere.

The occurrence of fungi (filamentous and yeasts) has been studied to a much greater degree in near-surface deep marine environments. The question of dominance of basidiomycetes vs. ascomycetes in deep marine sedimentary environments remains debatable and may depend on particular habitats and local carbon sources. Nagahama et al. (2001) reported that culturable fungal diversity in surface sediments at 2,000 m or greater was dominated by ascomycetous yeasts. In contrast to this, basidiomycetes dominated subsurface sediments at the same sediment depths and are also found in deep-sea clams, tubeworms, and mussels. The dominance of basidiomycetes was also observed by Takishita et al. (2006, 2007) in a molecular survey of sediments from deep methane seeps and by Bass et al. (2007) who found yeast forms to dominate fungal diversity in deep ocean environments. However, these studies have not elucidated the potential effects that fungi may have on deep-sea environments.

The boundaries between terrestrial and marine fungi are currently not clear, with the most current definition (circa 1979) of a marine fungus as being "those that grow and sporulate exclusively in a marine or estuarine habitat" (Kohlmeyer 1979). While fungi are increasingly found in deep-sea environments (Burgaud et al. 2009; Connell et al. 2009), a clear differentiation between marine and terrestrial fungi has been lost. Fungi thought to cause coral diseases are sometimes labeled as terrestrial invaders (Ravindran et al. 2001); however, a few have been shown to be marine species of a typically terrestrial group (Geiser et al. 1998). The presence of decaying algae and higher plants and animals in the deep sea may provide a clue as to how fungi come to dominate this environment. While the possibility remains that some fungi are truly marine, many terrestrial fungi may be delivered to the subsurface attached to detritus or may simply sediment as unattached spores or fungal filaments (Lorenz and Molitoris 1997). There is some evidence that terrestrial/surface-dwelling fungi may be capable of colonizing deep-sea habitats due to their ability to alter their membrane composition to accommodate high hydrostatic pressure (Simonato et al. 2006). The relationship between terrestrial and marine fungi is an evolving story in which molecular phylogeny will play a role as new species are discovered.

2. A Study of Peru Margin and Peru Trench Sediments: 18S rRNA Genes

An analysis of eukaryotic small-subunit ribosomal RNA (18S rRNA) from marine subsurface sediment samples from Peru Margin and Peru Trench was recently performed (Edgcomb et al. 2011). In that analysis, DNA and RNA were extracted from selected subsurface sediment samples collected during ODP Leg 201, currently the best characterized deep subsurface microbial ecosystem with extensive background data on bacteria and archaea (Biddle et al. 2006; Inagaki et al. 2006; Lipp et al. 2008; Parkes et al. 2005; Schippers et al. 2005; Sorensen and Teske 2006; Teske and Sorensen 2008; Webster et al. 2006) and on microbial biogeochemistry (D'Hondt et al. 2003, 2004; Meister et al. 2005). Prokaryotic cell counts (AODC) in the Peru Margin sites are generally in line with or above global averages for subsurface cell counts at comparable sediment depths (D'Hondt et al. 2004) and revealed near-surface maxima and local peaks in some sulfate/methane transition zones (SMTZ) (Parkes et al. 2005; Schippers et al. 2005). Since the depth distribution of eukaryotes could follow a similar dynamic, Edgcomb et al. (2011) analyzed a near-surface sample (3.91 mbsf) and a deeper sample from the SMTZ (35.9 mbsf) from ODP Site 1228 (Peru Margin, 262 m water), and a near-surface sample (1.75 mbsf) from ODP 1230 (Peru Trench, 5,086 m water). All sediments had a total organic carbon (TOC) (w%) of 3–10% (Meister et al. 2005). DNA and RNA were extracted from each core horizon, and cDNA was prepared from the RNA extracts. The 18S rRNA gene was amplified from both DNA and cDNA preparations for each horizon using a nested PCR approach. The universal eukaryote primers EukA and EukB (Medlin et al. 1988) were used in the first round of PCR, followed by Euk360F/Univ1492R (Medlin et al. 1988) in a second nested PCR reaction. Full-length clones were sequenced and analyzed (Table 1).

Basidiomycota were the dominant sequence types recovered in all cDNA-based clone libraries, suggesting that fungi dominate the eukaryotic microbial community in the deep marine subsurface. The greatest taxonomic diversity was observed in the shallowest sample from 1.75 mbsf (Peru Trench), which is not surprising since at this shallow sediment depth, a variety of organic substrates is likely available due to sedimentation and the mixing activities of burrowing metazoa. DNA- and cDNA-based libraries are roughly congruent; however, differences do exist, suggesting that working from extracted RNA may be more reliable for identifying active members of the community.

Some recovered sequences in the Edgcomb et al. (2011) study have nearly identical 18S rRNA gene sequences to their closest matches in GenBank from marine surface sediment or pelagic environments, whereas many have clearly

Table 1. Percentage of Peru Margin (PM) and Peru Trench (PT) clones from different taxonomic groups based on BLASTn analyses in clone libraries based on DNA and cDNA, 192 clones sequenced for each library (Adapted from Edgcomb et al. 2011).

Highest BLAST neighbor with a known taxonomy	1228 PM DNA 35.9 mbsf	1228 PM cDNA 35.9 mbsf	1228 PM DNA 3.91 mbsf	1228 PM cDNA 3.91 mbsf	1230 PT DNA 1.75 mbsf	1230 PT cDNA 1.75 mbsf
Fungi incertae sedis		3			1	16
Fungi basidiomycete	28	50	20	91	1	42
Fungi ustilaginomycete	13	47	73	9		10
Metazoa (Polychaeta)						16
Stramenopiles (Oikmonadaceae)	8				6	
Viridiplantae (Chlorophyta)					2	8
Rhodophyta						1
Alveolata, Apicomplexa						2
Alveolata Ciliophora	51					
Rhizaria (Gromiidae)					8	4
Unclass. Euk. (Cryomonadida)					8	1
Unclass. Euk. (BOLA471)					60	

different 18S sequences from their closest matches. Edgcomb et al. (2011) recovered sequences affiliated with the basidiomycetes, which include single-celled forms (yeasts) and asexual species (smuts, yeasts, rusts, jelly fungi) that are found in virtually all terrestrial ecosystems, as well as freshwater and marine habitats (Binder and Hibbett 2002; Hibbett and Binder 2002; Kohlmeyer 1979; Kohlmeyer and Kohlmeyer 1979). Some sequences affiliated with uncultured fungal sequences from Guaymas Basin hydrothermal sediments (Edgcomb et al. 2002; López-García et al. 2007), the Lucky Strike vents at the Mid-Atlantic Ridge (Le Calvez 2008), Cariaco Basin anoxic water column (Stoeck and Epstein 2003), and non-hydrothermal deep-sea sediments (Bass et al. 2007).

Bass et al. (2007) also showed that fungi dominate eukaryotic diversity in deep-sea surficial sediments. Their study found that fungi with yeast growth forms appear to be the dominant and most successful fungal form in the deep seas, and that the requirement for osmotrophy does not appear to be a problem under high hydrostatic pressure. Takishita et al. (2006) found in a eukaryote-wide SSU library that the basidiomycete yeast *Cryptococcus curvatus* may be the dominant microbial eukaryote at one deep-sea methane seep site. In agreement with their observation, 42% of the sequences from the RNA-based library from the Peru Trench 1.75 mbsf sample (Edgcomb et al. 2011) were closely related to

Cryptococcus, supporting the hypothesis that *Cryptococcus* species may indeed represent one of the most dominant eukaryotes in some marine sediments. Close relatives of the yeast *Cryptococcus* did not, however, appear in the clone libraries from 3.91 or 35.9 mbsf from Peru Margin, indicating a broader evolutionary diversity of fungi in the deep subsurface.

Marine yeasts are divided into obligate and facultative groups: obligate yeasts being those that have never been isolated from a nonmarine environment and facultative yeasts being those that are also described from terrestrial habitats (Kohlmeyer and Kohlmeyer 1979). Kohlmeyer and Kohlmeyer (1979) identified 25 out of 176 yeasts examined from diverse marine habitats as being obligate marine yeasts, represented principally by the genera *Metschnikowia*, *Rhodosporidium*, *Candida*, and *Torulopsis*. Similar to the findings of Bass et al. (2007), some fungal types in the Edgcomb et al. (2011) study branched near to known pathogens, suggesting that they may be opportunistic pathogens of deep-sea mammals. However, their recovery in RNA-based clone libraries from 35.9 mbsf suggests that further from the sediment surface, they may survive on buried organic matter. With the commonly accepted figure of 1.5 million fungal species (Hawksworth 2002), this means that we have described less than 5% of fungi in nature (mostly terrestrial), and there remains much to learn about marine fungi.

Additional studies are needed to determine whether the novel sequences recovered in the Edgcomb et al. (2011) study represent organisms specifically adapted to the deep marine subsurface environments. The stringent precautions taken to avoid contamination during sampling in that study make it unlikely that the fungi detected represent contaminants from surface environments recovered as a result of sampling methods. It is possible that fungi, whose rRNA gene sequences most closely affiliate with those of terrestrial fungi, represent such fungi that can adapt to life in the buried deep marine subsurface.

Independent of our 18S rRNA survey, fungi have been detected in deep Peru Margin sediments from ODP Site 1229 by other molecular and by cultivation-based approaches (Fig. 1). Fungal DNA was recovered from sediments as deep as 158 mbsf (Fig. 1a) and revived fungi from those sediments in culture (Fig. 1c).

Additionally, a pyrosequenced metagenome from those sediments suggests that both Ascomycota and Basidiomycota are present (Fig. 1b) (Biddle et al. 2008). Ascomycota were also predominant in the pyrosequenced metagenome compared to Basidiomycota, and only Ascomycota were cultivable (Fig. 1c; Biddle et al. 2005). These studies suggest that fungi, with different intergenic transcriber spacer fingerprints, metagenomic sequences, and varying responses to cultivation, are active in the subsurface, although at present, this is circumstantial proof.

These initial results demonstrate that eukaryotes are present in the subsurface, and we have an indication that fungi are the dominant eukaryotes in the deeply buried biosphere, based on the primer-independent pyrosequencing study (Biddle et al. 2008) and the primer-dependent PCR results presented by Edgcomb et al. (2011) and Bass et al. (2007). Other lines of evidence that point to fungi as major players in the marine deep subsurface are the discovery of novel groups of

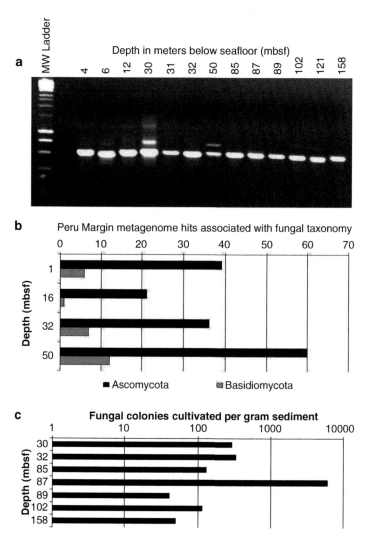

Figure 1. Selected results from studies done on ODP Site 1229 (Peru Margin): (**a**) Internal transcribed spacer (ITS) analysis of DNA extracted from sediments (depths 4–158 mbsf). Products were amplified by standard fungal ITS primers (ITS1F–ITS4R). *The black arrow* indicates the product usually associated with sequences of the *Ascomycota* (Biddle 2007). (**b**) Number of sequence reads related to fungal clades from 1, 16, 32, and 50 mbsf at Site 1229, from a pyrosequencing survey of sediment DNA without prior PCR amplification (Biddle et al. 2008). (**c**) Cultivation and enumeration of fungal colonies growing on heterotrophic medium, reanalyzed (Biddle et al. 2005).

fungi from a wide range of environments (for review see Xu 2006) and fossil evidence of filamentous structures highly suggestive of fungi within carbonate-filled vesicles within the upper oceanic crust in the North Pacific (ODP Site 1224) (Schumann et al. 2004).

In spite of these initial data, questions remain as to the function, activity, and diversity of subsurface fungi vs. other eukaryotes. The recovery of plant sequences in one DNA-based clone library indicates that preservation of DNA from deposited organisms confounds the interpretation of DNA-based subsurface data. For this reason, further RNA-based studies will provide valuable information.

3. Hydrothermal Vent and Cold Seep Sediments

There have been several investigations of microbial eukaryotes in deep-sea hydrothermal vent environments, although sediments sampled in these studies were primarily near-surface samples (e.g., Bass et al. 2007; Burgaud et al. 2009, 2010; Edgcomb et al. 2002; Gadanho and Sampaio 2005; López-García et al. 2003, 2007). Nonetheless, they provide valuable information about benthic eukaryotic communities that may help focus future investigations of the deeper marine subsurface. It should be noted that DNA-based surveys of diversity may not be reliable indicators of viability because it is possible that they include inputs from dead or encysted forms, such as dinoflagellate cysts, that have been detected in marine sediments (Nehring 1997). However, several groups of sequences detected in the study by Edgcomb et al. (2002), which was a DNA-based survey, affiliate with known specialists of anoxic environments such as the anaerobic ciliates *Metopus contortus* and *Trimyema compressum*. This is consistent with published accounts of viable flagellate protists from several hydrothermal vent sites including Guaymas Basin (Atkins et al. 2000). Rapid rates of reproduction allow these small flagellates to colonize temporally and spatially variable habitats such as these fine-particle, hydrothermally heated sediments. Atkins et al. (2000) isolated 18 strains of flagellated protists representing nine species from four Pacific deep-sea hydrothermal vents: Juan de Fuca Ridge (2,200 m water depth), Guaymas Basin (2,000 m water depth), 21° N (2,550 m water depth), and 9° N (2,000 m water depth). These vent flagellates belonged to six different taxonomic orders: the Ancyromonadida, Bicosoecida, Cercomonadida, Choanoflagellida, Chrysomonadida, and Kinetoplastida, and many are ubiquitous members of marine, freshwater, and terrestrial ecosystems worldwide. The most common kinetoplastids isolated from two vent sites (9°N and Juan de Fuca) were *Rhynchomonas nasuta* and *Bodo saliens*. The only cercomonad isolated was *Massisteria marina*, and this organism was recovered from all four vent sites. The bicosoecids *Cafeteria* sp. and *Caecitellus parvulus* were isolated from 9°N, and the ancyromonad *Ancyromonas sigmoides* was isolated from 9°N. The recovery of many of these organisms from diverse marine habitats suggests that their adaptability makes them relatively easy to grow in the laboratory, and hence, their recovery in culturing efforts cannot necessarily be interpreted as indicators of dominant active populations at these sites. However, many of these species are known to be tolerant of a wide range of environmental conditions including some extreme conditions (e.g., high hydrostatic pressure and high sulfate and heavy metal concentrations) common to hydrothermal vents (Atkins et al. 1998, 2002). The Guaymas SSU rRNA gene survey by Edgcomb et al. (2002) recovered sequences encompassing the majority of described lineages in the eukaryotic domain. In addition to anoxic specialists, there exists a diverse collection of eukaryotic microorganisms that is comparable in composition to benthic, pelagic, and near-surface water populations. These include certain fungi, radiolaria, acantharea, stramenopiles, foraminifera, and ciliates (Lee and Patterson 1998). Some portion of these taxa may be capable of adapting to selected hydrothermal vent sediment conditions. Bacterivorous heterotrophic flagellates such as the stramenopiles are known to be widely distributed in marine surface waters and to contribute a significant fraction (up to 35%) of heterotrophic flagellates in many regions (Massana et al. 2006), so it is not surprising that they were present in significant numbers in the Guaymas clone libraries from sediment samples spanning the oxic/anoxic interface (Edgcomb et al. 2002).

López-García et al. (2003) deployed colonization devices containing different organic, iron-rich, and porous mineral substrates that were exposed to a hydrothermal vent fluid source for 15 days at the Mid-Atlantic Ridge. The primary pioneers of the colonization process were bodonids and ciliates. Small-subunit ribosomal RNA gene sequences have been detected that probably belong to parasitic protist (alveolate) lineages including the Perkinsozoa, Apicomplexa, dinoflagellates, and ciliates in Pacific (Edgcomb et al. 2002) and Atlantic deep-sea vents (López-García et al. 2003). Moreira and Lopez-Garcia (2003) suggest that this diversity of parasitic protists could be hosted by the relatively abundant animal populations that are found in close proximity to these deep-sea vent sites. For example, *Perkinsus* ssp. are known to parasitize a variety of bivalves (Villalba et al. 2004), and one clone of *Perkinsus* spp. was recovered from a colonization substrate placed next to a colony of *Bathymodiolus azoricus*, making it likely that these mussels are parasitized by this organism. Additionally, abundant sequences from Apicomplexa, including Gregarines, which are known to be parasites mostly of the digestive tract or body cavities of invertebrates have been detected in these surveys (Edgcomb et al. 2002; López-García et al. 2003; Moreira and Lopez-Garcia 2003), suggesting that parasitization of different deep-sea hosts plays an important role in deep-sea vent ecology. High numbers of sequences from group I and group II marine alveolates in Mid-Atlantic Ridge (López-García et al. 2003), Pacific (Edgcomb et al. 2002), surface Atlantic, Pacific and Mediterranean waters (Diez et al. 2001; Moon-van der Staay et al. 2001), and deep-sea Antarctic plankton (López-García et al. 2003, 2001b) 18S clone libraries indicate that these organisms are likely to be ubiquitous in marine environments.

Broad molecular surveys of deep-sea hydrothermal vent sediments revealed a majority of sequences affiliating with the stramenopiles and the alveolates (Bass et al. 2007; Edgcomb et al. 2002; Gadanho and Sampaio 2005; López-García et al. 2003, 2007). The alveolates encompass the Apicomplexa, dinoflagellates, ciliates, and the taxonomically unassigned marine alveolates. The first two groups

(groups I and II) of uncultured marine alveolates were described by López-García et al. (2001a), and SSU rRNA sequences that affiliate with group I have since been recovered from a wide range of marine environments. The community structure of the oxic/anoxic interface at deep-sea hydrothermal vents likely includes both microaerophilic and facultative anaerobes; some of which may migrate into and out of the anoxic sediments. The active members of the protist community likely feed on the abundant bacteria sustained by hydrothermal activity and sedimentary input of organics from surface environments.

While these aforementioned studies did recover some novel fungal sequences, the proportion of fungal sequences recovered was not large. Using culture-dependent methods, Gadanho and Sampaio (2005) assessed yeast diversity in Mid-Atlantic Ridge (MAR) waters and found 12 phylotypes belonging to the *Ascomycota* and 7 to the *Basidiomycota*; 35% of which represented novel phylotypes. The isolation of some of the pink yeasts they cultured was only possible on media supplemented with sulfur, suggesting that these isolates were adapted to the ecosystems of the hydrothermal vents. Burgaud et al. (2010) theorized that the fungal community was being underestimated by previous approaches, and that yeasts likely have frequent interactions with endemic fauna. They investigated the diversity of culturable yeasts at the MAR, South Pacific Basins, and East Pacific Rise and recovered 32 yeast isolates from hydrothermal waters, mostly associated with animals. This was the first report of yeasts associated with deep-sea hydrothermal vent animals, and a phylogenetic analysis of 26 SSU rRNA gene sequences showed that these yeasts belonged to both the *Ascomycota* and *Basidiomycota* phyla. Burgaud et al. (2009) showed that the genera they recovered (*Rhodotorula, Rhodosporidium, Candida, Debaryomyces, and Cryptococcus*) are many of the same genera usually recovered from deep-sea environments. In a separate study of the same samples, Burgaud et al. (2009) isolated filamentous fungi from *Bathymodiolus azoricus* mussels and deep-sea shrimp, with only some filamentous fungi isolated from mineral substrates. As those authors note, the use of culture media rich in organic compounds may account in part for the low recovery of fungal isolates from sediments. More studies will be needed that incorporate a variety of culture approaches paired with culture-independent methods in order to determine in situ abundance. It is interesting to note that the taxa recovered from near-surface environments are roughly congruent with the results of the Edgcomb et al. (2011) study of sediments 1.75–35.9 mbsf, except that in the Edgcomb et al. study, members of the *Ascomycota* were not recovered.

A molecular survey of SSU rRNA gene amplicons present in the sediments (640 m water depth) of the cold methane seeps of the Kuroshima Knoll in the southern Ryukyu Arc revealed a eukaryotic community with extremely low diversity relative to other marine environments previously reported and indicated that the basidiomycetous fungus *Cryptococcus curvatus* was the dominant microbial eukaryote within this chemosynthetic ecosystem (Takishita et al. 2006). This fungus was the only sequence recovered in samples taken from the horizon just below the sediment surface to a depth of 9 cm. The possibility is mentioned by

Takishita et al. (2006) that this yeast may be an opportunistic pathogen of the local population of bivalves. Basidiomycetous yeasts are known to dominate yeast populations in oligotrophic oceanic waters and appear to be ubiquitously distributed in freshwater, marine, and deep-sea environments (Nagahama 2005). Basidiomycetes were also the dominant fungal signatures found in the deeper marine subsurface by Edgcomb et al. (2011). The dominance of this yeast in the microbial eukaryote community in the seep study by Takishita et al. (2006) may suggest a specific local association between this yeast and local seep fauna; however, it is too early to discount the possibility of a significant free-living basidiomycete population in both near-surface and deeper marine sediments.

4. Concluding Remarks

Marine eukaryotes are thought to play major roles in global carbon and mineral cycles through their contributions to primary production and through grazing of prokaryotes and small plankton (Fenchel 1984, 1987; Sherr and Sherr 2000). Initial studies of deep marine environments, including microoxic and anoxic habitats, support this notion. While reactions mediated by prokaryotes are undoubtedly crucial in redox processes, the role of eukaryotes in biotic redox reactions is also likely to be important. Although most eukaryotic taxa require O_2 to survive and hydrogen sulfide in micromolar concentrations is typically toxic to eukaryotes (e.g., National Research Council 1979), certain protists, fungi, and metazoans inhabit marine and freshwater redox zones (Fenchel and Finlay 1995) often in high densities (e.g., Bernhard et al. 2000) and may be important facilitators of redox reactions.

Recent studies indicate that fungi are likely the dominant eukaryotes in the marine subsurface. These studies were performed on continental margin sediment beneath an upwelling region off the coast of Peru, and as such, may not be representative of marine sediments as a whole. Additional studies are needed to examine the potential existence of eukaryotes in deep marine sediments.

If fungi continue to be seen as dominant eukaryotes in marine sediments, the significance of these eukaryotes should be examined further. It is important to know whether a deep subsurface fungal biosphere is seeded by highly adaptable, opportunistic, terrestrial organisms or whether these fungi are truly marine in origin and to gain insight into their levels of activity. The study of marine fungi at this point is in need of a detailed biodiversity study across many sediment conditions, using molecular, microscopic, and cultivation means, to determine which members of the fungi are active in the subsurface vs. organisms that are merely preserved (either dead or inactive). For future studies, it will be very important to be able to interpret accurately the potential significance of finding particular fungal groups in diverse marine subsurface samples, so the definition of a marine fungus should be revisited. If fungi and other eukaryotes can adapt and function in deeply buried sediments of the seafloor, global estimates of their activity and effect on geochemical cycles will also need to be reviewed.

5. Acknowledgements

Support for the deep subsurface investigation of fungi was provided by a grant from the Deep Ocean Exploration Institute, Woods Hole Oceanographic Institution (award number 32031109) and a NASA NPP fellowship administered by ORAU to JB.

6. References

Adl SM, Simpson AGB, Farmer MA, Andersen RA, Anderson OR, Barta JR et al (2005) The new higher level classification of eukaryotes with emphasis on the taxonomy of protists. J Eukaryot Microbiol 52:399–451

Atkins MS, Anderson OR, Wirsen CO (1998) Effect of hydrostatic pressure on the growth rate and encystment of flagellated protozoa isolated from a deep-sea hydrothermal vent and a deep shelf region. Mar Ecol Prog Ser 171:85–95

Atkins MS, Teske AP, Anderson OR (2000) A survey of flagellate diversity at four deep-sea hydrothermal vents in the eastern Pacific Ocean using structural and molecular approaches. J Eukaryot Microbiol 47:400–411

Atkins MS, Hanna MA, Kupetsky EA, Saito MA, Taylor CD, Wirsen CO (2002) Tolerance of flagellated protists to high sulfide and metal concentrations potentially encountered at deep-sea hydrothermal vents. Mar Ecol-Prog Ser 226:63–75

Bass D, Howe A, Brown N, Barton H, Demidova M, Michelle H et al (2007) Yeast forms dominate fungal diversity in the deep oceans. Proc R Soc B Biol Sci 274:3069–3077

Bernhard JM, Buck KR, Farmer MA, Bowser SS (2000) The Santa Barbara Basin is a symbiosis oasis. Nature 403:77–80

Biddle JF (2007) Microbial populations and processes in deep subseafloor sediments. PhD. Pennsylvania State University

Biddle JF, House CH, Brenchley JE (2005) Cultivation of deeply buried microbes shows influence of geochemistry. Geochimica Et Cosmochimica Acta 69:A228

Biddle JF, Lipp JS, Lever MA, Lloyd KG, Sorensen KB, Anderson R et al (2006) Heterotrophic Archaea dominate sedimentary subsurface ecosystems off Peru. Proc Natl Acad Sci USA 103:3846–3851

Biddle JF, Fitz-Gibbon S, Schuster SC, Brenchley JE, House CH (2008) Metagenomic signatures of the Peru Margin subseafloor biosphere show a genetically distinct environment. Proc Natl Acad Sci USA 105:10583–10588

Binder M, Hibbett DS (2002) Higher-level phylogenetic relationships of homobasidiomycetes (mushroom-forming fungi) inferred from four rDNA regions. Mol Phylogenet Evol 22:76–90

Burgaud G, Le Calvez T, Arzur D, Vandenkoornhuyse P, Barbier G (2009) Diversity of culturable marine filamentous fungi from deep-sea hydrothermal vents. Environ Microbiol 11:1588–1600

Burgaud G, Arzur D, Durand L, Cambon-Bonavita M-A, Barbier G (2010) Marine culturable yeasts in deep-sea hydrothermal vents: species richness and association with fauna. FEMS Microb Ecol 73:121–133

Connell LB, Barrett A, Templeton A, Staudigel H (2009) Fungal diversity associated with an active deep sea volcano: Vailulu'u Seamount, Samoa. Geomicrobiol J 26:597–605

Council NR (1979) Hydrogen sulfide. University Park Press, Baltimore, p 183

D'Hondt S, Jorgensen BB, Miller DJ (2003) Ocean Drilling Program, vol [CD-ROM]. Texas A&M University, College Station

D'Hondt S, Jorgensen BB, Miller DJ, Batzke A, Blake R, Cragg BA et al (2004) Distributions of microbial activities in deep subseafloor sediments. Science 306:2216–2221

Diez B, Pedros-Alio C, Marsh TL, Massana R (2001) Application of denaturing gradient gel electrophoresis (DGGE) to study the diversity of marine picoeukaryotic assemblages and comparison of DGGE with other molecular techniques. Appl Environ Microbiol 67:2942–2951

Edgcomb VP, Kysela DT, Teske A, Gomez AD, Sogin ML (2002) Benthic eukaryotic diversity in the Guaymas Basin hydrothermal vent environment. Proc Natl Acad Sci USA 99:7658–7662

Edgcomb VP, Beaudoin D, Gast R, Biddle JF, Teske A (2011) Marine subsurface eukaryotes: the fungal majority. Environ Microbiol 13(1):172–183

Fenchel T (ed) (1984) Suspended marine bacteria as a food source. Plenum, New York, pp 301–315

Fenchel T (1987) Ecology of protozoa: the biology of free-living phagotrophic protists. Science Tech/Springer, Berlin, p 197

Fenchel T, Finlay BJ (1995) Ecology and evolution in anoxic worlds. Oxford University Press, Oxford

Gadanho M, Sampaio JP (2005) Occurrence and diversity of yeasts in the Mid-Atlantic Ridge hydrothermal fields near the Azores Archipelago. Microb Ecol 50:408–417

Gadd GM (2007) Geomycology: biogeochemical transformations of rocks, minerals, metals and radionuclides by fungi, bioweathering and bioremediation. Mycol Res 111:3–49

Geiser AG, Zeng QQ, Sato M, Helvering LM, Hirano T, Turner CH (1998) Decreased bone mass and bone elasticity in mice lacking the transforming growth factor-beta1 gene. Bone 23:87–93

Hawksworth DL (2002) The magnitude of fungal diversity: the 1.5 million species estimate revisited. Mycol Res 105:1422–1432

Hibbett DS, Binder M (2002) Evolution of complex fruiting-body morphologies in homobasidiomycetes. P Roy Soc Lon B Biol Sci 269:1963–1969

Hyde KD, Jones EBG, Leao E, Pointing SB, Poonyth AD, Vrjmoed LLP (1998) Role of fungi in marine ecosystems. Biodivers Conserv 7:1147–1161

Inagaki F, Nunoura T, Nakagawa S, Teske A, Lever M, Lauer A et al (2006) Biogeographical distribution and diversity of microbes in methane hydrate-bearing deep marine sediments, on the Pacific Ocean Margin. Proc Natl Acad Sci USA 103:2815–2820

Jobard M, Rasconi S, Sime-Ngando T (2010) Diversity and functions of microscopic fungi: a missing component in pelagic food webs. Aquat Sci 72:255–268

Kohlmeyer J (1979) Marine fungal pathogens among Ascomycetes and Deuteromycetes. Experientia 35:437–439

Kohlmeyer J, Kohlmeyer E (1979) Marine mycology: the higher fungi. Academic, New York

Le Calvez T (2008) Third annual DOE Joint Genome Institute user meeting. U.S. Dept. of Energy, Office of Science, Walnut Creek

Lee WJ, Patterson DJ (1998) Diversity and geographic distribution of free-living heterotrophic flagellates – analysis by PRIMER. Protist 149:229–243

Lipp JS, Morono Y, Inagaki F, Hinrichs KU (2008) Significant contribution of Archaea to extant biomass in marine subsurface sediments. Nature 454:991–994

López-García P, Lopez-Lopez A, Moreira D, Rodriguez-Valera F (2001a) Diversity of free-living prokaryotes from a deep-sea site at the Antarctic Polar Front. FEMS Microbiol Ecol 36:193–202

López-García P, Rodriguez-Valera F, Pedros-Alio C, Moreira D (2001b) Unexpected diversity of small eukaryotes in deep-sea Antarctic plankton. Nature 409:603–607

López-García P, Philippe H, Gail F, Moreira D (2003) Autochthonous eukaryotic diversity in hydrothermal sediment and experimental microcolonizers at the Mid-Atlantic Ridge. Proc Natl Acad Sci USA 100:697–702

López-García P, Vereshchaka A, Moreira D (2007) Eukaryotic diversity associated with carbonates and fluid-seawater interface in Lost city hydrothermal field. Environ Microbiol 9:546–554

Lorenz R, Molitoris HP (1997) Cultivation of fungi under simulated deep sea conditions. Mycol Res 101:1355–1365

Massana R, Terrado R, Forn I, Lovejoy C, Pedros-Alio C (2006) Distribution and abundance of uncultured heterotrophic flagellates in the world oceans. Environ Microbiol 8:1515–1522

Medlin L, Elwood HJ, Stickel S, Sogin ML (1988) The characterization of enzymatically amplified eukaryotic 16 S-like rRNA-coding regions. Gene 71:491–499

Meister P, Prokopenko M, Skilbeck CG, Watson M, McKenzie JA (2005) In: Jorgensen BB, D'Hondt S, Miller DJ (eds) Proceeding ODP, scientific results, vol 201. ODP, College Station, pp 1–20

Moon-van der Staay SY, De Wachter R, Vaulot D (2001) Oceanic 18S rDNA sequences from picoplankton reveal unsuspected eukaryotic diversity. Nature 409:607–610

Moreira D, Lopez-Garcia P (2003) Are hydrothermal vents oases for parasitic protists? Trends Parasitol 19:556–558

Nagahama T (2005) Yeast biodiversity in freshwater, marine and deep-sea environments. In: The yeast handbook: biodiversity and ecophysiology of yeasts. Springer, Heidelberg

Nagahama T, Hamamoto M, Nakase T, Takami H, Horikoshi K (2001) Distribution and identification of red yeasts in deep-sea environments around the northwest Pacific Ocean. Antonie Van Leeuwenhoek 80:101–110

Nehring S (1997) Dinoflagellate resting cysts from recent German coastal sediments. Botanica Marina 40:307–324

Parkes RJ, Webster G, Cragg BA, Weightman AJ, Newberry CJ, Ferdelman T et al (2005) Deep sub-seafloor prokaryotes stimulated at interfaces over geological time. Nature 436:390–394

Ravindran J, Raghukumar C, Raghukumar S (2001) Fungi in Porites lutea: association with healthy and diseased corals. Dis Aquat Organ 47:219–228

Richards TA, Bass D (2005) Molecular screening of free-living microbial eukaryotes: diversity and distribution using a meta-analysis. Curr Opin Microbiol 8:240–252

Schippers A, Neretin LN, Kallmeyer J, Ferdelman TG, Cragg BA, Parkes RJ et al (2005) Prokaryotic cells of the deep sub-seafloor biosphere identified as living bacteria. Nature 433:861–864

Schumann G, Manz W, Reitner J, Lustrino M (2004) Ancient fungal life in North Pacific eocene oceanic crust. Geomicrobiol J 21:241–246

Sherr EB, Sherr BF (2000) Marine microbes. An overview. In: Kirchman DL (ed) Microbial ecology of the oceans. Wiley-Liss, New York, pp 13–46

Simonato F, Campanaro S, Lauro FM, Vezzi A, D'Angelo M, Vitulo N et al (2006) Piezophilic adaptation: a genomic point of view. J Biotechnol 126:11–25

Sørensen KB, Teske A (2006) Stratified communities of active Archaea in deep marine subsurface sediments. Appl Environ Microbiol 72:4596–4603

Stoeck T, Epstein S (2003) Novel eukaryotic lineages inferred from small-subunit rRNA analyses of oxygen-depleted marine environments. Appl Environ Microbiol 69:2657–2663

Takishita K, Tsuchiya M, Reimer JD, Maruyama T (2006) Molecular evidence demonstrating the basidiomycetous fungus Cryptococcus curvatus is the dominant microbial eukaryote in sediment at the Kuroshima Knoll methane seep. Extremophiles 10:165–169

Takishita K, Tsuchiya M, Kawato M, Oguri K, Kitazato H, Maruyama T (2007) Genetic Diversity of Microbial Eukaryotes in Anoxic Sediment of the Saline Meromictic Lake Namako-ike (Japan): On the Detection of Anaerobic or Anoxic-tolerant Lineages of Eukaryotes. Protist 158:51–64

Teske A, Sorensen KB (2008) Uncultured archaea in deep marine subsurface sediments: have we caught them all? ISME J 2:3–18

Villalba A, Reece KS, Ordas MC, Casas SM, Figueras A (2004) Perkinsosis in molluscs: a review. Aquat Living Resour 17:411–432

Webster G, Parkes RJ, Cragg BA, Newberry CJ, Weightman AJ, Fry JC (2006) Prokaryotic community composition and biogeochemical processes in deep subseafloor sediments from the Peru Margin. FEMS Microbiol Ecol 58:65–85

Whitman WB, Coleman DC, Wiebe WJ (1998) Prokaryotes: the unseen majority. P Natl Acad Sci USA 95:6578–6583

Xu J (2006) Fundamentals of fungal molecular population genetic analyses. Curr Issues Mol Biol 8:75–89

PART VI:
MODERN ANALOGS
AND TEMPLATES FOR EARTH HISTORY

Struck
Altenbach
Leiter
Mayr
Struck
Hiss
Radic
Glock
Schönfeld

Mallon
Almogi-Labin
Ashckenazi-Polivoda
Edelman-Furstenberg
Benjamini
Schieber
Gaulke
Altermann

Biodata of **Ulrich Struck**, author of the chapter "***On The Use of Stable Nitrogen Isotopes in Present and Past Anoxic Environments.***"

Dr. Ulrich Struck is head of the stable isotope facilities of the Museum für Naturkunde, Leibniz Institute for Research on Evolution and Biodiversity at the Humboldt University Berlin since 2006. His Diploma as well as his Ph.D. at Kiel University (obtained in 1992), occupied the sedimentology and paleo-ecology of benthic foraminifera in the European Arctic. More than 2 years were spent on research vessels and boats. Present focus in research and teaching is given to global organic matter cycles, trophic pathways, food webs, and induced isotopic fractionations.

E-mail: **Ulrich.struck@mfn-berlin.de**

ON THE USE OF STABLE NITROGEN ISOTOPES IN PRESENT AND PAST ANOXIC ENVIRONMENTS

ULRICH STRUCK[1,2]

[1]*GeoBio-CenterLMU, Richard-Wagner-Str. 10, 80333, Munich, Germany*
[2]*Museum für Naturkunde, Leibniz-Institut für Evolutions- und Biodiversitätsforschung, Humboldt-Universität zu Berlin, Invalidenstraße 43, 10115 Berlin, Germany*

1. Introduction

Knowledge about the biogeochemistry of nitrogen cycling in modern aquatic ecosystems and the associated fractionations of nitrogen isotopes ($\delta^{15}N$) has increased significantly during the past two decades. These insights also improved our ability to interpret $\delta^{15}N$ records recovered from ancient sediments. Specific environmental setups such as coastal upwelling areas, open pelagic realms, or stagnant basins are characterized by distinct biogenic processes and the formation of a typifying sedimentary record. We find growing evidence to recover these distinct biogenic processes in detail from $\delta^{15}N$ patterns observed in earth history. Sediments with elevated nitrogen contents (>0.2%) and low diagenetic offprint are most suitable for such investigations. The analysis of $\delta^{15}N$ in extracted biomarkers such as chlorines and porphyrines, or from nitrogen bearing hard parts of certain fossils offer valuable tools to assess the sample quality and a possible imprint derived from diagenesis.

2. The Modern Aquatic Nitrogen Cycle

Nitrogen often represents a limiting component for biological production in modern aquatic environments (Smith et al. 1986). Dissolved inorganic nitrogen (DIN) may occur in quite variable molecular states: as nitrate (NO_3^-), nitrite (NO_2^-), nitrous oxide (NO), di-nitrous oxide (N_2O), dinitrogen (N_2), or ammonia (NH_4^+). The oxidized forms tend to transform into reduced DIN forms in anoxic conditions, and vice versa (Fig. 1). As natural reduction/oxidation reactions are kinetically slow, their path relies mainly on biogenic catalysis (Grundl 1994).

The element nitrogen (N) has two stable (^{14}N, ^{15}N) and 15 instable isotopes (de Laeter et al. 2003). ^{14}N is by far more abundant (99,635%) in nature than ^{15}N (0,365%). Because nitrogen is a bio-reactive element, it is of great importance in

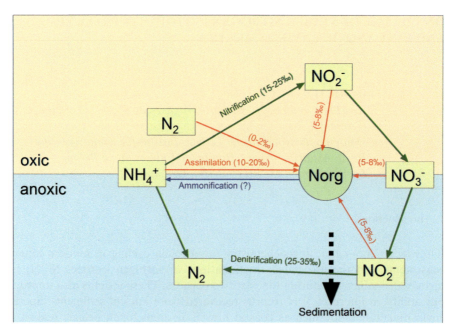

Figure 1. Simplified aquatic nitrogen cycle and associated processes. Estimates for process related isotope fractionation is shown in the fig as available from literature (for details see Table 1).

many processes related to biological production and degradation of organic matter. Many of these biological transformations are accompanied with a change of the isotopic composition from the respective nitrogen source to its product (the so-called isotope fractionation). The nitrogen-isotope composition of certain samples is typically presented as the ratio of $^{15}N/^{14}N$. Its variations are generally shown relative to the isotope ratio of air and nitrogen (Mariotti 1983) in per mill (‰), as given below.

$$\delta^{15}N_{Sample} = \left\{\frac{(^{15}N/^{14}N)_{Sample}}{(^{15}N/^{14}N)_{Standard}} - 1\right\} \times 1,000 \; (‰)$$

The nitrogen-isotope composition in sediments has often been utilized as an indicator of past changes in the nitrogen cycle and processes therein. But nitrogen in bulk sediments is not ultimately linked to local marine organic matter sources. Ammonia linked to clay minerals may dominate sedimentary nitrogen contents (Müller 1977; Scholten 1991). In addition, organic nitrogen compounds might undergo degrading processes which can therefore affect the isotope signature during sedimentation, diagenesis, or petrification. It is mandatory to select samples for the investigation of nitrogen isotopes with care.

2.1. SIGNATURES FROM THE SEDIMENTARY RECORD

Bulk sedimentary $\delta^{15}N$ has been attributed to distinct processes and their variability on global as well as regional scales of the nitrogen cycle (Altabet and François 1994; Holmes et al. 1999; Haug et al. 1998; Ganeshram et al. 1995; Struck et al. 2000, 2001, 2004). Most of these processes interact with each other, and thus cannot exclusively explain observed alterations in $\delta^{15}N$. In other words, variations in the $\delta^{15}N$ composition of bulk sediments will reflect changes in the dominance of specific processes in the nitrogen cycle rather than the presence or absence of those processes. Many observed changes in the $\delta^{15}N$ composition of bulk sediments can be attributed to alterations in the chemical setting of the surrounding ecosystem, including nutrient availability and oxygenation of the deeper water column. The following examples provide some cues to explain $\delta^{15}N$ variations recovered from bulk sediments.

2.2. MODERN EXAMPLES

Sedimentary records from the Holocene and specific Pleistocene periods are of great interest for studies related to modern climate change. Nitrogen isotopes offer very valuable proxy parameters for past environmental changes related to climate. The pioneering work of Francois et al. (1992) reported strong glacial to interglacial fluctuations in the bulk $\delta^{15}N$ composition (2–6‰) of sediment cores from the Antarctic frontal system. In a subsequent study, Altabet and François (1994) considered the sedimentary deposition of low $\delta^{15}N$ values as derived from enhanced nitrate supply in the euphotic zone, and decoded the N-isotope fractionation in phytoplankton during nitrate assimilation at excess nitrate concentrations (see Table 1). The concept of low sedimentary N-isotope as an indicator of marine surface-nitrate consumption has been applied in many other studies (Calvert et al. 1992; Holmes et al. 1999; Francois et al. 1992).

The first hype in the use of sedimentary $\delta^{15}N$ records was somewhat absorbed when potential diagenetic offprints were discussed, and contradictious interpretations isotopic patterns were offered. For example, Calvert et al. (1992) interpreted low $\delta^{15}N$ in the eastern Mediterranean sapropels as a result of meager nitrate consumption under a high productivity regime. More recent studies interpret such isotopic patterns with increased nitrogen supply from atmospheric sources ($\delta^{15}N = 0$‰), as derived from cyanobacterial N_2-fixation. This interpretation is supported by studies on the nitrogen cycle in the central Baltic Sea (Emeis et al. 2000; Struck et al. 1998; Struck et al. 2004); because the modern Baltic cycle is well comparable to the ancient sapropel formation of the eastern Mediterranean. The modern Baltic Sea and Black Sea are governed by basin-wide stagnation and deep-water anoxia, promoting the effective recycling of phosphorus into the oxygenated surface waters (Fry et al. 1991; Emeis et al. 2000). During such rather unusual settings, the available amount of phosphorus exceeds the amount of

Table 1. Average nitrogen-isotope fractionation ($\delta^{15}N$, ‰) of important processes in the aquatic nitrogen cycle and respective references.

Process	Source-Product	Fractionation	References
N_2-fixation	N_2-PON	0–2	Carpenter et al. (1997), Delwiche and Steyn (1970), Hoering and Ford (1960), Macko et al. (1987), and Montoya et al. 2002
Denitrification	NO_3-N_2	25–35	Barford et al. (1999), Cline and Kaplan (1975), Delwiche and Steyn (1970), Mariotti et al. (1981,1982), Miyake and Wada (1971), and Wada (1980)
Nitrification	NH_4-NO_3	15–25	Delwiche and Steyn (1970), Mariotti et al. (1981), Miyake and Wada (1971), and Yoshida (1988)
NO_3-assimilation	NO_3-PON	5–10	Montoya and McCarthy (1995), Needoba et al. (2003), Needoba and Harrison (2004), Pennock et al. (1996), Wada and Hattori (1978), and Waser et al. (1998a, b)
NO_2-assimilation	NO_2-PON	5–10	Cifuentes et al. (1989), Montoya et al. (1991), Pennock et al. (1988), Wada (1980), and Wada and Hattori (1978)
NH_4-assimilation	NH_4-PON	15–20	Waser et al. (1998a, b)

nitrogen required for plankton growth. The excess phosphorus gives rise to blooms of N_2-fixing cyanobacteria, which in result will introduce atmospheric nitrogen of low N-isotope composition ($\delta^{15}N = 0$‰) to the marine productivity cycle (Struck et al. 2004).

This scenario is likely to apply for many sapropels of the Eastern Mediterranean since about 2 Ma. Noteworthy, some of these sapropels show $\delta^{15}N$ compositions even below 0‰ (Struck et al. 2001; Arnaboldi and Meyers 2006; Milder et al. 1999). This cannot result from the supply of atmospheric nitrogen at 0‰ alone, because N_2-fixation generates a very low to negligible isotope fractionation. Accompanying processes, such as denitrification and nitrification, will outweigh their impact in the nitrogen cycle most probably (Table 1). Negative bulk sedimentary values can derive from phototrophic primary production under oxygen depletion. In the Black Sea or in meromictic lakes (Repeta et al. 1989; Halm et al. 2009), the anoxic water body can rise into the photic zone during long periods of stagnation. Under such conditions, green and purple phototrophic sulfur bacteria form dense populations in the uppermost part of the anoxic water body. Light may limit productivity here, but reduced inorganic nitrogen (NH_4^+) will be available in excess, so that strong isotope fractionation can cause exceptionally low $\delta^{15}N$ in the bacterial biomass (see Fig. 1). For instance, a dense population of purple bacteria of the small meromictic lake Alat of the Bavarian Alps (Struck et al. 2009) shows $\delta^{15}N$ of −5‰, whereas the organic matter in oxic surface waters has $\delta^{15}N$ clearly above 0‰. It is plausible under such conditions that a major contribution of nitrogen from anoxic production lowered the bulk sedimentary $\delta^{15}N$ signature to values below 0‰ (Struck et al. 2001). Independent evidence for such environmental conditions derived from specific biomarkers of phototrophic sulfur bacteria in sapropels from the Eastern Mediterranean Pliocene (Passier et al. 1999).

The modern Mediterranean is highly oligotrophic (Krom et al. 1991), and plankton productivity of the eastern parts of the Mediterranean nowadays is limited by phosphorus. N-nutrients remain unused during plankton blooms, which leads to low $\delta^{15}N$ (2–3‰) in the suspended matter and underlying surface sediments (Struck et al. 2001). However, other studies suggested that dominant N_2-fixation is responsible for low $\delta^{15}N$ in the modern eastern Mediterranean (Sachs et al. 1999).

As stated above, low $\delta^{15}N$ was initially considered for sediments formed under high productivity areas, such as upwelling centers and frontal zones (Altabet and François 1994). Detailed studies in the Peruvian, the Benguelan, and the Californian upwelling areas revealed more elevated $\delta^{15}N$ values at 6–12‰ (Ganeshram et al. 1995; Altabet et al. 2002; De Pol-Holz et al. 2006; Struck et al. 2002, Emeis et al. 2009; Pride et al. 1999).

The combination of elevated $\delta^{15}N$ and high sedimentary organic matter contents in upwelling areas are generally explained as resulting from high $\delta^{15}N$ in oceanic nitrate, which formed during extensive denitrification (Ganeshram et al. 1995; Pride et al. 1999; Voss et al. 2001). The so-called annamox ("anaerobic ammonia oxidation") is a second process which reduces the available amount of dissolved nitrogen. Similar to extensive denitrification, it leaves a reduced N-nutrient pool with an increased $\delta^{15}N$ signature (Kuypers et al. 2005). However, no stable isotope evidence for this scenario is available yet.

Nitrate-reducing processes, such as denitrification, show a strong isotope fractionation (Table 1), what leads to ^{15}N enrichment in residual nitrate (Voss et al. 2001). The entire food web of upwelling areas (Hückstädt et al. 2007) and the underlying sediments therefore are strongly enriched in ^{15}N when compared to open marine conditions. High nitrogen-isotope signatures in sediment cores underlying upwelling areas therefore were made reasonable from denitrification and associated processes (Ganeshram et al. 1995; Altabet et al. 2002; De Pol-Holz et al. 2006).

Another reason for high $\delta^{15}N$ in sediments accompanied by raised organic matter concentration is man-made eutrophication in coastal near marine and lacustrine environments (Struck et al. 2000; Voss and Struck 1997; Lu et al. 2010). This nitrogen-isotope enrichment is commonly due to increased loads of fertilizer or sewage-derived nutrients containing high $\delta^{15}N$ transported to the coastal sea mainly by river discharge (Deutsch et al. 2009; Dähnke et al. 2010).

Plant growth is commonly based on atmospheric nitrogen in natural terrestrial ecosystems and therefore shows relatively low $\delta^{15}N$ values around 0‰ (Peterson and Fry 1987). In areas of intensive agricultural land use and dense population, a mixture of manure or fertilizer-derived nitrogen enters the watersheds. The composition of this nitrogen is often strongly enriched in ^{15}N ($\delta^{15}N = 8–16‰$) due to several processes

Table 2. Rough estimates of typical ranges of $\delta^{15}N$ in bulk sediments in relation to the governing processes in the concomitant aquatic nitrogen cycle.

Governing process	$\delta^{15}N$ – Range in sediments	Environment	Range of TOC
Nitrate-based production	3–8‰	Open marine, oxic	<1%
Denitrification	8–12‰	Coastal upwelling with OMZ	>1–10%
Eutrophication	6–12‰	Coastal estuaries	>1–10%
N-fixation	0–3‰	Anoxia, stable water stratification	>2–10%
Sulfur bacteria	<0‰	Euphotic zone anoxia	0.5–5%

such as denitrification, ammonia volatization, and preferential uptake of ^{14}N during plant growth (see Dähnke et al. 2010).

This mixture of nitrogen sources commonly enters the ocean through rivers and leads to eutrophication of coastal-near ecosystems. However, most of these effects become negligible with increasing distance to the coast (e.g., Struck et al. 2000).

In summary, governing processes in the aquatic nitrogen cycle such as denitrification or N_2-fixation and associated isotope effects (see above) result in distinct nitrogen-isotope signatures in bulk sediments. In combination with additional sediment characteristics, such as organic matter contents, it is possible to reconstruct general modes of past nitrogen cycles based on $\delta^{15}N$ composition. Table 2 relates ranges of sedimentary $\delta^{15}N$ to dominant processes/modes in the concomitant nitrogen cycle.

2.3. EXAMPLES OF HOW TO VALIDATE THE QUALITY OF SEDIMENT- OR ROCK- $\delta^{15}N$

The preservation of the primary nitrogen-isotope composition in sedimentary rock depends on many processes, which start even before deposition and endure the final steps of petrogenesis or rock metamorphosis. Several studies have addressed the issue of early to late diagenesis to rock and sedimentary $\delta^{15}N$. Separate modes of diagenesis and alteration of $\delta^{15}N$ are evident from laboratory investigations for aerobe and anaerobe environments (Sachs and Repeta 1999; Lehmann et al. 2002; Freudenthal et al. 2001). Anoxic conditions tend to promote organic matter preservation as well as stability in isotope composition (Emeis et al. 1987). Different positive N-isotope offsets were found in aerobic environments of the Earth's oceans (Altabet et al. 1999; Kienast 2000; Struck et al. 2001), ranging from 0 to appr. +4‰ $\delta^{15}N$. However, there is a general trend towards a better preservation of the primary $\delta^{15}N$ signature with rising sedimentary organic matter contents (Altabet et al. 1999). Sedimentary nitrogen-isotope data from shallow oceans, shelves (i.e., Struck et al. 2004), and highly productive areas (Holmes et al. 1999; Altabet et al. 1999), therefore, are less different from expected isotope compositions in comparison to sedimentary $\delta^{15}N$ from oligotrophic deep-water realms

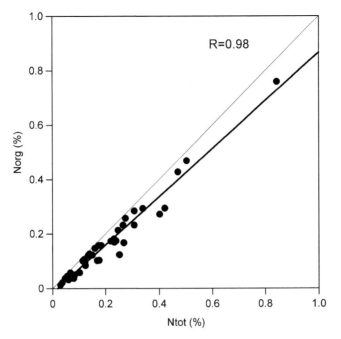

Figure 2. Bulk sedimentary nitrogen versus organic nitrogen in a selection of marine sediments (Scholten 1991). There is a clear and highly significant relation between the datasets pointing towards a primary organic origin of the nitrogen in most of the samples. Only samples with very small nitrogen concentrations (<0.1% Ntot) seem to be relatively enriched in inorganic (clay bond) nitrogen.

(Altabet et al. 1999). The relatively large difference between surface ocean $\delta^{15}N$ and sedimentary $\delta^{15}N$ in aerobic open-ocean sediments was often attributed to preferential bacterial degradation of settling particulate organic matter (Altabet and Deuser 1985) and deposition into oxygenated sediments (Freudenthal et al. 2001). However, direct observation of these processes remains difficult. An indirect approach was employed by Scholten (1991), who used a selection of modern and Mesozoic rock samples for comparison of their differing nitrogen fractions (organic N, clay bond N and bulk N). Figure 2 shows the relation of the concentration of organic N versus bulk N. On average, 90% of the bulk nitrogen in the samples is of organic origin, and only a minor part is derived from other sources (clay bond ammonia). Only samples with very low nitrogen content show a relatively high share of inorganic nitrogen. This is in agreement with Müller (1977), who recovered clay-bond nitrogen in high proportions from deep-sea sediments with very low bulk nitrogen concentrations.

For the validation of bulk rock $\delta^{15}N$, it is insufficient to solely compare relative shares of organic and inorganic nitrogen. Even if organic nitrogen obviously dominates, the nitrogen-isotope composition may have been modified during early diagenesis or later petrification. The comparison of compound specific

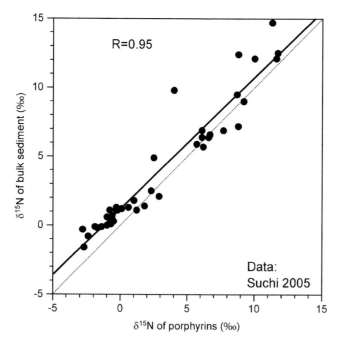

Figure 3. Comparison of nitrogen-isotope data from black shale samples ranging from the Silurian to Carboniferous (Suchi 2005) analyzed on bulk sediment and porphyrin extracts. This biomarker approach validates the origin of bulk sedimentary matter as a primary environmental tracer.

isotopes (biomarker isotopes) with concomitant bulk rock $\delta^{15}N$ can help to validate the primary origin of the signature. Time-consuming extraction process and well-equipped lab facilities are urgent for such studies (Chlorins: Sachs and Repeta 1999; Porphyrin: Suchi 2005). The work of Suchi (2005) showed promisingly that bulk rock $\delta^{15}N$ in black shales from Silurian to Carbonferous rocks are well comparable (Fig. 3). The small isotope enrichment between bulk and biomarker $\delta^{15}N$ was considered to result from the specifically lower $\delta^{15}N$ in chlorophyll as compared to total biomass (Sachs et al. 1999). A similar attempt for the validation of bulk $\delta^{15}N$ is provided by the $\delta^{15}N$ analysis of nitrogen-bearing remainders of fossil organisms, such as carapaces of crustaceans (Struck et al. 1998; Perga 2010) or planktic foraminifera (Ren et al. 2009).

3. Reading Past Changes in the Nitrogen Cycle From the $\delta^{15}N$ Composition Recorded in Bulk Sedimentary Rocks

Based on the knowledge of the distribution of $\delta^{15}N$ in modern aquatic ecosystems, it is basically possible to conclude on past fundamental variations in the nitrogen cycle related to climate change or other alterations. There is no continuous record

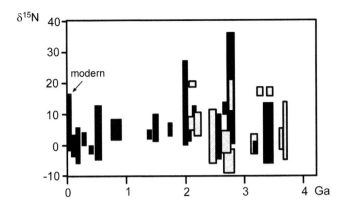

Figure 4. $\delta^{15}N$ in sedimentary rocks over the last four billion years (Data reported by Thomazo et al. 2009; Godfrey and Falkowsky 2009; Garvin et al. 2009; Papineau et al. 2009; LaPorte et al. 2009; Quan et al. 2008; Meyers et al. 2009; Jenkyns et al. 2001; Calvert and Pedersen 1996; Cremonese et al. submitted. For modern situation see references mentioned above).

of $\delta^{15}N$ available in literature covering larger parts of Earth's history comparable to Phanerozoic records of $\delta^{13}C$ and $\delta^{18}O$ showing secular trends in the carbon and oxygen cycles (Veizer et al. 1999). But a number of sedimentary sections from various episodes of Earth's history were studied to reconstruct general changes in the concomitant nitrogen cycle (Calvert et al. 1996, Rau et al. 1991, Schulz et al. 2004; Papineau et al. 2009; Pinti et al. 2001; Thomazo et al. 2009; Godfrey and Falkowsky 2009). Most of these studies rely on $\delta^{15}N$ from bulk rock or purified samples. Average $\delta^{15}N$ compositions in most cases vary between +15 and −5‰ (Fig. 4), which is quite comparable to the variability of $\delta^{15}N$ in sediments from modern environments (Fig. 4; Table 2).

One might conclude that the nitrogen cycle has undergone no major changes since the early stages of Earth's history (Godfrey and Falkowsky 2009). This is a reasonable interpretation because it is suggested that the governing huge nitrogen pool of the atmosphere has remained unchanged in its isotope composition and size since the very early days of the Earth (Godfrey and Falkowsky 2009). The relatively constant $\delta^{15}N$ composition in ancient sediments also implies that in most cases $\delta^{15}N$ of even very old sedimentary rocks shows little or no diagenetic change. This does not necessarily mean that there was no substantial removal of nitrogen from the source rocks during petrogenesis (Papineau et al. 2009). However, several studies indicate extraordinarily high average $\delta^{15}N$ in comparison to modern values (Papineau et al. 2009; Thomazo et al. 2009).

The study of Papineau et al. (2009) on the nitrogen cycle from Paleoproterozoic (~2.0 Ga) carbonaceous shales in India argues for a wide range of $\delta^{15}N$ values between −5 and +32‰. Although other authors report analytical studies indicating a lack of post depositional organic N alteration, these rocks, in our opinion, underwent metamorphic processes (as suggested by the same authors) responsible

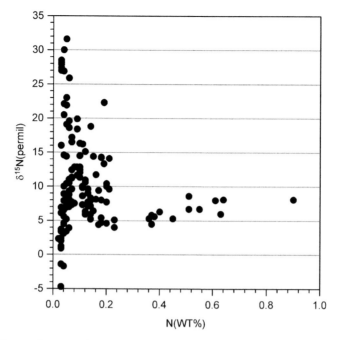

Figure 5. Nitrogen-isotope values vs. Nitrogen concentration from bulk rock samples (Papineau et al. 2009). $\delta^{15}N$ values <0 and >15‰ only occur in samples showing very low N-concentration, possibly linked to biased values from preferential loss of N during diagenesis. Samples with higher N-concentration show typical modern marine $\delta^{15}N$, indicating little or no diagenetic imprint (Graph modified from Cremonese et al. submitted).

for profound changes in the original N concentration. Specifically, nitrogen isotopic outliers are bound to samples with nitrogen contents below 0.1%, and C/N ratios above 100. This indicates a preferential nitrogen loss as reasonable for exceptional $\delta^{15}N$ values (Fig. 5). Taking into consideration only the $\delta^{15}N$ data above 0.1% N, the resulting $\delta^{15}N$ compositions are coherent with regular modern sedimentary $\delta^{15}N$ values.

4. Outlook

Nitrogen isotopes represent successful tools for the reconstruction of alterations in the biogenic nitrogen cycle. New concepts and methodologies in Paleoceanographic studies originate from this growing knowledge on modern cycles and dynamics of nitrogen isotopes in dissolved and particulate matter. Advancing research is focused on intervals in the Earth's history still not fully understood, such as the Cambrian explosion, or the mass extinction at the Permian-Triassic boundary. Anoxic conditions and euphotic zone anoxia find rapidly growing evidence during such events. The benefits of nitrogen-isotope studies for the reconstruction of aerobe and anaerobe environments have been demonstrated in numerous studies on modern aquatic realms. Even small meromictic lakes can help to decode historic $\delta^{15}N$ records. Nevertheless, it remains crucial to select samples for $\delta^{15}N$ analyses with great care, in order to avoid or at least control the influence of diagenesis and analytical pitfalls, such as too low nitrogen contents. The promising approach to investigate the nitrogen-isotope composition in biomarker extracts and/or fossil remains of organisms will significantly improve the quality of future bulk rock $\delta^{15}N$ analyses.

5. Acknowledgments

This book chapter greatly improved by the comments of one anonymous reviewer. This manuscript partly includes research, which was funded by the German Research Foundation (DFG) during Forschergruppe 736 (STR 356/4-1) and the Eu in the frame of HYPOX. I thank Alex Altenbach for the kind invitation to contribute to this book.

6. References

Altabet MA, Deuser W (1985) Seasonal variations in natural abundance of ^{15}N in particles sinking to the deep Sargasso Sea. Nature 315:218–219

Altabet MA, François R (1994) Sedimentary nitrogen isotopic ratio as a recorder for surface ocean nitrate utilization. Global Biogeochem Cycle 8:103–116

Altabet MA, Pilskaln C, Thunell R, Pride C, Sigman D, Chavez F, Francois R (1999) The nitrogen isotope biogeochemistry of sinking particles from the margin of the western North Pacific. Deep Sea Res I 46:655–679

Altabet MA, Higginson MJ, Murray DW (2002) The effect of millennial-scale changes in Arabian Sea denitrification on atmospheric CO2. Nature 415:159–162

Arnaboldi M, Meyers PA (2006) Patterns of organic carbon and nitrogen isotopic compositions of latest Pliocene sapropels from six locations across the Mediterranean Sea. Paleo Paleo Paleo 235(1–3):149–167

Barford CC, Montoya JP, Altabet MA, Mitchell R (1999) Steady state nitrogen isotope effects of N and NO production in *Paracoccus denitrificans*. Appl Environ Microbiol 65(3):989–994

Calvert SE, Pedersen TF (1996) Sedimentary geochemistry of manganese: implications for the environment of formation of manganiferous black shales. Econ Geol 91:36–47

Calvert S, Nielsen B, Fontugne MR (1992) Evidence from nitrogen isotope ratios for enhanced productivity during the formation of eastern Mediterranean Sapropels. Nature 359:223–225

Carpenter EJ, Harvey HR, Fry B, Capone DG (1997) Biogeochemical tracers of the marine cyanobacterium *Trichodesmium*. Deep Sea Res 44:27–38

Cifuentes LA, Fogel ML, Pennock JR, Sharp JH (1989) Biogeochemical factors that influence the stable nitrogen isotope ratio of dissolved ammonium in the Delaware Estuary. Geochim Cosmochim Acta 53:2713–2721

Cline JD, Kaplan IR (1975) Isotopic fractionation of dissolved nitrate during denitrification in the eastern tropical North Pacific Ocean. Mar Chem 3:271–299

Cremonese L, Shields G, Struck U (submitted) Nitrogen isotopes, from seawater to rocks through diagenesis – a review. Czech J of Geosci, spec vol subm

Dähnke K, Emeis K-C, Johannsen A, Nagel B (2010) Stable isotope composition and turnover of nitrate in the German Bight. Mar Ecol Prog Ser 408:7–U26

de Laeter JR, Böhlke JK, De Bièvre P, Hidaka H, Peiser HS, Rosman KJR, Taylo PDP (2003) Isotopic compositions and standard atomic masses from atomic weights of the elements. Review 2000 (IUPAC Technical Report). Pure Appl Chem 75(6):683–800

Delwiche CC, Steyn PL (1970) Nitrogen isotope fractionation in soils and microbial reactions. Environ Sci Technol 4:929–935

De Pol-Holz R, Ulloa O, Dezileau L, Kaiser J, Lamy F, Hebbeln D (2006) Melting of the Patagonian Ice Sheet and deglacial perturbations of the nitrogen cycle in the eastern South Pacific. Geophys Res Lett 33(4):L04704. doi:10.1029/2005GL024477

Deutsch B, Voss M, Fischer H (2009) Nitrogen transformation processes in the Elbe River: distinguishing between assimilation and denitrification by means of stable isotope ratios in nitrate. Aquat Sci 71:228–237. doi:10.1007/s00027-009-9147-9

Emeis K-C (1987) Cretaceous black shales of the South Atlantic Ocean: the role and origin of recycled organic matter. In: Degens ET, Und IE, Honjo S (eds) Particle flux in the Ocean. Mitt. Geol.-Paläont. Inst, Univ. Hamburg, Hamburg, pp 209–232

Emeis K-C, Struck U, Leipe T, Pollehne F, Kunzendorf H, Christiansen C (2000) Changes in the burial rates and C:N:P ratios in Baltic Sea sediments over the last 150 years – relevance to P regeneration rates and phosphorus cycle. Mar Geol 167(1–2):43–59

Emeis K-C, Struck U, Leipe T, Ferdelmann TG (2009) Variability in upwelling intensity and nutrient regime in the coastal upwelling system offshore Namibia: results from sediment archives. Int J of Earth Sci 89:309–326

Francois R, Altabet MA, Burkle LH (1992) Glacial to interglacial changes in surface nitrate utilization in the Indian sector of the Southern Ocean as recorded by sediment $\delta15N$. Paleoceanography 7:589–606

Freudenthal T, Wagner T, Wenzhöfer F, Zabel M, Wefer G (2001) Early diagenesis of organic matter from sediments of the eastern subtropical Atlantic: evidence from stable nitrogen and carbon isotopes. Geochimica Cosmochimica Ac 65:1795–1808

Fry B, Jannasch HW, Molyneaux SJ, Wirsen CO, Muramoto JA, King S (1991) Stable isotope studies of the carbon, nitrogen and sulfur cycles in the Black Sea and the Cariaco Trench. Deep Sea Res 38(suppl 2):S1003–S1019

Ganeshram RS, Pedersen TF, Calvert SE, Murray JW (1995) Large changes in oceanic nutrient inventories from glacial to interglacial periods. Nature 376:755–758

Garvin J, Buick R, Anbar AD, Arnold GL, Kaufman AJ (2009) Isotopic evidence for an aerobic nitrogen cycle in the latest Archean. Science 323:1045–1048

Godfrey LV, Falkowsky PG (2009) The cycling and redox state of nitrogen in the Archean ocean. Nat Geosci. doi:10.1038/NGE0633

Grundl T (1994) A review of the current understanding of redox capacity in natural, disequilibrium systems. Chemosphere 28:613–626

Halm H, Musat N, Lam P, Langlois R, Musat F, Peduzzi S, Lavik G, Schubert CJ, Sinha B, LaRoche J, Kuypers MMM (2009) Co-occurrence of denitrification and nitrogen fixation in a meromictic lake, Lake Cadagno (Switzerland). Environ Microbiol 11:1945–1958

Haug GH, Pedersen TF, Sigman DM, Calvert SE, Nielsen B, Peterson L (1998) Glacial/interglacial variations in production and nitrogen fixation in the Cariaco Basin during the last 580ka. Paleoceanography 13(5):427–432

Hoering TC, Ford HT (1960) The isotope effect in the fixation of nitrogen by *Azotobacter*. J Am Chem Soc 82:376–378

Holmes B, Eichner C, Struck U, Wefer G (1999) Reconstructions of surface ocean nitrate utilization using stable nitrogen isotopes in sinking particles and sediments. In: Fischer G, Wefer G (eds) Use of proxies in paleoceanography: examples from the South Atlantic. Springer, Berlin, pp 447–468

Hückstädt LA, Rojas CP, Antezana T (2007) Stable isotope analysis reveals pelagic foraging by the Southern sea lion in central Chile. J Exp Mar Biol Ecol 347:123–133

Jenkyns HC, Gröcke DR, Hesselbo SP (2001) Nitrogen isotope evidence for water mass denitrification during the early Toarcian (Jurassic) oceanic anoxic event. Paleoceanography 16(6):593–603

Kienast M (2000) Unchanged nitrogen isotopic composition of organic matter in the South China Sea during the last climatic cycle: global implications. Paleoceanography 15(2):244–253

Krom MD, Kress N, Brenner S, Gordon LI (1991) Phosphorus limitation of primary production in the eastern Mediterranean Sea. Limnol Oceanogr 36:424–432

Kuypers MMM, Lavik G, Woebken D, Schmid M, Fuchs BM, Amann R, Barker Jørgensen B, Jetten MSM (2005) Massive nitrogen loss from the Benguela upwelling system through anaerobic ammonium oxidation. Proc Natl Acad Sci 102:6478–6483

LaPorte DF, Holmden C, Patterson WP, Loxton JD, Melchin MJ, Mitchell CE, Finney SC, Sheets HD (2009) Local and global perspectives on carbon and nitrogen cycling during the Hirnantian glaciation. Palaeogeogr Palaeoclimatol Palaeoecol 276:182–195

Lehmann RF, Bernasconi SM, Barbieri A, McKenzie JA (2002) Preservation of organic matter and alteration of its carbon and nitrogen isotope composition during simulated and in situ early sedimentary diagenesis. Geochim Cosmochim Acta 66(3573–3584):2002

Lu YH, Meyers PA, Johengen PH, Eadie BJ, Robbins JA, Han HJ (2010) Delta N-15 values in Lake Erie sediments as indicators of nitrogen biogeochemical dynamics during cultural eutrophication. Chem Geol 273(1–2):1–7

Macko SA, Fogel ML, Hare PE, Hoering TC (1987) Isotopic fractionation of nitrogen and carbon in the synthesis of amino acids by microorganisms. Chem Geol Isot Geosci Section 65:79–92

Mariotti A (1983) Atmospheric nitrogen as a reliable standard for natural ^{15}N abundance measurements. Nature 303:685–687

Mariotti A, Germon JC, Hubert P, Kaiser P, Letolle R, Tardieux A, Tardieux P (1981) Experimental determination of nitrogen kinetic isotope fractionation: some principles; illustration for the denitrification and nitrification processes. Plant Soil 62:413–430

Mariotti A, Germon JC, LeClerc A, Catroux G, Letolle R (1982) Experimental determination of kinetic isotopic fractionation of nitrogen isotopes during denitrification. In: Schmidt HL, Forstel H, Heinzinger K (eds) Stable isotopes. Elsevier, Amsterdam, pp 459–464

Meyers PA, Bernasconi SM, Yum JG (2009) 20 My of nitrogen fixation during deposition of mid-cretaceous black shales on the demerara rise, equatorial Atlantic Ocean. Org Geochem 40:158–166

Milder JC, Montoya JP, Altabet MA (1999) Carbon and nitrogen stable isotope ratios at Sites 969 and 974: interpreting spatial gradients in sapropel properties. In: Zahn R, Comas MC, Klaus A (eds) Proc ODP Sci Res 161:401–411

Miyake Y, Wada E (1971) The isotope effect on the nitrogen in biochemical oxidation-reduction reactions. Rec Oceanogr Works Jpn 11:1–6

Montoya JP, McCarthy JJ (1995) Nitrogen isotope fractionation during nitrate uptake by marine phytoplankton in continuous culture. J Plankton Res 17(3):439–464

Montoya JP, Horrigan SG, McCarthy JJ (1991) Rapid, storm-induced changes in the natural abundance of 15 N in a planktonic ecosystem. Geochim Cosmochim Acta 55:3627–3638

Montoya JP, Carpenter EJ, Capone DG (2002) Nitrogen-fixation and nitrogen isotope abundances in zooplankton of the oligotrophic North Atlantic. Limnol Oceanogr 47:1617–1628

Müller PJ (1977) C/N ratios in Pacific deep-sea sediments: effect of inorganic ammonium and organic nitrogen compounds sorbed by clays. Geochim Cosmochim Acta 41:549–553

Needoba JA, Harrison PJ (2004) Influence of low light and a light:dark cycle on NO3 uptake, intracellular NO3, and nitrogen isotope fractionation by marine phytoplankton. J Phycol 40(3):505–516

Needoba JA, Waser NAD, Harrison PJ, Calvert SE (2003) Nitrogen isotope fractionation by 12 species of marine phytoplankton during growth on nitrate. Mar Ecol Prog Ser 255:81–91

Papineau D, Purohitb R, Goldberg T, Pi D, Shields G, Bhuf H, Steele A, Fogel ML (2009) High primary productivity and nitrogen cycling after the Paleoproterozoic phosphogenic event in the Aravalli Supergroup. India Precambrian Res 171:37–56

Passier HF, Bosch H-J, Nijenhuis IA, Lourens LJ, Böttcher ME, Leenders A, Sinninghe Damste JS, de Lange GJ, de Leeuw JW (1999) Sulphic Mediterranean surface waters during pliocene sapropel formation. Nature 397:146–149

Pennock JR, Sharp JH, Ludlam J, Velinsky DJ, Fogel ML (1988) Isotopic fractionation of nitrogen during the uptake of NH and NO by *Skeletonema costatum*. Eos 69(44):1098

Pennock JR, Velinsky DJ, Ludlam JM, Sharp JH, Fogel ML (1996) Isotopic fractionation of ammonium and nitrate during uptake by *Skeletonema costatum*: Implications for delta N-15 dynamics under bloom conditions. Limnol Oceanogr 41(3):451–459

Perga ME (2010) Potential of delta C-13 and delta N-15 of cladoceran subfossil exoskeletons for paleo-ecological studies. J Paleolimnol 44(2):387–395

Peterson BJ, Fry B (1987) Stable isotopes in ecosystem studies. Annu Rev Ecol Syst 18:293–320

Pinti DL, Hashizume K (2001) N-15-depleted nitrogen in early Archean kerogens: clues on ancient marine chemosynthetic-based ecosystems? A comment to V. Beaumont, F. Robert, Precambrian Res. 96 (1999) 62–82. Precambrian Res 105:85–88

Pride C, Thunell R, Sigman D, Keigwin L, Altabet M, Tappa E (1999) Nitrogen isotopic variations in the Gulf of California since the last deglaciation: Response to global climate change. Paleoceanography 14(3):397–409

Quan TM, van de Schootbrugge B, Field PM, Rosenthal Y, Falkowsky PG (2008) Nitrogen isotope and trace metal alalyses from the Mingolsheim core (Germany): Evidence for redox variations across the Triassic-Jurassic boundary. Global Biogeochem Cycl 22:GB2014. doi:10.1029/2007GB002981

Rau Gh, Takahashi T, Desmarais Dj (1991) Particulate organic-matter delta-c-13 variations across the drake passage. Journal of geophysical research-oceans 96(C8):15131–15135

Ren H, Sigman DM, Meckler AN, Plessen B, Robinson RS, Rosenthal Y, Haug GH (2009) Foraminiferal isotope evidence of reduced nitrogen fixation in the ice age Atlantic Ocean. Science 323(5911):244–248

Repeta DJ, Simpson DJ, Jørgensen BB, Jannasch HW (1989) Evidence for anoxygenic photosynthesis from the distribution of bacteriochlorophylls in the Black Sea. Nature 342:69–72

Sachs JP, Repeta DJ (1999) Oligotrophy and N-fixation during Eastern Mediterranean Sapropel Events. Science 286:2485–2488

Sachs JP, Repeta DJ, Goericke R (1999) Nitrogen and carbon isotopic ratios from marine phytoplankton. Geochim Cosmochim Acta 63(9):1431–1441

Scholten SO (1991) The distribution of nitrogen isotopes in sediments. Geologica Ultraiectina 81:101

Schulz HM, Bechtel A, Rainer T, Sachsenhofer RF, Struck U (2004) Paleoceanography of the Western Central paratethys during early oligocene nannoplankton Zone NP23 in the Austrian Molasse Basin. Geologica Carpathica 55(4):311–323

Smith SV, Kimmerer WJ, Walsh TW (1986) Vertical flux and biogeochemical turnover regulate nutrient limitation of net organic production in the North Pacific Gyre. Limnol Oceanogr 31:161–167

Struck U, Voss M, Mumm N, Bodungen VB (1998) Stable isotopes of nitrogen in fossil cladoceran exoskeletons: implications for nitrogen sources in the Baltic Sea during the last century. Naturwissenschaften 85:597–603

Struck U, Emeis KC, Voss M, Christiansen CC, Kunzendorf H (2000) Records of southern and central Baltic Sea eutrophication in $\delta^{13}C$ and $\delta^{15}N$ of sedimentary organic matter. Mar Geol 164(3–4):157–171

Struck U, Emeis KC, Voß M, Krom MD, Rau GH (2001) Biological productivity during sapropel S5 formation in the eastern Mediterranean Sea. Evidence from a stable isotopes of nitrogen and carbon. Geochim Cosmochim Acta 65(19):3249–3266

Struck U, Altenbach AV, Emeis KC (2002) Changes of the upwelling rates of nitrate preserved in the delta(15)N-signature of sediments and fish scales from the diatomaceous mud belt of Namibia. GEOBIOS 35(1):3–11

Struck U, Pollehne F, Bauerfeind E, Bodungen VB (2004) Sources of nitrogen for the vertical particle flux in the Gotland Sea (Baltic Proper) – results from sediment trap studies. J Mar Syst 45:91–101

Struck U, Falk M, Altenbach AV, Pollehne F, Schneider M (2009) Nitrogen and carbon isotope ratios in suspended matter and dissolved inorganic carbon in a meromictic lake of the northern Alps (Bavaria, Germany). In: ASLO Aquatic Sciences Meeting 2009, Nice, France, 25–30 Jan 2009

Suchi EC (2005) Rekonstruktion der Evolution des marinen Nähstoffangebots im Verlauf des Jungpaläozoikums unter besonderer Berücksichtigung der Stickstoffisotopen-zusammensetzung. PhD thesis, Westfälische Wilhelms-Universität, Münster,151 pp

Thomazo C, Pinti DL, Busigny B, Ader M, Hashizume K, Philippot P (2009) Biological activity and the Earth's surface evolution: insights from carbon, sulfur, nitrogen and iron stable isotopes in the rock record. C R Palevol 8:665–678

Veizer J, Ala D, Azmy K, Bruckschen P, Buhl D, Bruhn F, Carden GAF, Diener A, Ebneth S, Goddéris Y, Jasper T, Korte C, Pawellek F, Podlaha OG, Strauss H (1999) $^{87}Sr/^{86}Sr$, $\delta^{13}C$ and $\delta^{18}O$ evolution of Phanerozoic seawater. Chem Geol 161:59–88

Voss M, Struck U (1997) Stable nitrogen and carbon isotopes as indicator of eutrophication of the Oder river (Baltic sea), Marine Chemistry 59(1–2):35–49

Voss M, Dippner JW, Montoya JP (2001) Nitrogen isotope patterns in the oxygen deficient waters of the Eastern Tropical North Pacific Ocean. Deep Sea Res Pt I 48(8):1905–1921

Wada E (1980) Nitrogen isotope fractionation and its significance in biogeochemical processes occurring in marine environments. In: Goldberg ED, Horibe Y, Saruhashi K (eds) Isotope marine chemistry. Uchida Rokakuho, Tokyo, pp 375–398

Wada E, Hattori A (1978) Nitrogen isotope effects in the assimilation of inorganic nitrogenous compounds. Geomicrobiol J 1:85–101

Waser NAD, Kedong Y, Zhiming Y, Kuninano T, Harrison PJ, Turpin DH, Calvert SE (1998a) Nitrogen isotope fractionation during nitrate, ammonium and urea uptake by marine diatoms and coccolithophores under various conditions of N availability. Mar Ecol Prog Ser 169:29–41

Waser NAD, Harrison PJ, Nielsen B, Calvert SE, Turpin DH (1998b) Nitrogen isotope fractionation during the uptake and assimilation of nitrate, nitrite, ammonium, and urea by a marine diatom. Limnol Oceanogr 43(2):215–224

Yoshida N (1988) ^{15}N-depleted N_2O as a product of nitrification. Nature 335:528–529

Biodata of **Alexander Volker Altenbach, Carola Leiter, Christoph Mayr, Ulrich Struck, Martin Hiss**, and **Antonio Radic**, authors of the chapter "*Carbon and Nitrogen Isotopic Fractionation in Foraminifera: Possible Signatures from Anoxia.*"

Dr. Alexander Volker Altenbach is professor for micropaleontology at the Ludwig Maximilians University in Munich. For more information, see biodata of the editors.

E-mail: **a.altenbach@lrz.unimuenchen.de**

Dr. Carola Leiter is a free lance geologist in a Munich consulting company. She started working in geoconsulting companies during her studies. As a technical assistant at the isotopic lab of the GeoBio-Center^{LMU}, she learned a lot about Foraminifera, decided to stay, and earned her PhD on benthic Foraminifera living in suboxic sediments off Namibia in 2008 at Munich University. Current research activities concentrate on anoxic environments and cretaceous Foraminifera.

E-mail: **c.leiter@gmx.de**

Alexander Volker Altenbach   **Carola Leiter**

Dr. Christoph Mayr is head of the stable isotope laboratory at the Institute of Geography of the University of Erlangen (Germany). His fields of research are stable isotope applications in aquatic systems and in paleoclimatology. A main focus of his work is the investigation of natural archives of climate change, natural hazards, anthropogenic disturbance, and nutrient dynamics such as tree-rings and lacustrine sediments. He was investigating the potential to reconstruct Holocene climate from isotope dendrochronologies in his PhD. Since his post-doc time, he works on lake sediments in South America and is a co-initiator of the ICDP-project PASADO.

E-mail: **cmayr@geographie.uni-erlangen.de**

Dr. Ulrich Struck is head of the stable isotope facilities of the Museum für Naturkunde, Leibniz Institute for Research on Evolution and Biodiversity at the Humboldt University Berlin since 2006. His Diploma as well as his PhD at Kiel University (obtained in 1992) occupied the sedimentology and paleoecology of benthic Foraminifera in the European Arctic. More than 2 years were spent on research vessels and boats. Present focus in research and teaching is given to global organic matter cycles, trophic pathways and food webs, and resulting isotopic fractionations related to nitrogen isotopes.

E-mail: **ulrich.struck@mfn-berlin.de**

Christoph Mayr

Ulrich Struck

Dr. Martin Hiss is a micropaleontologist and the head of the laboratory department at the Geological Survey of North Rhine Westphalia in Krefeld (Germany). Primarily his field of research is foraminiferal biostratigraphy in Mesozoic and Cainozoic sediments of the Münsterland (Westphalia), the Lower Rhine Embayment, and adjacent areas.

E-mail: **hiss@gd.nrw.de**

Dipl. Geol. Antonio Radic was involved in measurements and analysis for his thesis at the Ludwig Maximilians University in Munich. At present he is working for a navigation technology company in Munich/Germany.

E-mail: **tony.radic@gmail.com**

Martin Hiss   **Antonio Radic**

CARBON AND NITROGEN ISOTOPIC FRACTIONATION IN FORAMINIFERA: POSSIBLE SIGNATURES FROM ANOXIA

ALEXANDER VOLKER ALTENBACH[1,2], CAROLA LEITER[1], CHRISTOPH MAYR[2,3], ULRICH STRUCK[2,4], MARTIN HISS[5], AND ANTONIO RADIC[1]

[1]*Department for Earth and Environmental Science, LMU Munich, Richard-Wagner-Str. 10, 80333 Munich, Germany*
[2]*GeoBio-CenterLMU, Richard-Wagner-Str. 10, 80333 Munich, Germany*
[3]*Inst. for Geography, FAU Nürnberg/Erlangen, Kochstr. 4/4, 91054 Erlangen, Germany*
[4]*Natural History Museum, Humboldt-University, 10099 Berlin, Germany*
[5]*Geological Survey NRW, 1080, 47710 Krefeld, Germany*

1. Introduction

Our study aims to record carbon and nitrogen isotopic fractionations in Foraminifera recovered from the sulfidic and long-term anoxic environment of the Namibian diatomaceous mud belt. Two species which can be considered facultative anaerobes were investigated, *Virgulinella fragilis* GRINDELL and COLLEN and *Nonionella stella* CUSHMAN and MOYER. The former species shelters presumably chemotrophic bacteria as endosymbionts and utilizes electron acceptors other than dissolved oxygen (Bernhard 2003; Tsuchiya et al. 2006), and the latter species accumulates large quantities of intracellular nitrate for denitrification (Risgaard-Petersen et al. 2006; Høgslund 2008; Høgslund et al. 2008) and sequesters chloroplasts (Bernhard and Bowser 1999; Grzymski et al. 2002). The carbon and nitrogen isotopic compositions of their biomass and their biomineralized carbonatic tests were analyzed and discussed with regard to environmental conditions, supposed metabolic pathways, and isotopic fractionations reported for Foraminifera elsewhere. Our discussion includes remarks on varying laboratory procedures in use for isotopic studies on foraminiferal biomass.

2. The Namibian Diatomaceous Mud Belt

The Benguela Current establishes annual primary productivity rates near 600 gC m^{-2}, mainly mediated by seasonal blooms of diatoms (Barlow et al. 2009). Calculated from the empirical function of Suess (1980), the annual fluxes

Table 1. Location and environmental parameters for stations sampled during cruise AHAB leg 5.

Sample number	Latitude (S)	Longitude (E)	Water depth (m)	O_2 (ml l^{-1})	Seafloor-arriving flux rate (g org. $C m^{-2} year^{-1}$)
266360	22°32.70'	14°14.95'	72	0.30	> 231
266630	23°30.03'	14°22.71'	65	0.31	> 231
266710	24°59.99'	14°25.10'	125	0.22	188
266750	25°29.83'	14°45.17'	75	0.33	> 231
266940	22°59.79'	14°00.01'	136	0.28	174

of organic matter (OM) to the seafloor range from 174 g C m^{-2} for our deepest station at 136 m water depth (Table 1) to >230 g C m^{-2} for stations shallower than 100 m water depth. Due to these high fluxes of OM and diatom frustules, a diatomaceous mud belt (DMB) accumulates on the inner Namibian shelf from 19°S to 25°S (van der Plas et al. 2007) at annual accumulation rates near 1 mm (Bremner 1983; Struck et al. 2002). Seismic investigations revealed a sediment thickness of up to 15 m, most pronounced between 21°S to 23°S (Bremner 1980; Emeis et al. 2004). Oxygen concentrations of bottom waters are below detectability in this area, suggesting denitrification and anaerobic ammonia oxidation (anammox) even in the water column (Emeis et al. 2009). The amount of oxygen advected by bottom near inflow is much smaller than OM fluxes would require for heterotrophic consumption (Mohrholz et al. 2008). Anoxic conditions may persist at the seafloor for several months, and dysoxic conditions (<10 μmol O_2) persist for more than a year (Brüchert et al. 2009). Mean redox levels near −200 mV are indicated by modern phosphorite formation (Baturin 2002). Gas accumulations of methane and H_2S were observed within the DMB in all areas exceeding 12 m sediment thickness and in 20–60% of locations with a sediment thickness of 6–10 m (Emeis et al. 2004). Massive eruptions of this gas occur irregularly, with facultatively catastrophic impacts for the environment (Weeks et al. 2002, 2004; Emeis et al. 2004, 2009).

Activity rates and biomass of chemotrophic bacteria vary in surface sediments and in the water column (Schulz et al. 1999; Tyrrell and Lucas 2002; Lavik et al. 2009), so consistent budgets for formation rates, effluxes, and oxidation rates of hydrogen sulfide and methane are not easily established. The availability of electron acceptors for the subsequent microbial oxidation of sulfide largely controls these dynamics in the DMB (Brüchert et al. 2009; Lavik et al. 2009). As methanogenic archea are outcompeted by bacteria that reduce sulfate or nitrate, anaerobic methane oxidation is estimated to have much less quantitative control in comparison to sulfide oxidation. Rather, the episodic ebullition of gaseous methane and hydrogen sulfide from deeper sediment layers was considered reasonable for the coexistence of methane and hydrogen sulfide in the bottom waters and for the coexistence of methane and high concentrations of sulfate in pore waters of the DMB (Brüchert et al. 2009). The oxidation of these additionally ascending gasses will steepen the gradients of electron acceptors in the surface sediments and elevate the methanogenic zone near to the sediment surface (Dale et al. 2009). The dynamics of advective processes, microbial turnover rates, and resulting isotopic fractionations in the DMB are still a matter of discussion, including the assumption of hitherto unobserved microbial loops (Dale et al. 2009). More information on local environmental settings, distributional patterns, and standing stock of benthic Foraminifera are reported by Leiter and Altenbach (2010).

3. Material and Methods

Surface sediments were sampled with multicorers from the Namibian shelf during cruise AHAB, leg 5, of R.V. ALEXANDER VON HUMBOLDT in 2004 (Table 1). Dissolved oxygen concentrations were measured 5–10 m above the seafloor with a ship-based CTD. Only a few Foraminifera could be recovered from the surface sediments onboard. They protruded their reticulopodia in Petri dishes filled with cold, supernatant water from the multicorer tubes. The sediment column was sampled at intervals of 1 cm and stained onboard with a solution of 2 g of Rose Bengal (C.I. 45440) per liter of ethanol immediately after recovery. About 150 cm^3 of the solution was added per 100 cm^3 of wet sediment. This solution takes 3–6 weeks for best staining results (Lutze and Altenbach 1991). Samples were wet sieved for the size fractions 63–250 μm and >250 μm at Munich University and dried in an oven at 40°C.

For the detection of the carbon and nitrogen isotopic composition of foraminiferal biomass, 400 Rose Bengal-stained tests of *V. fragilis* and 200 stained tests of *N. stella* were pooled from five stations (Table 1). For carbon isotopic analysis of test carbonates, 5–25 unstained tests were pooled for each analysis. Analyses were performed at LMU Munich with a Finnigan Mat Delta S at the GeoBio-Center. A Carlo Erba (EA 1108) elemental analyser was coupled to the Finnigan MAT by a ConFlo interface. Isotope ratios are given by elevational delta notation versus air ($\delta^{15}N$) and VPDB ($\delta^{13}C$). Mean deviations derived from repetitive measures of laboratory standards are below 0.03‰, thus significantly better than estimates on the mean accuracy expectable from natural samples (Mill et al. 2008; Bahlmann et al. 2009; Katz et al. 2010).

Foraminifera were not pretreated by acidification and rinsing in order to remove carbonates before isotopic analyses. This was done (a) to prevent loss of volatile compounds (Bunn et al. 1995), (b) to keep control on mass balances of empty as well as cytoplasm-filled tests, and (c) to compare our results with methodologies involving acidification. Mixtures of proportions with different isotopic composition were separated by the stepwise reconstruction of their respective share and specific isotopic composition. For stable proportions, such mixtures are intermediate between the compositions of the end-members (Faure 1986; Uhle et al. 1997). Thus, when two components with known fractions (f_1, f_2) and known isotopic compositions (dE_1, dE_2) join, the isotopic composition dE_{all} of the mixture M_{all} can be calculated by:

$$M_{all} = f_1 + f_2 \quad (1)$$

$$dE_{all} = dE_1 f_1 + dE_2 f_2 \quad (2)$$

These calculations were applied for subdividing the carbon isotopic compositions derived from the biomass, the carbonate of the test, and the carbon adhered by Rose Bengal staining. The software systems PAST version 1.91 (Hammer et al. 2001) and XACT version 7.5 (SciLab ltd., Germany) were used for statistical analysis and scatter plots.

4. Results

The isotopic measurement of complete, e.g., not decalcified, Rose Bengal-stained *V. fragilis* and *N. stella* reveals a nitrogen composition of 11.1 and 7.2‰ $\delta^{15}N$, respectively (Table 2). These measures integrate the nitrogen sources of the cell and of the tiny organic layers covering the calcite crystals of the carbonatic test. The latter amount can be neglected because it ranges around a few micromole per gram of test carbonate (King and Hare 1972; King 1977). Each specimen was scrutinized carefully for OM attached to the test during picking, thus we may assume negligible impact also for adhered matter. *N. stella* is depleted by 3.9‰ $\delta^{15}N$ in comparison to *V. fragilis*. Such interspecific differences, equivalent to about one or two trophic steps of a food chain, are commonly observed in benthic Foraminifera (Nomaki et al. 2008; Iken et al. 2001).

The mean carbon isotopic composition for complete specimens was detected at −18.5 and −15.7‰ $\delta^{13}C$ for *V. fragilis* and *N. stella*, respectively. These carbon fractionations comprise three different sources of carbon: (I) the cellular biomass, (II) the test carbonate, and (III) the Rose Bengal stain with an elementary composition of $C_{20}H_2Cl_4I_4Na_2O_5$. The test carbonate of empty (unstained) tests picked from the sample locations (Table 1) revealed a fractionation of −12.5

Table 2. Isotopic composition of Foraminifera.

Measurement	V. fragilis	N. stella
Total $\delta^{15}N$ ‰ (air)	11.1	7.2
Total $\delta^{13}C$ ‰ (VPDB)	−18.54	−15.66
Fraction of C derived from test carbonate	40%	26%
Test $\delta^{13}C$ ‰	−12.46	−2.39
Bengal Rose $\delta^{13}C$ ‰	−26.09	−26.09
$\delta^{13}C$ ‰ for Rose Bengal stained biomass	−22.59	−20.38
Corrected for 5% of Rose Bengal derived $\delta^{13}C$ ‰	−22.41	−20.08
Range of $\delta^{13}C$ ‰ for biomass	−22.6 to −22.4	−20.4 to −20.1

and −2.4‰ δ^{13}C for *V. fragilis* and *N. stella*, respectively. The δ^{13}C composition of Rose Bengal was detected at −26.09‰ δ^{13}C (Table 2), analyzed from the same packing unit in use for the onboard staining procedure. A similar fractionation of −25.32‰ δ^{13}C was recorded for Rose Bengal by Panieri (2006).

The definition of the specific fractions attributable to the different carbon sources is the most critical step for our chosen methodology. The quantity of carbon derived from test carbonate can only be reconstructed from a bulk of unstained tests with a size distribution most perfectly approaching the size distribution of the stained specimen. Otherwise, the amount of carbonate may significantly differ for both measurements, despite an identical number of specimens measured. Lower size classes, for example, may bias results by a factor of more than two. Correction values based on an estimated mean mass balance between cellular and carbonate carbon sources will be misleading as well, due to the fact that this ratio, e.g., the filling degree of internal test space with biomass, may vary by nearly one order of magnitude (Altenbach 1987). By carefully considering the size distribution, the number of specimens measured, and the mass dependent intensity signalized by the MS, we allocate 40% and 26% of carbonate-derived carbon to *V. fragilis* and *N. stella*, respectively. Calculating the given fraction and isotopic composition by formula (2), the carbon composition of the stained biomass revealed −22.6‰ δ^{13}C for *V. fragilis* and −20.4‰ δ^{13}C for *N. stella* (Table 2).

Control measures on the carbon content of living and Rose Bengal-stained Foraminifera were reported by Altenbach (1985). Comparing identical test size ranges, both series revealed identical variability patterns of their respective carbon contents (Snedecor-F-test passed at 1% level). In brief, the allometric growth curve for stained specimens ranged within the confidence interval of the allometric growth curve for unstained living specimens (Altenbach 1985). Despite the lack of statistical significance, however, Rose Bengal must definitely add some additional carbon. We calculated a worst-case scenario for a quarter of carbon derived from the stain and a more relaxed value of 5%. With respect to the given stained biomass composition, a proportion as large as one quarter would require correction values of 1.2‰ and 1.9‰ δ^{13}C for the unstained biomass of *V. fragilis* and *N. stella*, respectively. The more realistic estimate of about 5% or less delimits the influence of the stain at or below 0.3‰ (Table 2). As the amount of carbon introduced by Rose Bengal staining of foraminiferal tests does not significantly effect isotopic measurements of carbonate derived δ^{13}C and δ^{18}O (Bernhard et al. 2006; Serrano et al. 2008), an estimate of 5% or less seems justified. Panieri (2006) considered the influence of Rose Bengal for measurements on biomass-derived carbon isotopic fractionations as negligible.

However, the amount of Rose Bengal-derived carbon received by a mass spectrometer might perhaps even depend on the differing laboratory treatments in use for C/O analysis of test carbonates and C/N analysis of OM. Rose Bengal is a large organic molecule; its chemical reaction on sample preparation with acids for vaporizing carbon and oxygen from carbonates may be weaker in comparison to combustion methods in use for C/N analyses.

In sum, we consider a range between 5% and 0% impact of Rose Bengal-derived carbon as solid, resulting in −22.6‰ to −22.4‰ $\delta^{13}C$ for *V. fragilis*, and −20.4‰ to −20.1‰ $\delta^{13}C$ for *N. stella* (Table 2).

5. Discussion

5.1. TEST-DERIVED CARBON

Isotopic measurements of test-derived $\delta^{13}C$ in Foraminifera are broadly used in paleoclimatology, neontology, and environmental research. For some species, species-specific ranges are well documented and reproducible for distinct environmental settings, permitting the pursuit of environmental change within fluctuations of less than 1‰ $\delta^{13}C$ (Broecker and Peng 1982; Grossman 1984, 1987; Katz et al. 2010). Enhanced depletion in foraminiferal tests typically reveals the imprint of interstitial metabolic CO_2 production (Altenbach and Sarnthein 1988; McCorkle et al. 1990, 1997). This depletion is not considered to produce an offset of more than 4–5‰ $\delta^{13}C$ between bottom water and interstitial composition, even for the combined metabolic effects derived from dissolved oxygen consumption and from sulfate reduction (McCorkle and Emerson 1988).

The carbon fractionation of −12.5‰ $\delta^{13}C$ obtained for *V. fragilis* (Table 2) is the lightest value ever recorded for unaltered foraminiferal test carbonates of stained specimens. Similar depletions were recorded from empty tests at methane seeps but discussed with respect to some potential evidence for authigenic carbonate precipitation involved (Hill et al. 2004).

Within the surface sediments of the DMB, sulfur bacteria extensively oxidize hydrogen sulfide to sulfate using nitrate as the terminal electron acceptor at annual microbial sulfide oxidation rates exceeding 6 mol sulfur m^2 (Dale et al. 2009). A small fraction of the microbial sulfate production diffuses into deeper sediment layers, where it fuels microbial methane oxidation. Methane concentrations of more than 10 mM in deeper sediments thus vanish within the uppermost sediment column. The depth integrated $\sum CO_2$ efflux rate from the DMB is calculated to range near 12 mol m^{-2} year^{-1} (Dale et al. 2009). Marine biogenic methane is defined isotopically by $\delta^{13}C$ ratios of −60‰ to −110‰, exceeding the fractionation of thermogenic methane by far (Whiticar et al. 1986). Maximum depletions of dissolved inorganic carbon (DIC) result from enhanced methane consumption rather than methane production (Hinrichs et al. 1999). A substantial pool of CO_2 derived from methane oxidation may cause authigenic carbonate precipitations depleted by more than 30‰ $\delta^{13}C$ (Rathburn et al. 2000; von Rad et al. 1996). Therefore, the carbon depletion observed in *V. fragilis* may well result from enhanced amounts of oxidized methane in interstitial water (Wefer et al. 1994).

But the sympatric *N. stella* derived from the same samples is less depleted by around 10‰ $\delta^{13}C$ in comparison to *V. fragilis* (Table 2). In order to interpret this substantial interspecific divergence, we have to assume disparity in species-specific behavior. *N. stella* is found sympatric to *V. fragilis* frequently in the DMB and may penetrate even more deeply into the sediment column than *V. fragilis* (Leiter 2008). Therefore, *N. stella* must biomineralize its test during DIC compositions significantly different from *V. fragilis*. This may be either derived by suppressing test formation during enhanced oxygen depletion and flux rates of methane and hydrogen sulfide, or it is performed by active migration toward the sediment surface in the course of temporal advection of oxygenated bottom waters (Torres et al. 2003). The presence or absence of bacterial symbionts with differing physiologies, or species-specific differences in the ability to denitrify, may be influential as well. All Foraminifera sympatric to *V. fragilis* in the DMB show depletions around −2‰ to −5‰ $\delta^{13}C$ (Leiter 2008). This range agrees very well with the maximum carbon isotopic fractionation calculated for pore water DIC and foraminiferal test formation under oxygen depletion (McCorkle and Emerson 1988). Especially, facultative deep-dwelling species are adapted to such conditions, with respect to oxygen depletion as well as to metabolic DIC enrichment (McCorkle et al. 1997). Foraminiferal assemblages recovered from methane seeps commonly consist of such species, but their tests tend to show much less pronounced $\Delta\delta^{13}C$ values in comparison to local DIC (Rathburn et al. 2000, 2003; Torres et al. 2003; Mackensen et al. 2006).

Comparison of test and biomass-derived carbon isotopic compositions from methane seeps and non-seep sites did not reveal alterations above 3‰ $\delta^{13}C$ for biomass and 4‰ $\delta^{13}C$ for test-derived carbon compositions (Panieri 2006). The impact of metabolic CO_2 incorporated during test calcification by planktic foraminifera was observed to culminate at 15% (Spero and Lea 1996). In comparison to the test fractionation of the benthic *N. stella* at −2.4‰ $\delta^{13}C$, a metabolic amount of 15% would cause the fivefold test-derived carbon depletion observed in the sympatric *V. fragilis* only if the metabolic carbon transferred during test precipitation would range near −70‰ $\delta^{13}C$. This would require metabolic fractionations similar to biogenic methane formation (Whiticar et al. 1986). Carbon fractionation in the biomass of *V. fragilis* is only slightly elevated in comparison to the range commonly observed for many other benthic Foraminifera (Nomaki et al. 2008; Moodley et al. 2002; Iken et al. 2001), thus extreme metabolic fractionations are not indicated. Any species-specific specialization inducing severe metabolic carbon fractionation should lead to a distinctive carbon fractionation for this species. But the test carbonates secreted by *V. fragilis* revealed −2.6‰ $\delta^{13}C$ in sulfur-oxidizing bacterial mats off Pakistan (Erbacher and Nelskamp 2006), −6.4‰ $\delta^{13}C$ in the hypoxic to anoxic Cariaco Basin (Bernhard 2003), between −9 and −12‰ $\delta^{13}C$ in the years 2000–2003 in the DMB (Altenbach et al. 2002; Leiter 2008), and −12.5‰ $\delta^{13}C$ in our samples recovered from the DMB in 2004. In addition to this enormous range of intraspecific $\Delta\delta^{13}C$, *V. fragilis* reveals peak depletion in ^{13}C whenever compared to test carbonates of sympatric Foraminifera (Altenbach et al. 2002; Erbacher and Nelskamp 2006; Leiter 2008).

Primary productivity rates show considerable seasonal and inter-annual variability, but mean annual primary productivity can be estimated at 440 g C m^{-2} for the Arabian Sea (Barber et al. 2001) and between 370 and 570 gC m^{-2} for the Cariaco Basin (Goni et al. 2009). The water depths of the stations in the Cariaco Basin (375 m) and the Arabian Sea (576 m) are about three to nine times deeper in comparison to our DMB stations. Degradation of settling OM fluxes in the water column will be prolonged, and the amount of OM reaching the sea floor will be lowered. Calculated by the empiric function in Suess (1980), the annual seafloor-arriving fluxes of OM range at 32 gC m^{-2} for *V. fragilis* recovered from the Arabian Sea at 40–62 gC m^{-2} for the Cariaco Basin and above 173 gC m^{-2} for our stations. This provides a principal, albeit rough correlation of annual OM fluxes with observed depletions of ^{13}C in test carbonates of *V. fragilis*. However, methane seepage has to be considered an additional carbon source for all named areas.

The Arabian Sea station is placed at the upper edge of gas hydrate formation, which is induced by primary production in the surface waters (von Rad et al. 2000). Methane concentrations measured at the time of sampling were too low to support methane efflux, and dissolved oxygen was available in bottom waters at 0.3–0.6 ml l^1 (Erbacher and Nelskamp 2006). Therefore, the δ^{13}C composition of tests of *V. fragilis* revealed a range (−2.6‰) consistent with enhanced incorporation of metabolic carbon dioxide derived from present OM decomposition under low oxic conditions (McCorkle and Emerson 1988). In contrast, seepage of fossil methane dominates over microbial methane production in the Cariaco Basin. Anaerobic oxidation is the major sink for both sources, and resulting carbon isotopic fractionations of DIC are extensive (Kessler et al. 2006). This seems most reasonable for the observed fractionation of −6.4‰ δ^{13}C in the tests of *V. fragilis* (Bernhard 2003).

As already described for the DMB, the alternating prevalence of methane formation and oxidation may enhance carbon fractionation and cause significant local accumulation of DIC, especially within the topmost sediment column, where biogenic degradation and ascending methane sources intermix. A number of uncertainties are quoted for the biochemical pathway of methane oxidation in general (Valentine 2002; Wakeham et al. 2004). At least some of these ambiguities result from the episodic nature of methane fluxes, far from steady state and not obvious from short-term investigations (Scranton et al. 1993; Mackensen et al. 2006). The tests of *V. fragilis* offer a promising probe for the detection of such irregular and pulsed processes.

All these observations indicate that *V. fragilis* does not eschew test formation under substantially depleted DIC δ^{13}C concentrations. The test carbonate compositions of most other species are indicative for such avoidance (Rathburn et al. 2003; Torres et al. 2003). In fact, the large interspecific divergence in our measures and the published inter-location differences in δ^{13}C values for *V. fragilis* could hardly be explained otherwise. This modern species has an extremely thin and fully translucent test, thus secondary carbonate precipitation significantly affecting the test composition (Torres et al. 2003) is easily recognized. Tracing capabilities for depleted fractionation in DIC might have been initiated already by ancestral species of the genus. As noted by Altenbach et al. (2002), Miocene virgulinids generally show ^{13}C values more depleted in comparison to all sympatric Foraminifera.

5.2. CARBON AND NITROGEN ISOTOPIC COMPOSITION OF BIOMASS

Isotopic measurements of a single-celled amoeboid, e.g., a foraminiferan, faces a number of specific restrictions. Starvation, for example, can reduce the filling level of the internal test volume to just 10% (Altenbach 1987). As many reserve substances, such as fatty acids, are more depleted by 3‰–7‰ $\delta^{13}C$ than total foraminiferal biomass values, the loss of such specific compounds will lower observed bulk depletion (Uhle et al. 1997). The same authors speculate that the enrichment of lipids in well-fed Foraminifera, in turn, could lead to a bulk depletion in ^{13}C, even exceeding that of their food. Such nutritional influences can be considered reasonable for the species-specific seasonal alterations of up to 1.7‰ $\delta^{13}C$ and 2.7‰ $\delta^{15}N$ recorded by Nomaki et al. (2008). All data on the isotopic composition of N and C in foraminiferal biomass published hitherto are based on decalcification (Uhle et al. 1997, 1999; Iken et al. 2001; Moodley et al. 2002; Panieri 2006; Nomaki et al. 2008; Bergmann et al. 2009). This procedure will remove the test carbonate and also an unknown portion of volatile and soluble organic compounds (Bunn et al. 1995). For sediment samples, this loss may account for nearly half of the original contents of organic carbon (Froelich 1980) or nitrogen (Umezawa et al. 2008). Differing sample composition and applied laboratory methodologies can be highly influential for the loss of volatiles, thus even studies monitoring the influence of acidification and rinsing on isotopic compositions may offer contradictory evaluations (Kennedy et al. 2005). Divergent laboratory methodologies may bias measurements by more than 2‰ in general (Mill et al. 2008; Bahlmann et al. 2009).

Preservation with pure ethanol or a solution of ethanol and Rose Bengal leads to rapid dehydration via the cell membrane and massive cell shrinking, so the solutes might be lost. Freezing isolated metazoans are considered to have no effect on subsequent isotopic measures (Dannheim et al. 2007). Cryofracturing of a single cell, however, indeed may result in the loss of volatile compounds by membrane rupture. We rarely observed ruptured and spotted cytoplasm fragments in translucent tests of foraminifera frozen at −20°C to −30°C, but this proportion increases noticeably at freezing temperatures near −80°C (unpublished lab. observations). Only flash freezing techniques can suppress growth of large water crystals, and thus avoid damage to membranes completely (Ryan and Knoll 1994). Rupture of the outer cell membrane will definitely cause loss of volatiles; especially when rinsing is applied subsequent to defrosting. Therefore, the recommendation of Bunn et al. (1995) to measure without any acidification and rinsing can be followed by staining and preservation with Rose Bengal instead of deep freezing. Foraminiferal tests offer a natural container for the cytoplasm, albeit some control measures and calculations are needed for the separation of carbon fractions derived from test carbonate, stain, and biomass.

5.2.1. Nitrogen Isotopic Composition

The nitrogen isotopic composition of OM fluxes of the SE Atlantic seasonally fluctuates around 5.4‰ $\delta^{15}N$, between 4‰ and 7‰ (Loncaric et al. 2007).

As phytoplankton varies between 0‰ and 6‰ $\delta^{15}N$ (Struck et al. 2002), the 7.2‰ $\delta^{15}N$ detected for the biomass of *N. stella* may well infer a single nutritional step of 2–3‰ $\delta^{15}N$, based on herbivory. On the one hand, this corresponds to the observed attraction to fresh diatom fluxes for this species (Diz and Frances 2008); on the other hand, the diatoms from which chloroplasts are sequestered by this species are not ingested as a food source (Bernhard and Bowser 1999). Detrivory must be less influential because OM of surface sediment ranges near 7‰, and thus would induce a biomass value near 9–10‰ (Struck et al. 2002). Høgslund (2008) reported large amounts of intracellular nitrate stored by *N.* cf. *stella* for denitrification. Considering 6‰ $\delta^{15}N$ for dissolved inorganic nitrate (DIN) in the DMB (Struck et al. 2002), an intracellular amount near to 50% for this pool would infer a biomass value of 8.4‰ $\delta^{15}N$, as calculated from the bulk value of 7.2‰ $\delta^{15}N$. This shows that larger amounts of stored marine DIN will tend to lower bulk ^{15}N depletion. Denitrifying Foraminifera, such as *Globobulimina affinis* and *N.* cf. *stella* (Høgslund et al. 2008; Høgslund 2008; Pina-Ochoa et al. 2010), indeed show fractionation values below 8‰ $\delta^{15}N$, whereas many other species range within 8–12‰ $\delta^{15}N$ (Nomaki et al. 2008; Table 2). Høgslund (2008) speculated that the nitrate pool of *N. stella* might be derived from ingested diatoms. This seems questionable because the intracellular nitrate in diatoms may be enriched by some 6–13‰ $\delta^{15}N$, in comparison to the external pool of DIN (Needoba et al. 2004). The diatom-derived pool would rather tend to rise the enrichment of ^{15}N but not explain the comparably light nitrogen composition around 5–7‰ $\delta^{15}N$ detected for *N. stella* (Table 2) and *G. affinis* (Nomaki et al. 2008).

The nitrogen isotopic composition of 11.1‰ $\delta^{15}N$ requires a more pronounced fractionation for *V. fragilis*. Such a range was recorded before only for one benthic foraminiferal species, namely, *Bolivina spissa* (Nomaki et al. 2008). This species resides in large densities in surface sediments (0–5 cm) on the edge of the oxygen minimum zone near the Santa Barbara Basin; surface sediments are rich in OM (> 4%), and bottom near water masses are dysoxic (< 0.5 ml O_2 l^{-1}) even 10 m above the sea bottom (Shepherd et al. 2007). Environmental parameters and behavior were recorded quite similar for *B. spissa* and *B. pacifica*, both with a distinct increase in abundance in the presence of fresh phytodetritus, preferential growth and reproduction during surface bloom times, and similar food uptake rates (Shepherd et al. 2007). This all would suggest a common herbivorous behavior for *B. spissa*, albeit the enrichment of up to 5‰ $\delta^{15}N$ in comparison to POM (Nomaki et al. 2008) seems somewhat exaggerated. Detailed investigations of its sympatric congener, *B. pacifica*, reveal a much more complex adaptation to the oxygen-depleted environment, including specific internal organization of organelles and putative symbiosis with bacteria (Bernhard et al. 2010). As nearly identical conclusions were drawn for the ecology of *V. fragilis* (Tsuchiya et al. 2006; Bernhard and Bowser 2008), we tend to speculate that the heavy nitrogen compositions near 11‰ to 12‰ $\delta^{15}N$ are related to such specific ecological adaptations. Bacterivory, or the substantial uptake of protein-rich bacterial byproducts, may trigger trophic steps in nitrogen fractionation exceeding 3‰ $\delta^{15}N$ (McCutchan et al. 2003; Sweetman and Witte 2008).

For planktic Foraminifera, Uhle et al. (1999) assume a nitrogen isotopic enrichment of 4.1‰ $\delta^{15}N$ associated with each trophic interaction. This cannot be generalized for benthic Foraminifera because a number of species and morphotypes show biomass nitrogen isotopic compositions near or below POM fractionations derived from the appropriate water column or surface sediments (Iken et al. 2001; Nomaki et al. 2008; Bergmann et al. 2009). These observations do not support the assumption that these Foraminifera follow a simple, single-step strategy in nutrition based on detritus, bacteria, or algae. The C/N ratio of nutritional sources may highly influence nitrogen fractionation in trophic levels, but neither severe nitrogen limitation nor starvation should deplete the bulk ^{15}N of a heterotroph below its dietary $\delta^{15}N$ (Adams and Sterner 2000). Additional organic pools with significantly lighter ^{15}N compositions must be considered influential. Mainly, groups related to Xenophyophoracea (*Rhizammina* sp.; see Lecroq et al. 2009) or Komokiacea (termed "mudballs" and "soft-walled tests" in Iken et al. 2001) reveal nitrogen isotopic compositions between 0.5‰ and 2.8‰ $\delta^{15}N$, lighter in comparison to POM (Iken et al. 2001). Such depletion may be derived from the uptake of aged or dissolved OM (Smith et al. 2001) or from enhanced bacterial commensalism, gardening, and exchange of exudates. All these nutritional pools are accessible for Xenophyophoracea, Komokiacea, and other so-called "unfamiliar" or "allogromid-like" Foraminifera (Tendal 1979; DeLaca et al. 2002; Laureillard et al. 2004; Gooday et al. 2008). This group of simply agglutinated, single-chambered (monothalamiid) Foraminifera indicates that such nutritional sources are much more influential for the biology and evolution of Foraminifera than commonly estimated (Richardson and Cedhagen 2001; Altenbach and Gaulke 2011).

The most depleted stable isotopic composition in foraminiferal biomass was recorded by Bergman et al. (2009) at 4.1‰ $\delta^{15}N$ and −26.66‰ $\delta^{13}C$. This was lower than the surrounding POM and sediment bulk values, and much lower than any other taxon investigated in this study. Dual stable-isotope depletions below POM and sediment bulks tend to leave food webs based on photosynthetic carbon toward organic matter derived from sulfide oxidation. This was observed as a significant trend for the infauna analyzed from sulfur oxidizing bacterial mats and also considered for translocations from bacterial symbiosis (Demopoulos et al. 2010). Unfortunately, Bergmann et al. (2009) did not specify the foraminiferal taxa regarded.

Overall, if we compare this minimum value with the maximum reported by Iken et al. (2001) for benthic Foraminifera at 15.3‰ $\delta^{15}N$, this would span an interspecific range of more than 11‰ $\delta^{15}N$. This magnitude corresponds to three to four successive trophic transfers, each of which enriches the ^{15}N content by about 3‰ (Peterson and Fry 1987).

5.2.2. Carbon Isotopic Composition
Carbon isotopic fractionations of foraminiferal biomass tend to show a quite narrow and stable range between −18‰ and −23‰ $\delta^{13}C$ (Uhle et al. 1997; Iken et al. 2001; Moodley et al. 2002; Nomaki et al. 2008). One planktic species with

endosymbiotic phototrophs was found slightly less depleted at −17.4‰ $\delta^{13}C$ PDB (Uhle et al. 1997). For one specimen of *Rhizammina* sp., even −16.5‰ $\delta^{13}C$ was noted, but somewhat in conflict with two other congeneric specimens with −20.8‰ and −20.6‰ $\delta^{13}C$ (Iken et al. 2001). Our carbon isotopic measures on the biomass of *V. fragilis* and *N. stella* fully resemble the range commonly recorded between −18‰ and −23‰ $\delta^{13}C$ (Table 2).

In contrast, species recovered from active methane seeps reveal maximum fractionations of ∼−24‰ to −25.51‰ $\delta^{13}C$ (Panieri 2006). Non-seep stations chosen by Panieri (2006) for comparison still show depletions between −22.8‰ and −25.3‰ $\delta^{13}C$, so some minor, or perhaps pulsed, impact from methane seepage might have been influential there also. Large amounts of anaerobic methanotrophs cause overall biomass compositions near −70‰ $\delta^{13}C$ (Treude et al. 2007), with specific Archean lipids reaching −111‰ $\delta^{13}C$ (Blumenberg et al. 2004). Due to this severe depletion, even very small proportions ingested directly, or from trophic intermediates of such bacterial clusters, will explain the shift in carbon depletion in the biomass of Foraminifera observed by Panieri (2006). For the DMB, neither *V. fragilis* nor *N. stella* provides biomass carbon fractionations that indicate even subtle influence derived from such diets of severely depleted food.

6. Conclusions

Published isotopic carbon and nitrogen fractionations of foraminiferal biomass obtained from either pretreatment by acidification or from the mass balance calculations applied in our study reveal no obvious differences. We do not infer exactness for both methodologies but rather assume similar levels for possible biases. Flash freezing and direct measurement without decalcification appear as the most promising techniques for the detection of nitrogen isotopic compositions of Foraminifera with calcitic tests. Carbon isotopic measures require either decalcification or mass balance calculations, and both methodologies may cause substantial errors. Comparative application of both methodologies may permit bias tracking. Natural intraspecific variability ranges up to 2.7‰ $\delta^{15}N$ and 1.7‰ $\delta^{13}C$. In light of the relative small number of species investigated and data published on isotopic fractionation of foraminiferal biomass to date, it seems appropriate to interpret differences near or below these ranges with great care.

The total range of nitrogen fractionation in foraminiferal biomass recorded hitherto is amazingly large. More than 11‰ variance in interspecific $\delta^{15}N$, from below local POM to levels comparable to second and third level consumers, indicate an ambiguously broad range of trophic adjustments. Denitrifying Foraminifera can be influenced by large intracellular pools of stored nitrate, pending on origin and fractionation of this supply. Two species identified as denitrifiers range at lighter nitrogen compositions < 8‰ $\delta^{15}N$, probably due to stored marine DIN.

Biomass isotopic fractionations of carbon may indicate a phase transfer at $\delta^{13}C$ values lighter than −23‰, as enhanced depletion was observed only at active methane seeps. The number of observations is limited, however, and sometimes not fully explainable.

Carbon isotopic compositions of test carbonate reveal small variance patterns. Most Foraminifera seem to avoid biomineralization during periods of enhanced carbon isotopic depletion of bottom water or interstitial DIC; observed fractionations are consistent with the theoretical maximum of ^{13}C depletion derived from extensive heterotrophic consumption, oxygen depletion, and accumulation of metabolic CO_2. *Virgulinella fragilis* may be exceptional in mineralizing its carbonate test without regard to present DIC isotopic fractionation. Our observed interspecific difference of more than 10‰ $\delta^{13}C$ for tests of sympatric species fully corroborates the assumption of species-specific behavior as a ruling agent for the $\delta^{13}C$ composition of foraminiferal test carbonates.

7. References

Adams TS, Sterner RW (2000) The effect of dietary nitrogen content on trophic level 15N enrichment. Limnol Oceanogr 45:601–607

Altenbach AV (1985) Die Biomasse der benthischen Foraminiferen. – Auswertungen von Meteor-Expeditionen im östlichen Nordatlantik. PhD thesis, Math.-Nat. Faculty, Christian-Albrechts-Universität, Kiel

Altenbach AV (1987) The measurement of organic carbon in foraminifera. J Foraminifer Res 17:106–109

Altenbach AV, Gaulke M (2011) Did redox conditions trigger test templates in Proterozoic Foraminifera? In: Altenbach AV, Bernhard JM, Seckbach J (eds) Anoxia: Evidence for Eukaryote Survival and Paleontological Strategies. Springer, Dordrecht

Altenbach AV, Sarnthein M (1988) Productivity record in benthic Foraminifera. In: Berger WH, Smetacek V, Wefer G (eds) Productivity of the ocean: past and present. Wiley, Chichester, pp 255–269

Altenbach AV, Struck U, Graml M, Emeis K-C (2002) The genus *Virgulinella* in oxygen deficient, oligotrophic, or polluted sediments. In: Revets SA (ed) FORAMS 2002 international symposium on foraminifera, Volume of Abstracts, vol 1. The University of Western Australia, Perth, p 20

Bahlmann E, Bernasconi SM, Bouillon S, Houtekamer M, Korntheuer M, Langenberg F, Mayr C, Metzke M, Middelburg JJ, Nagel B, Struck U, Voss M, Emeis K (2009) Performance evaluation of nitrogen isotope ratio determination in marine and lacustrine sediments: an inter-laboratory comparison. Org Geochem 41:3–12

Barber RT, Marra J, Bidigare RC, Codispoti LA, Halpern D, Johnson Z, Latasa M, Goericke R, Smith SL (2001) Primary productivity and its regulation in the Arabian Sea during 1995. Deep-Sea Res II 48:1127–1172

Barlow R, Lamont T, Mitchell-Innes B, Lucas M, Thomalla S (2009) Primary production in the Benguela ecosystem, 1999–2002. Afr J Mar Sci 31:97–101

Baturin GN (2002) Manganese and molybdenium in phosphorites from the Ocean. Lithol Miner Resour 37:412–428

Bergmann M, Dannheim J, Bauerfeind E, Klages M (2009) Trophic relationships along a bathymetric gradient at the deep-sea observatory HAUSGARTEN. Deep-Sea Res I 56:408–424

Bernhard JM (2003) Potential symbionts in bathyal foraminifera. Science 299:861

Bernhard JM, Bowser SS (1999) Benthic Foraminifera of dysoxic sediments: chloroplast sequestration and functional morphology. Earth-Sci Rev 46:149–165

Bernhard JM, Bowser SS (2008) Peroxisome proliferation in foraminifera inhabiting the chemocline: an adaptation to reactive oxygen species exposure? J Eukaryot Microbiol 55:135–144

Bernhard JM, Ostermann DS, Williams DS, Blanks JK (2006) Comparison of two methods to identify live benthic foraminifera: a test between Rose Bengal and CellTracker Green with implications for stable isotope paleoreconstructions. Paleoceanography 21: doi:10.1029/2006PA001290

Bernhard JM, Goldstein ST, Bowser SS (2010) An ectobiont-bearing foraminiferan, *Bolivina pacifica*, that inhabits microxic pore waters: cell-biological and paleoceanographic insights. Environ Microbiol 12:2107–2119

Blumenberg M, Seifert R, Reitner J, Pape T, Michaelis W (2004) Membrane lipid patterns typify distinct anaerobic methanotrophic consortia. Proc Natl Acad Sci USA 101:11111–11116

Bremner JM (1980) Concretionary phosphorite from SW Africa. J Geol Soc London 137:773–786

Bremner JM (1983) Biogenic sediments on the South West African (Namibian) continental margin. In: Suess E, Thiede J (eds) Coastal upwelling. Plenum, New York, pp 73–103

Broecker WS, Peng T-H (1982) Tracers in the sea. Eldigio Press, Lamont Doherty Geological Observatory, USA

Brüchert V, Currie B, Peard KR (2009) Hydrogen sulphide and methane emissions on the central Namibian shelf. Prog Oceanogr 83:169–179

Bunn SE, Loneragan NR, Kempster MA (1995) Effects of acid washing on stable-isotope ratios of C and N in penaeid shrimp and seagrass – implications for food-web studies using multiple stable isotopes. Limnol Oceanogr 40:622–625

Dale AW, Brüchert V, Alperin M, Regnier P (2009) An integrated sulfur model for Namibian shelf sediments. Geochim Cosmochim Acta 73:1924–1944

Dannheim J, Struck U, Brey T (2007) Does sample bulk freezing affect stable isotope ratios of infaunal macrozoobenthos? J Exp Mar Biol Ecol 351:27–41

DeLaca TE, Bernhard JM, Reilly AA, Bowser SS (2002) *Notodendrodes hyalinosphaira* (sp. nov.): structure and autecology of an allogromiid-like agglutinated foraminifer. J Foram Res 32:177–187

Demopoulos AWJ, Gualtieri D, Kovacs K (2010) Food-web structure of seep sediment macrobenthos from the Gulf of Mexico. Deep-Sea Res II 57:1972–1981

Diz P, Frances G (2008) Distribution of live benthic foraminifera in the Ria de Vigo (NW Spain). Mar Micropaleontol 66:165–191

Emeis K-C, Brüchert V, Currie B, Endler R, Ferdelman T, Kiessling A, Leipe T, Noli-Peard K, Struck U, Vogt T (2004) Shallow gas in shelf sediments of the Namibian coastal upwelling ecosystem. Cont Shelf Res 24:627–642

Emeis K-C, Struck U, Leipe T, Ferdelman TG (2009) Variability in upwelling intensity and nutrient regime in the coastal upwelling system offshore Namibia: results from sediment archives. Int J Earth Sci 98:309–326

Erbacher J, Nelskamp S (2006) Comparison of benthic foraminifera inside and outside a sulphur-oxidizing bacterial mat from the present oxygen-minimum zone off Pakistan (NE Arabian Sea). Deep-Sea Res I 53:751–775

Faure G (1986) Principles of isotope geology, 2nd edn. Wiley, New York, 589 p

Froelich PN (1980) Analysis of organic carbon in marine sediments. Limnol Oceanogr 25:564–572

Goni M, Aceves H, Benitez-Nelson B, Tappa E, Thunell R, Black D, Muller-Karger F, Astor Y, Varela R (2009) Oceanographic and climatologic controls on the compositions and fluxes of biogenic materials in the water column and sediments of the Cariaco basin over the late Holocene. Deep-Sea Res I 56:614–640

Gooday AJ, Nomaki H, Kitazatao H (2008) Modern deep-sea foraminifera: a brief review of their morphology-based diversity and trophic diversity. In: Austin WEN, James RH (eds) Biogeochemical controls on palaeoceanographic environmental proxies. Geological Society London, Special Publication, 303, London, pp 97–119

Grossman EL (1984) Carbon fractionation in live benthic foraminifera – comparison with inorganic precipitate studies. Geochim Cosmochim Acta 48:1505–1512

Grossman EL (1987) Stable isotopes in modern benthic foraminifera: a study of vital effect. J Foraminifer Res 17:48–61

Grzymski J, Schofield OM, Falkowski PG, Bernhard JM (2002) The function of plastids in the deep-sea benthic foraminifer, *Nonionella stella*. Limnol Oceanogr 47:1569–1580

Hammer Ø, Harper DAT, Ryan PD (2001) PAST: Paleontological Statistics Software Package for Education and Data Analysis. Palaeontol Electron 4(1):9

Hill TM, Kennet JP, Valentine DL (2004) Isotopic evidence for the incorporation of methane-derived carbon into foraminifera from modern methane seeps, Hydrate Ridge, Northeast Pacific. Geochim Cosmochim Acta 68:4619–4627

Hinrichs K-W, Hayes JM, Sylva SP, Brewer PG, DeLong EF (1999) Methane-consuming archaebacteris in marine sediments. Nature 398:802–805

Høgslund S (2008) Nitrate storage as an adaptation to benthic life. PhD thesis, Dept. Biol. Science, University of Aarhus, Denmark

Høgslund S, Revsbech N, Cedhagen T, Nielsen L, Gallardo V (2008) Denitrification, nitrate turnover, and aerobic respiration by benthic foraminiferans in the oxygen minimum zone off Chile. J Exp Mar Biol Ecol 359:85–91

Iken K, Brey T, Wand U, Voigt J, Junghans P (2001) Food web structure of the benthic community at the Porcupine abyssal plain (NE Atlantic): a stable isotope analysis. Prog Oceanogr 50:383–405

Katz ME, Cramer BS, Franzese A, Honisch B, Miller KG, Rosenthal Y, Wright JD (2010) Traditional and emerging geochemical proxies in foraminifera. J Foram Res 40:165–192

Kennedy P, Kennedy H, Papadimitriou S (2005) The effect of acidification on the determination of organic carbon, total nitrogen and their stable isotopic composition in algae and marine sediment. Rapid Commun Mass Spectrom 19:1063–1068

Kessler JD, Reeburgh WS, Tyler SC (2006) Controls on methane concentration and stable isotope (delta H-2-CH4 and delta C-13-CH4) distributions in the water columns of the Black Sea and Cariaco Basin. Global Biogeochemical Cycles 20, GB4004: doi:10.1029/2005GB002571

King K (1977) Amino acid survey of recent calcareous and siliceous deep-sea microfossils. Micropaleontology 23:180–193

King KJ, Hare PE (1972) Amino acid composition of planctonic foraminifera: a paleobiochemical approach to evolution. Science 175:1461–1463

Laureillard J, Mejanelle L, Sibuet M (2004) Use of lipids to study the trophic ecology of deep-sea xenophyophores. Mar Ecol Progr Ser 270:129–140

Lavik G, Stührmann T, Brüchert V, Van der Plas A, Mohrholz V, Lam P, Mußmann M, Fuchs BM, Amann R, Lass U, Kuypers MMM (2009) Detoxification of sulphidic African shelf waters by blooming chemolithotrophs. Nature 457:581–585

Lecroq B, Gooday AJ, Tsuchiya M, Pawlowski J (2009) A new genus of xenophyophores (Foraminifera) from Japan Trench: morphological description, molecular phylogeny and elemental analysis. Zool J Linn Soc 156:455–464

Leiter C (2008) Benthos-Foraminiferen in Extremhabitaten: auswertung von METEOR-Expeditionen vor Namibia. PhD, Faculty of Geosciences, LMU Munich, Germany

Leiter C, Altenbach AV (2010) Benthic Foraminifera from the diatomaceous mud belt off Namibia: characteristic species for severe anoxia. Palaeontologica Electronica 13(2):11A, 19 p

Loncaric N, van Iperen J, Kroon D, Brummer G-JA (2007) Seasonal export and sediment preservation of diatomaceous, foraminiferal and organic matter mass fluxes in a trophic gradient across the SE Atlantic. Prog Oceanogr 73:27–59

Lutze GF, Altenbach AV (1991) Technik und Signifikanz der Lebendfärbung benthischer Foraminiferen mit Bengal-Rosa. Geol Jahrbuch Reihe A128:251–265

Mackensen A, Wollenburg J, Licari L (2006) Low delta C-13 in tests of live epibenthic and endobenthic foraminifera at a site of active methane seepage. Paleoceanography 21: PA2022. doi: 10.1029/2005PA001196

McCorkle D, Emerson SR (1988) The relationship between pore water carbon isotopic composition and bottom water oxygen concentration. Geochim Cosmochim Acta 52:1169–1178

McCorkle DC, Keigwin LD, Corliss BH, Emerson SR (1990) The influence of microhabitats on the carbon isotopic composition of deep-sea benthic foraminifera. Paleoceonography 5:161–185

McCorkle DC, Corliss BH, Farnham CA (1997) Vertical distributions and stable isotopic compositions of live (stained) benthic foraminifera from the North Carolina and California continental margin. Deep-Sea Res I 44:983–1024

McCutchan JH, Lewis WM, Kendall C, McGrath CC (2003) Variation in trophic shift for stable isotope ratios of carbon, nitrogen, and sulfur. Oikos 102:378–390

Mill AC, Sweeting CJ, Barnes C, Al-Habsi SH, MacNeil A (2008) Masspectrometer bias in stable isotope ecology. Limnol Oceanogr Methods 6:34–39

Mohrholz V, Bartholomae CH, van der Plas AK, Lass HU (2008) The seasonal variability of the northern Benguela undercurrent and its relation to the oxygen budget on the shelf. Cont Shelf Res 28:424–441

Moodley L, Middelburg JJ, Boschker HTS, Duineveld GCA, Pel R, Herman PMJ, Heip CHR (2002) Bacteria and Foraminifera: key players in a short-term deep-sea response to phytodetritus. Mar Ecol Prog Ser 236:23–29

Needoba JA, Sigman DM, Harrison PJ (2004) The mechanism of isotope fractionation during algal nitrate assimilation as illustrated by the 15N/14N of intracellular nitrate. J Phycology 40:517–522

Nomaki H, Ogawa N, Ohkouchi N, Suga H, Toyofuku T, Shimanaga M, Nakatsuka T, Kitazato H (2008) Benthic foraminifera as trophic links between phytodetritus and benthic metazoans: carbon and nitrogen isotopic evidence. Mar Ecol Prog Ser 357:153–164

Panieri G (2006) Foraminiferal response to an active methane seep environment: a case study from the Adriatic Sea. Mar Micropaleontol 61:116–130

Peterson BJ, Fry B (1987) Stable isotopes in ecosystem studies. Ann Rev Ecol Syst 18:293–320

Piña-Ochoa E, Høgslund S, Geslin E, Cedhagen T, Revsbech NP, Nielsen LP, Schweizer M, Jorissen F, Rysgaard S, Risgaard-Petersen N (2010) Widespread occurrence of nitrate storage and denitrification among Foraminifera and Gromiida. Proc Natl Acad Sci U S A 107:1148–1153

Rathburn AE, Levin LA, Held Z, Lohmann KC (2000) Benthic foraminifera associated with cold methane seeps on the northern California margin: ecology and stable isotopic composition. Mar Micropaleontol 38:247–266

Rathburn AE, Perez EM, Martin JB, Day SA, Mahn C, Gieskes J, Ziebis W (2003) Relationships between the distribution and stable isotopic composition of living benthic foraminifera and cold methane seep biogeochemistry in Monterey Bay, California. Geochem Geophys Geosys 4:1–28

Richardson K, Cedhagen T (2001) Quantifying pelagic-benthic coupling in the North Sea: are we asking the right questions? Senckenb marit 31:215–224

Risgaard-Petersen N, Langezaal AM, Ingvardsen S, Schmid MC, Jetten MSM, Op den Camp HJM, Derksen JWM, Pina-Ochoa E, Eriksson SP, Nielsen LP, Revsbech NP, Cedhagen T, van der Zwaan GJ (2006) Evidence for complete denitrification in a benthic foraminifer. Nature 443:93–96

Ryan KP, Knoll G (1994) Time-resolved cryofixation methods for the study of dynamic cellular events by electron-microscopy – a review. Scanning Microsc 8:259–288

Schulz HN, Brinkhoff T, Ferdelman TG, Hernandez Marine M, Teske A, Jorgensen BB (1999) Dense populations of a ginat sulfur bacterium in Namibian shelf sediments. Science 284:493–495

Scranton MI, Crill P, Deangelis MA, Donaghay PL, Sieburth JM (1993) The Importance of episodic events in controlling the flux of methane from an anoxic basin. Glob Biogeochem Cycles 7:491–507

Serrano O, Serrano L, Mateo MA (2008) Effects of sample pre-treatment on the $\delta 13C$ and $\delta 18O$ values of living benthic foraminifera. Chem Geol 257:221–223

Shepherd AS, Rathburn AE, Perez EM (2007) Living foraminiferal assemblages from the Southern California margin: a comparison of the >150, 63–150, and >63 µm fractions. Mar Micropal 65:54–77

Smith KL, Kaufmann RS, Baldwin RJ, Carlucci AF (2001) Pelagic-benthic coupling in the abyssal eastern North Pacific: an 8-year time-series study of food supply and demand. Limnol Oceanogr 46:543–556

Spero HJ, Lea DW (1996) Experimental determination of stable isotope variability in *Globigerina bulloides*: implications for paleoceanographic reconstructions. Mar Micropaleontol 28:231–246

Struck U, Altenbach AV, Emeis K-C, Alheit J, Eichner C, Schneider R (2002) Changes of the upwelling rates of nitrate preserved in the d15N signature of sediments and fish scales from the diatomaceous mud belt of Namibia. Geobios 35:3–11

Suess E (1980) Particulate organic carbon flux in the oceans-surface productivity and oxygen utilisation. Nature 288:260–263

Sweetman AK, Witte U (2008) Macrofaunal response to phytodetritus in a bathyal Norwegian fjord. Deep-Sea Res I 55:1503–1514

Tendal OS (1979) Aspects of the biology of Komokiacea and Xenophyophoria. Sarsia 64:13–17

Torres ME, Mix AC, Kinports K, Haley B, Klinkhammer GP, McManus J, de Angelis MA (2003) Is methane venting at the seafloor recorded by delta C-13 of benthic foraminifera shells? Paleoceanography 18:1062

Treude T, Orphan V, Knittel K, Gieseke A, House CH, Boetius A (2007) Consumption of methane and CO_2 by methanotrophic microbial mats from gas seeps of the anoxic Black Sea. Appl Environ Microbiol 73:2271–2283

Tsuchiya M, Toyofuku T, Takishita K, Yamamoto H, Collen J, Kitazato H (2006) Molecular characterization of bacteria and kleptoplast within *Virgulinella fragilis*. In: Kitazato H, Bernhard JM (ed) FORAMS 2006, vol 29. Anuário do Instituto de Geociências, pp 471–472

Tyrrell T, Lucas MI (2002) Geochemical evidence of denitrification in the Benguela upwelling system. Cont Shelf Res 22:2497–2511

Uhle ME, Macko SA, Spero HJ, Engel MH, Lea DW (1997) Sources of carbon and nitrogen in modern planktonic foraminifera: the role of algal symbionts as determined by bulk and compound specific stable isotopic analyses. Org Geochem 27:103–113

Uhle ME, Macko SA, Spero HJ, Lea DW, Ruddiman WF, Engel MH (1999) The fate of nitrogen in the *Orbulina universa* foraminifera-symbiont. Limnol Oceanogr 44:1968–1977

Umezawa Y, Miyajima T, Koike I (2008) Stable nitrogen isotope composition in sedimentary organic matter as a potential proxy of nitrogen sources for primary producers at a fringing coral reef. J Oceanogr 64:899–909

Valentine DL (2002) Biogeochemistry and microbial ecology of methane oxidation in anoxic environments: a review. Antonie Leeuwenhoek 81:271–282

van der Plas AK, Monteiro PMS, Pascall A (2007) Cross-shelf biogeochemical characteristics of sediments in the central Benguela and their relationship to overlying water column hypoxia. Afr J Mar Sci 29:37–47

von Rad U, Rösch H, Berner U, Geyh M, Marchig V, Schulz H (1996) Authigenic carbonates derived from oxidized methane vented from the Makran accretionary prism off Pakistan. Mar Geol 136:55–77

von Rad U, Berner U, Delisle G, Doose-Rolinski H, Fechner N, Linke P, Lücke A, Roeser HA, Schmaljohann R, Wiedicke M (2000) Gas and fluid venting at the Makran accretionary wedge off Pakistan. Geo-Mar Lett 20:10–19

Wakeham SG, Hopmans EC, Schouten S, Damste JSS (2004) Archaeal lipids and anaerobic oxidation of methane in euxinic water columns: a comparative study of the Black Sea and Cariaco Basin. Chem Geol 205:427–442

Weeks SJ, Currie B, Bakun A (2002) Massive emissions of toxic gas in the Atlantic. Nature 415:493–494

Weeks SJ, Currie B, Bakun A, Peard KR (2004) Hydrogen sulphide eruptions in the Atlantic Ocean off southern Africa: implications of a new view based on SeaWiFS satellite imagery. Deep-Sea Res I 51:153–172

Wefer G, Heinze P-M, Berger WH (1994) Clues to ancient methane release. Nature 369:282

Whiticar MJ, Faber E, Schoell M (1986) Biogenic methane formation in marine and freshwater environments: CO_2 reduction vs. acetate fermentation – isotope evidence. Geochim Cosmochim Acta 50:693–709

Biodata of **Nicolaas Glock**, **Joachim Schönfeld**, and **Jürgen Mallon**, authors of "*The Functionality of Pores in Benthic Foraminifera in View of Bottom Water Oxygenation: A Review.*"

Dipl. Chem. Nicolaas Glock is currently working on his Ph.D. thesis at the IFM-Geomar Leibnitz Institute for Marine Research. He obtained his diploma in chemistry at the Philipps University in Marburg in 2008. The topic of his Ph.D. thesis is the development of a set of geochemical and micropaleontological proxies for the reconstruction of past redox conditions in the Peruvian oxygen minimum zone.

E-mail: **nglock@ifm-geomar.de**

Dr. Joachim Schönfeld is currently staff scientist from the Research Unit Paleoceanography of Leibniz Institute for Marine Sciences IFM-GEOMAR, Kiel, Germany. He obtained his Dr. rer nat. from the University of Kiel in 1989 and continued research at the Federal Bureau of Geosciences and Natural Resources, Hannover; GEOMAR Research Center for Marine Geosciences, Kiel; at Department 5 – Geowissenschaften, Bremen University; and with Tethys Geoconsulting Ltd., Kiel, Germany. Doctor Schönfeld's scientific interests are in benthic foraminifera, their taxonomy, ecology, Mesozoic and Cenozoic paleoceanography, stratigraphy and climate history.

E-mail: **jschoenfeld@ifm-geomar.de**

Nicolaas Glock   **Joachim Schönfeld**

Jürgen Mallon is currently a PhD student in the Faculty of Paleoceanography at the IFM-GEOMAR at the University of Kiel, Germany. He studied Geology and Paleontology at the University of Hamburg, Germany, where he obtained his diploma. Now he works on benthic foraminifera of the Peruvian oxygen minimum zone regarding the collaboration project SFB754 "Climate–Biogeochemistry Interactions in the Tropical Ocean," funded by the DFG. The scientific interests are the ecology and distribution of low-oxygen benthic foraminifera and their applications in paleoceanographical and biogeochemical research.

E-mail: **jmallon@ifm-geomar.de**

THE FUNCTIONALITY OF PORES IN BENTHIC FORAMINIFERA IN VIEW OF BOTTOM WATER OXYGENATION: A REVIEW

NICOLAAS GLOCK[1], JOACHIM SCHÖNFELD[2], AND JÜRGEN MALLON[2]

[1]Marine Geosystems, IFM-GEOMAR, Wischhofstr. 1-3, 24148 Kiel, Germany
[2]Paleoceanography, IFM-GEOMAR, Wischhofstr. 1-3, 24148 Kiel, Germany

1. Introduction

Pores are developed in rotaliid calcareous foraminifera and are important morphological features. Their shape, size, and density are diagnostic for discerning between several species (Lutze 1986). But only few publications are discussing the function or the origin of these pores and whether they are or are not important for the survival of benthic foraminifera. With advances in the field of electron microscopy in the early 1950s, researchers started to describe the microstructure of the pores and discovered that these pores are often covered by some sieve-like microporous organic plates (Jahn 1953; Arnold 1954a, b). Until the late 1970s other workers (Le Calvez 1947; Angell 1967; Sliter 1974; Berthold 1976; Leutenegger 1977) observed that the pores in many benthic foraminifera are covered by one or more organic layers, but not all of them showed microperforations. Several terms have been given to these structures: "sieve plates," "pore diaphragms," "dark discs," "pore plugs," and "pore plates." Only very few experiments have been done to analyze the function of pores and their permeability to dissolved substances into the cytoplasm. It was demonstrated that *Patellina corrugata* is able to take up neutral red dye through the pores (Berthold 1976) and that *Amphistigina lobifera* takes up CO_2 in a similar way (Leutenegger and Hansen 1979). Additionally, some low-oxygen-tolerant species show that their mitochondria, i.e., cell organelles involved in respiration, were more abundant near the pores than in other species from well-oxygenated waters. This covariance implies an evolutionary linkage between pores and mitochondria (Leutenegger and Hansen 1979; Bernhard et al. 2010). These observations led to the most widespread interpretation in the literature that the pores in benthic foraminifera promote the uptake of oxygen and the release of metabolic CO_2, the uptake of CO_2 for symbiont-bearing foraminifera, and the osmoregulation and the intake and excretion of dissolved substances in general. On the other hand it has been suggested that the pores of *Rosalina floridana* are purely an ornamental feature because of the lack of micropores in the pore plates and the thick inner organic lining between cytoplasm and test walls which seals the pores (Angell 1967).

Another term of studies was the variability of morphological features like pore size and pore density among several benthic species. In the 1960s it has been found that *Bolivina spissa* from the Californian borderlands show a strong variation in the pore-free area of their test surface among different water depths (Lutze 1962). Homeomorphs of *Bolivina spissa* from different time periods at the Santa Barbara Basin, California, show a strong variability in pore density and shape (Harmann 1964). A connection between pore size and density and the oxygen concentration of their habitats has been documented for *Hanzawaia nitidula* from the oxygen minimum-zone (OMZ) in the Gulf of Tehuantepec (Perez-Cruz and Machain-Castillo 1990) and for laboratory cultures of *Ammonia beccarii* (Moodley and Hess 1992). Furthermore, species with high test-porosity in general may serve as indicator for oxygen-depleted environmental conditions (Sen-Gupta and Machain-Castillo 1993; Kaiho 1994). The same variation in the pore density of *Bolivina spissa* among different water depths that was described by Lutze (1962) was found again for *Bolivina spissa* from the oxygen minimum zone off Peru. But it appears that this variability in pore density is closely related to the nitrate-concentration in the bottom-water (Glock et al. 2011). So it is speculated whether the pores in *Bolivina spissa* are involved in the mechanism of nitrate respiration. The ability to store nitrate inside the cells and to switch to nitrate respiration in times when no or too less oxygen is available has been recently documented for several benthic foraminiferal species (Risgaard-Petersen et al. 2006; Høgslund et al. 2008; Glud et al. 2009; Piña-Ochoa et al. 2010).

The fact that a rod-shaped microbial ectobiont of unknown identity and physiology was found to inhabit the outer part of the pore void in *Bolivina pacifica* while mitochondria are clustered at the inner pore face (Bernhard et al. 2010) gives reason to speculate if and in how far these bacteria are involved into foraminiferal denitrification whether they are symbionts or parasites. The pores in amphisteginids and nummulites are quite obvious adapted for hosting algal symbionts (Hansen and Burchardt 1977; Lee and Anderson 1991): The inner surface of the test around the pores is excavated into cuplike pore rims. The symbiotic diatoms are concentrated along the surfaces of the cytoplasm in cytoplasmic bulges which fit into the pore rims. Besides all of these evidences, the function of pores still remains conjectural, and it is unclear in how far the pore function varies among the different benthic foraminiferal species.

2. Materials and Methods

Two different data sources were used for this review. First, for the major part, relevant publications related to the function of pores were compiled (Table 1). A few studies concerning the pores in planktonic foraminifera are also listed although they will not be discussed in the progress of this manuscript. Second, sampling material for the pictures in Fig. 1 was recovered during Meteor Cruise M77/1.

Table 1. Publications used as data source for this review.

Source	Results of the study
Doyle (1935)	Light microscopic observations of *Iridia diaphana* show that this species is able to move mitochondria through its pseudopodia.
Arnold (1954b)	Sieve-like plates are covering the pores of several benthic foraminiferal species. These so-called sieve plates or pore plugs contain a large number of micropores in a diameter range of 0.1–0.3 µm. These micropores might restrict the flow of smaller cytoplasmic elements like mitochondria into pore pseudopodia.
Lutze (1962)	*Bolivina spissa* from the Californian borderlands show a strong variation in the pore-free area of their test-surface among different water depths. For the explanation of this phenomenon, a temperature dependence of different chemical processes is proposed.
Harman (1964)	*Bolivinidae* from the Santa Barbara Basin, California, show morphological variations in response to environmental factors like oxygenation. Additionally, there are variations in recent and ancient homogenous sediments. Homeomorphs of *B. spissa* from different time periods show strong differences in pore density and shape.
Angell (1967)	The pores of *Rosalina floridana* are filled with organic "pore processes" which are anchored to the inner organic lining. These structures lack micropores and it is speculated that the pores in *R. floridana* are eliminated on purely morphological grounds.
Bé (1968)	Shell porosities of 22 planktonic foraminiferal species are relatively uniform for those co-occurring in same latitudinal belts. Because of this co-variation of porosity and temperature shell porosity in planktonic foraminifera is proposed as climatic index.
Sliter (1970)	Laboratory cultures of *Bolivina doniezi* show variations in pore morphology and pore density in the clone culture compared to the natural populations.
Hansen (1972)	Freeze-dried specimens of living *Amphistigina* show, in addition to the apertural pseudopodia, other pseudopodia closely connected with the pores.
Frerichs et al. (1972)	Pore density in *Globigerinoides sacculifer, Globorotalia tumida,* and *Neogloboquadrina dutertrei* decreases directly with distance from the equator, but *Globigerinella siphonifera* and *Globorotalia tumida* show no such relationship. The test porosity, however, decreases in all five species with distance from the equator. It is speculated that the test porosity is related to the water density, which in turn, is related to temperature.
Hottinger and Dreher (1974)	Pores in tests of *Operculina ammonoides* and *Heterostegina depressa* are not covered by pore plates. The inner organic lining is thickened at pore rims and thins out over the pore holes while the plasma membrane is differentiated by coarse granules below the pore holes. These observations and the position of the symbionts in the chamber plasma point to a physiological relationship between symbionts and pores.
Sliter (1974)	In contrast to many other foraminiferal taxa, *Bolivinitidae* and *Caucasinidae* appear to construct their tests in a monolamellar concept. The studied *Bolivinitidae* show double pore membranes between consecutive calcitic lamellae.
Berthold (1976)	Experiments on *Patellina corrugata* show that neutral red from ambient water is actively pumped into the cytoplasm through test pores even when the aperture is closed. It is speculated that the pore function is related to osmoregulation, gas exchange, or the intake and excretion of dissolved substances.

(continued)

Table 1. (continued)

Source	Results of the study
Hansen and Buchardt (1977)	The inner surface of the test around the pores in amphisteginids and nummulites is excavated into cuplike pore rims. The symbiotic diatoms are concentrated along the surfaces of the cytoplasm in cytoplasmic bulges which fit into the pore rims.
Leutenegger and Hansen (1979)	Mitochondria are clustered behind the pores of foraminiferal species from low-oxygen habitats. In several foraminiferal species from more oxygenated habitats, mitochondria are more uniformly distributed throughout the cytoplasm. Additionally, the inner organic lining is disrupted behind the pores of several species from oxygen-depleted habitats. It appears that the pores are related to gas exchange. This includes an uptake of O_2 and an elimination of CO_2, as well as an uptake of CO_2 for photosynthetic symbiont-bearing foraminifera like *Amphistigina lobifera* during day time.
Bé et al. (1980)	The mechanism of the formation of pores and pore plates in planktonic foraminifera is described. Pores are formed due to resorption of already precipitated material. There might be differences in pore formation and function between spinose and non-spinose species.
Bijma et al. (1990)	Laboratory cultures of *Globigerinoides sacculifer*, *Globigerinoides ruber*, *Globigerinoides siphonifera* and *Orbulina universa* show that changes in shell porosity are correlated with changes in salinity and temperature. The highest porosities are attained at higher temperatures and lower salinities.
Perez-Cruz and Machain-Castillo (1990)	*Hanzawaia nitidula* from the oxygen minimum zone (OMZ) in the Gulf of Tehuantepec shows more and larger pores than specimens from oxygenated waters.
Moodley and Hess (1992)	Laboratory cultures of *Ammonia beccarii* show an increase in pore size under low-oxygen conditions.
Sen Gupta and Machain-Castillo (1993), Kaiho (1994)	Benthic foraminiferal species with high test-porosity are postulated as an indicator for oxygen-depleted environmental conditions.
Risgaard-Petersen (2006)	First evidences that foraminiferal species from oxygen-depleted habitats switch to nitrate respiration in times when no oxygen is available are discovered.
Høgslund (2008)	Denitrification rates for benthic foraminifera from the Chilean OMZ are measured.
Allen et al. (2008)	Laboratory cultures of the planktic foraminifer *Orbulina universa* show a relationship of pore density and pore size to pH but no dependence of temperature.
Glud et al. (2009)	The contribution of foraminiferal denitrification to the nitrogen cycling at Sagami Bay, Japan, is quantified. The production of N_2 was attributed to foraminiferal denitrification in a total amount of 4%. Additionally, the nitrate storage in foraminiferal cells was measured for several species. It represented 80% of the total benthic nitrate pool.
Piña-Ochoa et al. (2010)	The nitrate storage among many benthic foraminifera from the Peruvian OMZ was measured.
Bernhard et al. (2010)	The outer part of the pore void of *Bolivina pacifica* in this study is inhabited by a rod-shaped microbial ectobiont of unknown identity and physiology. Again a clustering of mitochondria behind the pores is observed.

(continued)

Table 1. (continued)

Source	Results of the study
Glock et al. (2011)	The pore density in tests of *Bolivina spissa* from the Peruvian OMZ shows strong locational variations and a relationship to several environmental factors like oxygen or nitrate concentrations in the bottom waters. Because of the strong relationship to the bottom water nitrate concentrations gives a reason to speculate if the pores are related to nitrate respiration, either for the intracellular nitrate uptake or to act as "valve" for the release of waste products like N_2.

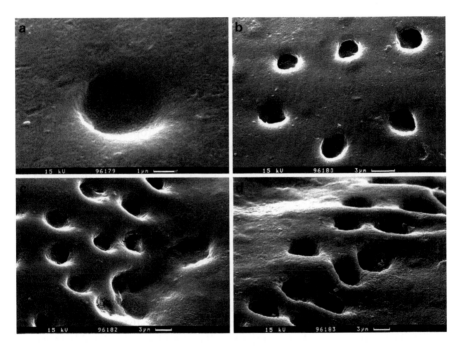

Figure 1. SEM pictures with close-ups of the pores of one specimen of *Bolivina spissa* (oversight shown in Fig. 2c). (**a**) A pore of the ultimate chamber with a well-preserved pore plate. (**b**) Several pores in the ultimate chamber covered by several layers of membranes (pore plates). Slits in some pore plates are probably deteriorations caused by drying process of samples. (**c**) The pore plates in this specimen are preserved until the middle chamber. (**d**) Pores near the proloculus are open to the surface.

A detailed description of the sampling locations and sampling procedure could be found elsewhere (Glock et al. 2011).

All specimens were mounted on aluminum stubs, sputter-coated with gold, and photographed with a CamScan CS-44 scanning electron microscope (SEM) at the Christian Albrecht University in Kiel.

3. The Pore Plates

The pores of many benthic foraminifera are sealed by one or more organic layers (Le Calvez 1947; Jahn 1953; Arnold 1954a, b; Angell 1967; Sliter 1974; Berthold 1976; Leutenegger 1977), while some species like *Operculina ammonoides* and *Heterostegina depressa* lack pore plates (Hottinger and Dreher 1974). In some species, these pore plates are additionally perforated by micropores with a diameter in a range of 0.05–0.3 µm, depending on the species. These micropores have been described in some unknown nonionid and camerinid species (Jahn 1953), in *Discorinopsis aguayoi* (Arnold 1954a), and in *Patellina corrugata* (Berthold 1976). The pores of *Rosalina floridana* are filled with organic "pore processes" anchored to an inner organic lining and are covered with an organic membrane (Angell 1967). All of these structures in *Rosalina* lack micropores. Specimens of *Bolivina* and *Coryphostoma* construct their tests in a monolamelar concept and show double pore-membranes between consecutive calcitic lamellae (Sliter 1974). The surface membrane seems to cover the ultimate chamber completely, while a progressive perforation in the pores of successively older chambers could be observed. This results in that the pores of the oldest chambers are open to the surface. If the surface membrane was intact, micropores could be observed only occasionally. These micropores became larger and more common in the penultimate and towards the older chambers. Similar structures exist in *Bolivina spissa*. Most of the pores in the ultimate chamber are covered with several layers of pore plates. The slits in some of the pore plates are probably deteriorations caused by the drying process of the samples or by the electron beam of the REM (Fig. 1a, b). In some specimens, these pore plates are preserved only in the ultimate chamber, while others show well-preserved pore plates among several other chambers (Fig. 1c). The pores in the earliest chambers near the proloculus are open to the surface (Fig. 1d).

It was speculated whether the micropores in some pore plates could serve as outlet for pore-pseudopodia and selectively control the flow of cytoplasmic elements into and back from the pseudopodia (Arnold 1954b). Because of the small size of the micropores, only very minute cell organelles would be able to pass the pore plates. At least mitochondria are able to move through the cytoplasm and flow into pseudopodia of *Iridia diaphana* (Doyle 1935). Indeed, some thin, threadlike structures emerging from pore plates were observed in freeze-dried specimens of living *Amphistigina* which have been interpreted as pore pseudopodia (Hansen 1972). This is not undisputed since it was discussed that the threadlike extrusions have more similarity to the hyphae of fungi than to granuloreticulose pseudopods (Berthold 1976). Observations with the light-microscope on *Patellina* first led to the conclusion that pores could serve as an outlet for pseudopodia, as does the aperture (Myers 1935). In contrast, later investigations found that it is not possible to show a correlation between pores and pseudopods in *Patellina* (Berthold 1971). Because most of the pores in the tests of the Bolivinitidae and Caucasinidae where sealed completely by a complex of imperforated pore plates, Sliter (1974) came to the conclusion that a free exchange of cytoplasm to the test surface is precluded through most of the pores. In summary, it could not be proven to date if benthic species indeed could move pseudopodia through some pores or not. But at least for some taxa (*Rosalina, Patellina,* Bolivinitidae, and Caucasinidae), due to the lack or the very minute size of micropores, such pseudopodial movements are not very probable (Angell 1967; Sliter 1974; Berthold 1976).

4. Permeability of Pores and the Previous Understanding of Pore Function

Only few papers describe experiments to test the permeability of pores of benthic foraminifera to dissolved substances in the ambient waters. *Patellina corrugata* could actively pump neutral red dye from surrounding water into the cell through the pores while the aperture was closed (Berthold 1976).

These results inferred that "the function of pores probably lies in the field of osmoregulation, gas exchange, or the intake and excretion of dissolved organic substances." The pores probably had a special importance during the reproductive phase of a foraminifer because the protoplasm is isolated from the medium and the pseudopodia cannot be extruded through the aperture. Another study on several benthic species from oxygen-depleted habitats showed that mitochondria are clustered behind the pores and the inner organic lining is interrupted behind the pores (Leutenegger and Hansen 1979). Additionally, the same study showed that the symbiont-bearing *Amphistigina lobifera* takes up C^{14} labeled CO_2 through the pores while the aperture is closed. Hottinger and Dreher (1974) showed that *Operculina ammonoides* and *Heterostegina depressa* lack pore plates, and the inner organic lining thins out over the pore holes while the plasma membrane is differentiated by coarse granules below the pore holes.

These observations and the position of the symbionts in the chamber plasma pointed to a physiological relationship between symbionts and pores. So it seemed obvious that the pores are related to gas exchange. But at least one publication mentions *Rosalina floridana*, where the pores might be built purely as ornamentation or just to provide organic continuity to the test exterior (Angell 1967). These conclusions were drawn because the pores of *R. floridana* are filled by organic "pore processes" anchored to the inner organic lining and these structures lack micropores. However, no experiments on the permeability of pores in *Rosalina* were performed.

The most widespread opinion today is that the pores in benthic foraminifera are related to the uptake of O_2 and the release of metabolic CO_2. These interpretations were based on observations that foraminiferal species from low-oxygen habitats show a high test porosity. These species were, in turn, used as an indicator for oxygen depleted environmental conditions (Sen-Gupta and Machain-Castillo 1993; Kaiho 1994). Furthermore, some species show a response in their pore size and pore density to variations in oxygen supply. For instance, *Hanzawaia nitidula* from the oxygen minimum zone (OMZ) in the Gulf of Tehuantepec show more and larger pores than specimens from oxygenated waters (Perez-Cruz and Machain-Castillo 1990). Laboratory cultures of *Ammonia beccarii* show an increase in pore-size under low-oxygen conditions (Moodley and Hess 1992). These experimental results were corroborated by field observations from Flensburg and Kiel Fjords. *Ammonia beccarii* showed large pores in Flensburg Fjord, where seasonal anoxia occurred (Nikulina and Dullo 2009; Polovodova et al. 2009), and small pores in Kiel Fjord which experiences only a moderate oxygen dropdown during summer (Nikulina et al. 2008). These observations denote the potential of using the pore size and pore density as a proxy for recent and past oxygen variations. A recent study shows a variability in the pore density of *Bolivina spissa* from the Peruvian OMZ, which might be related to oxygen supply (Glock et al. 2011). But this variability of pore density in *B. spissa* might be more related to the variations in nitrate availability. Three specimens of *B. spissa* from different oxygenated locations are shown in Fig. 2. At least some foraminiferal species are able to move their mitochondria into their pseudopodia (Doyle 1935). Because the pseudopodia could extend to ten times the test diameter of a foraminifer (Travis and Bowser 1991), it is possible that foraminifera which inhabit an environment

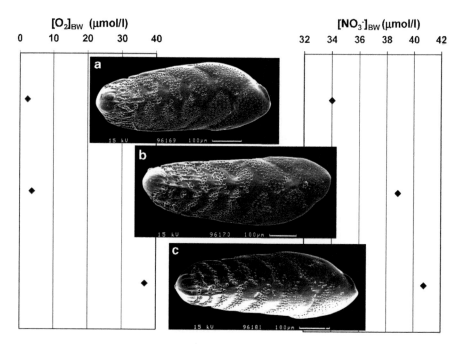

Figure 2. SEM pictures of three specimens of *Bolivina spissa* from different locations. The diagram on the right shows the different bottom water nitrate concentrations, and the diagram on the left, the bottom water oxygen concentrations from the different locations. A quantification of these relationships is presented by Glock et al. (2011) (A M77/1-445/MUC-21 (465 m) B M77/1-487/MUC-38 (579 m) C M77/1-445/MUC-15 (928-m water depth)).

with a steep oxygen gradient could use mitochondrial activity in their extended pseudopodia to maintain oxygen supply even when their tests are located in anoxic sediments (Bernhard and Sen Gupta 2003). In this case, an uptake of oxygen through test pores might not be very convincing. But it might be that different species follow different mechanisms of oxygen uptake.

5. Evidence of Pore Involvement in Nitrate Respiration Pathways

Recently, it has been shown that some benthic species from oxygen-depleted habitats respire nitrate via denitrification (Risgaard-Petersen et al. 2006). This fact, in combination with new results which show a strong relationship of the pore density in *Bolivina spissa* from the Peruvian continental margin to the bottom-water nitrate concentrations (Glock et al. 2011), gives a reason to speculate if and how far the pores might be related to the nitrate-respiration pathways in some benthic foraminifera.

If the pores of *B. spissa* are indeed related to the intracellular nitrate uptake, an increase of pore density would optimize their availabilities for nitrate uptake when nitrate is depleted. This would result in an advantage in the competition of nitrate uptake against other foraminiferal and prokaryotic species. These results might open a first window to carry modern knowledge on foraminiferal denitrification into a fossil record and earth history.

A study on the nitrogen cycle at Sagami Boy, Japan, which showed that *Bolivina spissa* have significant intracellular nitrate enrichment (Glud et al. 2009), supports this assumption. At Sagami Bay intracellular nitrate storage in foraminifera represented about 80% of the total benthic nitrate pool. The diagram on the right in Fig. 2 shows the bottom water nitrate concentrations in the different habitats of the three specimens of *B. spissa*.

The denitrification process in benthic foraminifera has not been attributed to a specific cell organelle yet (Høgslund et al. 2008). If it is assumed that the pores are directly related to intracellular nitrate uptake, the fact that mitochondria tend to cluster behind the pore plugs, which indeed has been observed from a *Bolivina* at low-oxygen conditions, might imply that mitochondria may also be involved in the mechanism of foraminiferal denitrification. Indeed, earlier studies showed that mitochondria are involved in nitrate respiration of the primitive eukaryote *Loxodes* (Finlay et al. 1983). In particular, the number of mitochondria became significantly enhanced when *Loxodes* switched from oxygen to nitrate respiration. The recent observation of microbial ectobionts of unknown identity and physiology inhabiting the pore void of *Bolivina pacifica* (Bernhard et al. 2010) provokes speculations if such ectobionts exist in the pore void of *Bolivina spissa* as well and are indeed denitrifiers. Another possibility would be that these ectobionts act more as parasites. Some Bolivinidae seem to produce N_2O as a product of denitrification instead of N_2 (Piña-Ochoa et al. 2010). Hence, it might be that these bacteria cluster in the pore void of some foraminifera for the uptake of N_2O which is released as a waste product through the pores. In a reaction catalyzed by the protein nitrous oxide reductase, N_2O is reduced to N_2 (Zumft 1997; Riester, Zumft and Kroneck 1989). The reaction:

$$N_2O + 2e^- + 2H^+ \rightarrow N_2 + H_2O \tag{1}$$

is highly exergonic ($\Delta G°' = 340$ kJ/mol) and thus a good energy source although N_2O is very inert at room temperature, and thus reaction (1) needs an efficient catalyst to occur (Haltia et al. 2003). Some rod-shaped bacteria like *Escherichia coli* or *Pseudomonas stutzeri* are known to have the ability to reduce N_2O to N_2 (Kaldorf et al. 1993; Lalucat et al. 2006).

A study of Lutze (1962) on *Bolivina spissa* from the Californian borderland shows that the pore-free area increases with water depth. This was assumed to reflect decreasing water temperatures. But later it was shown by comparison to nitrate data from the same study area (Boyer et al. 2009) that the variability in the pore-free area also reflects most probably the nitrate distribution in the water column (Glock et al. 2011). Nitrate is depleted at shallower water depth due to enhanced primary production on the ocean's surface. So the habitats of the analyzed specimens from shallower water depths most probably were more nitrate-depleted than those from the deeper waters.

6. Conclusions

Although the pores in tests of benthic foraminifera are an important morphological feature which even is used to distinguish several species, the understanding of the pore function is very fragmentary. This might be due to the fact that the pore function differs between the species. Symbiont-bearing foraminifera like *Amphistigina lobifera* seem to take up CO_2 through the pores during daytime for photosynthesis, while some species from oxygen-depleted habitats probably take up oxygen through the pores (Leutenegger and Hansen 1979). But also the uptake of dissolved organic substances is possible, as shown for *Patellina corrugata* (Berthold 1976). Recent findings even hint that at least one species, *Bolivina spissa*, takes up nitrate for nitrate respiration through the pores or uses these to release denitrification products like N_2O or N_2 and that they seem to adapt their pore densities to survive in extreme habitats (Glock et al. 2011). But for a lot of species, the pore function still remains conjectural, and it is even speculated that some species just build the pores on purely morphological grounds or just to provide organic continuity to the test exterior (Angell 1967). A better understanding of the pore function would help to understand which metabolic adaptations help foraminifera to survive extreme (even anoxic) habitats, because species from oxygen-depleted environments generally show a high porosity. Furthermore, it should be tested in how far pore densities in some benthic species might be used as proxy to reconstruct environmental factors like the oxygen or nitrate concentrations.

7. Faunal Reference List

Ammonia beccarii (Linné) = *Nautilus beccarii* Linnaeus 1758, p. 710, pl. 1, fig. 1.
Amphistegina lobifera (Larsen), 1976, p.11, pl. 3, figs. 1–5.
Bolivina doniezi (Cushman & Wickenden), 1929, p. 9, pl. 4, fig. 3a–b.
Bolivina pacifica (Cushman & McCulloch) = *Bolivina acerosa* Cushman var. *pacifica* Cushman and McCulloch, 1942, p. 185, pl. 21, figs. 2, 3.
Bolivina spissa (Cushman) = *Bolivina subadvena* Cushman var. *spissa*, Cushman 1926, p. 45, pl. 6, fig. 8a–b.
Discorinopsis aguayoi (Bermúdez) 1935, p. 204, pl. 15, figs. 10–14.
Globigerinella siphonifera (d'Orbigny) = *Globigerina siphonifera* d'Orbigny 1839, pl. 4, figs. 15–18.
Globigerinoides ruber (d'Orbigny), 1839, p. 82, pl. 4, figs. 12–14.
Globigerinoides sacculifer (Brady) 1877 [type fig. not given], Brady 1884: p. 604, pl. 80, figs. 11–17.
Globorotalia tumida (Brady) = *Pulvinulina menardii* (d'Orbigny) var. *tumida* Brady 1877 [type fig. not given], Brady 1884, pl. 103, figs. 4–6.
Hanzawaia nitidula (Bandy) = *Cibicidina basiloba* Cushman var. *nitidula* Bandy 1953, p. 178, pl. 22, figs. 3a–c.
Heterostegina depressa (d'Orbigny), 1826, p. 305, pl. 17, figs. 5–7.
Iridia diaphana (Heron-Allen & Earland), 1914, p. 371, pl. 36.
Neogloboquadrina dutertrei (d'Orbigny) = *Globoquadrina dutertrei* d'Orbigny 1839, pl. 4, figs. 19–21.
Operculina ammonoides (Gronovius) = *Nautilus ammonoides* Gronovius 1781, p. 282, pl. 19, figs. 5, 6.
Orbulina universa (d'Orbigny), 1839, p. 3, pl. 1, fig. 1.
Patellina corrugata (Williamson), 1858, p. 46, pl. 3, figs. 86–89, 89a.
Rosalina floridana (Cushman) = *Discorbis floridana* Cushman 1922, p. 39, pl. 5, figs. 11, 12.

8. Acknowledgments

We thank Anton Eisenhauer for his helpful general support and the providing of the Ph.D. position of the first author. Volker Liebetrau supported this work with helpful general discussions and organization. Support in the operation of the SEM at the Christian Albrecht University in Kiel came by Ute Schuldt. The scientific party on R/V METEOR cruise M77 is acknowledged for their general support and advice in multicorer operation and sampling. The "Deutsche Forschungsgemeinschaft, (DFG)" provided funding through SFB 754 "Climate–Biogeochemistry Interactions in the Tropical Ocean."

9. References

Allen KA, Hoenisch B, James KM, Eggins SM, Spero HJ (2008) Effects of pH and temperature on calcification of the planktonic foraminifer *O. universa*: insights from culture experiments: Eos Transactions, AGU, Fall Meeting Supplement, 89, Abstract PP51C-1519

Angell RW (1967) The test structure of the foraminifer *Rosalina floridana*. J Protozool 14:299

Arnold ZM (1954a) *Discorinopsis aguayoi* (Bermudez) and *Discorinopsis vadescens* Cushman and Brönnimann: a study of variation in cultures of living foraminifera. Contr Cushman Found Foram Res 5:4–13

Arnold ZM (1954b) A note on foraminiferan sieve plates. Contr Cushman Found Foram Res 5:77

Bandy OL (1953) Ecology and paleoecology of some California foraminifera, Part 1 – the frequency distribution of recent foraminifera off California. J Paleontol 27(2):161–182, Tulsa, Oklahoma

Bé AWH (1968) Shell porosity of recent planktonic foraminifera as a climatic index. Science 161:881–884

Bé AWH, Hemleben C, Andersen OR, Spindler M (1980) Pore structures in planktonic foraminifera. J Foraminifer Res 10:117–128

Bermudez PJ (1935) Foraminiferos de la Costa Norte de Cuba. Memorias de la Sociedad Cubana de Historia Natural 9:129–224, La Habana, Cuba

Bernhard JM, Bowser SS, Goldstein S, Sen Gupta BK (2003) Foraminifera of oxygen-depleted environments. In: Sen Gupta BK (ed.) Modern foraminifera. Kluwer Academic, New York/Boston/Dordrecht/London/Moscow, pp 201–216

Bernhard JM, Bowser SS, Goldstein S (2010) An ectobiont-bearing foraminiferan, *Bolivina pacifica*, that inhabits microxic pore waters: cell-biological and paleoceanographic insights. Environ Microbiol 12:2107–2119

Berthold W-U (1971) Untersuchungen über die sexuelle Differenzierung der Foraminifere *Patellina corrugata* Williamson mit einem Beitrag zum Entwicklungsgang und Schalenbau. Archiv für Protistenkunde 113:147–184

Berthold W-U (1976) Ultrastructure and function of wall perforations in *Patellina corrugata* Williamson, Foraminiferida. J Foraminifer Res 6:22–29

Bijma J, Faber WW, Hemleben C (1990) Temperature and salinity limits for growth and survival of some planktonic foraminifers in laboratory cultures. J Foraminifer Res 20:95–116

Boyer TP, Antonov JI, Garcia HE, Johnson DR, Locarnini RA, Mishonov AV, Pitcher MT, Baranova OK, Smolyar IV (2009) In: Levitus S (ed.) World ocean database 2009, NOAA atlas NESDIS 60. U.S. Government Printing Office, Washington, DC, 190 pp, DVDs

Brady HB (1877) Supplementary note on the foraminifera of the Chalk of the New Britain Group. Geol Mag, London, new series 4(12):534–536

Brady HB (1884) Report on the foraminifera dredged by H.M.S. Challenger during the years 1873–1876 (Reports of the scientific results of the voyage H.M.S. Challenger, 1873–1876). Zoology 9:1–814

Cushman JA (1922) Shallow-water foraminifera of the Tortugas Region. Publications of the Carnegie Institution of Washington, no. 311, vol 17. Department of Marine Biology, Papers, Washington, DC, pp 1–75

Cushman JA (1926) Some pliocene bolivinas from California. Contrib Cushman Lab Foram Res 2:40–46

Cushman JA, McCulloch I (1942) Some Virgulininae in the collections of the Allan Hancock Foundation, vol 6, no 4. Southern California University, Publications, Allan Hancock Pacific Expedition, Los Angeles, pp 179–230

Cushman JA, Wickenden RTD (1929) Recent foraminifera from off Juan Fernandez Islands. In: Proceedings of the United States National Museum, vol 75, no 2780, Washington, DC

D'Orbigny A (1826) Tableau méthodique de la classe des Céphalopodes. Annales des Sciences Naturelles Paris 7:245–314

D'Orbigny A (1839) Foraminiferes. In: de la Sagra R (ed.) Histoire physique et naturelle de l' Ile de Cuba. Arthus Bertrand, Paris, pp 1–224

Doyle WL (1935) Distribution of mitochondria in the foraminiferan, *Iridia diaphana*. Science 81:387

Finlay BJ, Span ASW, Harman JMP (1983) Nitrate respiration in primitive eukaryotes. Nature 303:333–335

Frerichs WE, Heiman ME, Borgman LE, Bé AWH (1972) Latitudinal variations in planktonic foraminiferal test porosity: Part 1. Optical studies. J Foram Res 2:6–13

Glock N, Eisenhauer A, Milker Y, Liebetrau V, Schönfeld J, Mallon J, Sommer S, Hensen C (2011) Environmental influences on the pore-density in tests of Bolivina spissa. J Foraminifer Res 41:22–32

Glud RN, Thamdrup B, Stahl H, Wenzhoefer F, Glud A, Nomaki H, Oguri K, Revsbech NP, Kitazato H (2009) Nitrogen cycling in a deep ocean margin sediment (Sagami Bay, Japan). Limnol Oceanogr 54:723–734

Gronovius LT (1781) Zoophylacii Gronoviani. Leyden: Theodorus Haak et Soc 3:241–380

Haltia T, Brown K, Tegoni M, Cambillau C, Saraste M, Mattila K, Djinovic-Carugo K (2003) Crystal structure of nitrous oxide reductase from Paracoccus denitrificans at 1.6 Å resolution. Biochem J 369:77–88

Hansen HJ (1972) Pore pseudopodia and sieve plates of *Amphistigina*. Micropaleontology 18:223–230

Hansen HJ, Buchardt B (1977) Depth distribution of Amphistigina in the Gulf of Elat. Utrecht Micropaleontol Bull 1:225–239

Harmann RA (1964) Distribution of foraminifera in the Santa Barbara basin, California. Micropaleontology 10:81–96

Heron-Allen E, Earland A (1914) The foraminifera of the Kerimba Archipelago (Portuguese East Africa); Part 1: The Transactions of the Zoological Society, London, vol 20, pp 363–390

Høgslund S, Revsbech NP, Cedhagen T, Nielsen LP, Gallardo VA (2008) Denitrification, nitrate turn over, and aerobic respiration by benthic foraminiferans in the oxygen minimum zone off Chile. J Exp Mar Biol Ecol 359:85–91

Hottinger L, Dreher D (1974) Differentiation of Protoplasm im Nummulitidae (Foraminifera) from Elat, Red Sea. Mar Biol 25:41–61

Jahn B (1953) Elektronenmikroskopische Untersuchungen an Foraminiferenschalen. Zeitschrift für Wissenschaftliche Mikroskopie und Mikrotechnik 61:294–297

Kaiho K (1994) Benthic foraminiferal dissolved-oxygen index and dissolved-oxygen levels in the modern ocean. Geology 22:719–722

Kaldorf M, Linne von Berg K-H, Maier U, Bothe H (1993) The reduction of nitrous oxide by *Escherichia coli*. Arch Microbiol 160:432–439

Lalucat J, Bennasar A, Bosch R, García- Valdéz E, Palleroni NJ (2006) Biology of *Pseudomonas stutzeri*. Microbiol Mol Biol Rev 70:510–547

Larsen AR (1976) Studies of recent Amphistegina, taxonomy and some ecological aspects. Israel J Earth-Sci 25:1–26

Le Calvez J (1947) Les perforations du test de *Discorbis erecta* (Foraminifère). Bull Lab Marit Dinard 29:1–4

Lee JJ, Anderson OR (1991) Symbiosis in foraminifera. In: Lee JJ, Anderson OR (eds.) Biology of foraminifera. Academic, London, pp 157–222

Leutenegger S (1977) Ultrastructure de foraminifères perforés et imperforés ainsi que de leurs symbiotes. Cahiers de Micropaléontologie 3:1–52

Leutenegger S, Hansen HJ (1979) Ultrastructural and radiotracer studies of pore function in foraminifera. Mar Biol 54:11–16

Linne C (1758) Systema naturae, vol 10, 10th edn. Typis Laurentii Salvii, Stockholm, 824 pp

Lutze GF (1962) Variationsstatistik und Ökologie bei rezenten Foraminiferen. Paläontologische Zeitschrift 36:252–264

Lutze GF (1986) *Uvigerina* species of the eastern North Atlantic. In: van der Zwaan GJ, Jorissen FJ, Verhallen PJJM, von Daniels CH (eds.) Atlantic-European Oligocene to recent *Uvigerina*: Utrecht Micropaleontological Bulletins, vol 35, pp 21–46

Moodley L, Hess C (1992) Tolerance of infaunal benthic foraminifera for low and high oxygen concentrations. Biol Bull 183:94–98

Myers EH (1935) Morphogenesis of the test and biological significance of dimorphism in the foraminifer *Patellina corrugata* Williamson. Bull Scripps Inst Oceanogr 3:393–404, Technical Series

Nikulina A, Polovodova I, Schönfeld J (2008) Foraminiferal response to environmental changes in Kiel Fjord, SW Baltic Sea. eEarth 3:37–49

Nikulina A, Dullo W-C (2009) Eutrophication and heavy metal pollution in Flensburg Fjord: a reassessment after 30 years. Mar Pollut Bull 58:905–915

Perez-Cruz LL, Machain-Castillo ML (1990) Benthic foraminifera of the oxygen minimum zone, continental shelf of the Gulf of Tehuantepec, Mexico. J Foraminifer Res 20:312–325

Piña-Ochoa E, Høgslund S, Geslin E, Cedhagen T, Revsbech NP, Nielsen LP, Schweizer M, Jorissen F, Rysgaard S, Risgaard-Petersen N (2010) Widespread occurence of nitrate storage and denitrification among Foraminifera and *Gromiida*. Proc Natl Acad Sci U S A 107:1148–1153

Polovodova I, Nikulina A, Schönfeld J, Dullo WC (2009) Recent benthic foraminifera from the Flensburg Fjord. J Micropaleontol 28:131–142

Riester J, Zumft WG, Kroneck PMH (1989) Nitrous oxide from *Pseudomonas stutzeri*. Eur J Biochem 178:751–762

Risgaard–Petersen N, Langezaal AM, Ingvardsen S, Schmid MC, Jetten MS, Op den Camp HJM, Derksen JWM, Pina–Ochoa E, Eriksson SP, Nielsen LP, Revsbech NP, Cedhagen T, van der Zwaan GJ (2006) Evidence for complete denitrification in a benthic foraminifer. Nature 443:93–96

Sen Gupta BK, Machain-Castillo ML (1993) Benthic foraminifera in oxygen-poor habitats. Mar Micropaleontol 20:183–201

Sliter WV (1970) *Bolivina doniezi* Cushman and Wickenden in clone culture. Contr Cushman Found Foram Res 21:87–100

Sliter WV (1974) Test ultrastructure of some living benthic foraminifers. Lethaia 7:5–16

Travis JL, Bowser SS (1991) The motility of foraminifera. In: Lee JJ, Anderson OR (eds.) Biology of the foraminifera. Academic, London, pp 91–155

Williamson WC (1858) On the recent foraminifera of Great Britain. Ray Society, London

Zumft WG (1997) Cell biology and molecular basis of denitrification. Microbiol Mol Biol Rev 61:533–616

Biodata of **Ahuva Almogi-Labin, Sarit Ashckenazi-Polivoda, Yael Edelman-Furstenberg**, and **Chaim Benjamini**, authors of *"Anoxia-Dysoxia at the Sediment-Water Interface of the Southern Tethys in the Late Cretaceous: Mishash Formation, Southern Israel."*

Dr. Ahuva Almogi-Labin is currently a senior scientist at the Geological Survey of Israel and head of the Micropaleontology laboratory. She obtained her Ph.D. from the Hebrew University of Jerusalem, Jerusalem, Israel in 1982. Dr. Ahuva Almogi-Labin's scientific interests are in the areas of: Micropaleontology of planktic and benthic foraminifera and pteropods, Paleoceanography of the Eastern Mediterranean, Red Sea, and Indian Ocean. Quaternary and late upper Cretaceous high productivity/oxygen depleted environments. Brackish foraminifera of inland water. Anthropogenic impact on the Israeli Mediterranean shelf.

E-mail: **almogi@gsi.gov.il**

Sarit Ashckenazi-Polivoda is currently a Ph.D. candidate in the Department of Geological and Environmental Sciences at Ben-Gurion University of the Negev, Beer-Sheva, Israel where she also obtained her M.Sc. degrees. Sarit Ashckenazi-Polivoda's scientific interests are in paleoenvironmental reconstruction of the Late Cretaceous southern Tethys high productivity setting by using planktic and benthic foraminifera assemblages.

E-mail: **ashcenaz@bgu.ac.il**

Ahuva Almogi-Labin   **Sarit Ashckenazi-Polivoda**

Dr. Yael Edelman-Furstenberg is currently a scientific researcher at the Geological Survey of Israel. She obtained her Ph.D. from the University of Chicago in 2004. Dr. Edelman-Furstenberg's scientific interests are in the areas of: paleoecology of bivalves and gastropods, taphonomy of bivalves, paleoceanography of planktic and benthic foraminifera and pteropods, and paleoecology of marine environments of differing nutrient levels.

E-mail: **yael@gsi.gov.il**

Prof. Chaim Benjamini is currently an Associate Professor in the Department of Geological and Environmental Sciences, Ben Gurion University of the Negev, Beer-Sheva, Israel. He obtained is Ph.D. from the Hebrew University of Jerusalem in 1980. Prof. Chaim Benjamini's scientific interests are in the areas of: Micropaleontology of planktonic and benthic calcareous microfossils. Genesis, stratigraphy, and microfacies of Mesozoic and Tertiary Tethyan carbonate platform- to pelagic facies. The Mediterranean continental shelf of Israel. Regional stratigraphy of Israel and the Near East.

E-mail: **chaim@bgu.ac.il**

Yael Edelman-Furstenberg   **Chaim Benjamini**

ANOXIA-DYSOXIA AT THE SEDIMENT-WATER INTERFACE OF THE SOUTHERN TETHYS IN THE LATE CRETACEOUS: MISHASH FORMATION, SOUTHERN ISRAEL

AHUVA ALMOGI-LABIN[1], SARIT ASHCKENAZI-POLIVODA[1,2], YAEL EDELMAN-FURSTENBERG[1], AND CHAIM BENJAMINI[2,3]
[1]*Geological Survey of Israel, 30 Malkhe Israel St., Jerusalem 95501, Israel*
[2]*Department of Geological and Environmental Sciences, Ben Gurion University of the Negev, Beer Sheva 84105, Israel*
[3]*Ramon Science Center, Mizpe Ramon 80600, Israel*

1. Geological and Paleoceanographic Setting

Oceanic upwelling prevailed offshore of the southern Tethyan margin during the Late Cretaceous and Early Tertiary (Parrish and Curtis 1982; Soudry et al. 2006). Nutrient excess and consequently, a high flux of organic matter on the seafloor resulted in widespread development of poorly oxygenated sediments (Bein et al. 1990; Almogi-Labin et al. 1990a, 1993). These features have been described from a band extending from the northern coast of South America and northwestern Africa to the west to Mesopotamia to the east, and temporally from the Santonian to the Eocene (Parrish and Curtis 1982; Soudry et al. 2006).

Upwelling-induced nutrient enrichment and high productivity develop under appropriate oceanographic conditions on continental shelves and slopes of modern oceans (e.g., Bremner 1983; Suess and Thiede 1983; Thiede and Suess 1983; Monteiro et al. 2005). The Cretaceous Tethyan upwelling system was situated a few hundred kilometers north of the Gondwanan shoreline, between 8°N and 15°N latitude (Parrish and Curtis 1982; Dercourt et al. 1993). Paleoposition above an outer shelf or slope is assumed, based on the broad regional distribution of the pelagic facies and the absence of indications of storm activity, wave, or current action (Reiss 1988; Almogi-Labin et al. 1990a, 1993; Dercourt et al. 1993). Abramovich et al. (2010) place the water depths of the Maastrichtian in this region in the outer slope setting. There is no direct noticeable effect in this region of the dramatic Tethyan orogenic events, such as ophiolite emplacement, tectonic thrusting, subduction, and island-arc volcanism, that were taking place on plate margins well to the north (e.g., Almogi-Labin et al. 1990b).

Recent studies in Israel have shown indications of nutrient excess and biotic response already from the Cenomanian and Turonian (Buchbinder et al. 2000), and subsequently, from the Maastrichtian onward until the Eocene. The acme of this system in Israel was in the Campanian, part of the Mount Scopus Group (Fig. 1a, b), when organic-rich carbonate, silica in the form of chert and porcelanite, and disseminated and sometimes concentrated phosphorite, formed from a highly fertile, nutrient-rich ocean (Almogi-Labin et al. 1990a, 1993; Eshet et al. 1994; Eshet and Almogi-Labin 1996; Soudry and Champetier 1983; Soudry et al. 1985; Bein et al. 1990; Kolodny and Garrison 1994; Edelman-Furstenberg 2008, 2009; Ashckenazi-Polivoda et al. 2010). These lithofacies are exceptional, as pelagic chalk or marl with little terrestrial input was deposited in this region in the Late Cretaceous (Dercourt et al. 1993).

These cherts, phosphorites, porcelanites, and organic-rich carbonates constitute the Mishash Formation of southern Israel. This formation is divided into a lower Chert Member and an upper Phosphate Member. The latter has been subdivided into the Phosphatic Carbonate unit at the base, the intervening Porcelanite unit, topped by the Phosphorite unit (Soudry et al. 1985). The biostratigraphic position of the upper Phosphate Member (Fig. 2) correlates to the *Globotruncana rosetta* and *Radotruncana calcarata* zones (Almogi-Labin et al. 1993). The Mishash Formation lies above the pelagic chalks of the Menuha Formation, Santonian-Early Campanian in age, and below pelagic chalky marls of the Ghareb Formation of Maastrichtian age. The distinctive lithofacies of the Mishash Formation were draped over the anticlinal ridge – synclinal basin topography of the Syrian Arc compressional tectonic regime in southern Israel (Figs. 1 and 2). Structural position on this fold belt has a significant effect on thickness

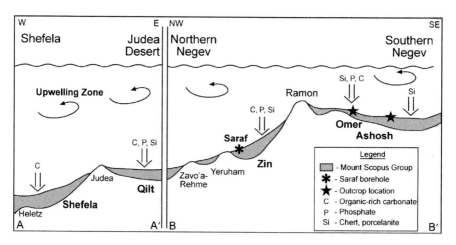

Figure 1. (a) Schematic cross-section (not to scale) showing two-part traverse from coast to Judean Desert (**a–a'**) and from the northern to the southern Negev (**b–b'**), across the Syrian Arc paleotopography, with Mount Scopus Group filling synclines. Upwelling zone extends across the entire traverse, with organic-rich carbonate (*C*), phosphorite (*P*), and siliceous sediment as chert and porcelanite (*Si*) deposited in varying proportions in the synclines. (b) Location map for two parts of traverse shown in Fig. 1a. Position of Ashosh, Omer, and Qilt sections, Saraf borehole, and Shefela syncline, are shown. Base topography is structural map showing Syrian Arc structures, main anticlinal axes (Modified after Gardosh et al. 2008).

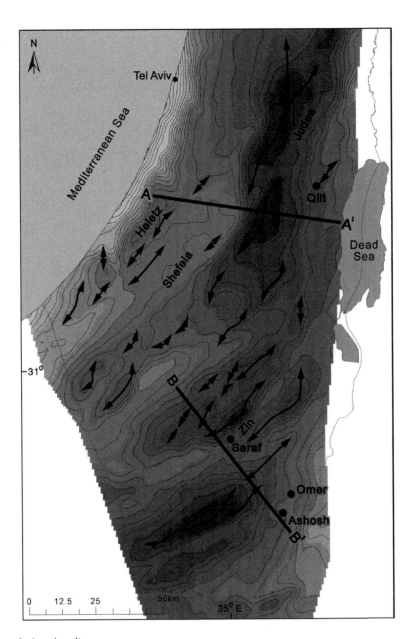

Figure 1. (continued).

and lithofacies. Beyond the Syrian Arc to the south the Mishash Formation becomes more uniform, predominantly of pelagic chalk interbedded with chert-porcelanite horizons and some continent-derived clastics (Fig. 2). Above Syrian Arc structures to the north and west of the Negev, the Qilt syncline in the Judean

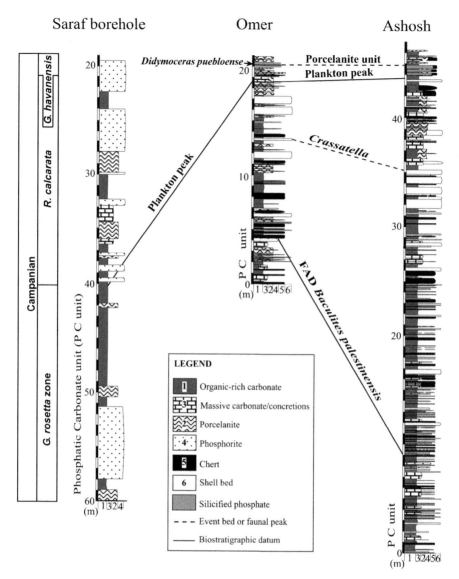

Figure 2. Columnar sections of phosphatic carbonate unit, Phosphate Member of the Mishash Formation. Saraf borehole, Omer, and Ashosh outcrop sections are shown, showing main lithologies and correlation using planktonic foraminifera and ammonites. First appearance datum (FAD) of *Baculites palestinensis* indicating Late Campanian is shown for the outcrop sections. Plankton peak is coeval with base *Radotruncana calcarata* Zone (Almogi-Labin et al. 1993). Locations of sections are given in Fig 1b.

Desert is poorer in chert and phosphorite (Edelman-Furstenberg 2009), while the Shefela syncline of the coastal belt lacks biogenic silica deposits (Bein et al. 1990) and is dominated by the organic-rich carbonates of the Ein Zeitim Formation (Flexer 1968; Reiss et al. 1985; Bein et al. 1990). The upwelling belt therefore extended in Israel across a paleoslope over 250 km wide compared to 140 km off Namibia (Monteiro et al. 2005).

The coincidence of the Tethyan high-productivity zone with restriction of circulation in synclinal basins led to severe oxygen depletion where organic-rich sediments accumulated. The Ramon structure, the most prominent anticline in the Negev, formed a submarine paleohigh in the Cretaceous (Fig. 1a, b), while the Zin syncline just to the north formed an organic-rich sediment trap. Just to the south of the Ramon axis, the Omer anticline and Ashosh syncline are smaller folds upon which organic-rich sediments also accumulated.

Therefore a variety of oxic, dysoxic, and anoxic environments are represented in the Mishash Formation over these structures (Edelman-Furstenberg 2008, 2009).

At the acme of this system, a gradient culminating in extreme anoxia developed, represented by the Phosphate Member of the Mishash Formation. We show here that populations of specially adapted foraminifera can thrive even under those conditions.

2. Microfaunal Assemblages Associated with Anoxia

Food flux and oxygen availability control benthic foraminiferal assemblages (Jorissen et al. 1995; van der Zwaan et al. 1999). In environments of high productivity, food is abundant, and assemblages are limited according to oxygen concentration (Almogi-Labin et al. 1990a, 1993; Kaiho 1994; Ashckenazi-Polivoda et al. 2010). Upwelling conditions adversely affect planktonic foraminifera that live within the water body, while benthic foraminifera on the seafloor can flourish due to the increased food supply. However, assemblages of the more dysaerobic conditions display characteristics of stressed systems, namely, low diversity and high dominance, and sometimes high numerical abundance (Sen Gupta et al. 1981; Sen Gupta and Machain-Castillo 1993; Kaiho 1994; Bernhard et al. 1997; Bernhard and Bowser 1999; Diz et al. 2006; Diz and Frances 2008; Leiter and Altenbach 2010). In the Cretaceous upwelling belt, benthic buliminid foraminifera become dominant, as diversity falls due to oxygen stress (Reiss 1962, 1988; Deutch 1986; Almogi-Labin et al. 1993; Ashckenazi-Polivoda et al. 2010).

Edelman-Furstenberg (2008, 2009) studied the molluscan assemblages of sedimentary cycles in the upper part of the Mishash Formation. Cycles of 1–9 m in thickness were constructed of organic-rich carbonate at the base, overlain by porcelanite, chert, carbonate, and by a molluscan shell bed cap (Fig. 2), sometimes with coarse phosphorite. Each part of the cycle has a characteristic macrobenthos, sedimentary structure, and taphonomy. The very low diversity of this primarily deposit-feeding molluscan assemblage indicates high organic loads, low oxygen levels, and a stressed environment at the base of the cycle. Increase in oxygen gradient upward in each cycle is accompanied by increasing species richness, body size, and abundance of these bivalves. At the base of each of these Mishash cycles, organic-rich carbonates devoid of bivalves (Edelman-Furstenberg 2008), or sometimes with only low-diversity assemblages, represent quasi-aerobic to dysaerobic facies according to the classification of Savrda and Bottjer (1987, 1991). They defined five inferred dissolved-oxygen zones below the sediment/water interface, namely, anaerobic, quasi-aerobic, and exaerobic, all indicative of 0.0 ml O_2/l in modern analogues, and dysaerobic (0.1–1.0 ml O_2/l) and aerobic (>1.0 ml O_2/l).

The poorly aerated environments at the base of the cycles, as defined by their sterility in molluscan fauna, imply oxygenation below 0.1 ml O_2/l (Edelman-Furstenberg 2008). These horizons are, however, populated by foraminiferal faunas, sometimes in substantial numbers (Reiss 1962). The gradient toward extreme anoxia is accompanied by development of increasingly specialized microfaunal assemblages (Ashckenazi-Polivoda et al. 2010). Presence of such assemblages can trace the acme of the upwelling system and concomitant extreme depletion of oxygen on the sea floor.

3. The Tethyan Anoxic Regime: Sections from Southern Israel

The strongest upwelling in the southern Tethyan Late Cretaceous is represented in Israel by the Phosphate Member, forming the upper part of the Mishash Formation. We here summarize data from southern Israel focusing on the extreme end of the dysaerobic-anaerobic continuum. Data are from the Phosphate Member in the Saraf borehole of the Zin Valley (Figs. 1 and 2), and from two sections limited to the Phosphatic Carbonate unit (Ashosh; Omer), from outcrops studied by Edelman-Furstenberg (2008, 2009). Focus is on the specialized foraminiferal associations from the organic-rich carbonate, porcelanite, and phosphate lithofacies. Foraminiferal assemblages of the poorly oxygenated organic-rich carbonates were studied by Ashckenazi-Polivoda et al. (2010). We have added comparison of foraminiferal assemblages from published records from the Judean Desert (Qilt, Ashckenazi-Polivoda et al. 2010) and Shefela foothills (Almogi-Labin et al. 1993), deposited more distally beneath the upwelling belt.

In the present study, we expand the data base to foraminiferal populations from all soft lithologies that could be disaggregated. Buliminid species used to define B-assemblages of the organic-rich carbonates by Ashckenazi-Polivoda et al. (2010) were used here to track the oxygenation gradient also in the siliceous and phosphatic carbonate lithofacies. The Saraf borehole has previously yielded data on the water column and seafloor of this part of the Mishash Formation, including geochemical and mineralogical properties of the sediment, and the foraminiferal, calcareous nannoplankton, and dinoflagellate populations (Bein et al. 1990; Almogi-Labin et al. 1993; Eshet et al. 1994; Eshet and Almogi-Labin 1996).

Chert, shell beds, carbonate concretions, and indurated phosphorite and porcelanite are generally unsuitable for foraminiferal studies, explaining the patchy distribution of samples in the sections. Preparation procedure was common for all samples. Approximately 20 g of the sample was disintegrated, sieved at 63 µm, and foraminifera (250–300 specimens/sample) were identified and counted at >63 µm. Total abundance (specimens per g sediment), relative abundance for buliminid species, and relative abundance of the dominant buliminid species, as well as raw diversity (species richness), were calculated. Benthic foraminiferal assemblages were defined by cluster analysis based on the relative abundance of the main (>3%) benthic species from all analyzed samples. The relative abundance of all specimens from all samples was used to construct a cluster dendrogram using PRIMER v6 software (PRIMER-E Ltd, Plymouth, UK). A series of "similarity profile" (SIMPROF) permutation tests were used to reveal statistically significant evidence for genuine clusters in the generated dendrogram, such that the group being subdivided had significant ($P<0.05$) internal structure.

4. The Characteristics of the Foraminiferal Assemblage

The Ashosh section (Figs. 1 and 2) is from a shallow syncline south of the Omer anticline, located on the flank of the larger Ramon structure. The Phosphate Member is 80 m thick, but the present study concentrates on the lower 40 m, from which 29 samples were taken. Foraminiferal numbers per g sediment are highly variable, ranging from 30 to 30,000 (Fig. 3). Peaks represent abundance well beyond values

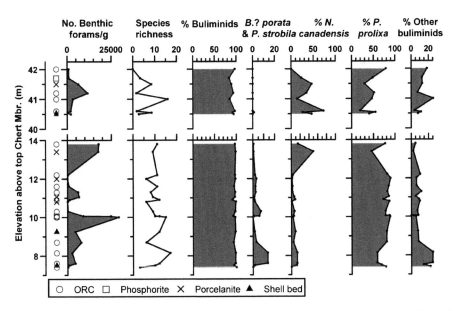

Figure 3. Benthic foraminiferal data for Ashosh section. Close sampling was from between 7–14 m and 40–42 m. No significant difference is found between foraminiferal populations of the different lithofacies: organic-rich carbonates (*ORC*), phosphorites, porcelanites, and shell beds, indicating that populations are representative of pore-waters within the sediments, rather than particle flux at sediment-water interface. Populations are highly variable in numbers but may reach extreme values. Number of species varies but does not exceed 17, indicating stress. Microfaunas are mostly buliminid foraminifera, with two species dominating. ORC-rich samples include data from Ashckenazi-Polivoda et al. (2010).

typical for pelagic carbonates. The number of species is low, however, ranging from 3 to 17, averaging around 10, much less than usual for pelagic carbonates below and above the Mishash Formation (Almogi-Labin et al. 1993).

Buliminid foraminifera are highly dominant, making up 90–100% of the benthic assemblage. Of these, nearly all belong to *Praebulimina prolix* (Cushman and Parker), *Neobulimina canadensis* (Cushman and Wickenden), *Bolivina? porata* and *Praebulimina strobila* (Marie), species used by Ashckenazi-Polivoda et al. (2010) to track the dysoxic – anoxic trajectory. Some other foraminiferal species, especially other buliminids, are also relatively abundant in this environment, as shown by Ashckenazi-Polivoda et al. (2010).

The Omer section is from the flank of the Omer anticline. The Phosphate Member is 40 m thick, and the 19 samples studied were from the lower 20 m. This section was richer in chert in the lower part and in thick shell beds in the upper part (Fig. 2). Foraminiferal abundance is variable, with several horizons in the lower half with numbers above 20,000 per g sediment (Fig. 4). Number of species ranges from 2 to 16, with an average of 11 in the very rich horizons. Buliminid foraminifera make up from 70% to 100% of the numbers, but peaks of *Gavelinella* sp. 1 and *Nonionella austinana* Cushman together with *Nonionella robusta* Plummer are responsible for up to 21% in some samples (Ashckenazi-Polivoda et al. 2010). The dominant buliminids are again *P. prolixa* and *N. canadensis*.

The Saraf borehole is located to the northwest of the Zin Valley synclinal axis, north of the Ramon structure (Fig. 1). 31 samples from the full 40 m of the Phosphate

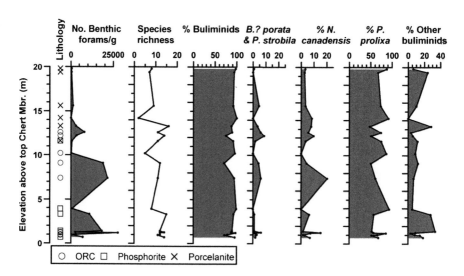

Figure 4. Benthic foraminiferal data for 20 m section at Omer. Foraminiferal populations of the different lithofacies are as at Ashosh but porcelanites appear to be somewhat poorer; lower part of section shows some extremely abundant foraminiferal populations. Number of species varies as at Ashosh with maximum at 16, indicating stress. Microfaunas are mostly buliminid foraminifera, with two species dominating. ORC-rich samples include data from Ashckenazi-Polivoda et al. (2010).

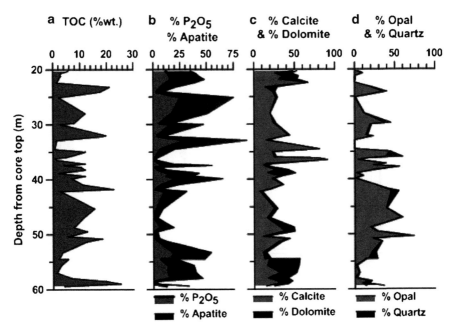

Figure 5. Lithological and geochemical data for Saraf borehole from Bein et al. (1990). (**a**) – TOC (total organic carbon); (**b**) – %P_2O_5 from laboratory analysis, % apatite from XRD; (**c**) – % calcite (*gray*) and cumulative % calcite and dolomite (*black*); (**d**) – % opal (*gray*) and cumulative % opal and quartz (*black*).

Member were sampled for foraminifera. The subdivision of this formation into members is not well marked in the Zin Valley. Total organic carbon (TOC) ranges from 1% to 25%, %P_2O_5 ranges from 1% to 31%, % apatite from 5% to 90% and SiO_2 (quartz and opal CT), 0–60% (Fig. 5, data from Bein et al. 1990); these values are particularly characteristic of an upwelling system (e.g., Bremner 1983; Suess and Thiede 1983). Carbonate as calcite and occasionally as dolomite rarely reaches 50%, considerably less than in the chalks under- and overlying the Mishash Formation.

Foraminiferal abundance peaks occasionally approach 60,000 specimens per g sediment (Fig. 6). On the other hand, number of species averages 6, rarely achieving 10 species, except at the top. Nearly all of the excess numbers are buliminid foraminifera belonging to the four species, *P. prolixa,* dominant in the lower 10 m, alternating with *N. canadensis* in the upper part, and with *B.? porata* and *P. strobila* occasionally forming peaks. The gradual transition to organic-rich oil shale, a more aerated facies (Almogi-Labin et al. 1993) occurs at the top.

The highly variable foraminiferal abundance is apparently lithology-independent. The lithological and mineralogical data from the Saraf borehole demonstrates that there is little correlation to foraminiferal abundance or diversity; the Omer and Ashosh outcrop data supports this conclusion.

Figure 7 shows the results of cluster analysis of all the 58 samples from the three sections. Five foraminiferal assemblages can be distinguished on a

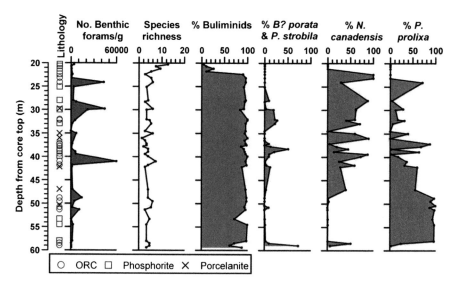

Figure 6. Foraminiferal data for Saraf borehole. Close sampling was from the full Phosphate Member, the Omer and Ashosh outcrops correlate only to the lower part, as shown in Fig. 2. Three population peaks of nearly 60,000 foraminifera per g sediment occur, but numbers are usually lower. Species richness average around 6, lower than the outcrops. Microfaunas are mostly buliminid foraminifera, with two species dominating. *Praebulimina prolixa* dominates the lower part, *N. canadaensis* dominates the upper part, and the middle has some more variable populations. Uppermost samples show transition to richer, more diverse populations with buliminids less dominant. ORC-rich samples from 40 to 60 m include data from Ashckenazi-Polivoda et al. (2010). Data from samples from 20 to 40 m are from Almogi-Labin et al. (1993).

multidimensional scaling (MDS) diagram that show how these clusters separate (Fig. 8). The assemblages joining clusters 1–3 are characterized by >75% *P. prolixa*. Assemblage 4 has c. 56% *P. prolixa*, 10% *N. canadensis*, and contributions of some other species, and Assemblage 5 in which both *P. prolixa* and *N. canadensis* each contribute some 41%. Figure 8 shows that in the MDS diagram, the nearly overlapping assemblages 1–3 occupy the upper right quadrant, assemblage 4 occupies the lower right quadrant, and assemblage 5, the entire left field of the diagram. There is almost no overlap between assemblages 1–3 and Assemblage 4 and Assemblage 5.

A comparison of these three sections, as well as the Qilt section of the Judean desert and the Shefela section, progressively more distal toward the open Tethys, is shown in Fig. 9. These sections exhibit the main lithological features and microfaunal assemblages of the seafloor beneath the upwelling system, along the dysoxic – anoxic gradient (Ashckenazi-Polivoda et al. 2010; Almogi-Labin et al. 1990a, 1993). Five foraminiferal assemblages occur along this gradient: 5 and 4 represent the most extreme conditions of near anoxia, while 1–3 occur along the dysoxic anoxic gradient under somewhat more aerated conditions.

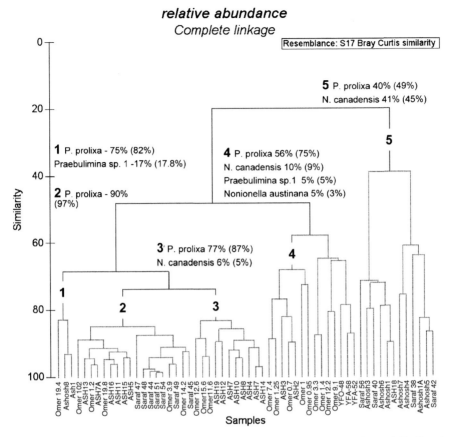

Figure 7. Bray-Curtis similarity clustering for all samples, including the depth interval 40–60 m in Saraf borehole which is equivalent of the Carbonate-Phosphate unit in the Zin borehole sequence. Assemblages 1–3 include the samples in which *P. prolixa* form 75–90% of the assemblage. Assemblage 4, *P. prolixa*, forms only 56% of the assemblage and *N. canadensis* forms 10%, and other species make significant contributions. For assemblage 5, *P. prolixa* and *N. canadensis* are subequal and make similar contributions.

5. Discussion

The same stressed foraminiferal assemblages occur in all lithologies, and all water depths, represented by the three Negev sections, as well as at Qilt (Fig. 9) (Ashckenazi-Polivoda et al. 2010). They occupy microenvironments largely independent of the nature of the inorganic sediment particles arriving at the seafloor, namely, siliceous opaline, carbonate, or precursors to phosphorite particles. We therefore conclude that the upwelling system in the upper levels of the water column was a regional feature oblivious to details of local bottom topography,

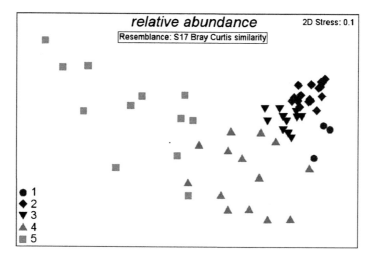

Figure 8. Multidimensional scaling (MDS) diagram for all samples showing good separation of assemblages from cluster analysis (Fig. 7). All samples of assemblage 5 plot to *left*, assemblage 4 to *lower right*, assemblages 1–3 plot in *upper right*.

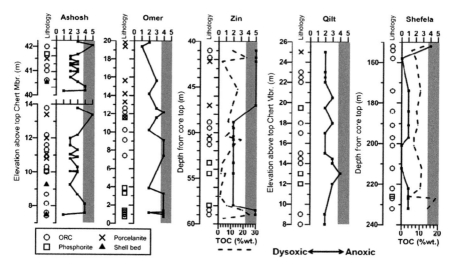

Figure 9. Plot of benthic foraminifera assemblages in the three studied sections; only the part of the Saraf borehole (*Zin*) equivalent to the Omer and Ashosh sections is given. Assemblages 1–3 are more aerated, 4, 5 are most depleted. Maximum depletion at the Saraf section (*Zin*) is in the *upper part* of the section; Omer and Ashosh are more aerated. The most depleted assemblages correspond to samples with high total organic carbon (TOC). The Qilt section (Fig. 1) is given for comparison. This more distal basin (Ashckenazi-Polivoda et al. 2010) shows the same range of variations in the assemblage characteristics as in the Negev sections. In the Shefela (location of syncline in Fig. 1), the benthic foraminiferal assemblage indicates a generally more aerated bottom water though organic carbon content is still high. Foraminiferal assemblage 0 represents foraminifera of more open marine conditions (Almogi-Labin et al. 1993).

e.g., that of the Syrian Arc system (Fig. 1). Local differences of thickness and lithology in this system are the product of near- or in-sediment sedimentary processes, such as bacterial generation of phosphate within the anoxic environment itself (e.g., Soudry and Champetier 1983; Goldhammer et al. 2010), mobilization and resedimentation of opaline silica as porcelanite or chert, and possibly, dolomitization of calcite. On the other hand, the Saraf section (Fig. 9) shows that foraminiferal assemblages can be sensitive to TOC levels.

Previous studies on mollusca indicate a strong connection between abundance and diversity of the assemblages and the suite of upwelling-sedimentary features (Edelman-Furstenburg 2008, 2009). Under upwelling conditions macrofaunas are limited by their tolerance to low oxygen conditions. The sediment arriving at the seafloor was composed of organic matter including dinoflagellates, siliceous microorganisms especially diatoms, carbonate in the form of pelagic nannofossils, and foraminifera (Moshkovitz et al. 1983; Almogi-Labin et al. 1993; Eshet et al. 1994; Eshet and Almogi-Labin 1996; Hoek et al. 1996). Molluscan assemblages are indicative of degree of water aeration and bottom-water energy conditions directly beneath the upwelling cell, reflected in textural attributes of the sediment. Molluscan faunas decline as oxygenation below 0.1 ml O_2/l (in modern analogues) is approached. At this point, microfaunas in the sediments sterile of mollusks remain abundant or may even increase.

At least five distinguishable foraminiferal assemblages were found under these conditions. Each decreased oxygenation level approaching anoxia is represented by specialized assemblages, each tolerant of ever more extreme conditions; these assemblages either fluctuate or replace each other in sequence. Ultimately, large populations can thrive even under the most extreme anoxic conditions. The displaced assemblages reappear in reverse order upon reoxygenation (Fig. 6, uppermost sample). A similar foraminiferal response to decreasing oxygenation of the seafloor, sometimes with high population numbers, was recorded in modern sediments of the Santa Barbara Basin (Bernhard et al. 1997).

The main factors known to influence foraminiferal numbers under these conditions are the availability of organic matter as a food resource (e.g., Diz et al. 2006; Diz and Francés 2008; Monteiro et al. 2005), absence of predation, and absence of toxic by-products of anoxia (cf. Leiter and Altenbach 2010). Foraminiferal abundance suggests that the dysoxia-anoxia gradients reflect changes taking place in the pore-water habitat deeper in the sediment. Special adaptation of different foraminiferal species to near-anoxic conditions is therefore the apparent reason for the changing assemblages.

Infaunal foraminiferal assemblages are known to be mobile within the sediment, inhabiting different levels in response to predation pressure or different resources at different levels. Predation pressure is unlikely under conditions of near anoxia. Upwelling conditions provide a continuous flux of organic particles to the sediment. Modern infaunal foraminifera do not necessarily directly scavenge food particles, as do macrofaunal filter or deposit feeders. Some are known to graze bacterial colonies that break down particulate organic matter, while others live within the denitrification zone under conditions of complete anoxia (cf. Risgaard-Petersen et al. 2006; Høgslund et al. 2008; Piña-Ochoa et al. 2010).

Buliminid foraminifera include many species that are common today in aerated environments (e.g., *Bulimina aculeata; Bulimina marginata;* Almogi-Labin et al. 2000; Hintz et al. 2004), but some species are known to be highly adapted to oxygen stress (Sen Gupta and Machain-Castillo 1993). Bernhard et al. (2010) reported on *Bolivina pacifica* (a modern buliminid species) that inhabits an extremely oxygen deficient microhabitat.

In the Mishash Formation material, two buliminid species, *Praebulimina prolixa* and *Neobulimina canadensis*, are responsible for most of the variability in the five low-oxygen assemblages as defined above (Figs. 8 and 9). These two buliminid species are morphologically and ecologically quite different (Fig. 10).

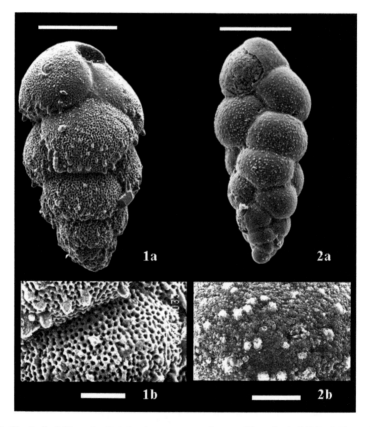

Figure 10. Two buliminid species that dominate near-anoxic assemblages in the Mishash Formation: (**1a**) *Praebulimina prolixa* – lateral view (bar = 100 μm), and (**1b**) Close up (bar = 10 μm) showing dense distribution of 1.5–2 μm in diameter pores (Saraf borehole sample 16, 24 m depth); (**2a**) *Neobulimina canadensis* – lateral view (bar = 100 μm); (**2b**) Pore types – close up (bar = 10 μm) showing combination of scattered pore cones and high density of micropores with a diameter of ~0.3 μm (Saraf borehole sample 16, 24 m depth).

Praebulimina prolixa has a more robust, ornamented test, perforated all over by large pores of 1.5–2 µm in diameter. It has been considered an opportunist species, proliferating in tandem with increased organic flux as long as the minimal oxygenation conditions remain tolerable (Ashckenazi-Polivoda et al. 2010). Notably, Bernhard et al. (2010) showed how similar test pore morphology in *Bolivina pacifica* was an adaptation for supporting symbiotic ectobionts that were crucial to survival in pore waters that are extremely depleted in oxygen. *Neobulimina canadensis*, on the other hand, has an unusual smooth wall microstructure, with numerous small micropores c. 0.3 µm in diameter and a second pore type that occurs in lower numbers centered on cones. This wall microstructure also implies a deep infaunal low-oxygen habitat (cf. Corliss 1985; Corliss and Chen 1988; Bernhard 1986; Kaiho 1994). This species, like related Cenomanian forms with similar wall microstructure, have been positively correlated with low-oxygen concentrations (Eicher and Worstell 1970; Bernhard 1986; Gebhardt et al. 2004).

Although functions of test features in foraminifera are generally unknown, these wall microstructures are assumed to represent different functional adaptations in different species. Both species belong to the buliminid group, among which some species are known to proliferate in environments of oxygenation stress. The co-occurrence of these species is attributed to their partition of resources in the in-sediment space. This partition may be due to tiering within the sediment, with each species living at a different depth and grazing on different bacterial populations (e.g., Goldhammer et al. 2010). Under conditions of high organic matter flux due to upwelling, vertical partitioning would take place in pore waters within the uppermost millimeters of the sediment (e.g., Leiter and Altenbach 2010). Tiering corresponding to adaptation to food resources at different levels would correspond to differing but extremely low levels of O_2 concentrations generated by these reduction processes, all below the 0.1 ml/l oxygen dysoxia cutoff.

In the Omer section, *Nonionella austinana* and *Nonionella robusta* sporadically peak as high as 21% (Ashckenazi-Polivoda et al. 2010). This section is the most aerated of the Negev Mishash sections. Notably, *Nonionella stella* is the most stress-tolerant species of the dysoxic foraminiferal assemblage of the Santa Barbara Basin, also located beneath an upwelling system (Bernhard et al. 1997). Modern and Cretaceous stressed assemblages are therefore highly analogous.

Although upwelling all across the southern Tethys has been described from the mid-Cretaceous until the Eocene, its acme was in the Campanian. In the sections of the Negev and Judean desert, the interval of extreme upwelling, represented by the phosphate-carbonate unit of the Mishash Formation, was never interrupted by aerated intervals with populations of benthic foraminifera with high diversity and dominance of rotaliids. Such populations represented the open marine system and oxygenated seafloor that prevailed before and after deposition of the Mishash Formation (Almogi-Labin et al. 1993; Ashckenazi-Polivoda et al. 2010). In the Shefela, the correlative section does not have anoxic assemblages, and the section was relatively aerated (with rotaliids occurring in foraminiferal assemblage 0, Fig. 9).

Nevertheless, TOC there ranges between 2% and 19%, indicating the high organic flux of the upwelling system. Thus, the Shefela was positioned at the seaward edge of the upwelling zone, that intensified to the south and east across the traverse described here, continuing well to the south, across at least 250 km of the shelf-slope system of the Gondwanan margin of the southern Tethys.

6. References

Abramovich S, Yovel-Corem S, Almogi-Labin A, Benjamini C (2010) Global climate change and planktic foraminiferal response in the Maastrichtian. Paleoceanography 25:PA2201. doi:10.1029/2009PA001843

Almogi-Labin A, Bein A, Sass E (1990a) Agglutinated foraminifera in organic-rich neritic carbonates (Upper Cretaceous, Israel) and their use in identifying oxygen levels in oxygen-poor environments. In: Hemleben C, Kaminski MA, Kuhnt W, Scott DB (eds) Paleoecology, biostratigraphy, paleoceanography and taxonomy of agglutinated foraminifera. Kluwer Academic, Dordrecht, pp 565–585

Almogi-Labin A, Flexer A, Honigstein A, Rosenfeld A, Rosenthal E (1990b) Biostratigraphy and tectonically controlled sedimentation of the Maastrichtian in Israel and adjacent countries. Rev Esp Paleontol 5:41–52

Almogi-Labin A, Bein A, Sass E (1993) Late Cretaceous upwelling system along the southern Tethys margin (Israel): interrelationship between productivity, bottom water environments and organic matter preservation. Paleoceanography 8:671–690

Almogi-Labin A, Schmiedl G, Hemleben C, Siman-Tov R, Segl M, Meischner D (2000) The influence of the NE winter monsoon on productivity changes in the Gulf of Aden, NW Arabian Sea, during the last 530 ka as recorded by foraminifera. Mar Micropaleontol 40:295–319

Ashckenazi-Polivoda S, Edelman-Furstenberg Y, Almogi-Labin A, Benjamini C (2010) Characterization of lowest oxygen environments within ancient upwelling environments: benthic foraminifera assemblages. Palaeogeogr Palaeoclimatol Palaeoecol 289:134–144

Bein A, Almogi-Labin A, Sass E (1990) Sulfur sinks and organic carbon relationships in Cretaceous organic-rich carbonates: implications for evaluation of oxygen-poor depositional environments. Am J Sci 290:882–911

Bernhard JM (1986) Characteristic assemblages and morphologies of benthic foraminifera from anoxic, organic-rich deposits: Jurassic through Holocene. J Foramin Res 16:207–215

Bernhard JM, Bowser SS (1999) Benthic foraminifera of dysoxic sediments: chloroplast sequestering and functional morphology. Earth-Sci Rev 46:149–165

Bernhard JM, Sen Gupta BK, Borne PF (1997) Benthic foraminiferal proxy to estimate dysoxic bottom-water oxygen concentrations: Santa Barbara Basin, U.S. Pacific continental margin. J Foraminiferal Res 27:301–310

Bernhard JM, Goldstein ST, Bowser SS (2010) An ectobiont-bearing foraminiferan, *Bolivina pacifica*, that inhabits microxic pore waters: cell-biological and paleoceanographic insights. Environ Microbiol 12:2007–2019

Bremner JM (1983) Biogenic sediments on the South West African (Namibian) continental margin. In: Theide J, Suess E (eds) Coastal upwelling, its sediment record, part B, sedimentary record of ancient coastal upwelling. NATO conference series, IV, marine sciences 10(b):73–104

Buchbinder B, Benjamini C, Lipson-Benitah S (2000) Sequence development of Late Cenomanian–Turonian carbonate ramps, platforms and basins in Israel. Cretaceous Res 21:813–843

Corliss BH (1985) Microhabitats of benthic foraminifera within deep-sea sediments. Nature 314:435–438

Corliss BH, Chen C (1988) Morphotype patterns of Norwegian Sea deep-sea benthic foraminifera and ecological implications. Geology 16:716–719

Dercourt J, Ricou JE, Vrielynck B (1993) Atlas Tethys paleoenvironmental maps. Gauthier-Villars, Paris

Deutch S (1986) Biostratigraphy and paleoecology of buliminids (benthic foraminifera) in the Upper Cretaceous in Israel. M.Sc. thesis, Hebrew University, Jerusalem 160 pp (in Hebrew, abstract and taxonomy in English)

Diz P, Frances G (2008) Distribution of live benthic foraminifera in the Ria de Vigo (NW Spain). Mar Micropaleontol 66:165–191

Diz P, Francés G, Rosón G (2006) Effects of contrasting upwelling–downwelling on benthic foraminiferal distribution in the Ría de Vigo (NW Spain). J Mar Syst 60:1–18

Edelman-Furstenberg Y (2008) Macrobenthic community structure in a high-productivity region: Upper Campanian Mishash Formation (Israel). Palaeogeogr Palaeoclimatol Palaeoecol 261:58–77

Edelman-Furstenberg Y (2009) Cyclic upwelling facies along the Late Cretaceous southern Tethys (Israel): taphonomic and ichnofacies evidence of a high-productivity mosaic. Cretaceous Res 30:847–863

Eicher DL, Worstell P (1970) Cenomanian and Turonian foraminifera from the Great Plains, United States. Micropaleontology 16:269–324

Eshet Y, Almogi-Labin A (1996) Calcareous nannofossils as paleoproductivity indicators in Upper Cretaceous organic-rich sequences in Israel. Mar Micropaleontol 29:37–61

Eshet Y, Almogi-Labin A, Bein A (1994) Dinoflagellate cysts, paleoproductivity and upwelling systems: A Late Cretaceous example from Israel. Mar Micropaleontol 23:231–240

Flexer A (1968) Stratigraphy and facies development of Mount Scopus group (Senonian-Paleocene) in Israel and adjacent countries. Israel J Earth Sci 17:85–114

Gardosh M, Druckman Y, Buchbinder B, Rybakov M (2008) The Levant basin offshore Israel, stratigraphy, structure, tectonic evolution and implications for hydrocarbon exploration. Geological Survey of Israel Report GSI/4/2008, 118 p

Gebhardt H, Kuhnt W, Holbourn A (2004) Foraminiferal response to sea level change, organic flux and oxygen deficiency in the Cenomanian of the Tarfaya Basin, southern Morocco. Mar Micropaleontol 53:133–157

Goldhammer T, Brüchert V, Ferdelman TG, Zabel M (2010) Microbial sequestration of phosphorous in anoxic upwelling sediments. Nature Geosci 3:557–561

Hintz CJ, Chandler GT, Bernhard JM, McCorkle DC, Havach SM, Blanks JK, Shaw TJ (2004) A physicochemically constrained seawater culturing system for production of benthic foraminifera. Limnol Oceanogr Methods 2:160–170

Hoek RP, Eshet Y, Almogi-Labin A (1996) Dinoflagellate cyst zonation of Campanian- Maastrichtian sequences in Israel. Micropaleontology 42:125–150

Høgslund S, Revsbech NP, Cedhagen T, Nielsen LP, Gallardo VA (2008) Denitrification, nitrate turnover, and aerobic respiration by benthic foraminiferans in the oxygen minimum zone off Chile. J Exp Mar Biol Ecol 359:85–91

Jorissen FJ, de Stigter HC, Widmark GV (1995) A conceptual model explaining benthic foraminiferal microhabitats. Mar Micropaleontol 26:3–16

Kaiho K (1994) Benthic foraminiferal dissolved-oxygen index and dissolved-oxygen levels in modern ocean. Geology 22:719–722

Kolodny Y, Garrison RE (1994) Sedimentation and diagenesis in paleo-upwelling zones of epeiric sea and basinal settings: a comparison of the Cretaceous Mishash Formation of Israel and the Miocene Monterey Formation of California. In: Proceedings of the 29th International Geological Congress, part C, California, pp 133–158

Leiter C, Altenbach AV (2010) Benthic Foraminifera from the diatomaceous mud belt off Namibia: characteristic species for severe anoxia. Paleontol Electron 13: 13.2.11A

Monteiro PMS, van der Nelson G, Plas A, Mabille E, Bailey GW, Klingelhoeffer E (2005) Internal tide—shelf topography interactions as a forcing factor governing the large-scale distribution and burial fluxes of particulate organic matter (POM) in the Benguela upwelling system. Cont Shelf Res 25:1864–1876

Moshkovitz S, Ehrlich A, Soudry D (1983) Siliceous microfossils of the Upper Cretaceous Mishash Formation, Central Negev, Israel. Cretaceous Res 4:173–194

Parrish JT, Curtis RL (1982) Atmospheric circulation, upwelling, and organic-rich rocks in the Mesozoic and Cenozoic eras. Palaeogeogr Palaeoclimatol Palaeoecol 40:31–66

Piña-Ochoa E, Høgslund S, Geslin E, Cedhagen T, Revsbech NP, Nielsen LP, Schweizer M, Jorissen F, Rysgaard S, Risgaard-Petersen N (2010) Widespread occurrence of nitrate storage and denitrification among Foraminifera and Gromiida. PNAS 107:1148–1153

Reiss Z (1962) Stratigraphy of phosphate deposits in Israel. Israel Geo Surv Bull 34:1–23

Reiss Z (1988) Assemblages from a Senonian high-productivity sea. Rev Paléobiol 2:323–332, Special volume

Reiss Z, Almogi-Labin A, Honigstein A, Lewy Z, Lipson-Benitah S, Moshkovitz S, Zak Y (1985) Late Cretaceous multiple stratigraphic framework of Israel. Israel J Earth Sci 34:147–166

Risgaard-Petersen N, Langezaal AM, Ingvardsen S et al (2006) Evidence for complete denitrification in a benthic foraminifer. Nature 443:93–96

Savrda CE, Bottjer DJ (1987) The exaerobic zone, a new oxygen-deficient marine biofacies. Nature 327:54–56

Savrda CE, Bottjer DJ (1991) Oxygen-related biofacies in marina strata: an overview and update. In: Tyson RV, Pearson TH (eds) Modern and ancient continental shelf anoxia. Geological Society, London, pp 201–219, Special Publication 58

Sen Gupta BK, Lee RF, May MS (1981) Upwelling and an unusual assemblage of benthic foraminifera on the northern Florida continental slope. J Paleontol 55:853–857

Sen Gupta BK, Machain-Castillo ML (1993) Benthic foraminifera in oxygen-poor habitats. Mar Micropaleontol 20:183–201

Soudry D, Champetier Y (1983) Microbial processes in the Negev phosphorites (southern Israel). Sedimentology 30:411–423

Soudry D, Nathan Y, Roded R (1985) The Ashosh-Haroz facies and their significance for the Mishash palaeogeography and phosphorite accumulation in northern and central Negev, southern Israel. Israel J Earth Sci 34:211–220

Soudry D, Glenn CR, Nathan Y, Segal I, VonderHaar D (2006) Evolution of Tethyan phosphogenesis along the northern edges of the Arabian–African shield during the Cretaceous–Eocene as deduced from temporal variations of Ca and Nd isotopes and rates of P accumulation. Earth-Sci Rev 78:27–57

Suess E, Thiede J (eds) (1983) Coastal upwelling, its sediment records; Part A, responses of the sedimentary regime to present coastal upwelling. NATO conference series IV, marine sciences 10a, 604 pp

Thiede J, Suess E (eds) (1983) Coastal upwelling, its sediment records; Part B, sedimentary records of ancient coastal upwelling. NATO conference series IV, marine sciences 10b, 610 pp

van der Zwaan GJ, Duijnstee IAP, den Dulk M, Ernst SR, Jannink NT, Kouwenhoven TJ (1999) Benthic foraminifera: proxies or problems? A review of paleoecological concepts. Earth-Sci Rev 46:213–236

Biodata of **Jürgen Schieber**, author of *"Styles of Agglutination in Benthic Foraminifera from Modern Santa Barbara Basin Sediments and the Implications of Finding Fossil Analogs in Devonian and Mississippian Black Shales."*

Jürgen Schieber received his Ph.D. in Geology from the University of Oregon in 1985. He is currently a Professor of Geology at Indiana University in Bloomington, Indiana. His research focus is the study of fine-grained sediments (muds) and sedimentary rocks (shales and mudstones), and in particular, the conditions under which shales and mudstones of the past have accumulated. He takes a holistic approach to shale studies and integrates field studies (facies, stratigraphy), the study of modern muds (seafloor coring), and experimental approaches (flume studies). His work on agglutinated foraminifera grew out of their importance for constraining oxygen availability in the bottom waters of ancient ocean basins. He has published extensively on the sedimentology of shales and mudstones and has lectured on the subject across the globe at universities, research organizations, industry short courses, and symposia.

E-mail: **jschiebe@indiana.edu**

STYLES OF AGGLUTINATION IN BENTHIC FORAMINIFERA FROM MODERN SANTA BARBARA BASIN SEDIMENTS AND THE IMPLICATIONS OF FINDING FOSSIL ANALOGS IN DEVONIAN AND MISSISSIPPIAN BLACK SHALES

JÜRGEN SCHIEBER
Department of Geological Sciences, Indiana University,
1001 E. 10th Str., Bloomington, IN 47405, USA

1. Introduction

Oxygen-deficient environments, such as the central portions of the Santa Barbara Basin, have long been studied for the particular role of benthic foraminifera in such settings (e.g., Bernhard and Reimers 1991; Pike et al. 2001; Bernhard et al. 2003). Benthic foraminiferal assemblages are sensitive to bottom water oxygen levels and have been used to estimate paleo-oxygen concentrations in such settings (Bernhard et al. 1997). Remains of benthic foraminifera have also been discovered in ancient black shale successions of Devonian and Mississippian age (Milliken et al. 2007; Schieber 2009), but in both instances, the only recovered foraminiferal remains were of agglutinated benthic foraminifera that consisted of well sorted, fine detrital quartz and were cemented with diagenetic silica. These silica-cemented foraminiferal remains are easily recognized in polished thin sections, and their detrital origin is confirmed by application of scanned cathodoluminescence (Schieber 2009).

When looking at modern sediments, however, one realizes that acquisition of small quartz grains is not the only mode of chamber agglutination for benthic foraminifera. Looking at resin impregnated thin sections of Santa Barbara Basin sediments, one quickly finds that foraminifera also use poorly sorted and polymineralic materials in the construction of their chamber walls, and they may also construct chamber walls almost exclusively from clay and mica flakes. The obvious question that arises is whether these other styles of chamber construction arose in more recent earth history, or whether they simply were missed because they are less obvious than silicified quartzose remains (Milliken et al. 2007; Schieber 2009).

New observations from the Mississippian Barnett Shale (Milliken et al. 2007) and the Devonian Cleveland Shale (Schieber 2009) show that other styles of foraminiferal chamber agglutination are indeed preserved in these ancient shales. By comparison with the Santa Barbara Basin, this suggests that similarly diverse agglutinated benthic foraminiferal assemblages existed at the seabed of several Paleozoic black shale basins. By extension, this suggests that in spite of claims to the contrary (e.g., Rimmer 2004; Loucks and Ruppel 2007), the bottom waters of these basins were not persistently anoxic, but instead may have been much more comparable to modern suboxic sea bottoms (oxygen content ~0.2–0.0 ml/l; Tyson and Pearson 1991).

Careful consideration of the many factors that control accumulation of organic matter in sediments shows that organic matter preservation is not a black vs. white issue (i.e., productivity vs. preservation). A thoughtful appraisal of the issue by Bohacs et al. (2005) delineates multiple scenarios that can give rise to organic-rich sediments that fundamentally reflect the complex interactions between production, destruction, and dilution. Anoxic conditions alone, although they do imply relatively low rates of destruction, are not always sufficient to ensure the deposition of an organic-rich sediment.

Modern suboxic sea bottoms are accessible for study in various locations and include the Santa Barbara Basin (Bernhard et al. 2003), the Cariaco Basin (Hughen et al. 1996), the Baltic (Bonsdorff et al. 1996), the Arabian Sea (Schulz et al. 1996), and the Saanich Inlet (Russel and Morford 2001). Studying ancient black shales from a suboxic basin perspective may lead us to a much more realistic appraisal of the boundary conditions of Phanerozoic black shale accumulation.

2. Methods and Materials

The examined materials for this study were polished thin sections from the Late Devonian Cleveland Shale of Kentucky and from the Mississippian Barnett Shale of Texas. In addition, samples of Santa Barbara Basin (SBB) sediments, collected from the central portion of the basin, were provided by A. Schimmelmann. The latter were embedded in Spurr Resin and polished with diamond lapping film.

Polished thin sections were prepared by a commercial lab, Petrographic International, in Choiceland, Saskatchewan. Initial screening of thin sections was done with a petrographic microscope (Zeiss Photo III) in transmitted and reflected light. Petrographic microscope images shown here were acquired with a Pixera Pro 600ES digital camera with 5.8 megapixel resolution. A subset of these samples was then examined by scanned color cathodoluminescence in order to confirm the agglutinated character of features identified as agglutinated foraminifera. The used instrumentation was a FEI Quanta FEG 400 ESEM equipped with an energy dispersive x-ray microanalysis (EDS) system and a Gatan Chroma CL cathodoluminescence (CL) detector. High-resolution CL scans ($4,000 \times 4,000$ pixels, 1,000 μs beam dwell time) were run at 10 kV with a narrow lens aperture (aperture 4) and spot size 5. The thin sections were carbon coated and the observations were made under high vacuum. Working conditions for SEM backscatter imaging (BSE) were 15 kV, high vacuum, aperture 4, and spot size 3.

3. Observations

3.1. SANTA BARBARA BASIN

The SBB sediments contain a range of benthic foraminifera (e.g., Bernhard et al. 1997), but in this study the main effort was devoted to agglutinated foraminifera because those are not in danger of dissolution during diagenesis. No attempt was made to identify genera or species. The focus was on chamber wall composition and wall construction because those were considered features most readily recognized in ancient shales.

With regard to composition and wall structure, there appear to be three main types that occur side by side in SBB sediments (Fig. 1a). The first and apparently most common type in the examined SBB material has chamber walls that are lined with a thin layer of small silicate grains. The silicates are predominantly angular quartz grains that range in size from a few to ten microns, and the wall is typically no thicker than two grains (Fig. 1b, e). Whereas in some foraminifera the chamber walls seem almost exclusively constructed of quartz grains, in others feldspar grains and mica flakes are also incorporated into the wall structure. The second type of wall structure is of the same general composition as the first type, but the size range of incorporated grains is broader (a few microns to several ten microns) and the walls typically are three or more grains in thickness (Fig. 1c, f). In the third type of wall structure (Fig. 1d), the chamber walls are covered predominantly with platy minerals (mica flakes, clay). The latter type of wall structure is the thinnest and may be constructed from single layers of overlapping platelets (Fig. 1g) or may consist of several layers of overlapping platelets (Fig. 2d, e).

Whereas the outlines of agglutinated foraminifera are largely undistorted in samples that have only experienced shallow burial (Figs. 1a and 2a), once samples from greater depths (1–2 m core depth) are examined, an increasing number shows partial chamber collapse (Fig. 2b) and flattening (Fig. 2c). Comparing, for example, the well-preserved specimen in Fig. 1b with a collapsed counterpart in Fig. 2c, one can readily imagine that all that may be left after complete compaction is a streak of quartz grains in an otherwise more clay-rich heterogeneous matrix.

Thus, what we can expect to find in fully compacted ancient muds is not the chambered structure of the original foraminifera, but rather whatever remains after the chambers have completely collapsed. There is a potential that in places a pre-compaction chamber fill forms a thin separation between chamber walls after compaction, something referred to as a medial suture (Schieber 2009). Without such an infill, the chamber walls can no longer be resolved, and all we are left with is a thin lens of agglutinated material. Collapsed remains of agglutinated foraminifera are illustrated with the following examples from the rock record.

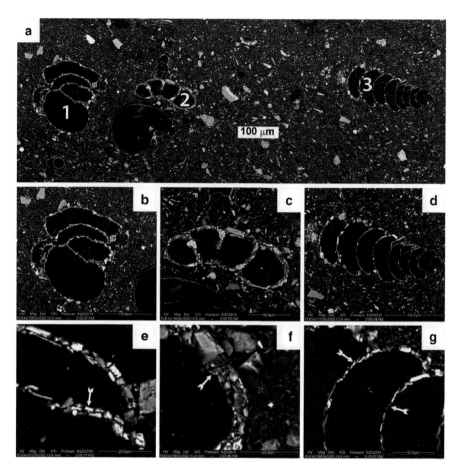

Figure 1. SEM images of common types of agglutinated foraminifera in sediments from the Santa Barbara Basin. (**a**) Image with several multi-chambered agglutinated foraminifera (marked 1 through 3) in a lamina. (**b**) Close-up of specimen 1, showing the agglutinated nature of chamber walls. The agglutinated material consists of silicate grains (mostly quartz) of a narrow size range (<10 microns). (**c**) Close-up of specimen 2. The chamber walls consist of a thicker layer (in comparison to specimen 1) of agglutinated silicate grains, and the grain size range of agglutinated grains is broader (up to 30 microns). (**d**) Close-up of specimen 3. In this foraminifera a large portion of the chamber walls is covered with platy minerals, mica flakes and clay, although other portions are also covered with quartz and some feldspar grains. The specimen is an example of mixed mineralogy wall construction. (**e**) Detail of chamber wall (arrow) of specimen 1. (**f**) Detail of chamber wall (arrow) of specimen 2. Note the poorer sorting and thicker agglutinate layer when compared to specimen 1. (**g**) Detail of chamber wall (arrows) of specimen 3. Note dominance of platy minerals.

Figure 2. (continued) agglutinated foraminifera from the Santa Barbara Basin. The thinness of the wall reflects the fact that the agglutinate consists exclusively of platy minerals (clay and mica flakes). Bright grains in the interior of chambers are pyrite framboids and framboid clusters. (**e**) Close-up of the wall structure (between *arrows*) from (**d**) shows that the wall consists of overlapping clay and mica platelets. Sorting of platy particles in the foraminifera wall is significantly better than in the shale matrix to the left. Bright grains to the right of wall are pyrite framboids.

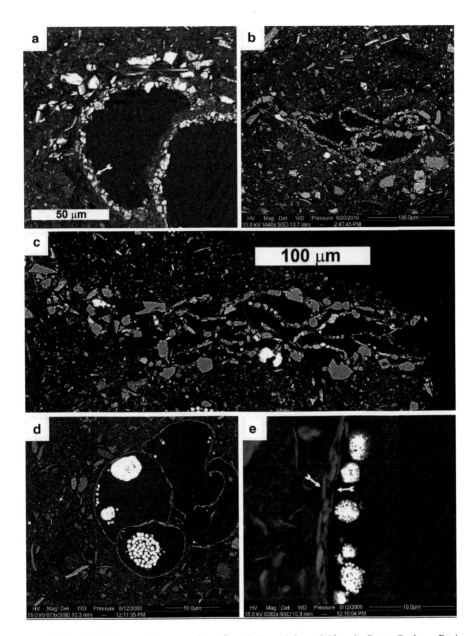

Figure 2. (**a–c**) Collapse and compaction of agglutinated foraminifera in Santa Barbara Basin sediments (SEM images). (**a**) Close-up of undeformed chamber of thick-walled (*arrow*) agglutinated foraminifera (similar to specimen 2 in Fig. 1). (**b**) Partially collapsed foraminifera with deformation of chamber walls, but chambers still partially preserved. The image also highlights size selective behavior of foraminifera. The material (mostly quartz) that has been incorporated into the chamber walls is finer and much better sorted than quartz in the surrounding mud matrix. (**c**) Almost completely collapsed multi-chambered foraminifera. When compaction is complete, the chamber interiors will no longer be visible and the foraminifer will be a horizontal streak of quartz grains. (**d**) A thin-walled multi-chambered

3.2. CLEVELAND SHALE, DEVONIAN

Lenticular bodies of fine-crystalline silica that have been interpreted as foraminiferal remains (Schieber 2009) occur, for example, in numerous petrographic thin sections of Devonian black shales (Fig. 3a). They are up to 500 μm in length and up to 50 μm thick, have a fine-crystalline cherty appearance (Fig. 3a, b), and were initially considered silicified organic remains. Their granular appearance in SEM

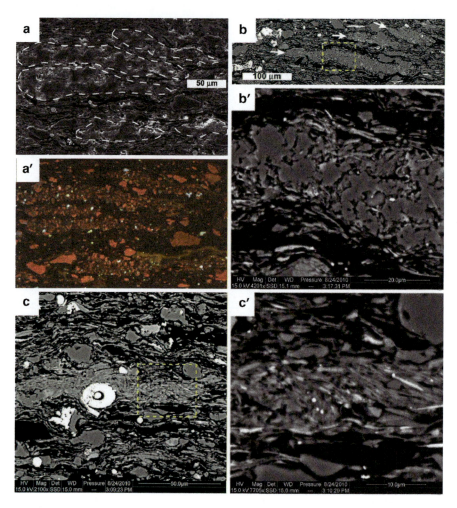

Figure 3. Petrographic features (SEM images) in the Devonian Cleveland Shale of Kentucky that are interpreted as remains of agglutinated benthic foraminifera (Schieber, 2009). (**a**) Multiple siliceous streaks in a shale sample, interpreted as agglutinated benthic foraminifera, are marked with *dashed lines*. (**a'**) A cathodoluminescence scan of the same area as in (**a**). Detrital quartz appears as *reddish to bluish* angular particles. The areas occupied by foraminiferal remains are invariably much finer grained

images also suggested that they might have originated as tiny ripples of fine silt that migrated over the muddy seafloor due to bottom currents (Schieber 2009).

Charge contrast imaging (CCI; Watt et al. 2000) and scanned color cathodoluminescence (CL) reveal that these features consist of discrete quartz grains set in a matrix of quartz cement (Fig. 3a'). Typically, only a few microns in size, the quartz grains are well sorted, angular, and distinctly finer-grained than the quartz silt in the surrounding shale matrix (Fig. 3a, a').

Whereas a strong argument has already been made that above cherty streaks are foraminiferal remains (Schieber 2009), new petrographic observations on the Late Devonian Cleveland Shale of Kentucky suggest that additional features in these rocks could be attributed to agglutinated foraminifera as well.

For example, elongate lenticular features (Fig. 3b) that consist of variable mixtures of quartz silt, clay, and mica flakes with quartz cement (Fig. 3b'), and of a size and morphology comparable or identical to cherty streaks are common in these shales. They occur in the same sediment layers as cherty streaks, and their morphology suggests a common origin. We also observe in these same layers bedding parallel streaks that consist almost exclusively of clays and mica flakes (Fig. 3c') and are again morphologically very similar to cherty streaks (Fig. 3c).

3.3. BARNETT SHALE, MISSISSIPPIAN

Cherty streaks like those seen in the Cleveland Shale also occur abundantly in the Mississippian Barnett Shale of Texas (Fig. 4a, b) and have likewise been interpreted as foraminiferal remains (Milliken et al. 2007). A re-examination of Barnett Shale samples from that study has revealed additional features in these rocks that are close analogs to potential foraminiferal remains seen in the Cleveland Shale (Fig. 3) and Santa Barbara Basin sediments.

For example, in the Barnett Shale, not all siliceous streaks consist entirely of quartz. There is a portion of them where part of the wall consists of a mixture of quartz, clay, and mica flakes (Fig. 4c). Other structures of comparable shape and size, and with medial suture, may consist entirely of clay minerals and mica flakes (Fig. 4d–f).

Figure 3. (continued) and better sorted than the surrounding shale matrix. (**b**) Another area in the same sample that shows morphological features (arrows) identical to the siliceous streaks seen in (**a**). These features differ in composition and consist of a mixture of quartz grains, clay and mica flakes. (**b'**) A close-up (dashed box) of the features in (**b**). Shows the textural and compositional difference of this feature to the shale below and above. Shows that the feature consists of a mixture of quartz grains with platy clay and mica flakes. (**c**) Another area in the same sample that shows morphological features (arrows) identical to the siliceous streaks seen in (**a**). These features consist entirely of clay and mica flakes. (**c'**) Enlargement of an area in (**c**) that is outlined by dashed box. The image shows clear textural contrast between feature and surrounding shale. The feature itself consists of aligned/shingled clay and mica flakes.

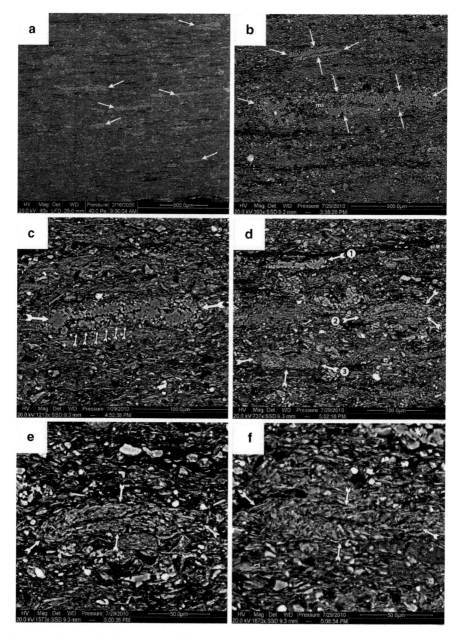

Figure 4. Petrographic features in the Mississippian Barnett Shale of Texas that are interpreted as remains of agglutinated benthic foraminifera. (**a**) Low magnification SEM image of Barnett Shale, bedding is close to horizontal. *Dark streaks* are rich in organic matter; *light streaks* (some marked with *arrow*) are silica-rich and consist of fine detrital quartz with quartz cement. (**b**) A close-up of two of these siliceous streaks (marked with *arrows*). Both streaks show a medial clay-rich portion (marked ms, "medial suture" in lower specimen). These features are interpreted as collapsed agglutinated foraminifera that

4. Implications

In modern suboxic seafloor settings, multiple species of benthic agglutinated foraminifera are able to persist (Bernhard et al. 1997), and thus finding several different strategies of chamber wall agglutination preserved in thin sections of SBB sediments (Fig. 1) is to be expected. Well-developed grain selectivity has been documented for agglutinated benthic foraminifera in such settings (Pike and Kemp 1996), and grain selectivity can therefore be considered a strong criterion for recognition in older sediments. In grain selection, foraminifera first accumulate random grains and then incorporate desired grain types and sizes into new chambers. The unused material is left behind as a "detritic heap" (Pike and Kemp 1996). Agglutinated grains are held together by organic cement, and in the samples studied by Pike and Kemp (1996), a preference for fine-grained quartz silt was noted. Pike and Kemp (1996) suggested that well-sorted streaks of quartz grains in the rock record may be interpreted as remains of agglutinated benthic foraminifera, and that poorly sorted streaks represent "detrital heaps" of rejected material. Yet, as we can see from the multiple styles of agglutination in modern SBB sediments (Fig. 1), some types of benthic foraminifera are "sloppy" with regard to grain size selectivity (Fig. 1c). Thus, in the rock record, poorly sorted streaks may at times also represent collapsed foraminiferal tests.

Fecal pellets of benthic organisms that ingest sediment and digest parts of the organic component, such as polychaete worms (Fauchald and Jumars 1979), are in the same size range as mineral streaks produced by collapse of agglutinated foraminifera in the SBB (Cuomo and Rhoads 1987; Cuomo and Bartholomew 1991), and in cross section would have the same lenticular morphology as collapsed agglutinated foraminifera (Fig. 5). Fecal pellets, however, lack selectivity with regard to grain size and mineralogy and have a randomized interior texture (Fig. 5). Also, unlike collapsed foraminifera, they would not show any evidence of medial sutures filled with a contrasting mineral matter (Fig. 4b, e, f).

In ancient shales, such as the Cleveland Shale and the Barnett Shale, lenticular bodies of fine-crystalline silica (Figs. 3 and 4) have been identified as the likely remains of benthic agglutinated foraminifera in recent studies (Milliken et al. 2007; Schieber 2009). The identification was made using criteria developed by Pike and Kemp (1996), such as morphology, and selectivity with regard to mineralogy and

---

Figure 4. (continued) armored their chamber walls with small quartz grains. (**c**) A close-up of a siliceous streak/agglutinated foraminifera (between *large white arrows*) with medial suture visible. In this case, parts of the wall structure are dominated by clay minerals and mica flakes; the section pointed out with small white *arrows*. (**d**) Overview image that shows a siliceous streak (*arrow* 1) as seen in (**b**) and also two comparable features (*arrows* 2 and 3) that consist of mica and clay mineral flakes. (**e**) Close-up SEM image of feature marked by *arrow* 2 in (**d**) shows a shingled structure of clays and micas and a medial suture. (**f**) Close-up SEM image of feature marked by *arrow* 3 in (**d**) shows a shingled structure of clays and micas and a medial suture. The features shown in (**e**) and (**f**) are interpreted as collapsed benthic foraminifera that armored their chamber walls with clay and mica flakes.

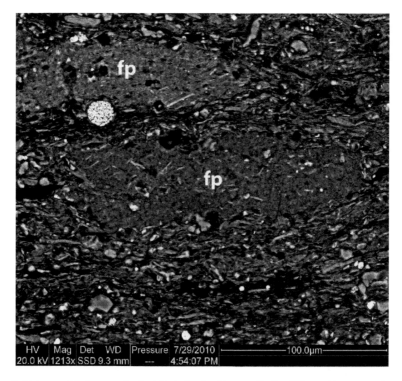

Figure 5. Petrographic features in the Mississippian Barnett Shale of Texas. For comparison with collapsed agglutinated foraminifera, two elongate lumps of material (marked *fp*) that are similar in morphology to siliceous streaks. Based on their random internal structure, these are interpreted as fecal pellets of benthic organisms.

grain size. That approach, however, because it limited itself to agglutinated foraminifera that constructed their chambers of quartz grains, has its limitations.

Obviously, agglutinated benthic foraminifera that coat their chamber walls with grains of mixed mineralogy do exist in modern sediments (e.g., Fig. 1d). In the rock record, as long as there is good morphological correspondence to the more readily identified remains of agglutinated foraminifera with well-sorted monomineralic tests (Figs. 3a and 4a), one can have some confidence that mixed mineral streaks (e.g., Figs. 3b, b' and 4 c) are foraminiferal remains as well.

Likewise, benthic agglutinated foraminifera that are selective toward clays and micas and incorporate platy grains of a narrow size range when compared with the size range available in the mud that surrounds them are evident in SBB sediments as well (e.g., Bernhard et al. 2006; and Figs. 1d and 2d), and such forms have also been reported from other modern seafloor sediments (e.g., Gooday et al. 2008). Thin-walled varieties like the one shown in Fig. 2d would of course yield a very thin mineral streak once compacted, but when high quality polishes (e.g., via argon ion milling) are used, the textural contrast and the better sorting may still alert the observer to the likely presence of a collapsed agglutinated foraminifera. The clay and mica dominated streaks from the Cleveland Shale (Fig. 3c, c') and the Barnett Shale (Fig. 4d–f) resemble the earlier described mineral streaks (quartz dominated and mixed mineralogy) in outward morphology, and that resemblance should probably be taken as a first hint that we may be dealing with yet another example of collapsed agglutinated foraminifera.

The other visual clue that indicates that these clay-mica streaks are unlikely to simply be squashed fecal pellets of benthic sediment ingesting organisms is the texture displayed by the platy minerals (Figs. 3c' and 4e, f) and the medial sutures in some examples from the Barnett Shale (Fig. 4). Sediment ingestion would randomize the particles (Fig. 5), and there is no known alternative mechanism that could produce the observed particle arrangements and the medial sutures.

The suspected clay-mica agglutinated foraminifera from the Cleveland and Barnett Shales are unlike the clay-agglutinated examples from the SBB in two respects; the coating of platy minerals must have been quite thick when compared to the SBB example in Fig. 2d, and the minerals show an overlapping shingled arrangement (Figs. 3c' and 4e, f) that is unlike anything reported from modern equivalents. It may represent a style of wall agglutination that is no longer used by modern agglutinated foraminifera, or it is a style that has thus far escaped notice in modern settings.

That mixed mineralogy (Fig. 1d) and clay-dominated (Fig. 2d) foraminiferal remains were not discussed by Pike and Kemp (1996) when they reported on agglutinated foraminifera from the SBB probably reflects the fact that these types tend to "blend" in more successfully with the shale matrix, and thus probably escaped further scrutiny in these earlier studies. Another factor that matters in that regard is the quality of the polished surface that is examined under the SEM. Unless the polish is of high quality, the necessary textural details tend to be less obvious.

In the SBB, bottom water oxygen levels are too small to permit more than meiofaunal bioturbators (Pike et al. 2001). Because of the absence of macroscopic bioturbation, the sediments retain a laminated appearance at the mm-scale (Pike and Kemp 1996; Schimmelmann and Lange 1996). Once these muds become part of the rock record, they will be laminated carbonaceous shales that contain compacted tests of agglutinated benthic foraminifera. Given the observations reported here, we should expect not only to find remains of agglutinated foraminifera constructed of quartz grains, but also collapsed grain assemblages that remain from mixed mineralogy foraminifera tests and those that are composed of clays and micas.

The SBB samples used in this study come from the deepest portions of the Santa Barbara Basin, where bottom waters are suboxic (oxygen content ~0.05 ml/l) and, through intermittent influx of oxygenated bottom waters, experience renewal events every few years (Bograd et al. 2002). The redox interface is located just beneath or at the sediment–water interface and marked by a mat of sulfide oxidizing *Beggiatoa* (Reimers et al. 1996). Intermittently, the redox interface may creep upward into the bottom millimeters to centimeters of the water column (Reimers et al. 1996). Other than that, the water column is never entirely devoid of oxygen. That this was the likely state of affairs for most of the recent as well as the more distant geological past is indicated by long-term monitoring (Bograd et al. 2002; Reimers et al. 1996), and the remains of agglutinated foraminifera that are found throughout the laminated deposits (Pike and Kemp 1996; personal core observations). Brief interludes of anoxic bottom waters do not preclude seafloor colonization by benthic agglutinated foraminifera, because a number of modern agglutinated benthic foraminifera can endure brief (a few months) spells of bottom water anoxia. Some oxygen, however, is required so that they can persist on the seafloor (Bernhard and Reimers 1991; Bernhard et al. 2003).

There have been recent reports of benthic foraminifera that can tolerate anoxic and even sulfidic conditions via foraminiferal denitrification (e.g., Risgaard-Petersen et al. 2006; Leiter and Altenbach 2010; Piña-Ochoa et al. 2010). The latter authors found intracellular nitrate pools in a number of benthic foraminifera (including agglutinated forms) that are considered to be facultative anaerobes. Their oxygen respiration rates were considerably higher than their denitrification rates. In light of the higher energy yield from aerobic respiration, this observation suggests that denitrification represents an auxiliary metabolic mode that supports the organism during temporary stays in oxygen-free environments, and that oxygen respiration may be required for growth and reproduction (Piña-Ochoa et al. 2010). In none of these studies has it been documented that any of the described species can survive anoxia through an entire life cycle, much less persist in such an environment for geologically relevant time spans (decades to millennia). Thus, although these observations indicate that certain agglutinated foraminifera are capable of intracellular denitrification, there is at present no evidence to suggest that this alternative metabolic pathway could have sustained them for multiple life cycles.

Identification of siliceous lenses (Figs. 2a and 3a), mixed-mineral lenses (Figs. 3b and 4c), and clay-mica lenses (Figs. 3c and 4d) in laminated Devonian and Mississippian black shales provides prima facie evidence that at the time of deposition, an assemblage of agglutinated benthic foraminifera lived at the seafloor. By comparison with modern sediments of the SBB, we may further presume that the respective bottom waters must have contained at least some oxygen (Pike and Kemp 1996; Bernhard and Reimers 1991; Bernhard et al. 2003). The preservation of laminae and the absence of macroscopic burrows on the other hand suggest that oxygen levels were too low to allow colonization by macrobenthos. As such, the situation with regard to bottom water oxygenation appears to be oxygen-depleted but not wholly anoxic. The observation of multiple types of relict agglutinated benthic foraminifera in laminated black shales of the Cleveland Shale and the Barnett Shale also indicates that in the Paleozoic suboxic sea-bottom environments were widespread in epicontinental seas. Analogous to what we observe in the SBB, brief anoxic interludes (a few months) probably occurred, because we know that benthic foraminifera from the Santa Barbara Basin can survive brief anoxia (Pike and Kemp 1996; Bernhard and Reimers 1991; Bernhard et al. 2003).

Whereas laminated black shales in the Illinois and Appalachian Basins have been interpreted as deposited beneath persistently anoxic and even euxinic bottom waters on the basis of geochemical proxies (Sageman et al. 2003; Rimmer 2004; Algeo 2004; Werne et al. 2002), observation of siliceous streaks (Milliken et al. 2007; Schieber 2009) served to indicate that other factors than persistent bottom water anoxia must have been at play to produce extensive black shales during the Late Devonian and the Mississippian. Finding a larger variety of agglutinated benthic foraminifera in these sediments further strengthens that argument and gives more credence to an alternative scenario where short-lived anoxia were produced by seasonal thermoclines, in conjunction with seasonal mixing (Van Cappellen and Ingall 1994). This "productivity-anoxia feedback" scenario allows high burial fluxes of organic carbon and is not in conflict with continued seafloor colonization by agglutinated benthic foraminifera (Van Cappellen and Ingall 1994; Sageman et al. 2003).

5. Conclusions

Through careful petrography and comparison to modern analogs, multiple types of agglutinated benthic foraminifera have been recognized in two ancient black shale successions. Given that the siliceous streak type of relict agglutinated foraminifera has already been observed in a wide variety of other black shales (Schieber 2009), it is likely that evidence of ancient assemblages of benthic agglutinated foraminifera will be uncovered in many other black shale successions as well.

Expanding our knowledge of benthic agglutinated foraminifera in ancient black shale successions does on one hand force re-examination of current concepts of organic carbon preservation in the past. On the other hand, by looking at these rocks from the perspective of apparently quite comparable modern analogs, we no longer need to postulate specialized non-uniformitarian conditions (such as ocean-wide anoxia) to explain their existence. Combinations of variables that we can observe in modern oceans (such as in the SBB) may have sufficed to produce widespread deposition of carbonaceous sediments at certain times in the past.

The described approach for the detection of agglutinated benthic foraminifera in black shales can readily be expanded to other examples from the rock record.

6. Acknowledgments

Research on Paleozoic black shales was supported by the Petroleum Research Fund administered by the American Chemical Society (grant # 38523-AC8). ESEM and color SEM-CL analyses were made possible through a NSF equipment grant (grant EAR 0318769). I also would like to thank David Valentine (UCSB) and Alex Sessions (Caltech) for their invitation to participate in a Santa Barbara Basin coring cruise.

7. References

Algeo TJ (2004) Can marine anoxic events draw down the trace element inventory of seawater? Geology 32:1057–1060

Bernhard JM, Reimers CE (1991) Benthic foraminiferal population fluctuations related to anoxia: Santa Barbara Basin. Biogeochemistry 15:127–149

Bernhard JM, Sen Gupta BK, Borne PF (1997) Benthic foraminiferal proxy to estimate dysoxic bottom-water oxygen concentrations: Santa Barbara Basin, U.S. Pacific continental margin. J Foraminifer Res 21:301–310

Bernhard JM, Visscher PT, Bowser SS (2003) Sub-millimeter life positions of bacteria, protists, and metazoans in laminated sediments of the Santa Barbara Basin. Limnol Oceanogr 48:813–828

Bernhard JM, Habura A, Bowser SS (2006) An endobiont-bearing allogromiid from the Santa Barbara Basin: implications for the early diversification of foraminifera. J Geophys Res Biogeosci 111:G03002, 10 pp

Bograd SJ, Schwing FB, Castro CG, Timothy DA (2002) Bottom water renewal in the Santa Barbara Basin. J Geophys Res 107:3216–3224

Bohacs KM, Grabowski GJ, Carroll AR, Mankiewicz PJ, Miskell-Gerhardt KJ, Schwalbach JR, Wegner MB, Simo JA (2005) Production, destruction, dilution, and accommodation – the many paths to source-rock development. Soc Sediment Geol 82:61–101, Special Publication

Bonsdorff E, Diaz RJ, Rosenberg R, Norkko A, Cutter GR (1996) Characterization of soft-bottom benthic habitats of the Åland Islands, Northern Baltic Sea. Mar Ecol Prog Ser 142:235–245

Cuomo MC, Bartholomew PR (1991) Pelletal black shale fabrics: their origin and significance. In: Tyson RV, Pearson TH (eds.) Modern and ancient continental shelf anoxia. Geological Society, London, pp 221–232, Special Publication No. 58

Cuomo MC, Rhoads DC (1987) Biogenic sedimentary fabrics associated with pioneering polychaete assemblages: modern and ancient. J Sediment Petrol 57:537–543

Fauchald K, Jumars P (1979) The diet of worms: a study of polychaete feeding guilds. Oceanogr Mar Biol Annu Rev 17:193–284

Gooday AJ, Todo Y, Uematsu K, Kitazato H (2008) New organic-walled foraminifera (Protista) from the ocean's deepest point, the Challenger Deep (Western Pacific Ocean). Zool J Linn Soc 153: 399–423

Hughen K, Overpeck JT, Peterson LC, Anderson R (1996) Varve analysis and palaeoclimate from sediments of the Cariaco Basin, Venezuela. In: Kemp AES (ed.) Palaeoclimatology and palaeoceanography from laminated sediments. Geological Society, London, pp 171–183, Special Publication No. 116

Leiter C, Altenbach AV (2010) Benthic foraminifera from the diatomaceous mud-belt off Namibia: characteristic species for severe anoxia. Palaeontol Electron 13(2):1–19

Loucks RG, Ruppel SC (2007) Mississippian Barnett Shale: lithofacies and depositional setting of a deepwater shale-gas succession in the Fort Worth Basin, Texas. Am Assoc Pet Geol Bull 91:579–601

Milliken KL, Choh S-J, Papazis P, Schieber J (2007) "Cherty" stringers in the Barnett Shale are agglutinated foraminifera. Sediment Geol 198:221–232

Pike J, Kemp AES (1996) Silt aggregates in laminated marine sediment produced by agglutinated foraminifera. J Sediment Res 66:625–631

Pike J, Bernhard JM, Moreton SG, Butler IB (2001) Microbioirrigation of marine sediments in dysoxic environments: implications for early sediment fabric formation and diagenetic processes. Geology 29:923–926

Piña-Ochoa E, Høgslund S, Geslin E, Cedhagen T, Revsbech NP, Nielsen LP, Schweizer M, Jorissen F, Rysgaard S, Risgaard-Petersen N (2010) Widespread occurrence of nitrate storage and denitrification among foraminifera and Gromiida. PNAS Proc Natl Acad Sci 107:1148–1153

Reimers CE, Ruttenberg KC, Canfield DE, Christiansen MB, Martin JB (1996) Porewater pH and authigenic phases formed in the uppermost sediments of the Santa Barbara Basin. Geochim et Cosmochim Acta 60:4037–4057

Rimmer SM (2004) Geochemical paleoredox indicators in Devonian-Mississippian black shales, central Appalachian Basin (U.S.A.). Chem Geol 206:373–391

Risgaard-Petersen N, Langezaal AM, Ingvardsen S, Schmid MC, Jetten MSM, Op den Camp HJM, Derksen JWM, Pina-Ochoa E, Eriksson SP, Nielsen LP, Revsbech NP, Cedhagen T, van der Zwaan GJ (2006) Evidence for complete denitrification in a benthic foraminifer. Nature 443:93–96

Russel AD, Morford JL (2001) The behavior of redox-sensitive metals across a laminated-massive-laminated transition in Saanich Inlet, British Columbia. Mar Geol 174:341–354

Sageman BB, Murphy AE, Werne JP, Ver Straeten CA, Hollander DJ, Lyons TW (2003) A tale of shales: the relative roles of production, decomposition, and dilution in the accumulation of organic-rich strata, Middle-Upper Devonian, Appalachian basin. Chem Geol 195:229–273

Schieber J (2009) Discovery of agglutinated benthic foraminifera in Devonian black shales and their relevance for the redox state of ancient seas. Paleogeogr Paleoclimatol Paleoecol 271:292–300

Schimmelmann A, Lange CB (1996) Tales of 1001 varves: a review of Santa Barbara Basin sediment studies. In: Kemp AES (ed.) Palaeoclimatology and palaeoceanography from laminated sediments. Geological Society, London, pp 121–141, Special Publication 116

Schulz H, von Rad U, von Stackelberg U (1996) Laminated sediments from the oxygen-minimum zone of the northeastern Arabian Sea. In: Kemp AES (ed.) Palaeoclimatology and palaeoceanography from laminated sediments. Geological Society, London, pp 185–207, Special Publication No. 116

Tyson RV, Pearson TH (1991) Modern and ancient continental shelf anoxia: an overview. In: Tyson RV, Pearson TH (eds.) Modern and ancient continental shelf anoxia. Geological Society, London, pp 1–24, Special Publication No. 58

Van Cappellen P, Ingall ED (1994) Benthic phosphorus regeneration, net primary production, and ocean anoxia; a model of the coupled marine biogeochemical cycles of carbon and phosphorus. Paleoceanography 9:677–692

Watt GR, Gruffin BJ, Kinny PD (2000) Charge contrast imaging of geological materials in the environmental scanning electron microscope. Am Mineral 85:1784–1794

Werne JP, Sageman BB, Lyons TW, Hollander DJ (2002) An Integrated assessment of a "type euxinic" deposit: evidence for multiple controls on black shale deposition in the Middle Devonian Oatka Creek Formation. Am J Sci 302:110–143

Biodata of **Alexander Volker Altenbach** and **Maren Gaulke**, authors of the chapter *"Did Redox Conditions Trigger Test Templates in Proterozoic Foraminifera?"*

Dr. Alexander Volker Altenbach is Professor for micropalaeontology at the Dept. of Earth and Environmental Science of the Ludwig-Maximilians-University in Munich (Germany) since 1994. He is one of the editors of this C.O.L.E. Volume; for more information see biodata of the editors.

E-mail: **a.altenbach@lrz.uni-muenchen.de**

Dr. Maren Gaulke studied geo- and bio-sciences and earned her Ph.D. at Kiel University in 1988, in cooperation with the Senckenberg Research Institute (Frankfurt/Main, Germany). She spent many years of field research, mainly in SE-Asia and on the Philippine Islands. Her biological studies revealed a number of new insights into the ecology, ethology, and trophic dependencies of tropic animals, including the description of new vertebrate and invertebrate species. At present, she cooperates as a consultant with national and international institutions in the status classification of endangered species and the initiation of protective areas.

E-mail: **mgaulke@web.de**

Alexander Volker Altenbach   **Maren Gaulke**

DID REDOX CONDITIONS TRIGGER TEST TEMPLATES IN PROTEROZOIC FORAMINIFERA?

ALEXANDER VOLKER ALTENBACH[1,2]
AND MAREN GAULKE[1]
[1] GeoBio-Center, Ludwig-Maximilians-University,
Richard-Wagner-Str. 10, 80333 Munich, Germany
[2] Department for Earth and Environmental Science,
Ludwig-Maximilians-Universität Munich,
Richard-Wagner-Str. 10, 80333 Munich, Germany

1. The Stem Group of Foraminifera

The genetically controlled formation of shells ("tests") in the protistan first-rank Foraminifera [Rhizaria] offers a considerable advantage in evolutionary and phylogenetic studies on their development and diversification. Fossilized tests reveal the evolutionary patterns for some thousand genera, from the Cambrian to the Holocene. The progressive complexity of internal and external morphologies and test-wall structures provides a diversification scheme, basic for their taxonomic grouping (Loeblich and Tappan 1964, 1988). The monothalamous (single-chambered), organic, membranous, or firm proteinaceous tests of the Allogromina (Precambrian to Recent) seemed to offer an appropriate stem group as the ancestors of the Textulariina (Cambrian to Recent) with an agglutinated test and the Fusulinina (Silurian to Permian) with calcitic test walls (Tappan and Loeblich 1988). But the fossil record of allogromiids is extremely scarce due to the low preservation potential of the purely organic test composition. Knowledge on allogromiid ecology, diversity, and taxonomy, and thus their test morphological evolution, is highly fragmentary as well (Gooday 2002; Habura et al. 2008). Therefore, phylogenetic considerations deduce an unknown, organic-walled or athalamous (naked) Precambrian ancestor. This impasse was overcome by molecular phylogenetic studies, revealing distinct genetic relationships for formerly divergent taxonomic groupings of primitive monothalamous Foraminifera, and the convergence of their organic and agglutinated wall structures. Xenophyophoreans, astrorhizids, allogromiids, and an athalamous freshwater foraminifer were combined in the "Monothalamida," replacing the allogromiids as the ultimate stem group for Phanerozoic clades (Pawlowski et al. 2003a; Flakowski et al. 2005).

Genetic clocks date the early radiation of Foraminifera between 690 and 1,150 Ma (million years before Present) by rRNA sequences (Pawlowski et al. 2003a), with a diversification branch in the actin family tree by 814 Ma

(Flakowski et al. 2006). Athalamous, allogromiid, and more or less agglutinated, monothalamous test constructions branch near the stem of the foraminiferal tree (Pawlowski et al. 2002). All multichambered lineages evolved from either naked, or proteinaceaous, or agglutinating Monothalamida (Longet and Pawlowski 2007). In sum, genetic approaches increase the time span for the evolution of Foraminifera by about one hundred to several hundred million years, in comparison to preceding considerations based on the fossil record.

Analysis of small subunit rDNA (SSU rDNA) ranges among the most common and valuable tracer for phylogenetic studies, but still carries quite a number of contradictions (Roger and Hug 2006; Yoon et al. 2008). Genetic clocks often reveal minimum estimates at best (Graur and Martin 2004). The pronounced heterogeneity of substitution rates common in amoeboid protists leads to difficult interpretations, a serious problem for the definition of the position of Foraminifera and their ultimate stem lineage (Pawlowski and Burki 2009). Despite some uncertainties, basic phylogenetic results were corroborated by recoveries from the fossil record. Morphological evidence for rigid, probably mineralized or agglutinated, mid-Neoproterozoic testate amoebae was supported by Porter and Knoll (2000). The authors point to the possibility of unrecognized Foraminifera among their remarkably diverse test collections. The recovery of monothalamid tests from late Neoproterozoic sediments is under discussion, although definitive taxonomic assignments are uncertain (Gaucher and Sprechmann 1999; Seilacher et al. 2003; Gaucher et al. 2005; Li et al. 2008). Further, because complex, multichambered textularids seem to appear in the lower to middle Cambrian (Scott et al. 2003), the phylogenetic calibration point for the branch of textularids from Monothalamia would drift from the Ordovician (Pawlowski et al. 2003a; Flakowski et al. 2006) toward the Cambrian/Ediacaran boundary. The oldest fossil structures under discussion for the foraminiferal lineage reach back deep into the Proterozoic (Dong et al. 2008; Matz et al. 2008), but these trace fossils are a matter of debate (Pawlowski and Gooday 2009).

A quite uncertain Meso- to a better-justified Neoproterozoic origin of early foraminiferan taxa seems indicated. To identify the structures and strategies that branched the Foraminifera from their Cercozoan ancestors (Pawlowski et al. 2003a), specific environmental controls of the upper Proterozoic must be considered.

2. Constraints for Proterozoic Environments

During the Proterozoic eon, the atmospheric pO_2 ranged somewhere between 1% and 10% of modern values (Canfield 2005; Kump 2008). Reconstructions of Proterozoic marine-oxygenation stages differ somewhat (Holland 2006; Komiya et al. 2008), but most agree in the subdivision of slightly oxygenated surface waters, underlain by sulfidic to anoxic deeper waters (e.g., Anbar and Knoll 2002). Redox (reduction–oxidation reaction)-dependent behavior of nutrients and trace metals subdivided the oligotrophic open-ocean surface waters from the enhanced nutrient availability of shallow coastal regions, a key factor for the evolution of green- and red-algal lineages (Anbar and Knoll 2002; Katz et al. 2004; Anbar 2008). Biogenic oxygen production was the source for both atmospheric and marine pO_2. Thus, local biogenic dynamics may have been more influential than implied by large-scale reconstructions obtained from geochemical tracers (Butterfield 2009). Microbial mats with phototrophs offered enhanced oxygen concentrations before diffusional processes to surface waters and to the atmosphere. In general, it is agreed that prolonged retention of oxygen existed in estuarine and coastal waters due to lowered salinity or temperature (Rothman et al. 2003; Knauth 2005). Knauth and Kennedy (2009) even assume an explosion of continental photosynthesizing communities by the Neoproterozoic. All these factors raise doubts as to whether present knowledge of Precambrian paleoenvironmental oxygen conditions allows confident inferences of the ecology of earliest eukaryotes.

Whenever local productivity or flux rates of organic matter from the water column reached the benthic boundary layer at pO_2 levels above the Pasteur point, aerobic respiration offered elevated metabolic conversion rates for benthic life. On the other hand, such immediate oxidation of reduced organic matter enhanced nutrient recycling and the pelagic-benthic coupling. Nitrate and sulfate concentrations reached peak concentrations in shallow coastal waters (Anbar and Knoll 2002; Katz et al. 2004). On the other hand, the substantial oxygen demand for aerobic respiration constantly impacted the meager oxygen levels (Butterfield 2009). The balance of oxygen availability is not easily deciphered from biogenic production and consumption, as oxidation of inorganic compounds from weathering and the regeneration of aged biogenic carbon pools inhibits a closed biogenic cycling of oxygen (Gaidos et al. 2007). These parameters have to be considered unstable during most of the late Proterozoic (Rothman et al. 2003); thus, the shallow benthic boundary layer most probably was attributed to severe fluctuations in redox conditions (Komiya et al. 2008). Anoxia may have been induced when the underlain sulfidic waters ascended, or when storm- and tidal-induced hydrodynamics resuspended deeper sulfidic sediment layers. Organic flux rates and pelagic-benthic-coupled food webs, somewhat comparable to modern dynamics, occur at the end of the Neoproterozoic. The accelerated removal of organic matter from surface waters was bolstered by the spread of pelagic fecal pellet fluxes, algal production of resistant biopolymers, and biomineralization (Rothman et al. 2003). Far-spread cyanobacterial (photosynthetic) microbial mats may have offered additional support for the propagation of benthic eukaryotes (Gaidos et al. 2007). Excess oxygen produced by phototrophic activity within bacterial mats will accumulate in bubbles when captured by the bacterial mucus. This attractively rich source of oxygen accompanied the food offered by phototrophic and heterotrophic microbes for the eukaryotes.

Many natural redox reactions are very slow in reaching equilibrium; thus, different redox levels may co-occur and retain prolonged in the same locale (Stumm 1984). When there are no larger organisms to bioturbate, the path to stationary redox conditions relies first on the mediating impact of microorganisms

(Frevert 1984; Grundl 1994). When phototrophic oxygen production and diffusion of hydrogen sulfide from deeper sediment layers intermix at the benthic interface, the redox cline shifts from the sediment column into the overlying waters and back into sediments in diurnal phases (Schulz and Jørgensen 2001). Generalized "oxic" or "anoxic" conditions can be attributed over larger distances, but not in terms of microhabitats. Centers of bacterial clusters with differing respiratory demands, inputs and outputs, and redox conditions can easily establish and persist in neighboring micro-compartments. As a result, the intra-sedimentary redox cline is not an even surface. It more resembles a three-dimensional, interwoven, submillimeter-scaled intercalation of differing redox conditions, even in laminated sediments (Bernhard et al. 2003). Such compartmentalization must be considered ubiquitous for the benthic boundary layer in Proterozoic surface waters. Benthic shallow-water environments must have been continuously exposed to alterations in redox conditions in both space and time, and thus were not anoxic or oxic per se.

Even oxygen concentrations below the Pasteur point would not necessarily negate the possibility that all basal radiations developed from mitochondrial eukaryotes (Cavalier-Smith 2002), if mitochondria, hydrogenosomes, and mitosomes were derived from anoxic metabolisms (Mentel and Martin 2008; Falkowski and Godfrey 2009). Recently, a modified anaerobic mitochondrion was described from a deep-branching anoxic eukaryote, casting new light on the ancestral position of mitochondriate and amitochondriate organisms (Minge et al. 2009). Other authors, in contrast, consider all anaerobic function of mitochondria and modern amitochondriates as secondarily derived from aerobe ancestors (Embley et al. 2003; van Hellemond et al. 2003; Embley 2006). Within the last decade, modern anoxic environments revealed a burst of eukaryote biodiversity, including enigmatic kingdom levels and new cercozoan groups (Dawson and Pace 2002; Stoeck and Epstein 2003; Takishita et al. 2005a, 2007; Bass et al. 2009; Edgcomb et al. 2009; Fenchel and Finlay 2009; Stoeck et al. 2009; Danovaro et al. 2010). Even a new biological pathway for oxygen production was discovered recently, based on nitrite-driven methane oxidation (Ettwig et al. 2010). At present, the general discussion on metabolic processes, pathways, and specifically anoxic or oxic eukaryote ancestral lineages, remains open. Regarding foraminifera, we have to focus on the questions to what extent foraminifera are able to cope with the oxygen-depleted to anoxic and sulfidic conditions of benthic Proterozoic environments, and to assess which strategies likely were employed by these basal ancestors.

3. Foraminifera and Anoxia

Many microaerophilic protists grow in anoxic conditions based on fermentative metabolism or nitrate respiration, all considered "facultative anaerobes" (Fenchel and Finlay 2008). Only during the last decade, this behavior was revealed in more detail for Foraminifera. Their autogenic and complete denitrification was observed in the laboratory, and subsequently corroborated by field studies (Risgaard-Petersen et al. 2006; Høgslund et al. 2008; Glud et al. 2009). Denitrification offers independence from supply of other electron acceptors (i.e., oxygen); thus, all denitrifying foraminifera were considered facultative anaerobes by Fenchel and Finlay (2008). Foraminifera were also mentioned among the relevant unknowns for biogenic ammonia oxidation (Francis et al. 2007). But observed species-specific intracellular nitrate concentrations span several orders of magnitude, raising the question whether this defines a complete alternative metabolic path, or an auxiliary pathway to bridge longer periods of anoxia (Høgslund 2008). Foraminifera may be responsible for more than 50% (Høgslund et al. 2008) or just 4% (Glud et al. 2009) of local N_2 production derived from denitrification. Both ranges would indicate a noteworthy share, rather than an aberrant biochemical path with negligible impact.

Detailed microstructural investigations of complexes of peroxisomes, endoplasmic reticulae, and mitochondria lead to the definition of a metabolic model, by assumption of their collaboration in the oxidative phosphorylization of hydrogen peroxide (Bernhard and Bowser 2008). This process would yield free O_2 for the metabolic demand of the foraminifer, not despite, but because of oxygen depletion and the accumulation of free radicals. Chloroplasts sequestered from phototrophs can remain active in the foraminiferal cytoplasm for several months to more than 1 year (Correia and Lee 2002; Grzymski et al. 2002). Such kleptoplastic endobionts frequently occur in taxa recovered from dysoxic to anoxic pore-water conditions, in the photic zone as well as in the abyss. Under sufficient light conditions, photosynthesis may be functional (Cedhagen 1991; Bernhard and Bowser 1999). Below the photic zone, other biochemical paths must be considered (Grzymski et al. 2002). The most striking one is the use of free radicals, congruent to the model of Bernhard and Bowser (2008).

The first proof for an anaerobic foraminifer was given by Bernhard et al. (2006). This deep-sea allogromiid thrives in anoxic, sulfidic environments and harbors prokaryotic endosymbionts. Due to the lack of complex buildings or kleptoplasts in the cytoplasm of this hitherto undescribed species, either the performance of the symbionts or an unknown supplemental mechanism must be elemental for its metabolism. Allogromiids are well adapted to marine extreme levels in acidity and pCO_2 (Bernhard et al. 2009a).

Leiter (2008) was able to trace four calcitic foraminiferal taxa down to a 10–19 cm sediment depth in sulfidic and anoxic sediments on the Namibian shelf (*Virgulinella fragilis, Nonionella stella, Fursenkoina fusiformis,* and *Bolivina pacifica*). Sulfate reduction rates reach maxima at the surface or within the uppermost centimeters of the sediment column, providing a constant net flux of hydrogen sulfide out of the sediment column into the bottom water (Brüchert et al. 2003). When this sulfide is metabolized by chemolithotrophic bacteria in the benthic boundary layer, anoxic bottom waters extend some meters above the sea floor (Lavik et al. 2009). Intracellular nitrate concentrations of 35 millimole (mM) infer *N.* cf. *stella* as a candidate for denitrification (Piña-Ochoa et al. 2010).

Other taxa of the genera *Fursenkoina* (regarded *Stainforthia* by some authors) and *Bolivina* range up to 180 mM (Piña-Ochoa et al. 2010). Endo- and ectobiotic bacteria were observed in *Virgulinella fragilis* and *Bolivina pacifica* respectively, which may support specific metabolic paths (Bernhard 2003; Bernhard et al. 2009b) under long-term, sulfidic oxygen deficiency (Leiter and Altenbach 2010).

Denitrification and oxidative phosphorylation of free radicals offer an alternative electron donator when dissolved oxygen is a limiting source for foraminifera. Interactions with endo- or ectobiotic bacteria and/or sequestered chloroplasts seem to play important roles for this specific metabolic path (Bernhard and Bowser 1999, 2008; Bernhard 2003; Tsuchiya et al. 2006, 2009; Høgslund et al. 2008; Bernhard et al. 2009b). It is worthwhile to note that such specific adaptations seemingly remain rare and exceptional in modern oxygenated oceans. But they must be considered key factors for the early evolution of foraminifera during the much less oxygenated Precambrian (Richardson and Rutzler 1999; Bernhard et al. 2006).

4. Functional Morphology and the Evolution of Precambrian Foraminifera

Morphologically, monothalamous Foraminifera appear somewhat simple, irregular, and/or "primitive." For the functionality of their tests, early researchers considered nothing but heavy weight-preventing buoyancy (Marszalek 1969). But after closer examination, their approach to access the laminated flow in order to reach elevated food sources from increasing hydrodynamic shear can reveal structural and technical optimizations (Altenbach et al. 1987). Anchoring in the sediment, flexibility of reticulopodial mountings or shape, size, and rigidity of tests often correspond with known environmental parameters and trophic behavior (Nyholm 1957; DeLaca 1982; Cedhagen 1988, 1993; Levin and Thomas 1988; Tendal and Thomsen 1988; DeLaca et al. 2002). A number of specimens were observed to leave the test by complete detachment, moving away as a naked amoeboid (Marszalek 1969; Cedhagen and Tendal 1989). The former author cites other colleagues who observed the return and resettlement by the test constructor, and the latter authors point to the erratic descriptions of amoeboid species from naked foraminifers sampled in the field (see also Bowser et al. 1995; Heinz et al. 2005). With regard to the limited number of long-term cultivations and observations published on monothalamous foraminifera, test detachment might not be a rare abnormality. The tests of monothalamous foraminifera offer useful and appreciated functionality for their constructors, rather than an indispensable prerequisite for survival. Foraminiferal tests differ from shells of other testate amoeboids by an ontogenetic growth, confluent with a tendency toward increasing internal subdivision and compartmentalization. Even strait tubular forms may show internal constrictions or cavities within the agglutinated wall. The combining or separating usage of such subsections can be considered a task for the reticulopodia.

A set of five primary and five derived structures of the reticulopodia offer diagnostic value (Bowser and Travis 2002), and a molecular synapomorphy could be issued from the foraminiferan small subunit rDNA (Habura et al. 2004). The more conservative term "granuloreticulosean protist" used in both elaborations was meanwhile specified for the rank Foraminifera within the monophyletic supergroup Rhizaria (Adl et al. 2005; Pawlowski 2008) and reassigned for the naked reticulose Cercozoa (Bass et al. 2009). The first polyubiquitin insertion after the development of reticulopodia discussed in Bass et al. (2005) tends to develop into a foraminiferal synapomorphy against all other rhizarian groupings. The next branching subphylum considered in Bass et al. (2005), Endomyxa, was meanwhile emended and extensively revised (Bass et al. 2009). The new order Reticulosida now comprises all naked and truly reticulose Cercozoa. Families of this order can be easily separated by ultrastructural and genetic analysis from Foraminifera, albeit one exception: the family Rhizoplasmidae. For this family, however, Bass et al. (2009) consider that their members ultimately may be determined as naked foraminifera. The genetic and morphological definition of the uniqueness of the foraminiferal reticulopodial system is currently investigated. Bowser and Travis (2002) raised the question whether test or reticulopodia derived first. Most of such "chicken or egg problems" seem funny in first view, but often point to crucial problems in comparative physiology.

4.1. THE CELLULAR VIEW

The bidirectional flow of organelles and membranes in the densely anastomosing network of foraminiferal granular reticulopodia offers a perfect tool for handling internal and external micro-compartments with differing chemical and biochemical composition. Foraminiferal pseudopodial networks have three extraordinary features that optimize this usage: (1) their forceful and far reaching extension into distant and multiple microenvironments; (2) the logarithmic range control of the cellular surface-volume ratio by extension and retraction of reticulopodia; (3) the rapid internal and external bidirectional flow (Bowser and Travis 2002). As observed by light microscopy, reticulopodial extensions may reach into the environment with a multiple of the individual test size (Jepps 1942; Linke and Lutze 1993). Extensions of 15 mm even might be prolonged by reticolopodia hidden to light microscopy due to diameters of much less than 1 µm (Jahn and Rinaldi 1959; Rinaldi and Jahn 1964). Reticulopodia can be bundled, bent, or turned into spirals exhibiting large tensional forces. They are sufficient for the breakage of large opaline diatom tests (Austin et al. 2005), or the capture and dissection of larger invertebrates (Bowser et al. 1992). Foreign particles continuously stick to the bidirectional flow and therefore may pass each other going in opposite directions. Reticulopodia sectioned at both ends to a length of only 40 µm still show continued bidirectional flow (Jahn and Rinaldi 1959). Therefore, the distal and proximal bidirectional flow in reticulopodia follows a continuous automatic mechanism of attraction and adhesion or ingestion, rather than a simple strategy of search and hunt (Travis et al. 2002).

Globular cell size reaches universal thresholds derived from diffusional processes, such as pO_2 (Catling et al. 2005). Reticulopodia combine bacterial surface to volume ratios with meiofaunal biovolumes (Richardson and Cedhagen 2001; Cedhagen and Frimanson 2002). This enhancement of surface area for metabolic exchange and nutrient procurement was speculated a central driving force for the evolution of reticulopodia, in combination with their far reaching extension into distant, nutrient-laden microenvironments (Bowser and Travis 2002). Perhaps the most efficient usage of all the capabilities foraminiferal reticulopodia offer was considered by Høgslund (2008): the transport of electron donors and acceptors from distant microenvironments. Her speculation about "drinking nitrate through a straw" enlarges our assumptions on what purposes are reasonable for their gradual evolution. A first laboratory attempt to prove this, however, failed (Koho et al. 2011). Nevertheless, most likely due to such enhanced transport capabilities for all kinds of electron donors and acceptors, Foraminifera may metabolize some ten times more rapidly than naked amoebae of equivalent size (Hannah et al. 1994). In addition, one should not underestimate the sensory capabilities of reticulopodia required for all respiratory and mechanical procedures mentioned. The communication with external phototrophs and the logistic management of biochemical inputs and outputs of several thousand phototrophic symbionts provide a vivid picture of their abilities (Anderson and Lee 1991; Lee et al. 1991).

Two cytoskeletal elements are basic for pseudopodial movement: actin and tubulin. These proteins can reversibly polymerize into filaments and microtubuli, providing fundamentals for eukaryotic cell dynamics (Doolittle 1995). Their functional connection to numerous other proteins providing counter bearings and ATP-driven movement is so essential that their rate of evolutionary modification is conservative, roughly 10% in one billion years. Therefore, these proteins, ubiquitous in eukaryotes, should allow judgment about common ancestry by simple sequence inspection (Doolittle 1995). Actin and tubulin analysis on foraminifera succeeded in ending a decade's lasting taxonomic debate on *Miliammina fusca* as either of miliolid or textularid origin (Fahrni et al. 1997; Habura et al. 2006). Bidirectional high-speed movement and continuous reorganization in reticulopodia (Koonce and Schliwa 1985; Bowser and DeLaca 1985; Rupp et al. 1986; Welnhofer and Travis 1997; Travis et al. 2002) revealed distinct specialization, quoted as the "aberrant tubulin evolution" in foraminifera (Takishita et al. 2005b). Such unique and phylogenetically distinct characters were revealed from the actin paralogs ACT1 and ACT2 (Flakowski et al. 2005), and from spliceseosomal introns in foraminiferal actin sequences, derived from internal, parallel gains (Flakowski et al. 2006).

4.2. EARLY TEST COMPOSITIONS

Foraminiferan tests may show organic, agglutinated, high-magnesium calcitic, calcitic, aragonitic, or opaline wall compositions. The biomineralization of calcareous tests may have arisen independently in various lineages (Tappan and Loeblich 1988). Indeed, convergent evolution can be considered a common phenomenon in Foraminifera, due to the repeated and even multiple appearances of test structures and modifications in unrelated foraminiferal lineages (Loeblich and Tappan 1988). But two traits of test formation are replicated in major lineages: the formation of temporal or permanent organic linings, and the sporadic or temporal agglutination of foreign particles. Some extant monothalamous taxa still resemble a provisional position in-between the organic-walled allogromiids and the agglutinated, early textularid wall structures (Bowser et al. 1995; DeLaca et al. 2002; Gooday et al. 2008a). Inner organic linings (IOL) occur in all modern agglutinated Foraminifera investigated (Bender 1995), as well as in lowermost Cambrian records (McIlroy et al. 2001; Winchester-Seeto and McIlroy 2006). They are common in many fossil and extant taxa from all major taxa (Tappan and Loeblich 1988). The creation of temporary IOL is often related to reproduction, chamber formation, or specific ontogenetic stages, whereas outer organic linings seem restricted to multichambered foraminiferans (Bender 1995; Goldstein 1999). The agglutinated miliolid, *Miliammina fusca*, lost its calcareous test by secondary loss of calcification (Habura et al. 2006; see also Flakowski et al. 2006). Despite this massive and unusual phylogenetic alteration in test formation, the organic cement of this species still resembles structures identical to allogromiid and agglutinated tests (Gooday et al. 2008a). The large-sized miliolid *Miliolinella subrotunda* is capable of producing an even larger flexible organic tube, attached outside of the test (Altenbach et al. 1993). Equivalence and homology of organic test walls in allogromiids and the inner organic linings of agglutinated tests is discussed in Gooday et al. (2008a, b). At present, no contradiction is known for considering IOL and allogromid walls as homologies in Monothalamids. This is corroborated by the conspicuous similarities observed by several authors, and the fact that foraminiferal secretion of organic material as well as the production of organic linings and organic cementations are likely caused by one intracellular mechanism (Langer 1992). Protein-containing matrices cover biomineralizations in virtually all eukaryotic microfossil groups (King 1977), and thus most probably point to plesiomorphy of organic lining production from far ancestral lineages. The diverse composition of these proteins, however, is considered group specific, and thus might reveal phylogenetic lineages (King 1977).

Benthic foraminifera from the early branching taxa to most recent lineages retain the ability to rapidly agglutinate detrital canopies covering their test partially or completely. The lumps or cysts are produced facultatively, built and left within hours, or inhabited for longer periods (Heinz et al. 2005). They may be loosely consolidated (Linke and Lutze 1993: Plate III, Fig. F) more uniformly and rigidly (Licari and Mackensen 2005: Plate II, Fig. 10), or show combinations of globular and branching tubular shapes (Gross 2002: Plate 1, Fig. 2). Heinz et al. (2005) provide an extensive and thoughtful review of foraminiferal cyst production. They discuss the possible synapomorphy of cysts, but consider plesiomorphy derived from ancestral lineages. Trophic adaptations and bacterial gardening were often invoked as possible explanations for the production of temporal cysts and envelopes (Gooday 1986; Linke and Lutze 1993; Goldstein and Corliss 1994; Richardson and Cedhagen 2001; Heinz et al. 2005). But cyst building associated with reproduction (Goldstein and Moodley 1993: Plate 1, Fig. 4), chamber formation, and several other environmental constraints were also observed (see Heinz et al. 2005 for additional references).

Athalamous foraminifera have not yet been recovered from the fossil record, but linked to the stem group by similarity of DNA sequences (Pawlowski et al. 2003a; Li et al. 2008). However, the former authors note that the ancestor of the stem group might have been testate, and thus athalamous Foraminifera would result from complete test reduction. Moreira et al. (2007) consider that even the ancestral rhizarian might have been endowed with a test during its trophic phase. Therefore, the recovery of a primitive globular test from Proterozoic sediments will not reveal the ancestral foraminiferan irrefutably.

4.3. THE SEDIMENTARY SURROUNDING

Microbial mats may explain most of the life styles observed in benthic organisms of the shallow Late Proterozoic ocean; larger motile benthic life was confined to the lower and upper border of the erosion-resistant mats, due to the denseness of mat construction (Seilacher 1999). These mats offered rich sources of oxygen and food assessable for the reticulopodia, but a rigid test could severely hinder infiltration. This scenario infers that food acquisition and early test formation occurred in separate places. But which conditions were sufficiently beneficial from the start for an ancestral foraminifer to collect, rearrange, and agglutinate foreign particles around its naked or testate cell? We may neglect protective shelter or complex functionalities observed in modern Foraminifera. The former reason might have been induced by testate ancestors already, and both constructs would require rigidity, and thus enhance potential fossilization. In contrast, specific trophic reasons confer benefits from the start of a softly arranged agglomeration, and convincingly fit with inferred Precambrian environmental conditions.

Exopolymers are commonly exuded by single- and multicellular organisms for attachment, as a flotation aid, supporting locomotion and feeding, for building biofilms, for protection against harsh environmental conditions, as a barrier against attack by pathogens, parasitic organisms, and predators, and even for communication. Their ubiquitous appearance is considered essential for the biochemical functioning of all aquatic ecosystems (Wotton 2005). Mucus tracks produced by epiphytic foraminifera are a fertile substrate for bacteria and fungi. Harvesting the overgrowth on their formerly produced traces was observed for foraminifera from different orders; thus, a specific farming strategy was proposed by Langer and Gehring (1993). Even naked freshwater foraminifera produce such tracks (Koonce and Schliwa 1985). Extracellular digestion was observed in a number of monothalamous genera (DeLaca et al. 1981, 2002; DeLaca 1982; Richardson and Cedhagen 2001). The hydrolytic activities of the monothalamous taxa, *Hyperammina* sp. and *Reophax* sp., may be so significant that surrounding bacteria directly benefit from the foraminiferal metabolism (Meyer-Reil and Köster 1991; Köster et al. 1991). Therefore, the excretion of exoenzymes, the establishment of affiliated bacterial clusters, and the uptake of dissolved nutrients are considered common phenomena in Foraminifera (Richardson and Cedhagen 2001). Bacteria often obtain metabolites from the excretions of other microbes. This process can enhance the effectivity and growth of the community, and thus may rapidly favor evolutionary specialization and adaptation of the partners (Harcombe 2010). A spatially structured environment that centers the beneficial byproducts for reciprocation is an essential prerequisite. It assures conditions sufficient for the origin of such cooperation from the start (Harcombe 2010).

Comparable to foraminiferal mucus, the tracks of nematodes significantly support bacterial growth and remineralization rates (Leduc and Probert 2009). The mucus triggers specific patterns of bacterial colonization and succession, leading to substantial effects in food-web interactions (Moens et al. 2005). Riemann and Helmke (2002) assume that nematodes produce agglutinated detrital lumps or burrows in order to participate in these enzymatic peak activities. Both parties invest their differing enzyme pools, cyst constructor as well as the microbes settling in the compartments, and cavities of the detrital agglomeration. Resulting solutes of detrital organic matter can overcome the lack of specific proteolytic enzymes of one party. Therefore, gathering one's own exudates and incompletely degraded organic matter that is transferred to higher levels of the food chain by bacterial counterparts seems quite effective. Parallel to the increasing number of bacteria available as food for the producer of the agglomeration, it enhances its metabolic pathways by enzyme sharing.

Xenophyophoreans provide some hybrid of the functional and trophic concepts discussed above. Knowledge on modern xenophyophors is still restricted, but at least some of them are definitely foraminifers (Pawlowski et al. 2003b; Pawlowski 2008). Commonly found in modern bathyal to hadal environments, they produce comparably fragile, but large and voluminous tests, in contrast to their much smaller biovolume (Gooday and Nott 1982; Gooday and Tendal 1988). The test interior can show strong compartmentalization (Tendal 1994), a trait that offers functionality comparable to numerous agglutinated taxa with alveolar or canaliculate wall structures (Loeblich and Tappan 1989). Foraminiferal tests can act as an environmental buffer against external perturbations (Marszalek 1969; Marszalek et al. 1969). Internal compartmentalization, on the contrary, supports the establishment of chemically differing microenvironments within the test interior. Such biochemical management culminates in larger Foraminifera with thousands of phototrophic symbionts. The spatial arrangement of the symbionts, the intake of nutrients supporting photosynthesis, and the consumption or elimination of exudates are controlled by the foraminiferal host (Anderson and Lee 1991).

Microscopic barite crystals in the endoplasm of xenophyophores were considered a proof for their biogenic barite precipitation (Hopwood et al. 1997;

Bertram and Cowen 1997). However, these rhizarians are typical members of modern oligotrophic deep-sea environments. They ingest considerable amounts of highly refractory organic matter in an oligotrophic environment, and produce large amounts of waste pellets (stercomata). This offers an exceptional food source for deep-ocean microbial communities (Nozawa et al. 2006). Gardening of a specialized bacterial flora growing on the stercomata was postulated by Tendal (1979), including suggestions regarding exudate exchange. The existence of a specific bacterial community degrading the stercomata within the test was shown by Laureillard et al. (2004), despite remarkable interspecific and even interoceanic differences. Such affiliated microbial communities may alter the redox conditions when degrading the stercomata within the test compartments, a factor that should contribute to the observed intracellular agglomeration of barite crystals. Barite precipitation is restricted to marine microenvironments with sulfate and large amounts of decaying organic matter (Sternberg et al. 2008). Authigenic precipitation only occurs in heavily oxygen-depleted or anoxic ocean realms, such as the Black Sea (Monnin et al. 1999). These authors mentioned that biogenic barite production would follow an energetically unfavorable pathway. However, barite crystals were also observed in the facultatively anaerobic, denitrifying ciliate *Loxodes* (Finlay et al. 1983).

5. The "Envelope Hypothesis"

In the absence of agglutinated cysts or tests older than the Neoproterozoic to Cambrian boundary strata (Lipps 1992), any further assumptions on ancestral cyst and test building must of course take into account known ecological or structural concepts and reasons. And we have to admit that oxygen concentrations, as well as related processes and their timing, are highly debatable (Canfield 2005; Wille et al. 2008, 2009; Jiang et al. 2009). We can summarize the most important conclusions as follows: (1) during the oxygen-deficient Precambrian, the adaptation to negative redox conditions was a key factor for the early evolution of foraminifera; (2) several extant Foraminifera adapt to long-term anoxic and sulfidic conditions, and a number of extant species from several lineages use electron acceptors other than dissolved oxygen; (3) bidirectional streaming in reticulopodia offers a perfect tool for the concurrent utilization and linkage of diffusional gradients, redox-dependent electron acceptors, and nutritional offers dispersed in environmental microhabitats, as well as for internal test compartmentalization and endosymbiosis; (4) the functional relationships with sequestered organelles, ecto- and/or endosymbionts, and enzyme sharing is considered much more influential for Precambrian times than hitherto expected; (5) the formation of agglutinated structures surrounding the rhizarian most probably is derived from trophic and symbiotic interactions, due to resulting beneficial conditions from the earliest formation on.

We propose four distinct steps suitable for the development of cyst and test production in foraminifera (Fig. 1):

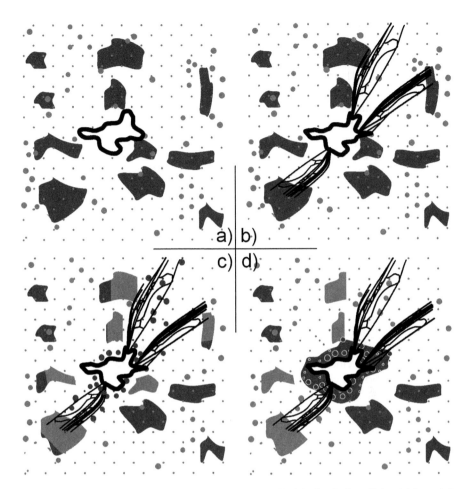

Figure 1. The four steps which combine the development and habit of reticulopodial activities and the evolvement of agglutinated envelopes in Foraminifera.

(a) A rhizarian with short lobopodia or filopodia comes into contact with only a small range of microenvironments providing electron acceptors and donors appropriate for their metabolism. This organism depends on a strategy of search and hunt. At the border between anoxic to microxic conditions (i.e., redoxcline, anoxic–oxic interface), motility is an enforced key strategy for many facultative anaerobic prokaryotes and eukaryotes (Schulz and Jørgensen 2001; Fenchel and Finlay 2008). In case of a testate amoeboid, this task is considerably energy consuming (Gross 2002) and reduces the overall metabolic efficiency.

(b) Far reaching reticulopodia allow concomitant utilization of multiple sources from distant microenvironments. High tensional forces and rapid bidirectional flow of organelles in reticulopodia are perfectly suitable for this strategy.

This concept is identical to the behavior of large sulfur bacteria near the redox cline, when shuttling between sediment spots providing reduced sulfur in one location, and an appropriate electron acceptor in another (Schulz and Jørgensen 2001; Høgslund 2008).

(c) Adhesion and dislocation of sediment particles is an automated process in reticulopodia, especially when multiple bundles are stretched and tightened toward an attractive source. The sedimentary structure of the surrounding becomes reorganized. Near the redox cline, species were observed to either arrange a free space directly in front of the aperture, or agglomerate numerous particles around their test (Linke and Lutze 1993). This process gives a starting phase for the selective arrangement of an extracellular space surrounding the naked or testate foraminifer. Substrate and porosity can be controlled, what alters diffusional gradients. Biogenic matter can be added, altered, or removed, and thus reticulopodian activities can channel the biochemistry within the cyst as well as in the enlarged outer pore volume (Heinz et al. 2005).

(d) The formation of this biogenic confined space favors the growth of bacteria exploiting the foraminiferal exudates. The onset for bacterial gardening and an "enzyme tank" sensu Riemann and Helmke (2002) is reached. Its formation needs particle agglomeration and reorganization of the surrounding sediment and an agglutinant or a cement, thus all prerequisites for monothalamid test formation. Such early test constructions might have been left occasionally, however. The evolvement of a more rigid and genetically fixed test structure pursued specialization by channeling chemical gradients and functional morphology.

6. Conclusions

Prior assumptions on allogromiids, monothalamids, or naked or testate cercocoans as the ancestors of foraminiferan rigid test constructions do not answer a crucial question: Why and how did foraminiferan tests appear? Functional constraints from modern assemblages with predominant heterotrophic uptake of detritus, bacteria, and algae would be misleading. Modern coupled pelagic-benthic food chains depend on upwelling and downwelling, thus alternating water mass densities, deep water formation, and paleoclimate play a crucial role. Today's scenarios had their onset at the Eocene–Oligocene boundary and evolved into the deep water formation of the glacial Quaternary (Thomas 2007). We have to dissect, if any, maintained structures and strategies which are appropriate for Precambrian environments. Our review of present knowledge derived from genetic studies, phylogenetic constraints, ultrastructural investigations, and palaeoenvironmental data generates a novel hypothesis. Dispersed microenvironments with alternating redox partners shaped the oxygen-deficient Precambrian. The adaptability to negative redox conditions was a key factor for the early evolution of foraminifera. Extant Foraminifera from several lineages still adapt to long-term anoxic and sulfidic conditions, and a number of species can handle electron acceptors other than dissolved oxygen. The bidirectional streaming in reticulopodia offers a perfect tool for the concurrent utilization and linkage of diffusion gradients, redox-dependent electron acceptors, and potential prey dispersed in the microenvironments presumed to occur in the Precambrian. The synchronous exploitation of disparate and restricted resources offered a general metabolic gain. The encapsulation of gathered resources and subsequent metabolic exudates in a spatially structured environment generates two additional advantages: it offers resources not accessible to one's own metabolic pathways to affiliated microbes, and it assures the reciprocation of their growth and byproducts. Relationships with exo- and endosymbionts, sequestered organelles, and enzyme sharing must be considered highly effective during the pronounced instability of redox conditions from the Neoproterozoic to the lowermost Phanerozoic. The early formation of agglutinated structures surrounding the ancestral foraminifer is most probably derived from such trophic and symbiotic interactions. It paved the way for monothalamid and textularid test structures. The resurgence of these capabilities is maintained in the temporary agglomerated cysts produced by many modern benthic taxa.

Phanerozoic test constructions followed functional morphologies convergent to metazoans: suspension filterers, grazers, burrowers, raptors, parasites. Symbiosis and mutualism were employed in the evolutionary path of foraminifera repeatedly, predominantly where or whenever resources were limited.

7. References

Adl SM, Simpson AGB, Farmer MA, Andersen RA, Anderson OR, Barta JR, Bowser SS, Brugerolle G, Fensome RA, Fredericq S, James TY, Karpov S, Kugrens P, Krug J, Lane CE, Lewis LA, Lodge J, Lynn DH, Mann DG, McCourt RM, Mendoza L, Moestrup Ø, Mozley-Standridge SE, Nerad TA, Shearer CA, Smirnov AV, Spiegel FW, Taylor MFJR (2005) The new higher level classification of eucaryotes with emphasis on the taxonomy of protists. J Eukaryot Microbiol 52:399–451

Altenbach AV, Unsöld G, Walger E (1987) The hydrodynamic environment of *Saccorhiza ramosa* (Brady). Meyniana 40:119–132

Altenbach AV, Heeger T, Linke P, Spindler M, Thies A (1993) *Miliolinella subrotunda* (Montagu), a miliolid foraminifer building large detritic tubes for a temporary epibenthic life-style. Mar Micropaleontol 20:293–301

Anbar AD (2008) Elements and evolution. Science 322:1481–1483

Anbar AD, Knoll AH (2002) Proterozoic ocean chemistry and evolution: a bioinorganic bridge? Science 297:1137–1142

Anderson OR, Lee JJ (1991) Symbiosis in foraminifera. In: Lee JJ, Anderson OR (eds.) Biology of foraminifera. Academic, London, pp 157–220

Austin HA, Austin WE, Paterson DM (2005) Extracellular cracking and content removal of the benthic diatom *Pleurosigma angulatum* (Quekett) by the benthic foraminifera *Haynesina germanica* (Ehrenberg). Mar Micropaleontol 57:68–73

Bass D, Moreira D, Lopez-Garcia P, Polet S, Chao EE, von der Heyden S, Pawlowski J, Cavalier-Smith T (2005) Polyubiquitin insertions and the phylogeny of Cercozoa and Rhizaria. Protist 156:149–161

Bass D, Chao EEY, Nikolaev S, Yabuki A, Ishida KI, Berney C, Pakzad U, Wylezich C, Cavalier-Smith T (2009) Phylogeny of novel naked filose and reticulose Cercozoa: Granofilosea cl. n. and Proteomyxidea revised. Protist 160:75–109

Bender H (1995) Test structure and classification in agglutinated foraminifera. In: Kaminski MA, Geroch S, Gasinski MA (eds.) Proceedings of the fourth international workshop on agglutinated foraminifera. Grzybowski Foundation, Krakow, pp 27–70

Bernhard JM (2003) Potential symbionts in bathyal foraminifera. Science 299:861

Bernhard JM, Bowser SS (1999) Benthic foraminifera of dysoxic sediments: chloroplast sequestration and functional morphology. Earth-Sci Rev 46:149–165

Bernhard JM, Bowser SS (2008) Peroxisome proliferation in foraminifera inhabiting the chemocline: an adaptation to reactive oxygen species exposure? J Eukaryot Microbiol 55:135–144

Bernhard JM, Visscher PT, Bowser SS (2003) Submillimeter life positions of bacteria, protists, and metazoans in laminated sediments of the Santa Barbara Basin. Limnol Oceanogr 48:813–828

Bernhard JM, Habura A, Bowser SS (2006) An endobiont-bearing allogromiid from the Santa Barbara Basin: implications for the early diversification of foraminifera. J Geophys Res Biogeosci 111:G03002. doi:10.1029/2005JG000158

Bernhard JM, Mollo-Christensen E, Eisenkolb N, Starczak VR (2009a) Tolerance of allogromiid foraminifera to severely elevated carbon dioxide concentrations: implications to future ecosystem functioning and paleoceanographic interpretations. Global Planet Change 65:107–114

Bernhard JM, Goldstein ST, Bowser SS (2009b) An ectobiont-bearing foraminiferan, *Bolivina pacifica*, that inhabits microxic pore waters: cell-biological and paleoceanographic insights. Environ Microbiol. doi:10.1111/j.1462-2920.2009.02073.x

Bertram MA, Cowen JP (1997) Morphological and compositional evidence for biotic precipitation of marine barite. J Mar Res 55:577–593

Bowser SS, DeLaca TE (1985) Rapid intracellular motility and dynamic membrane events in an Antarctic foraminifer. Cell Biol Int Rep 9:901–910

Bowser SS, Travis JL (2002) Reticulopodia: structural and behavioral basis for the suprageneric placement of granuloreticulosan protists. J Foraminifer Res 32:440–447

Bowser SS, Alexander SP, Stockton WL, DeLaca TE (1992) Extracellular matrix augments mechanical properties of pseudopodia in the carnivorous foraminiferan *Astrammina rara*: role in prey capture. J Protozool 39:724–732

Bowser SS, Gooday AJ, Alexander SP, Bernhard JM (1995) Larger agglutinated foraminifera of McMurdo Sound, Antarctica: are *Astrammina rara* and *Notodendrodes antarctikos* allogromiids incognito? Mar Micropaleontol 26:75–88

Brüchert V, Jørgensen BB, Neumann K, Riechmann D, Schlösser M, Schulz HN (2003) Regulation of bacterial sulfate reduction and hydrogen sulfide fluxes in the central Namibian coastal upwelling zone. Geochim Cosmochim Acta 67:4505–4518

Butterfield NJ (2009) Oxygen, animals and oceanic ventilation: an alternative view. Geobiology 7:1–7

Canfield DE (2005) The early history of atmospheric oxygen: homage to Robert M. Garrels. Annu Rev Earth Pl Sci 33:136

Catling DC, Glein CR, Zahnle KJ, Mckay CP (2005) Why O_2 is required by complex life on habitable planets and the concept of planetary "Oxygenation Time". Astrobiology 5:415–438

Cavalier-Smith T (2002) The phagotrophic origin of eukaryotes and phylogenetic classification of protozoa. Int J Syst Evol Microbiol 52:297–354

Cedhagen T (1988) Position in the sediment and feeding of *Astrorhiza limicola* Sandahl, 1857 (Foraminiferida). Sarsia 73:43–47

Cedhagen T (1991) Retention of chloroplasts and bathymetric distribution in the sublittoral foraminiferan *Nonionella labradorica*. Ophelia 33:17–30

Cedhagen T (1993) Taxonomy and biology of *Pelosina arborescens* with comparative notes on *Astrorhiza limicola* (Foraminiferida). Ophelia 37:143–162

Cedhagen T, Frimanson H (2002) Temperature dependence of pseudopodial organelle transport in seven species of foraminifera and its functional consequences. J Foraminifer Res 32:434–439

Cedhagen T, Tendal OS (1989) Evidence of test detachment in *Astrorhiza limicola*, and two consequential synonyms: *Amoeba gigantea* and *Megaamoebomyxa argillobia* (Foraminiferida). Sarsia 74: 195–200

Correia MJ, Lee JJ (2002) How long do the plastids retained by *Elphidium excavatum* (Terquem) last in their host? Symbiosis 32:27–37

Danovaro R, Dell'Anno A, Pusceddu A, Gambi C, Heiner I, Kristensen RM (2010) The first metazoa living in permanently anoxic conditions. BMC Biol 8:30. doi:10.1186/1741-7007-8-30

Dawson SC, Pace NR (2002) Novel kingdom-level eukaryotic diversity in anoxic environments. Proc Natl Acad Sci 99:8324–8329

DeLaca TE (1982) Use of dissolved amino-acids by the foraminifer *Notodendrodes antarcticos*. Am Zool 22:683–690

DeLaca TE, Karl DM, Lipps JH (1981) Direct use of dissolved organic carbon by agglutinated benthic foraminifera. Nature 289:287–289

DeLaca TE, Bernhard JM, Reilly AA, Bowser SS (2002) *Notodendrodes hyalinosphaira* (sp. nov.): structure and autecology of an allogromiid-like agglutinated foraminifer. J Foraminifer Res 32:177–187

Dong L, Xiao S, Shen B, Zhou C (2008) Silicified *Horodyskia* and *Palaeopascichnus* from upper Ediacaran cherts in South China: tentative phylogenetic interpretation and implications for evolutionary stasis. J Geol Soc UK 165:367–378

Doolittle RF (1995) The origins and evolution of eukaryotic proteins. Philos T Roy Soc B 349: 235–240

Edgcomb V, Orsi W, Leslin C, Epstein SS, Bunge J, Jeon S, Yakimov MM, Behnke A, Stoeck T (2009) Protistan community patterns within the brine and halocline of deep hypersaline anoxic basins in the Eastern Mediterranean Sea. Extremophiles 13:151–167

Embley TM (2006) Multiple secondary origins of the anaerobic lifestyle in eukaryotes. Philos T Roy Soc B 361:1055–1067

Embley M, van der Giezen M, Horner DS, Dyal PL, Foster P (2003) Mitochondria and hydrogenosomes are two forms of the same fundamental organelle. Philos T Roy Soc B 358:191–203

Ettwig KF, Butler MK, Le Paslier D, Pelletier E (2010) Nitrate-driven anaerobic methane oxidation by oxygenic bacteria. Nature 464:543–548

Fahrni JF, Pawlowski J, Richardson S, Debenay J-P, Zanetti L (1997) Actin suggests *Miliammina fusca* (Brady) is related to porcellaneous rather than to agglutinated foraminifera. Micropaleontology 43:211–214

Falkowski PG, Godfrey LV (2009) Electrons, life and the evolution of Earth' oxygen cycle. Philos T Roy Soc B 363:2705–2716

Fenchel T, Finlay B (2008) Oxygen and the spatial structure of microbial communities. Biol Rev 83:553–569

Fenchel T, Finlay BJ (2009) The diversity of microbes: resurgence of the phenotype. Philos T Roy Soc B 361:1965–1973

Finlay BJ, Span ASW, Harman JMP (1983) Nitrate respiration in primitive eukaryotes. Nature 303:333–336

Flakowski J, Bolivar I, Fahrni J, Pawlowski J (2005) Actin phylogeny of foraminifera. J Foraminifer Res 35:93–102

Flakowski J, Bolivar I, Fahrni J, Pawlowski J (2006) Tempo and mode of spliceosomal intron evolution in actin of foraminifera. J Mol Evol 63:30–41

Francis CA, Beman JM, Kuypers MMM (2007) New processes and players in the nitrogen cycle: the microbial ecology of anaerobic and archaeal ammonia oxidation. ISME J 1:19–27

Frevert T (1984) Can the redox conditions in natural waters be predicted by a single parameter? Aquat Sci 46:270–290

Gaidos E, Diubuc T, Dunford M, McAndrew P, Padilla-Gamino J, Studer B, Weersing K, Stanley S (2007) The Precambrian emergence of animal life: a geobiological perspective. Geobiology 5:351–373

Gaucher C, Sprechmann P (1999) Upper Vendian skeletal fauna of the Arroyo del Soldado Group, Uruguay. Beringeria 23:55–91

Gaucher C, Frimmel HE, Germs GJB (2005) Organic-walled microfossils and biostratigraphy of the upper Port Nolloth Group (Namibia): implications for latest Neoproterozoic glaciations. Geol Mag 142:539–559

Glud RN, Thamdrup B, Stahl H, Wenzhoefer F, Glud A, Nomaki H, Oguri K, Revsbech NP, Kitazato H (2009) Nitrogen cycling in a deep ocean margin sediment (Sagami Bay, Japan). Limnol Oceanogr 54:723–734

Goldstein SL (1999) Foraminifera: a biological overview. In: Sen Gupta BK (ed.) Modern Foraminifera. Kluwer Academic, Dordrecht, pp 37–55

Goldstein ST, Corliss BH (1994) Deposit feeding in selected deep-sea and shallow-water benthic foraminifera. Deep-Sea Res 41:229–241

Goldstein ST, Moodley L (1993) Gametogenesis and the life-cycle of the foraminifer *Ammonia beccarii* (Linne) forma Tepida (Cushman). J Foraminifer Res 23:213–220

Gooday AJ (1986) Meiofaunal foraminiferans from the bathyal Porcupine Seabight (north-east Atlantic): size structure, standing stock, taxonomic composition, species diversity and vertical distribution in the sediment. Deep-Sea Res 33:1345–1373

Gooday AJ (2002) Biological response to seasonally varying fluxes of organic matter to the sea floor: a review. J Oceanogr 58:305–322

Gooday AJ, Nott AJ (1982) Intracellular barite crystals in two xenophyophores, *Aschemonella ramuliformis* and *Galatheammina* sp. (Protozoa, Rhiziopoda) with comments on the taxonomy of *A. ramuliformis*. J Mar Biol Assoc UK 62:595–605

Gooday AJ, Tendal OS (1988) New xenophyophores (Protista) from the bathyal and abyssal north-east Atlantic Ocean. J Nat Hist 22:413–434

Gooday AJ, Todo Y, Uematsu K, Kitazato H (2008a) New organic-walled foraminifera (Protista) from the ocean's deepest point, the Challenger Deep (Western Pacific Ocean). Zool J Linn Soc 153: 399–423

Gooday AJ, Kamenskaya OE, Kitazato H (2008b) The enigmatic, deep-sea, organic-walled genera Chitinosiphon, Nodellum and Resigella (Protista, Foraminifera): a taxonomic re-evaluation. Syst Biodivers 6:385–404

Graur D, Martin W (2004) Reading the entrails of chickens: molecular timescales of evolution and the illusion of precision. Trends Genet 20:80–86

Gross O (2002) Sediment interactions of foraminifera: implications for food degradation and bioturbation processes. J Foraminifer Res 32:414–424

Grundl T (1994) A review of the current understanding of redox capacity in natural, disequilibrium systems. Chemosphere 28:613–626

Grzymski J, Schofield OM, Falkowski PG, Bernhard JM (2002) The function of plastids in the deep-sea benthic foraminifer, *Nonionella stella*. Limnol Oceanogr 47:1569–1580

Habura A, Rosen DR, Bowser SS (2004) Predicted secondary structure of the foraminiferal SSU 3 'major domain reveals a molecular synapomorphy for granuloreticulosean protists. J Eukaryot Microbiol 51:464–471

Habura A, Goldstein ST, Parfrey LW, Bowser SS (2006) Phylogeny and ultrastructure of *Miliammina fusca*: evidence for secondary loss of calcification in a miliolid foraminifer. J Eukaryot Microbiol 53:204–210

Habura A, Goldstein ST, Broderick S, Bowser SS (2008) A bush, not a tree: the extraordinary diversity of cold-water basal foraminiferans extends to warm-water environments. Limnol Oceanogr 53:1339–1351

Hannah F, Rogerson A, Laybournparry J (1994) Respiration rates and biovolumes of common benthic foraminifera (Protozoa). J Mar Biol Assoc UK 74:301–312

Harcombe W (2010) Novel cooperation experimentally evolved between species. Evolution. doi:10.1111/j.1558-5646.2010.00959.x

Heinz P, Geslin E, Hemleben C (2005) Laboratory observations of benthic foraminiferal cysts. Mar Biol Res 1:149–159

Høgslund S (2008) Nitrate storage as an adaptation to benthic life. Ph.D. thesis, Department of Biological Science, University of Aarhus, Denmark

Høgslund S, Revsbech N, Cedhagen T, Nielsen L, Gallardo V (2008) Denitrification, nitrate turnover, and aerobic respiration by benthic foraminiferans in the oxygen minimum zone off Chile. J Exp Mar Biol Ecol 359:85–91

Holland HD (2006) The oxygenation of the atmosphere and oceans. Philos T Roy Soc B 361:903–915

Hopwood JD, Mann S, Gooday AJ (1997) The crystallography and possible origin of barium sulphate in deep sea rhizopod protists (Xenophyophorea). J Mar Biol Assoc UK 77:969–987

Jahn TL, Rinaldi RA (1959) Protoplasmic movement in the foraminiferan, *Allogromia laticollaris*, and a theory of its mechanism. Biol Bull 117:100–118

Jepps MW (1942) Studies on *Polystomella* Lamarck (Foraminifera). J Mar Biol Assoc UK 25:607–666

Jiang S-Y, Pi D-H, Heubeck C, Frimmel H, Liu Y-P, Deng H-L, Ling H-F, Yang J-H (2009) Early Cambrian ocean anoxia in South China. Nature 459:E5–E6

Katz ME, Finkel ZV, Grzebyk D, Knoll AH, Falkowski PG (2004) Evolutionary trajectories and biogeochemical impacts of marine eucaryotic phytoplankton. Ann Rev Ecol Syst 35:523–556

King K (1977) Amino acid survey of recent calcareous and siliceous deep-sea microfossils. Micropaleontology 23:180–193

Knauth LP (2005) Temperature and salinity history of the Precambrian ocean: implications for the curse of microbial evolution. Palaeogeogr Palaeoecol 219:53–69

Knauth LP, Kennedy MJ (2009) The late Precambrian greening of the earth. Nature. doi:10.1038/nature08213

Koho KA, Pina-Ochoa E, Geslin E, Risgaard-Petersen N (2011) Vertical migration, nitrate uptake and denitrification: survival mechanisms of foraminifers (*Globobulimina turgida*) under low oxygen conditions. FEMS Microbiol Ecol 75:273–283

Komiya T, Hirata T, Kitajima K, Yamamoto S, Shibuya T, Sawaki Y, Ishikawa T, Shu D, Li Y, Han J (2008) Evolution of the composition of seawater through geologic time, and its influence on the evolution of life. Gondwana Res 14:159–174

Koonce MP, Schliwa M (1985) Bidirectional organelle transport can occur in cell processes that contain single microtubules. J Cell Biol 100:322–326

Köster M, Jensen P, Meyer-Reil L (1991) Hydrolytic activities of organisms and biogenic structures in deep-sea sediments. In: Chrost RJ (ed.) Microbial enzymes in aquatic environments. Springer, New York, pp 298–310

Kump LR (2008) The rise of atmospheric oxygen. Nature 451:277–278

Langer MR (1992) Biosynthesis of glucosaminoglycans in foraminifera: a review. Mar Micropaleontol 19:245–255

Langer MR, Gehring C (1993) Bacteria farming: a possible feeding strategy of some smaller, motile foraminifera. J Foraminifer Res 23:40–46

Laureillard J, Mejanelle L, Sibuet M (2004) Use of lipids to study the trophic ecology of deep-sea xenophyophores. Mar Ecol Prog Ser 270:129–140

Lavik G, Stührmann T, Brüchert V, Van der Plas A, Mohrholz V, Lam P, Mußmann M, Fuchs BM, Amann R, Lass U, Kuypers MM (2009) Detoxification of sulphidic African shelf waters by blooming chemolithotrophs. Nature 457:581–585

Leduc D, Probert PK (2009) The effect of bacterivorous nematodes on detritus incorporation by macrofaunal detritivores: a study using stable isotope and fatty acid analyses. J Exp Mar Biol Ecol 371:130–139

Lee JJ, Faber WW, Lee RE (1991) Antigranulocytes reticulopodal digestion – a possible preadaptation to benthic foraminiferal symbiosis. Symbiosis 10:47–61

Leiter C (2008) Benthos-Foraminiferen in Extremhabitaten: Auswertung von METEOR-Expeditionen vor Namibia. Ph.D., Faculty of Geosciences, LMU Munich, Germany, p 103 [in German, free access: http://edoc.ub.uni-muenchen.de/view/subjects/fak20.html]

Leiter C, Altenbach AV (2010) Benthic foraminifera from the diatomaceous mud-belt off Namibia: characteristic species for severe anoxia. Palaeontol Electron 13(2):1–19

Levin LA, Thomas CL (1988) The ecology of xenophyophores (Protista) on Eastern Pacific seamounts. Deep-Sea Res 35:2003–2027

Li Y, Guo JF, Zhang XL, Zhang WQ, Liu YH, Yang WX, Li YY, Liu LQ, Shu DG (2008) Vase-shaped microfossils from the Ediacaran Weng'an biota, Guizhou, South China. Gondwana Res 14: 263–268

Licari L, Mackensen A (2005) Benthic foraminifera off West Africa (1 N to 32 S): do live assemblages from the topmost sediment reliably record environmental variability? Mar Micropaleontol 55: 205–233

Linke P, Lutze GF (1993) Microhabitat preferences of benthic foraminifera – a static concept or a dynamic adaption to optimize food acquisition? Mar Micropaleontol 20:215–234

Lipps JH (1992) Proterozoic and Cambrian skeletonized protists. In: Schopf JW, Klein C (eds.) The Proterozoic biosphere: a multidisciplinary study. Cambridge University Press, Cambridge, UK, pp 237–240

Loeblich AR, Tappan H (1964) Sarcodina chiefly "Thecamoebians" and Foraminiferida. In: Moore RC (ed.) Treatise on invertebrate paleontology, part C, protista 2. The University of Kansas Press and The Geological Society of America, Boulder, USA

Loeblich AR, Tappan H (1988) Foraminiferal genera and their classification. Van Nostrand Reinhold, New York

Loeblich AR, Tappan H (1989) Implications of wall composition and structure in agglutinated foraminifers. J Paleontol 63:769–777

Longet D, Pawlowski J (2007) Higher-level phylogeny of foraminifera inferred from the RNA polymerase II (RPBI) gene. Eur J Protistol 43:171–177

Marszalek DS (1969) Observations on *Iridia diaphana*, a marine foraminifer. J Protozool 16:599–611

Marszalek DS, Wright RC, Hay WW (1969) Function of the test in foraminifera. Trans Gulf Coast Assoc Geol Soc 19:341–352

Matz MV, Frank TM, Marshall NJ, Widder EA, Johnsen S (2008) Giant deep-sea protist produces bilaterian-like traces. Curr Biol 18:1849–1854

McIlroy D, Green OR, Brasier MD (2001) Palaeobiology and evolution of the earliest agglutinated foraminifera: Platysolenites, Spirosolenites and related forms. Lethaia 34:13–29

Mentel M, Martin W (2008) Energy metabolism among eukaryotic anaerobes in light of Proterozoic ocean chemistry. Philos T Roy Soc B 363:2717–2729

Meyer-Reil L, Köster M (1991) Fine-scale distribution of hydrolytic activity associated with foraminiferans and bacteria in deep-sea sediments of the Norwegian-Greenland Sea. Kieler Meeresforschungen 8:121–126, sp.vol

Minge MA, Silberman JD, Orr RJS, Cavalier-Smith T, Shalchian-Tabrizi K, Burki F, Skjaeveland A, Jakobsen KS (2009) Evolutionary position of breviate amoebae and the primary eukaryote divergence. P Roy Soc UK Bio 276:597–604

Moens T, dos Santos GAP, Thompson F, Swings J, Fonseca-Genevois V, Vincx M, De Mesel I (2005) Do nematode mucus secretions affect bacterial growth? Aquat Microb Ecol 40:77–83

Monnin C, Jeandel C, Cattaldo T, Dehairs F (1999) The marine barite saturation state of the world's oceans. Mar Chem 65:253–261

Moreira D, von der Heyden S, Bass D, López-García P, Chao E, Cavalier-Smith T (2007) Global eukaryote phylogeny: combined small- and large-subunit ribosomal DNA trees support monophyly of Rhizaria, Retaria and Excavata. Mol Phylogenet Evol 44:255–266

Nozawa F, Kitazato H, Tsuchiya M, Gooday AJ (2006) 'Live' benthic foraminifera at a abyssal site in the equatorial Pacific nodule province: abundance, diversity and taxonomic composition. Deep-Sea Res 53:1406–1422

Nyholm K-G (1957) Orientation and binding power of recent monothalamous foraminifera in soft sediments. Micropaleontology 3:75–76

Pawlowski J (2008) The twilight of Sarcodina: a molecular perspective on the polyphyletic origin of amoeboid protists. Protistology 5:281–302

Pawlowski J, Burki F (2009) Untangling the phylogeny of amoeboid protists. J Eukaryot Microbiol 56:16–25

Pawlowski J, Gooday AJ (2009) Precambrian biota: protistan origin of trace fossils? Curr Biol 19: R28–R30

Pawlowski J, Holzmann M, Fahrni J, Cedhagen T, Bowser SS (2002) Phylogeny of allogromiid foraminifera inferred from SSU rRNA gene sequences. J Foraminifer Res 32:334–343

Pawlowski J, Holzmann M, Berney C, Fahrni J, Gooday AJ, Cedhagen T, Habura A, Bowser SS (2003a) The evolution of early foraminifera. Proc Natl Acad Sci USA 100:11494–11498

Pawlowski J, Holzmann M, Fahrni J, Richardson SL (2003b) Small subunit ribosomal DNA suggests that the xenophyophorean *Syringammina corbicula* is a foraminiferan. J Eukaryot Microbiol 50:483–487

Piña-Ochoa E, Høgslund S, Geslin E, Cedhagen T, Revsbech NP, Nielsen LP, Schweizer M, Jorissen F, Rysgaard S, Risgaard-Petersen N (2010) Widespread occurrence of nitrate storage and denitrification among foraminifera and Gromiida. Proc Natl Acad Sci USA 107:1148–1153

Porter SM, Knoll AH (2000) Testate amoebae in the Neoproterozoic Era: evidence from vase-shaped microfossils in the Chuar Group, Grand Canyon. Paleobiology 26:360–385

Richardson K, Cedhagen T (2001) Quantifying pelagic-benthic coupling in the North Sea: are we asking the right questions? Senckenbergiana maritima 31:215–224

Richardson S, Rutzler K (1999) Bacterial endosymbionts in the agglutinating foraminiferan *Spiculidendron corallicolum* Rutzler and Richardson 1996. Symbiosis 26:299–312

Riemann F, Helmke E (2002) Symbiotic relations of sediment-agglutinating nematodes and bacteria in detrital habitats: the enzyme sharing concept. Mar Ecol 23:93–113

Rinaldi RA, Jahn TL (1964) Shadowgraphs of protoplasmic movement in *Allogromia-laticollaris* and a correlation of this movement to striated muscle contraction. Protoplasma 58:369–390

Risgaard-Petersen N, Langezaal AM, Ingvardsen S, Schmid MC, Jetten MSM, Op den Camp HJM, Derksen JWM, Pina-Ochoa E, Eriksson SP, Nielsen LP, Revsbech NP, Cedhagen T, van der Zwaan GJ (2006) Evidence for complete denitrification in a benthic foraminifer. Nature 443:93–96

Roger AJ, Hug LA (2006) The origin and diversification of eukaryotes: problems with molecular phylogenetics and molecular clock estimation. Philos T Roy Soc B 361:1039–1054

Rothman DH, Haynes JM, Summons RE (2003) Dynamics of the Neoproterozoic carbon cycle. Proc Natl Acad Sci USA 100:8124–8129

Rupp G, Bowser SS, Manella CA, Rieder CL (1986) Naturally occuring tubulin-containing paracrystals in *Allogromia*: immunocytochemical identification and functional significance. Cell Motil Cytoskeleton 6:363–375

Schulz H, Jørgensen BB (2001) Big bacteria. Annu Rev Microbiol 55:105–137

Scott DB, Medioli F, Braund R (2003) Foraminifera from the Cambrian of Nova Scotia: the oldest multichambered foraminifera. Micropaleontology 49:109–126

Seilacher A (1999) Biomat-related lifestyles in the Precambrian. Palaios 14:86–93

Seilacher A, Grazhdankin D, Legouta A (2003) Ediacaran biota: the dawn of animal life in the shadow of giant protists. Paleontol Res 7:43–54

Sternberg E, Jeandel C, Robin E, Souhaut M (2008) Seasonal cycle of suspended barite in the mediterranean sea. Geochim Cosmochim Acta 72:4020–4034

Stoeck T, Epstein S (2003) Novel eukaryotic lineages inferred from small-subunit rRNA analyses of oxygen-depleted marine environments. Appl Environ Microb 69:2657–2663

Stoeck T, Behnke A, Christen R, Amaral-Zettler L, Rodriguez-Mora MJ, Chistoserdov A, Orsi W, Edgcomb VP (2009) Massively parallel tag sequencing reveals the complexity of anaerobic marine protistan communities. BMC Biol 7:72

Stumm W (1984) Interpretation and measurement of redox intensity in natural waters. Aquat Sci 46:291–296

Takishita K, Miyake H, Kawato M (2005a) Genetic diversity of microbial eukaryotes in anoxic sediment around fumaroles on a submarine caldera floor on the small-subunit rDNA phylogeny. Extremophiles 9:185–196

Takishita K, Inagaki Y, Tsuchiya M, Sakaguchi M, Maruyama T (2005b) A close relationship between Cercozoa and foraminifera supported by phylogenetic analyses based on combined amino acid sequences of three cytoskeletal proteins (actin, α-tubulin, and β-tubulin). Gene 362:153–160

Takishita K, Tsuchiya M, Kawato M, Oguri K, Kitazato H, Maruyama T (2007) Genetic diversity of microbial eukaryotes in anoxic sediment of the saline meromictic Lake Namako-ike (Japan): on the detection of anaerobic or anoxic-tolerant lineages of eukaryotes. Protist 158:51–64

Tappan H, Loeblich AR (1988) Foraminiferal evolution, diversification, and extinction. J Paleontol 62:695–714

Tendal OS (1979) Aspects of the biology of Komokiacea and Xenophyophoria. Sarsia 64:13–17

Tendal OS (1994) Protozoa Xenophyophorea Granuloreticulosa: *Psammina zonaria* sp. nov. from the West Pacific and some aspects of the growth of xenophyophores. Mem Mus Nat d'Histoire Naturelle 161:49–54

Tendal OS, Thomsen E (1988) Observation on the life position and size of the large foraminifer *Astrorhiza arenaria* Norman, 1876 from the shelf off Northern Norway. Sarsia 20:39–42

Thomas E (2007) Cenozoic mass extinctions in the deep sea: what perturbs the largest habitat on Earth? In: Monechi S, Coccioni R, Rampino M (eds.) Large ecosystem perturbations: causes and consequences. Geological Society of America special papers, 424:1–23

Travis JL, Welnhofer EA, Orokos DD (2002) Autonomous reorganization of foraminiferan reticulopodia. J Foraminifer Res 32:425–433

Tsuchiya M, Toyofuku T, Takishita K, Yamamoto H, Collen J, Kitazato, H (2006) Molecular characterization of bacteria and kleptoplast within *Virgulinella fragilis*. In: Kitazato H, Bernhard JM (eds.) FORAMS 2006, Anuário do Instituto de Geociências, 29(6):471–472. http://www.anuario. igeo.ufrj.br/anuario_2006_1/Anuario_2006_1_471_472.pdf. Last access 9 June 2009

Tsuchiya M, Grimm GW, Heinz P, Stögerer K, Ertan KT, Collen J, Brüchert V, Hemleben C, Hemleben V, Kitazato H (2009) Ribosomal DNA shows extremely low genetic divergence in a world-wide distributed, but disjunct and highly adapted marine protozoan (*Virgulinella fragilis*, Foraminiferida). Mar Micropaleontol 70:8–19

van Hellemond JJ, van der Klei A, van Weelden SWH, Tielens AGM (2003) Biochemical and evolutionary aspects of anaerobically functioning mitochondria. Philos T Roy Soc B 358:205–215

Welnhofer EA, Travis JL (1997) Evidence for a direct conversion between two tubulin polymers – microtubules and helical filaments – in the foraminiferan, *Allogromia laticollaris*. Cell Motil Cytoskeleton 41:107–116

Wille M, Nägler TF, Lehmann B, Schröder S, Kramers JD (2008) Hydrogen sulfide release to surface waters at the Precambrian/Cambrian boundary. Nature 453:767–769

Wille M, Nägler TF, Lehmann B, Schröder S, Kramers JD (2009) Wille et al. reply. Nature 459:E6

Winchester-Seeto TM, McIlroy D (2006) Lower Cambrian melanosclerites and foraminiferal linings from the Lontova Formation, St. Petersburg, Russia. Rev Palaeobot Palynol 139:71–79

Wotton RS (2005) The essential role of exopolymers (EPS) in aquatic systems. Oceanogr Mar Biol 42:57–94

Yoon HS, Grant J, Tekele YI, Wu M, Chaon BC, Cole JC, Logsdon JM, Patterson DJ, Bhattacharya D, Katz LA (2008) Broadly sampled multigene trees of eukaryotes. BMC Evol Biol 8. doi:10.1186/1471-2148-8-14

Biodata of **Wladyslaw Altermann**, author of *"The Relevance of Anoxic and Agglutinated Benthic Foraminifera to the Possible Archean Evolution of Eukaryotes"* (with co-authors **Alexander V. Altenbach**, co-editor of this volume, and **Carola Leiter**, co-author of *"Carbon and Nitrogen Isotopic Fractionation in Foraminifera: Possible Signatures from Anoxia"*).

Dr. Wladyslaw (Wlady) Altermann serves as the Kumba-Exxaro Chair in Geodynamics of Ore Deposits at the Department of Geology, University of Pretoria, South Africa. He is specialized in Precambrian geology and sedimentary mineral deposits with emphasis on the Archean life habitats and microfossils. Wlady Altermann was active for many years as professor at the Ludwig Maximilians University of Munich (LMU), Germany, as Commissary Chair in Historical Geology, Geological Remote Sensing, and in Geology (1998–2007 with interruptions). He obtained Diploma in Geology and Paleontology in 1983 and doctoral degree (Dr. *rer. nat.*) in 1988 from the Free University of Berlin and the Dr. *rer. nat. habil.* degree in 1998, from the LMU. He spent several years of research at the University of Stellenbosch, South Africa; at the Center for the Studies of the Evolution of Life, University of California, Los Angeles; and at the Centre Biophysique Moléculaire, CNRS, Orléans, France.

Dr. Altermann is interested in all aspects of Precambrian sedimentology and biogeology and of bio-sedimentary processes, next to many other geological disciplines, reaching as far as Carbon Capture and Storage (CCS) techniques and Enhanced Oil and Gas Recovery methods (EOR). He has participated in research projects and fieldwork around the world from Australia to South America, Southeast Asia, India and China. He was Honorary Professor at the Shandong University of Science and Technology in Huangdao, Quing Dao, P.R. China (2005–2008) and at the University of Pretoria, R.S.A. (1998–2009).

E-mail: **wlady.altermann@up.ac.za**

Dr. Alexander Volker Altenbach is professor for micropaleontology at the Ludwig Maximilians University in Munich. For more information, see biodata of the editors.

E-mail: **a.altenbach@lrz.unimuenchen.de**

Dr. Carola Leiter is a free lance geologist in a Munich consulting company. She started working in geoconsulting companies during her studies. As a technical assistant at the isotopic lab of the GeoBio-Center^{LMU}, she learned a lot about Foraminifera, decided to stay, and earned her PhD on benthic Foraminifera living in suboxic sediments off Namibia in 2008 at Munich University. Current research activities concentrate on anoxic environments and cretaceous Foraminifera.

E-mail: **c.leiter@gmx.de**

Alexander Volker Altenbach

Carola Leiter

THE RELEVANCE OF ANOXIC AND AGGLUTINATED BENTHIC FORAMINIFERA TO THE POSSIBLE ARCHEAN EVOLUTION OF EUKARYOTES

WLADYSLAW ALTERMANN[1], ALEXANDER VOLKER ALTENBACH[2,3], AND CAROLA LEITER[2]
[1]*Department of Geology, University of Pretoria, Pretoria 0002, South Africa*
[2]*GeoBioCenter, Ludwig-Maximilians-University, Richard-Wagner-Str. 10, 80333 Munich, Germany*
[3]*Department for Earth and Environmental Science, Ludwig-Maximilians-Universität Munich, Richard-Wagner-Str. 10, 80333 Munich, Germany*

1. Introduction to the Enigmas of Precambrian Earth

The history of the Earth, its lithosphere, hydrosphere, atmosphere, and biosphere is closely interlinked in many disciplines (geology, paleoclimatology, paleoceanography, and geobiology or paleobiology). This relationship is particularly complicated for the Precambrian constituting c. 90% of this history. When stepping backward in time, the history is increasingly obliterated by alteration processes connected to plate tectonics, metamorphism, diagenesis, biodegradation and taphonomy, and simply by increasing coverage by younger rocks. The biggest problems in deciphering this history are imposed by the often diametrically controversial interpretation of lithological, biological, and geochemical signatures preserved from these times, but especially from the Precambrian far past, the Archean (4.6–2.5 billion years ago). The most intriguing problems in our understanding of the earliest c. 60% of the Earth's history are: the geochemistry of the primordial oceans and atmosphere, their pH and oxidation state, and mineralization processes, e.g., the origin of banded iron formations; periods of global glaciations (Snowball Earth scenarios); and the appearance and early evolution of life, including photosynthesis, cyanobacteria, prokaryotes, and eukaryotes. A comprehensive overview of these problems and the discussed solutions can be gained from the series of state-of-the-art articles treating the entire Precambrian period, in Eriksson et al. (2004). Moreover, an all-embracing view on the close relationships of mineralogy and biology and the general evolution of the Earth and the lithosphere and biosphere can be detracted from Hazen et al. (2008). The present contribution is partly based on these discussions, perhaps not familiar to a wide audience of microbiologists.

We elaborate on the possible appearance of eukaryotes in the Archean and discuss herein the implications resulting from the ability of some foraminifera to survive and thrive in anoxia, a stage of minimum oxygen fugacity, generally attributed to the time of 4.6–2.5 Ga. The timing of the rise of eukaryotes in the Earth's history is habitually debated based on bodily microfossil record, reaching back to middle Proterozoic and on uncertain eukaryotic biomarker occurrence in rocks of Neoarchean age (3.0–2.7 Ga), at oxygen levels generally regarded too low to support oxygen-dependent metabolism (comp. Zimmer 2009). Here, we deviate from this custom and attempt to discuss a possibility of the evolution of eukaryotes without strictly adhering to the fossil record, but with support of biomarkers record from the Archean. This endeavor allows for scenarios less restricted by the sometimes erratic, fragmentary fossil record and our poor understanding of the taphonomy of unicellular Precambrian life.

Recent reports of diverse unicellular eukaryotes flourishing in anoxic or suboxic conditions (references in this book), but especially reports of naked, allogromiid foraminifera thriving under anoxic conditions and of Paleozoic agglutinating foraminifera from black, anoxic shales (Schieber 2009), open a possibility for earliest eukaryotes to have been able to use similar metabolic pathways as foraminifera in anoxia. We are assuming herein that this is not a newly acquired ability of some foraminifera and thus a modern evolutionary step, but, according to the conservative and hypobradytelic evolution of early life on Earth (Schopf 1995), an ancient metabolic pathway.

Foraminifera belong to the early eukaryotic lineages and may have inherited the primordial ability of life in the absence of free oxygen from their ancestors (Altenbach and Gaulke, this volume). Eukaryotes and their ancestors may have originated through capturing of a purple sulfur bacterium (endosymbiosis), long before the "great oxygenation event" in the Paleoproterozoic, and may have gradually acquired the ability of oxygenic respiration, with the increase of oxygen in the atmosphere during the Proterozoic. The rapid oxygenation of the hydrosphere and atmosphere at c. 2.3–2.4 Ga, dubbed the "great oxidation event" of the Paleoproterozoic, becomes strongly disputed (Ohmoto 2004) with increasing evidence found in the geological record for low levels of free oxygen, even during the Archean (Watanabe et al. 2009; Duan et al. 2010). This possibility becomes an option that should be seriously considered, especially in the lights of findings of eukaryotic biomarkers and large microfossils in Archean rocks (Buick 2010; Javaux et al. 2010).

The occurrence of benthic foraminifera in Devonian black shales may not necessarily evidence local and periodic oxygen oases in these environments, but may perhaps hint to a primordial ability of anoxygenic respiration for such foraminifera. For example, Schieber (2009) argues that black shales must have been at least in part deposited during periods of higher oxygen levels, and thus not be entirely euxinic (comp. Schieber, this volume). It seems, however, worthwhile to consider yet another possibility of foraminifera thriving in anoxic conditions and environments of shale sedimentation, already in the Proterozoic and throughout the Phanerozoic, but escaping fossilization due to lack of tests and *postmortem* degradation. Such a scenario becomes particularly attractive in the light of increasingly abundant reports of Archean eukaryotic biomarkers (Brocks et al. 1999, 2003; Brocks and Banfield 2009; Waldbauer et al. 2009). They are often considered contaminants (Brocks et al. 2008; Altermann 2007, 2009) because of assumed oxygen deficiency of these times and sediments (black shales) in which such biomarkers are found. On the other hand, Archean environments may have been moderately oxidized and not entirely anoxic, as argued against the "great oxidation event" and the "Cloud–Walker–Holland–Casting" model for the oxygenation of the Earth's surface by Ohmoto (2004). Alternatively and simultaneously, early eukaryotes may have been able to thrive in anoxia.

2. General Evolutionary Considerations

It is generally accepted in the discussion of the evolution of eukaryotes that all eukaryotes require a minimum of free oxygen concentration in the environment, which has only been achieved in the Proterozoic with c. 1–2% of the present atmospheric level (PAL). Such considerations are consistent with the findings of eukaryotic microfossils in 1.7 to 1.8 rocks (Canfield 2005; Knoll et al. 2006), post-dating the alleged great oxygenation event by c. 500 million years, but contradict the presence of eukaryotic biomarkers in rocks of Archean age when virtually no free oxygen was present (Brocks et al. 2003). Nevertheless, in recent years, several lines of evidence have been presented that atmospheric oxygen might have reached levels of 1–2% PAL already during or at the end of the Archean (e.g., Ohmoto 2004; Kump and Barley 2007; Sleep and Bird 2008; Kato et al. 2009). Early eukaryotes, however, might have not needed such relatively high free oxygen levels. A solution to the debate of the early rise of eukaryotes under Archean low oxygen or oxygen-free atmospheric conditions may be offered by the multiple reports of flourishing foraminifera (eukaryotes) in recent anoxic environments and from reports of fossilized remains of benthic foraminifera in Paleozoic black shales.

The exciting findings of agglutinated foraminifera in black shales of Devonian age (Schieber 2009; see Schieber, this volume) may bear important impact on the interpretation of the paleoenvironment and ecology of black shales in general. The discussion on periodic oxygenation stages during the deposition of black shales is seemingly supported by findings of cyanobacterial fossils in, e.g., Silurian and Devonian black shales and radiolarites (e.g., Kremer and Kazmierczak 2005). The apparently compulsory conclusion that such black shales must have been weakly oxygenated at least for some time intervals, in order to support eukaryotic life, might nevertheless be misleading. Instead, we propose a different interpretation with far reaching consequences for the above portrayed debate.

The findings of agglutinated foraminifera in euxinic Paleozoic environments are in accordance with recent reports of extant foraminifera thriving under anoxic conditions. Furthermore, the evolutionary implications of these findings, combined with new reconnaissance on extant benthic foraminifera found in anoxic sediments, may also imply dramatic consequences for the general interpretation of the evolutionary history of eukaryotes.

Foraminifera have been described from sulfuric deep sea environments and are known to colonialize deep sea manganese nodules (Frieling and Mrazek 2007), where bacteria are considered sulfur oxidizers. Sulfur must have been abundant in the Archean, although the sulfur cycle of the Archean ocean and biosphere is basically unknown and controversially debated. The sulfur content of the ocean, from the Archean to recent, strongly depends on the oxygenation stage of the hydrosphere (and atmosphere). Under anoxia, the SO_2^{-4} content of the ocean should be expected to be significantly lower than the present value of 28 mM, corresponding to 900 ppm S ("Cloud–Walker–Holland–Casting" model for the oxygenation of the Earth's surface). Habicht et al. (2002) suggested that the oceanic SO_2^{-4} content was at c. 200 µM and increased gradually to 10 mM only at c. 800 Ma, in the Neoproterozoic. Following the arguments of Ohmoto (2004), however, the SO_2^{-4} content of the oceans was constantly similar to present day values, above 10 mM since c. 4.0 Ga. With the involvement of sulfur reducing bacteria (SRB) in the Neoarchean (Ohmoto et al. 2006) and with abundant volcanic exhalations of the hotter Precambrian Earth, no shortage of sulfur in both, reduced and oxidized forms is thus implied for the Archean sedimentary environments.

3. Anoxic Foraminifera and Environments

We hypothesize that the recovery of benthic foraminifera in various Devonian shales (Schieber 2009) reflects the ancient ability of some foraminifera to thrive in anoxia, which descends from at least Neoarchean times. Bernhard and Reimers (1991) and Bernhard et al. (2003) have already considered findings of living foraminifera in sulfidic and anoxic microhabitats, as of far reaching and yet unforeseeable consequences in eukaryotes research. Particularly, the interwoven, submillimetric structures of oxic and anoxic microhabitats according to these authors, exclude the definition of a horizontal anoxic border above or within the benthic boundary layer, as well as a significant definition for the survival range of the foraminifera involved (Bernhard et al. 2003). The earlier considerations of Bernhard and Reimers (1991) that such communities require at least a temporal higher oxygenation level were based on the overwhelming number of biological and neontological observations from oxic to oxygen-depleted environments at that time, in absence of reports from long-term sulfidic and anoxic habitats.

In the intervening time since 1991 (opt. cit.), chemotrophical endosymbionts in *Virgulinella fragilis* were recovered from the hypoxic to sulfidic Cariaco Basin (Bernhard 2003), and an hitherto undescribed allogromiid from sulfidic deep sea environments (Bernhard et al. 2006). In both cases, the endosymbiotic bacteria were considered "sulfur oxidizers." A constant flow of reduced sulfur is necessary for keeping the metabolism of these bacteria active, and any oxygen source must be considered for the explanation of the survivability of the host.

Within the last decade therefore, our common opinion on the principally aerobe and heterotrophic habit of foraminifera capsized dramatically. The first recovery of a foraminifer from long-term anoxic environments can be considered an anaerobe species, as it never was reported from oxygenated sediments before (Bernhard et al. 2006). In the same year, nitrate storage and full denitrification was proven for these eukaryotes (Risgaard-Petersen et al. 2006). It is a widespread metabolic pathway for members of all major foraminiferal tribes, and for nearly every second species investigated (Piña-Ochoa et al. 2010a). Large intracellular nitrate pools definitely allow beneficial metabolic activity under complete absence of dissolved oxygen, limited only by the availability of nitrate (Piña-Ochoa et al. 2010b). Other foraminiferal species not involved in denitrification were found in deep sediment layers which generate a constant efflux of sulfide into the overlaying water masses (Leiter and Altenbach 2010).

Within the cytoplasm of *V. fragilis*, chloroplasts sequestered from ingested phototrophes occur in distinct arrangements, in conjunction with peroxisome proliferation. As chloroplasts can degrade hydroxyl radicals, they may offer a source of oxygen for the host by H_2O_2 breakdown (Bernhard and Bowser 2008). This hypothesis appears convincing, because many foraminiferal species thriving at redox boundary conditions shelter such kleptoplasts (Bernhard and Bowser 2008). They are often sequestered at water depth, where photosynthetic activities are excluded for energetic reasons. As they may retain functionality for at least 1 year in the cytoplasm, also other purposes have to be considered. Reaching redox boundary conditions, foraminifera are enforced to adapt to heavily toxic stressors, such as free sulfides and different radicals (Grzymski et al. 2002). Under anoxic conditions, this provides a functional task for the kleptoplasts as well (Bernhard and Bowser 2008). The recently discovered authigenic and complete denitrification by foraminifera at redox boundary conditions (Risgaard-Petersen et al. 2006; Høgslund et al. 2008) gives proof for such extensive self-determination of the foraminiferal biochemistry, managing the intracellular redox boundary conditions appropriate for the reduction of dissolved organic nitrogen to N_2. Such biochemical activities of benthic foraminifera are speculated to contribute significantly to the global nutrient cycle, rather than providing rare abnormalities of negligible impact (Høgslund et al. 2008).

Our observations in the diatomaceous mud belt off Namibia (Leiter and Altenbach 2010) fully corroborate the long-term survival of benthic foraminifera under anoxic and sulfidic conditions. The offshore Namibia mud belt is one of the most inhospitable, oxygen-depleted, and sulfidic open shelf environments on the Earth (Baturin 2002; van der Plas et al. 2007). Oxygen concentrations of bottom waters are below detectability in the central area, suggesting denitrification and anaerobic ammonia oxidation (anammox) even in the water column (Emeis et al. 2007). Methane and hydrogen sulfide accumulates within the sediment column over large areas (Emeis et al. 2004), and blowouts of these toxic gasses occur frequently (Weeks et al. 2004). We have repeatedly recovered living *Virgulinella fragilis* and three syntopic taxa of the genera *Fursenkoina, Nonionella,* and *Bolivina* from anoxic surface sediments (Altenbach et al. 2002; Altenbach and Struck 2006; Leiter 2008). Within the uppermost sediment column, sulfate-reducing bacteria were detected in all cores sampled by Brüchert et al. (2003) on this shelf. Their sulfate reduction rates culminate near the sediment surface, or within the uppermost centimeters of the sediment column, providing a net flux of hydrogen sulfide into the bottom water (Brüchert et al. 2003). This implies sulfidic environments for the benthic boundary layer, and especially for the deeper sediment portions, despite residual traces of dissolved oxygen (0–0.3 ml l^{-1}) detected by CTD measurements (Leiter 2008). Enhanced advection of slightly oxygenated water masses provides temporal oxygen availability near the sea floor (Chapman and Shannon 1987), but the benthic consumption rates on the inner shelf are sufficient to reduce the pool of bottom water oxygen completely (van der Plas et al. 2007; Mohrholz et al. 2008). Living (Rose Bengal stained) foraminifera were recovered deep infaunal, down to 19 cm sediment depth. The standing stock of these populations even increased downcore, i.e., toward the diffusion coefficient of sulfide, by nearly one order of magnitude (Leiter 2008). Mean foraminiferal migratory speeds of close to 10–100 μm per minute, exclude any rapid movement toward sporadic oxygen availability at the sediment surface, not to mention the drastically enhanced energy demand required (comp. Gross 2000).

Two of the species recovered from the deepest layers, *Nonionella stella* and *Fursenkoina fusiformis* (referred to the genus *Stainforthia* by many authors), were observed to thrive better under anoxia as compared to oxic conditions (Moodley et al. 1997). For both species, authigenic denitrification is considered (Risgaard-Petterson et al. 2006; Høgslund et al. 2008), and for both species functional kleptoplasts were reported (Bernhard and Bowser 2008). *Stainforthia* will even flourish under anoxia when large pools of organic matter are available (Ernst et al. 2005). Under such conditions, *Stainforthia* most probably relies on organic metabolites enabling denitrification.

Summarizing these findings, future foraminiferal research must draw attention to anoxic and sulfidic habitats much more than in the past (Bernhard 2006). Even more importantly, at present we have to repeal the dogma that all benthic foraminifera are irrefutably bound to habitats with at least some, or periodical, dissolved oxygen availability. Foraminifera have to be ranked among the facultatively anaerobe eukaryotes (Fenchel and Finlay 2008; Koho et al. 2011). Skeptic minds may argue that modern species in anoxic environments were never proven to thrive under fully anoxic conditions for at least one generation cycle. In response, we have to state that the opposite was never proven either. The autecology of ancient Precambrian eukaryotes, as of the benthic foraminifera recovered from Devonian black shales (Schieber 2009) may have resembled *Virgulinella fragilis*. This species thrives in all oceans, always in conjunction with oxygen depletion, sulfur-oxidizing bacteria, or seeps of methane and hydrogen sulfide (Altenbach et al. 2002; Bernhard 2002, 2003; Takata et al. 2005; Erbacher and

Nelskamp 2006; Diz and Frances 2008). It avoids all habitats with oxygen values above 0.3 ml l^{-1} in the near bottom waters (Altenbach and Struck 2006; Leiter 2008). However, leaning on the now repealed dogma, many authors considered it a regular inhabitant of oxic environments; either due to the lack of own oxygen measures or based on dead or unstained assemblages, a clear methodological mismatch by itself (Murray 2000).

At present, the Janus face of modern foraminiferal metabolic capabilities increasingly turns from an obligate aerobic view toward multiple adaptations for anaerobiosis (Altenbach and Gaulke, this volume). For more than a century, the appearance of foraminifera was considered a tracer for at least sporadic availability of dissolved oxygen. Now we know that this assumption is no more an irrefutable matter of course.

For the Devonian shales, enhanced oxygenation, as indicated by increasing bioturbation, reduces the amount of benthic foraminifera. This was interpreted by Schieber (2009) as a result of mechanical disturbance and complete destruction of the agglutinated tests. Under high loads of organic matter, agglutinated tests are most resistible to bioturbation, but total destruction still may occur in exceptional cases. However, it is most probable that some layers will still carry increasing numbers of broken tests, sporadic new invaders introduced by the ecological change, or relict faunas still recording the ecological gradients (Debenay et al. 2004). The same plausibility for the vanishing of benthic foraminifera is given by the oxygen itself, similar to the avoidance of dissolved oxygen by *V. fragilis*.

In the Archean, of course, despite the lack of any bioturbation, no signs of foraminifera were observed and perhaps never will be. It is entirely speculative whether the ancestors of foraminifera existed in the Archean and represent the carriers of eukaryotic biomarkers found in Archean anoxic shales. Agglutination of sediment grains to eukaryotic cells might have been superfluous in the Archean and of no evolutionary or life advantage. Archean eukaryotes might have been benthic and members of microbial-cyanobacterial mat communities freely planctonic, as at the generally low oxygen levels no special ecological niches were required. Our assumptions and considerations may be speculative. But as we deal with the Archean to Proterozoic environments and evolution, it is the most ancestral, sulfidic, and most oxygen-independent evolvement of Foraminifera to focus on (Altenbach and Gaulke, this volume). We have to rebalance some formerly well-established considerations with quite opposing modern findings.

4. Evolution of Eukaryotes as Recorded in Black Shales

It is most fascinating to conclude for the evolution of unicellular eukaryotes that the doctrine of utterly aerobic metabolic pathways is not exclusively valid. Survival of short-time anoxia is known even in the animal kingdom (Menuz et al. 2009; Borgonie 2011), although the strategies are entirely different than in single-celled eukaryotes. The question arises how far in the past such metabolic pathways were developed? Does the ability of surviving under anoxia represent a relatively newly acquired or perhaps a primordial capacity of eukaryotes in general? If the first case is valid, the question remains whether this new development descends from the ancient memories of the metabolic pathways programmed in the genetic code of foraminifera. During the early evolution of foraminifera, the majority of benthic environmental setups, trophic structures, and available bacterial counterparts for gardening or symbiosis, were fated to hypoxic to sulfidic conditions (Martin et al. 2008). Organizing the intracellular chemistry under anoxic conditions, activating functional kleptoplasts derived from ingested algae, or bacterial gardening or symbiosis with chemotrophs might have been demanded most frequently for the naked foraminiferal ancestors, and their Proterozoic successors, the organic-walled allogromiids. For their rigid test constructing successors in the Palaeozoic, it is worth to note that hard-shelled foraminifera were found more resistant to prolonged anoxia than allogromiids (Moodley et al. 1997). The early formation of agglutinated (Lower Cambrian) and calcitic (Lower Silurian) tests thus may have had a functional correspondence with anoxia. Toward the late Paleozoic, Mesozoic, and Cainozoic, such necessity will have declined stepwise, in conjunction with the shifting marine oxygen budgets. But the facultative ability to thrive under anoxic condition is still carried by the most basal extant clade (allogromiids) as well as by modern taxa developed in the Cainozoic (*V. fragilis*). This may provide a key factor for the evolution of foraminifera (and eukaryotes), much more influential, and reaching further back in time than hitherto suspected (Richardson and Rutzler 1999; Bernhard et al. 2006). The Devonian agglutinated foraminifera may have been much better adapted to anoxia than any modern species considered for comparison. The geochemical proxies derived from Devonian black shales are thus no longer at odds with the foraminiferal fauna (Schieber 2009). Nevertheless, we support Schieber's (2009) call for global reexamination of black shales, and extend it into the Precambrian, in benefit for a better understanding of ancient anoxia and foraminiferal and eukaryotic evolution.

Concluding from the above, considerations are proposed herein to investigate whether the recent controversy on Neoarchean and Palaeo- to Mesoproterozoic eukaryotic biomarkers and microfossils and the rise of atmospheric oxygen can be resolved under the auspices of the above discussion. The ensured eukaryotic fossil record extends back to the Mesoproterozoic, at c. 1.6 Ga (Javaux et al. 2004), but may reach further back to at least 1.87 Ga, based on fossil algal remains (Grypania described by Han and Runnegar 1992). Typical eukaryotic biomarkers as C28–30 steranes, however, reach back into 2.7 Ga black shales of the Mount Bruce Supergroup from the Pilbara Craton, Western Australia (Brocks et al. 1999, 2003), where the oxygen levels where agreeably below the eukaryotic life support threshold (but see Ohmoto 2004). Despite the fact that the findings of steranes from the Neoarchean were often criticized as possible contaminants (e.g., Altermann 2004, 2007) and reconsidered as post-metamorphic contaminants by Rasmussen et al. (2008) and Brocks et al. (2008), no clear picture of the early evolution of eukaryotes emerges from these discussions (Fischer 2008).

An independent team of researchers reported eukaryotic (methylhopane) biomarkers from Neoarchean rocks of the Pilbara Craton (Eigenbrode et al. 2008). At the same time, Rasmussen et al. (2008) and Brocks et al. (2008) basically excluded the possibility that the eukaryotic biomarkers found in 2.7 Ga Mount Bruce Supergroup rocks are in situ, indigenous, and syngenetic. Surprisingly, also the cyanobacterial biomarkers in these rocks were assigned contaminants and thus the existence of cyanobacteria in the Neoarchean was refuted, despite firm, multiple microfossil evidence implies that cyanobacteria and oxygen-producing photosynthesis existed already at 2.6 Ga (Altermann and Schopf 1995; Kazmierczak and Altermann 2002; Kazmierczak et al. 2004, 2009; Waldbauer et al. 2009).

Oxygenic photosynthesis dominates increasingly strongly the ecology of the Earth since at least the Neoarchean, as evidenced by the sequestration of carbon in widespread, thick carbonate platforms (Altermann et al. 2006) and in black shales (Sleep and Bird 2007, 2008). It seems thus worth considering, whether the earliest eukaryotic ancestors could not have utilized symbionts or kleptoplasts to be able to survive under anoxic conditions. Allogromiid foraminifera reach back in origin about a billion years (Pawlowski et al. 2003), based on molecular clock calculations, which have often been shown to be minimum age estimates at best (e.g., Graur and Martin 2004). Moreira et al. (2007) consider foraminifera among the early lineages within the supergroup Rhizaria, and suggest that even the ancestral rhizarian might have been endowed with a test during the trophic phase. It is not unlikely that such structures have been overlooked in older sedimentary records. However, relatively large putative microfossils, with astonishingly complicated and enigmatic morphology, are found in Archean cherts (Ueno et al. 2006; Sugitani et al. 2007, 2010).

Canopies of detritic covers, thickly laminated envelopes, or coarse-grained cysts are commonly gathered by benthic foraminifera from very basal, organic-walled clades as well as from modern agglutinated and calcareous taxa (Heinz et al. 2005; see also plate III, Fig. F, in Linke and Lutze 1993; Plate II, Fig. 10, in Licari and Mackensen 2005). Such cysts may be built and left in the sediment within hours or inhabited for weeks (Heinz et al. 2005). Their usage is discussed in connection with protection from environmental perturbations, reproduction, or as a provider of additional food sources from enhanced bacterial growth within the cyst (Linke and Lutze 1993; Heinz et al. 2005; Licari and Mackensen 2005). The preservation potential of such primitive cysts is negligible in sedimentary record.

Algal eukaryotes that certainly predate the development of foraminifera may have used similar strategies to the "hydroxyl-theory" in Neoarchean, oxygen-deficient environments. Another interesting group to investigate could be fermentative, anaerobically living eukaryotes, such as wine yeast (J. Brocks, pers. communication). The ancestors of aerobic eukaryotes may have left their biomarkers, e.g., sterols, in rocks, which is as old as the 2.7 Ga Mount Bruce black shales of the Pilbara Craton, Western Australia, when the pO_2 of the atmosphere can be estimated to have been 1.5% of the present atmospheric level (Kato et al. 2009). The evolution of the oxygen consuming eukaryotic kingdom must not need to have post-dated the "great oxidation event," which in turn, must not have been as dramatic as usually assumed (i.e., Anabar et al. 2007; Kato et al. 2009; Duan et al. 2010). The evolution may rather have gradually followed the rise of oxygen, toxic to most prokaryotes, during the Neoarchean and Paleoproterozoic. The developing eukaryotes adapted to oxygen, to their energetic advantage, as oxygen more and more competed with CO_2 for binding sites in the Rubisco molecule (Knoll 2003).

According to the endosymbiosis theory, mitochondria are the descendants of organelles that once acted as an anoxygenic photosynthetic "chloroplast." Eukaryotes or their ancestors originated through capturing of a purple sulfur bacterium by the last eukaryotic common ancestor. If the phylogenetic tree is correct, stem group eukaryotes must have been anaerobes (Woese and Fox 1977). The combined organism must have, at least initially, preserved the ability of anoxygenic respiration. Only with the increase of oxygen in the atmosphere, latest at c. 2.45 Ga (Canfield 2005), the eukaryotes may have gradually switched to oxygen consumers. Some, however, may have preserved their anoxic heritage, perhaps even until today. Evolutionary inventions usually do not get lost and can be reactivated when conditions change accordingly, at least in "simple," extremely slowly evolving (hypobradytelic) organisms (Schopf 1995).

5. Acknowledgments

We are grateful to Jochen Brocks and two anonymous reviewers for comments and critical assessment of an earlier version of our manuscript. Funding by the Deutsche Forschungsgemeinschaft (Al 331/7 and Al 331/14) for our research on modern Foraminifera in anoxic environments is gratefully acknowledged. WA is grateful for support by the University of Pretoria (RDP) and by the NRF.

6. References

Altenbach AV, Struck U (2006) Some remarks on Namibia's shelf environments, and a possible teleconnection to the hinterland. In: Leser H (ed.) The changing culture and nature of Namibia: case studies. The sixth Namibia workshop Basel 2005. In Honour of Dr. h.c. Carl Schlettwein (1925–2005), Basler Afrika Bibliographien, Basel, pp 109–124

Altenbach AV, Struck U, Graml M, Emeis K (2002) The genus *Virgulinella* in oxygen deficient, oligotrophic, or polluted sediments. In: Revets SA (ed.) FORAMS 2002 international symposium on foraminifera, volume of abstracts, The University of Western Australia 1:20

Altermann W (2004) Evolving life and its effect on Precambrian sedimentation. In: Eriksson PG, Altermann W, Nelson DR, Mueller W, Catuneanu O (eds.) The Precambrian Earth: Tempos and Events, vol 12, Developments in Precambrian Geology. Elsevier, Amsterdam, pp 539–545

Altermann W (2007) The early Earth's record of enigmatic cyanobacteria and supposed extremophilic bacteria at 3.8 to 2.5 Ga. In: Seckbach J (ed.) Algae and cyanobacteria in extreme environments, vol 11, Cellular origin, life in extreme habitats and astrobiology (COLE). Springer, Dordrecht, pp 759–778

Altermann W (2009) Introduction to from fossils to astrobiology – a roadmap to a Fata Morgana? In: Seckbach J, Walsh M (eds.) From fossils to astrobiology (COLE), vol 12. Springer, Dordrecht, pp Xv–xxvii

Altermann W, Schopf JW (1995) Microfossils from the Neoarchean Campbell Group, Griqualand West Sequence of the Transvaal Supergroup, and their paleoenvironmental and evolutionary implications. Precambrian Res 75:65–90

Altermann W, Kazmierczak J, Oren A, Wright D (2006) Microbial calcification and its impact on the sedimentary rock record during 3.5 billion years of Earth history. Geobiology 4:147–166

Anbar AD, Duan Y, Lyons TW, Arnold GL, Kendall B, Creaser RA, Kaufman AJ, Gordon GW, Scott C, Garvin J, Buick R (2007) A whiff of oxygen before the Great Oxidation Event? Science 317:1903–1906

Baturin GN (2002) Nodular fraction of phosphatic sand from the Namibia Shelf. Lithol Miner Resour 37:1–17

Bernhard JM (2002) The anoxic Carioca Basin has benthic foraminifers: preliminary observations on the ecology and ultrastructure of *Virgulinella fragilis*. In: Revets SA (ed.) FORAMS 2002 international symposium on foraminifera, volume of abstracts, The University of Western Australia 1:24–25

Bernhard JM (2003) Potential symbionts in bathyal foraminifera. Science 299:861

Bernhard JM (2006) Foraminifera living in sulfidic environments: biology, ecology, and geological implications. FORAMS 2006, international symposium on foraminifera. Natal, Brasil. Abstracts, [URL: http://www.fgel.uerj.br/forams2006/joan.htm]

Bernhard JM, Bowser SS (2008) Peroxisome proliferation in foraminifera inhabiting the chemocline: an adaptation to reactive oxygen species exposure? J Eukaryot Microbiol 55:135–144

Bernhard JM, Reimers CE (1991) Benthic foraminiferal population fluctuations related to anoxia: Santa Barbara Basin. Biogeochemistry 15:127–149

Bernhard JM, Visscher PT, Boewser SS (2003) Submillimeter life positions of bacteria, protists, and metazoans in laminated sediments of the Santa Barbara Basin. Limnol Oceanogr 48:813–828

Bernhard JM, Habura A, Bowser SS (2006) An endobiont-bearing allogromiid from the Santa Barbara Basin: implications for the early diversification of foraminifera. J Geophys Res 111:G03002. doi:10.1029/2005JG000158

Borgonie G (2011) Worms from Hell: Nematoda from the terrestrial deep subsurface of South Africa. New Horizons for international investigations into carbon cycling in the deep crustal biosphere, Abstract Vol. University of the Free State, Bloemfontein, SA. O6

Brocks JJ, Banfield J (2009) Unravelling ancient microbial history with community petrogenomics and lipid geochemistry. Nat Rev Microbiol 7:601–609

Brocks JJ, Logan GA, Buick R, Summons RE (1999) Archean molecular fossils and the early rise of eukaryotes. Science 285:1033–1036

Brocks JJ, Buick R, Summons RE, Logan GA (2003) A reconstruction of Archean biological diversity based on molecular fossils from the 2.78 to 2.45 billion-year-old Mount Bruce Supergroup, Pilbara Craton, Western Australia. Geochim Cosmochim Acta 67:289–4319

Brocks JJ, Grosjean E, Logan GA (2008) Assessing biomarker syngeneity using branched alkanes with quarternary carbon (BAQCs) and other plastic contaminants. Geochim Cosmochim Acta 72:871–888

Brüchert V, Jorgensen BB, Neumann K, Riechmann D, Schloesser M, Schulz H (2003) Regulation of bacterial sulfate reduction and hydrogen sulfide fluxes in the central Namibian coastal upwelling zone. Geochim et Cosmochim Acta 67:4505–4518

Buick R (2010) Ancient acritarchs. Nature 463:885–886

Canfield DE (2005) The early history of atmospheric oxygen: homage to Robert M. Garrels. Annu Rev Earth Planet Sci 33:1–36

Chapman P, Shannon L (1987) Seasonality in the oxygen minimum layers at the extremities of the Benguela system. S Afr J Mar Sci 5:11–34

Debenay J, Guiral D, Paarra M (2004) Behaviour and taphonomic loss in foraminiferal assemblages of mangrove swamps of French Guiana. Mar Geol 208:295–314

Diz P, Frances G (2008) Distribution of live benthic foraminifera in the Ria de Vigo (NW Spain). Mar Micropaleontol 66:165–191

Duan Y, Anbar AD, Arnold GL, Lyons TW, Gordon GW, Kendall B (2010) Molybdenum isotope evidence for mild environmental oxygenation before the Great Oxidation Event. Geochim et Cosmochim Acta 74:6655–6668

Eigenbrode JL, Freeman KH, Summons RE (2008) Methylhopane biomarker hydrocarbons in Hamersley Province sediments provide evidence for Neoarchean aerobiosis. Earth Planet Sci Lett 273:323–331

Emeis K, Brüchert V, Currie B, Endler R, Ferdelman T, Kiessling A, Leipe T, Noli-Peard K, Struck U, Vogt T (2004) Shallow gas in shelf sediments of the Namibian coastal upwelling ecosystem. Cont Shelf Res 24:627–642

Emeis K, Struck U, Leipe T, Ferdelman TG (2007) Variability in upwelling intensity and nutrient regime in the coastal upwelling system offshore Namibia: results from sediment archives. Int J Earth Sci (Geol. Rundsch). doi: 10.1007/s00531-007-0236-5

Erbacher J, Nelskamp S (2006) Comparison of benthic foraminifera inside and outside a sulphur-oxidizing bacterial mat from the present oxygen-minimum zone off Pakistan (NE Arabian Sea). Deep-Sea Res Part I 53:751–775

Eriksson PG, Altermann W, Nelson DR, Mueller W, Catuneanu O (eds.) (2004) The Precambrian Earth: tempos and events, vol 12, Developments in Precambrian geology. Elsevier, Amsterdam, 941 pp

Ernst S, Bours R, Duijnstee I, van der Zwaan B (2005) Experimental effects of an organic matter pulse and oxygen depletion on a benthic foraminiferal shelf community. J Foraminifer Res 35:177–197

Fenchel T, Finlay B (2008) Oxygen and the spatial structure of microbial communities. Biol Rev 83:553–569

Fischer WW (2008) Life before the rise of oxygen. Nature 455:1051–1052

Frieling D, Mrazek J (2007) Sind Manganknollen Tiefwasser-Onkoide? Senckenberg maritima 37(2):93–128

Graur D, Martin W (2004) Reading the entrails of chickens: molecular timescales of evolution and the illusion of precision. Trends Genet 20:80–86

Gross O (2000) Influence of temperature, oxygen and food availability on the migrational activity of bathyal benthic foraminifera: evidence by microcosm experiments. Hydrobiologia 426:123–137

Grzymski J, Schofield OM, Falkowski PG, Bernhard JM (2002) The function of plastids in the deepsea benthic foraminifer, *Nonionella stella*. Limnol Oceanogr 47:1569–1580

Habicht KS, Gade M, Thamdrup B, Berg P, Canfield DE (2002) Calibration of sulfate levels in the Archean Ocean. Science 298:2372–2374

Han T-M, Runnegar B (1992) Megascopic eukaryotic algae from the 2.1-billion-year-old Negaunee Iron-Formation, Michigan. Science 257:232–235

Hazen RM, Papineau D, Bleeker W, Downs RT, Ferry JM, Mccoy TJ, Sverjensky DA, Yang H (2008) Mineral evolution. Am Mineral 93:1693–1720

Heinz P, Geslin E, Hemleben C (2005) Laboratory observations of benthic foraminiferal cysts. Mar Biol Res 1:149–159

Høgslund S, Revsbech N, Cedhagen T, Nielsen L, Gallardo V (2008) Denitrification, nitrate turnover, and aerobic respiration by benthic foraminiferans in the oxygen minimum zone off Chile. J Exp Mar Biol Ecol 359:85–91

Javaux EJ, Knoll AH, Walter MR (2004) TEM evidence for eukaryotic diversity in mid-Proterozoic oceans. Geobiology 2:121–132

Javaux EJ, Marshall CP, Bekker A (2010) Organic-walled microfossils in 3.2-billion-years-old shallow-marine siliciclastic deposits. Nature 463:934–938

Kato Y, Suzuki K, Nakamura K, Hickman AH, Nedachi M, Kusakabe M, Bevacqua DC, Ohmoto H (2009) Hematite formation by oxygenated groundwater more than 2.76 billion years ago. Earth Planet Sci Lett 278:40–49

Kazmierczak J, Altermann W (2002) Neoarchean biomineralisation by benthic cyanobacteria. Science 298:2351

Kazmierczak J, Kempe S, Altermann W (2004) Microbial origin of Precambrian carbonates: lessons from modern analogues. In: Eriksson PG, Altermann W, Nelson DR, Mueller W, Catuneanu O (eds.) The Precambrian Earth: Tempos and Events, vol 12, Developments in Precambrian Geology. Elsevier, Amsterdam, pp 545–563

Kazmierczak J, Altermann W, Kremer B, Kempe S, Eriksson PG (2009) Mass occurrence of benthic coccoid cyanobacteria and their role in the production of carbonates in Neoarchean of South Africa. Precambrian Res 173:79–92

Knoll AH (2003) The geologic consequences of evolution. Geobiology 1:3–14

Knoll AH, Javaux EJ, Hewitt D, Cohen P (2006) Eukaryotic organisms in Proterozoic oceans. Philos Trans R Soc B 361:1023–1038

Koho KA, Pina-Ochoa E, Geslin E, Risgaard-Petersen N (2011) Vertical migration, nitrate uptake and denitrification: survival mechanisms of foraminifers (*Globobulimina turgida*) under low oxygen conditions. FEMS Microbiol Ecol 75:273–283

Kremer B, Kazmierczak J (2005) Cyanobacterial mats from Silurian black radiolarian cherts: phototrophic life at the edge of darkness? J. Sediment Res 75:897–906. doi:10.2110/jsr.2005.069

Kump LR, Barley ME (2007) Increased subaerial volcanism and the rise of atmospheric oxygen 2.5 billion years ago. Nature 48:1033–1036

Leiter C (2008) Benthos-Foraminiferen in Extremhabitaten: Auswertung von Meteor-Expeditionen vor Namibia. Ph.D. thesis, Ludwig-Maximilians-Universitaet, Munich, Germany, URL http://edoc.ub.uni-muenchen.de/view/subjects/fak20.html (In German)

Leiter C, Altenbach AV (2010) Benthic foraminifera from the diatomaceous mud belt off Namibia: characteristic species for severe anoxia. Palaeontol Electron, vol 13(2); 11A:19p; http://palaeo-electronica.org/2010_2/188/index.html

Licari L, Mackensen A (2005) Benthic foraminifera off West Africa (1N to 32S): Do live assemblages from the topmost sediment reliably record environmental variability? Mar Micropaleontol 55:205–233

Linke P, Lutze GF (1993) Microhabitat preferences of benthic foraminifera – a static concept or a dynamic adaption to optimize food acquisition? Mar Micropaleontol 20:215–234

Martin RE, Quigg A, Podkovyrov V (2008) Marine Biodiversification in response to evolving phytoplankton stoichiometry. Palaeogeogr Palaeoclimatol Palaeoecol 258:277–291

Menuz V, Howell KS, Gentina S, Epstein S, Riezman I, Fornallaz-Mulhauser M, Hengartner MO, Gomez M, Riezman H, Martinou J-C (2009) Protection of *C. elegans* from Anoxia by HYL-2 Ceramide Synthase. Science 324:381–384

Mohrholz V, Bartholomae CH, van der Plas AK, Lass HU (2008) The seasonal variability of the Northern Benguela undercurrent and its relation to the oxygen budget on the shelf. Cont Shelf Res 28:424–441

Moodley L, van der Zwaan GJ, Herman PMJ, Kempers L, van Breugel P (1997) Differential response of benthic meiofauna to anoxia with special reference to Foraminifera (Protista: Sarcodina). Mar Ecol Prog Ser 158:151–163

Moreira D, von der Heyden S, Bass D, López-García P, Chao E, Cavalier-Smith T (2007) Global eukaryote phylogeny: combined small- and large-subunit ribosomal DNA trees support monophyly of Rhizaria, Retaria and Excavata. Mol Phylogenet Evol 44:255–266

Murray JW (2000) The enigma of the continued use of total assemblages in ecological studies of benthic foraminifera. J Foraminifer Res 30:244–245

Ohmoto H (2004) Archean atmosphere, hydrosphere and biosphere. In: Eriksson PG, Altermann W, Nelson DR, Mueller W, Catuneanu O (eds.) The Precambrian Earth: tempos and events, vol 12, Developments in Precambrian geology. Elsevier, Amsterdam, pp 361–387

Ohmoto H, Watanabe Y, Ikemi H, Poulson SR, Taylor BE (2006) Sulphur isotope evidence for an oxic Archaean atmosphere. Nature 424:908–911

Pawlowski J, Holzmann M, Berney C, Fahrni J, Gooday AJ, Cedhagen T, Habura A, Bowser SS (2003) The evolution of early foraminifera. PNAS 100:11494–11498

Pina-Ochoa E, Koho KA, Geslin E, Risgaard-Petersen N (2010) Survival and life strategy of the foraminiferan *Globobulimina turgida* through nitrate storage and denitrification. Mar Ecol Prog Ser 417:39–49

Piña-Ochoa E, Høgslund S, Geslin E, Cedhagen T, Revsbech NP, Nielsen LP, Schweizer M, Jorissen F, Rysgaard S, Risgaard-Petersen N (2010) Widespread occurrence of nitrate storage and denitrification among Foraminifera and Gromiida. PNAS 107:1148–1153

Rasmussen B, Fletcher IR, Brocks JJ, Kilburn MR (2008) Reassessing of the first appearance of eukaryotes and cyanobacteria. Nature 455:1101–1104

Richardson S, Rutzler K (1999) Bacterial endosymbionts in the agglutinating foraminiferan *Spiculidendron corallicolum* Rutzler and Richardson 1996. Symbiosis 26:299–312

Risgaard-Petersen N, Langezaal AM, Ingvardsen S, Schmid MC, Jetten MSM, Op den Camp HJM, Derksen JWM, Pina-Ochoa E, Eriksson SP, Nielsen LP, Revsbech NP, Cedhagen T, van der Zwaan GJ (2006) Evidence for complete denitrification in a benthic foraminifer. Nature 443:93–96

Schieber J (2009) Discovery of agglutinated benthic foraminifera in Devonian black shales and their relevance for the redox state of ancient seas. Palaeogeo Palaeoclim Palaeoecol 271:292–300

Schopf JW (1995) Metabolic memories of the Earth's earliest biosphere. In: Marshall C, Schopf JW (eds.) Evolution and molecular revolution. Jones and Bartlett, Boston, pp 73–107

Sleep NH, Bird DK (2007) Niches of the pre-photosynthetic biosphere and geologic preservation of Earth's earliest ecology. Geobiology 5:101–117

Sleep NH, Bird DK (2008) Evolutionary ecology during the rise of dioxygen in the Earth's atmosphere. Philos Trans R Soc B. doi:10.1098/rstb.2008.0018

Sugitani K, Grey K, Allwood A, Nagaoka T, Mimura K, Minami M, Marshall CP, van Kranendonk MJ, Walter MR (2007) Diverse microstructures from c. 3.4 Ga Strelley Pool Chert, Pilabra Craton, Western Australia: microfossils, dubiofossil, or pseudofossils. Precambrian Res 158:228–262

Sugitani K, Lepot K, Nagaoka T, Mimura K, Van Kranendonk M, Oehler DZ, Walter MR (2010) Biogenicity of morphologically diverse carbonaceous microstructures from the *ca.* 3400 Ma Strelley Pool Formation, in the Pilbara Craton, Western Australia. Astrobiology 10(9):899–920

Takata H, Seto K, Sakai S, Tanaka S, Takayasu K (2005) Correlation of *Virgulinella fragilis* Grindell & Collen (benthic foraminiferid) with near-anoxia in Aso-kai Lagoon, central Japan. J Micropalaeontol 24:159–167

Ueno Y, Isozaki Y, McNamara KJ (2006) Coccoid-like microstructures in a 3.0 Ga chert from Western Australia. Int Geol Rev 48:78–88

van der Plas AK, Monteiro PMS, Pascall A (2007) Cross-shelf biogeochemical characteristics of sediments in the central Benguela and their relationship to overlying water column hypoxia. Afr J Mar Sci 29:37–47

Waldbauer JR, Sherman LS, Sumner DY, Summons RE (2009) Late Archean molecular fossils from the Transvaal Supergroup record the antiquity of microbial diversity and aerobiosis. Precambrian Res 169:28–47

Watanabe Y, Farquhar J, Ohmoto H (2009) Anomalous fractionations of sulfur isotopes during thermochemical sulfate reduction. Science 324:370–373

Weeks SJ, Currie B, Bakun A, Peard KR (2004) Hydrogen sulphide eruptions in the Atlantic Ocean off Southern Africa: implications of a new view based on SeaWiFS satellite imagery. Deep-Sea Res Part I 51:153–172

Woese CR, Fox GE (1977) The concept of cellular evolution. J Mol Evol 10:1–6

Zimmer C (2009) On the origin of eukaryotes. Science 325:666–668

ORGANISM INDEX

A

Acanthamoeba castellanii, 60, 66
Acer saccharum, 228
Acetabularia acetabulum, 61
Acorus calamus, 223, 224, 226, 227, 230
Actuariola framvarensis, 426
Adercotryma glomeratum, 259
Agelastica alni, 168
Ageneotettix deorum, 167, 169
Akrav israchanani, 452
Allochromatium sp., 442
   A. vinosum, 440
Alvinella, 457
Amara alpina, 168
Amblycheila sp., 177
   A. cylindriformis, 167, 170
Ameira sp., 380
   A. parvula, 380
Ammodiscus catinus, 282
Ammonia sp.
   A. beccarii, 251, 518, 520, 524
   A. Elphidium, 243
   A. parkinsoniana, 243, 251
   A. tepida, 251, 287
Amoebidium parasiticum, 61, 83
Amoebobacter, 440
Amphascella subdebilis, 380
Amphistigina sp., 519, 522
   A. lobifera, 517, 520, 523, 526, 527
Ancyromonas sigmoides, 470
Andalucia incarcerata, 51, 60, 66
Angulogerina angulosa, 300, 302, 304–307
Anisonema platysomum, 129
Anobium punctatum, 174
Antherenus sp.
   A. verbasci, 175
   A. vorax, 175
Antonospora sp., 98
A. locustae, 56, 97–98
A. loustae, 88
Anurophorus laricis, 168
Aphelenchus avenae, 188
   A. fossor, 166
Aponema, 376
Archesola typhlops, 380
Arenicola marina, 409
Arphia xanthoptera, 167, 169
Artemia sp., 186–193
   A. franciscana, 69, 183–193
   A. salina, 200, 201
Ascaris, 5
Ascomycota, 468, 469, 472
Askenasia, 382
Astasia longa, 61
Attagenus unicolor, 175
Ayyalonia dimentmani, 452

B

Baculites palestinensis, 536
Bangiomorpha sp., 411
Barbulanympha, 155
Basidiomycota, 466, 468, 472
Bathyallogromia, 373
Bathymodiolus azoricus, 471, 472
Bathysiphon, 319
Beggiatoa, 25, 261, 296, 323, 448, 450, 562
Bemisia, 176
Betula sp., 228
   B. pubescens, 210
Bigenerina cylindrica, 249, 251
Blastocystis sp., 51, 53–55, 60, 61, 66, 68, 69, 81, 84, 87, 89, 92–93, 95, 100, 101
   B. hominis, 409
Blattella germanica, 174
Bledius spectabilis, 167, 168, 171

Bodo saliens, 470
Bolboschoenus maritimus, 223
Bolivina sp., 522, 525, 540, 597
   B. acerosa, 527
   B. alata, 251, 301, 305–307
   B. costata, 300–302, 304–307, 317, 319, 321, 322, 324
   B. dilatata, 244, 251
   B. doniezi, 519, 527
   B. humilis, 317, 319, 321, 322, 324
   B. interjuncta, 300, 301, 304–307
   B. minuta, 300, 302, 304–307
   B. pacifica, 244, 246, 252, 319, 518, 520, 525, 527, 546, 547, 572, 573
   B. plicata, 251, 300, 301, 304–307
   B. seminuda, 246, 247, 252, 300, 301, 303–307, 315, 319, 321, 322
   B. serrata, 301, 307
   B. skagerrakensis, 251
   B. spiss, 246, 252, 261, 301, 307, 507, 518, 519, 521, 522, 524–526
   B. spissa, 246, 252, 261, 301, 307, 507, 518, 519, 521, 522, 524–527
   B. striatula, 249
   B. subadvena, 527
   B. subaenariensis, 252
Bolivinellina sp.
   B. humilis, 322
   B. pseudopunctata, 260
Bolivinida, 301
Bolivinidae, 519, 525
Bolivinita minuta, 300, 302, 306
Bolivinitidae, 300, 303, 519, 522, 523
Bradysia, 176
Bulimina sp., 285
   B. aculeata, 244, 252, 257, 546
   B. costata, 246
   B. elongata, 252
   B. exilis, 244, 252
   B. marginata, 252, 257, 259, 279, 283, 546
   B. morgani, 243
Buliminella sp., 243
   B. morgani, 242, 253
   B. subfusiformis, 319
   B. tenuata, 264, 319, 321, 324
Byrrhus pilula, 168

C

C. parvulus, 470
Caenomorpha levanderi, 11
Cafeteria, 470
Calamites, 219
Calkinsia sp., 336
   C. aureus, 424
Calliactis parasitica, 347
Callitroga macellaria, 167
Camponotus andersoni, 171
Campylaimus, 376
Cancris sp.
   C. auriculus, 318
   C. carmenensis, 300, 302, 304, 306, 307
Candida, 468, 472
Candidatus, 43, 44, 135, 156
Carex sp.
   C. bigelowii, 217
   C. papyrus, 222
   C. rostrata, 222, 225
Carpediemonas membranifer, 89
Cassidulina crassa, 302, 306, 307
Caucasinidae, 519, 522, 523
Celosia argentea, 217
Cereus pedunculatus, 347
Chilostomella sp., 250, 258
   C. oolina, 253
   C. ovoidea, 246, 247, 253
Chironomus sp.
   C. anthracinus, 163
   C. plumosus, 163
   C. thummi, 177
Chlamydomonas sp., 5, 83, 100
   C. reinhardtii, 60–62, 65, 67, 68, 71, 409
Chlamys varia, 349
Chlorella, 8, 61, 384
Chlorobium phaeovibrioides, 397
Chorisia speciosa, 216, 217
Chortophaga viridifasciata, 167, 169
Chromatium sp., 442
   C. okenii, 439, 442, 443
Cibicides wuellestrofi, 258
Cibicidina basiloba, 526
Cibicidoides, 304
Cicer arietinum, 211, 220, 221, 230
Cicindela sp., 170
   C. denverensis, 167, 168

C. formosa, 167, 168, 171
C. hirticollis, 167, 168, 170, 171
C. nevadica, 167, 168
C. punctulata, 167, 168
C. repanda, 167, 168
C. togata, 167, 170, 177
C. tranquebarica, 167, 168
Clavulina cylindirica, 249
Cloeon dipterum, 163
Cobbionema, 376
Coptotermes, 151
  C. formosanus, 154, 156
Coronympha sp., 157
  C. clevelandi, 157
  C. octonaria, 157
Coryphostoma, 522
Cosmopolites sordidus, 168, 172
Cryptocercus sp., 151, 155, 156
Cryptococcus sp., 468, 472
  C. curvatus, 467, 472
Cryptopygus antarcticus, 168
Cryptosporidium sp., 51, 54, 92, 99
  C. parvum, 56, 61, 88, 98–99, 409
Cryptotermes sp., 151
  C. brevis, 175
Cucumis sativus, 217
Cyanophora paradoxa, 61
Cyclammina cancellata, 246, 253, 305, 307
Cyclidium sp., 131, 132
  C. porcatum, 13, 409

D

Dactylis glomerata, 225
Dasytricha sp., 40, 93, 94
  D. ruminantium, 94
Daucus carota, 212
Debaryomyces, 472
Deinococcus radiodurans, 202
Dermestes lardarius, 175
Deschampsia beringensis, 223
Desulfobulbaceae, 146
Devescovina striata, 150
Diabrotica sp., 171
  D. balteata, 167
  D. undecimpunctata, 167
  D. virgifera, 167
Diplodinium, 40

Discorbis sp.
  D. aguayoi, 522, 527
  D. floridana, 527
Drosophila sp., 173, 177, 178
  D. melanogaster, 79, 165, 169, 173, 175

E

Echiniscus japonicus, 203
Echinochloa sp., 215
Eggerella sp., 243
  E. scabra, 285
Ehrenbergina trigona, 244
Eleocharis palustris, 222
Elphidium sp.
  E. excavatum, 243
  E. magellanicum, 242, 253
Encephalitozoon cuniculi, 56, 88, 97–98, 409
Enhydrosoma, 380
Entamoeba sp., 6, 7, 54, 57, 59, 66, 71, 88, 96, 97
  E. histolytica, 51, 54, 56, 58–60, 66–68, 88, 96–97, 407–409
Enterolobium contortisiliquum, 212
Epiblema scudderiana, 167
Epidinium, 93
Epistominella sp., 243
  E. exigua, 244
  E. vitrea, 242, 243, 260
Equisetum sp., 219
  E. telmateia, 219
Eriophorum sp.
  E. angustifolium, 222
  E. vaginatum, 222
Erythrina caffra, 217
Escherichia coli, 56, 437, 526
Eucalyptus citriodora, 212
Eudiplodinium, 93
Eudorylaimus andrassyi, 456
Euglena sp., 61, 67
  E. gracilis, 61, 67, 92, 409
Euilyodrilus heuscheri, 456
Euplotes sp., 8, 84, 93
  E. crassus, 84
  E. minuta, 84
  E. vannus, 135
Eurosta solidaginis, 167

F

Fasciola hepatica, 60, 68
Ferroplasma acidiphilum, 146
Festuca vivipara, 222
Filipendula ulmaria, 222
Folsomia sp.
   F. candida, 169
   F. quadrioculata, 169
Frankliniella occidentalis, 176
Fraxinus pennsylvanica, 228
Frontonia, 8
Fursenkoina sp., 250, 573, 597
   F. bradyi, 250
   F. fusiformis, 300, 302, 304, 306, 307, 572, 597
   F. mexicana, 250, 253
Fusarium oxysporum, 409

G

Gastrophilus intestinalis, 173
Gavelinella, 540
Geotrupes, 180
Giardia sp., 7, 51, 54, 66, 71, 99
   G. intestinalis, 56, 60, 88, 99
   G. lamblia, 66, 68, 409
Ginkgo biloba, 219
Globigerina sp.
   G. siphonifera, 527
Globigerinella siphonifera, 527
Globigerinoides sp.
   G. ruber, 527
   G. sacculifer, 519, 520, 527
   G. siphonifera, 520
Globobulimina sp., 244, 250, 253, 258, 260, 285
   G. affinis, 246, 253, 261, 285, 507
   G. auriculata f. arctica, 253
   G. pacifica, 246, 253, 261, 302, 304, 307
   G. turgida, 254, 260–263, 283, 285, 287
Globocassidulina sp.
   G. cf. biora, 264
   G. subglobosa, 281
Globoquadrina dutertrei, 519, 527
Globorotalia tumida, 519, 527
Globotruncana rosetta, 534
Glyceria maxima, 222, 223, 225

Gobius niger, 349
Goesella flintii, 254
Goniomonas, 421
Goodayia rostellatum, 373
Gromia sp., 261, 365, 367, 409
Gromiida, 369, 409
Gyrodinium aureolum, 416

H

Haematococcus pluvialis, 61
Haloshizopera pontarchis, 380
Hanzawaia nitidula, 518, 520, 523, 527
Haynesina germanica, 254
Heterostegina depressa, 519, 522, 523, 527
Hexagenia limbata, 164
Hexaplex trunculus, 344, 349
Hicanonectes teleskopos, 100
Histomonas meleagridis, 66, 84
Hoeglundina elegans, 258
Holomastigotoides mirabile, 154
Holosticha, 145, 147
Hordeum vulgare, 212
Hyalinea balthica, 254
Hydromedion sparsutum, 168
Hylotrupes bajulus, 175
Hyperammina, 578
Hypogastrura sp.
   H. tullbergi, 169
   H. viatica, 169
Hypothenemus obscures, 176

I

Incisitermes sp., 151, 157
   I. harbors, 157
   I. marginipennis, 156
   I. minor, 152, 175
Iridia diaphana, 519, 522, 527
Iris sp., 223, 226, 230
   I. germanica, 222, 229
   I. pseudacorus, 223, 226, 227, 230
Isotoma sp.
   I. anglicana, 169
   I. tschernovi, 169
   I. violacea, 169
Isotricha, 40, 93

J
Joenia annectens, 156
Juncus sp., 225
   J. conglomeratus, 222, 225
   J. effusus, 222, 225

K
Kalotermes flavicollis, 156

L
Lamprocystis sp.
   L. purpurea, 440–443
   L. purpureus, 440, 442
Laophonte sp., 380
   L. setosa, 380
Lasioderma serricorne, 174, 176
Lepyrus arcticus, 168
Linhomoeus, 376
Loxodes sp., 8, 409, 525, 579
Luzula parviflora, 217
Lyctus brunneus, 175

M
Macrobiotus sp.
   M. areolatus, 201
   M. hufelandi, 199, 203
Ma. occidentalis, 203
Masselina secans, 282, 286
Massisteria marina, 470
Mastigamoeba sp., 54, 57, 59
   M. balamuthi, 51, 60, 61, 66, 87, 88, 97, 409
Mastigella, 11
M. darwiniensis, 151
Medicago sativa, 225
Melanoplus sp.
   M. confusus, 169
   M. femurrubrum, 167, 169
Melasoma collaris, 168
Melonis barleeanus, 249, 254, 260
Mentha aquatica, 222
Mesochra, 380
Metacoronympha, 157
Metacoronympha sp.
   M. senta, 157
Metacyclops sp., 453
   M. longimaxillis, 450, 453
   M. subdolus, 453

Metalinhomoeus, 376
Metopus sp., 141, 397
   M. contortus, 11, 394, 470
Metschnikowia, 468
Microcosmus, 344, 349
Microlaimus, 376
Micromonas pusilla, 61
Miliammina fusca, 575, 576
Miliolinella subrotunda, 576
Milnesium tardigradum, 200–204
Mirocaris sp., 457
Modiolula phaseolina, 362
Monhystera, 376
Monocercomonas, 84
Musca domestica, 179
Myzus persicae, 176

N
Naegleria sp., 71, 86
   N. fowleri, 91
   N. gruberi, 66, 91, 101
Nautilus sp.
   N. ammonoides, 527
   N. beccarii, 527
Neagleria, 101
Necallimastix, 61
Nelumbo nucifera, 217
Neobulimina canadensis, 540, 546, 547
Neocallimastix sp., 54, 84, 88, 91–92
   N. frontalis, 65, 69, 409
   N. patriciarum, 53, 60
Neogloboquadrina dutertrei, 519, 521
Nicobium sp., 174
Nonionella sp., 259, 283, 597
   N. auris, 246, 319, 321–323, 325
   N. austinana, 540, 547
   N. canadaensis, 542
   N. canadensis, 540–543
   N. labradorica, 260
   N. miocenica, 304
   N. robusta, 540, 547
   N. stella, 246, 247, 254, 260, 264, 300, 302, 304–307, 322, 498, 500–504, 507, 509, 547, 572, 597
   N. turgida, 243, 249, 285
Nonion sp., 283
   N. fabum, 249
   N. scaphum, 249, 250, 254

Normanella serrata, 380
Nouria polymorphinoides, 285
Noviholosticha, 145, 147
  N. fasciola, 145
Nyctotherus, 6, 53, 55, 87, 92–94
  N. ovalis, 54, 63, 66, 68, 80, 84, 87, 89, 91, 93, 101, 409

O

Ocnus planci, 349
Onychiurus sp.
  O. arcticus, 169
  O. groenlandicus, 169
  O. vontoernei, 169
Oopterus soledadinus, 168
Operculina ammonoides, 519, 522, 523, 527
Ophiothrix quinquemaculata, 344, 347–349
Orbulina universa, 520, 527
Orthosoma brunnem, 173
Oryza sativa, 234
Ostreococcus lucimarinus, 61
Ostreococcus tauri, 61
Otiorrhynchus dubius, 168
Oxymonas sp., 152, 153
  O. pediculosa, 150
Oxyria digyna, 222
Oxyrrhis marina, 60
Oxytricha sp., 93

P

Parablepharisma, 397
Parablepharsma, 11
Paracoccus denitrificans, 490, 529
Paralinhomoeus, 376
Paramacrobiotus richtersi, 200, 204
Paramecium sp., 84, 93, 157
  P. aurelia, 84
Paramphiascopsis longirostris, 380
Pardalophora haldemani, 169
Parisotoma octoculata, 168
Patellina sp., 522, 523
  P. corrugata, 517, 519, 522, 523, 526, 527
Pelomyxa, 11
Pelophila borealis, 168, 172
Pemphigus trehernei, 169
Peranema trichophorum, 60
Perimylops antarcticus, 168

Periplaneta americana, 174
Perkinsus sp., 471
  P. marinus, 60
Phaeoxantha klugii, 167, 168
Phalaris arundinacea, 222
Phallusia mammilata, 349
Phaseolus vulgaris, 212
Phleum pratense, 225
Phragmites australis, 223, 232
Pinus, 219
Piromyces sp., 88
  P. E2, 54, 61, 84, 91–92
Plagiopyla frontata, 10, 394
Planctomycetes, 42
Planulina, 304
Plasmodium sp., 83
  P. falciparum, 56, 68, 83, 100
Pleuronema marinum, 382
Pliciloricus, 40
Podospora anserina, 83
Polypedilum vanderplanki, 200, 201, 203
Polyrhachis sokolova, 169
Porphyra haitanensis, 61
Postgaardi, 11
Praebulimina sp.
  P. prolixa, 540–543, 546, 547
  P. strobila, 540, 541
Protodrilus, 377, 378, 381, 384
Prymnesium parvum, 61
Psalteriomonas sp., 6, 11, 101
  P. lanterna, 51, 53, 63, 66, 69, 84, 89, 91, 100
Psammechinus microtuberculatus, 349
Psammosphaera bouwmanni, 259
Pseudoeponides falsobeccarii, 285
Pseudokeronopsidae, 145, 147
  P. carna, 145
  P. flava, 145
  P. rubra, 145
Pseudomonas stutzeri, 526
Pseudononion sp., 243
  P. atlanticum, 243, 254
Pseudopenilia bathyalis, 381
Pseudotrichonympha grassii, 66, 156
Pterotermes occidentis, 153
Pulvinulina menardii, 527
Pyrsonympha, 152–154, 156

Q

Quadricoma, 376
Quinqueloculina lamarckiana, 285

R

Radotruncana calcarata, 534, 536
Ragactis pulchra, 344
Ramazzottius varieornatus, 199, 201–205
Ranunculus repens, 222
Ra. oberhauseri, 202, 203
Ra. varioeornatus, 203
Reclinomonas americana, 83
Rectuvigerina phlegeri, 255
Reniera, 344
Reophax sp., 578
   R. dentaliniformis, 255, 283
Reticulitermes sp., 151, 154, 156
   R. flavipes, 151, 156
   R. speratus, 156
Rhizammina, 508, 509
Rhizopus oryzae, 61
Rhodnius prolixus, 166
Rhodopseudomonas, 12
Rhodosporidium, 468, 472
Rhodotorula, 472
Rhynchomonas nasuta, 470
Richtersius coronifer, 201–204
Rimicaris, 457
Rosalina sp., 522, 523
   R. floridana, 517, 519, 522, 523, 527
Rotaliatinopsis semiinvoluta, 244
Rugiloricus, 40

S

Sabatieria sp., 376
   S. pulchra, 376
Saccharomyces cerevisiae, 53, 69
Sawyeria marylandensis, 51, 53, 63, 66
Saxifraga sp.
   S. caespitosa, 222
   S. cernua, 222
   S. hieracifolia, 222
   S. oppositifolia, 213, 222
Scenedesmus sp., 65, 86
   S. obliquus, 61
Schistocerca gregaria, 168, 177
Schizaster sp.
S. canaliferus, 349
S. chiajei, 344
Schoenoplectus lacustris, 223
Shistocerca gregaria, 169, 177
Sialis velata, 164
Sitophilus sp.
   S. granarius, 176
   S. oryzae, 175
Skeletonema costatum, 416
Sminthurides malmgreni, 169
Solanum tuberosum, 222
Sonderia, 11, 397
Spartina anglica, 223, 232
Spizellomyces punctutatus, 83
Spiculodendron corallicolun, 264
Spinoloricus sp., 40, 409
Spirinia, 376
Spironucleus sp., 88
   S. barkhanus, 60, 66, 100
   S. salmonicida, 99
Squilla mantis, 344
Stainforthia sp., 246, 259, 260, 283, 573, 597
   S. fusiformis, 242, 243, 255, 259, 260
Stainfortia fusiformis, 259
Stegodium paniceum, 174
Strombidium purpureum, 12, 396
Supella longipalpa, 174

T

Temnoschoita nigroplagiata, 168, 172
Tenebrio molitor, 167
Tethysbaena sp., 453, 456
   T. ophelica, 451, 457
   T. ophelicola, 450, 451, 454
   T. relicta, 450, 457
Tetracanthella sp.
   T. afurcata, 169
   T. arctica, 169
   T. wahlgreni, 169
Tetrahymena sp., 84, 93
   T. malaccensis, 84
   T. paravorax, 84
   T. pigmentosa, 84
Tetrahymena, 89, 98
Textularia sp., 261
   T. earlandi, 247
   T. kattegatensis, 246
   T. tenuissima, 255

Thalassiosira pseudonana, 61, 62
Theristus, 376
Thermobia domestica, 175
Thermo electron, 316
Thermosbaena mirabilis, 457
Thiocystis, 442
Thiohalocapsa halophila, 457
Thiomargarita, 25
Thioploca, 25, 261, 296, 318, 323
Tineola bisselliella, 175
Tinogullmia sp., 372, 373
　T. cf. riemanni, 373
Torulopsis, 468
Trachipleistophora hominis, 57, 59, 97–98, 409
Treponema sp., 154
　T. primitia, 154
Tribolium sp.
　T. castaneum, 176
　T. confusum, 175, 176
Trichonympha agilis, 156
Trichomnas sp.
　T. vaginalis, 88
Trichomonas sp., 6, 55, 62, 63, 71, 84–87, 91–95, 101, 153
　T. vaginalis, 51, 56, 60, 62, 63, 65–67, 69, 80, 84, 85, 91, 100, 154, 409
Trichonympha, 153, 154, 156
Trifolium pratense, 225
Trimastix sp., 87
　T. priformis, 89
　T. pyriformis, 51, 66, 84, 86–91
Trimerotropis maritima, 169
Trimyema compressum, 94, 470
Tritrichomonas foetus, 63, 84, 85
Troglopedetes, 451, 453
　T. somala, 457
Tulumella sp., 456
Tussilago farfara, 222
Typha latifolia, 223
Typhlocaris, 456

T. ayyaloni, 451–453
T. galilea, 449, 451

U
Utricularia, 211
Uvigerina sp., 244, 246, 249, 255, 256, 261, 300, 302, 305–307
　U. akitaensis, 246, 255, 261
　U. elongatastriata, 249, 255
　U. ex. gr semionata, 244, 255
　U. mediterranea, 256
　U. peregrina, 244, 256, 300, 302, 304–307
　U. peregrine, 306

V
Vaccinium sp.
　V. macrocarpon, 214, 223
　V. macrocarpum, 214
Valvulineria sp.
　V. bradyana, 256, 409
　V. glabra, 300, 306, 307
　V. laevigata, 256
Vigtorniella zaikai, 377, 378, 384
Virgulinella fragilis, 247, 256, 287, 319, 323–325, 498, 510, 572, 573, 597
Virgulinella glabra, 321, 323, 324, 500–505, 507, 509, 596, 598, 599
Volvox carteri, 61

X
Xenylla maritima, 168
Xestobium rufovillosum, 174

Z
Zizania sp., 215
　Z. texana, 215
Zoobenthos, 362
Zootermopsis, 151, 156
Zostera sp., 232
　Z. marina, 228

SUBJECT INDEX

A

Acetate production, 41–42, 69, 71, 100
Acetate:succinate CoA transferase (ASCT), 70, 71, 92–95, 97, 100
Acetyl-CoA, 56, 62–65, 69–73, 91, 97, 99, 100, 103
Acetyl-CoA generation, 62, 64
Acetyl-CoA synthetase (ACS), 70, 93–95, 102
Acetyl-phosphate, 64, 70
Actin, 593, 600
Adenosine triphosphate assay (ATP), 291
Adriatic Sea, 253, 254, 355–365
Aerobic respiration, 24, 31, 42, 53, 54, 56, 179, 230, 274, 299, 586, 595
Aerotaxis, 141, 296–298
Agglutinated foraminifera, 332, 573, 576–579, 581–587, 601, 619, 624
Alat lake, 451–459, 502
Allogromiid, 258, 298, 299, 332, 336, 338, 377, 384–386, 593, 594, 597, 601, 606, 618, 620, 624, 625
Allogromina, 593
Altitude, 115, 179, 405
American cranberry, 225–226, 234
Ammonia monooxygenase, 42, 43
Ammonium, 20, 28, 29, 33, 373
Amoebae, 6, 14, 382, 594, 600
Anaerobes, 5–14, 45, 63, 67, 69, 85, 153, 224, 272, 273, 275, 289, 295, 299, 407, 423, 425, 426, 434, 442, 489, 504, 509, 519, 586, 596, 597, 621, 622, 626
Anaerobic ATP production, 68
Anaerobic flagellates, 8, 10
Anaerobic metabolism, 42–46, 73, 87, 93, 94, 106, 142, 178, 184, 236, 239, 299, 408–411, 426, 427
Anaerobic mitochondria, 71, 87, 100, 106, 409, 425, 426

Anaerobic syntrophy, 409
Anammox, 43, 520, 621
Anhydrobiosis, 207–210, 212
Anoxia longevity, 191, 194
Anoxia metabolism, 194
Anoxygenic photosynthesis, 42, 414, 430
Apicomplexa, 88, 93, 103, 425, 484, 488
Aquatic insects, 169–171
Archaea, 42, 152, 271, 406, 483
Archaic, 365
Archean eukaryotes, 623
Archezoa, 85, 423–425
Arctic anoxia, 223–224
Artemia (brine shrimp) embryos, 191–199
Astrobiology, 207–214
ATP/ADP carrier, 58
ATP generation, 8, 10, 55, 69–73
ATP production, 55, 56, 68, 74, 87, 473

B

Bacteria, ix, xiii, 8, 26, 41, 55, 115, 133, 152, 157, 207, 308, 327, 343, 374, 406, 426, 449, 451–459, 466, 502, 519, 540, 597, 620
Bacterial diversity, 452
Baltic Sea, 27, 195, 405, 428, 501
Banded iron formations (BIFs), 33, 427, 617
Bathyal, 147–154, 262–265, 267, 316, 334, 336, 344, 603
Beetle, 173, 177–180, 183, 185
Benthic community, 147, 257, 260, 271, 357
Benthic foraminifera, 9, 46, 249, 251–276, 287–300, 305–319, 323, 327, 328, 332, 334, 336, 338, 369, 497, 515, 516, 521, 522, 525, 529, 539–549, 575–587, 601, 615, 617–626
Benthic foraminiferal assemblages, 328, 331–335, 337, 559, 561, 575

639

Biodiversity pattern, 437, 438
Biogeochemistry, 19–24, 305, 370, 483, 499, 538
Biomass, 347–351, 357, 375, 410–413, 437, 470, 481, 482, 502, 506, 519–523, 525, 527–530
Biomineralization, 131, 132, 531, 595, 600
Biomineralized test, 519
Bioturbation, 23, 24, 27, 30, 147, 268, 292, 298, 309, 364, 585, 623
Black Sea, 27, 28, 43, 147, 258, 369, 373–397, 405, 427, 429, 430, 501, 502, 604
Black shales, 506, 573, 575–587, 618, 619, 622–626
Buliminids, 562, 564

C

Campanian, 556, 558, 569
Carbonatic test, 519, 522
CellTracker Green (CTG), 291, 292
Chemically stratified, 134, 135, 141, 403, 405–406, 414
Chemocline, 9, 145, 344, 406, 407, 413, 414, 437, 451, 452, 455, 458, 459, 470, 473, 474
Chemolithoautotrophy, 160
Chironomidae, 170
Chytridiomycete fungi (chytrids), 6, 89, 96, 99, 409
Ciliates, ix, xii, xiv, 6, 41, 147, 161, 274, 347, 407, 434, 451, 487
C-layer, xiv
Clone libraries, 148, 439, 483–485, 488
Colonization of land, 170, 476
CORR hypothesis, 105, 106
Cultural control, 178, 179

D

Dead zone, 343, 345–351, 356
Delta^{15}N, 521
Denitrification, 271–273, 502, 504, 542
Desiccation tolerance, 205, 208
Devonian, 29, 229, 573, 575–587, 618–620, 622–624
Diagenesis, 499, 500, 504, 505, 508, 509, 577, 617
Diet, 530
Discontinuous gas exchange, 171–173

Dissolved inorganic carbon, 259, 524
Diversity, 32, 39, 41–46, 53, 71, 73, 86–87, 146, 157, 159–164, 169, 172, 185, 235, 253, 255, 257, 258, 270, 289, 307, 313, 314, 317, 338, 346, 350, 351, 353, 371, 375, 388
δ^{15}N. *See* Nitrogen isotopes (δ^{15}N)
Dung beetles, 172, 180, 183, 185
Dysoxia, 553, 555–570

E

Ecology of anoxia tolerance, 243
Ectobionts, 149, 153, 412, 547, 569
Ectosymbionts, 11, 307
Edge effects, 397
Electron transport chain (ETC), 54, 69, 85, 86, 98, 100, 103, 106
Endosymbionts, 9, 11–13, 160, 163, 275, 299, 348, 409, 411, 519, 597, 604, 607, 620
Endosymbiosis, 85, 87, 604, 618, 626
Endosymbiotic bacteria, 11, 162, 274, 620
Energy metabolism, 5–8, 13, 54–56, 100, 101, 103, 169, 226, 426
Eukaryotic tree of life, 424–426
Eutrophication, 253, 254, 273, 343, 353, 355, 364, 365, 503, 504
Euxinia, 414
Evolution, xi, xx, xxvi, xxx, 12, 14, 31, 45, 51, 52, 73, 83, 86, 89, 106, 131, 146, 147, 157, 158, 162, 163, 169–170, 173, 228–230, 408, 410, 423–442, 475–476, 528, 598–604, 617–626
Evolution of anoxia tolerance, 228–230
Evolution of eukaryotes, 421, 425, 615, 617–626
Extremophiles, 451

F

Facultative anaerobes, 6, 14, 63, 272, 289, 295, 299, 434, 489, 519, 586, 596, 597
[FeFe]-hydrogenase, 55, 56, 59, 65–69, 73
[FeFe]-hydrogenase maturases, 68
Fermentation processess, 7, 8, 11, 20, 41, 42, 46, 86, 99, 157, 159, 184, 221, 226, 230
Ferredoxin, 7, 8, 55, 60–65, 67, 91, 92, 96, 97, 100, 102, 104
Fe–S cluster synthesis, 56, 101, 102

SUBJECT INDEX

Flagellates, 347, 348, 350
Flooding, 177–178, 223, 228–230
Flowering plants, 219, 221–243
Fluorescein diacetate (FDA), 291, 292
Flux rates, 520
Foraminifer, 258, 271, 293, 294, 298–300, 331, 425, 542, 545, 546, 579, 593, 597, 602, 606, 607, 621
Fractionation, 99, 269, 500–503, 515, 519–531, 615, 616
Framvaren Fjord, 395, 421, 423–442
Free energy rule, 195, 196, 199
Freezing, 173, 179, 184, 405, 527, 530
Fumarate respiration, 106
Fungi, 6, 14, 41, 55, 88, 96, 207, 409, 425, 437, 481–490, 544

G

Genetics, 83, 133, 252, 275, 449
Geochemistry, 131, 252, 324, 370, 617
Geomicrobiology, 249
Glass electrodes, 292
Glycine cleavage system (GCS), 92–94, 102
Gp_4G, 194
Grain pests, 183
Grazing, 140, 141, 243, 409, 410, 474, 481, 490, 569
Greigite, 133, 135, 137, 140, 141
Gromiids, 9, 275, 369, 377, 382, 383, 385
Groundwater, 463, 465–476
Growth efficiency, 410–412
Grypania, 426, 624
Gycolysis end-products, 99, 236–239

H

Harpacticoids, 377, 387, 391–393, 397
H-clusters, 68
Hemocyanin, 472, 475
Human R antigen (HuR), 119–121, 126
Humidity, 183
Hybrid organelles, 56
Hydrogenases, 66–69, 86, 90–92, 101
Hydrogen hypothesis, 67, 68
Hydrogenosomes, x, xi, 6–12, 41, 42, 55, 69, 83, 85–106, 113, 409, 424, 425, 596
Hydrogen production, 64, 68, 69, 74, 96, 101
Hydrogen sulfide, 26, 33, 135, 252, 258, 259, 270, 373, 375, 384, 385, 388, 390, 391, 394, 395, 397, 406, 426, 470, 473, 490, 520, 524, 525, 596, 597, 621, 622
Hypersaline, 39, 41, 146, 192, 193, 394, 406, 426, 427, 479
Hypoxia, ix, xi–xv, 24, 115–126, 170, 221, 253, 295, 345, 355, 375, 472–474
Hypoxia inducible factor (HIF), 116–119, 122, 123, 125, 126
Hypoxic environments, 172, 180, 343, 346, 351, 412, 475

I

Immersion, 173, 177–179, 185
Internal ribosome entry site (IRES), 122–124, 126
Iron, 19, 20, 27, 32, 33, 58–61, 67, 118, 132, 133, 140–141, 198, 229, 240, 370, 425, 427, 488, 617
Iron–sulfur (Fe–S) cluster assembly, 55, 58–62, 425
Iron–sulfur (Fe–S) clusters, 55, 58
Iron–sulfur (Fe–S) proteins, 58–61
ISC system, 59–61

K

Kleptoplast, 336, 597, 621, 622, 624, 625

L

Labile organic matter, 261, 271, 297, 318, 329, 335, 336, 338
Laboratory experiments, 269, 270, 295, 297, 298, 358, 442
L'Atalante basin, 41, 46
Late Cretaceous, 29, 553, 555–570
Late embryogenesis abundant (LEA) proteins, 198–199, 209
Long-term anoxia tolerance, 219, 221–243
Loricifera, 6, 41, 46, 55, 426

M

Magnetite, 133–135, 137, 139–141
Magnetotactic bacteria, 131–137, 139–141
Magnetotactic protists, 131, 133–142
Magnetotaxis, 132, 133, 136
Malate decarboxylation, 64–65
Manganese, 19, 20, 27, 229, 259, 269, 373, 620

Marine sediments, 9, 17, 19–33, 44, 146, 151, 195, 259, 370, 479, 481, 485, 487, 490, 505
Mass mortality, 357, 362
Meiobenthic communities, 374, 397
Meiofauna, 6, 41, 46, 251, 258, 295, 298, 345, 346, 349, 350, 377, 380, 397
Meromictic lakes, 406, 407, 451, 458, 502, 509
Meromixis, 406, 451–452
Metabolic depression, 184, 273
Metabolic rate depression (MRD), 191, 195, 196
Metabolism, 5–8, 42–46, 53, 86, 142, 169, 191, 208, 221, 270, 299–300, 408–410, 427, 452, 596, 618
Metagenomics, 450
Metazoa, x, xii, 41, 147, 375, 386–393, 413–414, 425, 483, 484
Methane, 17, 20, 24, 26, 28, 29, 31, 42–45, 258, 369, 370, 375, 377, 394, 397, 467, 482, 483, 489, 520, 524–526, 530, 531, 596, 621, 622
Methane monooxygenase, 42–44
Methanogenesis, 12, 24, 26, 31, 44
Methanogens, 5, 10–12, 20, 159, 274, 409
Microbial food web, 403, 409
Microbial mats, 595, 602
Microbial zonation, 406–407
Microhabitat, 208, 257, 259, 261–270, 275, 296, 297, 314, 323, 327–338, 568
Micropaleontology, xviii, xxx, 287, 288, 323, 515, 553, 554, 616
MicroRNAs, 124
Microsporidia, 53, 56, 74, 89, 93, 102, 103, 423–425
Migration, 268, 272, 297, 316, 405, 525
Mishash Formation, 553, 555–570
Mississippian, 573, 575–587
Mitochondria, x, xi, xiv, 6–9, 14, 41, 53–64, 67–71, 73, 85–90, 92, 96–98, 100, 103–106, 157, 193, 242, 274, 299, 307, 409, 423–426, 539–542, 545–547, 596, 597, 626
Mitochondrial carrier family (MCF), 55, 58, 104, 125
Mitochondrial complex I, 91, 98, 100, 106
Mitochondrial complex III, 97, 98, 105
Mitochondrial complex IV, 97, 98, 105
Mitochondrial complex V, 97, 98, 105
Mitochondrial protein import, 57, 101
Mitochondrial proteins, 55–57, 95, 99, 102, 423, 424
Mitochondrion-related organelles, 51–74, 87
Mitosomes, xi, 7, 8, 53, 55–59, 61, 74, 83, 85–106, 113, 424, 425, 596
Mixolimnion, 406, 413, 414, 451
Models for iron-sulfur cluster biosynthesis, 60
Modified atmospheres, 180–183
Molecular chaperones, 118, 197–198
Molecular chaperones artemin, 197, 198
Molecular chaperones hsp70, 60, 197
Molecular chaperones p26, 197–199
Molybdenum, 32, 33, 427
Monimolimnion, 406, 413, 451, 458
Monothalamous, 371, 382, 384, 385, 395, 593, 594, 598, 601, 602
mRNA stabilization, 119, 120, 126
mRNA transcription, 126
mRNA translation, 121, 126
Mucus tracks, 602

N

N_2, 621
Nematodes, 6, 195, 207, 209, 308, 336, 347, 349–351, 369, 371, 377, 387–390, 395, 397, 473, 603
N_2-fixation, 43, 59, 501–504
NIF system, 59, 61
Nitrate, xiv, xv, 9, 19, 42, 86, 252, 271–273, 290, 307, 335, 394, 409, 499, 519, 540, 586, 595–597, 621
Nitric oxide dismutase, 44
Nitrification, 502
Nitrite, 42, 43, 45, 242, 409, 499, 596
Nitrogen, 31, 43, 115, 242, 255, 292, 329, 455, 482, 497, 499–509, 515, 519–531, 547, 621
Nitrogen cycle, 33, 43, 249, 499–508, 547
Nitrogen isotopes ($\delta^{15}N$), 497, 499–509, 516
Non-photosynthetic aerobes, 69
Northern Adriatic Sea, 254, 353, 355–365
Norway, 224, 395, 428, 429, 442

SUBJECT INDEX

Nuclear prelaminin-A Recognition Factor (NARF/Nar) proteins, 68

O

Obligate anaerobes, 5, 6, 153, 434
Ophel, 463, 465–476
Optodes, 292, 293, 297, 298
Organellar (mitochondrial) genome, 55, 56, 86, 89, 98, 105, 106
Organic geochemistry, 324
Organic matter, ix, 19–22, 253, 257, 259, 261, 268, 292, 296, 297, 307, 309, 314, 316, 318, 329, 335, 336, 338, 356, 374, 465, 481, 485, 497, 500, 502–505, 516, 520, 529, 555, 567, 569, 576, 582, 595, 603, 604, 622, 623
Organic-rich carbonates, 556, 558–561
Orientation, 138, 296–298
Oxic–anoxic transition zone, 133, 258, 375, 394, 395
Oxygen, ix–xvi, 6, 19, 42, 98, 115, 141, 147, 159, 169, 191, 221, 249, 251–276, 288–300, 305, 307–319, 327, 345, 356, 373, 405, 425, 452, 519, 539, 559, 575, 595, 618
Oxygen deficient, xii, xv, 223, 225, 226, 317, 327, 357, 360, 414, 598, 619
Oxygen depletion, ix, xiii, xv, 19, 252–254, 294, 295, 297, 298, 307, 308, 334–336, 360, 361, 502, 525, 531, 559, 560, 597, 622
Oxygen minimum zone (OMZ), 26, 249, 257, 305, 307–319, 345, 414, 428, 528, 537, 538, 540, 542, 545
Oxygen reconstruction, 147, 595
Oxygen sensitivity, 62, 63
Oxygen sensors, 118, 292

P

Parabasalia, 160, 425
Peru, x, xxii, xxiv, xxvi, 25, 26, 308–314, 316, 323, 324, 327–338, 345, 481, 483–487, 490, 540
Peruvian margin, 316, 327, 328
Phagotrophic protozoa, 409, 410, 412, 413
Phanerozoic, 33, 414, 507, 576, 593, 607, 619
Phosphorites, 556, 561

Phosphorous, 27, 29
Phosphotransacetylase, 64, 70
Photosynthetic sulfur bacteria, 407, 412–414
Plant survival strategies, 231, 241
Polychaetes, 357, 387, 390–392, 394, 397
Polymerase chain reaction (PCR), 148, 149, 152, 453, 483, 485, 486
Population density, 307, 312, 314, 317
Porcelanites, 556, 561, 562
Pores, 274, 295, 315, 537, 539–549, 568, 569
Post-anoxic injury, 239–241
Precambrian foraminifera, 598
Prolyl hydroxylase (PHD), 117, 118
Proteobacteria, 101
α–Proteobacteria, 61, 85, 87, 88, 423
γ–Proteobacteria, 452, 456–458
Proterozoic, 5, 33, 414, 426, 427, 430, 475, 591, 593–607, 618, 619, 623, 624
Protists, ix, xii, xix, 6–14, 56, 65, 70, 74, 86–87, 99, 133–142, 145–147, 152, 155, 157–164, 289, 343–345, 377, 394, 407–410, 412, 422–427, 430–440, 479, 488, 489, 594, 599
Protozoa, ix, 7, 41, 46, 55–59, 62, 137, 162, 375, 382–386, 403, 408–413, 431
Prox(y)ies, xix, 228, 249, 305, 308, 317–319, 328, 329, 337, 501, 537, 546, 548, 587, 624
Purple non-sulfur bacteria, 11, 411
Pyrosequencing, 408, 439, 485, 486
Pyruvate dehydrogenase (PDH), 93, 94, 96, 97, 99, 100, 102, 103
Pyruvate dehydrogenase complex (PDC), 54, 62–65, 69, 73, 100
Pyruvate:ferredoxin oxidoreductase (PFO), 55, 59, 61–65, 68, 69, 73, 91–97, 99, 100, 102–104
Pyruvate formate lyase (PFL), 56, 62–65, 73, 93, 96, 97, 99, 100
Pyruvate formate lyase-activating enzyme (PFLA), 64, 65
Pyruvate metabolism, 56, 62–65, 71, 73, 100
Pyruvate:NADP$^+$ oxidoreductase (PNO), 62–65, 70, 73, 93, 97, 103

R

Radiation tolerance, 205, 209–211
Redox, xiii, xv, 20, 21, 31, 32, 63, 64, 105,

106, 239, 241, 242, 252, 259–261, 268–271, 273, 275, 296, 305, 316, 335, 336, 338, 373, 388, 406, 427, 435–438, 467, 490, 520, 537, 585, 591, 593–607, 621
Redoxcline, xi, 33, 407, 605
Respiration rate, 299, 473
Reticulopodia, 521, 598–600, 602, 604–607
Reverse methanogenesis, 44
Romer's Gap, 169
Rose Bengal, 291, 292, 308, 310, 328, 374, 377, 394, 397, 521–524, 527, 622
Rumen, ix, 14, 41, 46, 55, 98, 99, 151, 161, 162

S
SAM complex, 57
Santa Barbara Basin, x, xv, 147, 257, 258, 274, 275, 318, 334, 345, 346, 348–351, 528, 540, 541, 567, 569, 573, 575–587
Sapropels, 29, 501–503
Sedimentary nitrogen, 500, 504, 505
Sediments, x, xiv, 5, 17, 19–33, 41, 53, 146–153, 170, 227, 273, 289, 308, 327, 377, 405, 440, 481, 499, 520, 573, 575–587, 594, 619
Seeds and anoxia, 226–228
Sensing anoxia-plants, 241–243
Shales, xv, 426, 506, 507, 573, 575–587, 618–620, 622–626
Short-term anoxia tolerance, 230–231
Small subunit ribosomal RNA, 431, 481, 483, 488
Southern Tethys, 553, 555–570
Speciation, 432, 459
Speleology, 465
Spirotrichea, 151
16S rDNA, 152, 453, 455
18S rRNA, 148, 481, 483–487
Stable isotopes, xv, xviii, 252, 497, 503, 516, 529
Staining, 328, 374, 394, 521–523, 527
Stercomata, 604
Sterols, 6, 10, 625
Stratification, 504
Stratified aquatic environments, 141, 403, 405–407
Stress behaviors, 357

Suboxic, xiv, xv, xvi, 24, 252, 289, 308, 327, 335, 437, 442, 515, 576, 583, 585, 586, 616, 618
Subsurface biosphere, 481, 482
Subsurface marine eukaryotes, 479, 481–490
Succinate thiokinase (STK), 70, 71, 93, 98
SUF mobilization, 59, 62
Sulfate reducers, 5, 11, 12, 20, 412
Sulfate reduction, 24, 26, 30–32, 45, 261, 524, 597, 622
Sulfide, xiv, 299, 376
Sulfur bacteria, 26, 33, 272, 395, 407, 412–414, 451–459, 466, 467, 473, 502, 503, 606
Sulfuretum, 458–459
Supersulfidic, 430
Symbiosis, 9, 157, 162, 528, 529, 607, 624
Synapomorphy, 599, 601
Syntrophic hydrogen transfer, 11

T
Tardigrades, x, xii, 205, 207–214, 450
TCA cycle, 6, 7, 54, 62, 65, 69, 71, 73, 92, 96, 98, 100, 102, 103, 106, 226
Thermosbaenacea, 467, 473, 475, 476
Thresholds, xiii, xvi, 252, 269, 359, 363, 365, 600
Tiger beetle, 173, 177, 178, 185
TOM complex. *See* Translocase of the Outer Membrane (TOM) complex
Tracheal system, 169, 171, 172
Translational arrest, 123
Translocase of the Outer Membrane (TOM) complex, 57
Trehalose, 184, 185, 199, 209, 210
TRophic OXygen (TROX) model, 259, 296, 307, 308, 316, 336
Tubulin, 600

U
Ultrastructure, 53, 96, 193, 196, 273–275, 377, 423, 430, 440–442
Unfolded protein response (UPR), 118, 124
Upstream open reading frame (uORF), 122, 124
Upwelling, xv, 23, 24, 33, 254, 307, 308, 317, 325, 327, 345,

428, 490, 499, 503, 504, 555, 556, 558–560, 563–565, 567, 569, 570, 606

V

Vascular endothelial growth factor (VEGF), 113, 116–126

Vertical distribution, 252, 288, 296, 297, 307, 327, 328, 331–338, 380, 385, 389, 390, 414, 428

von Hippel–Lindau protein (VHL), 117, 125

X

Xenophyophores, 369, 603

AUTHOR INDEX

A
Almogi-Labin, A., xxi, 553, 555–570
Altenbach, A.V., xvii, xviii, xxi, 515, 519–531, 591, 593–607, 616–626
Altermann, W., xxi, 615–626
Anikeeva, O.V., xxi, 371, 373–397
Ashckenazi-Polivoda, S., xxi, 553, 555–570

B
Barnhart, M.C., xxix
Barry, J.P., xxi, 344–351
Bazylinski, D.A., xxi, 131, 133–142
Beaudoin, D.J., xxi, 145, 147–153
Behnke, A., xxii, 422–442
Benjamini, C., xxii, 554–570
Bernhard, J., xvii, xix, xxii, 145, 147–153, 344–351
Biddle, J.F., xxii, 479, 481–490
Bonsdorff, E., xxix
Brümmer, F., xxii, 450–459
Brust, M.L., xxix
Buck, K.R., xxii, 343, 345–351
Bunge, J.A., xxix

C
Cardich, J., xxii, 323, 327–338
Clegg, J.S., xxii, 189, 191–200
Crawford, R.M.M., xxiii, 219, 221–243

D
de Graaf, R.M., xxiii, 83, 85–106
De tullio, M.C., xxix
Dolan, M.F., xxiii, 155, 157–164
Dyer, B.D., xxix

E
Edelman-furstenberg, Y., xxiii
Edgcomb, V.P., xxiii, 146–153, 479, 481–490

F
Fenchel, T., xxiii, 3, 5–14
Frankel, R.B., xxiii, 132–142
Frenzel, P., xxix
Fritz, G.B., xxiii, 449, 451–459

G
Gaulke, M., xxiii, 591, 593–607
Geslin, E., xxiii, 288–300
Glock, N., xxiv, 305, 307–319, 537, 539–549
Gooday, A.J., xxiv, 369, 373–397
Gutiérrez, D., xxiv, 325, 327–338

H
Hackstein, J.H.P., xxiv, 83, 85–106
Hasemann, C., xxix
Heidelberg, K., xxx
Heinz, P., xxiv, 287, 289–300
Hengherr, S., xxiv, 450–459
Hiss, M., xxiv, 517, 519–531
Hoback, W.W., xxiv, 167, 169–185
Horikawa, D.D., xxiv, 205, 207–214
Hromic, T., xxx

K
Koho, K.A., xxv, 249, 251–276
Kolesnikova, E.A., xxv, 370, 373–397
Kormas, K.AR., xxx
Kosheleva, T.N., xxv, 371, 373–397
Kuhnt, W., xxx

L
Leadbetter, J., xxx
Lefèvre, C.T., xxv, 131, 133–142
Leger, M.M., xxv, 51, 53–74
Leiter, C., xxv, 515, 517–531, 616–626
Levy, A.P., 113, xxv, 113, 115–125

Levy, N.S., xxv, 113, 115–125
Lichtschlag, A., xxv, 370, 373–397
Lopez-garcia, P., xxx

M

Mallon, J., xxvi, 305, 307–319, 538–549
Mayr, C., xxvi, 516, 519–531
Mazlumyan, S.A., xxvi, 370, 373–397
Mclennan, A., xxx
Meysman, F., xxx
Morales, M., xxvi, 323, 327–338

N

Nielsen, L.P., xxx

O

Oren, A., xxvi, 39, 41–46

P

Pawlowski, J., xxx
Pfannkuchen, M., 449, 451–459
Piña-Ochoa, E., xxvi, 249, 251–276
Pollehne, F., xxx
Por, F.D., xxvi, 463, 465–476

Q

Quipúzcoa, L., xxvi, 324, 327–338

R

Rabalais, N.N., xxvii, 343, 345–351
Radic, A., xxvii, 517, 519–531
Riedel, B., xxvii, 353, 355–365
Roger, A.J., xxvii, 52–74

S

Saccà, A., xxvii, 403, 405–414
Schieber, J., xxvii, 573, 575–587
Schönfeld, J., xxvii, 306–319, xxvii, 537, 539–549
Schüler, D., xxxi
Schweikert, M., xxxi
Seckbach, J., xvii, xx, xxvii
Sergeeva, N.G., xxvii, 369, 373–397
Sifeddine, A., xxvii, 324, 327–338
Simpson, A., xxxi
Stachowitsch, M., xxviii, 353, 355–365
Stairs, A.W., xxviii, 52–74
Stoeck, T., xxviii, 421, 423–442
Strohmeier, S., xxviii, 450–459
Struck, U., xxviii, 497, 499–509, 516, 519–531

T

Tielens, L., xxxi
Treude, T., xxviii, 17, 19–33
Tsaousis, A.D., xxviii, 51, 53–74

V

van der Giezen, M., xxix

W

Wetzel, M., xxxi

Z

Zuschin, M., xxviii, 353, 355–365